NanoScience and Technology

The series NanoScience and Technology is focused on the fascinating nano-world, mesoscopic physics, analysis with atomic resolution, nano and quantum-effect devices, nanomechanics and atomic-scale processes. All the basic aspects and technology-oriented developments in this emerging discipline are covered by comprehensive and timely books. The series constitutes a survey of the relevant special topics, which are presented by leading experts in the field. These books will appeal to researchers, engineers, and advanced students.

More information about this series at http://www.springer.com/series/3705

Quan Li
Editor

Nanomaterials for Sustainable Energy

Springer

Editor
Quan Li
Liquid Crystal Institute and Chemical
Physics Interdisciplinary Program
Kent State University
Kent, OH
USA

ISSN 1434-4904 ISSN 2197-7127 (electronic)
NanoScience and Technology
ISBN 978-3-319-32021-2 ISBN 978-3-319-32023-6 (eBook)
DOI 10.1007/978-3-319-32023-6

Library of Congress Control Number: 2016935971

Printed on acid-free paper

This Springer imprint is published by Springer Nature
The registered company is Springer International Publishing AG Switzerland

Preface

Each generation of human civilization on our planet has confronted with new challenges and opportunities for survival and prosperity. Availability of consumable energy in different forms has greatly influenced the living standards of humankind on earth. The quality of life and prosperity of our future generations will primarily be defined by how we live up to the current and future energy demand challenges. Energy has apparently been the cornerstone of sustainable development of our economy and society. Some of the basic needs of the modern society such as food production and processing, water, health, clean environment, and education are critically dependent on energy supply and sufficiency. Developing materials and systems toward sustainable energy is a crucial challenge for the twenty-first century. Most of the energy needs of the current world are met through fossil fuels. As the energy demands are substantially increasing, the limited reserves of fossil fuels are progressively depleting and their continuous use is adversely affecting both the human health and environment. Therefore, research and development worldwide focus on diminishing the negative health and environmental consequences of fossil fuel energy consumption and to explore new materials paradigm for sustainable energy production and applications.

Nanomaterials and nanostructures provide unique electrical, optical, and mechanical properties and have played an important role in the recent advances of energy-related applications. Different nanomaterials have been employed in energy harvest, generation, saving, storage, transport, and conversion processes very effectively and efficiently. This book covers recent progresses on the preparation, characterization, and usage of semiconducting, dielectric, 1D and 2D nanomaterials and hybrid architectures for energy-related applications, technologies and devices, such as solar cells, piezoelectronics, fuel cells, supercapacitors, thermoelectronics, solar water-splitting for hydrogen production, batteries, and biofuels. Moreover, the topics in the book also highlights novel approaches in nanomaterials design and synthesis and evaluate materials characteristics with regard to sustainability issues. Contributions from active and leading experts regarding important aspects like the synthesis, assembly, and properties of nanomaterials for energy-related applications are compiled into a reference book. As evident from the diverse topics, the book is expected to be useful to researchers working in the interface of physics, chemistry, biology, materials science, and engineering. It is anticipated that the work would stimulate meaningful future developments in this rapidly emerging and critical frontier of nanomaterials for sustainable energy.

This book is not an attempt to exhaustively cover all the interesting and relevant topics on nanomaterials for sustainable energy; apparently it is simply not possible to do so within a single book in concise manner. Instead, the book focuses on the recent developments of fascinating themes for sustainable energy. The various chapters cover the following trendy topics: nanostructured materials for high-efficiency perovskite solar cells (Chap. 1), dielectric nanomaterials for silicon solar cells (Chap. 2), nanostructured cathode buffer layers for inverted polymer solar cells (Chap. 3), nanomaterials for stretchable energy storage and conversion devices (Chap. 4), piezoelectric nanomaterials for mechanical energy harvesting (Chap. 5), liquid crystals for self-organizing photovoltaics (Chap. 6), aligned carbon nanotubes for energy conversion and storage (Chap. 7), graphene-based electrochemical microsupercapacitors for miniaturized energy storage applications (Chap. 8), graphene in fuel cells (Chap. 9), mesoporous materials for fuel cells (Chap. 10), thermoelectric nanocomposites for thermal energy conversion (Chap. 11), nanomaterials for hydrogen generation from solar water-splitting (Chap. 12), nanomaterials for rechargeable lithium batteries (Chap. 13), self-organized chiral liquid crystalline nanostructures for energy-saving devices (Chap. 14), and nanomaterials for the production of biofuels (Chap. 15). In each chapter, the state of the art, along with future prospects in the respective fields, is discussed and highlighted by the experts.

Finally, I would like to express my gratitude to Claus Ascheron at Springer for inviting us to bring this exciting and essential field of research to a wide audience, and to all our distinguished contributors for their dedicated efforts. Also, I am indebted to my wife Changshu, my sons Daniel and Songqiao for their great support and encouragement.

Kent, OH, USA Quan Li

Contents

Contributors

Craig E. Banks Faculty of Science and Engineering, Manchester Metropolitan University, Manchester, UK

Jiming Bao Institute of Fundamental and Frontier Sciences, University of Electronic Science and Technology of China, Chengdu, China; Department of Electrical and Computer Engineering, University of Houston, Houston, TX, USA

Hari Krishna Bisoyi Liquid Crystal Institute and Chemical Physics Interdisciplinary Program, Kent State University, Kent, OH, USA

Guozhong Cao Department of Materials Science and Engineering, University of Washington, Seattle, WA, USA

Liming Dai Department of Macromolecular Science and Engineering, Center of Advanced Science and Engineering for Carbon (Case4Carbon), Case Western Reserve University, Cleveland, OH, USA

Quanbin Dai Department of Macromolecular Science and Engineering, Center of Advanced Science and Engineering for Carbon (Case4Carbon), Case Western Reserve University, Cleveland, OH, USA

Sudipta De Departamento de Quimica Organica, Universidad de Cordoba, Cordoba, Spain

Ingo Dirnstorfer NaMLab gGmbH, Dresden, Germany

Feng Du Department of Macromolecular Science and Engineering, Center of Advanced Science and Engineering for Carbon (Case4Carbon), Case Western Reserve University, Cleveland, OH, USA

Levi Elston Air Force Research Laboratory (AFRL), WPAFB, Fairborn, OH, USA

James Iocozzia School of Materials Science and Engineering, Georgia Institute of Technology, Atlanta, GA, USA

Kory Jenkins Department of Mechanical Engineering, University of Minnesota, Minneapolis, MN, USA

San Ping Jiang Department of Chemical Engineering, Fuels and Energy Technology Institute, Curtin University, Perth, WA6102, Australia

Yucheng Lan Department of Physics and Engineering Physics, Morgan State University, Baltimore, MD, USA

Quan Li Liquid Crystal Institute and Chemical Physics Interdisciplinary Program, Kent State University, Kent, OH, USA

Zhiqiang Liang Institute of Functional Nano and Soft Materials Laboratory (FUNSOM), Jiangsu Key Laboratory for Carbon-Based Functional Materials and Devices, Collaborative Innovation Center of Suzhou Nano Science and Technology, Soochow University, Suzhou, Jiangsu, China; Department of Materials Science and Engineering, University of Washington, Seattle, WA, USA

Zhiqun Lin School of Materials Science and Engineering, Georgia Institute of Technology, Atlanta, GA, USA

Xiangyang Liu Fujian Provincial Key Laboratory for Soft Functional Materials Research, Department of Physics, Research Institute for Biomimetics and Soft Matter, College of Physical Science and Technology, Xiamen University, Xiamen, China

Xueqin Liu School of Materials Science and Engineering, Georgia Institute of Technology, Atlanta, GA, USA

Wu Lu Department of Electrical and Computer Engineering, The Ohio State University, Columbus, OH, USA

Rafael Luque Departamento de Quimica Organica, Universidad de Cordoba, Cordoba, Spain

Sebastien Martinet Laboratoire d'Innovation pour les Technologies des Energies Nouvelles et les nanomateriaux (LITEN), French Alternative Energies and Atomic Energy Commission (CEA), Grenoble Cedex, France

Thomas Mikolajick NaMLab gGmbH, Dresden, Germany; Institute of Semiconductor and Microsystems, TU Dresden, Dresden, Germany; Center for Advancing Electronics Dresden (cfaed), TU Dresden, Dresden, Germany

Edward P. Randviir Faculty of Science and Engineering, Manchester Metropolitan University, Manchester, UK

Thomas Reitz Air Force Research Laboratory (AFRL), WPAFB, Fairborn, OH, USA

Zhifeng Ren Department of Physics and TcSUH, University of Houston, Houston, TX, USA

Zhiming Wang Institute of Fundamental and Frontier Sciences, University of Electronic Science and Technology of China, Chengdu, China; State Key Laboratory of Electronic Thin Films and Integrated Devices, University of Electronic Science and Technology of China, Chengdu, China

Bingqing Wei State Key Laboratory of Solidification Processing and School of Materials Science and Engineering, Center for Nano Energy Materials, Northwestern Polytechnical University, Xi'an, China; Department of Mechanical Engineering, University of Delaware, Newark, DE, USA

Keyu Xie State Key Laboratory of Solidification Processing and School of Materials Science and Engineering, Center for Nano Energy Materials, Northwestern Polytechnical University, Xi'an, China

Hao Yang Department of Electrical and Computer Engineering, The Ohio State University, Columbus, OH, USA

Rusen Yang Department of Mechanical Engineering, University of Minnesota, Minneapolis, MN, USA

Meidan Ye Fujian Provincial Key Laboratory for Soft Functional Materials Research, Department of Physics, Research Institute for Biomimetics and Soft Matter, College of Physical Science and Technology, Xiamen University, Xiamen, China

Jin Zhang Fuels and Energy Technology Institute and Department of Chemical Engineering, Curtin University, Perth, WA6102, Australia

Qiuhong Zhang Air Force Research Laboratory (AFRL), WPAFB, Fairborn, OH, USA

Zhenhuan Zhao Institute of Fundamental and Frontier Sciences, University of Electronic Science and Technology of China, Chengdu, China; Department of Electrical and Computer Engineering, University of Houston, Houston, TX, USA

Zhigang Zheng Liquid Crystal Institute and Chemical Physics Interdisciplinary Program, Kent State University, Kent, OH, USA

Chapter 1
Nanostructured Materials for High Efficiency Perovskite Solar Cells

Meidan Ye, Xueqin Liu, James Iocozzia, Xiangyang Liu and Zhiqun Lin

Abstract Capturing renewable energy has become an increasingly urgent task due to concerns over fossil fuel supplies and accessability in the future. Photovoltaic technology represents the most promising strategy to address this problem. However to date, no solar cell module has been able to satisfy the simultaneous requirements of high efficiency, low cost and long lifetime necessary for large scale production and application. Recent development in organic-inorganic halide perovskite solar cells (PSCs) has shown great potential for large-scale application owing to the superb power conversion efficiency (PCE), facile device fabrication and low material cost. Since the first report of methyl ammonium lead halide, $CH_3NH_3PbX_3$ (X = Br, I), sensitized liquid solar cells in 2009, the PCEs of PSCs have jumped from 3.8 to 20.2 %. Research efforts on new preparation processes and materials design for PSCs still remain highly active. This chapter will concentrate on summarizing the recently reported nanostructured materials (i.e. TiO_2, Al_2O_3, ZnO, carbon materials and others) applied in PSCs. This chapter strives to present a comprehensive overview and provide a deep understanding of nanostructured materials in PSC devices.

1.1 Introduction

Increasing energy demands, diminishing amounts of easily-accessible fossil fuel sources, and concerns of global warming have motivated a great deal of effort into the exploration and development of clean, inexpensive and renewable energy

M. Ye · X. Liu
Research Institute for Biomimetics and Soft Matter,
Fujian Provincial Key Laboratory for Soft Functional Materials Research,
Department of Physics, College of Physical Science and Technology,
Xiamen University, Xiamen 361005, China

X. Liu · J. Iocozzia · Z. Lin (✉)
School of Materials Science and Engineering, Georgia Institute of Technology,
Atlanta, GA 30332, USA
e-mail: zhiqun.lin@mse.gatech.edu

Q. Li (ed.), *Nanomaterials for Sustainable Energy*,
NanoScience and Technology, DOI 10.1007/978-3-319-32023-6_1

sources [1]. It is important to note that research into alternative and renewable energy resources is motivated not only by concerns of planetary damage and economics but also a desire to conserve more and waste less as a point of basic societal betterment. Solar energy is commonly acknowledged as a promising renewable energy source to support the future of humanity [1, 2]. To date, besides the conventional silicon solar cells (single crystal and mesoporous) and thin film solar cells, several new photovoltaic technologies for converting solar photons into electricity have been developed. These include organic photovoltaic cells (OPVs), dye-sensitized solar cells (DSSCs) and quantum dot-sensitized solar cells (QDSCs). Such devices have inspired much research interest due to their potential to serve as low-cost alternatives to the traditional solar cell modules [3–7].

Recently, a new type of solar cell based on mixed organic-inorganic halide perovskites ABX_3 (Fig. 1.1a, A = CH_3NH_3 (MA) or NH_2CHNH_2 (FA), B = Pb or Sn, X = Cl, Br or I) has quickly attracted much attention and gradually become one of the most important all-solid-state solar cells materials in what are called perovskite solar cells (PSCs) [8–15]. As the core of PSCs, hybrid organic-inorganic halide perovskites ABX_3 possess some fascinating properties such as an appropriate band gap, a high absorption coefficient, a long charge diffusion length and solution processability. Taken together, this gives PSCs some of the highest efficiencies with values as large as 20 % [16]. Perovskite, named after the Russian mineralogist Lew A. Perovski, is defined as a class of compounds by the general formula ABX_3 in which X is an anion and A and B are cations of different sizes (where A > B) [17]. Ideally, a perovskite structure consists of corner sharing BX_6 octahedra with the A ion placed in the cuboctahedral interstices which belong to the cubic $Pm\bar{3}m$ crystal structure, as shown in Fig. 1.1a [18]. In the hybrid organometal halide perovskites, A is an organic cation (i.e., MA^+ or FA^+), B is a metal cation (i.e., Sn^{2+} or Pb^{2+}), and X is a halide anion (i.e., Cl^-, Br^- or I^-) [19, 20]. In particular, hybrid organometal halide perovskites $ABX_{3-x}Y_x$ with mixed halides, for example, $MAPbI_{3-x}Cl_x$ and $MAPbI_{3-x}Br_x$ (Fig. 1.1b, c) have also attracted special attention due to their tunable optical properties which leads to improved performance in PSCs [21–23].

As indicated in Fig. 1.2, early in 2009, hybrid organometal halide perovskites were initially used in conventional liquid electrolyte-based DSSCs as light absorbers, yielding PCEs of 3.13 % ($MAPbBr_3$) and 3.81 % ($MAPbI_3$), respectively [24]. In 2011, Park and coworkers achieved a PCE of 6.54 % in a perovskite quantum dot-sensitized 3.6 μm-thick TiO_2 film using the iodide/iodine redox couple [25]. However, these solar cells are usually unstable and the performance degraded rapidly due to the dissolution or decomposition of perovskites in the liquid electrolyte.

The real breakthrough for PSCs came in 2012, as shown in Fig. 1.2 and Table 1.1, when Grätzel's group employed a solid hole-transporting material (HTM) 2,2',7,7'-tetrakis(N,N-di-p-methoxyphenylamine)-9,9'-spirobifluorene (spiro-MeOTAD) instead of the conventional liquid redox electrolyte. In their PSCs, PCEs of 9.7 % were achieved under AM 1.5 G illumination along with excellent long term stability for over 500 h. Charge separation proceeded via electron transfer to a submicron-thick mesoporous TiO_2 film and hole injection from the excited $MAPbI_3$ into

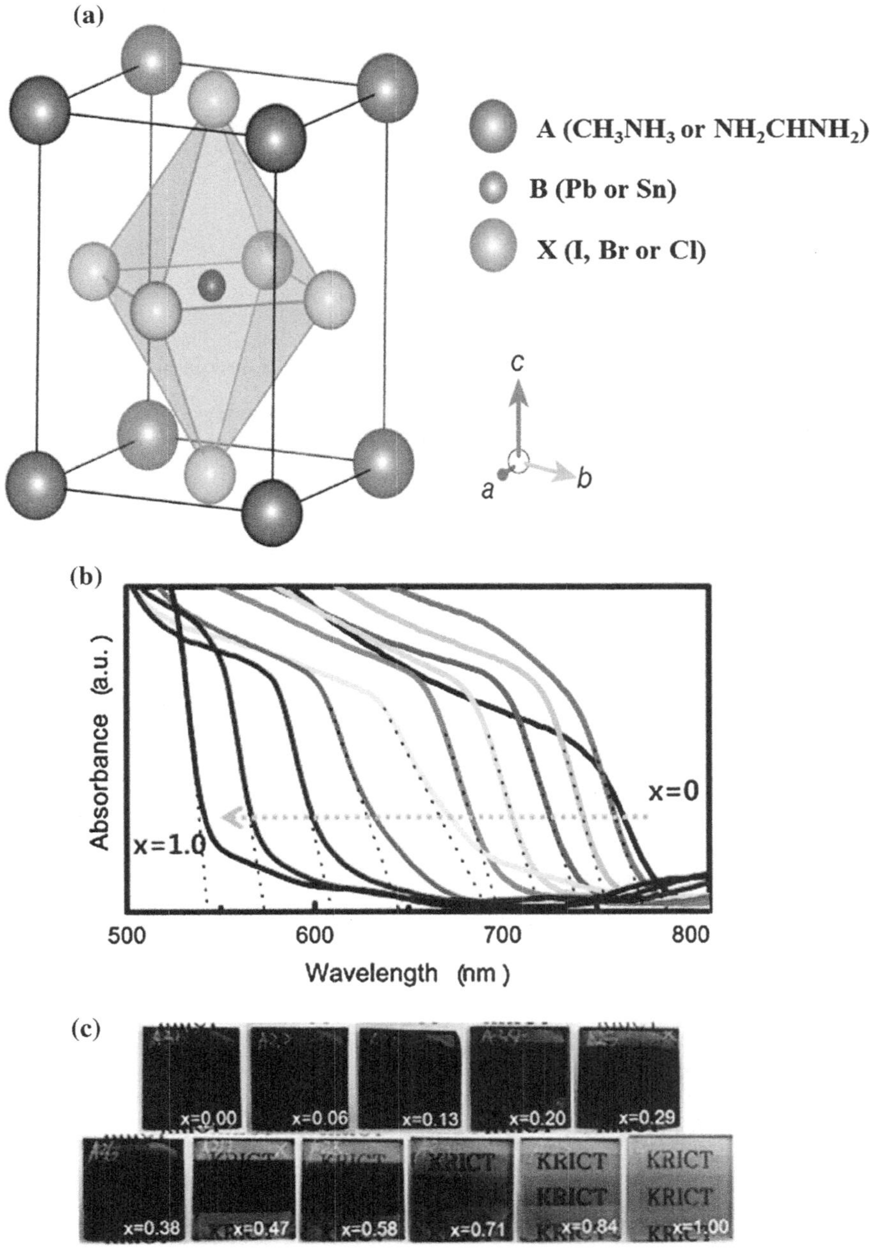

Fig. 1.1 **a** Crystal structure of ABX_3, **b** UV-vis absorption spectra and **c** photographs of $MAPbI_{3-x}Br_x$ (reprinted with permission from [18] **a** M. Liu et al., Nature 2013, 501, 395–398. [21] **b–c** J.H. Noh et al., Nano Lett. 2013, 13, 1764–1769. Copyright © Nature Publishing Group and American Chemical Society)

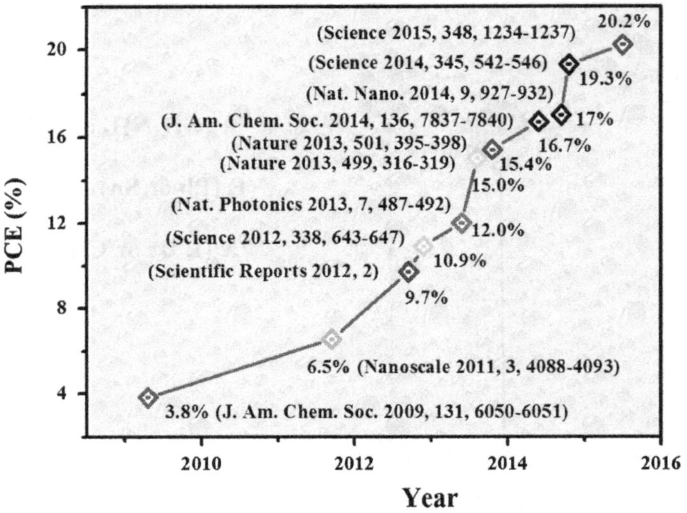

Fig. 1.2 Progress in perovskite solar cells demonstrating the amazingly sharp rise in PCE over the past 6 years

Table 1.1 Photovoltaic performance of PSCs depending on materials and cell configurations

Cell configuration[e]	J_{sc} (mA/cm^2)[a]	V_{oc} (V)[b]	FF[c]	PCE (%)[d]	References
FTO/bl-TiO_2/mp-TiO_2/$MAPbI_3$/Spiro-MeOTAD/Au	17.6	0.888	0.62	9.7	[26]
FTO/bl-TiO_2/mp-Al_2O_3/$MAPbI_2Cl$/Spiro-MeOTAD/Ag	17.8	0.98	0.63	10.9	[27]
FTO/bl-TiO_2/mp-TiO_2/$MAPbI_3$/PTAA/Au	16.5	0.997	0.727	12.0	[28]
FTO/bl-TiO_2/mp-Al_2O_3/$MAPbI_{3-x}Br_x$/PTAA/Ag	18.0	1.02	0.67	12.3	[29]
FTO/bl-TiO_2/mp-TiO_2/$MAPbI_3$/spiro-MeOTAD/Au	20.0	0.993	0.73	15.0	[30]
FTO/bl-TiO_2/$MAPbI_{3-x}Cl_x$/spiro-MeOTAD/Ag	21.5	1.07	0.67	15.4	[18]
FTO/bl-TiO_2/mp-TiO_2/$MAPbI_3$/PTAA/Au	21.3	1.04	0.73	16.2	[31]
FTO/bl-TiO_2/mp-TiO_2/$MAPbI_3$/pp-spiro-MeOTAD/Au	21.2	1.02	0.776	16.7	[32]
FTO/bl-TiO_2/mp-TiO_2/$MAPbI_3$ cuboid/spiro-MeOTAD/Au	21.64	1.056	0.741	17.01	[33]
ITO-PEIT/bl-Y doped-TiO_2/M $APbI_3$ $_{-x}Cl_x$/spiro-MeOTAD/Au	22.75	1.13	0.75	19.3	[34]
FTO/bl-TiO_2/mp-TiO_2/$FAPbI_3$/PTAA/Au	24.7	1.06	0.775	20.2	[35]

[a]Short-circuit voltage
[b]Open-circuit voltage
[c]Fill factor
[d]Power conversion efficiency
[e]*bl* stands for blocking layer and *mp* stands for mesoporous

the spiro-MeOTAD [26]. Almost simultaneously in 2012 (Fig. 1.2 and Table 1.1), Snaith et al. replaced the *n*-type mesoporous TiO_2 with insulating mesoporous Al_2O_3 in PSCs. The FTO/blocking layer (bl)-TiO_2/mesoporous (mp)-Al_2O_3/$MAPbI_2Cl$/spiro-MeOTAD/Ag-based device, termed the meso-superstructured solar cell (MSSC), boosted the reported PCEs to 10.9 %. Moreover, it was also revealed that perovskites can be used not only as sensitizers but also as an electron and hole transport layer between cell terminals [27].

In early 2013 (Fig. 1.2 and Table 1.1), a further jump to a PCE of 12.0 % was obtained by Seok, Grätzel et al. by using a bicontinuous three-dimensional nanocomposite of mp-TiO_2 with $MAPbI_3$ perovskites as a light harvester and poly-triarylamine (PTAA) as a hole transport material (HTM) [28]. Subsequently, Seok et al. further improved the PCE to 12.3 % using similar structures but with mixed-halide $MAPbI_{3-x}Br_x$ perovskites [29]. In late 2013 (Fig. 1.2 and Table 1.1), further progress was reported by two research groups with PCEs above 15 % [18, 30]. First, Grätzel et al. developed a sequential deposition process to fabricate $MAPbI_3$ films with improved morphology on the TiO_2 scaffold layers and measured a remarkable PCE of 15 % in a FTO/bl-TiO_2/mp-TiO_2/$MAPbI_3$/spiro-MeOTAD/Au-based PSC [30]. Subsequently, Snaith et al. deposited a high quality $MAPbI_{3-x}Cl_x$ film via two-source thermal evaporation in a new planar heterojunction (PHJ) perovskite solar cell (FTO/bl-TiO_2/$MAPbI_{3-x}Cl_x$/spiro-MeOTAD/Ag) without the mp-TiO_2 scaffold and achieved a PCE of 15.4 % [18].

In early 2014 (Fig. 1.2 and Table 1.1), Seok et al. further improved the PCEs of FTO/bl-TiO_2/mp-TiO_2/$MAPbI_3$/PTAA/Au-based PSCs to 16.2 % [31]. At the same time, Lee and Seok achieved a 16.7 % PCE in PSCs with a similar solar cell structure by instead using spiro-MeOTAD derivatives as the HTM [32]. Later, Park and Grätzel applied a two-step spin coating procedure to control the size of the $MAPbI_3$ cuboids during their growth and achieved a PCE of 17.01 % [33]. This was increased to a confirmed efficiency of 17.9 % in early 2014 by Seok [8]. Then, Yang et al. further raised the PCE to 19.3 % in a planar geometry by using an ITO-PEIT/bl-Yttruim doped-TiO_2/$MAPbI_{3-x}Cl_x$/spiro-MeOTAD/Au device structure [34]. In 2015, Seok et al. fabricated $FAPbI_3$-based PSCs with the highest PCE to date (up to 20.2 %) using a FTO-glass/bl-TiO_2/mp-TiO_2/$FAPbI_3$/PTAA/Au cell structure [35].

During the past few years, extensive investigation into the morphology and crystal optimization of perovskites [36–44], the material selection and deposition modulation of blocking and scaffold layers [33, 45–48], HTM modification and interface/band-gap engineering [28, 49–54] has led to an accelerated boost in the reported device PCEs from less than 4 % to over 20 %. This trend has opened up a new way to develop highly efficient solid-state solar cells at low cost and high stability.

The cell configuration of PSCs can be generally classified into two types: porous and planar architectures [55]. As shown in Fig. 1.3, three cell models can be found in the porous configuration [56, 57]: (1) sensitization: the perovskite with nanodot morphology is produced in one step by spin-coating onto the mp-TiO_2 surface where the amount of deposited perovskite is relatively low and the HTM is

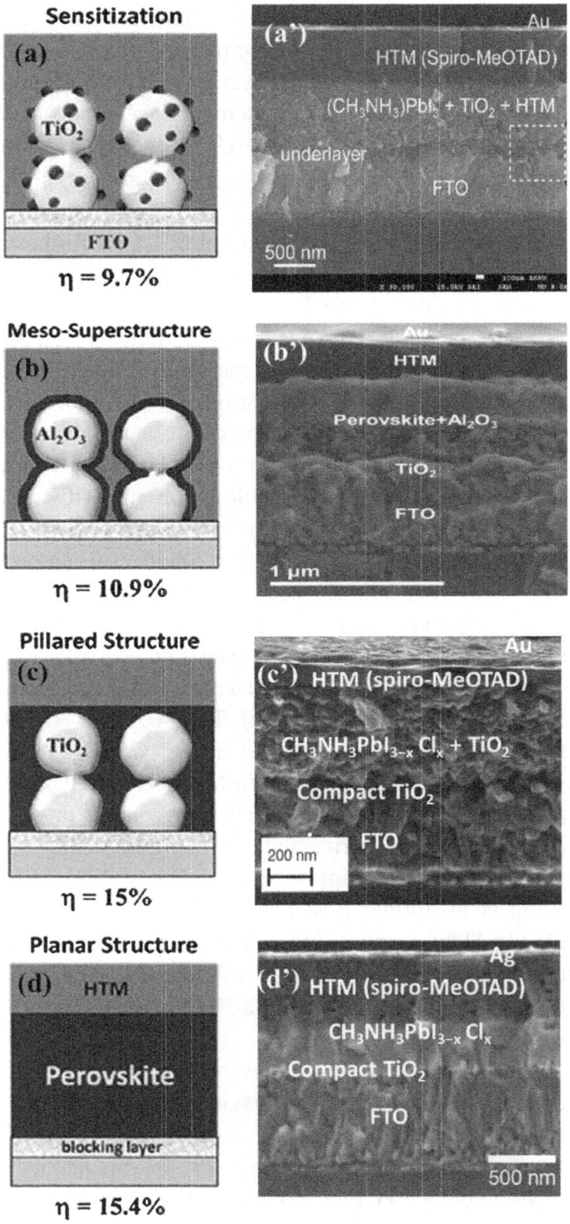

Fig. 1.3 The evolution of device structures for perovskite solar cells: **a–d** schematic diagrams, and **a′–d′** cross-sectional SEM images of different cell modules (reprinted with permission from [56] **a–d** H.-S. Kim et al., J. Phys. Chem. C 2014, 118, 5615–5625. [26] **a′** H.-S. Kim et al., Scientific Reports 2012, 2. [150] **b′** S. Guarnera et al., J. Phys. Chem. Lett. 2015, 6, 432–437. [30] **c′** J. Burschka et al., Nature 2013, 499, 316–320. [18] **d′** M. Liu et al., Nature 2013, 501, 395–398. Copyright © Nature Publishing Group and American Chemical Society)

infiltrated into the mp-TiO_2 pores [26]; (2) meso-superstructure: a thin perovskite layer is loaded on the mp-Al_2O_3 surface and the HTM is infiltrated into the mp-Al_2O_3 pores [27]; (3) pillared structure: the mp-TiO_2 pores are filled with perovskite instead of HTM and a thin capping layer (over layer) of perovskite is deposited to contact the HTM [30].

In the planar configuration, a mesoporous oxide film is not necessary because perovskite is capable of transporting electrons and holes. However, a thin blocking layer is still required to prevent direct contact between the conductive substrate and the HTM [58]. It is worth noting that the mesoporous scaffold layers can not only be composed of semiconductor materials (e.g. TiO_2, ZnO, NiO, SnO_2, WO_3, metal sulfides) [59–64], but also can contain metal oxide insulators (e.g. Al_2O_3, ZrO_2, SiO_2) owing to the ability of perovskites to transport electrons as well as holes [65–67]. In the following section, a comprehensive summary of the performance of such metal oxide semiconductors and insulators, as well as other materials (e.g., carbon materials) used in PSCs, is presented.

1.2 Nanostructured Scaffold Layers in PSCs

For nanostructured scaffolds, significant progress in the deposition technology, material type, film morphology, and layer thickness has been made as a consequence of the rapid development of PSCs in order to accommodate a variety of device architectures [47, 68–72]. In the following section, some representative examples will be presented and discussed.

1.2.1 Nanostructured TiO_2 Layers

It has been observed that the morphology, thickness and crystallinity of the TiO_2 electron-blocking layer and mesostructured layer play a crucial role in the efficiency of PSCs [73–75]. Next, a general overview and some detailed description are presented on the recent development of nanostructured TiO_2 as applied in PSCs.

1.2.1.1 Compact TiO_2 Blocking Layers

For typical porous and planar PSCs, the fabrication processes starts with the deposition of a compact TiO_2 layer onto a pre-cleaned conductive substrate which acts as a blocking layer to prevent direct contact between the conductive substrate and the infiltrated HTM layer [18, 26]. At present, four strategies have been developed to deposited such a compact TiO_2 layer [76], including (1) spin-coating a colloidal dispersion of TiO_2 nanoparticles (NPs) followed by a thermal treatment (titanium source: $TiCl_4$ [77], titanium tetraisopropoxide (TTIP) [78], and tetra-*n*-butyl-titanate [79]); (2) spin-coating the titanium precursor solutions followed by a thermal treatment (titanium source: TTIP [80], tetra-*n*-butyl-titanate [81, 82], and titanium diisopropoxidebis(acetylacetonate) [33]); (3) spray pyrolysis deposition (titanium source: TTIP [83], titanium diisopropoxidebis(acetylacetonate) [33, 84]); (4) atomic layer deposition (ALD) (titanium source: $TiCl_4$ [85], cyclopentadienyl alkylamido titanium [86], tetrakis(dimethylamino) titanium [87]). Other methods such as thermal oxidation of a sputtered Ti film [88] and magnetron sputtering TiO_2 NPs on FTO substrates [89, 90] have been reported as well. In particular, the ALD method enables the deposition of higher quality films, as defined by lower pinhole density, in the range of temperatures compatible with conductive plastic substrates [86].

For the fabrication of flexible PSCs, several solutions have been exploited to avoid the high temperature heating treatment: (1) low temperature methods (e.g., spin coating followed by low temperature annealing [77], ultrasonic spray coating [91], and dip-coating [92]) to prepare the TiO_2 compact layers without high temperature heating treatment; (2) TiO_2 layer deposition through RF magnetic sputtering as an efficient barrier layer to replace the TiO_2 compact layer and create a PET-ITO/Ti/$MAPbI_3$/spiro-MeOTAD/Ag based flexible PSC with a PCE of 8.39 % (Fig. 1.4a–c) [93]; and (3) Ti metal as the conductive substrate for flexible PSCs with a PCE of 6.15 % by using a silver thin film as a semi-transparent top electrode on the Ti substrate (Fig. 1.4d–f) [94].

Furthermore, in order to improve the electrical properties at the interface between the perovskite layer and the compact TiO_2 layer, surface modification techniques such as $TiCl_4$- and UV(O_3)-treatment were employed to increase the PCEs of planar PSCs from 13.7 to 16.3 and 16.9 %, respectively [95]. Also, thin dense Mg-doped TiO_2 was used as the blocking layer instead of pure TiO_2 because of its better properties including improved optical transmission properties, up shifted conduction band minimum (CBM) and downshifted valence band maximum (VBM), better hole-blocking effect, and higher electron lifetime. These properties led to enhanced performance (PCE: pure TiO_2-9.16 % vs. Mg-doped TiO_2-12.28 %) in porous PSCs [96].

1.2.1.2 Mesoporous TiO_2 Scaffold Layers

As the most common device configuration for PSCs, porous PSCs based on perovskite sensitized mesoporous TiO_2 scaffolds have achieved PCEs from 9.7 % to over 20 % [26, 35]. Mesoporous TiO_2 films not only transport photo-generated electrons from perovskite sensitizers but also assist in the perovskite crystal transformation when the perovskite layer is deposited by the sequential deposition process [76]. In general, a mesoporous TiO_2 layer is deposited by spin-coating, screen-printing or doctor blading the TiO_2 nanoparticle (NP) paste on the top of the compact TiO_2 blocking layer followed by a sintering process [97–101]. Importantly, efficient infiltration of the perovskite and the subsequent HTM into the pores of the mesoporous scaffold is necessary for high performance PSCs. For mesoporous TiO_2 scaffolds, several aspects have been investigated to study their effect on the porous PSC device performance.

First, Kim et al. showed that the increase of the TiO_2 film thickness reduced the open circuit voltage (V_{OC}) and fill factor (FF) of cell devices. This was attributed to an increase in dark current and electron transport resistance according to impedance spectroscopic studies (Fig. 1.5a–b). Meanwhile, the short circuit photocurrent

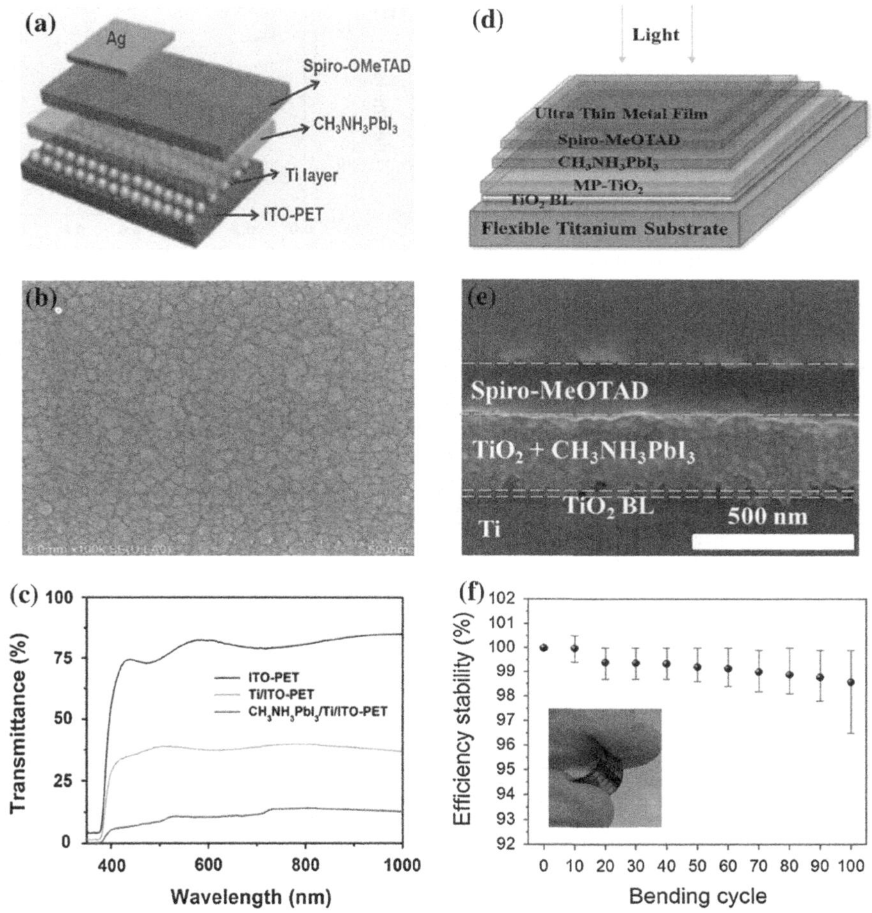

Fig. 1.4 **a** Schematic representation of the fabricated PET-ITO/Ti/$MAPbI_3$/spiro-MeOTAD/Ag flexible PSC, **b** SEM image of a Ti thin film with a thickness of ~100 nm, **c** UV-vis spectra of different samples, **d** device structure, **e** Cross-sectional image, and **f** efficiency stability as a function of bending cycle in a PSC using a Ti substrate with a semi-transparent Ag layer. Inset image is an optical image taken during a bending test (reprinted with permission from [93] **a–c** S. Ameen et al., Dalton Trans. 2015, 44, 6439–6448. [94] **d–f** M. Lee et al., J. Mater. Chem. A 2015, 3, 4129–4133. Copyright © The Royal Society of Chemistry)

density (J_{SC}) was not strongly influenced by a change in thickness of the TiO_2 layer from 0.6 to 1.4 μm; maintaining a high level of 16–17 mA cm^{-2} due to the large optical absorption [26]. Since the electron diffusion length in $MAPbI_3$ is about 100 nm [102], studies have revealed that when the thickness of mesoporous TiO_2 exceeds 600 nm, the electron/hole transport encounters larger resistances. This results in a significant loss of V_{OC} and therefore FF [103]. As a result, the optimized thickness for mesoporous TiO_2 should be controlled in the range of 400–600 nm [30, 104–107].

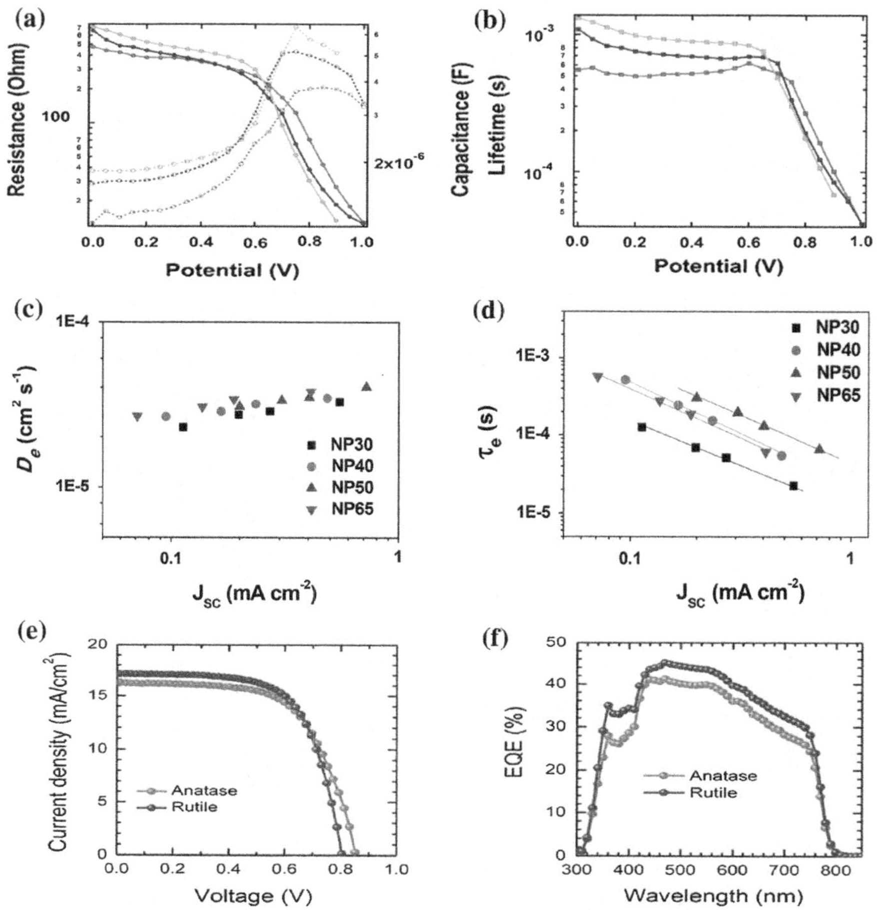

Fig. 1.5 **a** Recombination resistance (*solid lines*) and accompanying capacitance (*dashed lines*), and **b** electron lifetime under illumination of TiO_2, thickness: *red* 0.6 μm, *blue* 1.15 μm and *green* 1.4 μm. Plots of **c** the electron diffusion coefficient (D_e) versus J_{SC} and **d** the electron lifetime (τ_e) versus J_{SC} for the PSCs employing TiO_2 NPs with different sizes. **e** Current density-voltage (J-V) curves and **f** EQE spectra for the PSCs based on anatase (*red*) and rutile (*blue*) TiO_2 films (reprinted with permission from [26] **a, b** H.-S. Kim et al., Scientific Reports 2012, 2. [70] **c, d** S.D. Sung et al., Nanoscale 2015, 7, 8898–8906. [75] **e, f** J.-W. Lee et al., J. Mater. Chem. A 2014, 2, 9251–9259. Copyright © Nature Publishing Group and the Royal Society of Chemistry) (Color figure online)

Second, the size of TiO_2 NPs also impacts the device performance of PSCs. Han et al. used anatase TiO_2 scaffold layers with different particle sizes (15–30 nm) to investigate their performance in hole-conductor-free fully printable mesoscopic PSCs based on a carbon counter electrode and $(5\text{-}AVA)_x(MA)_{1-x}PbI_3$ perovskite, and found that the size of TiO_2 particles not only affected the infiltration of the precursor and the contact between the perovskite crystal and TiO_2, but also significantly influenced the charge transfer kinetics at the perovskite/TiO_2 interface. The device based on TiO_2 NPs with a diameter of 25 nm exhibited the best PCE of 13.41 % [108]. Meanwhile, Sung et al. prepared larger anatase TiO_2 NPs with diameter sizes from 30 to 65 nm and found that the spherical TiO_2 NP of 50 nm size exhibited the longest electron lifetime although the electron injection from perovskite to TiO_2 was less efficient than the devices with smaller TiO_2 NPs (Fig. 1.5c, d) and offered the highest PCE of 17.19 % (J_{SC}: 21.58 mA cm^{-2}, V_{OC}: 1049 mV, FF: 0.759) when it was employed in porous PSCs [70].

Finally, Yella, Grätzel et al. found that the nanocrystalline rutile TiO_2 prepared via hydrolysis of $TiCl_4$ at 70 °C achieved a much better performance (PCE: 13.7 %) than a planar TiO_2 (anatase) film (PCE: 3.7 %) prepared by high temperature spin coating of $TiCl_4$. This is likely attributed to the formation of an intimate junction of large interfacial area between the nanocrystalline rutile TiO_2 and the $MAPbI_3$ layer. This leads to effective extraction of the photo-generated electrons [62]. Similarly, Park et al. reported that the device performance of PSCs can be influenced by the crystal phase of TiO_2. In terms of infiltration of the $MAPbI_3$ perovskite and the smooth surface of the perovskite capping layer, the rutile TiO_2 film was found to be better than the anatase TiO_2 film. Rutile TiO_2-based devices (PCE: 14.46 %) exhibited better photovoltaic performance than the anatase TiO_2-based devices (PCE:13.99 %) along with lower standard deviation (Fig. 1.5e, f) [75].

1.2.1.3 One-Dimensional (1D) TiO_2 Scaffold Layers

Mesoscopic particle films as scaffolds in PSCs frequently suffer from complications such as inefficient electron transport in the nanocrystalline films and incomplete filling by organic HTMs in the mesopores of the films. Herein, one-dimensional (1D) nanostructures are considered as one promising alternative because they can provide a direct path for photo-generated electron transport. Moreover, the straight channels encountered in 1D architectures could benefit from the filling of the perovskites and the HTMs thus enhancing the hole transport efficiency [104, 109–111]. At present, several kinds of 1D nanostructures have been applied as scaffolds in PSCs including nanowires (NWs)/nanorods (NRs) [104, 109, 112–116], nanotubes (NTs) [110, 117], nanofibers (NFs) [111], and nanocones (NCs) [118].

In early 2013, Yang et al. organized PSC devices based on 1.5 μm-thick rutile TiO_2 NW arrays grown via a hydrothermal method, and obtained PCEs of 4.29 % and 4.87 % for $MAPbI_3$ and $MAPbI_2Br$ (Fig. 1.6a, b), respectively [113]. Subsequently, Park et al. raised the PCE to 9.4 % based on a sub-micrometer (~0.6 μm) rutile TiO_2 NR sensitized with $MAPbI_3$ perovskite nanodots (Fig. 1.6c, d) [112]. In 2014, Jiang et al. reported a PCE of 11.7 % for $MAPbI_3$ PSCs using optimized 900 nm-thick rutile TiO_2 NWs with an open face for effective material filling (Fig. 1.6e, f) [104]. Cai et al. synthesized rutile TiO_2 NR arrays on FTO substrates using an acid-free hydrothermal method in an aqueous Na_2EDTA solution with glycerol modified TTIP as the precursor. A PCE of 11.1 % was

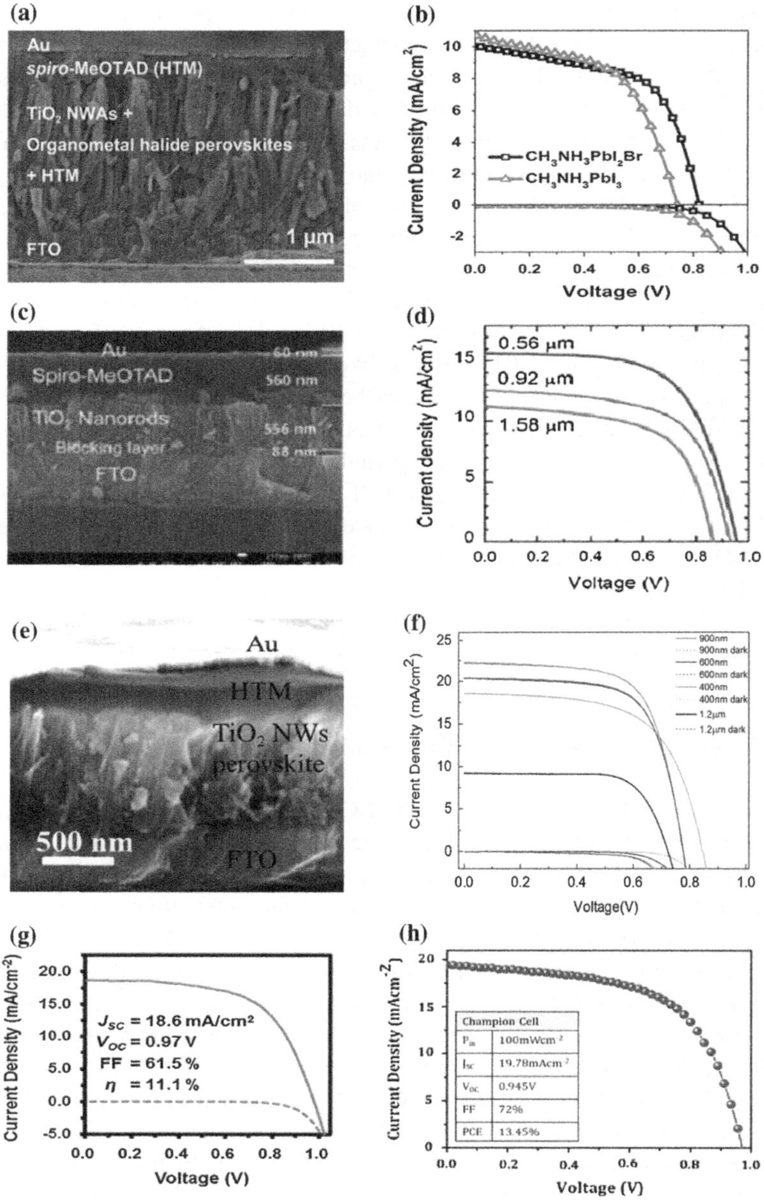

Fig. 1.6 **a**, **c**, **e** Cross-sectional SEM images and **b**, **d**, **f–h** J-V characteristics of the TiO_2 NW/NR-based PSCs (reprinted with permission from [113] **a, b** J. Qiu et al., Nanoscale 2013, 5, 3245–3248. [112] **c**, **d** H.-S. Kim et al., Nano Lett. 2013, 13, 2412–2417. [104] **e, f** Q. Jiang et al., Chem. Commun. 2014, 50, 14720–14723. [114] **g** B. Cai et al., J. Mater. Chem. C 2015, 3, 729–733. [116] **h** S.S. Mali et al., Chem. Mater. 2015, 27, 1541–1551. Copyright © The Royal Society of Chemistry and American Chemical Society)

obtained for the 0.7 μm-thick NR-based $MAPbI_3$ PSCs (Fig. 1.6g) [114]. In 2015, Mali et al. deposited an ultrathin TiO_2 passivation layer on the hydrothermally-grown TiO_2 NRs via ALD and achieved a significantly enhanced PCE of 13.45 % (pure NRs: 5.03 %, Fig. 1.6h) [116].

For 1D TiO_2 NTs, several groups have reported their application in PSCs [110, 117, 119, 120]. In 2014, Gao et al. first applied them as scaffold materials in $MAPbI_3$ PSCs. They fabricated freestanding TiO_2 NT array films with varied thicknesses by a two-step anodization process. The NTs were subsequently detached them from the Ti substrate by in situ field-assisted chemical dissolution and then transferred to the bl-TiO_2 layer pre-covered FTO substrates followed by the deposition of perovskite, HTM and novel metal films. The NT-based devices demonstrated improved light absorption and a reduced charge recombination rate as compared with TiO_2 NP-based systems using the same thickness of 4.8 μm (Fig. 1.7a, c). A peak PCE of 6.52 % was achieved with an optimized thickness of 2.3 μm [110]. Subsequently, Wang et al. prepared flexible PSCs based on Ti foil/TiO_2 NTs in conjunction with transparent carbon nanotubes (CNTs). These materials are deposited on top of perovskite to function as a hole collector as well as transparent electrode for light illumination. Such devices exhibited a PCE of 8.31 % and showed little deterioration after 100 mechanical bending cycles. This durability suggests potential applications in building wearable photovoltaic devices [117].

In addition, Dharani et al. employed electrospun TiO_2 NFs as scaffolds for PSCs yielding a PCE of 9.8 % (11.79 % at 0.1 sun). The open and porous nature of the electrospun NF network, which varied with the film thicknesses and fiber diameters, determined the device performance (Fig. 1.7d–f) [111]. Moreover, nanocone (NC) nanostructures were constructed under a nearly neutral hydrothermal process by using Na_2ETAD. A PCE of 11.9 % was achieved for such TiO_2 NC-based PSC devices. It was further found that electron transfer from $MAPbI_3$ to TiO_2 NCs was significantly faster than to TiO_2 NRs. This is an important factor to suppress charge recombination and improve device performance [118].

1.2.1.4 Two-Dimensional (2D) TiO2 Scaffold Layers

2D TiO_2 nanostructures have been widely used in DSSCs, and many studies have proved that the (001) facets of anatase TiO_2 are more reactive than the (101) facets, and it was proposed that the higher ionic charge of the exposed (001) facets strengthens the attachment of the light absorbers to the TiO_2 surface, thereby facilitating the electron injection into the conduction band of the oxide [121–123]. In 2014, Han et al. employed a double layer of TiO_2 NSs containing high levels of exposed (001) facets and ZrO_2 as a scaffold in a hole-conductor-free fully printable porous PSC (Fig. 1.8a). The high reactivity of (001) facets in TiO_2 NSs improved the interfacial properties between the perovskite and the electron collector (Fig. 1.8b, c). As a result, the PCE of this HTM-free PSC reached 10.64 % [124]. Subsequently, Dar et al. developed a TiO_2 scaffold composed of (001)-oriented nanoplatelets of anatase TiO_2 for PSCs (Fig. 1.8d, e). By adjusting

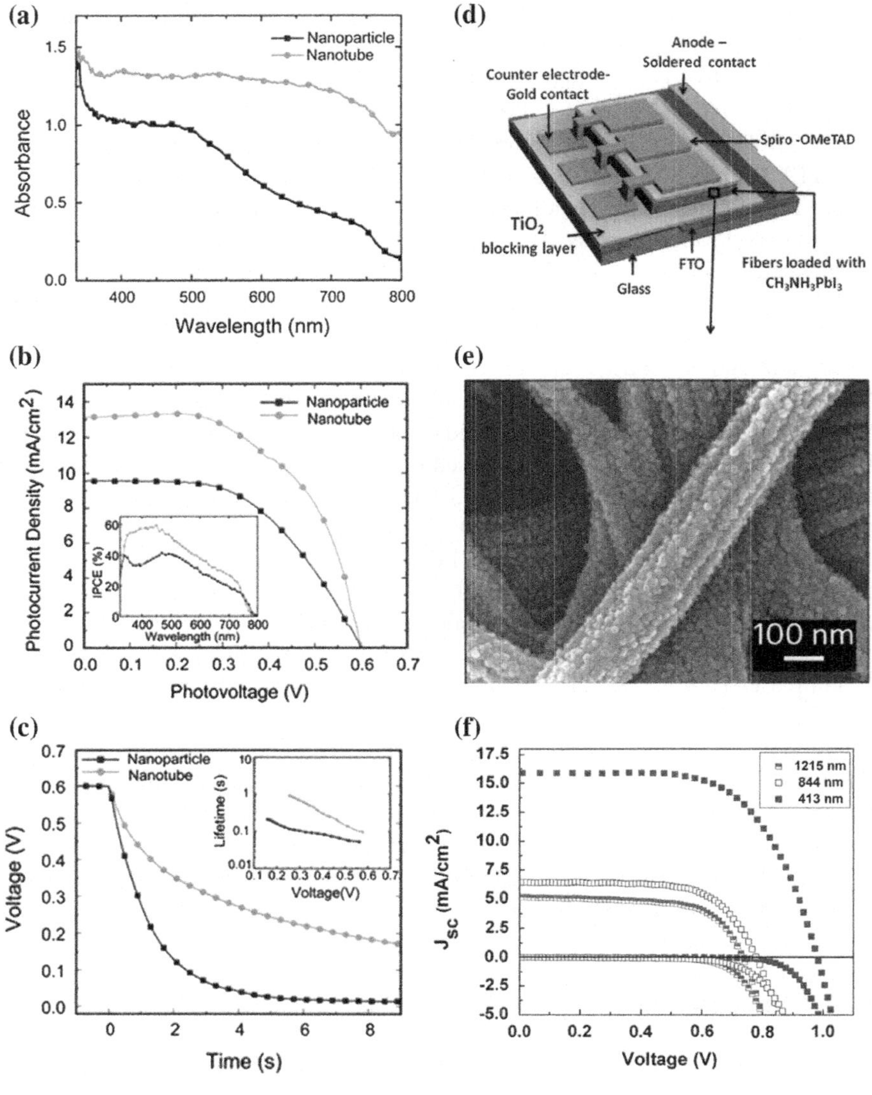

Fig. 1.7 **a** UV-Vis absorption of $MAPbI_3$ sensitized TiO_2 NP and NT electrodes, **b** J-V curves, IPCE (*Inset* of image b), and **c** open-circuit voltage decay and electron lifetimes (*Inset* of image c) of the PSCs with TiO_2 NP and NT electrodes. **d** Schematic illustration of the TiO_2 NF-based PSCs, **e** SEM of the $TiCl_4$ treated rough TiO_2 nanofibers, and **f** J-V curves of the PSCs based on TiO_2 NFs with different thicknesses (reprinted with permission from [110] **a–c** X. Gao et al., Chem. Commun. 2014, 50, 6368–6371. [111] **d**, **e** S. Dharani et al., Nanoscale 2014, 6, 1675–1679. Copyright © The Royal Society of Chemistry)

the thickness of the TiO_2 nanoplatelet and $MAPbI_3$/HTM layers, a PCE of 12.30 % was realized in the device using 100 nm-thick TiO_2 nanoplatelets and 300 nm-thick $MAPbI_3$/HTM layers (Fig. 1.8f) [125].

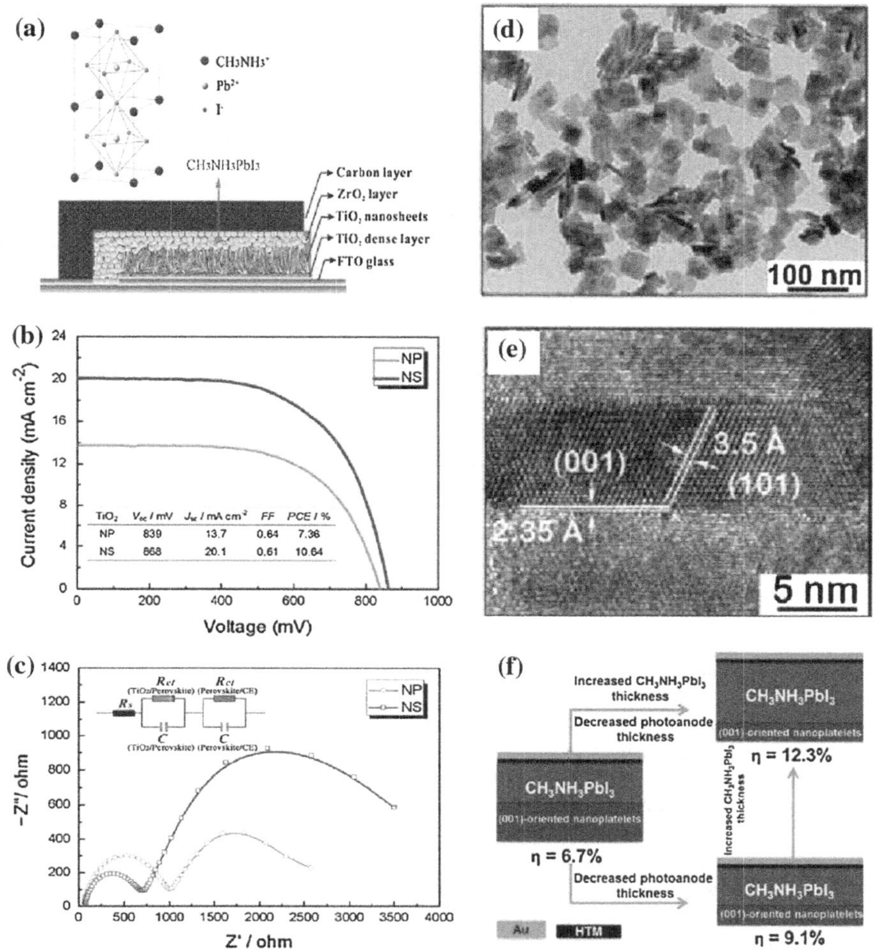

Fig. 1.8 **a** Schematic diagram, **b** J-V curves, and **c** Nyquist plots of a hole-conductor-free mesoscopic TiO_2/$MAPbI_3$ heterojunction solar cell based on P25 NPs/anatase TiO_2 NSs and carbon counter electrodes. **d** TEM and **e** HRTEM images for (001) anatase TiO_2 nanoplatelets. **f** Sketch map of the thickness optimization of the TiO_2 nanoplatelet-based PSCs (reprinted with permission from [124] **a–c** Y. Rong et al., J. Phys. Chem. Lett. 2014, 5, 2160–2164. [125] **d–f** M.I. Dar, et al., Chem. Mater. 2014, 26, 4675–4678. Copyright © American Chemical Society)

1.2.1.5 Three-Dimensional (3D) TiO_2 Scaffold Layers

Besides the above-mentioned nanostructures, some complicated 3D TiO_2 materials have been used in PSCs owing to their special properties such as high-speed conduction pathways and large surface areas for sufficient chemical reaction and/or dye loading [126–129]. For example, Kim, Park et al. made TiO_2 NRs into well-aligned helical arrays by the oblique-angle electron beam evaporation method

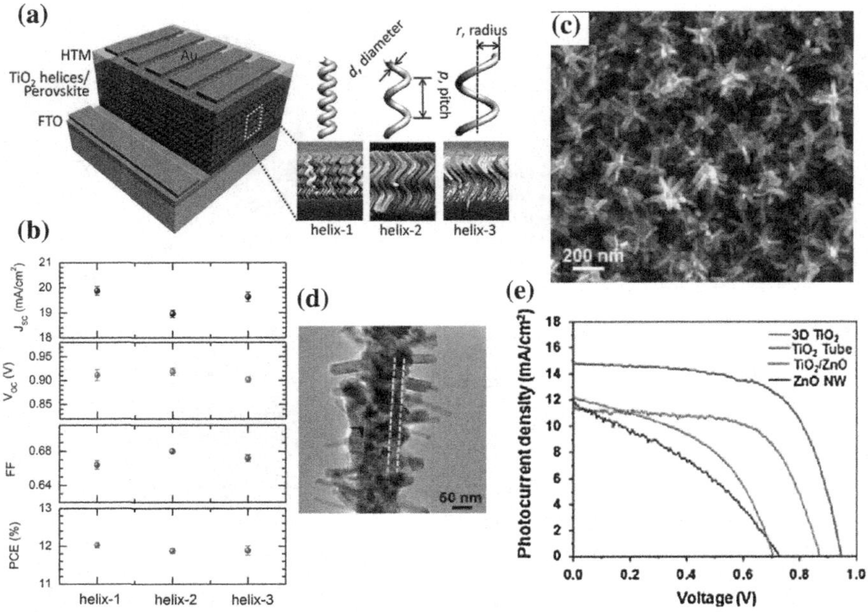

Fig. 1.9 **a** Schematic description of the PSC consisting of TiO_2 helices, and cross-sectional SEM images of three different TiO_2 helices grown on a silicon wafer. **b** Device performance of the TiO_2 helix-based PSCs. **c** *Top view* SEM, and **d** TEM images of the 3D TiO_2 NW architecture. **e** J-V curves of PSCs made from 3D TiO_2 nanostructures, TiO_2 NTs, TiO_2-coated ZnO NWs, and ZnO NWs with an identical film thickness of ~600 nm (reprinted with permission from [126] **a**, **b** J.-W. Lee et al., J. Mater. Chem. A 2015, 3, 9179–9186. [130] **c–e** Y. Yu et al., ACS Nano 2015, 9, 564–572. Copyright © The Royal Society of Chemistry and American Chemical Society)

(Fig. 1.9a). When this helical scaffold was used in PSCs, a best PCE of 12.03 % was measured (Fig. 1.9b). It was found that the contact area between TiO_2 and perovskite was critical to achieve a high performance PSC [126]. In particular, 3D hyper-branched NW/NR/NF architectures with high charge transport rates, sufficient surface area, and a 3D interconnected network for electron extraction were intensively studied and applied in solar cells. Yu et al. reported a 3D TiO_2 NW-based PSC with a 3D hierarchical TiO_2 architecture possessing a mixed anatase and rutile phases was constructed via a surface-reaction-limited pulsed chemical vapor deposition (SPCVD) technique (Fig. 1.9c, d). The highest PCE measured from such a device was 9.0 % (Fig. 3.9e). This was substantially higher than the PSCs made from ZnO NWs (PCE: 3.0 %), TiO_2-coated ZnO NWs (PCE: 4.0 %), and TiO_2 NTs (PCE: 6.2 %) [130].

Similarly, highly branched anatase TiO_2 NWs with varied orientation were grown via a facile one-step hydrothermal process on a FTO substrate. The multi-functionality of this dendritic TiO_2 thin film was found in PSCs: (1) its dense surface coverage on the transparent conductive oxide (TCO) substrate made it an effective electron transport layer; (2) its porosity enabled good perovskite anchoring and TiO_2/perovskite interconnection; (3) its 3D configuration can efficiently capture and confine the incident light to yield improved light harvesting efficiency and electron transfer from the perovskite layer; and (4) the TiO_2 film can function as a good hole-blocking layer due to the varied orientation of the wires that would effectively block the direct contact between spiro-MeOTAD and FTO or between the back electrode and FTO even in the absence of a TiO_2 blocking layer. Consequently, the cell device achieved a high PCE of 14.21 % [127]. Moreover, Mahmood et al. introduced a 3D hyper-branched anatase TiO_2 NR-NF array prepared via a combination of electrospinning and hydrothermal processes. The hyper-branched scaffold with optimal electron transport and carrier lifetime led to highly efficient mesostructured PSCs with an average PCE of 15.03 % and a maximum PCE of 15.50 % [129].

1.2.1.6 Modified TiO_2 Scaffold Layers

A lot of work suggests that the interface between TiO_2 and $MAPbX_3$ is a crucial factor in determining the crystal growth of perovskite and the effectiveness of charge separation in devices. The interfacial treatment and modification of TiO_2 to adjust the electronic structure have been widely investigated to enhance the charge injection and collection and simultaneously reduce the carrier recombination in PSCs [34, 131–135]. Roughly, the strategies used to modify the TiO_2 scaffolds can be divided into three directions as follows.

(1) Surface treatment has frequently been used to improve the quality at the interface. A common approach is to introduce a carboxylate monolayer on the mesoporous TiO_2 surfaces to passivate the interface and then suppress the charge recombination in the solar cells (Fig. 1.10a–c) [136]. Another process involves chemical etching to tailor the inner space of TiO_2 electrodes in order to provide an optimized space for perovskite sensitizers and infiltration of a hole transport layer without sacrificing their original electron transport ability [137].
(2) Elemental doping has also been employed. Typical elements used include Y [34, 138, 139], Zr [140], Al [131], Sn [141], Nb [132], and Mg [142] among others. Yttrium was poorly soluble in TiO_2 and the segregation of Y_2O_3 at the TiO_2 surface facilitated an increased perovskite loading on TiO_2 (Fig. 1.10d, e) [138]. The Y-TiO_2-based devices exhibited faster photocurrent decay (Fig. 1.10f) suggesting a more efficient carrier extraction in the Y-TiO_2 devices which probably resulted from the improved interface at the Y-TiO_2 layer. Moreover, the Y-TiO_2 devices exhibited a slight upward shift of the Fermi level in the Y-TiO_2 layer. This indicates an increased donor concentration in response to the enhanced conductivity [34].
(3) Incorporation of cooperative materials (e.g., graphene [107, 143], Sb_2S_3 [144], ZrO_2 [67, 145, 146], MgO [134, 147], WO_3 [135], and Au [133]). For example, MgO surface modification can reduce the perovskite degradation

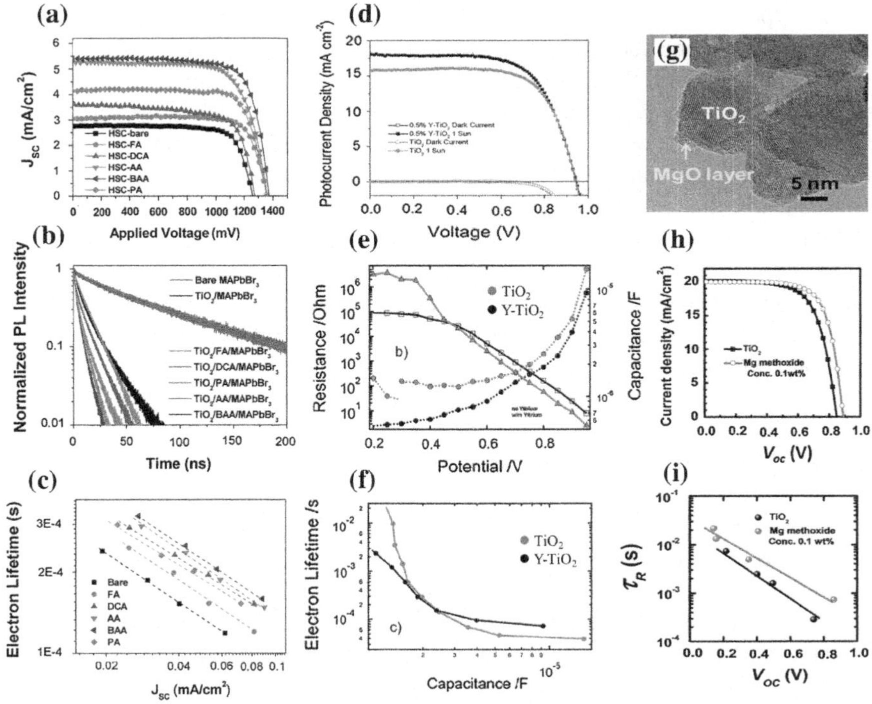

Fig. 1.10 **a** J-V curves, **b** PL decays, and **c** electron lifetime (τ_e) versus J_{SC} of PSCs employing different carboxylic acids. **d** J-V curves, **e** Recombination resistance (*solid lines*) and accompanying capacitance (*dashed lines*), and **f** evolution of the election lifetime as a function of capacitance of devices based on TiO_2 and Y-TiO_2. **g** HRTEM image of MgO/TiO_2 NPs. **h** J-V curves and (i) Electron lifetime of PSCs based on MgO/TiO_2 core-shell NPs and TiO_2 NPs (reprinted with permission from [136] **a–c** H.B. Kim et al., J. Mater. Chem. A 2015, 3, 9264–9270. [138] **d–f** P. Qin et al., Nanoscale 2014, 6, 1508–1514. [134] **g–i** G.S. Han et al., J. Mater. Chem. A 2015, 3, 9160–9164. Copyright © The Royal Society of Chemistry)

under H_2O and UV conditions by reducing the H_2O adsorbed on the mp-TiO_2 and blocking direct contact between TiO_2 and perovskite (Fig. 1.10g–j) [134, 147]. It also can reduce the electron/hole recombination at the TiO_2/$CH_3NH_3PbI_3$ surface thereby extending the carrier lifetime and yielding enhanced cell performance (Fig. 1.10k, l) [134, 147]. In addition, the PSC device incorporated with a unique sandwiched TiO_x-Au-NPs-TiO_x composite showed a 20–30 % performance enhancement. This was mainly attributed to the enhanced conductivity of the TiO_x layers to match the hole transporting capability, the decreased space-limited charges and the decreased surface potential of TiO_x film to lead to larger built-in potential in the device with improved V_{OC}. Both effects were ascribed to the plasmon-mediated hot carrier injection from the Au-NPs to TiO_x [133].

1.2.2 Nanostructured Al_2O_3 Layers

Since 2012 when Snaith et al. employed a mesoporous Al_2O_3 insulating layer as a scaffold in meso-superstructured PSC which gave a high PCE of 10.9 % and a high V_{OC} of 1.1 V [27], insulating Al_2O_3 materials have been extensively used in PSCs [36, 65, 68, 148–154]. Researchers find that the nature of the mesoporous scaffold impacts the final properties of perovskites and a high degree of preferential orientation has been detected for alumina scaffolds [155]. Furthermore, PSCs based on mesoporous Al_2O_3 electrodes present lower charge recombination rates than devices prepared on mesoporous TiO_2 [47].

It is worth noting that the PSCs based on the co-deposition of a combined Al_2O_3-perovskite layer, where the alumina NPs were suspended in the perovskite precursor solution and the Al_2O_3-perovskite layer was co-deposited by spin coating in a single deposition process followed by a low temperature heating step at T = 110 °C, showed an average PCE of 7.2 % on a non-sintered Al_2O_3 scaffold. [148] In addition, an ultrathin AlO_x layer was deposited onto a $MAPbI_3$ film using ALD, to construct a metal-insulator-semiconductor (MIS) back contact for the HTM-free PSC (Fig. 1.11a, b). This yielded a high PCE of 11.10 % after optimizing the ALD deposition cycles. It was revealed that this MIS back contact contributed to the enhancement in charge collection resulting from the electron blocking effect of the AlO_x layer (Fig. 1.11c) [154].

Moreover, when Al_2O_3 was coated on TiO_2 and $MAPbI_3$ layers, it could act as an insulating barrier to protect $MAPbI_3$ from degradation by moisture and suppress electron recombination between TiO_2 and spiro-MeOTAD. This leads to an enhancement of device stability when exposed to moisture [149, 150, 156]. In a similar study, an ultrathin Al_2O_3 film loaded on the HTM layer via an ALD method can act as a waterproofing agent to isolate the $MAPbI_3$ layers from moisture and thus realize enhanced device stability (Fig. 1.11d, e). Remarkably, the PCE of the PSC with ALD-Al_2O_3 retained ~90 % of its initial value after 24 days of storage in air (Fig. 1.11f) [68]. In addition, an insulating buffer layer between the perovskite and the metal was introduced into the PSC devices (Fig. 1.11g, h). This Al_2O_3 buffer layer significantly reduced the shunting degradation, inhibited the formation of a direct contact between the metal electrode and the perovskite, and improved the device stability with nearly no degradation in the first 350 h under simulated standard full sunlight (Fig. 1.11i). This Al_2O_3 buffer layer enabled the regulation of the thickness of the hole transporter layer which lead to lower device series resistance and higher PCEs [152].

1.2.3 Nanostructured ZnO Layers

Compared to TiO_2, ZnO is known to have a higher electron mobility which makes it an ideal choice for an electron-selective contact. Additionally, ZnO layers can be

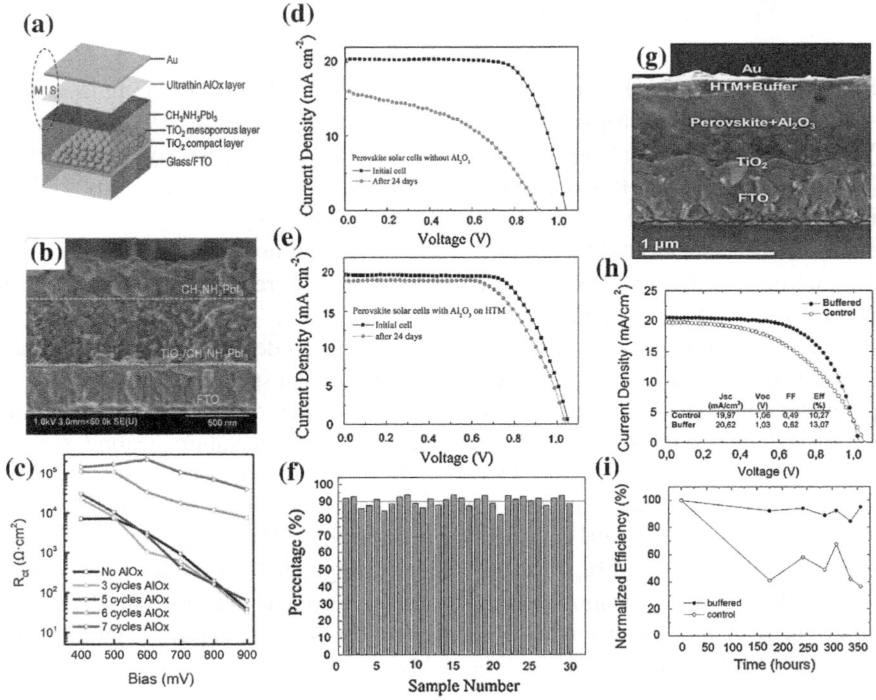

Fig. 1.11 **a** The scheme of the PSC with the MIS back contact, **b** cross-sectional SEM image of the FTO/TiO_2-$MAPbI_3$-AlO_x film, and **c** plots of charge transfer resistance (R_{ct}) for PSCs with different ALD cycles at various bias voltages. J-V characteristics for **d** FTO/bl-TiO_2/$MAPbI_3$/HTM/Ag and **e** FTO/bl-TiO_2/$MAPbI_3$/HTM/Al_2O_3/Ag at room temperature in 50 % humidity for 0 and 24 days. **f** The histogram of the PCE of thirty PSCs after 24 days in air compared with the original efficiency. **g** False color SEM image of a PSC with an Al_2O_3 buffer layer, **h** J-V curves and (i) normalized efficiency of the best solar cells realized with and without a buffer layer (reprinted with permission from [154] **a–c** H. Wei et al., Phys. Chem. Chem. Phys. 2015, 17, 4937–4944. [68] **d–f** X. Dong et al., J. Mater. Chem. A 2015, 3, 5360–5367. [152] **g–i** S. Guarnera et al., J. Phys. Chem. Lett. 2015, 6, 432–437. Copyright © The Royal Society of Chemistry and American Chemical Society)

easily deposited by spin coating and, depending on the use, require either no sintering step or merely a low-temperature heating treatment. This makes them ideal for the fabrication of flexible solar cells [157]. Devices based on ZnO NP scaffolds gave PCEs reaching 15.7 % [158], while ZnO-based flexible PSCs displayed PCEs over 13.14 % [157]. The solution-processing of the ZnO layers is simple, economical and proceeds at room temperature. This shows the remarkable advantages over TiO_2 scaffolds [64, 159, 160]. In addition to sol-gel methods [159], hydrothermal processes [161], chemical bath deposition [162], electrodeposition [64], and ALD methods [163] have also been used to grow ZnO structures for applications in PSCs. In addition to ZnO NP layers [164, 165], ZnO nanostructures in the form of NRs (Fig. 2.12a, b) [161, 162], NSs (Fig. 1.12c, d) [166] and quantum dots (Fig. 1.12e, f)

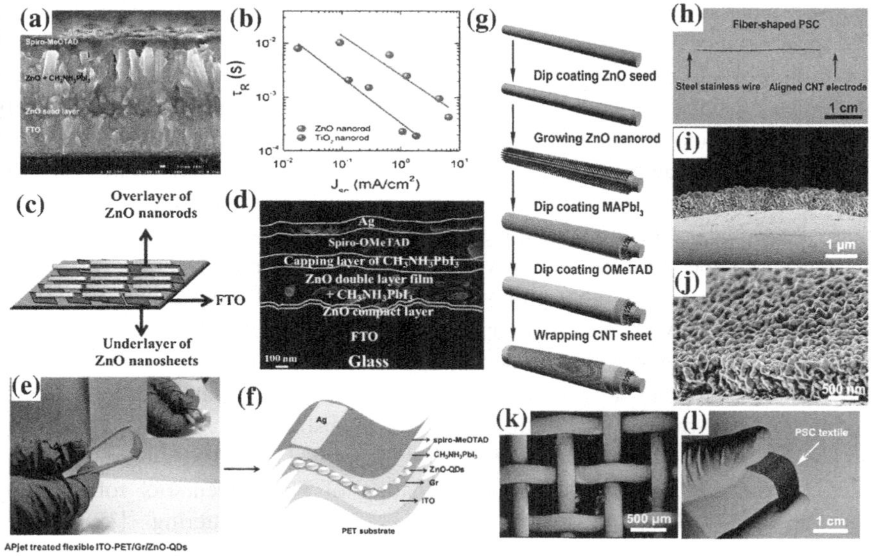

Fig. 1.12 **a** Cross-sectional SEM image of a ZnO NR-based PSC, **b** charge recombination τ_R as a function of light intensity, represented by photocurrent density, for the ZnO and TiO_2 nanorod-based PSCs. **c** Schematic illustration of ZnO film-based double-layer structure, and **d** cross-sectional SEM image of the corresponding PSC. **e** Photograph of ITO-PET/Graphene (Gr) and **f** schematic illustration of the flexible ZnO QD-based PSC. **g–j** Fiber-shaped PSC based on the aligned ZnO nano-obelisks: **g** Schematic illustration showing the fabrication process. **h** Photograph of a fiber-shaped PSC. SEM images of **i**, **j** a perovskite layer on the ZnO nano-obelisk array, and **k** the ZnO nano-obelisk array grown on a stainless steel fabric. **l** Photograph of the PSC fabric (reprinted with permission from [162] **a**, **b** D.Y. Son et al., J. Phys. Chem. C 2014, 118, 16567–16573. [166] **c**, **d** K. Mahmood et al., Nanoscale 2014, 6, 14674–14678. [167] **e**, **f** S. Ameen et al., J. Phys. Chem. C 2015, 119, 10379–10390. [173] **g–l** S. He et al., J. Mater. Chem. A 2015, 3, 9406–9410. Copyright © The Royal Society of Chemistry and American Chemical Society)

[167] have been used to fabricate PSC devices. Furthermore, doping treatments (e.g., Al, N and F) have been employed to modify the performance of ZnO nanostructures used in PSCs [168–171]. For large scale roll-to-roll production of flexible PSCs, a slot-die coating method was applied to realize the fabrication of fully printed ZnO-based PSCs yielding a peak PCE of 11.96 % [172]. Toward wearable solar cell device, PSCs in both fiber (Fig. 1.12g–j) and fabric (Fig. 1.12k, l) formats were built based on a ZnO nano-obelisk array, showing a peak PCE of 2.61 % [173].

1.2.4 Nanostructured NiO Layers

NiO is a cubic *p*-type semiconductor material with a large band gap that is widely applied in photovoltaic devices as a hole-conducting and collecting layer. When used in PSCs, it has several advantages [174–176]: (1) NiO can prevent the degradation of PSC performance which is common in TiO_2-based PSCs because of the light induced desorption of surface-adsorbed oxygen on TiO_2 when it is subjected to long-term sunlight exposure; (2) unlike another p-type organic material, PEDOT:PSS (poly (3,4-ethylene dioxythiophene):poly(4-styrenesulfonate)), NiO is inert and does not corrode ITO substrates; (3) NiO_x has a tunable work function from 5.0 to 5.6 eV by using different O_2-plasma surface treatments. A better alignment of the work function of the NiO_x electrode interlayer with the VB (valence band) edge level of perovskite can be realized to improve the transfer of holes and increase the device photovoltage as well as other photovoltaic parameters; (4) NiO is suitable for the integration of PSC with silicon or copper indium gallium selenide (CIGS) solar cells to build tandem devices for higher device efficiency due to the inverse device architecture with holes flowing to the ITO substrate instead of the traditional device involving TiO_2 where electrons flow to the ITO substrate.

The NiO films used as hole-transporting layers (HTL) in PSCs have been prepared in various ways including spin coating of NiO precursors followed by annealing [174–176], sol-gel methods [177], RF sputtering [178, 179], screen-printing [180], spray pyrolysis [181], and pulsed laser deposition [60]. Both planar and mesoporous cell structures have been applied in NiO-based PSCs and their highest PCEs are 15.40 % [182] and 17.3 % [60], respectively. In addition, some special designs are also presented for NiO-based PSCs [182–185]. For example, Cu-doped NiO_x was used in planar PSCs based on a device configuration of ITO/Cu doped-NiO_x/$MAPbI_3$/[6,6]-phenyl-C_{61}-butyric acid methyl ester ($PC_{61}BM$)/C_{60}-bis surfactant/Ag. This device showed an impressive PCE up to 15.40 % and fair environmental stability [182]. A PSC device with a cell configuration of FTO/bl-TiO_2/mp-TiO_2/mp-NiO ($MAPbI_3$)/carbon achieved a PCE of 11.4 % by using NiO as an electron blocker and a hole conductor (Fig. 1.13a–c) [183]. An ultrathin NiO compact layer (10–20 nm) cooperating with an inert mesoporous Al_2O_3 (meso-Al_2O_3) scaffold was prepared in an inverted PSC with a cell configuration of FTO/NiO/meso-Al_2O_3/$MAPbI_3$/PCBM/bathocuproine(BCP)/Ag (Fig. 1.13d) and exhibited a PCE of 13.5 % owing to the minimal light absorption loss and interfacial recombination loss (Fig. 1.13e, f) [184].

1.2.5 Nanostructured Carbon Materials

Carbon materials are inexpensive and readily-available on an industrial scale. They are well-known, versatile and have been extensively applied in many fields [186, 187]. As to PSCs, carbon materials (e.g., grapheme, carbon, CNTs, and C_{60}) also play an important role and have been employed in almost all the functional layers [188–204].

For instance,

(1) Graphene: graphene nanoflakes were added into the electron collection layer of TiO_2 NPs in PSCs to reduce the device series resistance and recombination

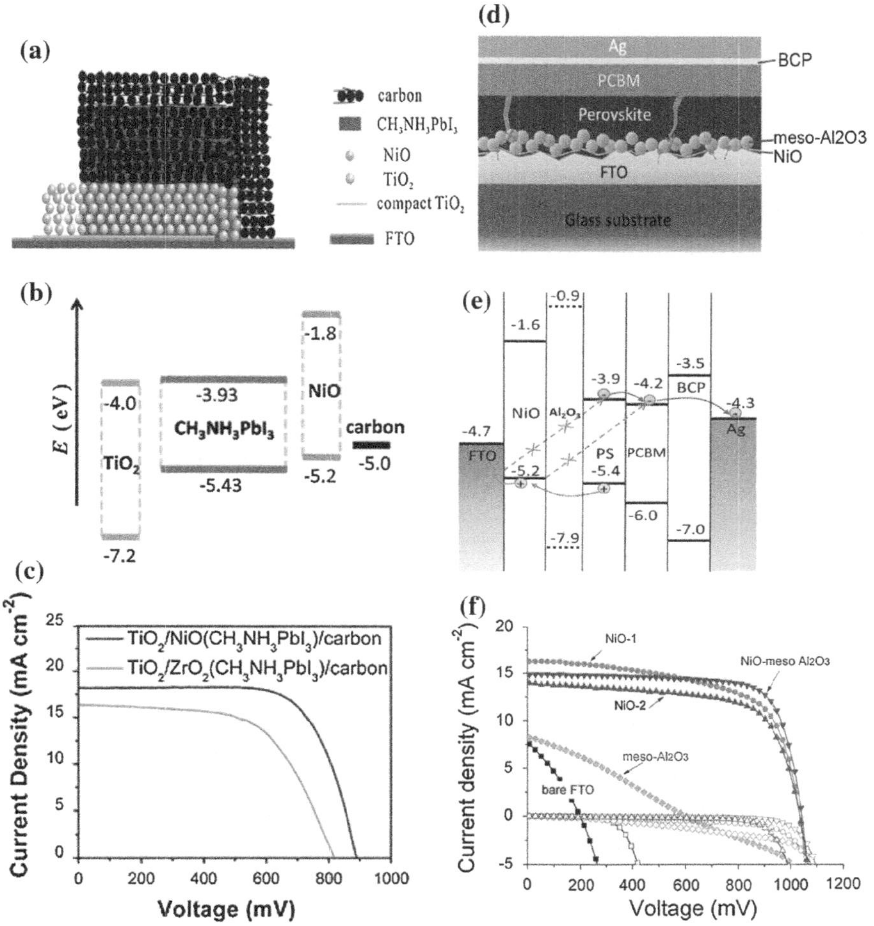

Fig. 1.13 **a**, **b** Cell configuration and energy level (versus vacuum) diagram of the TiO_2/NiO ($MAPbI_3$)/carbon device and **d**, **e** the inverted NiO/Al_2O_3-based PSCs. **c**, **f** J-V curves of the different PSCs (reprinted with permission from [183] **a–c** Z. Liu et al., Dalton Trans. 2015, 44, 3967–3973. [184] **d**, **e** W. Chen et al., *Energy Environ. Sci.* 2015, 8, 629–640. Copyright © The Royal Society of Chemistry)

losses (Fig. 1.14a, b) [143]; graphene quantum dots were inserted between the perovskite and the mesoporous TiO_2 to facilitate the electron transfer from the perovskite to the current collector (Fig. 1.14c, d) [107]; amphiphilic graphene oxide (GO) was used to form an insulating buffer layer between the perovskite and the HTL to prevent charge recombination [188]; GO and reduced graphene oxide (RGO) can act as hole conductors in inverted planar heterojunction PSCs (Fig. 1.14e, f) [189–191]; single-layered graphene (SG) and multilayered graphene (MG) function as hole extraction materials in PSCs with a device structure FTO/TiO_2/$MAPbI_3$/SG (or MG) [192]; multilayer

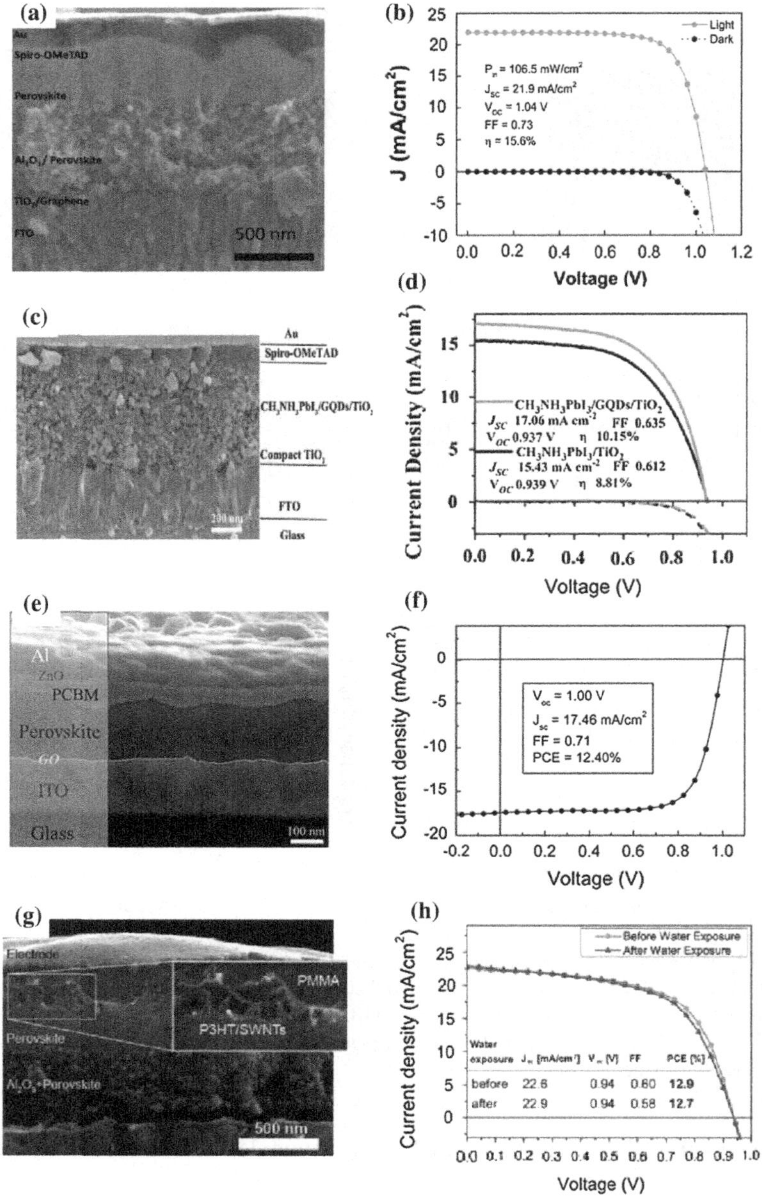

Fig. 1.14 Cross-sectional SEM images and I-V curves of the PSCs modified by **a–f** graphene materials and **g–h** carbon nanotubes (reprinted with permission from [143] **a, b**: J.T.-W. Wang et al., Nano Lett. 2014, 14, 724–730. [107] **c, d**: Z. Zhu et al., J. Am. Chem. Soc. 2014, 136, 3760–3763. [189] **e, f** Z. Wu et al., Nanoscale 2014, 6, 10505–10510. [202] **g, h** S.N. Habisreutinger et al., Nano Lett. 2014, 14, 5561–5568. Copyright © The Royal Society of Chemistry and American Chemical Society)

graphene prepared by CVD methods and modified by PEDOT:PSS was applied as a transparent top electrode on the HTL, which is promising for the fabrication of flexible PSCs by printing or roll-to-roll processes [193]; Graphene sheets were mixed into a functionalized nanographene (perthiolated trisulfur-annulated hexa-peri-hexabenzocoronene, TSHBC) which was employed as a HTM in PSCs [194].

(2) Carbon: carbon is typically used as the counter electrode in PSCs, especially in a hole-conductor-free cell configuration, by a low cost and fully-printable process [195–198]. Han's group developed a hole-conductor-free, fully printable mesoscopic PSC with carbon black counter electrodes (Fig. 2.8a) [67, 124, 197, 199, 200]. The resulting device gave a PCE of 12.8 % and a high environmental stability (>1000 h) in ambient air under full sunlight [200].

(3) Carbon nanotubes (CNTs): CNT network films can serve as hole collectors for PSCs without an organic HTM and Au electrode [201]; functionalized single-walled CNTs were embedded in an inert polymer matrix and achieved long-term stability in high-efficiency PSCs (Fig. 1.14g, h) [202]; transparent CNT electrodes were prepared for flexible PSCs based on a device structure of Ti foil/TiO_2 NTs/$MAPbI_3$/spiro-MeOTAD/CNTs [117]; CNT fiber electrodes were used to fabricate highly flexible, double-twisted PSCs with a maximum PCE of 3.03 % and performance stability longer than 96 h in open air as well as flexural stability for more than 1000 bending cycles [203].

1.2.6 Other Nanostructured Layers

Many other nanostructured materials, such as SnO_2 [61, 205–207], SiO_2 [66, 208, 209], Zn_2SnO_4 [210, 211], WO_x [63], $SrTiO_3$ [212], Nb_2O_5 [213], metal sulfides (i.e., PbS, $CuInS_2$, ZnS and CdS) [59, 214, 215], Cs_2CO_3 [216], MoO_x [217, 218], CuI [219], and CuSCN [220–222] have also been employed in the construction of PSC devices. Some representative examples are shown in the following section.

(1) SnO_2:Compared to TiO_2 and ZnO, SnO_2 has a similar band-gap, high transparency, and large electron mobility. However, SnO_2 has sufficient conductivity without a high temperature sintering process which is beneficial for the fabrication of flexible PSCs. Also, SnO_2 has a deeper conduction band (CB) than TiO_2 which facilitates a more efficient transfer of photo-generated electrons from the perovskite absorber to the SnO_2 CB. Furthermore, bulk SnO_2 has an electron mobility of up to 240 $cm^2 V^{-1} s^{-1}$ which is 100 times higher than that of TiO_2 and promising for highly efficient solar cells [61, 205–207]. Thus, low-temperature solution-processed nanocrystalline SnO_2 films have been employed as the electron transporting layers (ETLs) in PSCs. Resulting devices achieved a peak PCE of 17.21 % (Fig. 1.15a–c) [61].

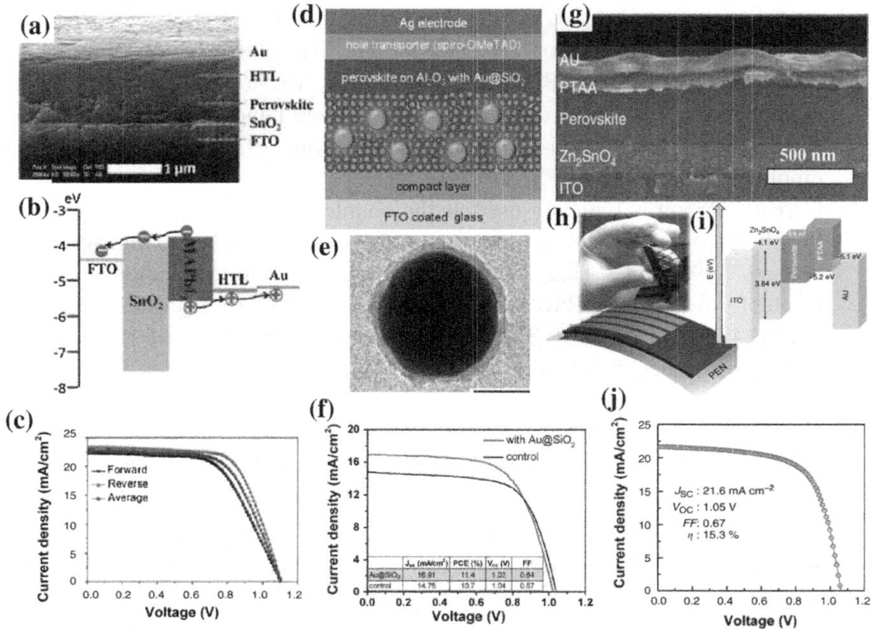

Fig. 1.15 Cross-sectional SEM images and energy band diagrams of the PSC devices based on **a**, **b** SnO_2 and **g**, **i** $ZnSnO_4$. I-V curves of the PSCs based on **c** SnO_2, **f** Au@SiO_2, and **j** $ZnSnO_4$. **e** TEM image of Au@SiO_2 NPs. **h** Photograph and of the $ZnSnO_4$-based flexible PSC (reprinted with permission from [61] **a–c** W. Ke et al., J. Am. Chem. Soc. 2015, 137, 6730–6733. [66] **d–f** W. Zhang et al., Nano Lett. 2013, 13, 4505–4510. [210] **g–j** S.S. Shin et al., Nature communications 2015, 6, 7410. Copyright © American Chemical Society and Nature Publishing Group)

(2) SiO_2: Similar to the insulating mesoporous Al_2O_3 scaffold, SiO_2 NPs were also employed as scaffold materials in PSCs. Resulting devices showed a PCE of up to 12.4 % [208, 209]. In particular, core-shell Au@SiO_2 NPs were incorporated into the Al_2O_3-based PSCs (Fig. 1.15d, e) and delivered a PCE of up to 11.4 % with enhanced J_{SC} (Fig. 1.15f). This is attributed to a reduced exciton binding energy and hence enhanced generation of free charge carriers with the incorporation of the metal NPs [66].

(3) Zn_2SnO_4: ZSO is an *n*-type semiconductor with a small electron effective mass of 0.23 m_e, a high electron Hall mobility of 10–30 $cm^2\ Vs^{-1}$, a wide optical band gap of 3.8 eV and a relatively low refractive index of $\sim$2.0 in the visible spectrum. Its conduction band edge is similar to TiO_2 and ZnO, making it an excellent electrode material in solar cell applications. More importantly, it has chemical stability with respect to acid/base solutions and polar organic solvents [210, 211]. Recently, dispersed ZSO NPs prepared at low-temperatures (<100 °C) were introduced into the development of flexible PSCs with a peak PCE of 15.3 % (Fig. 1.15g–j). This makes ZSO a promising candidate as an electron-conducting material for high efficiency flexible PSC applications [210].

1.3 Summary and Outlook

In this chapter, an overview of the historical progress in PSCs is presented. Next a focused review of the nanostructured materials used in the fabrication of PSCs is detailed. A summary of the development and prospects for several popular nanostructured materials; such as TiO_2, Al_2O_3 and ZnO, as well as other promising nanostructured materials, including NiO, carbon materials, metal sulfides, SnO_2, Zn_2SnO_4 is also outlined. These nanostructured materials can not only serve as effective blocking layers and scaffolds but can also play key roles in the HTM and novel metal electrodes in PSCs. In addition, these nanostructured inorganic materials can be used to modify the surface of perovskites and HTMs as well as the interface between them in order to improve the device efficiency and stability.

The high-efficiency, low material cost and simple assembly processes of PSCs make them economically viable for commercialization. More importantly, PSCs show promise for use in all-solid-state and flexible solar cells which are attractive power solutions for future wearable device technology. However, PSCs currently still face some challenges that hinder large-scale commercialization: (1) the toxicity of Pb atoms, (2) long-term stability, and (3) cost-effectiveness. In this regard, future research will aim at finding Pb-free perovskite materials, exploiting effective surface modification methods and external encapsulation technologies, and developing low cost materials and processes. It is probable that continued investigation of the nanostructured materials in PSCs will improve device stability and reduce the manufacturing costs of PSCs. If one thing is clear about PSCs, it is that the research and development of them in the last 6 years has been some of the most aggressive and effective of any technology. With ongoing challenges to push the PCE even higher, the trend in research will likely shift to one less focused on pushing the upper efficiency limit and more on the industrialization, scaling, fine tuning and repeatability of an already well-understood technology.

Acknowledgements This work was supported by the National Science Foundation (NSF ECCS-1305087), the National Nature Science Foundation of China (No.21503177), the Fundamental Research Funds for the Central Universities of China (No. 20720150031), and the 111 Project of China (B16029).

References

1. E. Serrano, G. Rus, J. García-Martínez, Nanotechnology for sustainable energy. Renew. Sustain. Energy Rev. **13**, 2373–2384 (2009)
2. N.S. Lewis, Toward cost-effective solar energy use. Science **315**, 798–801 (2007)
3. L. El Chaar, N. El Zein, Review of photovoltaic technologies. Renew. Sustain. Energy Rev. **15**, 2165–2175 (2011)
4. M. Grätzel, Recent advances in sensitized mesoscopic solar cells. Acc. Chem. Res. **42**, 1788–1798 (2009)
5. M.A. Green, Third generation photovoltaics: ultra-high conversion efficiency at low cost. Prog. Photovoltaics Res. Appl. **9**, 123–135 (2001)

6. Y.L. Lee, Y.S. Lo, Highly efficient quantum-dot-sensitized solar cell based on Co-sensitization of CdS/CdSe. Adv. Funct. Mater. **19**, 604–609 (2009)
7. S. Zhang, X. Yang, Y. Numata, L. Han, Highly efficient dye-sensitized solar cells: progress and future challenges. Energy Environ. Sci. **6**, 1443–1464 (2013)
8. N.-G. Park, Perovskite solar cells: an emerging photovoltaic technology. Mater. Today **18**, 65–72 (2014)
9. M. Gratzel, The light and shade of perovskite solar cells. Nat. Mater. **13**, 838–842 (2014)
10. M.D. McGehee, Perovskite solar cells: continuing to soar. Nat. Mater. **13**, 845–846 (2014)
11. G. Hodes, D. Cahen, Photovoltaics: Perovskite cells roll forward. Nat. Photon. **8**, 87–88 (2014)
12. M.A. Green, T. Bein, Photovoltaics Perovskite cells charge forward. Nat. Mater. **14**, 559–561 (2015)
13. N.-G. Park, Perovskite solar cells switchable photovoltaics. Nat. Mater. **14**, 140–141 (2015)
14. M.A. Green, A. Ho-Baillie, H.J. Snaith, The emergence of perovskite solar cells. Nat. Photonics **8**, 506–514 (2014)
15. G. Hodes, Perovskite-based solar cells. Science **342**, 317–318 (2013)
16. T.C. Sum, N. Mathews, Advancements in perovskite solar cells: photophysics behind the photovoltaics. Energy Environ. Sci. **7**, 2518–2534 (2014)
17. B.V. Lotsch, New light on an old story: Perovskites go solar. Angew. Chem. Int. Ed. **53**, 635–637 (2014)
18. M. Liu, M.B. Johnston, H.J. Snaith, Efficient planar heterojunction perovskite solar cells by vapour deposition. Nature **501**, 395–398 (2013)
19. M. Hu, L. Liu, A. Mei, Y. Yang, T. Liu, H. Han, Efficient hole-conductor-free, fully printable mesoscopic perovskite solar cells with a broad light harvester $NH_2CH{=}NH_2PbI_3$. J. Mater. Chem. A **2**, 17115–17121 (2014)
20. C.C. Stoumpos, C.D. Malliakas, M.G. Kanatzidis, Semiconducting tin and lead iodide perovskites with organic cations: phase transitions, high mobilities, and near-infrared photoluminescent properties. Inorg. Chem. **52**, 9019–9038 (2013)
21. J.H. Noh, S.H. Im, J.H. Heo, T.N. Mandal, S.I. Seok, Chemical management for colorful, efficient, and stable inorganic-organic hybrid nanostructured solar cells. Nano Lett. **13**, 1764–1769 (2013)
22. Y. Zhao, A.M. Nardes, K. Zhu, Mesoporous perovskite solar cells: material composition, charge-carrier dynamics, and device characteristics. Faraday Discuss. **176**, 301–312 (2014)
23. M. Jiang, J. Wu, F. Lan, Q. Tao, D. Gao, G. Li, Enhancing the performance of planar organo-lead halide perovskite solar cells by using a mixed halide source. J. Mater. Chem. A **3**, 963–967 (2015)
24. A. Kojima, K. Teshima, Y. Shirai, T. Miyasaka, Organometal halide perovskites as visible-light sensitizers for photovoltaic cells. J. Am. Chem. Soc. **131**, 6050–6051 (2009)
25. J.-H. Im, C.-R. Lee, J.-W. Lee, S.-W. Park, N.-G. Park, 6.5 % Efficient perovskite quantum-dot-sensitized solar cell. Nanoscale, **3**, 4088–4093 (2011)
26. H.-S. Kim, C.-R. Lee, J.-H. Im, K.-B. Lee, T. Moehl, A. Marchioro, S.-J. Moon, R. Humphry-Baker, J.-H. Yum, J.E. Moser, M. Graetzel, N.-G. Park. Lead iodide perovskite sensitized all-solid-state submicron thin film mesoscopic solar cell with efficiency exceeding 9 %. Sci. Rep., **2** (2012)
27. M.M. Lee, J. Teuscher, T. Miyasaka, T.N. Murakami, H.J. Snaith, Efficient hybrid solar cells based on meso-superstructured organometal halide perovskites. Science **338**, 643–647 (2012)
28. J.H. Heo, S.H. Im, J.H. Noh, T.N. Mandal, C.-S. Lim, J.A. Chang, Y.H. Lee, H.-J. Kim, A. Sarkar, M.K. Nazeeruddin, M. Graetzel, S.I. Seok, Efficient inorganic-organic hybrid heterojunction solar cells containing perovskite compound and polymeric hole conductors. Nat. Photonics **7**, 487–492 (2013)
29. J.M. Ball, M.M. Lee, A. Hey, H.J. Snaith, Low-temperature processed meso-superstructured to thin-film perovskite solar cells. Energy Environ. Sci. **6**, 1739–1743 (2013)

30. J. Burschka, N. Pellet, S.-J. Moon, R. Humphry-Baker, P. Gao, M.K. Nazeeruddin, M. Graetzel, Sequential deposition as a route to high-performance perovskite-sensitized solar cells. Nature **499**, 316–320 (2013)
31. S. Ryu, J.H. Noh, N.J. Jeon, Y. Chan Kim, W.S. Yang, J. Seo, S.I. Seok, Voltage output of efficient perovskite solar cells with high open-circuit voltage and fill factor. Energy Environ. Sci. **7**, 2614–2618 (2014)
32. N.J. Jeon, H.G. Lee, Y.C. Kim, J. Seo, J.H. Noh, J. Lee, S.I. Seok, o-Methoxy substituents in spiro-OMeTAD for efficient inorganic–organic hybrid perovskite solar cells. J. Am. Chem. Soc. **136**, 7837–7840 (2014)
33. T. Moehl, J.H. Im, Y.H. Lee, K. Domanski, F. Giordano, S.M. Zakeeruddin, M.I. Dar, L.-P. Heiniger, M.K. Nazeeruddin, N.-G. Park, M. Graetzel, Strong photocurrent amplification in perovskite solar cells with a porous TiO_2 blocking layer under reverse bias. J. Phys. Chem. Lett. **5**, 3931–3936 (2014)
34. H. Zhou, Q. Chen, G. Li, S. Luo, T.-B. Song, H.-S. Duan, Z. Hong, J. You, Y. Liu, Y. Yang, Interface engineering of highly efficient perovskite solar cells. Science **345**, 542–546 (2014)
35. W.S. Yang, J.H. Noh, N.J. Jeon, Y.C. Kim, S. Ryu, J. Seo, S.I. Seok, High-performance photovoltaic perovskite layers fabricated through intramolecular exchange. Science **348**, 1234–1237 (2015)
36. K.W. Tan, D.T. Moore, M. Saliba, H. Sai, L.A. Estroff, T. Hanrath, H.J. Snaith, U. Wiesner, Thermally induced structural evolution and performance of mesoporous block copolymer-directed alumina perovskite solar cells. ACS Nano **8**, 4730–4739 (2014)
37. Y. Deng, E. Peng, Y. Shao, Z. Xiao, Q. Dong, J. Huang, Scalable fabrication of efficient organolead trihalide perovskite solar cells with doctor-bladed active layers. Energy Environ. Sci. **8**, 1544–1550 (2015)
38. C. Law, L. Miseikis, S. Dimitrov, P. Shakya-Tuladhar, X. Li, P.R.F. Barnes, J. Durrant, B.C. O'Regan, Performance and stability of lead perovskite/TiO_2, polymer/PCBM, and dye sensitized solar cells at light intensities up to 70 Suns. Adv. Mater. **26**, 6268–6273 (2014)
39. B.-w. Park, B. Philippe, T. Gustafsson, K. Sveinbjornsson, A. Hagfeldt, E.M.J. Johansson, G. Boschloo, Enhanced crystallinity in organic-inorganic lead halide perovskites on mesoporous TiO_2 via disorder-order phase transition. Chem. Mater., **26**, 4466–4471 (2014)
40. J.-H. Im, J. Luo, M. Franckevicius, N. Pellet, P. Gao, T. Moehl, S.M. Zakeeruddin, M.K. Nazeeruddin, M. Graetzel, N.-G. Park, Nanowire perovskite solar cell. Nano Lett. **15**, 2120–2126 (2015)
41. N. Yantara, D. Sabba, F. Yanan, J.M. Kadro, T. Moehl, P.P. Boix, S. Mhaisalkar, M. Graetzel, C. Graetzel, Loading of mesoporous titania films by $CH_3NH_3PbI_3$ perovskite, single step vs. sequential deposition. Chem. Commun. **51**, 4603–4606 (2015)
42. W.C. Choy, Vacuum-assisted thermal annealing of $CH_3NH_3PbI_3$ for highly stable and efficient perovskite solar cells. ACS Nano **9**, 639–646 (2015)
43. D. Bi, A.M. El-Zohry, A. Hagfeldt, G. Boschloo, Unraveling the effect of PbI_2 concentration on charge recombination kinetics in perovskite solar cells. ACS Photonics **2**, 589–594 (2015)
44. T. Zhang, M. Yang, Y. Zhao, K. Zhu, Controllable sequential deposition of planar $CH_3NH_3PbI_3$ perovskite films via adjustable volume expansion. Nano Lett. **15**, 3959–3963 (2015)
45. T. Leijtens, G.E. Eperon, S. Pathak, A. Abate, M.M. Lee, H.J. Snaith, Overcoming ultraviolet light instability of sensitized TiO_2 with meso-superstructured organometal tri-halide perovskite solar cells. Nat. Commum. **4**, 2885 (2013)
46. A.K. Jena, H.-W. Chen, A. Kogo, Y. Sanehira, M. Ikegami, T. Miyasaka, The interface between FTO and the TiO_2 compact layer can be one of the origins to hysteresis in planar heterojunction perovskite solar cells. ACS Appl. Mater. Interfaces **7**, 9817–9823 (2015)
47. B. Suarez, V. Gonzalez-Pedro, T.S. Ripolles, R.S. Sanchez, L. Otero, I. Mora-Sero, Recombination study of combined halides (Cl, Br, I) perovskite solar cells. J. Phys. Chem. Lett. **5**, 1628–1635 (2014)

48. D. Bi, S.-J. Moon, L. Haggman, G. Boschloo, L. Yang, E.M.J. Johansson, M.K. Nazeeruddin, M. Graetzel, A. Hagfeldt, Using a two-step deposition technique to prepare perovskite ($CH_3NH_3PbI_3$) for thin film solar cells based on ZrO_2 and TiO_2 mesostructures. RSC Adv. **3**, 18762–18766 (2013)
49. J. Xiao, J. Shi, H. Liu, Y. Xu, S. Lv, Y. Luo, D. Li, Q. Meng, Y. Li, Efficient $CH_3NH_3PbI_3$ perovskite solar cells based on graphdiyne (GD)-modified P3HT hole-transporting material. Adv. Energy Mater. **5**, 1401943 (2015)
50. A. Abrusci, S.D. Stranks, P. Docampo, H.-L. Yip, A.K.Y. Jen, H.J. Snaith, High-performance perovskite-polymer hybrid solar cells via electronic coupling with fullerene monolayers. Nano Lett. **13**, 3124–3128 (2013)
51. J.-Y. Jeng, Y.-F. Chiang, M.-H. Lee, S.-R. Peng, T.-F. Guo, P. Chen, T.-C. Wen, $CH_3NH_3PbI_3$ Perovskite/fullerene planar-heterojunction hybrid solar cells. Adv. Mater. **25**, 3727–3732 (2013)
52. C. Kuang, G. Tang, T. Jiu, H. Yang, H. Liu, B. Li, W. Luo, X. Li, W. Zhang, F. Lu, J. Fang, Y. Li, Highly efficient electron transport obtained by doping PCBM with graphdiyne in planar-heterojunction perovskite solar cells. Nano Lett. **15**, 2756–2762 (2015)
53. N. Marinova, W. Tress, R. Humphry-Baker, M.I. Dar, V. Bojinov, S.M. Zakeeruddin, M.K. Nazeeruddin, M. Grätzel, Light harvesting and charge recombination in $CH_3NH_3PbI_3$ perovskite solar cells studied by hole transport layer thickness variation. ACS Nano **9**, 4200–4209 (2015)
54. H. Azimi, T. Ameri, H. Zhang, Y. Hou, C.O.R. Quiroz, J. Min, M. Hu, Z.G. Zhang, T. Przybilla, G.J. Matt, A universal interface layer based on an amine-functionalized fullerene derivative with dual functionality for efficient solution processed organic and perovskite solar cells. Adv. Energy Mater. **5**, 1401692 (2015)
55. S. Gamliel, L. Etgar, Organo-metal perovskite based solar cells: sensitized versus planar architecture. RSC Adv. **4**, 29012–29021 (2014)
56. H.-S. Kim, S.H. Im, N.-G. Park, Organolead halide perovskite: new horizons in solar cell research. J. Phys. Chem. C **118**, 5615–5625 (2014)
57. S. Luo, W.A. Daound, Recent progress in organic–inorganic halide perovskite solar cells: mechanisms and material design. J. Mater. Chem. A **3**, 8992–9010 (2015)
58. C.-W. Chen, H.-W. Kang, S.-Y. Hsiao, P.-F. Yang, K.-M. Chiang, H.-W. Lin, Efficient and uniform planar-type perovskite solar cells by simple sequential vacuum deposition. Adv. Mater. **26**, 6647–6652 (2014)
59. J. Liu, C. Gao, L. Luo, Q. Ye, X. He, L. Ouyang, X. Guo, D. Zhuang, C. Liao, J. Mei, W. Lau, Low-temperature, solution processed metal sulfide as an electron transport layer for efficient planar perovskite solar cells. J. Mater. Chem. A **3**, 11750–11755 (2015)
60. J.H. Park, J. Seo, S. Park, S.S. Shin, Y.C. Kim, N.J. Jeon, H.W. Shin, T.K. Ahn, J.H. Noh, S. C. Yoon, Efficient CH3NH3PbI3 perovskite solar cells employing nanostructured p-type NiO electrode formed by a pulsed laser deposition. Adv. Mater. **27**, 4013–4019 (2015)
61. W. Ke, G. Fang, Q. Liu, L. Xiong, P. Qin, H. Tao, J. Wang, H. Lei, B. Li, J. Wan, G. Yang, Y. Yan, Low-temperature solution-processed tin oxide as an alternative electron transporting layer for efficient perovskite solar cells. J. Am. Chem. Soc. **137**, 6730–6733 (2015)
62. A. Yella, L.-P. Heiniger, P. Gao, M.K. Nazeeruddin, M. Graetzel, Nanocrystalline rutile electron extraction layer enables low-temperature solution processed perovskite photovoltaics with 13.7 % efficiency. Nano Lett. **14**, 2591–2596 (2014)
63. K. Wang, Y. Shi, Q. Dong, Y. Li, S. Wang, X. Yu, M. Wu, T. Ma, Low-temperature and solution-processed amorphous WOX as electron-selective layer for perovskite solar cells. J. Phys. Chem. Lett. **6**, 755–759 (2015)
64. J. Zhang, P. Barboux, T. Pauporte, Electrochemical design of nanostructured ZnO charge carrier layers for efficient solid-state perovskite-sensitized solar cells. Adv. Energy Mater. **4**, 1400932 (2014)

65. H.-Y. Hsu, C.-Y. Wang, A. Fathi, J.-W. Shiu, C.-C. Chung, P.-S. Shen, T.-F. Guo, P. Chen, Y.-P. Lee, E.W.-G. Diau, Femtosecond excitonic relaxation dynamics of perovskite on mesoporous films of Al_2O_3 and NiO nanoparticles. Angew. Chem. Int. Ed. **53**, 9339–9342 (2014)
66. W. Zhang, M. Saliba, S.D. Stranks, Y. Sun, X. Shi, U. Wiesner, H.J. Snaith, Enhancement of perovskite-based solar cells employing core-shell metal nanoparticles. Nano Lett. **13**, 4505–4510 (2013)
67. L. Liu, A. Mei, T. Liu, P. Jiang, Y. Sheng, L. Zhang, H. Han, Fully printable mesoscopic perovskite solar cells with organic silane self-assembled monolayer. J. Am. Chem. Soc. **137**, 1790–1793 (2015)
68. X. Dong, X. Fang, M. Lv, B. Lin, S. Zhang, J. Ding, N. Yuan, Improvement of the humidity stability of organic-inorganic perovskite solar cells using ultrathin Al_2O_3 layers prepared by atomic layer deposition. J. Mater. Chem. A **3**, 5360–5367 (2015)
69. G. Murugadoss, G. Mizuta, S. Tanaka, H. Nishino, T. Umeyama, H. Imahori, S. Ito, Double functions of porous TiO_2 electrodes on $CH_3NH_3PbI_3$ perovskite solar cells: enhancement of perovskite crystal transformation and prohibition of short circuiting. APL Mater. **2**, 081511 (2014)
70. S.D. Sung, D.P. Ojha, J.S. You, J. Lee, J. Kim, W.I. Lee, 50 nm sized spherical TiO_2 nanocrystals for highly efficient mesoscopic perovskite solar cells. Nanoscale **7**, 8898–8906 (2015)
71. J. Song, J. Bian, E. Zheng, X.-F. Wang, W. Tian, T. Miyasaka, Efficient and environmentally stable perovskite solar cells based on ZnO electron collection layer. Chem. Lett. **44**, 610–612 (2015)
72. D.-Y. Son, K.-H. Bae, H.-S. Kim, N.-G. Park, Effects of seed layer on growth of ZnO nanorod and performance of perovskite solar cell. J. Phys. Chem. C **119**, 10321–10328 (2015)
73. S. Aharon, S. Gamliel, B.E. Cohen, L. Etgar, Depletion region effect of highly efficient hole conductor free $CH_3NH_3PbI_3$ perovskite solar cells. Phys. Chem. Chem. Phys. **16**, 10512–10518 (2014)
74. T. Leijtens, B. Lauber, G.E. Eperon, S.D. Stranks, H.J. Snaith, The importance of perovskite pore filling in organometal mixed halide sensitized TiO_2-based solar cells. J. Phys. Chem. Lett. **5**, 1096–1102 (2014)
75. J.-W. Lee, T.-Y. Lee, P.J. Yoo, M. Graetzel, S. Mhaisalkar, N.-G. Park, Rutile TiO_2-based perovskite solar cells. J. Mater. Chem. A **2**, 9251–9259 (2014)
76. S. Shi, Y. Li, X. Li, H. Wang, Advancements in all-solid-state hybrid solar cells based on organometal halide perovskites. Mater. Horiz. **2**, 378–405 (2015)
77. K. Wojciechowski, M. Saliba, T. Leijtens, A. Abate, H.J. Snaith, Sub-150 °C processed meso-superstructured perovskite solar cells with enhanced efficiency. Energy Environ. Sci. **7**, 1142–1147 (2014)
78. B. Conings, L. Baeten, T. Jacobs, R. Dera, J. D'Haen, J. Manca, H.-G. Boyen, An easy-to-fabricate low-temperature TiO_2 electron collection layer for high efficiency planar heterojunction perovskite solar cells. APL Mater. **2**, 081505 (2014)
79. W. Liu, Y. Zhang, Electrical characterization of $TiO_2/CH_3NH_3PbI_3$ heterojunction solar cells. J. Mater. Chem. A **2**, 10244–10249 (2014)
80. N.K. Noel, A. Abate, S.D. Stranks, E.S. Parrott, V.M. Burlakov, A. Goriely, H.J. Snaith, Enhanced photoluminescence and solar cell performance via Lewis base passivation of organic–inorganic lead halide perovskites. ACS Nano **8**, 9815–9821 (2014)
81. Y. Xiao, G. Han, Y. Li, M. Li, Y. Chang, J. Wu, Preparation of high performance perovskite-sensitized nanoporous titanium dioxide photoanodes by in situ method for use in perovskite solar cells. J. Mater. Chem. A **2**, 16531–16537 (2014)
82. Y. Xiao, G. Han, Y. Chang, Y. Zhang, Y. Li, M. Li, Investigation of perovskite-sensitized nanoporous titanium dioxide photoanodes with different thicknesses in perovskite solar cells. J. Power Sour. **286**, 118–123 (2015)

83. X. Wang, Y. Fang, L. He, Q. Wang, T. Wu, Influence of compact TiO_2 layer on the photovoltaic characteristics of the organometal halide perovskite-based solar cells. Mater. Sci. Semicond. Process. **27**, 569–576 (2014)
84. P. Qin, S. Tanaka, S. Ito, N. Tetreault, K. Manabe, H. Nishino, M.K. Nazeeruddin, M. Grätzel, Inorganic hole conductor-based lead halide perovskite solar cells with 12.4 % conversion efficiency. Nat. Commun., **5** (2014)
85. W. Li, H. Dong, X. Guo, N. Li, J. Li, G. Niu, L. Wang, Graphene oxide as dual functional interface modifier for improving wettability and retarding recombination in hybrid perovskite solar cells. J. Mater. Chem. A (2014)
86. F. Di Giacomo, V. Zardetto, A. D'Epifanio, S. Pescetelli, F. Matteocci, S. Razza, A. Di Carlo, S. Licoccia, W.M.M. Kessels, M. Creatore, T.M. Brown, Flexible perovskite photovoltaic modules and solar cells based on atomic layer deposited compact layers and UV-irradiated TiO_2 scaffolds on plastic substrates. Adv. Energy Mater. **5**, 1401808 (2015)
87. Y. Wu, X. Yang, H. Chen, K. Zhang, C. Qin, J. Liu, W. Peng, A. Islam, E. Bi, F. Ye, M. Yin, P. Zhang, L. Han, Highly compact TiO_2 layer for efficient hole-blocking in perovskite solar cells. Appl. Phys. Express **7**, 052301 (2014)
88. W. Ke, G. Fang, J. Wang, P. Qin, H. Tao, H. Lei, Q. Liu, X. Dai, X. Zhao, Perovskite solar cell with an efficient TiO_2 compact film. ACS Appl. Mater. Interfaces **6**, 15959–15965 (2014)
89. Q. Gao, S. Yang, L. Lei, S. Zhang, Q. Cao, J. Xie, J. Li, Y. Liu, An effective TiO_2 blocking layer for perovskite solar cells with enhanced performance. Chem. Lett. **44**, 624–626 (2015)
90. S.D. Sung, M.S. Kang, I.T. Choi, H.M. Kim, H. Kim, M. Hong, H.K. Kim, W.I. Lee, 14.8 % perovskite solar cells employing carbazole derivatives as hole transporting materials. Chem. Commun. **50**, 14161–14163 (2014)
91. S. Das, B. Yang, G. Gu, P.C. Joshi, I.N. Ivanov, C.M. Rouleau, T. Aytug, D.B. Geohegan, K. Xiao, High-performance flexible perovskite solar cells by using a combination of ultrasonic spray-coating and low thermal budget photonic curing. ACS Photonics **2**, 680–686 (2015)
92. S. Hong, A. Han, E.C. Lee, K.-W. Ko, J.-H. Park, H.-J. Song, M.-H. Han, C.-H. Han, A facile and low-cost fabrication of TiO_2 compact layer for efficient perovskite solar cells. Curr. Appl. Phys. **15**, 574–579 (2015)
93. S. Ameen, M.S. Akhtar, H.-K. Seo, M.K. Nazeeruddin, H.-S. Shin, Exclusion of metal oxide by an RF sputtered Ti layer in flexible perovskite solar cells: energetic interface between a Ti layer and an organic charge transporting layer. Dalton Trans. **44**, 6439–6448 (2015)
94. M. Lee, Y. Jo, D.S. Kim, Y. Jun, Flexible organo-metal halide perovskite solar cells on a Ti metal substrate. J. Mater. Chem. A **3**, 4129–4133 (2015)
95. L. Cojocaru, S. Uchida, Y. Sanehira, J. Nakazaki, T. Kubo, H. Segawa, Surface treatment of the compact TiO_2 layer for efficient planar heterojunction peroyskite solar cells. Chem. Lett. **44**, 674–676 (2015)
96. J. Wang, M. Qin, H. Tao, W. Ke, Z. Chen, J. Wan, P. Qin, L. Xiong, H. Lei, H. Yu, G. Fang, Performance enhancement of perovskite solar cells with Mg-doped TiO_2 compact film as the hole-blocking layer. Appl. Phys. Lett. **106**, 121104 (2015)
97. T. Krishnamoorthy, F. Kunwu, P.P. Boix, H. Li, T.M. Koh, W.L. Leong, S. Powar, A. Grimsdale, M. Gratzel, N. Mathews, S.G. Mhaisalkar, A swivel-cruciform thiophene based hole-transporting material for efficient perovskite solar cells. J. Mater. Chem. A **2**, 6305–6309 (2014)
98. K. Do, H. Choi, K. Lim, H. Jo, J.W. Cho, M.K. Nazeeruddin, J. Ko, Star-shaped hole transporting materials with a triazine unit for efficient perovskite solar cells. Chem. Commun. **50**, 10971–10974 (2014)
99. P. Qin, N. Tetreault, M.I. Dar, P. Gao, K.L. McCall, S.R. Rutter, S.D. Ogier, N.D. Forrest, J.S. Bissett, M.J. Simms, A.J. Page, R. Fisher, M. Grätzel, M.K. Nazeeruddin, A novel oligomer as a hole transporting material for efficient perovskite solar cells. Adv. Energy Mater. **5**, 1400980 (2014)

100. Z. Zhu, Y. Bai, H.K.H. Lee, C. Mu, T. Zhang, L. Zhang, J. Wang, H. Yan, S.K. So, S. Yang, Polyfluorene derivatives are high-performance organic hole-transporting materials for inorganic–organic hybrid perovskite solar cells. Adv. Funct. Mater. **24**, 7357–7365 (2014)
101. N. Fu, C. Huang, Y. Liu, X. Li, W. Lu, L. Zhou, F. Peng, Y. Liu, H. Huang, Organic-free anatase TiO_2 paste for efficient plastic dye-sensitized solar cells and low temperature processed perovskite solar cells. ACS Appl. Mater. Interfaces **7**, 19431–19438 (2015)
102. S.D. Stranks, G.E. Eperon, G. Grancini, C. Menelaou, M.J.P. Alcocer, T. Leijtens, L.M. Herz, A. Petrozza, H.J. Snaith, Electron-hole diffusion lengths exceeding 1 micrometer in an organometal trihalide perovskite absorber. Science **342**, 341–344 (2013)
103. S. Kazim, M.K. Nazeeruddin, M. Grätzel, S. Ahmad, Perovskite as light harvester: a game changer in photovoltaics. Angew. Chem. Int. Ed. **53**, 2812–2824 (2014)
104. Q. Jiang, X. Sheng, Y. Li, X. Feng, T. Xu, Rutile TiO_2 nanowire-based perovskite solar cells. Chem. Commun. **50**, 14720–14723 (2014)
105. F. Hao, C.C. Stoumpos, D.H. Cao, R.P.H. Chang, M.G. Kanatzidis, Lead-free solid-state organic-inorganic halide perovskite solar cells. Nat. Photon. **8**, 489–494 (2014)
106. P. Qin, S. Paek, M.I. Dar, N. Pellet, J. Ko, M. Grätzel, M.K. Nazeeruddin, Perovskite solar cells with 12.8 % efficiency by using conjugated quinolizino acridine based hole transporting material. J. Am. Chem. Soc. **136**, 8516–8519 (2014)
107. Z. Zhu, J. Ma, Z. Wang, C. Mu, Z. Fan, L. Du, Y. Bai, L. Fan, H. Yan, D.L. Phillips, S. Yang, Efficiency enhancement of perovskite solar cells through fast electron extraction: the role of graphene quantum dots. J. Am. Chem. Soc. **136**, 3760–3763 (2014)
108. Y. Yang, K. Ri, A. Mei, L. Liu, M. Hu, T. Liu, X. Li, H. Han, The size effect of TiO_2 nanoparticles on a printable mesoscopic perovskite solar cell. J. Mater. Chem. A **3**, 9103–9107 (2015)
109. C.B. Zhong D, Wang X, et al., Synthesis of oriented TiO_2 nanocones with fast charge transfer for perovskite solar cells. Nano Energy, **11**, 409–418 (2015)
110. X. Gao, J. Li, J. Baker, Y. Hou, D. Guan, J. Chen, C. Yuan, Enhanced photovoltaic performance of perovskite $CH_3NH_3PbI_3$ solar cells with freestanding TiO_2 nanotube array films. Chem. Commun. **50**, 6368–6371 (2014)
111. S. Dharani, H.K. Mulmudi, N. Yantara, T. Pham Thi Thu, N.G. Park, M. Graetzel, S. Mhaisalkar, N. Mathews, P.P. Boix, High efficiency electrospun TiO_2 nanofiber based hybrid organic-inorganic perovskite solar cell. Nanoscale **6**, 1675–1679 (2014)
112. H.-S. Kim, J.-W. Lee, N. Yantara, P.P. Boix, S.A. Kulkarni, S. Mhaisalkar, M. Graetzel, N.-G. Park, High efficiency solid-state sensitized solar cell-based on submicrometer rutile TiO_2 nanorod and $CH_3NH_3PbI_3$ perovskite sensitizer. Nano Lett. **13**, 2412–2417 (2013)
113. J. Qiu, Y. Qiu, K. Yan, M. Zhong, C. Mu, H. Yan, S. Yang, All-solid-state hybrid solar cells based on a new organometal halide perovskite sensitizer and one-dimensional TiO_2 nanowire arrays. Nanoscale **5**, 3245–3248 (2013)
114. D. Zhong, B. Cai, Z. Yang, B. Huang, S. Miao, W-H. Zhang, J. Qiu, C. Li, An acid-free medium growth of rutile TiO_2 nanorods arrays and their application in perovskite solar cells. J. Mater. Chem. C, **3**, 729–733 (2015)
115. A. Fakharuddin, F. Di Giacomo, I. Ahmed, Q. Wali, T.M. Brown, R. Jose, Role of morphology and crystallinity of nanorod and planar electron transport layers on the performance and long term durability of perovskite solar cells. J. Power Sour. **283**, 61–67 (2015)
116. S.S. Mali, C.S. Shim, H.K. Park, J. Heo, P.S. Patil, C.K. Hong, Ultrathin atomic layer deposited TiO_2 for surface passivation of hydrothermally grown 1D TiO_2 nanorod arrays for efficient solid-state perovskite solar cells. Chem. Mater. **27**, 1541–1551 (2015)
117. X. Wang, Z. Li, W. Xu, S.A. Kulkarni, S.K. Batabyal, S. Zhang, A. Cao, L.H. Wong, TiO_2 nanotube arrays based flexible perovskite solar cells with transparent carbon nanotube electrode. Nano Energy **11**, 728–735 (2015)
118. D. Zhong, B. Cai, X. Wang, Z. Yang, Y. Xing, S. Miao, W.-H. Zhang, C. Li, Synthesis of oriented TiO_2 nanocones with fast charge transfer for perovskite solar cells. Nano Energy **11**, 409–418 (2015)

119. R. Salazar, M. Altomare, K. Lee, J. Tripathy, R. Kirchgeorg, N.T. Nguyen, M. Mokhtar, A. Alshehri, S.A. Al-Thabaiti, P. Schmuki, Use of anodic TiO_2 nanotube layers as mesoporous scaffolds for fabricating $CH_3NH_3PbI_3$ perovskite-based solid-state solar cells. ChemElectroChem (2015)
120. J. Zhang, T. Pauporté, One-dimensional self-standing TiO_2 nanotube array layers designed for perovskite solar cell applications. ChemPhysChem **16**, 2836–2841 (2015)
121. H. Yang, W. Fang, X. Yang, H. Zhu, Z. Li, H. Zhao, X. Yao, Yolk@ shell anatase TiO_2 hierarchical microspheres with exposed 001 facets for high-performance dye sensitized solar cells. J. Mater. Chem. **22**, 22082–22089 (2012)
122. W. Yang, J. Li, Y. Wang, F. Zhu, W. Shi, F. Wan, D. Xu, A facile synthesis of anatase TiO_2 nanosheets-based hierarchical spheres with over 90 %{001} facets for dye-sensitized solar cells. Chem. Commun. **47**, 1809–1811 (2011)
123. J.G. Yu, J.J. Fan, K.L. Lv, Anatase TiO_2 nanosheets with exposed (001) facets: improved photoelectric conversion efficiency in dye-sensitized solar cells. Nanoscale **2**, 2144–2149 (2010)
124. Y. Rong, Z. Ku, A. Mei, T. Liu, M. Xu, S. Ko, X. Li, H. Han, Hole-conductor-free mesoscopic $TiO_2/CH_3NH_3PbI_3$ heterojunction solar cells based on anatase nanosheets and carbon counter electrodes. J. Phys. Chem. Lett. **5**, 2160–2164 (2014)
125. M.I. Dar, F.J. Ramos, Z. Xue, B. Liu, S. Ahmad, S.A. Shivashankar, M.K. Nazeeruddin, M. Graetzel, Photoanode based on (001)-oriented anatase nanoplatelets for organic-inorganic lead iodide perovskite solar cell. Chem. Mater. **26**, 4675–4678 (2014)
126. J.-W. Lee, S.H. Lee, H.-S. Ko, J. Kwon, J.H. Park, S.M. Kang, N. Ahn, M. Choi, J.K. Kim, N.-G. Park, Opto-electronic properties of TiO_2 nanohelices with embedded $HC(NH_2)_2PbI_3$ perovskite solar cells. J. Mater. Chem. A **3**, 9179–9186 (2015)
127. W.Q. Wu, F. Huang, D. Chen, Y.B. Cheng, R.A. Caruso, Thin films of dendritic anatase titania nanowires enable effective hole-blocking and efficient light-harvesting for high-performance mesoscopic perovskite solar cells. Adv. Funct. Mater. **25**, 3264–3272 (2015)
128. H. Sun, P. Ruan, Z. Bao, L. Chen, X. Zhou, Shell-in-shell TiO_2 hollow microspheres and optimized application in light-trapping perovskite solar cells. Solid State Sci. **40**, 60–66 (2015)
129. K. Mahmood, B.S. Swain, A. Amassian, Highly efficient hybrid photovoltaics based on hyperbranched three-dimensional TiO_2 electron transporting materials. Adv. Mater. **27**, 2859–2865 (2015)
130. Y. Yu, J. Li, D. Geng, J. Wang, L. Zhang, T.L. Andrew, M.S. Arnold, X. Wang, Development of lead iodide perovskite solar cells using three-dimensional titanium dioxide nanowire architectures. ACS Nano **9**, 564–572 (2015)
131. S.K. Pathak, A. Abate, P. Ruckdeschel, B. Roose, K.C. Goedel, Y. Vaynzof, A. Santhala, S.-I. Watanabe, D.J. Hollman, N. Noel, A. Sepe, U. Wiesner, R. Friend, H.J. Snaith, U. Steiner, Performance and stability enhancement of dye-sensitized and perovskite solar cells by Al doping of TiO_2. Adv. Funct. Mater. **24**, 6046–6055 (2014)
132. M. Yang, R. Guo, K. Kadel, Y. Liu, K. O'Shea, R. Bone, X. Wang, J. He, W. Li, Improved charge transport of Nb-doped TiO_2 nanorods in methylammonium lead iodide bromide perovskite solar cells. J. Mater. Chem. A **2**, 19616–19622 (2014)
133. Z. Yuan, Z. Wu, S. Bai, Z. Xia, W. Xu, T. Song, H. Wu, L. Xu, J. Si, Y. Jin, B. Sun, Hot-electron injection in a sandwiched TiOx-Au-TiOx structure for high-performance planar perovskite solar cells. Adv. Energy Mater. **5**, 1500038 (2015)
134. G.S. Han, H.S. Chung, B.J. Kim, D.H. Kim, J.W. Lee, B.S. Swain, K. Mahmood, J.S. Yoo, N.-G. Park, J.H. Lee, Retarding charge recombination in perovskite solar cells using ultrathin MgO-coated TiO_2 nanoparticulate films. J. Mater. Chem. A **3**, 9160–9164 (2015)
135. B.S. Swain, K. Mahmood, A.R. Kirmani et al., Highly efficient perovskite solar cells based on a nanostructured WO_3–TiO_2 core–shell electron transporting material. J. Mater. Chem. A, **3**, 9051–9057 (2015)

136. H.B. Kim, I. Im, Y. Yoon, S.D. Sung, E. Kim, J. Kim, W.I. Lee, Enhancement of photovoltaic properties of CH3NH3PbBr 3 heterojunction solar cells by modifying mesoporous TiO_2 surfaces with carboxyl groups. J. Mater. Chem. A **3**, 9264–9270 (2015)
137. J. Kwon, S.J. Kim, J.H. Park, The tailored inner space of TiO_2 electrodes via a 30 second wet etching process: high efficiency solid-state perovskite solar cells. Nanoscale **7**, 10745–10751 (2015)
138. P. Qin, A.L. Domanski, A.K. Chandiran, R. Berger, H.-J. Butt, M.I. Dar, T. Moehl, N. Tetreault, P. Gao, S. Ahmad, M.K. Nazeeruddin, M. Gratzel, Yttrium-substituted nanocrystalline TiO_2 photoanodes for perovskite based heterojunction solar cells. Nanoscale **6**, 1508–1514 (2014)
139. H.-H. Wang, Q. Chen, H. Zhou, L. Song, Z. St Louis, N. De Marco, Y. Fang, P. Sun, T.-B. Song, H. Chen, Y. Yang, Improving the TiO_2 electron transport layer in perovskite solar cells using acetylacetonate-based additives. J. Mater. Chem. A, **3**, 9108–9115 (2015)
140. H. Nagaoka, F. Ma, D.W. deQuilettes, S.M. Vorpahl, M.S. Glaz, A.E. Colbert, M.E. Ziffer, D.S. Ginger, Zr incorporation into TiO_2 electrodes reduces hysteresis and improves performance in hybrid perovskite solar cells while increasing carrier lifetimes. J. Phys. Chem. Lett. **6**, 669–675 (2015)
141. X. Zhang, Z. Bao, X. Tao, H. Sun, W. Chen, X. Zhou, Sn-doped TiO_2 nanorod arrays and application in perovskite solar cells. RSC Adv. **4**, 64001–64005 (2014)
142. K. Manseki, T. Ikeya, A. Tamura, T. Ban, T. Sugiura, T. Yoshida, Mg-doped TiO_2 nanorods improving open-circuit voltages of ammonium lead halide perovskite solar cells. RSC Adv. **4**, 9652–9655 (2014)
143. J.T.-W. Wang, J.M. Ball, E.M. Barea, A. Abate, J.A. Alexander-Webber, J. Huang, M. Saliba, I. Mora-Sero, J. Bisquert, H.J. Snaith, R.J. Nicholas, Low-temperature processed electron collection layers of graphene/TiO_2 nanocomposites in thin film perovskite solar cells. Nano Lett. **14**, 724–730 (2014)
144. S. Ito, S. Tanaka, K. Manabe, H. Nishino, Effects of surface blocking layer of Sb_2S_3 on nanocrystalline TiO_2 for $CH_3NH_3PbI_3$ perovskite solar cells. J. Phys. Chem. C **118**, 16995–17000 (2014)
145. X. Xu, Z. Liu, Z. Zuo, M. Zhang, Z. Zhao, Y. Shen, H. Zhou, Q. Chen, Y. Yang, M. Wang, Hole selective NiO contact for efficient perovskite solar cells with carbon electrode. Nano Lett. **15**, 2402–2408 (2015)
146. L. Zhang, T. Liu, L. Liu, M. Hu, Y. Yang, A. Mei, H. Han, The effect of carbon counter electrodes on fully printable mesoscopic perovskite solar cells. J. Mater. Chem. A **3**, 9165–9170 (2015)
147. X. Guo, H. Dong, W. Li, N. Li, L. Wang, Multifunctional MgO layer in perovskite solar cells. ChemPhysChem **16**, 1727–1732 (2015)
148. M.J. Carnie, C. Charbonneau, M.L. Davies, J. Troughton, T.M. Watson, K. Wojciechowski, H. Snaith, D.A. Worsley, A one-step low temperature processing route for organolead halide perovskite solar cells. Chem. Commun. **49**, 7893–7895 (2013)
149. W. Li, J. Li, L. Wang, G. Niu, R. Gao, Y. Qiu, Post modification of perovskite sensitized solar cells by aluminum oxide for enhanced performance. J. Mater. Chem. A **1**, 11735–11740 (2013)
150. G. Niu, W. Li, F. Meng, L. Wang, H. Dong, Y. Qiu, Study on the stability of $CH_3NH_3PbI_3$ films and the effect of post-modification by aluminum oxide in all-solid-state hybrid solar cells. J. Mater. Chem. A **2**, 705–710 (2014)
151. S. Guarnera, A. Abate, W. Zhang, J.M. Foster, G. Richardson, A. Petrozza, H.J. Snaith, Improving the long-term stability of perovskite solar cells with a porous Al_2O_3 buffer-layer. J. Phys. Chem. Lett. (2015)
152. S. Guarnera, A. Abate, W. Zhang, J.M. Foster, G. Richardson, A. Petrozza, H.J. Snaith, Improving the long-term stability of perovskite solar cells with a porous Al_2O_3 buffer layer. J. Phys. Chem. Lett. **6**, 432–437 (2015)

153. S.N. Habisreutinger, T. Leijtens, G.E. Eperon, S.D. Stranks, R.J. Nicholas, H.J. Snaith, Carbon nanotube/polymer composites as a highly stable hole collection layer in perovskite solar cells. Nano Lett. **14**, 5561–5568 (2014)
154. H. Wei, J. Shi, X. Xu, J. Xiao, J. Luo, J. Dong, S. Lv, L. Zhu, H. Wu, D. Li, Y. Luo, Q. Meng, Q. Chen, Enhanced charge collection with ultrathin AlO_x electron blocking layer for hole-transporting material-free perovskite solar cell. Phys. Chem. Chem. Phys. **17**, 4937–4944 (2015)
155. A. Listorti, E.J. Juarez-Perez, C. Frontera, V. Roiati, L. Garcia-Andrade, S. Colella, A. Rizzo, P. Ortiz, I. Mora-Sero, Effect of mesostructured layer upon crystalline properties and device performance on perovskite solar cells. J. Phys. Chem. Lett. **6**, 1628–1637 (2015)
156. J.-J. Shi, W. Dong, Y.-Z. Xu, C.-H. Li, S.-T. Lv, L.-F. Zhu, J. Dong, Y.-H. Luo, D.-M. Li, Q.-B. Meng, Q. Chen, Enhanced performance in perovskite organic lead iodide heterojunction solar cells with metal-insulator-semiconductor back contact. Chin. Phys. Lett. **30**, 128402 (2013)
157. M.M. Tavakoli, K.-H. Tsui, Q. Zhang, J. He, Y. Yao, D. Li, Z. Fan, Highly efficient flexible perovskite solar cells with antireflection and self-cleaning nanostructures. ACS Nano **9**, 10287–10295 (2015)
158. D. Liu, T.L. Kelly, Perovskite solar cells with a planar heterojunction structure prepared using room-temperature solution processing techniques. Nat. Photonics **8**, 133–138 (2014)
159. J. Kim, G. Kim, T.K. Kim, S. Kwon, H. Back, J. Lee, S.H. Lee, H. Kang, K. Lee, Efficient planar-heterojunction perovskite solar cells achieved via interfacial modification of a sol-gel ZnO electron collection layer. J. Mater. Chem. A **2**, 17291–17296 (2014)
160. H. Zhou, Y. Shi, K. Wang, Q. Dong, X. Bai, Y. Xing, Y. Du, T. Ma, Low-temperature processed and carbon-based ZnO/CH(3)NH(3)Pbl(3)/C planar heterojunction perovskite solar cells. J. Phys. Chem. C **119**, 4600–4605 (2015)
161. D. Bi, G. Boschloo, S. Schwarzmueller, L. Yang, E.M.J. Johansson, A. Hagfeldt, Efficient and stable $CH_3NH_3PbI_3$-sensitized ZnO nanorod array solid-state solar cells. Nanoscale **5**, 11686–11691 (2013)
162. D.-Y. Son, J.-H. Im, H.-S. Kim, N.-G. Park, 11% efficient perovskite solar cell based on ZnO nanorods: an effective charge collection system. J. Phys. Chem. C **118**, 16567–16573 (2014)
163. X. Dong, H. Hu, B. Lin, J. Ding, N. Yuan, The effect of ALD-Zno layers on the formation of $CH_3NH_3PbI_3$ with different perovskite precursors and sintering temperatures. Chem. Commun. **50**, 14405–14408 (2014)
164. D. Liu, J. Yang, T.L. Kelly, Compact layer free perovskite solar cells with 13.5 % efficiency. J. Am. Chem. Soc. **136**, 17116–17122 (2014)
165. Y. Cheng, Q.-D. Yang, J. Xiao, Q. Xue, H.-W. Li, Z. Guan, H.-L. Yip, S.-W. Tsang, Decomposition of organometal halide perovskite films on zinc oxide nanoparticles. ACS Appl. Mater. Interfaces **7**, 19986–19993 (2015)
166. K. Mahmood, B.S. Swain, A. Amassian, Double-layered ZnO nanostructures for efficient perovskite solar cells. Nanoscale **6**, 14674–14678 (2014)
167. S. Ameen, M.S. Akhtar, H.-K. Seo, M.K. Nazeeruddin, H.-S. Shin, An insight into atmospheric plasma jet modified ZnO quantum dots thin film for flexible perovskite solar cell: optoelectronic transient and charge trapping studies. J. Phys. Chem. C **119**, 10379–10390 (2015)
168. K. Mahmood, B.S. Swain, H.S. Jung, Controlling the surface nanostructure of ZnO and Al-doped ZnO thin films using electrostatic spraying for their application in 12 % efficient perovskite solar cells. Nanoscale **6**, 9127–9138 (2014)
169. J. Han, S. Yuan, L. Liu, X. Qiu, H. Gong, X. Yang, C. Li, Y. Hao, B. Cao, Fully indium-free flexible Ag nanowires/ZnO: F composite transparent conductive electrodes with high haze. J. Mater. Chem. A **3**, 5375–5384 (2015)
170. J. Dong, Y. Zhao, J. Shi, H. Wei, J. Xiao, X. Xu, J. Luo, J. Xu, D. Li, Y. Luo, Q. Meng, Impressive enhancement in the cell performance of ZnO nanorod-based perovskite solar cells with Al-doped ZnO interfacial modification. Chem. Commun. **50**, 13381–13384 (2014)

171. K. Mahmood, B. S. Swain, A. Amassian, 16.1 % efficient hysteresis-free mesostructured perovskite solar cells based on synergistically improved ZnO nanorod arrays. Adv. Energy Mater., **5**, 1500568 (2015)
172. K. Hwang, Y.S. Jung, Y.J. Heo, F.H. Scholes, S.E. Watkins, J. Subbiah, D.J. Jones, D.Y. Kim, D. Vak, Toward large scale roll-to-roll production of fully printed perovskite solar cells. Adv. Mater. **27**, 1241–1247 (2015)
173. S. He, L. Qiu, X. Fang, G. Guan, P. Chen, Z. Zhang, H. Peng, Radically grown obelisk-like ZnO arrays for perovskite solar cell fibers and fabrics through a mild solution process. J. Mater. Chem. A **3**, 9406–9410 (2015)
174. J.-Y. Jeng, K.-C. Chen, T.-Y. Chiang, P.-Y. Lin, T.-D. Tsai, Y.-C. Chang, T.-F. Guo, P. Chen, T.-C. Wen, Y.-J. Hsu, Nickel oxide electrode interlayer in $CH_3NH_3PbI_3$ perovskite/PCBM planar-heterojunction hybrid solar cells. Adv. Mater. **26**, 4107–4113 (2014)
175. K.-C. Wang, J.-Y. Jeng, P.-S. Shen, Y.-C. Chang, E.W.-G. Diau, C.-H. Tsai, T.-Y. Chao, H.-C. Hsu, P.-Y. Lin, P. Chen, T.-F. Guo, T.-C. Wen. *p*-Type mesoscopic nickel oxide/organometallic perovskite heterojunction solar cells. Sci. Rep., **4** (2014)
176. L. Hu, J. Peng, W. Wang, Z. Xia, J. Yuan, J. Lu, X. Huang, W. Ma, H. Song, W. Chen, Y.-B. Cheng, J. Tang, Sequential deposition of $CH_3NH_3PbI_3$ on planar NiO film for efficient planar perovskite solar cells. ACS Photonics **1**, 547–553 (2014)
177. Z. Zhu, Y. Bai, T. Zhang, Z. Liu, X. Long, Z. Wei, Z. Wang, L. Zhang, J. Wang, F. Yan, S. Yang, High-performance hole-extraction layer of sol-gel-processed NiO nanocrystals for inverted planar perovskite solar cells. Angew. Chem. Int. Ed. **53**, 12571–12575 (2014)
178. J. Cui, F. Meng, H. Zhang, K. Cao, H. Yuan, Y. Cheng, F. Huang, M. Wang, $CH_3NH_3PbI_3$-based planar solar cells with magnetron-sputtered nickel oxide. ACS Appl. Mater. Interfaces **6**, 22862–22870 (2014)
179. K.-C. Wang, P.-S. Shen, M.-H. Li, S. Chen, M.-W. Lin, P. Chen, T.-F. Guo, Low-temperature sputtered nickel oxide compact thin film as effective electron blocking layer for mesoscopic NiO/$CH_3NH_3PbI_3$ perovskite heterojunction solar cells. ACS Appl. Mater. Interfaces **6**, 11851–11858 (2014)
180. H. Tian, B. Xu, H. Chen, E.M.J. Johansson, G. Boschloo, Solid-state perovskite-sensitized *p*-type mesoporous nickel oxide solar cells. ChemSusChem **7**, 2150–2153 (2014)
181. H. Wang, X. Zeng, Z. Huang, W. Zhang, X. Qiao, B. Hu, X. Zou, M. Wang, Y.-B. Cheng, W. Chen, Boosting the photocurrent density of *p*-type solar cells based on organometal halide perovskite-sensitized mesoporous NiO photocathodes. ACS Appl. Mater. Interfaces **6**, 12609–12617 (2014)
182. J.H. Kim, P.-W. Liang, S.T. Williams, N. Cho, C.-C. Chueh, M.S. Glaz, D.S. Ginger, A.K.Y. Jen, High-performance and environmentally stable planar heterojunction perovskite solar cells based on a solution-processed copper-doped nickel oxide hole-transporting layer. Adv. Mater. **27**, 695–701 (2015)
183. Z. Liu, M. Zhang, X. Xu, L. Bu, W. Zhang, W. Li, Z. Zhao, M. Wang, Y.-B. Cheng, H. He, *p*-Type mesoscopic NiO as an active interfacial layer for carbon counter electrode based perovskite solar cells. Dalton Trans. **44**, 3967–3973 (2015)
184. W. Chen, Y. Wu, J. Liu, C. Qin, X. Yang, A. Islam, Y.-B. Cheng, L. Han, Hybrid interfacial layer leads to solid performance improvement of inverted perovskite solar cells. Energy Environ. Sci. **8**, 629–640 (2015)
185. Y. Bai, H. Yu, Z. Zhu, K. Jiang, T. Zhang, N. Zhao, S. Yang, H. Yan, High performance inverted structure perovskite solar cells based on a PCBM: polystyrene blend electron transport layer. J. Mater. Chem. A **3**, 9098–9102 (2015)
186. M.F. De Volder, S.H. Tawfick, R.H. Baughman, A.J. Hart, Carbon nanotubes: present and future commercial applications. Science **339**, 535–539 (2013)
187. D.S. Hecht, L. Hu, G. Irvin, Emerging transparent electrodes based on thin films of carbon nanotubes, graphene, and metallic nanostructures. Adv. Mater. **23**, 1482–1513 (2011)

188. W. Li, H. Dong, X. Guo, N. Li, J. Li, G. Niu, L. Wang, Graphene oxide as dual functional interface modifier for improving wettability and retarding recombination in hybrid perovskite solar cells. J. Mater. Chem. A **2**, 20105–20111 (2014)
189. Z. Wu, S. Bai, J. Xiang, Z. Yuan, Y. Yang, W. Cui, X. Gao, Z. Liu, Y. Jin, B. Sun, Efficient planar heterojunction perovskite solar cells employing graphene oxide as hole conductor. Nanoscale **6**, 10505–10510 (2014)
190. J.-S. Yeo, R. Kang, S. Lee, Y.-J. Jeon, N. Myoung, C.-L. Lee, D.-Y. Kim, J.-M. Yun, Y.-H. Seo, S.-S. Kim, S.-I. Na, Highly efficient and stable planar perovskite solar cells with reduced graphene oxide nanosheets as electrode interlayer. Nano Energy **12**, 96–104 (2015)
191. T. Liu, D. Kim, H. Han, A.R.B. Mohd Yusoff, J. Jang, Fine-tuning optical and electronic properties of graphene oxide for highly efficient perovskite solar cells. Nanoscale **7**, 10708–10718 (2015)
192. K. Yan, Z. Wei, J. Li, H. Chen, Y. Yi, X. Zheng, X. Long, Z. Wang, J. Wang, J. Xu, S. Yang, High-performance graphene-based hole conductor-free perovskite solar cells: Schottky junction enhanced hole extraction and electron blocking. Small **11**, 2269–2274 (2015)
193. P. You, Z. Liu, Q. Tai, S. Liu, F. Yan, Efficient semitransparent perovskite solar cells with graphene electrodes. Adv. Mater. **27**, 3632–3638 (2015)
194. J. Cao, Y.-M. Liu, X. Jing, J. Yin, J. Li, B. Xu, Y.-Z. Tan, N. Zheng, Well-defined thiolated nanographene as hole-transporting material for efficient and stable perovskite solar cells. J. Am. Chem. Soc. **137**, 10914–10917 (2015)
195. Y. Yang, J. Xiao, H. Wei, L. Zhu, D. Li, Y. Luo, H. Wu, Q. Meng, An all-carbon counter electrode for highly efficient hole-conductor-free organo-metal perovskite solar cells. RSC Adv. **4**, 52825–52830 (2014)
196. F. Zhang, X. Yang, H. Wang, M. Cheng, J. Zhao, L. Sun, Structure engineering of hole-conductor free perovskite-based solar cells with low-temperature-processed commercial carbon paste as cathode. ACS Appl. Mater. Interfaces **6**, 16140–16146 (2014)
197. Z. Ku, Y. Rong, M. Xu, T. Liu, H. Han, Full printable processed mesoscopic $CH_3NH_3PbI_3$/TiO_2 heterojunction solar cells with carbon counter electrode. Sci. Rep., **3** (2013)
198. H. Wang, X. Hu, H. Chen, The effect of carbon black in carbon counter electrode for $CH_3NH_3PbI_3/TiO_2$ heterojunction solar cells. RSC Adv. **5**, 30192–30196 (2015)
199. M. Xu, Y. Rong, Z. Ku, A. Mei, T. Liu, L. Zhang, X. Li, H. Han, Highly ordered mesoporous carbon for mesoscopic $CH_3NH_3PbI_3/TiO_2$ heterojunction solar cell. J. Mater. Chem. A **2**, 8607–8611 (2014)
200. A. Mei, X. Li, L. Liu, Z. Ku, T. Liu, Y. Rong, M. Xu, M. Hu, J. Chen, Y. Yang, A hole-conductor–free, fully printable mesoscopic perovskite solar cell with high stability. Science **345**, 295–298 (2014)
201. Z. Li, S.A. Kulkarni, P.P. Boix, E. Shi, A. Cao, K. Fu, S.K. Batabyal, J. Zhang, Q. Xiong, L.H. Wong, N. Mathews, S.G. Mhaisalkar, Laminated carbon nanotube networks for metal electrode-free efficient perovskite solar cells. ACS Nano **8**, 6797–6804 (2014)
202. S.N. Habisreutinger, T. Leijtens, G.E. Eperon, S.D. Stranks, R.J. Nicholas, H.J. Snaith, Carbon nanotube/polymer composites as a highly stable hole collection layer in perovskite solar cells. Nano Lett. **14**, 5561–5568 (2014)
203. R. Li, X. Xiang, X. Tong, J. Zou, Q. Li, Wearable double-twisted fibrous perovskite solar cell. Adv. Mater. **27**, 3831–3835 (2015)
204. K. Wojciechowski, T. Leijtens, S. Spirova, C. Schlueter, M. Hoerantner, J.T.-W. Wang, C.-Z. Li, A.K.-Y. Jen, T.-L. Lee, H.J. Snaith, C_{60} as an efficient N-type compact layer in perovskite solar cells. J. Phys. Chem. Lett. **6**, 2399–2405 (2015)
205. Y. Li, J. Zhu, Y. Huang, F. Liu, M. Lv, S. Chen, L. Hu, J. Tang, J. Yao, S. Dai, Mesoporous SnO_2 nanoparticle films as electron-transporting material in perovskite solar cells. Rsc Adv. **5**, 28424–28429 (2015)
206. J. Song, E. Zheng, J. Bian, X.-F. Wang, W. Tian, Y. Sanehira, T. Miyasaka, Low-temperature SnO_2-based electron selective contact for efficient and stable perovskite solar cells. J. Mater. Chem. A **3**, 10837–10844 (2015)

207. Q. Dong, Y. Shi, K. Wang, Y. Li, S. Wang, H. Zhang, Y. Xing, Y. Du, X. Bai, T. Ma, Insight into Perovskite solar cells based on SnO_2 compact electron-selective layer. J. Phys. Chem. C **119**, 10212–10217 (2015)
208. M.J. Carnie, C. Charbonneau, M.L. Davies, B.O. Regan, D.A. Worsley, T.M. Watson, Performance enhancement of solution processed perovskite solar cells incorporating functionalized silica nanoparticles. J. Mater. Chem. A **2**, 17077–17084 (2014)
209. S.H. Hwang, J. Roh, J. Lee, J. Ryu, J. Yun, J. Jang, Size-controlled SiO_2 nanoparticles as scaffold Layers in thin-film perovskite solar cells. J. Mater. Chem. A **2**, 16429–16433 (2014)
210. S.S. Shin, W.S. Yang, J.H. Noh, J.H. Suk, N.J. Jeon, J.H. Park, J.S. Kim, W.M. Seong, S.I. Seok, High-performance flexible perovskite solar cells exploiting Zn_2SnO_4 prepared in solution below 100 °C. Nat. Commum. **6**, 7410 (2015)
211. L.S. Oh, D.H. Kim, J.A. Lee, S.S. Shin, J.-W. Lee, I.J. Park, M.J. Ko, N.-G. Park, S.G. Pyo, K.S. Hong, J.Y. Kim, Zn_2SnO_4-based photoelectrodes for organolead halide perovskite solar cells. J. Phys. Chem. C **118**, 22991–22994 (2014)
212. A. Bera, K. Wu, A. Sheikh, E. Alarousu, O.F. Mohammed, T. Wu, Perovskite oxide $SrTiO_3$ as an efficient electron transporter for hybrid perovskite solar cells. J. Phys. Chem. C **118**, 28494–28501 (2014)
213. A. Kogo, Y. Numata, M. Ikegami, T. Miyasaka, Nb_2O_5 blocking layer for high open-circuit voltage perovskite solar cells. Chem. Lett. **44**, 829–830 (2015)
214. L. Hu, W. Wang, H. Liu, J. Peng, H. Cao, G. Shao, Z. Xia, W. Ma, J. Tang, PbS colloidal quantum dots as an effective hole transporter for planar heterojunction perovskite solar cells. J. Mater. Chem. A **3**, 515–518 (2015)
215. M. Lv, J. Zhu, Y. Huang, Y. Li, Z. Shao, Y. Xu, S. Dai, Colloidal $CuInS_2$ quantum dots as inorganic hole-transporting material in perovskite solar cells. ACS Appl. Mater. Interfaces **7**, 17482–17488 (2015)
216. Q. Hu, J. Wu, C. Jiang, T. Liu, X. Que, R. Zhu, Q. Gong, Engineering of electron-selective contact for perovskite solar cells with efficiency exceeding 15 %. ACS Nano **8**, 10161–10167 (2014)
217. Y. Zhao, A.M. Nardes, K. Zhu, Effective hole extraction using MoO_x-Al contact in perovskite $CH_3NH_3PbI_3$ solar cells. Appl. Phys. Lett., **104**, (2014)
218. F. Hou, Z. Su, F. Jin, X. Yan, L. Wang, H. Zhao, J. Zhu, B. Chu, W. Li, Efficient and stable planar heterojunction perovskite solar cells with an MoO_3/PEDOT:PSS hole transporting layer. Nanoscale **7**, 9427–9432 (2015)
219. J.A. Christians, R.C.M. Fung, P.V. Kamat, An inorganic hole conductor for organo-lead halide perovskite solar cells. Improved hole conductivity with copper iodide. J. Am. Chem. Soc. **136**, 758–764 (2014)
220. S. Chavhan, O. Miguel, H.-J. Grande, V. Gonzalez-Pedro, R.S. Sanchez, E.M. Barea, I. Mora-Sero, R. Tena-Zaera, Organo-metal halide perovskite-based solar cells with CuSCN as the inorganic hole selective contact. J. Mater. Chem. A **2**, 12754–12760 (2014)
221. S. Ye, W. Sun, Y. Li, W. Yan, H. Peng, Z. Bian, Z. Liu, C. Huang, CuSCN-based inverted planar perovskite solar cell with an average PCE of 15.6. Nano Lett. **15**, 3723–3728 (2015)
222. S. Ito, S. Tanaka, H. Vahlman, H. Nishino, K. Manabe, P. Lund, Carbon-double-bond-free printed solar cells from $TiO_2/CH_3NH_3PbI_3$/CuSCN/Au: structural control and photoaging effects. ChemPhysChem **15**, 1194–1200 (2014)

Chapter 2
Dielectric Nanomaterials for Silicon Solar Cells

Ingo Dirnstorfer and Thomas Mikolajick

Abstract Dielectric nanomaterials are emerging as key components in today's highly efficient silicon solar cells. The most successful materials are SiO_2, SiN_x:H and Al_2O_3 due to their excellent material properties for surface passivation and light management. Dielectric passivation layers ensure high level of chemical passivation by effectively reducing the density of silicon surface states, which could act as recombination centers for photo-generated charge carriers. Additionally, these materials provide strong field-effect passivation by a high density of intrinsic fixed charges. However, novel solar cell concepts demand for dielectric nanomaterials with additional functionalities, such as electrical conductivity, simultaneous applicability on *n*- and *p*-type substrates or compatibility with low-thermal budget processing. Recent developments show that these functionalities can be realized in multi-oxide nanolaminates including HfO_2 or TiO_2 sublayers. Additionally, alternative deposition techniques and innovative processes, such as flash light annealing, become available to reduce process complexity and thermal budget. This chapter reviews the theory and application of dielectric nanomaterials in today's solar cells and gives an outlook on promising solutions in future solar cells.

I. Dirnstorfer (✉) · T. Mikolajick
NaMLab gGmbH, Nöthnitzer Strasse 64, 01187 Dresden, Germany
e-mail: Ingo.Dirnstorfer@namlab.com

T. Mikolajick
e-mail: Thomas.Mikolajick@namlab.com

T. Mikolajick
Institute of Semiconductor and Microsystems, TU Dresden,
Nöthnitzer Strasse 64, 01187 Dresden, Germany

T. Mikolajick
Center for Advancing Electronics Dresden (cfaed), TU Dresden,
01062 Dresden, Germany

Q. Li (ed.), *Nanomaterials for Sustainable Energy*,
NanoScience and Technology, DOI 10.1007/978-3-319-32023-6_2

2.1 Dielectric Nanomaterials in Today's and Future Silicon Solar Cells

The introduction of dielectric nanomaterials was one of the major advances in silicon solar cell development. Thermal oxidation of the Si wafer resulted in a substantial improvement of the surface passivation quality and this improvement was essential for the first solar cells surpassing the 20 % efficiency mark in the 1980s [1, 2]. SiO_2 effectively suppresses surface recombination of photo-generated electron hole pairs. Additionally, SiO_2 provides excellent optical properties for controlling light reflection losses in the solar cell. In the subsequent years, additional dielectric nanomaterials emerged for solar cell passivation and these materials gradually developed from lab scale to industrial standard. SiO_2 is still used for high performance solar cells [3]. However, SiN_x:H became the most common dielectric material in solar cell manufacturing [4, 5]. Recently, Al_2O_3 also proved excellent passivation performance and established itself as an additional dielectric material in silicon solar cells [6, 7].

Figure 2.1a sketches a passivated emitter and rear cell (PERC), which is currently implemented in industrial manufacturing. The PERC concept is anticipated to become the dominant commercial cell technology by 2020 [8]. In the PERC solar cell, the front and rear sides are passivated with dielectric nanomaterials [9]. In most concepts, SiN_x:H is used to passivate the front side of the cell, whereas Al_2O_3 in combination with a SiN_x:H capping layer is applied at the rear side.

Since dielectric materials are insulators, the passivation layers are locally opened to facilitate the electrical contact between silicon and metallization. The state-of-the-art technology for contacting through the SiN_x:H layer is the application of a screen printing paste containing a lead borosilicate glass frit. During fast firing annealing at temperatures around 800 °C, this frit etches SiN_x:H and promotes the electrical contact between the front side metal paste and the Si surface [9]. The rear side metallization is contacted by local contact opening (LCO) [10] or laser fired contacts (LFC) [9]. These processes locally contact the Si wafer though point contacts, while the surface passivation layer is maintained in large parts.

Today's PERC solar cells benefit from considerable R&D effort during the last decades resulting in dielectric nanomaterials with excellent performance, stability and scalability in industrial environment. However, solar cell development moves forward and novel solar cell concepts emerge with higher conversion efficiencies and lower manufacturing costs. Selected cell concepts are displayed in Fig. 2.1(b–f). These novel concepts demand for advanced dielectric nanomaterials, which will be discussed in the following.

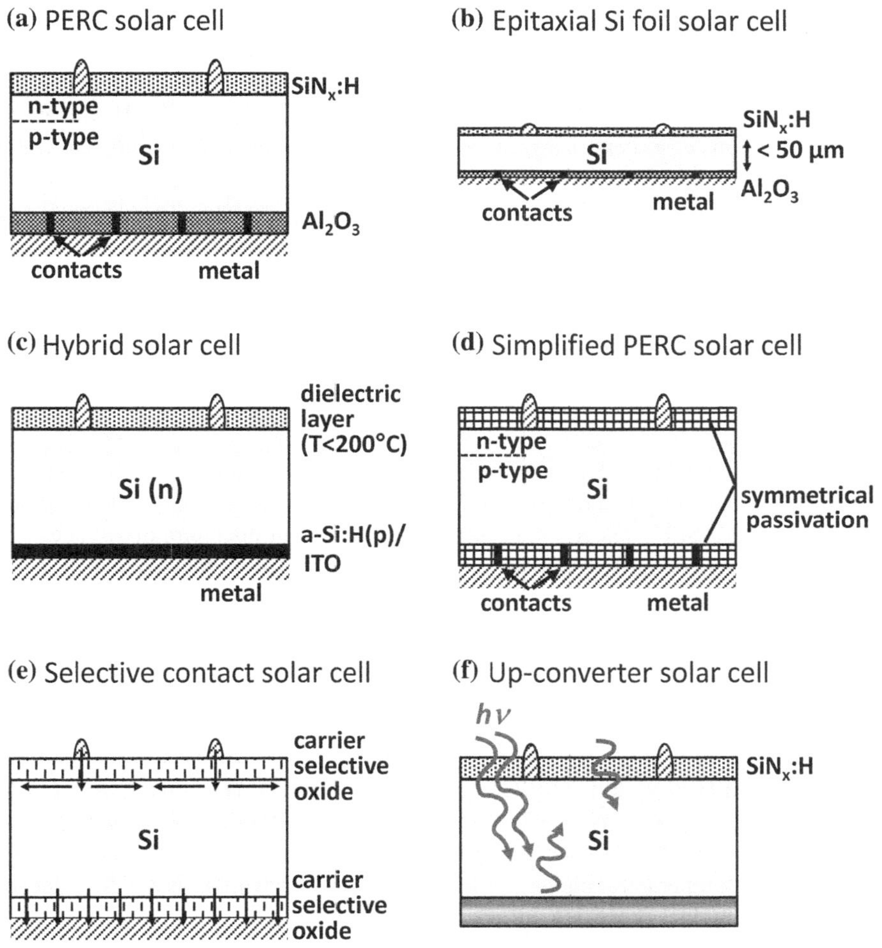

Fig. 2.1 Schematic drawing of the state-of-the-art PERC solar cell (**a**) and emerging solar cell concepts, which require further development of dielectric nanomaterials: epitaxial Si foil-based solar cell (**b**), hybrid silicon heterojunction solar cell with dielectric surface passivation (**c**), simplified PERC solar cell with symmetrical passivation (**d**), selective contact solar cell (**e**) and up-converter solar cell (**f**). The concepts are discussed in the text

2.1.1 Epitaxial Si Foil-Based Solar Cell

Thin, epitaxial crystalline Si foil-based solar cells are realized by CVD growth on porosified substrates. After deposition, the Si films are detached from the parent wafer and bonded to a foreign carrier, while the parent wafer is re-used [11]. A PERC-like epitaxial Si foil solar cell is shown in Fig. 2.1b. Epitaxial silicon solar cells with sub-50 μm absorber layers reach efficiencies of about 20 % today [12–14]. Thinning of Si absorbers is a very promising route to reduce material consumption. Additionally, wafer thinning is motivated by conversion efficiency enhancement because the intrinsic solar cell efficiency limit imposed by Auger recombination shifts towards higher values when the wafer thickness is decreased from today's standard ($\sim$180 μm) to values below 100 μm [15]. However, in thinned silicon absorbers, surface recombination losses and incomplete light absorption arise as critical limitations [16, 17], imposing higher requirements on the dielectric nanomaterials in the solar cells. The critical parameters for electrical and optical losses are discussed in Sects. 2.2 and 2.5, respectively.

2.1.2 Heterojunction Solar Cell with Dielectric Front Side Layer

Silicon heterojunction solar cells are based on a *pn*-junction with hydrogenated amorphous Si (a-Si:H) and crystalline Si (c-Si). Due to the very good surface passivation of a-Si:H, this cell concept reaches very high open circuit voltages (up to 750 mV) and efficiencies up to 24.7 % in double sided contacted cells [18]. Though excellent efficiencies are achieved, the short circuit current is limited by the very low quantum efficiency of the front side layers, i.e. a-Si:H and ITO, due to their poor photoelectric properties. To reduce the parasitic light absorption in these layers, hybrid solar cells are suggested. In hybrid solar cells, the *pn*-junction is located at the rear side and a highly transparent dielectric material is used at the front side (Fig. 2.1c) [19–21]. One obstacle for this concept is the low temperature budget of a-Si:H ($T < 200$ °C) [22] compared to the relatively high thermal activation energy required for the dielectric passivation layers ($T \sim 400$ °C). The integration of hybrid SHJ solar cells with dielectric front side layers can be facilitated if the temperature budget of the dielectric passivation layer is reduced. Low-thermal budget processes are discussed in Sect. 2.3.4.

2.1.3 Solar Cell with Symmetrical Passivation

The state-of-the-art PERC concept is realized with two different passivation materials for the front and rear side, respectively. Two different materials are required, since charged dielectric passivation layers perform best on substrates with oppositely charged majority carriers [23–25]. In a simplified PERC process flow, the same passivation material is deposited in a single process step on both sides of the solar cell. A solar cell with symmetrical passivation materials for *p*- and *n*-type Si is shown in Fig. 2.1d. Symmetrical passivation is also applied in interdigitated back contact (IBC) solar cells for the *p*- and *n*-doped areas at the rear side [26, 27]. Symmetrical passivation can be realized by materials, which are free of intrinsic charges. Zero-fixed-charge multi-oxide nanolaminates are discussed in Sect. 2.4.1.

2.1.4 Solar Cell with Carrier Selective Contacts

PERC solar cells employ insulating dielectric passivation layers with local interconnects between silicon and metallization. Though necessary for the electrical contact, the metal-semiconductor interconnects partly degrade the solar cell performance, because of their high density of interface states and a very high local surface recombination rate. Therefore, the contact area requires thorough optimization to achieve both low recombination and ohmic losses [10]. An alternative to local point contacts are full area, carrier selective contacts, which provide a high level of surface passivation and electrical conductivity (Fig. 2.1e). In general, these contacts separate the absorber material, i.e. Si, and the metal contact by a medium with different conductivities for majority and minority carriers [28]. Dielectric passivation layers can be applied as carrier selective contacts if they provide sufficient electrical conductivity in addition to high surface passivation. Multi-oxide nanolaminates employing only dielectric materials are very interesting candidates due to their synergies with materials already used in PERC manufacturing. Dielectric carrier selective contacts are discussed in Sect. 2.4.2.

2.1.5 Up-Converter Solar Cell

In single junction solar cells, the dominant loss processes are sub-band gap transmission and thermalization losses of electron-hole pairs generated by photons exceeding the band gap energy. These losses can be significantly reduced if the solar spectrum is matched to the silicon band gap [29]. Figure 2.1f shows a solar cell concept with an up-conversion layer attached to the rear side. Within this layer two sub-band gap photons are up-converted into one photon, which can be absorbed in Si. As the up-conversion process utilizes sub-band gap photons, this process overcomes the fundamental Shockley-Queisser efficiency limit of 30 % [30]. Up-conversion shifts the theoretical efficiency limit of a single junction Si solar cell towards 40 % [31]. In a similar way, down-conversion converts high energy photons into two or more photons of lower energies. Down-conversion also shifts the efficiency limit due to the better utilization of the photon energy. Concepts for spectral conversion are discussed in Sect. 1.5.3.

2.2 Theory of Surface Recombination and Surface Passivation

2.2.1 Surface Recombination Model

Open bonds at the surface of a crystal result in additional energy levels, which could be located within the band gap and form recombination centers for electrons and holes. Free carriers are first captured in these deep traps and then recombine in a second step. The liberated energy is converted to lattice vibrations and phonons. The Shockley-Read-Hall (SRH) model describes the recombination rate U_{SRH} via deep traps by [32–34]:

$$U_{\mathrm{SRH}} = \sigma_n \sigma_p v_{\mathrm{th}} \left(n_s p_s - n_i^2\right) \int \frac{D_{it} dE}{\sigma_n (n_s + n_1) + \sigma_p (p_s + p_1)} \approx \frac{N_T v_{th} \sigma_n \sigma_p n_s p_s}{\sigma_n n_s + \sigma_p p_s} \quad (2.1)$$

with the thermal velocity v_{th} and the electron and hole concentrations n_{s} and p_{s} at the surface (unit: cm^{-3}). The free carrier concentrations are a function of substrate doping and the excess carrier concentration, which depends on the optical generation rate in an illuminated Si wafer. Furthermore, n_1 and p_1 are defined as

$$n_1 = n_i \exp\left(\frac{E - E_i}{kT}\right) \text{ and } p_1 = n_i \exp\left(-\frac{E - E_i}{kT}\right) \quad (2.2)$$

with the intrinsic carrier concentration n_{i} and the intrinsic energy level E_{i}. The surface traps are characterized by their density D_{it} (unit: $\mathrm{eV}^{-1}\ \mathrm{cm}^{-2}$) and the capture cross sections σ_{n} and σ_{p} for electron and holes, respectively. In general, trap levels are distributed throughout the band gap and their densities and capture cross sections are functions of trap energy. Therefore, the recombination rate is determined by an integral from the conduction to valence band energy level.

The SRH formula simplifies when assuming relevant illumination (n_{s}, $p_{\mathrm{s}} \gg n_{\mathrm{i}}$) and solely mid-gap defects at E_{i}. Under these assumptions the recombination rate only depends on the defect density N_{T} (unit: cm^{-2}), the free carrier concentrations at the surface and the capture cross sections of the defect (2.1). In this simplified formula the surface recombination is a linear function of N_{T}. When differentiating this expression with respect to the carrier concentration, a maximum is found at $p_{\mathrm{s}}/n_{\mathrm{s}} = \sigma_{\mathrm{n}}/\sigma_{\mathrm{p}}$, i.e. the highest recombination rate is reached when the ratio of holes and electrons is equal to the inverse ratio of their capture cross sections. The capture cross section ratio of mid-gap surface defects is asymmetrical. Typical $\sigma_{\mathrm{n}}/\sigma_{\mathrm{p}}$-values of Si surface defects are 1000 [23] and 20–1000 [35–37], when the surface is passivated with SiO_2 and Al_2O_3, respectively. Consequently, the highest recombination occurs when the surface concentration of holes exceeds the concentration of electrons.

The quality of the surface passivation is measured by the surface recombination velocity, which is defined as [34]

$$S_{\mathrm{eff}} = \frac{U_{\mathrm{SRH}}}{\Delta n} \tag{2.3}$$

with the injection level Δn. At a planar surface, the upper limit of surface recombination velocity is constituted by the thermal velocity of free carriers in the order of 10^7 cm/s [38]. In today's PERC solar cells, S_{eff} is typically below 100 cm/s [7, 39]. Very low surface recombination velocities in the range of 1 cm/s and even below are achieved with high quality SiO_2 [24], Al_2O_3 [7, 40] and SiN_x:H [41] passivation layers.

The surface recombination velocity is usually not measured directly but deduced from the effective carrier lifetime (τ_{eff}) in a Si substrate. Several mature techniques are available to determine the carrier lifetime in Si, such as quasi-steady-state photoconductance (QSSPC) [42, 43] microwave-detected photoconductivity (MDP) [44] or photoluminescence [45]. In a both side passivated Si wafer, the correlation between carrier lifetime and surface recombination velocity is [46]:

$$\frac{1}{\tau_{\mathrm{eff}}} = \frac{1}{\tau_{\mathrm{bulk}}} + \frac{2S_{\mathrm{eff}}}{W} \tag{2.4}$$

with the wafer thickness W and the carrier lifetime of the bulk material τ_{bulk}. When the surface is poorly passivated and recombination rate becomes very high, an additional term describing the minority carrier diffusion towards the surface has to be taken into account [46]. The measured effective carrier lifetime is a function of surface and bulk recombination and both recombination channels cannot be separated with this measurement. Therefore, the evaluation of passivation materials is usually done on high quality float zone (fz) Si substrates. In fz Si wafers, the impurity concentration is very low and SRH recombination at bulk defects is negligible. The bulk lifetime is only limited by Auger and radiative recombination, where the excess energy is transferred to a free carrier or a photon. The recombination rate of these intrinsic processes depends on doping concentrations and injection level [47]. Considering a typical injection level of non-concentrating solar cells ($\Delta n = 10^{14}$ cm^{-3}) and a typical substrate resistivity of 2 Ωcm, the intrinsic carrier lifetime can be calculated to be 15 and 25 ms for *p*- and *n*-type Si, respectively. When the measured effective carrier lifetime is well below the calculated intrinsic limit, (2.4) simplifies to $S_{\mathrm{eff}} \approx 0.5\ W/\tau_{\mathrm{eff}}$. Based on this reasoning the measured carrier lifetime is commonly used as a measure for the surface passivation quality.

2.2.2 Dielectric Charges and Near Surface Recombination

According to (2.1), the surface SRH recombination is a function of the electron and hole concentrations at the surface. These concentrations are strongly influenced by a dielectric surface passivation layer if it contains intrinsic fixed charges. A dielectric layer with positive fixed charges attracts electrons, whereas a negatively charged layer attracts holes to the Si surface. The surface charge densities can be calculated by simultaneously solving the Poisson and continuity equations with the boundary conditions given by surface SRH recombination and optical generation. A numerical scheme for solving these equations was described by Girisch et al. [48]. Nowadays, fast calculation of carrier distributions and recombination rates is provided by several shareware Poisson solvers, such as PC1D [49] or AFORS-HET [50].

Figure 2.2a shows the simulated electron and hole surface concentrations as a function of the fixed charges density (Q_f). Densities are always positive; however, Q_f often ranges from negative to positive values, indicating the polarity of the charges. Fixed charges with densities in the order of 10^{12} cm^{-2} can heavily influence the surface carrier concentration. When the wafer is illuminated, the surface minority carrier concentration strongly increases whereas the relative change of majority carriers is small. As a consequence, the asymmetry in free carrier concentrations reduces with increasing optical generation. However, this asymmetry is still sufficient to effectively suppress surface recombination.

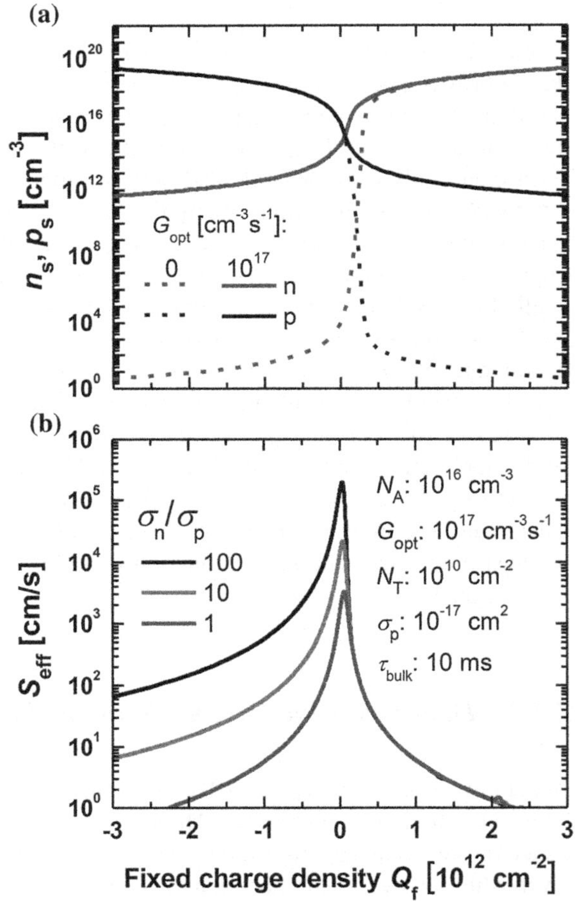

Fig. 2.2 AFORS-HET simulation of surface electron and hole concentrations with and without illumination (**a**) and surface recombination velocity (**b**) as a function of the fixed charge density in the dielectric. The input parameters for the simulation are displayed in the figure. Both, negative and positive fixed charges suppress surface recombination

The effect of fixed charges on the surface recombination is shown in Fig. 2.2b. For simulation, a constant generation rate is used as input parameter, e.g. caused by sun light illumination. The injection level is not constant as it depends on the effective carrier lifetime via $\Delta n = G_{\mathrm{opt}}/\tau_{\mathrm{eff}}$ and thus on variable surface recombination. The highest recombination is found at the point where the fixed charges result in $p_s/n_s = \sigma_n/\sigma_p$. A deviation from this condition reduces the surface recombination as the recombination becomes limited by either holes or electrons. Consequently, negative and positive fixed charges suppress surface recombination equally. At low injection conditions the relation between surface recombination velocity and fixed charges can be expressed analytically [51, 52]:

$$S_{\mathrm{eff}} \sim \frac{N_{\mathrm{dop}}}{Q_f^2} \tag{2.5}$$

with N_{dop} being the acceptor or donor concentration in *p*- or *n*-type Si, respectively. When the σ_n/σ_p-ratio increases, the recombination maximum slightly shifts towards higher surface hole concentrations, i.e. towards negative fixed charges. However, this shift is very small. More significant is the fact that the surface recombination curve becomes asymmetrical due to the higher capture rate for one carrier type.

The influence of charge polarity on the Si surface band bending is shown in Fig. 2.3. In *p*-type Si, negative fixed charges create an accumulation layer, where majority carriers are accumulated at the surface. Positive fixed charges form in inversion layer with increased minority carrier concentration. Both charge polarities suppress surface recombination as shown in Fig. 2.2. However, an additional near surface recombination (NSR) channel opens in the inversion layer. At a distinct point within the inversion layer, the carrier concentrations fulfill the requirement for highest recombination rate, i.e. $p_s/n_s = \sigma_n/\sigma_p$. For typical solar cell doping concentrations ($\sim 10^{16}\ \mathrm{cm}^{-3}$) and fixed charge densities ($\sim 10^{12}\ \mathrm{cm}^{-2}$), this point is approximately 100 nm below the surface. At this position, the recombination at deep

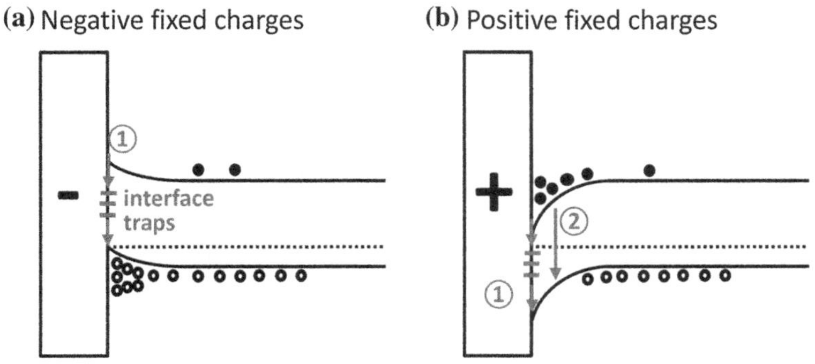

Fig. 2.3 Schematic band diagram of *p*-type Si with a passivation layer containing negative (**a**) and positive (**b**) fixed charges. *Arrows* indicate the recombination channels through interface traps at the surface (*1*) and in the near surface region (*2*)

bulk defects is strongly enhanced. This effect was first observed with positively charged SiO_2 [23–25] and SiN_x:H [53, 54] passivation layers on *p*-type Si. Later, this effect was also measured with negatively charged Al_2O_3 passivation layers on *n*-type Si [40, 55]. The effect was analytically described for the first time by Glunz et al. using an empirical formula, which was deduced from the ideal diode equation [56]:

$$S_{\text{NSR}} = \frac{J_0}{q\Delta n}\left(\sqrt{\Delta n/n_{0,\min} + 1} - 1\right) \tag{2.6}$$

with the equilibrium minority carrier concentration $n_{0,\text{min}}$ and the saturation current J_0 as fit parameter. In this equation, the near surface recombination velocity S_{NSR} increases with decreasing Δn. Consequently, this effect dominates the lifetime at low injection levels. Though the near surface recombination was described already more than one decade ago, the origin of the involved recombination centers is still under discussion. Steingrube et al. suggested the existence of a surface damage region containing additional defects, which are potentially formed by hydrogen incorporation during the deposition of the dielectric layer or the post-deposition annealing step [51, 57–59]. Alternatively, it was shown that homogenously distributed bulk defects could significantly contribute to near surface recombination even if their concentration is very small. In the inversion region of p-type Si, "harmless" bulk defects turn into highly efficient recombination centers when their σ_n/σ_p–ratio is large and the energy level of the defect is located in the lower half of the band gap [60]. The reduced lifetime in the low injection level range can also be consistently explained by transport of carriers towards recombination centers at the edge of the sample [61, 62].

Figure 2.4 shows the effective carrier lifetime in *p*- and *n*-type Si with an ideally passivated surface ($S_{\text{eff}} = 0$). A passivation layer with negative fixed charges is assumed resulting in an inversion layer in *n*-type Si. The carrier lifetime is limited

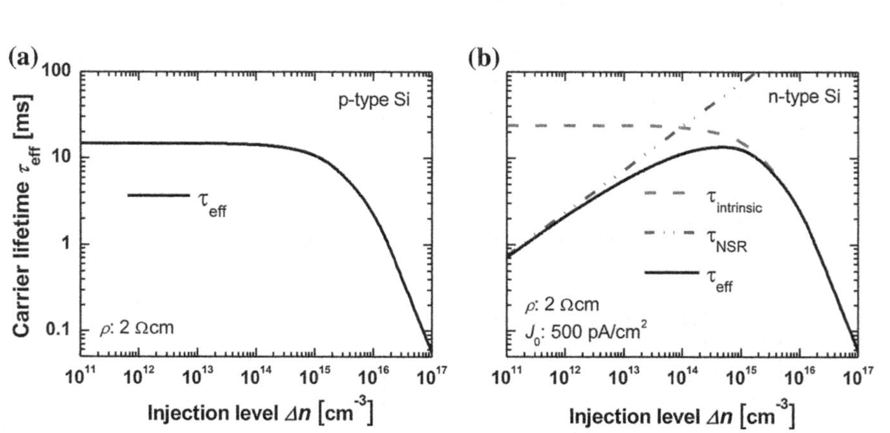

Fig. 2.4 Calculated carrier lifetime as a function of injection level for *p*- (**a**) and *n*-type (**b**) Si. A passivation layer with negative fixed charges is assumed ($S_{\text{eff}} = 0$). The intrinsic bulk lifetime ($\tau_{\text{intrinsic}}$) and near surface recombination are calculated according to [47] and (2.6)

by the intrinsic bulk lifetime and, in *n*-type Si, also by the near surface recombination. In the injection level range of 10^{13}–10^{14} cm^{-3}, which is typical in non-concentrator solar cells, the near surface recombination constitutes a critical limitation for carrier lifetime. As a consequence, *p*/*n*-type Si is usually combined with negatively/positively charged passivation layers.

2.2.3 *Surface Passivation*

Dielectric nanomaterials provide chemical and field-effect passivation. Chemical passivation reduces surface recombination centers and results in low D_{it} values. Field-effect passivation is caused by intrinsic charges in the dielectric layer, which produce an electric field altering the electron and hole concentrations at the surface. These two passivation effects control the key parameters of the SRH surface recombination process (2.1).

2.2.3.1 Chemical Passivation

The termination of a regular crystalline lattice results in a large number of unsaturated chemical bonds at the surface. For Si, thermal surface oxidation provides a very effective saturation of chemical bonds resulting in a low density of interface states. The high electrical quality of the Si/SiO_2 interface is one of the key advantage of Si as a semiconductor [63]. Deal and Grove studied the thermal oxidation kinetics of Si in detail. Based on their model, oxygen diffuses from the gas phase through the amorphous oxide towards the substrate where it reacts with silicon to SiO_2. The oxide formation occurs at an buried interface [64, 65]. Despite of a significant bond density mismatch of crystalline Si and amorphous SiO_2, the interface was found to be rather abrupt within very few atomic monolayers. Within this interface the Si atoms are either bound in Si or in SiO_2 leaving a very low density of open bonds [66].

When high-*k* oxides, such as Al_2O_3, TiO_2 or HfO_2, are deposited on Si, the interface is not abrupt but the materials are separated by an interfacial SiO_x layer [63]. The interfacial SiO_x layer thickness is usually in the order of a few monolayers [52, 67]. This layer is formed during the oxide deposition process and it is even found on H-terminated substrates where the native SiO_2 layer was removed by HF etching prior to deposition [40]. However, the SiO_x interface thickness increases when deposition is done on substrates covered by native SiO_2 [68]. Very thick interfacial SiO_x layers of up to 8 nm were found in samples produced by reactive sputtering of Al_2O_3 [69]. The SiO_x thickness also slightly increases during post-deposition annealing at 400 °C [40, 70]. It is suggested that residual –OH groups react with the Si surface [71]. Interfacial SiO_x is also found in HfO_2 passivation layers [72, 73]. In SiN_x:H, a thin SiN_xO_y layer is found at the interface to silicon. This oxynitride is suggested to be formed in a reaction with native SiO_2

during SiN_x:H deposition [74]. As a consequence, the interfaces between Si and the passivation layers are basically Si/SiO_2-like. The high electrical quality of interfacial SiO_x is suggested to be the key to the excellent chemical passivation of dielectric materials [52, 75].

The surface passivation is significantly enhanced by hydrogenation of the Si/SiO_x interface. Most dielectrics used for surface passivation already contain a high quantity of hydrogen, as hydrogen is part of the deposition chemistry. Typical Al_2O_3 passivation layers contain 2–7.5 at.% hydrogen, mainly incorporated as –OH groups [70], with increasing concentration at lower deposition temperatures [76, 77]. In SiN_x:H, the hydrogen concentration is in the range of 10 and 20 at.% [41, 78]. SiO_2 contains about 8 at.% of hydrogen [79] when grown by ALD, whereas SiO_2 grown by thermal oxidation of Si is free of hydrogen.

Hydrogenation of the interface is a temperature activated process, which involves the transport of hydrogen towards the interface and the reaction with open bonds at the Si surface. The elevated hydrogen concentration at the Si/SiO_x interface after annealing can be detected with SIMS measurements [80]. The dynamics of this process depends on the annealing temperature and the microstructure of the dielectric layer [81, 82].

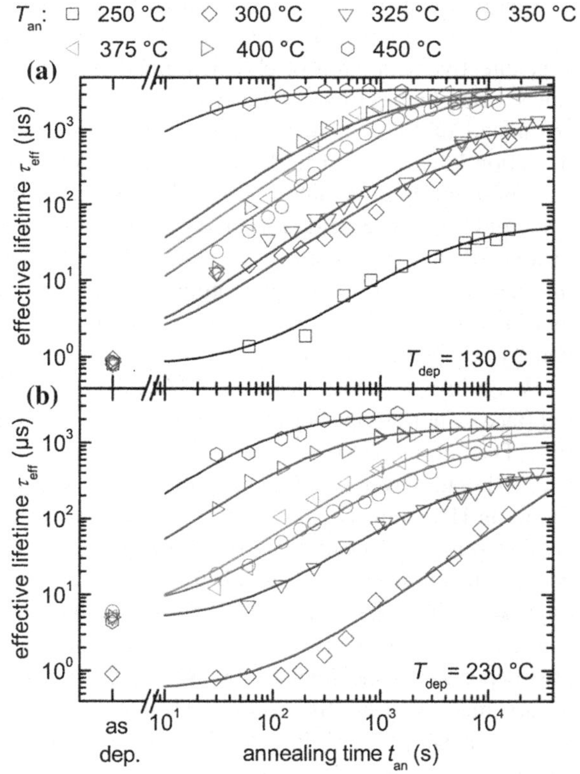

Fig. 2.5 Effective carrier lifetimes (at $\Delta n = 10^{15}$ cm^{-3}) as a function of cumulative annealing time t_{an} and annealing temperature T_{an} (in N_2 atmosphere). The deposition temperatures T_{dep} of the Al_2O_3 passivation layers are 130 °C (**a**) and 230 °C (**b**). The "as dep." annealing time refers to the as-grown layer. The *solid lines* are fits using (2.7). Reprinted with permission from Applied Physics Letters **104**, 061606. Copyright 2014, AIP Publishing LLC [82]

Figure 2.5 shows the activation dynamics of an Al_2O_3 layer. The activation of the passivation layer ($\tau_{eff} > 1$ ms) requires an annealing temperature between 350 and 450 °C and an annealing time of about 10 min. The evolution of carrier lifetime as a function of the annealing time t_{an} is described by [82]:

$$\frac{1}{\tau_{eff}(t_{an})} = \frac{1}{\tau_{bulk}} + \frac{2}{W}\left(S_{\infty} + \frac{S_0}{1 + R_{act}t_{an}}\right) \tag{2.7}$$

$$R_{act} = C \cdot \exp\left(-\frac{E_A}{k_B T_{an}}\right) \tag{2.8}$$

with S_0 and S_∞ being the surface recombination velocities prior to and after long-time annealing, respectively. These two parameters and the reaction rate R_{act} were fitted to the experimental data and it was found that the reaction rate exhibits Arrhenius behavior described by (2.8). Thus, the hydrogenation of interface defects is a temperature activated process with a characteristic activation energy. Though the applied fit functions varied in different investigations, similar activation energies of 1.5–1.6 eV [82] and 0.9–1.2 eV [81] were found for ALD-grown Al_2O_3 and SiO_2/Al_2O_3 layers after annealing in N_2 atmosphere. PVD-grown Al_2O_3 has an activation energy of 1.1 eV during H_2/N_2 annealing. In PVD-grown layers, the reaction rate is about four orders of magnitude lower due to a lack of mobile atomic hydrogen [83]. For thermally grown SiO_2, values of 1.5–1.7 eV are measured during annealing in H_2 atmosphere [84]. Since the activation energies of different materials match very well and exceed the activation energies reported for hydrogen diffusion, it is concluded that hydrogenation of the interface is a reaction limited process [81, 82, 84, 85].

Layers with high hydrogen content can be activated without external hydrogen source. For activation of Al_2O_3, a simple hot plate annealing in lab environment is virtually sufficient. The application of H_2-containing atmospheres hardly improves the passivation level of Al_2O_3 [81, 86] and TiO_2 [87]. Therefore, most groups perform annealing in a temperature controlled oven with N_2 atmosphere. However, annealing in H_2-containing gas atmosphere is required for passivation layers with low or zero H-content, such as Al_2O_3 deposited by PVD [83] or thermally grown SiO_2 [84].

The reaction rate is also influenced by the microstructure of the material, i.e. the structural material properties determine the constant C in (2.8). Figure 2.5 shows that activation of Al_2O_3 is significantly faster when the layer is grown at lower deposition temperatures. At low ALD and PECVD process temperatures, the mass density of films reduces due to enhanced incorporation of carbon and hydrogen contaminations [70, 76, 77]. The larger hydrogen reservoir supports accelerated hydrogenation during post-deposition annealing [81, 82].

Annealing supports hydrogenation of the interface, however, at the same time it also leads to outdiffusion and hydrogen loss. When the outdiffusion process prevails, the level of passivation degrades. This degradation is observed when the Al_2O_3 layer thickness descends below a value of about 10 nm (Fig. 2.6a). The reduction of carrier lifetime is correlated to a strong increase of D_{it} values (Fig. 2.6b), which is consistent with the creation of dangling bonds due to hydrogen release [67, 88].

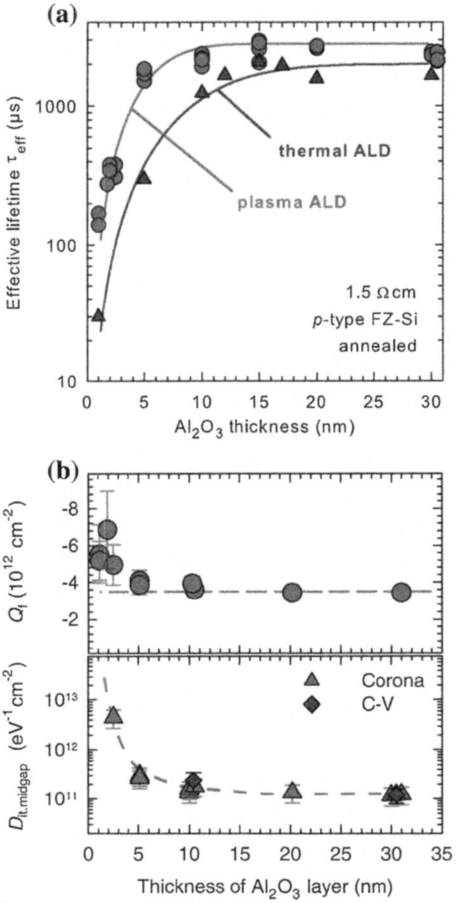

Fig. 2.6 Effective carrier lifetime (**a**) and Q_f and D_{it} values (**b**) as a function of Al_2O_3 layer thickness after annealing at 425 °C. At a layer thickness below of 10 nm, the carrier lifetime drops due to increased D_{it} values. Reprinted with permission from Journal of Applied Physics, **109**, 113701. Copyright 2011, AIP Publishing LLC [67]

The loss of hydrogen becomes more critical at higher annealing temperatures. In thermal effusion experiments on ALD-grown Al_2O_3, the hydrogen signal significantly rose at temperatures above 400 °C [81]. The peak temperature of H_2 and H_2O effusion depends on the microstructure of the films. Films deposited at temperatures between 200 and 400 °C are relatively dense and the maximum effusion shifts towards 700 °C, whereas the effusion maximum already appears at 400 °C when the films are grown at only 50 °C. Thus, denser films provide better stability against hydrogen loss during high temperature annealing [81]. Furthermore, higher layer thickness provides better temperature stability [89]. Hydrogen loss and film degradation is also observed in SiN_x:H films at temperatures above 500 °C [78].

Temperature stability of passivation layers is crucial as the PERC process sequence contains a short-term fast firing step at 800–900 °C. To increase the firing stability of thin Al_2O_3 layers, the films are usually capped with a thick (~100 nm), hydrogen-containing SiN_x:H layer [89–91]. Nuclear reaction analysis measurements confirm that this capping prevents H outdiffusion and maintains a high hydrogen concentration at the interface [92] resulting in a high chemical passivation after firing.

The available bonds per unit area of an unpassivated Si surface depend on the crystal orientation. In the Si (111) plane, the available bond density equals 1.2×10^{15} cm^{-2}, whereas the density is only 7×10^{14} cm^{-2} in the (100) plane [93]. After thermal oxidation of the surface, the available bond density at the silicon surface is found to correlate to the interface state density. As interface states below the gate dielectric are critical for the transistor functionally, todays CMOS manufacturing is based on (100) substrates [65].

The surface plane orientation is more complex in solar cells. Standard mono-crystalline solar cells are produced on (100) wafers where at least the front surface is textured and the surface plane orientations become (111). However, several studies revealed that the correlation between surface plane orientation and surface recombination is low. Texturing is found to increase the emitter saturation current density with a textured-to-planar ratio of 1.5–2. However, the degradation was mainly attributed to the larger surface area whereas the influence of the crystal orientation was small [94]. Liang et al. investigated the Al_2O_3 passivation of different surface orientations and measured a better passivation of the (100) surface, however, the improvement was correlated to an increase in fixed charges rather than a reduction of interface defects [95]. Black et al. also reported an influence of the different surface orientations on the performance of as-grown Al_2O_3 layers. However, the differences vanished after annealing [96]. On multi-crystalline wafers with well-passivated surfaces, the carrier lifetime was limited by recombination through crystal defects and the influence of surface planes orientation was negligable [97].

2.2.3.2 Field-Effect Passivation

Most dielectric layers contain intrinsic fixed charges, which provide field-effect passivation. Two methods are commonly applied to determine the fixed charge densities, the capacitance-voltage (CV) measurement on a metal-insulator-semiconductor test structure [34, 93] and the corona charge measurement directly applied on the passivated substrate [52, 56]. Additionally, lateral 2D mapping of the fixed charge distribution was demonstrated using a passivated substrate with a full area rear side electrode and photoluminescence imaging [98] or carrier lifetime mapping [99, 100]. Figure 2.7 shows a corona charge measurement on three different passivation materials. In this measurement the surface recombination velocity peaks when the deposited corona charges compensate the field-effect passivation. The measurement reveals positive fixed charges in SiN_x:H and SiO_2 layers and negative charges in Al_2O_3.

Fixed charges are formed during film deposition and during post-deposition annealing. In SiN_x:H, the intrinsic charges are related to K centers ($\cdot Si \equiv N_3$), which are formed by Si atoms backbonded to nitrogen [101]. With increasing [N]/[Si] ratio, the density of K centers increases at the interface, although these

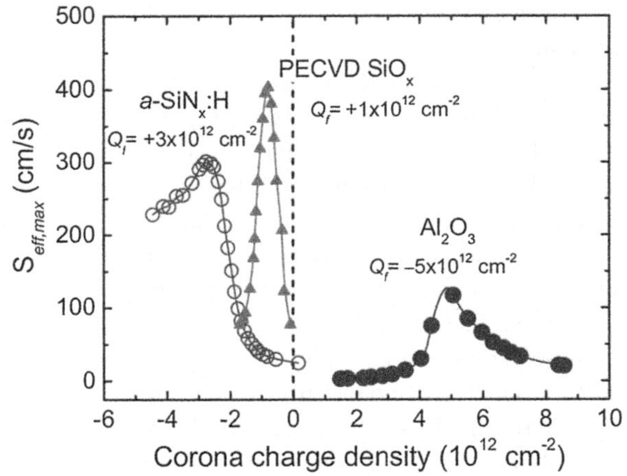

Fig. 2.7 Surface recombination velocity versus deposited corona charge density for SiN_x:H, SiO_2 and Al_2O_3 layers after annealing at 400 °C. Reprinted with permission from Journal of Vacuum Science & Technology A, **30**, 040802. Copyright 2012, American Vacuum Society [6]

centers are to some extent passivated by the presence of H [102]. These centers have amphoteric character with a positive, neutral or negative charge state. In stoichiometric and N-rich SiN_x:H films, the dangling bond of the K center is the dominant deep defect level, which becomes positively charged after electron emission. When the [N]/[Si] ratio is reduced, the density of K centers decreases and the films converge to amorphous Si, which is free of intrinsic charges [41, 103, 104].

In Al_2O_3, the fixed charge density depends on the applied process parameters during layer deposition. Plasma enhanced deposition methods tend to produce higher fixed charge densities than thermal processes [105–107], presumably due to the influence of highly reactive gas species in the plasma. Post-deposition annealing further supports the fixed charge formation and the charge density increases with annealing temperature up to about 500 °C [86, 108]. At higher annealing temperatures, the effect on charge formation reverses and charge densities decrease.

Oxide fixed charges are located at the interface to Si. When the Al_2O_3 thickness is reduced to 1 nm the fixed charge density hardly changes [67, 88] (Fig. 2.6b). This result is confirmed by CV-measurements on slant-etched oxides (Fig. 2.8). For these measurements, ALD-grown Al_2O_3 and SiO_2 layers are capped with an insulating HfO_2 top layer to block the leakage current even in very thin oxides. With this strategy it is possible to measure the flat band voltage of oxides as a function of thickness down to the sub-nanometer range [109]. The measured flat band voltage linearly increases with the effective oxide thickness (EOT), which suggests an oxide with solely interface charges [34, 93]. The densities are 2×10^{11}, 3×10^{12} (negative charges) and 1×10^{12} cm^{-2} (positive charges) in HfO_2, Al_2O_3 and SiO_2, respectively. A significant deviation from the linear function only appears when EOT approaches the value of the capping layer (about 8 nm EOT), i.e. the oxide thickness falls below 1 nm (grey line in Fig. 2.8). For Al_2O_3, a steep V_{fb} drop of about 1 V appears at an Al_2O_3 thickness below 1 nm. In the SiO_2 layer,

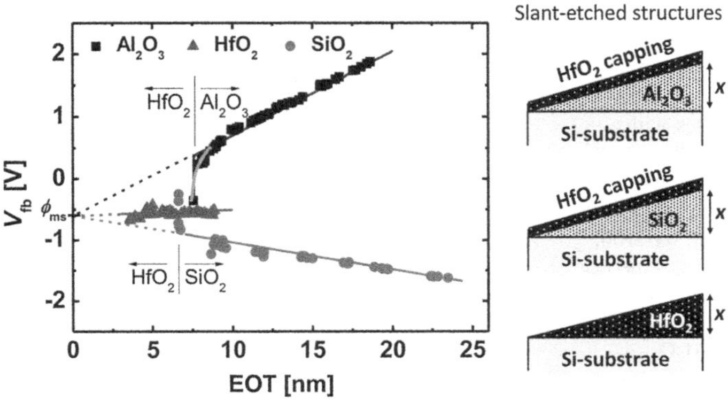

Fig. 2.8 Flat band voltages as a function of EOT determined on slant-etched Al_2O_3 and SiO_2 layers with HfO_2 capping and on slant-etched HfO_2 (stacks plotted on *right side*). The data follow a linear function (*solid lines*) down to sub-nanometer Al_2O_3 and SiO_2 layers (capping layer EOT: ~8 nm). This linearity suggests that charges are located at the interface to Si. Reprinted with permission from ACS Appl. Mater. Interfaces, **7**, 28215–28222. Copyright 2015, American Chemical Society [109]

the voltage drop is less pronounced. The flat band voltage drop correlates to the built-up of fixed charges within the first nanometer of the layer.

Bulk charges would result in a parabolic dependency on EOT, which is not visible in Fig. 2.8. Based on the V_{fb} function and a measurement error of ±100 mV, the bulk charge density is below 2×10^{17} cm^{-3} in ALD-grown SiO_2 and Al_2O_3 [109]. Assuming an EOT of 20 nm, the influence of bulk charges on the flat band voltage is comparable to the effect of interface charges with a density of 2×10^{9} cm^{-2}. This result indicates that the influence of bulk charges is negligible. However, earlier investigations on slant-etched Al_2O_3 layers indicated a significant density of negative [110] or positive [111, 112] bulk charges in the range of 10^{19} cm^{-3}. The large spread of values might be explained by the difficulty to exclude systematic error sources in the thickness series, such as charge trapping during CV measurements.

2.2.3.3 Formation of Fixed Charges in Al_2O_3

The origin of the interface charges in Al_2O_3 is still under discussion. It is found that Al_2O_3 has different structural properties at the interface to Si, where the fixed charges are located. In situ x-ray photon spectroscopy (XPS) revealed a strongly increased [O]/[Al] ratio at the interface to Si [67, 113]. Figure 2.9a shows the excess O fraction as a function of the layer thickness. At the interface, about 50 % of oxygen is neither stoichiometrically bound to Si as SiO_2 or to Al as Al_2O_3 [113]. When moving away from the interface, the O-excess rapidly drops and the [O]/[Al] ratio finally approaches the nominal value of 1.5. Post-deposition annealing at 400 °C only slightly affects the [O]/[Al] ratio suggesting that the O-rich layer is

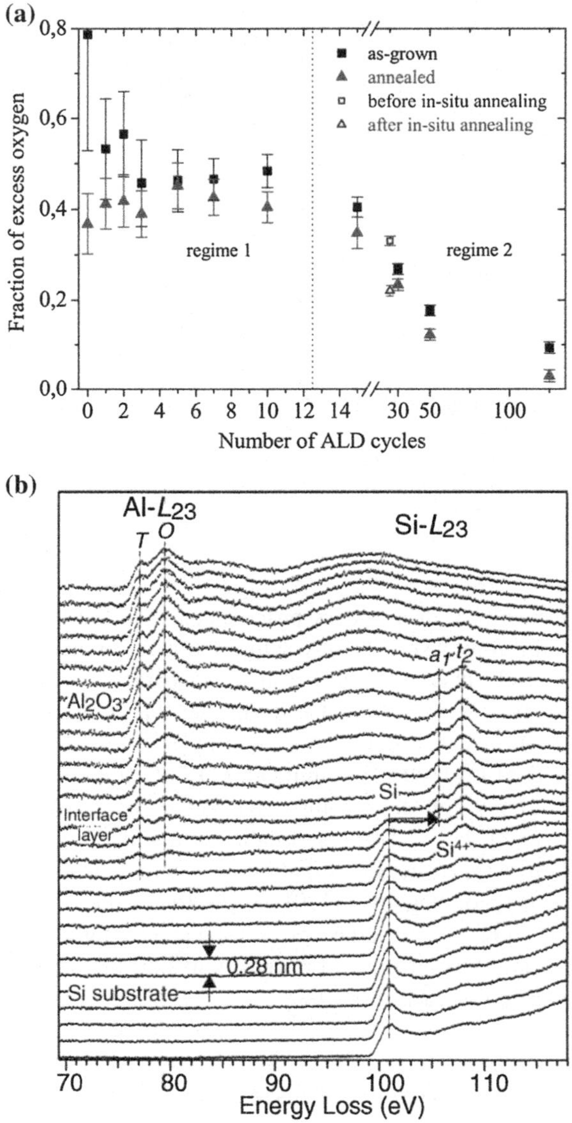

Fig. 2.9 XPS (**a**) and EELS (**b**) measurements of the initial growth region in Al_2O_3. XPS data reveal an O-excess, which decreases with increasing Al_2O_3 layer thickness. Reprinted with permission from J. Vac. Sci. Technol. A **30**, 04D106. Copyright 2012, American Vacuum Society [113]. Spatially resolved EELS data of a $Si/SiO_x/Al_2O_3$ cross-section show an enhanced tetrahedral coordination of Al atoms (Al-L_{23} T peak) close to the interface to Si. Reprinted with permission from Appl. Phys. Lett. **83**, 4306. Copyright 2003, AIP Publishing LLC [114]

already formed during layer deposition. Apparently, the stoichiometry deviation is caused by the initial growth of Al_2O_3 on the Si substrate.

Kimoto et al. investigated the Al coordination within Al_2O_3 by spatial resolved electron energy loss spectroscopy (EELS) [114]. Figure 2.9b shows the energy-loss near-edge structures (ELNES), which are sensitive to the valence and the coordination of the specific elements. The authors investigated the Al-L_{23} ELNES and found the T and O peaks throughout the layer, corresponding to tetrahedrally and octahedrally coordinated Al in Al_2O_3, respectively. However, the intensity of the T peak, i.e. the tetrahedral coordination, is dominant near the interface. Additionally, it is found that the Si-L_{23} fine structure does not differ within the amorphous oxide layer at the interface. It is suggested that an aluminum silicate interface layer is formed, where Si atoms exist as SiO_4 tetrahedra. This tetrahedral coordination of cations in aluminum silicate influences the initial growth of Al_2O_3 resulting in different Al coordination at the interface compared to the bulk layer [114].

Lucovsky et al. investigated the local atomic structure in binary SiO_2–Al_2O_3 glasses and found that Al atoms were incorporated in both Al^{3+} ions and in $AlO_{2/4}^-$ tetrahedral groups due to the different electronegativity of the cations. The Al^{3+} ions are incorporated in a site, which has an octahedral coordination by oxygen atoms. Its positive charge is autocompensated by the negative charge of an $AlO_{2/4}^-$ network group. To ensure charge neutrality, the ratio of tetrahedral and octahedral Al-coordination is 3:1 [115]:

$$2Al_2O_3 \equiv 3AlO_{2/4}^- + Al^{3+} \tag{2.9}$$

As the $AlO_{2/4}^-$ network group is O-rich, an excess of tetrahedral coordination is consistent with the O-excess observed with XPS at the interface. Since the structural deviations spatially correlate to the location of fixed charges it is suggested that the tetrahedral coordination is the root cause for the negative charges in Al_2O_3 [6, 52]. Another approach for understanding the intrinsic charge formation is based on charged point defects. First principles calculations of defect formation energies reveal that oxygen interstitials (O_i) and aluminum vacancies (V_{Al}) have the lowest formation energies and are the most stable defects in O-rich crystalline Al_2O_3 [116]. Both defects are negatively charged and form energy levels close to the valence band of Si according to simulation. Therefore, O_i and V_{Al} point defects are suggested to act as fixed charge centers in crystalline Al_2O_3 [117, 118]. Al_2O_3 passivation layers are amorphous since crystallization does not occur at the applied process temperatures. Nevertheless, point defects are also suggested to be the origin for the negative charges in amorphous Al_2O_3 [113].

Charges in passivation layers are commonly referred to as ‘fixed’ charges. However, several studies indicate that the charge density at least partly depends on external forces such as illumination and electric field. Gielis et al. observed a continuous increase of charge density in Al_2O_3 after laser illumination during second harmonic generation measurements [119]. Liao et al. illuminated the Al_2O_3 passivation with sun light for several 100 h. During light soaking the carrier lifetime increased from 1 to 1.5 ms, which was correlated to an increase of negative charges in Al_2O_3 [120]. Illumination also reduces the density of positive charge in SiN_x:H films [54]. The authors explain this effect by a photo-induced electron injection into the dielectric. When the samples are stored at dark the charge injection at least partly reverses and the measured charge density returns to the initial value.

Charging of Al_2O_3 is also observed as hysteresis in CV measurements [68, 121, 122]. Hysteresis becomes visible when the charge density changes during the

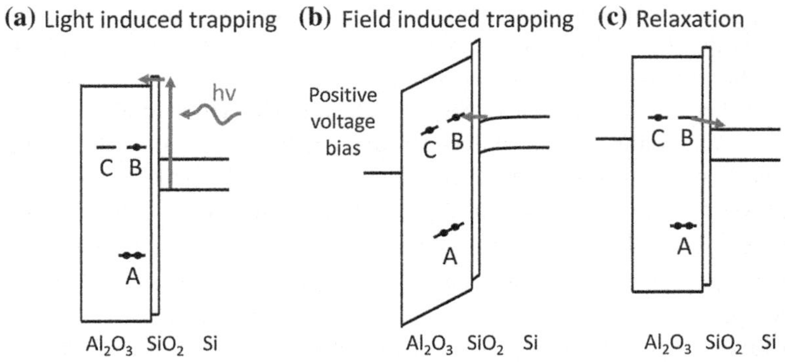

Fig. 2.10 Simplified energy diagram of Al_2O_3 to illustrate the role of fixed and trapped charges. Defect site A is negatively charged due to its low energy level. Defect sites B and C can trap electrons from Si during illumination (**a**) or application of electric field (**b**). After the external stress pulse, the electrons partly relax to the Si conduction band (**c**)

voltage sweep due to electron trapping at positive voltages and electron detrapping at negative voltages applied to the dielectric.

Figure 2.10 shows a simplified energy diagram of trap sites in Al_2O_3 to illustrate the difference between fixed and trapped charges. Structural defects create different energy levels in the Al_2O_3 band, which could interact with the Si substrate. Three different defect sites are distinguished as suggested in [123]. Defect site A is associated with a deep energy level, which is located close to the Si interface. Assuming an acceptor or amphoteric character of this defect, it will be intrinsically charged by electron capture from Si. Hence, this site is negatively charged and the charges are fixed due to the high activation energy. Defect site A could be related to tetrahedrally coordinated $AlO_{2/4}^-$ units or to other defects caused by oxygen excess. In crystalline Al_2O_3, intrinsic O_i and V_{Al} point defects are calculated to form deep traps with energy levels close to that of defect site A drawn in Fig. 2.10 [117, 118]. Similar structural defects might be responsible for deep electron traps in amorphous Al_2O_3.

Defect sites B and C are located above the Si Fermi level and charge trapping requires an external force such as illumination or electric field. Light absorption could lead to electron injection into the dielectric layer where the electrons are captured into trap sites (Fig. 2.10a). The application of an electric field lowers the trap level energy and the barrier height for electrons tunneling towards the defect (Fig. 2.10b). As soon as the layer returns in an unstressed state, i.e. light or electric field are switched off, the trapped charges tend to relax to the energetically favorable Si conduction band (Fig. 2.10c).

The dynamics of charge trapping and detrapping has been extensively investigated for nonvolatile memory devices [124–126]. Charge trapping is described by tunneling through the interfacial SiO_x layer into the high-k dielectric. As the tunneling probability depends on the energy barrier height and the tunneling distance [93], states close to the interface are occupied first. The same applies for detrapping via tunneling. States close to the interface relax the fastest. Using the simplified drawings in Fig. 2.10, charges in defect site B have poor retention and relaxation quickly starts after the external stress is removed. This defect could cause hysteresis effects during measurements. Defect site C has higher retention and it remains charged due to the lower tunneling probability for trapped charges. If charges are trapped at this defect, the layer remains charged even after the external stress is removed. As the trapped charges have to pass through the interfacial SiO_x layer, the thickness of this layer essentially determines the tunneling probability. Consequently, the trapped charge density in Al_2O_3 reduces when this barrier is enforced by additional SiO_2 deposition [68, 127].

Trapping states in Al_2O_3 were characterized by the trap spectroscopy by charge injection and sensing (TSCIS) method [128, 129]. It was found that crystalline Al_2O_3 contains a rather narrow ($\sim$0.2 eV) trap band with a high density of trapping states. This band is located at 2.16 eV below the Al_2O_3 conduction band, which is roughly the energy position of defect sites B and C in Fig. 2.10. However, in amorphous Al_2O_3 with post-deposition annealing temperatures below 1000 °C, a homogeneous trap distribution was found [128]. Therefore, defects sites B and C are representatives of a continuous defect distribution in amorphous Al_2O_3 layers.

As trapped charges could significantly contribute to the field-effect passivation it was suggested to engineer their density. Werner et al. could increase the charge density up to 10^{13} cm^{-2} after applying an external positive voltage to a metal electrode on top of Al_2O_3. Unless the samples are annealed at 350 °C the charge retention is very good and the trapped charges support the field-effect passivation [123]. A substantial increase of trapped charges is achieved by Ti-doping of Al_2O_3, which also results in improved carrier lifetimes [122]. Charge trapping is also observed in SiN_x:H films, which are positively charged. The charge density could be increased above 10^{13} cm^{-2} after applying an external voltage [130]. The net charge polarity could also be turned from positive to negative after electron trapping in SiN_x:H films [131].

2.3 Deposition Methods

In solar cell manufacturing, plasma enhanced chemical vapor deposition (PECVD) is the dominant deposition technique for front and rear side passivation layers. Besides PECVD, only the atomic layer deposition (ALD) technique reached a significant market share of 10–20 % for processing the Al_2O_3 rear side passivation. Other methods have been developed as well. However, according to the ITRPV forecast of 2015 they are not expected to gain significant market share within the next decade [132].

2.3.1 Plasma Enhanced Chemical Vapor Deposition

PECVD is used to deposit thin films from the gas phase. A plasma generator produces a glow discharge (plasma) in which precursor gases are (partly) transformed into radicals, ions or other highly excited species. These atomic and molecular species react with the substrate resulting in material deposition. Since the excitation of the precursor gases occurs by collision in the gas phase, the substrate can be maintained at low temperature. The low deposition temperature is one of the main advantages of PECVD over the conventional CVD process.

The film properties strongly depend on the deposition parameters and the reactor design [4, 5, 74]. Common reactor designs are parallel-plate ('direct' plasma) and 'remote' plasma systems. In parallel-plate reactors, the silicon wafer is placed between the two electrodes and is hence in direct contact with the plasma. In this configuration, the plasma frequency becomes a crucial parameter. To reduce the plasma damage by accelerated ions, high-frequency (>13.56 MHz) plasma systems are used [41, 78, 133]. In remote PECVD systems, the process gas is excited outside of the deposition chamber and the silicon wafer is not in direct contact with the plasma and ion bombardment damage is largely avoided. The widespread industrial high-throughput MAiA® system of Meyer Burger AG employs a 2.45 GHz linear microwave plasma source where a quartz tube with inner copper antennas generates the plasma outside of the tube [134].

Table 2.1 gives an overview of applied precursor gases and process conditions for PECVD of different passivation materials. For SiN_x:H deposition, ammonia and silane are commonly used. However, experiments were also done with additional N_2 and H_2 gas supply to grow layers close to ideal stoichiometry [78] and to enhance the hydrogen incorporation [41], respectively. The stoichiometry of SiN_x:H films can be widely adjusted by the ratio of precursor gases and this flexibility is used to tune the films for best performance. The [Si]/[N]-ratio controls the refractive index and absorption coefficient, which determine the antireflection properties of the SiN_x:H layer (Sect. 2.5.1). Additionally, the surface passivation properties depend on stoichiometry, whereby a general trend towards better performance is found in more Si-rich SiN_x:H films [41, 134, 135]. Al_2O_3 layers reach best passivation properties in nominal or slightly O-rich stoichiometry ([O]/[Al] $\geq$ 1.5) [70, 105].

Table 2.1 Typical process settings in PECVD of silicon surface passivation layers achieving S_{eff} < 10 cm/s after annealing. The abbreviation TMA stands for trimethylaluminium $Al(CH_3)_3$

Material	Precursor gases	Process ambient	References
SiN_x:H	$SiH_4 + NH_3$	350–450 °C	[134–137]
SiO_2 (SiN_x)	$SiH_4 + N_2O$	250–350 °C	[138, 139]
Al_2O_3	TMA + CO_2 (+H_2)	200 °C	[133, 140]
	TMA + O_2/Ar	200–300 °C	[76, 141]
	TMA + N_2O	350 °C	[94, 142–144]

2.3.2 Atomic Layer Deposition

ALD processes for oxides are based on a sequence with two surface reactions (Fig. 2.11). These sequential gas-surface reactions are called 'half-reactions'. During the first half-reaction, the precursor gas is pulsed into the reactor chamber until the substrate surface is fully adsorbed by the precursor gas. Subsequently, the chamber is purged with an inert carrier gas (typically N_2 or Ar) to remove any unreacted precursor or by-products of the reaction. This is then followed by an oxidizing agent pulse, which reacts with the adsorbed precursor molecules to the oxide. Finally, the chamber is purged to remove the reactants. Subsequently, this process is cycled until the thickness target is reached. As the process is self-limited by the gas adsorption to the substrate, the growth can be controlled on a single layer level. The growth per cycle (GPC) values depends on reactant gases and process temperature. Typical values are 1 Å/cycle for Al_2O_3, SiO_2 and HfO_2 and 0.3 Å/cycle for TiO_2.

Because the surface reactions are performed sequentially, the two gas phase reactants are not in contact in the gas phase. This separation of the two reactions limits possible gas phase reactions that can form particles and produce granular films after deposition on the surface. As a result, ALD films grow extremely smooth and conformal. Because of the homogeneous coverage with reactants no surface sites are left behind during film growth and the films tend to be very continuous and pinhole-free [146]. These factors result in excellent surface passivation properties of ALD-grown layers.

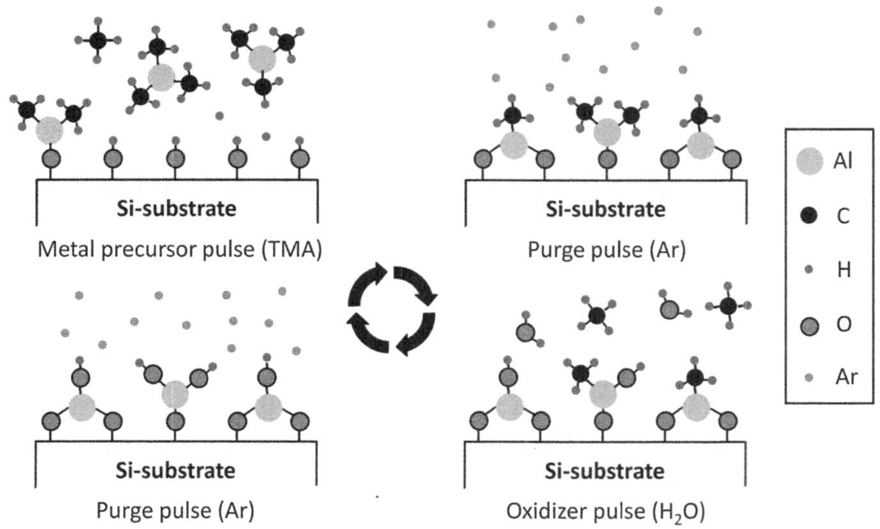

Fig. 2.11 Schematic drawing of the Al_2O_3 ALD process with TMA and H_2O precursors. The ALD cycle comprises of two half reactions with a sequence of metal precursor and oxidizing agent separated by purge pulses (according to [145])

The first investigations on ALD-grown Al_2O_3 passivation layers were published only one decade ago [40, 147]. Triggered by the excellent material properties of Al_2O_3, intensive research followed on new materials and processes. Today, excellent passivation performance has been demonstrated for numerous ALD-grown materials (Table 2.2). Interestingly enough, the most common passivation material, i.e. SiN_x:H, cannot be found in this list. The plasma enhanced ALD process of silicon nitride is developed for the application in microelectronic devices [148, 149]. However, reasonable surface passivation of this material has not been reported yet. One of the barriers is the stoichiometric nature of ALD-grown silicon nitride, i.e. Si_3N_4, whereas good passivation performance is achieved Si-rich films.

In the conventional temporal ALD the precursors are sequentially dosed into the reaction chamber and separated in time by purge pulses. ALD tools in research and microelectronic manufacturing are usually based on temporal processing. To meet the demand for high-throughput and low-cost equipment for solar cell manufacturing, spatial ALD concepts were introduced recently. In spatial ALD, the precursors are supplied continuously but the two reactions are spatially separated. The reactor contains two zones where the half-reactions take place. The ALD cycles are realized by moving the substrates between these two zones [159]. Spatial ALD systems are designed for industrial throughput above 1000 wafers/h. Several studies reveal the excellent passivation quality of spatially ALD-grown Al_2O_3 on wafers [150] and solar cells [160, 161].

Table 2.2 Typical process settings in ALD of silicon surface passivation layers achieving $S_{eff} < 10$ cm/s after annealing. The abbreviations BDEAS stands for bis(diethylamino)silan $H_2Si[N(C_2H_5)_2]_2$, TEMAHf for tetrakis(ethylmethylamino)hafnium $Hf[N(CH_3)C_2H_5]_4$ and TTIP for titanium tetraisopropoxide $Ti(C_{12}H_{28}O_4)$, respectively

Material	Precursor 1 (metal precursor)	Precursor 2 (oxidizing agent)	Process ambient	References
Al_2O_3	TMA	H_2O	100–300 °C	[6, 76, 105, 150, 151, 153]
		O_2 plasma	100–300 (500) °C	[6, 40, 76, 105, 119]
		O_3		[6, 105, 152]
SiO_2	BDEAS	O_2 plasma	150–200 °C	[79, 153]
HfO_2	TEMAHf	O_2 plasma	200 °C	[154]
		H_2O	150–200 °C	[109, 151 153]
	Trimethyl-hafnium	H_2O	300 °C	[72]
TiO_2	TTIP	H_2O	200–250 °C	[87, 151]
	$TiCl_4$	H_2O	100–200 °C	[155, 156]
Ta_2O_5 (SiN)	Tantalum-ethoxide	H_2O	250 °C	[157]
Ga_2O_3	Trimethyl-gallium	O_2 plasma	75 °C	[158]

2.3.3 Alternative Deposition Methods

Though PECVD and ALD tools dominate the market today, novel techniques emerge due to their potential to reduce production costs and to replace the pyrophoric TMA process gas. Atmospheric pressure chemical vapor deposition (APCVD) is very interesting for industrial high-throughput applications because of the elimination of vacuum and consequently reduced equipment costs. Additionally, Al_2O_3 deposition is possible with the non-pyrophoric triethyldialuminum tri-(sec-butoxide) precursor [162]. Black et al. reached excellent surface recombination velocities of below 3 cm/s with APCVD-grown Al_2O_3. Similar to films grown by PECVD or ALD, APCVD-grown layers also contain negative fixed charges in the order of 10^{12} cm^{-2} providing high field-effect passivation. Additionally, films capped with SiN_x:H feature very good thermal stability during fast firing [163, 164]. APCVD was also used to deposit TiO_2 for surface passivation and antireflection coating [165, 166].

Another promising deposition technology is sputtering. Sputtering is a mature technique widely used in electronics and thin-film industries and it is available for large-scale processing. Sputtering targets consist of non-toxic materials. However, surface passivation properties of sputtered layers lag behind the performance achieved with established methods. Most publications on sputtered passivation layers report carrier lifetimes in the range of 100 μs or even below. One of the obvious drawbacks of sputtering is that the process does not involve an obvious source of hydrogen. Contrary to metal-organic precursor gasses, metallic or ceramic sputtering targets are free of hydrogen. It was shown that reactive sputtering with additional H_2 gas supply significantly improved the passivation performance of SiN_x:H [167] and Al_2O_3 [168]. Zhang et al. demonstrated carrier lifetimes up to 5 ms with hydrogen-sputtered Al_2O_3 [168]. Despite of this excellent value, the role of hydrogen is still under discussion. This discussion is inspired by the fact that hydrogen was also detected in high concentrations in sputtered Al_2O_3 layers [169] potentially caused by remaining water vapor in the vacuum chamber. Additionally, it is shown that process parameters and film stoichiometry are at least as important and lifetimes in the millisecond range are feasible even without additional hydrogen supply [83, 170].

Though sputtered passivation layers have not yet reached the high quality of films grown by the established methods, the achieved level is already sufficient for the application in highly efficient PERC solar cells. Schmidt et al. demonstrated an cell efficiency above 20 % with sputtered rear side Al_2O_3 passivation [171].

2.3.4 Low-Thermal Budget Processing

For some solar cell concepts the thermal budget is restricted (Sect. 2.1.2). For example, concepts involving a-Si:H require process temperatures below 200 °C.

ALD-grown Al_2O_3 is a potential candidate for low-thermal budget passivation, as an excellent level of passivation is already achieved at low deposition temperatures (Sect. 2.2.3.1). However, films grown at low temperatures still require a thermal activation at about 400 °C. Considering hydrogenation of interface defects as a thermally activated process with associated activation energies (2.7) and (2.8), it appears difficult to by-pass this process at low temperatures. Lower temperatures might be compensated by longer annealing periods; however, long term annealing potentially also harms the temperature sensitive substrate.

The thermal budget of the annealing process could be reduced by optimized process parameters. Vandana et al. investigated reduced process times during annealing in N_2 atmosphere at 400 °C. The authors found carrier lifetimes above 1 ms already after 100 s of annealing. The passivation level increased with thickness and the best values were found for 100 nm thick Al_2O_3 layers [172]. Seguini et al. investigated Al_2O_3 annealing at 200 °C and found carrier lifetimes up to 1 ms when using layers, which were deposited at ALD process temperatures below 150 °C [173]. Due to enhanced incorporation of hydrogen during layer growth, low deposition temperatures increase the reaction rate during the hydrogenation process (Sect. 2.2.3.1). Therefore, the combination of low deposition and low annealing temperatures appears very promising [106, 107, 173]. However, the reduction of the thermal budget often results in a lower level of passivation. Electrical characterization reveals a reduced field-effect passivation and partly lower chemical passivation [172, 173] indicating that the passivation process is not fully completed.

Another approach to reduce the thermal budget is flash light annealing (FLA). This technique is known from various applications in microelectronics, e.g. doping activation [174], thermal treatment of high-*k* dielectrics [175] or flash-enhanced deposition techniques [176]. FLA employs a short light flash, typically in the millisecond range, to induce a short-term heating process [177]. During the flash, the uppermost region of the substrate is rapidly heated to high temperatures while the backside only experiences moderate heating [178]. Furthermore, the substrate acts as heat sink resulting in rapid cooling after the flash. A schematic drawing of the flash light setup and the surface temperature as a function of time are shown in Fig. 2.12.

Simon et al. applied the FLA technique to Al_2O_3 passivation layers [106]. Figure 2.13 shows the carrier lifetime as a function of flash iteration, annealing atmosphere and deposition process. The best result is achieved for Al_2O_3 grown by thermal ALD after FLA in H_2 atmosphere at 200 °C (Fig. 2.13a). In the as-grown state, the carrier lifetime is poor (3 μs) but it quickly increases to 455 μs after a single FLA cycle and reaches 5 ms after 100 cycles, which is comparable to values reached after reference annealing at 350 °C. FLA of samples grown by plasma enhanced ALD also results in a significant increase of carrier lifetime; however, the values hardly exceed 1 ms and clearly remain below the result of the standard anneal (Fig. 2.13b). Processing in Ar instead of H_2 atmosphere also results in lower carrier lifetimes.

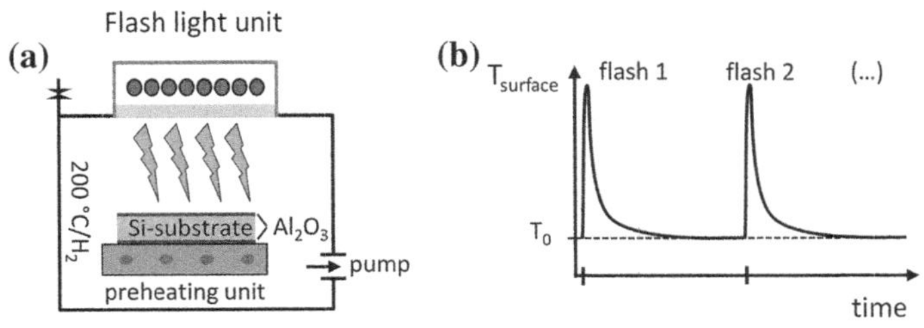

Fig. 2.12 Schematic drawing of a FLA setup (**a**) and surface temperature ($T_{surface}$) versus process time (**b**). The flash light induces a short-term temperature peak at the surface. After the flash, the surface rapidly cools down to the substrate temperature (T_0)

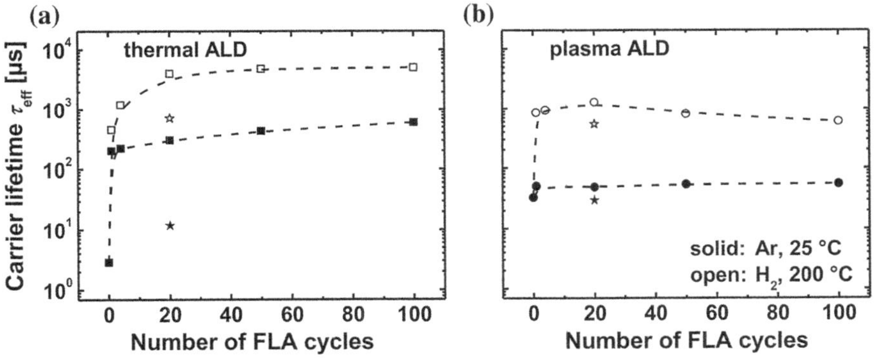

Fig. 2.13 Carrier lifetime as a function of FLA cycle numbers. The Al_2O_3 layers are deposited by thermal (**a**) and plasma enhanced (**b**) ALD. Reference samples with single-side FLA are plotted as *star symbols*. Reprinted with permission from Phys. Status Solidi RRL **9**, 631–635. Copyright 2015, John Wiley and Sons [106]

The different results after FLA are related to the effectiveness of interface hydrogenation at low process temperatures. In Al_2O_3 grown by thermal ALD, a high hydrogen incorporation facilitates the hydrogenation of the interface [105]. Contrary, plasma ALD-grown Al_2O_3 suffers from a higher interface state density in the as-grown state due to increased surface damage by exposure to oxygen radicals. Therefore, the thermal process provides a more suitable microstructure with a higher reaction rate for the hydrogenation process. Additionally to the influence of the microstructure, the surface passivation process is supported by H_2 atmosphere and enhanced temperature (200 °C). When further increasing the annealing temperature, the microstructure influence reduces due to the higher thermal budget involved. After reference annealing at 350 °C, samples grown by plasma and thermal ALD reach comparable carrier lifetimes of about 5 ms.

Flash light annealing is surface selective. When applying single-side FLA (Fig. 2.13, star symbols), the carrier lifetimes remain on a low level as single-side passivation could not improve the carrier lifetime by more than the factor of two, when lifetime is limited by surface recombination [46]. The small increase of carrier lifetime corresponds to the pure effect of process ambient, i.e. substrate temperature and process atmosphere. To activate the passivation layers on both wafer sides, FLA has to be applied on the front and rear side of the substrate. This surface selectivity of FLA provides high thermal activation on one wafer side, whereas the temperature budget remains low at other side. Therefore, FLA opens the possibility to combine dielectric surface passivation with low-thermal budget solar cell concepts.

2.4 Dielectric Multi-oxide Nanolaminates

The introduction of multi-oxide nanolaminates opens the possibility to tailor material properties and functionalities for novel passivation layers. Multi-oxide nanolaminates are used to realize symmetrical (Sect. 2.4.1) and conductive (Sect. 2.4.2) passivation layers. These nanolaminates are grown by ALD, due to the accurate thickness control and the possibility to combine many different materials in one process step.

2.4.1 Zero-Fixed-Charge Passivation Layers

Dielectric multi-oxide nanolaminates make it possible to control the fixed charges and even reduce their density to zero. Zero-fixed-charge nanolaminates are realized in Al_2O_3 passivation layers with a very thin interface layer of SiO_2 [55, 127], Al-doped SiO_2 ($Al:SiO_2$) [73] or HfO_2 [153] on moderately doped silicon. Additionally, the concept was demonstrated with SiO_2/Al_2O_3 stacks on p^+ and n^+-doped Si substrates [27]. Figure 2.14 displays the density of fixed charges in these stacks as a function of interface layer thickness, which is plotted as ALD cycle number. At zero ALD cycle, i.e. in pure Al_2O_3, a high density of negative fixed charges ($3\text{–}4 \times 10^{12}$ cm^{-2}) is found. By increasing the ALD cycle number, the negative fixed charges gradually disappear. The required interface layer thickness for charge annihilation is about 10 cycles for SiO_2 and 5 cycles for HfO_2. Considering a steady state growth rate of about 1.2 Å/cycle for both materials, the interface layer thickness corresponds to a few monolayers only. The $Al:SiO_2$ interface layer was introduced because of the lower passivation level of ALD-grown SiO_2 compared to the level reached with Al_2O_3 [73, 79]. The doping was realized by 1:1 ALD super-cycles of SiO_2 and Al_2O_3. Zero fixed charge density is achieved after about five super-cycles (steady state growth rate: $\sim$2.5 Å/super-cycle).

When the interface layer thickness is further increased, the properties of this layer determine the charge formation. For HfO_2, the density remains zero as HfO_2 is almost free of charges under the applied ALD process settings [151, 153]. For SiO_2, the charge polarity turns from negative to positive above 10 ALD cycles and the

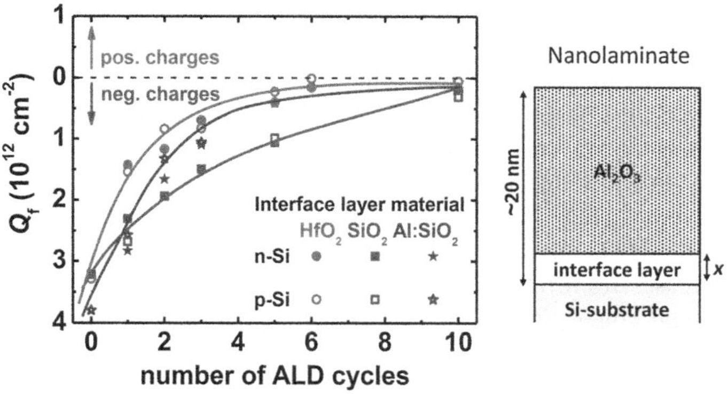

Fig. 2.14 Fixed charge density as a function of the interface layer material and thickness plotted as ALD (super-)cycle number. The charge density decreases with increasing interface thickness. The lines are guides to the eye. The nanolaminate layout is shown at the *right side*. Reprinted with permission from IEEE 42th Photovoltaic Specialists Conference (PVSC), 1–6. Copyright 2015, IEEE [73]

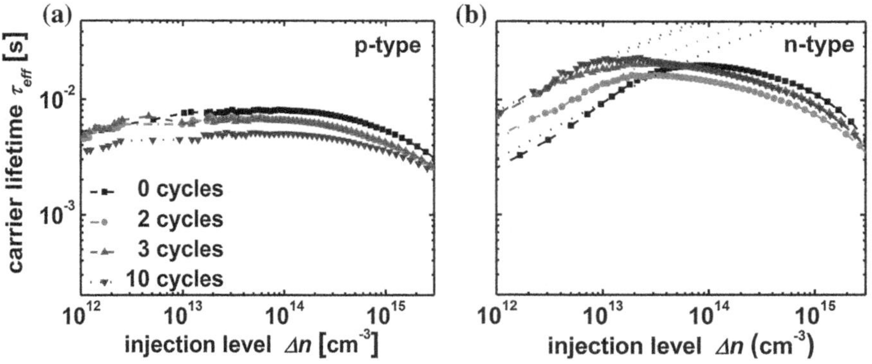

Fig. 2.15 QSSPC measurements of carrier lifetime as a function of the injection level for HfO_2/Al_2O_3 nanolaminates on *p*-type (**a**) and *n*-type (**b**) Si. The thickness of the HfO_2 interface layer was varied. *Dotted lines* are calculated using (2.4) and (2.6). Reprinted from Sol. Energy Mater. Sol. Cells **131**, 72–76, Copyright (2014), with permission from Elsevier [153]

density increases to 1–2 $\times$ 10^{12} cm^{-2} for thick SiO_2 layers [109, 179]. The substrate doping has no influence on the fixed charge formation, i.e. the same charge density is formed on *p*- and *n*-type Si.

The influence of fixed charge density on the carrier lifetime is different on *p*- and *n*-type Si substrates (Fig. 2.15). Both types of substrates are passivated with a nanolaminate comprising 20 nm Al_2O_3 and an HfO_2 interface layer with different ALD cycle numbers. On *p*-type Si, the carrier lifetime reaches values up to 8.0 ms and the values are almost independent on the interface layer thickness in the low thickness regime (Fig. 2.15a). Only at 10 cycles of HfO_2, the carrier lifetime slightly decreases (5.5 ms). The measured carrier lifetimes are also independent of the injection level for $\Delta n < 1 \times 10^{15}$ cm^{-3}.

On *n*-type Si, the observation is different because the negative charges in Al_2O_3 invoke a surface inversion layer resulting in near surface recombination (Sect. 2.2.2). For low injection levels ($\Delta n < 5 \times 10^{13}$ cm^{-3}), the carrier lifetime strongly decreases (Fig. 2.15b). However, this degradation could be inhibited by introducing the HfO_2 interface. At an injection level of 5×10^{12} cm^{-3}, the carrier lifetime increases from 6.0 ms for pure Al_2O_3 to 9.5 ms and 15.0 ms for Al_2O_3 with two and three ALD cycles of HfO_2 interface layer, respectively. Additional ALD cycles do not result in higher carrier lifetimes and the effect saturated. At high injection level ($\Delta n = 1 \times 10^{14}$ cm^{-3}), all samples show comparable values of about 10.0 ms, independent of their HfO_2 interface layer thickness.

The reduction of fixed charges results in symmetrical passivation as the near surface recombination is suppressed in *p*- and *n*-type Si substrates. However, because of the absence of field-effect passivation the chemical passivation becomes more critical. Symmetrical passivation layers only rely on chemical passivation, i.e. a very low density of interface defects. The results of this section show that this high level of chemical passivation could be realized with multi-oxide nanolaminates.

An "inert" spacer model was proposed for interpretation of the fixed charge modification in Al_2O_3 [109]. Fixed charges in Al_2O_3 are linked to structural defects and the stoichiometry deviation in the initial growth regime of Al_2O_3 (Sect. 2.2.3.3). An additional interface layer interferes with this initial growth process as it interrupts the contact to the crystalline silicon substrate. The domination of tetrahedral coordination of Al in $AlO_{4/2}^{-}$ units requires a physical contact to the tetrahedrally oriented Si/SiO_x interface [114, 115]. This contact is interrupted by an interface layer. In HfO_2, the Hf-atom is 8-fold coordinated [180] and this different structure efficiently suppresses the growth of tetrahedrally coordinated Al in Al_2O_3. The SiO_2 spacer is less effective than the HfO_2 layer in a sense that the required layer thickness for charge annihilation is twice as large. ALD-grown SiO_2 does not suppress the tetrahedral coordination of the SiO_x/Si interface as efficiently due to its similar chemical structure.

2.4.2 *Carrier Selective Contacts*

2.4.2.1 Concepts of Carrier Selective Contacts

In general, carrier selective contacts consist of a medium with different conductivities for majority and minority carriers [28]. The medium is transparent to majority carriers, which are passed through to the electrical contact. However, minority carrier transport is blocked and no recombination occurs within this medium or at its surfaces. Such carrier selective contacts have been realized in

- a-Si:H/c-Si heterocontacts (Fig. 2.16a), which are applied in silicon heterojunction solar cells [22]. In this concept, the carrier selectivity is implied by the band offsets between a-Si:H and c-Si. The majority carrier transport through the heterocontact is constituted by thermionic emission or tunneling in the presence of band bending. Owing to the excellent carrier selectivity, this concept reaches highest open circuit voltages up to 750 mV and efficiencies up to 24.7 % for both-side contacted solar cells [18].
- tunneling contacts (Fig. 2.16b), which comprise a high band gap tunneling oxide and a transport matrix. SiO_2 [181–183] and Al_2O_3 [184, 185] are common tunneling oxide materials due to their high band gap and low defect density at the interface to silicon. For the transport matrix, microcrystalline or amorphous silicon are used. Alternatively, transparent conductive oxides are investigated such as ITO or ZnO for contacts on n-type Si [184, 186] and WO_x or MoO_x for contacts on p-type Si [186]. The electrical conductivity is optimized by tailoring the tunneling probability, which is mainly determined by oxide thickness and transport matrix parameters including work function and band gap energy [187, 188]. Though the development of tunneling contacts gained momentum only a few years ago, several groups already integrated this concept into solar cells and demonstrated very high efficiencies [185, 189] of up to 24.4 % [190].

Carrier selective contacts can also be realized with dielectric multi-oxide nanolaminates (Fig. 2.16c). Promising candidates are Al_2O_3–TiO_2 nanolaminates, which combine the excellent surface passivation properties of Al_2O_3 with the higher conductivity of TiO_2 [73, 87]. Al_2O_3–TiO_2 passivation laminates are shown to feature superior properties compared to pure Al_2O_3 layers in terms of surface recombination [151] and optical properties [191, 192]. These nanolaminates have

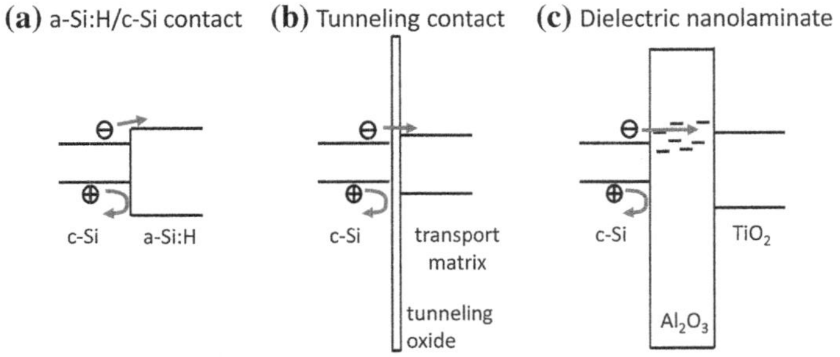

Fig. 2.16 Different concepts for carrier selective contacts on n-type silicon. The carrier selectivity is achieved by the band offsets at the a-Si:H/c-Si heterocontact (**a**), the tunneling probability at the tunneling contact (**b**) and the specific electrical transport properties at the dielectric nanolaminate (**c**)

also been intensively investigated for the application as gate dielectrics in microelectronic devices. The incorporation of TiO_2 is found to increase the *k*-value of the gate stack; however, it also strongly enhances the leakage current, which degrades the device performance. Both parameters can be controlled by the stacking sequence and sublayer thicknesses in the Al_2O_3–TiO_2 nanolaminates [193, 194]. For carrier selective contacts, the Al_2O_3–TiO_2 stack layout is optimized for high conductivity and high level of surface passivation. The main advantage of dielectric Al_2O_3–TiO_2 nanolaminates over other concepts is the strong synergy with today's PERC technology:

- Dielectric layers act as very effective optical reflectors at the rear side of the solar cell due to their relatively low refractive index of below 2 (Sect. 2.5.1). Carrier selective contacts with Si-based transport matrices often feature attenuated rear side reflection and require additional measures for light management to achieve the same level of absorption than realized in PERC solar cells.
- Dielectric materials provide a reasonable level of temperature stability, which is required during the fast firing step in the PERC process sequence. Carrier selective contact concepts partly have a restricted thermal budget, especially when a-Si:H is used as transport matrix. An integration of these concepts into the PERC solar cell would require a thermal budget reduction with major consequences, especially for the metallization strategy.
- Dielectric nanolaminates can be deposited by ALD and in principle also by PECVD. As the required equipment is already implemented in today's PERC manufacturing, the conversion from a point contact to a full area carrier selective contact scheme could be realized with low equipment invest.

2.4.2.2 Al_2O_3–TiO_2 Carrier Selective Contacts

The electrical properties of Al_2O_3–TiO_2 nanolaminates are essentially determined by the stack layout. Figure 2.17 shows the results of double and multilayer stacks on *p*-type Si substrates. Both stacks comprise an Al_2O_3 interface layer due to its better surface passivation. TiO_2 serves as a capping layer in the double layer stack. Al_2O_3–TiO_2 multilayers consist of alternating Al_2O_3 and TiO_2 sublayers of equal thickness and an adjusted number of iterations in order to maintain a total nominal thickness of 20 nm. The measured carrier lifetime as a function of the Al_2O_3 interface layer thickness x shows that pure TiO_2 passivation ($x = 0$ nm) results in poor carrier lifetime ($\tau_{eff} < 100$ µs). This result is not surprising as the passivation level of TiO_2 is known to be inferior to the one of Al_2O_3 [122, 154, 155]. When increasing the interface layer thickness, the carrier lifetime continuously improves. The highest value of 5 ms is measured with pure Al_2O_3 ($x = 20$ nm) passivation. This correlation is observed for double and multilayer stacks, i.e. the passivation performance is independent of the stack above the first TiO_2 (sub)layer. Essentially, the lifetime is determined by the distance of the first TiO_2 layer to the silicon surface [151].

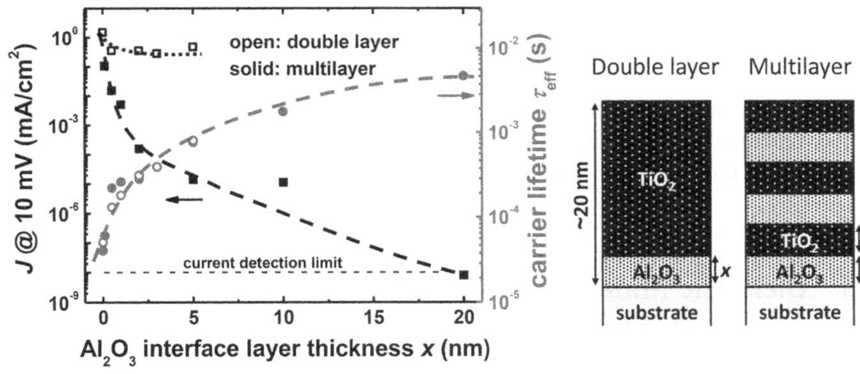

Fig. 2.17 Current density (at 10 mV) and carrier lifetime of different double layers (*open dots*) and multilayers (*solid dots*) as a function of the Al_2O_3 interface thickness x. Substrate material is *p*-type Si. The nanolaminate layout is shown at the *right side*. Reprinted with permission from IEEE 42th Photovoltaic Specialists Conference (PVSC), 1–6, Copyright 2015, IEEE [73]

Figure 2.17 also shows the measured current densities J at 10 mV. This voltage drop is chosen as an acceptable voltage loss at the contact. The electrical data are plotted as current densities rather than contact resistance because of the non-ohmic behavior of a part of the investigated stacks. The conductivity of pure Al_2O_3 is very low and the current density is below the measurement limit (10^{-8} mA/cm^2). In pure TiO_2, the current density is at least eight orders of magnitude higher. The influence of the Al_2O_3 interface layer thickness is different for both stack types. For multilayer stacks, the current density drastically drops with increasing thickness. In the double layer stack, the reduction is comparatively low. A 5 nm Al_2O_3 interface layer still results in a current density of about 0.5 mA/cm^2. This value is about four orders of magnitude higher than the current density of a multilayer stack with the same sublayer thickness.

In double layers, the current density is almost independent of the Al_2O_3 interface layer thickness, suggesting the electrical transport is not constituted by electron tunneling. The electrical properties of the nanolaminates are suggested to be determined by a TiO_2 phase transformation from amorphous to anatase and an interaction of TiO_2 and Al_2O_3. The phase transformation occurs when a critical layer thickness of about 10 nm is exceeded in ALD-grown TiO_2 layers [195, 196]. As the different conductivities of both stack types correlate with the appearance of anatase TiO_2 [87], the crystallization of the TiO_2 capping layer is suggested to weaken the insulating properties of Al_2O_3 resulting in good conductivity even through 5 nm thick layers. Potential transport processes through the Al_2O_3 interface layer are hopping through impurities levels (as shown in Fig. 2.16c) or pin-hole transport [197].

The Al_2O_3–TiO_2 double layer with $x = 5$ nm features the best balance of electrical transport and surface passivation with 0.5 mA/cm^2 current density (at 10 mV) and 15 cm/s surface recombination velocity. Compared to the requirements of a highly efficient solar cell [7], this nanolaminate features sufficient passivation performance. However, the current density is still about two orders of magnitude below target (40 mA/cm^2), although the Al_2O_3–TiO_2 nanolaminates strongly enhanced the conductivity compared to standard Al_2O_3 passivation layers. In conclusion, the electrical properties of Al_2O_3–TiO_2 nanolaminates can be controlled by stack layout and sublayer thicknesses within a wide range. With further improvements of conductivity, dielectric Al_2O_3–TiO_2 nanolaminates could provide powerful solutions for future carrier selective contacts.

2.5 Dielectric Materials and Light Management

Dielectric passivation layers are also part of the solar cell light management. Therefore, the optical properties of these layers have to be optimized. In the state-of-the-art PERC solar cell, dielectric nanomaterials control the surface reflections at the front and rear side (Fig. 2.18a). Additionally, dielectric materials are applied in novel light trapping and spectral conversion concepts.

2.5.1 Dielectric Layers for Surface Reflection Control

A planar Si surface appears like a mirror plane. Owing to the relatively high refractive index of Si (n_{Si} = 3.8 at 633 nm), the surface reflection of polished silicon is about 34 % making reflection control and the application of anti-reflection coating (ARC) essential.

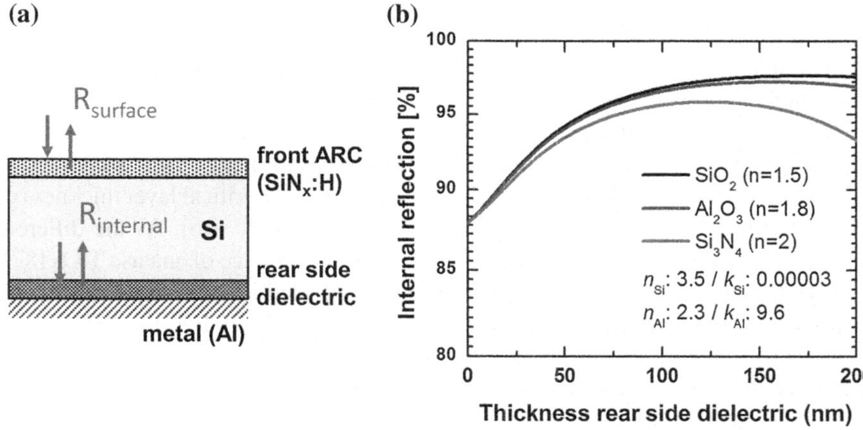

Fig. 2.18 Optical reflections at the front and rear surfaces of a PERC solar cell (**a**) and simulation of the internal reflection (λ: 1100 nm, angle of incident: 0°) as a function of thickness and material of the rear side dielectric layer (**b**). Optical parameters used for simulation are stated in the plot

For a single-layer, non-absorbing ARC, the reflection at perpendicular angle of incident becomes zero when the ARC thickness is optimized and the refractive indices of the materials are matched [198]:

$$d_{\mathrm{ARC}} = \frac{\lambda}{4n_{\mathrm{ARC}}} \text{ and } n_{\mathrm{ARC}} = \sqrt{n_{\mathrm{Si}} n_{\mathrm{ambient}}} \tag{2.10}$$

with d_{ARC} the ARC thickness and n the refractive indices of the materials. A system optimized for the peak intensity of sun light (633 nm) requires an ARC with refractive index of $n_{\mathrm{ARC}} = 1.95$ when the ambient is air ($n_{\mathrm{ambient}} = 1$, $n_{\mathrm{Si}} = 3.8$). The resulting ARC thickness is 80 nm. In solar cell manufacturing, minimum reflection losses are targeted within the module stack, where the ambient is defined by the packaging material consisting of lamination foil (e.g. EVA) and glass cover with a refractive index of about 1.5. Considering the packing material in (2.10), the optimum refractive index of the ARC shifts to a higher value. Consequently, the ARC material properties have to be optimized for either encapsulated or non-encapsulated solar cells. The calculation of optical reflection becomes more complex when considering the full relevant wavelength range, different angles of incident, the spectral response of the solar cell and the absorption coefficient of the ARC layer. A thorough optimization of the antireflection strategy usually employs numerical simulation [136, 199, 200].

The standard front side ARC material is SiN_x:H and its main advantage is the possibility to continuously tune the refractive index within a wide range. In the PECVD process, the stoichiometry of the film is controlled by the process gas ratio, which allows to continuously increase the refractive index from 1.9 (a-Si_3N_4:H) to 3.3 (a-Si:H) [201]. The optimum SiN_x:H ARC layers are thus Si-rich with [Si]/[N] > 3/4. The tunability by a gradual replacement of reactant gases can also be used to create graded index ARC layers with minimized surface reflection [202].

At the rear side of the solar cell, high internal reflection is targeted for the near band gap spectral range (1000–1100 nm), which is weakly absorbed in Si. As band-to-band absorption increases with photon energy, the shorter wavelength part of the spectrum is mainly absorbed within the Si wafer and hardly reaches the rear side of the wafer. Figure 2.18b shows the effect of a rear side dielectric layer on the internal reflection at a wavelength of 1100 nm. The reflection is about 88 %, when the Si rear side is terminated by an Al electrode, which is the contact scheme in a back surface field (BSF) solar cell. This relatively low reflection seems surprising since metals are known to be highly reflective in the IR spectral range. However, at the Si–Al interface the high refractive index of Si leads to considerably suppressed reflection compared to the reflection at the air-Al interface [8]. When Si and Al are separated by a low refractive index dielectric layer, the optical contrast of the interfaces is enhanced and the internal reflection significantly increases. The simulated internal reflection is a function of the stack parameters with highest values above 97 %, which are reached for dielectric materials with low refractive index and a matched layer thickness in the range of 100 and 200 nm.

Based on these simple optical simulations, the optimum dielectric layer thickness is larger at the rear side of the solar cell. Therefore, the SiN_x:H capping layer is not only required to improve the temperature stability of the rear side passivation layer but also to enhance the internal reflection. Beside of the reduction of surface recombination, the improved rear side light management is one of the main advantages of the PERC solar cell concept [8, 9]. This improvement can be measured as reduced reflection at the solar cell surface in the wavelength range between 1000 and 1200 nm [203, 204]. Reduced reflection consequently leads to higher quantum efficiencies of the PERC solar cells compared to cells with direct Si–Al contact (BSF solar cells) [9, 205].

2.5.2 Concepts for Light Trapping

Reflection control alone has only limited potential for absorption enhancement, especially when substrates become thinner and when high quantum efficiencies are targeted for the near band gap spectrum. Figure 2.19a shows a planar wafer with an incident light ray at perpendicular angle of incident. Considering a perfect reflection control on both surfaces, the light path through the system is twice the substrate thickness. After passing the substrate, the non-absorbed photons escape through the front surface. Therefore, light scattering features are introduced to enhance the light path and to trap the light within the absorber layer.

The effect of light scattering can be theoretically described by ideal Lambertian surfaces, which create a diffuse light spectrum with a statistical phase-space intensity distribution. When sunlight passes a Lambertian scatterer, the light path is enhanced and the light is then totally reflected within the surfaces until it escapes. As a consequence, the internal light intensity manifold exceeds the intensity of incident light in the sustrate. Based on statistical ray optics, Yablonovitch calculated this intensity enhancement factor as $2n^2$ with n the refractive index of the substrate under the assumption that the rear side is perfectly reflecting [206]. The resulting optical absorption enhancement is $4n^2$. A Lambertian scattering surface could thus enhance the absorption up to a factor of 50 in Si.

In crystalline silicon based solar cells, the surface is structured with randomized pyramids in a typical length scale of 10 μm (Fig. 2.19b) [207]. The texturing

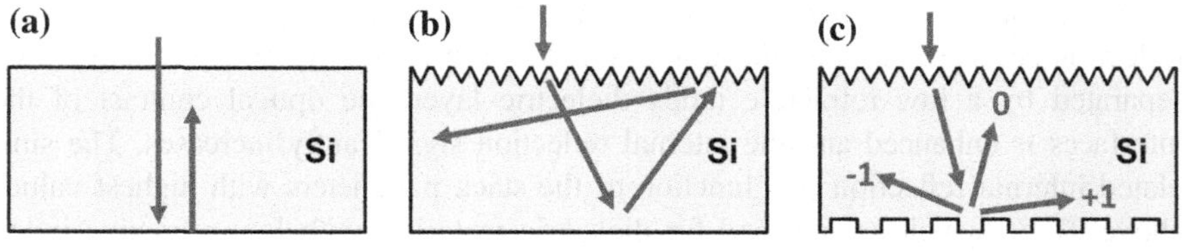

Fig. 2.19 Light path enhancement strategies in solar cells employing planar surfaces (**a**) a textured front surface (**b**) and an additional periodic structures at the rear surface (**c**)

process is based on anisotropic (i.e. orientation dependent) wet etching with an alkali (KOH or NaOH) etch solution. Alkaline etchants remove the (100) plane of Si much faster than the (111) plane and the result is a 3D pyramidal structure [208]. For crystalline wafers it was shown that randomized pyramids with optimized geometry get close to the $4n^2$ absorption limit of Yablonovitch [207, 209]. For multi-crystalline wafers with random crystal orientations this technique is much less effective since only grains with close to (100) orientation are well textured while grains with (111) orientation still remain highly reflective. Multi-crystalline wafers are usually textured with acid solutions resulting in a rough scattering surface. The resulting absorption enhancement is below the enhancement achieved on textured Si surfaces [210, 211].

Further enhancement of light trapping is possible with advanced photonic features, which exploit the wave nature of the light (Fig. 2.19c) [212, 213]. For periodic patterns the reflection is described by the diffraction equation [198]:

$$\sin(\theta_m) - \sin(\theta_i) = \frac{m\lambda}{nP} \tag{2.11}$$

with the wavelength λ, the order m, the grating period P, the refractive index n and the angles of incident Θ_i and diffraction Θ_m defined w.r.t. the surface nominal. Periodic gratings provide very powerful absorption enhancement when the diffracted light intensity is concentrated at high angles and the light transverse a long path through the solar cell before it escapes through a surface. Using (2.11), high first order ($m = \pm 1$) diffraction angles are reached when $\lambda \leq nP$. Thus, light absorption enhancement of the near band gap spectrum requires scattering features on micrometer scale. Periodic structures could significantly increase light trapping even beyond the limit of Lambertian light scattering [214–216].

Several techniques are applied to fabricate photonic nanostructures including nanoimprint [217, 218], laser interference [213], e-beam [219] and holographic [220] lithography. As an example, one concept generating the periodic optical contrast with dielectric materials is shown in Fig. 2.20. Eisenlohr et al. deposited a hexagonally ordered monodisperse SiO_2 sphere grating by spin coating on the rear side of a solar cell (Fig. 2.20a). These low refractive index spheres were embedded in a high refractive index matrix of polycrystalline Si deposited by APCVD [221, 222]. These spheres significantly improved the light absorption in the IR region (Fig. 2.20b) resulting in a higher short circuit current. Planar solar cells with back side scattering spheres reach an efficiency of 22.1 %.

2.5.3 Spectral Conversion of Light

In a single junction solar cell the sub-band gap light spectrum is not utilized. With photon up-conversion low-energy photons are converted into ‘useful’ photons ($h\nu > E_g$) having sufficient energy to be absorbed in Si. The up-conversion process

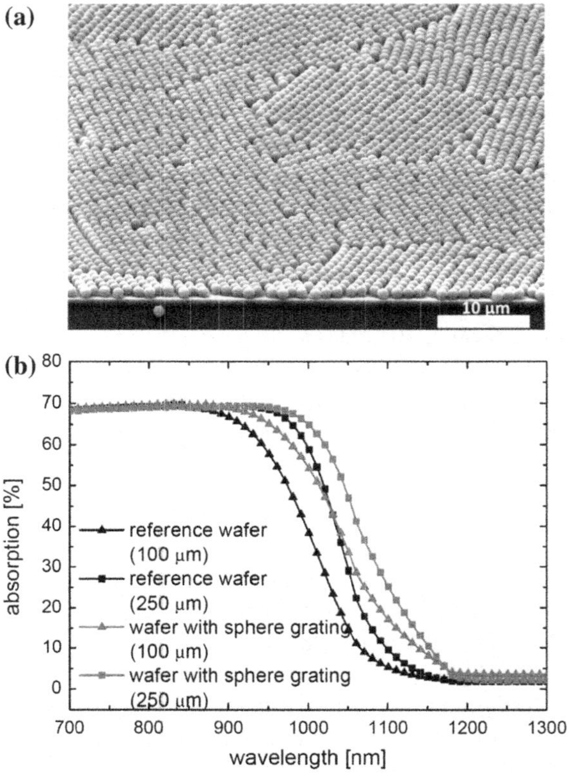

Fig. 2.20 SEM picture of a dielectric rear side grating consisting of SiO_2 spheres after spin coating (**a**) and absorption spectra measured with and without sphere grating on 100 and 250 μm thick wafers. The improved light trapping significantly increases the light absorption in the IR region. Reprinted with permission from Optics Express **22**, A111–A119, Copyright 2014, Optical Society of America [222]

is possible in trivalent lanthanide ions with metastable and long-lived intermediate levels. These ladder-like levels act as intermediate levels for the sequential excitation. The principal transitions in common lanthanide are summarized in [223]. Suitable ions with energy levels in the Si sub-band gap energy range are Er^{3+}, Tm^{3+} and Ho^{3+}, whereas the Er^{3+} ion is the most investigated material in combination with Si [29, 224, 225]. The energy levels of Er^{3+} are depicted in Fig. 2.21. Er^{3+} has a rather narrow absorption band at 1.56 μm (0.8 eV), which is capable of absorbing sub-band gap photons for up-conversion. To broaden the utilized spectral range, Lahoz et al. combined Er^{3+} and Ho^{3+} up-conversion layers. Ho^{3+} ions have an absorption band at 1.17 μm (1.06 eV), where the solar irradiation intensity is about twice as high as the intensity at the absorption band of Er^{3+} [226].

These lanthanide-based conversion ions are hosted in a matrix, which requires a low lattice mismatch to the dopant ions and low phonon energies. The low phonon energy suppresses non-radiative transitions between closely spaced energy levels due to multiphonon relaxation. The most commonly used matrix material for Er^{3+} is β-$NaYF_4$, however, several other materials such as β-BaY_2F_8 or fluoride glasses are investigated as well [29, 223].

Several mechanisms of up-conversion have been identified. The simplest up-conversion mechanism involves sequential photon absorption from the ground

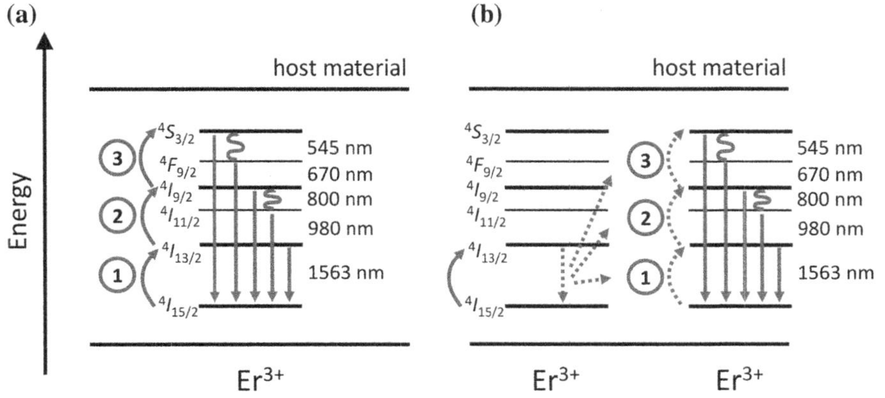

Fig. 2.21 Principal up-conversion mechanisms in Er^{3+} ions. The excited state absorption (**a**) involves a single ion. The energy transfer up-conversion (**b**) is based on the non-radiative energy exchange between two neighboring ions. Absorption/emission and energy transfer are indicated by *solid* and *dotted lines*, respectively. The sequential absorption steps are denoted by numbers. Energy levels are taken from [225, 229]

state to elevated excited states. This excited state absorption (ESA) process transfers a single ion to a higher energy level as illustrated in Fig. 2.21a. In the energy transfer up-conversion (ETU) two neighboring ions are involved. After both ions are excited into a meta-stable state, the ions then non-radiatively exchange energy. The activator ion is excited to a higher energy level and the sensitizer ion relaxes to the ground state (Fig. 2.21b). At high excitation powers also the photon avalanche mechanism results in up-conversion [223, 224]. After a two-step or multiple-step absorption process, the excited ion can emit one photon exceeding the Si band gap energy. This photon generates electron-hole pairs in the solar cell and thus increases the spectral response at an energy level corresponding to the absorption band of the converter material.

Up-conversion is a non-linear process. When the process involves two or more photons for up-converted emission, theory predicts that the occupation of the uppermost excited state correlates to P^{n}, where P is the power density and n the number of photons involved. This non-linearity results in an external quantum efficiency (EQE) proportional to $P^{(\mathrm{n}-1)}$, i.e. the EQE values linearly increases with illumination intensity for a two-photon absorption process. In experiments the exponent n was found to be close to the theoretical value of two at low irradiance. At higher power densities the exponent decreases as other processes, like non-radiative recombination, occur [225, 227, 228]. As a consequence of the power dependency of up-conversion, reasonable up-conversion efficiencies are only reached for high illumination intensities, which are typically only reached in concentrated sun light.

Up-conversion layers can be easily integrated in bifacial solar cells with a transparent rear side [228, 230]. The up-conversion layer is attached at the rear side of the solar cell and covered by an additional back reflector to reflect the anisotropic light emission back into the Si substrate. In principle, the up-conversion layer could also be integrated at the front side. However, this configuration is less attractive due to parasitic absorption of photons exceeding the band gap energy and due to the fact that half of the up-converted anisotropic emission is directed away from the solar cell. Figure 2.22a depicts a bifacial solar cell with an Er-doped $NaYF_4$ up-conversion layer. This conversion stack reaches an enhanced quantum efficiency yield of up to 3.4 % at the absorption band of Er^{3+} under an illumination of 1000 suns (Fig. 2.22b) [227]. In this early work, up-conversion hardly enhanced solar cell efficiency at lower illumination intensities. More recent studies demonstrate photocurrent gains of 0.55 % at 94 suns [230] and 3.89 mA/cm^2 at 50 suns, which corresponds to about 0.2 % enhancement [231]. These photocurrent gains prove a significant progress during the last years; however, the low conversion efficiency is still an obstacle for commercialization of up-converter solar cells.

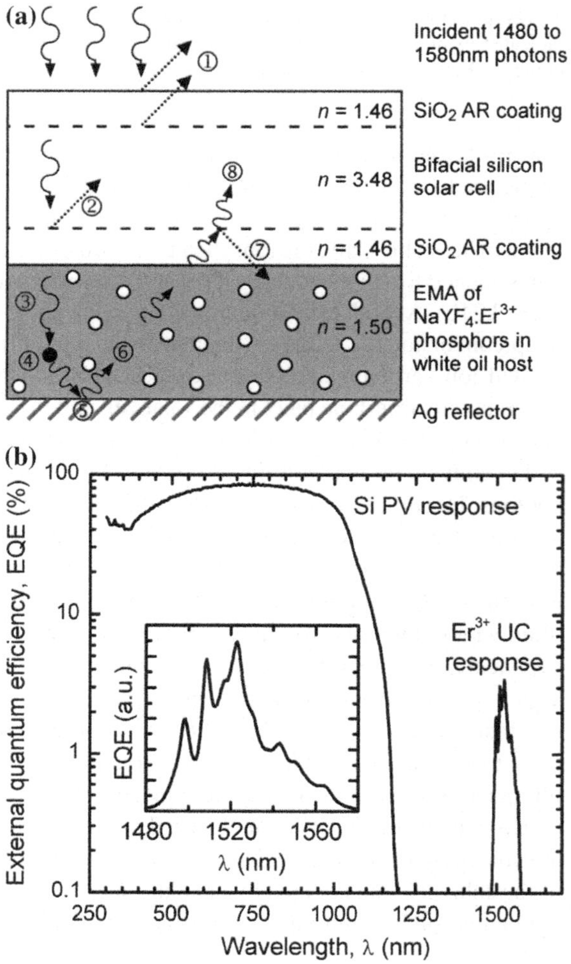

Fig. 2.22 Solar cell comprising a $NaYF_4$:Er^{3+} up-conversion layer (**a**) and measured spectral quantum efficiency (**b**). The up-conversion process results in an increased photo-response centered at 1540 nm. Reprinted with permission from IEEE Transactions on Electron Devices **54**, 2679–2684, Copyright (2007), IEEE [227]

In most concepts, the up-conversion layer is attached to a functional solar cell. Alternatively, up-conversion layers could also be used for surface passivation and integrated into the solar cell concept, provided that the electrical quality of the host matrix is sufficient. Lanthanide oxides used for up-conversion can be deposited by ALD within a temperature range of 200–400 °C [232], which coincides with the temperature window of common passivation materials (Table 2.2). Dingemans et al. realized Er-doped Al_2O_3 layers and found up-conversion under 1480 nm wavelength illumination [233]. Al_2O_3 is found to be a suitable host matrix for Er^{3+} on condition that a post-deposition annealing above 900 °C is applied. During annealing, OH residuals are removed, which cause quenching of the Er^{3+} luminescence [234]. However, the high thermal budget required for optical activation of the Er^{3+} ions also degrades the passivation quality of Al_2O_3. Solar cells based on passivating conversion layers (Fig. 2.1f) have not been experimentally realized yet.

Down-conversion transfers a high energy photon (>$2E_g$) into two photons of lower energy [29, 235, 236]. Down-conversion layers have to be integrated at the front side of the solar cell, where high energy photons are not already absorbed within the Si substrate. This is a fundamental restriction for down-conversion, since down-converted photons are isotropically emitted and partly directed away from the solar cell. This escape loss strongly reduces the conversion efficiency. Therefore, research activity mainly focuses on up-conversion, though the theoretical potential of up- and down-conversion is similar.

2.6 Conclusions and Outlook

Dielectric nanomaterials are an essential element in today's silicon PERC solar cells. SiO_2, SiN_x:H and Al_2O_3 nanolayers were introduced due to their excellent properties for surface passivation and light management. The excellent surface passivation is achieved by chemical and field-effect passivation. The chemical passivation reduces the density of silicon surface states, which could act as recombination centers for photo-generated charge carriers. The field-effect passivation results from a high density of intrinsic fixed charges, which are located at structural defect sites at the interface of the dielectric to silicon. These fixed charges produce an electric field, which causes a strong asymmetry of the electron and hole concentrations at the surface and thus reduces charge carrier recombination.

The front side SiN_x:H passivation layer also serves as optical ARC. As the refractive index of PECVD-grown SiN_x:H can be continuously tuned within a wide range, the optical properties of the material can be optimized for lowest reflection losses at the front surface. Additionally, the dielectric passivation stack enhances the internal light reflection at the rear side of the PERC solar cell resulting in a significant improvement of the quantum efficiency in the IR spectral range.

However, novel solar cell concepts with higher efficiencies and lower production costs emerge and these concepts require further development of materials and processes. Flash light annealing appears as a promising solution to reduce the temperature of the passivation process below 200 °C. A low-thermal budget process enables the combination of dielectric materials with low-temperature concepts, such as a-Si:H/c-Si heterojunction solar cells. Novel functionalities can be realized in multi-oxide nanolaminates employing materials, such as HfO_2 and TiO_2, which are not common in solar cell manufacturing today. Al_2O_3 nanolaminates with SiO_2 or HfO_2 interface layers provide symmetrical passivation with similar performance on both *n*- and *p*-type Si substrates. Al_2O_3–TiO_2 nanolaminates enhance the electrical conductivity of the passivation layer by several orders of magnitude without significantly deteriorating the passivation performance. These nanolaminates are very promising candidates for future carrier selective contact materials due to the strong synergy with today's PERC technology. Dielectric materials could also support future light management concepts as photonic scattering material or as host material for up-conversion layers.

Dielectric nanomaterials have been extensively studied for the application as high-*k* materials in microelectronic devices and this resulted in a profound understanding of material properties and mature deposition and characterization methods. This great knowledge base and the strong synergy between photovoltaics and microelectronics facilitated the integration of dielectric nanomaterials into Si solar cells in the past. However, this synergy is by far not exploited yet. Target of future R&D effort should be to further exploit this synergy and to realize novel functionalities for the next generation of highly efficient solar cells.

References

1. A.G. Aberle, Surface passivation of crystalline silicon solar cells: a review. Prog. Photovolt. Res. Appl. **8**, 473–487 (2000)
2. A.W. Blakers, M.A. Green, 20 % Efficiency silicon solar cells. Appl. Phys. Lett. **48**, 215 (1986)
3. J. Zhao, A. Wang, M.A. Green, 24.5 % efficiency silicon PERT cells on MCZ substrates and 24.7 % efficiency PERL cells on FZ substrates. Prog. Photovolt. Res. Appl. **7**, 471–474 (1999)
4. A.G. Aberle, R. Hezel, Progress in low-temperature surface passivation of silicon solar cells using remote-plasma silicon nitride. Prog. Photovolt. Res. Appl. **5**, 29–50 (1997)
5. J. Schmidt, J.D. Moschner, J. Henze, S. Dauwe, R. Hezel, Recent progress in the surface passivation of silicon solar cells using silicon nitride, in *Proc. 19th PVSEC* (2004), pp. 391–396
6. G. Dingemans, W.M.M. Kessels, Status and prospects of Al_2O_3-based surface passivation schemes for silicon solar cells. J. Vac. Sci. Technol. A **30**, 040802 (2012)
7. J. Schmidt, A. Merkle, R. Brendel, B. Hoex, M.C.M. van de Sanden, W.M.M. Kessels, Surface passivation of high-efficiency silicon solar cells by atomic-layer-deposited Al_2O_3. Prog. Photovolt. Res. Appl. **16**, 461–466 (2008)
8. M.A. Green, The passivated emitter and rear cell (PERC): from conception to mass production. Sol. Energy Mater. Sol. Cells **143**, 190–197 (2015)
9. S.W. Glunz, R. Preu, D. Biro, Crystalline silicon solar cells—state-of-the-art and future developments. Compr. Renew. Energy, Elsevier **1**, 353–387 (2012)
10. K. Wijekoon, F. Yan, Y. Zheng, D. Wang, H. Mungekar, L. Zhang, H. Ponnekanti, Optimization of rear local contacts on high efficiency PERC solar cells structures. Int. J. Photoenergy **2013**, 1–8 (2013)

11. F. Dross, K. Baert, T. Bearda, J. Deckers, V. Depauw, O. El Daif, I. Gordon, A. Gougam, J. Govaerts, S. Granata, R. Labie, X. Loozen, R. Martini, A. Masolin, B. O'Sullivan, Y. Qiu, J. Vaes, D. Van Gestel, J. Van Hoeymissen, A. Vanleenhove, K. Van Nieuwenhuysen, S. Venkatachalam, M. Meuris, J. Poortmans, Crystalline thin-foil silicon solar cells: where crystalline quality meets thin-film processing: crystalline quality meets thin-film processing. Prog. Photovolt. Res. Appl. **20**, 770–784 (2012)
12. P. Kapur, M. Moslehi, A. Deshpande, V. Rana, J. Kramer, S. Seutter, H. Deshazer, S. Coutant, A. Calcaterra, S. Kommera, S. Su, D. Grupp, S. Tamilmani, D. Dutton, T. Stalcup, T. Du, M. Wingert, A manufacturable, non-plated, non-Ag metallization based 20.44 % efficient, 243 cm^2 area, back contacted solar cell on 40 μm thick mono-crystalline silicon, in *Proc. 28th PVSEC* (2013), pp. 2228–2231
13. J.H. Petermann, D. Zielke, J. Schmidt, F. Haase, E.G. Rojas, R. Brendel, 19 %-Efficient and 43 μm-thick crystalline Si solar cell from layer transfer using porous silicon. Prog. Photovolt. Res. Appl. **20**, 1–5 (2012)
14. J. Govaerts, C. Trompoukis, H.S. Radhakrishnan, L. Tous, S.N. Granata, E.G. Carnemolla, R. Martini, A. Marchegiani, M. Karim, I. Sharlandziev, T. Bearda, V. Depauw, K. Van Nieuwenhuysen, I. Gordon, J. Szlufczik, J. Poortmans, Solar cells from epitaxial foils: an epifoil epiphany. Energy Procedia **77**, 871–880 (2015)
15. M.J. Kerr, P. Campbell, A. Cuevas, Lifetime and efficiency limits of crystalline silicon solar cells, in *Photovoltaic Specialists Conference (PVSC), 29th IEEE*, (2002), pp. 438–441
16. M.A. Green, Limiting efficiency of bulk and thin-film silicon solar cells in the presence of surface recombination. Prog. Photovolt. Res. Appl. **7**, 327–330 (1999)
17. A. Bozzola, P. Kowalczewski, L.C. Andreani, Towards high efficiency thin-film crystalline silicon solar cells: the roles of light trapping and non-radiative recombinations. J. Appl. Phys. **115**, 094501 (2014)
18. M. Taguchi, A. Yano, S. Tohoda, K. Matsuyama, Y. Nakamura, T. Nishiwaki, K. Fujita, E. Maruyama, 24.7 % Record efficiency HIT solar cell on thin silicon wafer. IEEE J. Photovolt. **4**, 96–99 (2014)
19. M. Bivour, C. Meinhardt, D. Pysch, C. Reichel, K.-U. Ritzau, M. Hermle, and S. W. Glunz, *n*-Type silicon solar cells with amorphous/crystalline silicon heterojunction rear emitter, in *Photovoltaic Specialists Conference (PVSC), 35th IEEE* (2010), pp. 001304–001308
20. L. Tous, S.N. Granata, P. Choulat, T. Bearda, A. Michel, A. Uruena, E. Cornagliotti, M. Aleman, R. Gehlhaar, R. Russell, F. Duerinckx, J. Szlufcik, Process simplifications in large area hybrid silicon heterojunction solar cells. Sol. Energy Mater. Sol. Cells **142**, 66–74 (2015)
21. I. Dirnstorfer, F. Benner, D. K. Simon, T. Mikolajick, N. Schilling, U. Klotzbach, Feasibility study for silicon heterojunction metal wrap through solar cells. *Proc. 28th PVSEC* (2013), pp. 1108–1112
22. S. De Wolf, A. Descoeudres, Z.C. Holman, C. Ballif, High-efficiency silicon heterojunction solar cells: a review. Green **2**, 7–24 (2012)
23. A.G. Aberle, S. Glunz, W. Warta, Impact of illumination level and oxide parameters on Shockley–Read–Hall recombination at the Si-SiO_2 interface. J. Appl. Phys. **71**, 4422 (1992)
24. M.J. Kerr, A. Cuevas, Very low bulk and surface recombination in oxidized silicon wafers. Semicond. Sci. Technol. **17**, 35–38 (2002)
25. S.W. Glunz, A.B. Sproul, W. Warta, W. Wettling, Injection-level-dependent recombination velocities at the Si-SiO_2 interface for various dopant concentrations. J. Appl. Phys. **75**, 1611 (1994)
26. K. Carstens, M. Dahlinger, E. Hoffmann, J.R. Köhler, R. Zapf-Gottwick, J.H. Werner, Universal passivation for p++ and n++ areas on IBC solar cells. Energy Procedia **77**, 779–785 (2015)
27. B.W.H. van de Loo, H.C.M. Knoops, G. Dingemans, G.J.M. Janssen, M.W.P.E. Lamers, I.G. Romijn, A.W. Weeber, W.M.M. Kessels, 'Zero-charge' SiO_2/Al_2O_3 stacks for the simultaneous passivation of n+ and p+ doped silicon surfaces by atomic layer deposition. Sol. Energy Mater. Sol. Cells **143**, 450–456 (2015)

28. U. Wurfel, A. Cuevas, P. Wurfel, Charge carrier separation in solar cells. IEEE J. Photovolt. **5**, 461–469 (2015)
29. C. Strümpel, M. McCann, G. Beaucarne, V. Arkhipov, A. Slaoui, V. Švrček, C. del Cañizo, I. Tobias, Modifying the solar spectrum to enhance silicon solar cell efficiency—an overview of available materials. Sol. Energy Mater. Sol. Cells **91**, 238–249 (2007)
30. W. Shockley, H.J. Queisser, Detailed balance limit of efficiency of *p-n* junction solar cells. J. Appl. Phys. **32**, 510 (1961)
31. T. Trupke, A. Shalav, B.S. Richards, P. Würfel, M.A. Green, Efficiency enhancement of solar cells by luminescent up-conversion of sunlight. Sol. Energy Mater. Sol. Cells **90**, 3327–3338 (2006)
32. W. Shockley, W.T. Read, Statistics of the recombinations of holes and electrons. Phys. Rev. **87**, 835–842 (1952)
33. R.N. Hall, Electron-hole recombination in germanium. Phys. Rev. **87**, 387 (1952)
34. D.K. Schroder, *Semiconductor material and device characterization*, 3rd edn. (N.J. IEEE Press. Wiley, Hoboken, 2006)
35. F. Werner, A. Cosceev, J. Schmidt, Interface recombination parameters of atomic-layer-deposited Al_2O_3 on crystalline silicon. J. Appl. Phys. **111**, 073710 (2012)
36. P. Saint-Cast, Y.-H. Heo, E. Billot, P. Olwal, M. Hofmann, J. Rentsch, S.W. Glunz, R. Preu, Variation of the layer thickness to study the electrical property of PECVD Al_2O_3/c-Si interface. Energy Procedia **8**, 642–647 (2011)
37. L.E. Black, K.R. McIntosh, Modeling recombination at the Si-Al_2O_3 interface. IEEE J. Photovolt. **3**, 936–943 (2013)
38. W.G.J.H.M. van Sark, L. Korte, F. Roca (eds.), *Physics and Technology of Amorphous-Crystalline Heterostructure Silicon Solar Cells* (Springer, Heidelberg, 2012)
39. T. Dullweber, M. Siebert, B. Veith, C. Kranz, J. Schmidt, R. Brendel, B. F. P. Roos, T. Dippell, A. Schwabedissen, and S. Peters, High-efficiency industrial-type PERC solar cells applying ICP AlO_x as rear passivation layer. *Proc. 27th PVSEC* (2012), pp. 672–675
40. B. Hoex, S.B.S. Heil, E. Langereis, M.C.M. van de Sanden, W.M.M. Kessels, Ultralow surface recombination of c-Si substrates passivated by plasma-assisted atomic layer deposited Al_2O_3. Appl. Phys. Lett. **89**, 042112 (2006)
41. H. Mäckel, R. Lüdemann, Detailed study of the composition of hydrogenated SiN_x layers for high-quality silicon surface passivation. J. Appl. Phys. **92**, 2602 (2002)
42. R.A. Sinton, A. Cuevas, M. Stuckings, Quasi-steady-state photoconductance, a new method for solar cell material and device characterization. in *Photovoltaic Specialists Conference (PVSC), 25th IEEE* (1996), pp. 457–460
43. A. Cuevas, R.A. Sinton, Prediction of the open-circuit voltage of solar cells from the steady-state photoconductance. Prog. Photovolt. Res. Appl. **5**, 79–90 (1997)
44. K. Dornich, T. Hahn, J.R. Niklas, Non destructive electrical defect characterisation and topography of silicon wafers and epitaxial layers. MRS Proc. **864**(E11), 2 (2005)
45. D. Kiliani, G. Micard, B. Steuer, B. Raabe, A. Herguth, G. Hahn, Minority charge carrier lifetime mapping of crystalline silicon wafers by time-resolved photoluminescence imaging. J. Appl. Phys. **110**, 054508 (2011)
46. A.B. Sproul, Dimensionless solution of the equation describing the effect of surface recombination on carrier decay in semiconductors. J. Appl. Phys. **76**, 2851 (1994)
47. A. Richter, S.W. Glunz, F. Werner, J. Schmidt, A. Cuevas, Improved quantitative description of Auger recombination in crystalline silicon. Phys. Rev. B **86**, 165202 (2012)
48. R.B.M. Girisch, R.P. Mertens, R.F. De Keersmaecker, Determination of Si-SiO_2 interface recombination parameters using a gate-controlled point-junction diode under illumination. IEEE Trans. Electron Devices **35**, 203–222 (1988)
49. D.A. Clugston, P.A. Basore, PC1D version 5: 32-bit solar cell modeling on personal computers, in *Proc. PVSC, 26th IEEE* (1997), pp. 207–210
50. R. Stangl, C. Leendertz, J. Haschke, Numerical simulation of solar cells and solar cell characterization methods: the open-source on demand program AFORS-HET, in *InTech e-book: "SolarEnergy"* (2010)

51. S. Steingrube, P.P. Altermatt, D.S. Steingrube, J. Schmidt, R. Brendel, Interpretation of recombination at c-Si/SiN_x interfaces by surface damage. J. Appl. Phys. **108**, 014506 (2010)
52. B. Hoex, J.J.H. Gielis, M.C.M. van de Sanden, W.M.M. Kessels, On the c-Si surface passivation mechanism by the negative-charge-dielectric Al_2O_3. J. Appl. Phys. **104**, 113703 (2008)
53. J.R. Elmiger, Recombination at the silicon nitride/silicon interface. J. Vac. Sci. Technol. A **15**, 2418 (1997)
54. J. Schmidt, A.G. Aberle, Carrier recombination at silicon-silicon nitride interfaces fabricated by plasma-enhanced chemical vapor deposition. J. Appl. Phys. **85**, 3626 (1999)
55. G. Dingemans, N.M. Terlinden, M.A. Verheijen, M.C.M. van de Sanden, W.M.M. Kessels, Controlling the fixed charge and passivation properties of Si(100)/Al_2O_3 interfaces using ultrathin SiO_2 interlayers synthesized by atomic layer deposition. J. Appl. Phys. **110**, 093715 (2011)
56. S.W. Glunz, D. Biro, S. Rein, W. Warta, Field-effect passivation of the SiO_2-Si interface. J. Appl. Phys. **86**, 683 (1999)
57. S. Steingrube, P.P. Altermatt, D. Zielke, F. Werner, J. Schmidt, R. Brendel, Reduced passivation of silicon surfaces at low injection densities caused by H-induced defects, in *Proc. 15th PVSEC* (2010), pp. 1748–1754
58. F.-J. Ma, G.G. Samudra, M. Peters, A.G. Aberle, F. Werner, J. Schmidt, B. Hoex, Advanced modeling of the effective minority carrier lifetime of passivated crystalline silicon wafers. J. Appl. Phys. **112**, 054508 (2012)
59. Z. Hameiri, Fa-Jun Ma, K.R. McIntosh, Investigation of low injection effects using the local ideality factor obtained from effective lifetime measurements. *Photovoltaic Specialists Conference (PVSC), 2014 IEEE 40th*, pp. 1842–1847
60. I. Dirnstorfer, D.K. Simon, P.M. Jordan, T. Mikolajick, Near surface inversion layer recombination in Al_2O_3 passivated *n*-type silicon. J. Appl. Phys. **116**, 044112 (2014)
61. M. Kessler, T. Ohrdes, P.P. Altermatt, R. Brendel, The effect of sample edge recombination on the averaged injection-dependent carrier lifetime in silicon. J. Appl. Phys. **111**, 054508 (2012)
62. B. Veith, T. Ohrdes, F. Werner, R. Brendel, P.P. Altermatt, N.-P. Harder, J. Schmidt, Injection dependence of the effective lifetime of n-type Si passivated by Al_2O_3: An edge effect? Sol. Energy Mater. Sol. Cells **120**, 436–440 (2014)
63. J. Robertson, R.M. Wallace, High-K materials and metal gates for CMOS applications. Mater. Sci. Eng. R Rep. **88**, 1–41 (2015)
64. B.E. Deal, A.S. Grove, General relationship for the thermal oxidation of silicon. J. Appl. Phys. **36**, 3770 (1965)
65. B.E. Deal, M. Sklar, A.S. Grove, E.H. Snow, Characteristics of the surface-state charge (Q_{ss}) of thermally oxidized silicon. J. Electrochem. Soc. **114**, 266 (1967)
66. F.J. Himpsel, F.R. McFeely, A. Taleb-Ibrahimi, J.A. Yarmoff, G. Hollinger, Microscopic structure of the SiO_2/Si interface. Phys. Rev. B **38**, 6084–6096 (1988)
67. F. Werner, B. Veith, D. Zielke, L. Kühnemund, C. Tegenkamp, M. Seibt, R. Brendel, J. Schmidt, Electronic and chemical properties of the c-Si/Al_2O_3 interface. J. Appl. Phys. **109**, 113701 (2011)
68. F. Benner, P.M. Jordan, M. Knaut, I. Dirnstorfer, J.W. Bartha, T. Mikojajick, Investigation of the c-Si/Al_2O_3 interface for silicon surface passivation, in *Proc 27th PVSEC* (2012), pp. 1793–1796
69. M. Bhaisare, A. Misra, A. Kottantharayil, Aluminum oxide deposited by pulsed-DC reactive sputtering for crystalline slicon surface passivation. IEEE J. Photovolt. **3**, 930–935 (2013)
70. V. Verlaan, L.R.J.G. van den Elzen, G. Dingemans, M.C.M. van de Sanden, W.M.M. Kessels, Composition and bonding structure of plasma-assisted ALD Al_2O_3 films. Phys. Status Solidi C **7**, 976–979 (2010)
71. O. Renault, L.G. Gosset, D. Rouchon, A. Ermolieff, Angle-resolved x-ray photoelectron spectroscopy of ultrathin Al_2O_3 films grown by atomic layer deposition. J. Vac. Sci. Technol. A **20**, 1867 (2002)

72. F. Lin, B. Hoex, Y.H. Koh, J. Lin, A.G. Aberle, Low-temperature surface passivation of moderately doped crystalline silicon by atomic-layer-deposited hafnium oxide films. ECS J. Solid State Sci. Technol. **2**, N11–N14 (2012)
73. D.K. Simon, P.M. Jordan, M. Knaut, T. Chohan, T. Mikolajick, I. Dirnstorfer, ALD Al_2O_3 based Nanolaminates for Solar Cell Applications, in *Photovoltaic Specialists Conference (PVSC), IEEE 42th* (2015), pp. 1–6
74. A.G. Aberle, Overview on SiN surface passivation of crystalline silicon solar cells. Sol. Energy Mater. Sol. Cells **65**, 239–248 (2001)
75. A. Stesmans, V.V. Afanas'ev, Si dangling-bond-type defects at the interface of (100)Si with ultrathin layers of SiO_x, Al_2O_3, and ZrO_2. *Appl. Phys. Lett.* **80**, 1957 (2002)
76. G. Dingemans, M.C.M. van de Sanden, W.M.M. Kessels, Influence of the deposition temperature on the c-Si surface passivation by Al_2O_3 films synthesized by ALD and PECVD. Electrochem. Solid-State Lett. **13**, H76–H79 (2010)
77. M.D. Groner, F.H. Fabreguette, J.W. Elam, S.M. George, Low-temperature Al_2O_3 atomic layer deposition. Chem. Mater. **16**, 639–645 (2004)
78. J. Schmidt, M. Kerr, A. Cuevas, Surface passivation of silicon solar cells using plasma-enhanced chemical-vapour-deposited SiN films and thin thermal SiO_2/plasma SiN stacks. Semicond. Sci. Technol. **16**, 164–170 (2001)
79. G. Dingemans, C. Van Helvoirt, M.C.M. Van de Sanden, W.M. Kessels, Plasma-assisted atomic layer deposition of low temperature SiO_2. ECS Trans. **35**, 191–204 (2011)
80. G. Dingemans, W. Beyer, M.C.M. van de Sanden, W.M.M. Kessels, Hydrogen induced passivation of Si interfaces by Al_2O_3 films and SiO_2/Al_2O_3 stacks. Appl. Phys. Lett. **97**, 152106 (2010)
81. G. Dingemans, F. Einsele, W. Beyer, M.C.M. van de Sanden, W.M.M. Kessels, Influence of annealing and Al_2O_3 properties on the hydrogen-induced passivation of the Si/SiO_2 interface. J. Appl. Phys. **111**, 093713 (2012)
82. A. Richter, J. Benick, M. Hermle, S.W. Glunz, Reaction kinetics during the thermal activation of the silicon surface passivation with atomic layer deposited Al_2O_3. Appl. Phys. Lett. **104**, 061606 (2014)
83. X. Zhang, A. Thomson, A. Cuevas, Silicon surface passivation by sputtered aluminium oxide: influence of annealing temperature and ambient gas. ECS Solid State Lett. **3**, N37–N39 (2014)
84. A. Stesmans, Passivation of P_{b0} and P_{b1} interface defects in thermal (100) Si/SiO_2 with molecular hydrogen. Appl. Phys. Lett. **68**, 2076 (1996)
85. L. Tsetseris, S.T. Pantelides, Migration, incorporation, and passivation reactions of molecular hydrogen at the Si-SiO_2 interface. Phys. Rev. B **70**, 245320 (2004)
86. F. Kersten, A. Schmid, S. Bordihn, J.W. Müller, J. Heitmann, Role of annealing conditions on surface passivation properties of ALD Al_2O_3 films. Energy Procedia **38**, 843–848 (2013)
87. I. Dirnstorfer, T. Chohan, P.M. Jordan, M. Knaut, D.K. Simon, J.W. Bartha, T. Mikojajick, Al_2O_3–TiO_2 nanolaminates for conductive silicon surface passivation. IEEE J. Photovolt. **6**, 86–91 (2015)
88. N.M. Terlinden, G. Dingemans, M.C.M. van de Sanden, W.M.M. Kessels, Role of field-effect on c-Si surface passivation by ultrathin (2–20 nm) atomic layer deposited Al_2O_3. Appl. Phys. Lett. **96**, 112101 (2010)
89. B. Veith, F. Werner, D. Zielke, R. Brendel, J. Schmidt, Comparison of the thermal stability of single Al_2O_3 layers and Al_2O_3/SiN_x stacks for the surface passiviation of silicon. Energy Procedia **8**, 307–312 (2011)
90. W. Liang, K.J. Weber, A.F. Thomson, Effective SiN_x: H capping layers on 1-nm Al_2O_3 for p^+ Surface Passivation. IEEE J. Photovolt. **4**, 1405–1412 (2014)
91. D. Schuldis, A. Richter, J. Benick, P. Saint-Cast, M. Hermle, S.W. Glunz, Properties of the c-Si/Al_2O_3 interface of ultrathin atomic layer deposited Al_2O_3 layers capped by SiN_x for c-Si surface passivation. Appl. Phys. Lett. **105**, 231601 (2014)

92. P. Saint-Cast, D. Kania, R. Heller, S. Kuehnhold, M. Hofmann, J. Rentsch, R. Preu, High-temperature stability of c-Si surface passivation by thick PECVD Al_2O_3 with and without hydrogenated capping layers. Appl. Surf. Sci. **258**, 8371–8376 (2012)
93. S.M. Sze, K.K. Ng, *Physics of Semiconductor Devices*, 3rd edn (2007)
94. S. Duttagupta, F. Lin, K.D. Shetty, A.G. Aberle, B. Hoex, Excellent boron emitter passivation for high-efficiency Si wafer solar cells using AlO_x/SiN_x dielectric stacks deposited in an industrial inline plasma reactor: Excellent boron emitter passivation for high-efficiency Si wafer solar cells. Prog. Photovolt. Res. Appl. **21**, 760–764 (2012)
95. W. Liang, K.J. Weber, D. Suh, S.P. Phang, J. Yu, A.K. McAuley, B.R. Legg, Surface passivation of boron-diffused *p*-type silicon surfaces with (100) and (111) orientations by ALD Al_2O_3 layers. IEEE J. Photovolt. **3**, 678–683 (2013)
96. L.E. Black, T.C. Kho, K.R. McIntosh, A. Cuevas, The influence of orientation and morphology on the passivation of crystalline silicon surfaces by Al_2O_3. Energy Procedia **55**, 750–756 (2014)
97. H.C. Sio, S.P. Phang, Y. Wan, W. Liang, T. Trupke, S. Cao, D. Hu, Y. Wan, D. Macdonald, The influence of crystal orientation on surface passivation in multi-crystalline silicon, in *Photovoltaic Specialists Conference (PVSC), IEEE 39th* (2013), pp. 1770–1775
98. H. Haug, Ø. Nordseth, E.V. Monakhov, E.S. Marstein, Photoluminescence imaging under applied bias for characterization of Si surface passivation layers. Sol. Energy Mater. Sol. Cells **106**, 60–65 (2012)
99. P.M. Jordan, D.K. Simon, F.P.G. Fengler, T. Mikolajick, I. Dirnstorfer, 2D mapping of chemical and field effect passivation of Al_2O_3 on silicon substrates. Energy Procedia **77**, 91–98 (2015)
100. P.M. Jordan, D.K. Simon, T. Mikolajick, I. Dirnstorfer, BiasMDP: carrier lifetime characterization technique with applied bias voltage. Appl. Phys. Lett. **106**, 061602 (2015)
101. W.L. Warren, F.C. Rong, E.H. Poindexter, G.J. Gerardi, J. Kanicki, Structural identification of the silicon and nitrogen dangling-bond centers in amorphous silicon nitride. J. Appl. Phys. **70**, 346 (1991)
102. K.T. Butler, M.P.W.E. Lamers, A.W. Weeber, J.H. Harding, Molecular dynamics studies of the bonding properties of amorphous silicon nitride coatings on crystalline silicon. J. Appl. Phys. **110**, 124905 (2011)
103. S. Garcia, I. Martil, G. Gonzalez Diaz, E. Castan, S. Dueñas, M. Fernandez, Deposition of SiNx: H thin films by the electron cyclotron resonance and its application to Al/SiN_x:H/Si structures. J. Appl. Phys. **83**, 332 (1998)
104. G. Lucovsky, Defect properties of Si-, O-, N-, and H-atoms at Si—SiO_2 interfaces. J. Vac. Sci. Technol. B **14**, 2832 (1996)
105. G. Dingemans, N.M. Terlinden, D. Pierreux, H.B. Profijt, M.C.M. van de Sanden, W.M.M. Kessels, Influence of the oxidant on the chemical and field-effect passivation of Si by ALD Al_2O_3. Electrochem. Solid-State Lett. **14**, H1 (2011)
106. D.K. Simon, T. Henke, P.M. Jordan, F.P.G. Fengler, T. Mikolajick, J.W. Bartha, I. Dirnstorfer, Low-thermal budget flash light annealing for Al_2O_3 surface passivation: low-thermal budget flash light annealing for Al_2O_3 surface passivation. Phys. Status Solidi RRL—Rapid Res. Lett. **9**, 631–635 (2015)
107. D.K. Simon, Realization and characterization of advanced nanolayers for photovoltaic applications, PhD thesis, TU Dresden (2016)
108. J. Benick, A. Richter, T.-T.A. Li, N.E. Grant, K.R. McIntosh, Y. Ren, K.J. Weber, M. Hermle, S.W. Glunz, Effect of a post-deposition anneal on Al_2O_3/Si interface properties, in *Photovoltaic Specialists Conference (PVSC), IEEE 35th* (2010), pp. 000891–000896
109. D.K. Simon, P.M. Jordan, T. Mikolajick, I. Dirnstorfer, On the control of the fixed charge densities in Al_2O_3 based silicon surface passivation schemes. ACS Appl. Mater. Interfaces **7**, 28215–28222 (2015)
110. A. Rothschild, B. Vermang, X. Loozen, B.J. O'Sullivan, J. John, J. Poortmans, ALD-Al_2O_3 passivation for solar cells: charge investigation, in *Proc 25th PVSEC* (2010), pp. 1382–1385

111. B. Shin, J.R. Weber, R.D. Long, P.K. Hurley, C.G. Van de Walle, P.C. McIntyre, Origin and passivation of fixed charge in atomic layer deposited aluminum oxide gate insulators on chemically treated InGaAs substrates. Appl. Phys. Lett. **96**, 152908 (2010)
112. Y. Ahn, S.H. Choudhury, D. Lee, S.M. Sadaf, M. Siddik, M. Jo, S. Park, Y.D. Kim, D.H. Kim, H. Hwang, Estimation of interfacial fixed charge at Al_2O_3/SiO_2 using Slant-Etched wafer for solar cell application. Jpn. J. Appl. Phys. **50**, 071503 (2011)
113. V. Naumann, M. Otto, R.B. Wehrspohn, C. Hagendorf, Chemical and structural study of electrically passivating Al_2O_3/Si interfaces prepared by atomic layer deposition. J. Vac. Sci. Technol. A **30**, 04D106 (2012)
114. K. Kimoto, Y. Matsui, T. Nabatame, T. Yasuda, T. Mizoguchi, I. Tanaka, A. Toriumi, Coordination and interface analysis of atomic-layer-deposition Al_2O_3 on Si(001) using energy-loss near-edge structures. Appl. Phys. Lett. **83**, 4306 (2003)
115. G. Lucovsky, A chemical bonding model for the native oxides of the III–V compound semiconductors. J. Vac. Sci. Technol. **19**, 456 (1981)
116. K. Matsunaga, T. Tanaka, T. Yamamoto, Y. Ikuhara, First-principles calculations of intrinsic defects in Al_2O_3. Phys. Rev. B **68**, 085110 (2003)
117. J.R. Weber, A. Janotti, C.G. Van de Walle, Native defects in Al_2O_3 and their impact on III-V/Al_2O_3 metal-oxide-semiconductor-based devices. J. Appl. Phys. **109**, 033715 (2011)
118. M. Choi, A. Janotti, C.G. Van de Walle, Native point defects and dangling bonds in α-Al_2O_3. J. Appl. Phys. **113**, 044501 (2013)
119. J.J.H. Gielis, B. Hoex, M.C.M. van de Sanden, W.M.M. Kessels, Negative charge and charging dynamics in Al_2O_3 films on Si characterized by second-harmonic generation. J. Appl. Phys. **104**, 073701 (2008)
120. B. Liao, R. Stangl, T. Mueller, F. Lin, C.S. Bhatia, B. Hoex, The effect of light soaking on crystalline silicon surface passivation by atomic layer deposited Al_2O_3. J. Appl. Phys. **113**, 024509 (2013)
121. J.A. Töfflinger, A. Laades, C. Leendertz, L.M. Montañez, L. Korte, U. Stürzebecher, H.-P. Sperlich, B. Rech, PECVD-AlO_x/SiN_x passivation stacks on silicon: effective charge dynamics and interface defect state spectroscopy. Energy Procedia **55**, 845–854 (2014)
122. F. Benner, M. Knaut, P.M. Jordan, I. Dirnstorfer, J. Ocker, J.W. Bartha, T. Mikolajick, Improvement of Al_2O_3 passivation by Ti-doping, in *Proc 28th PVSEC* (2013), pp. 1156–1161
123. F. Werner, J. Schmidt, Manipulating the negative fixed charge density at the c-Si/Al_2O_3 interface. Appl. Phys. Lett. **104**, 091604 (2014)
124. M. Cho, R. Degraeve, P. Roussel, B. Govoreanu, B. Kaczer, M.B. Zahid, E. Simoen, A. Arreghini, M. Jurczak, J.V. Houdt, G. Groeseneken, A consistent model for oxide trap profiling with the trap spectroscopy by charge injection and sensing (TSCIS) technique. Solid-State Electron. **54**, 1384–1391 (2010)
125. E. Vianello, F. Driussi, A. Arreghini, P. Palestri, D. Esseni, L. Selmi, N. Akil, M.J. van Duuren, D.S. Golubovic, Experimental and simulation analysis of program/retention transients in silicon nitride-based NVM cells. IEEE Trans. Electron Devices **56**, 1980–1990 (2009)
126. T. Mikolajick, M. Specht, N. Nagel, T. Mueller, S. Riedel, F. Beug, T. Melde, K.-H. Kusters, The future of charge trapping memories. Int. Symp. VLSI-TSA **2007**, 1–4 (2007)
127. N.M. Terlinden, G. Dingemans, V. Vandalon, R.H.E.C. Bosch, W.M.M. Kessels, Influence of the SiO_2 interlayer thickness on the density and polarity of charges in Si/SiO_2/Al_2O_3 stacks as studied by optical second-harmonic generation. J. Appl. Phys. **115**, 033708 (2014)
128. R. Degraeve, M. Cho, B. Govoreanu, B. Kaczer, M.B. Zahid, J. Van Houdt, M. Jurczak, G. Groeseneken, Trap spectroscopy by charge injection and sensing (TSCIS): a quantitative electrical technique for studying defects in dielectric stacks (IEDM, IEEE Int. Electron Devices Meet, 2008), pp. 775–778
129. M. Toledano-Luque, R. Degraeve, M.B. Zahid, B. Kaczer, J. Kittl, M. Jurczak, G. Groeseneken, J. Van Houdt, Resolving fast VTH transients after program/erase of flash memory stacks and their relation to electron and hole defects (IEDM, IEEE Int. Electron Devices Meet, 2009), pp. 749–752

130. M. Bazilchuk, H. Haug, E.S. Marstein, Modulating the fixed charge density in silicon nitride films while monitoring the surface recombination velocity by photoluminescence imaging. Appl. Phys. Lett. **106**, 143505 (2015)
131. Y. Ren, K. Weber, F. Karouta, K. Vora, W. Liang, Charge trapping and storage in SiN_x thin films deposited with Oxford PlasmaLab 100 system, in *Photovoltaic Specialists Conference (PVSC), 38th IEEE* (2012), pp. 001094–001097
132. www.itrpv.net, "International Technology Roadmap for Photovoltaic (ITRPV), 2014 Results," 01-Jul-2015
133. S. Miyajima, J. Irikawa, A. Yamada, M. Konagai, High quality aluminum oxide passivation layer for crystalline silicon solar cells deposited by parallel-plate plasma-enhanced chemical vapor deposition. Appl. Phys. Express **3**, 012301 (2010)
134. T. Lauinger, Optimization and characterization of remote plasma-enhanced chemical vapor deposition silicon nitride for the passivation of *p*-type crystalline silicon surfaces. J. Vac. Sci. Technol. A **16**, 530 (1998)
135. Y. Wan, K.R. McIntosh, A.F. Thomson, A. Cuevas, Low surface recombination velocity by low-absorption silicon nitride on c-Si. IEEE J. Photovolt. **3**, 554–559 (2013)
136. S. Duttagupta, F. Ma, B. Hoex, T. Mueller, A.G. Aberle, Optimised antireflection coatings using silicon nitride on textured silicon surfaces based on measurements and multidimensional modelling. Energy Procedia **15**, 78–83 (2012)
137. Z. Hameiri, L. Mai, N. Borojevic, S. Javid, B. Tjahjono, S. Wang, A. Sproul, S. Wenham, The influence of silicon nitride layer parameters on the implied Voc of CZ silicon wafers after annealing, in *Photovoltaic Specialists Conference (PVSC), 34th IEEE* (2009), pp. 1795–1800
138. M. Hofmann, S. Kambor, C. Schmidt, D. Grambole, J. Rentsch, S. W. Glunz, R. Preu, PECVD-ONO: a new deposited firing stable rear surface passivation layer system for crystalline silicon solar cells. Adv. Optoelectron., 1–10 (2008)
139. Z. Chen, S.K. Pang, K. Yasutake, A. Rohatgi, Plasma-enhanced chemical-vapor-deposited oxide for low surface recombination velocity and high effective lifetime in silicon. J. Appl. Phys. **74**, 2856 (1993)
140. J. Irikawa, S. Miyajima, S. Kida, T. Watahiki, M. Konagai, Effects of annealing and atomic hydrogen treatment on aluminum oxide passivation layers for crystalline silicon solar cells. Jpn. J. Appl. Phys. **50**, 012301 (2011)
141. G. Dingemans, P. Engelhart, R. Seguin, M.M. Mandoc, M.C.M. van de Sanden, W.M.M. Kessels, Comparison between aluminum oxide surface passivation films deposited with thermal ALD, plasma ALD and PECVD, in *Photovoltaic Specialists Conference (PVSC), 35th IEEE* (2010), pp. 003118–003121
142. H.-P. Sperlich, D. Decker, P. Saint-Cast, E. Erben, L. Peters, High productive solar cell passivation on Roth&Rau MAiA® MW-PECVD Inline machine—a comparison of Al_2O_3, SiO_2 and SiN_x-H process conditions and performance. in *Proc 25th PVSEC* (2010), pp. 1352–1357
143. P. Saint-Cast, D. Kania, M. Hofmann, J. Benick, J. Rentsch, R. Preu, Very low surface recombination velocity on *p*-type c-Si by high-rate plasma-deposited aluminum oxide. Appl. Phys. Lett. **95**, 151502 (2009)
144. A. Laades, H.-P. Sperlich, M. Bähr, U. Stürzebecher, C.A. Diaz Alvarez, M. Burkhardt, H. Angermann, M. Blech, A. Lawerenz, On the impact of interfacial SiO_x-layer on the passivation properties of PECVD synthesized aluminum oxide: on the impact of interfacial SiO_x-layer on the passivation properties of PECVD synthesized aluminum oxide. Phys. Status Solidi C **9**, 2120–2123 (2012)
145. M.D. Groner, S.M. George, High-k dielectrics grown by atomic layer deposition. *Interlayer Dielectrics for Semiconductor Technologies* (Elsevier, Amsterdam, 2003), pp. 327–348
146. S.M. George, Atomic layer deposition: an overview. Chem. Rev. **110**, 111–131 (2010)
147. G. Agostinelli, A. Delabie, P. Vitanov, Z. Alexieva, H.F.W. Dekkers, S. De Wolf, G. Beaucarne, Very low surface recombination velocities on *p*-type silicon wafers passivated with a dielectric with fixed negative charge. Sol. Energy Mater. Sol. Cells **90**, 3438–3443 (2006)

148. H.C.M. Knoops, E.M.J. Braeken, K. de Peuter, S.E. Potts, S. Haukka, V. Pore, W.M.M. Kessels, Atomic layer deposition of silicon nitride from bis(tert-butylamino)silane and N_2 Plasma. ACS Appl. Mater. Interfaces **7**, 19857–19862 (2015)
149. S.W. King, Plasma enhanced atomic layer deposition of SiN_x: H and SiO_2. J. Vac. Sci. Technol. A **29**, 041501 (2011)
150. F. Werner, B. Veith, V. Tiba, P. Poodt, F. Roozeboom, R. Brendel, J. Schmidt, Very low surface recombination velocities on *p*- and *n*-type c-Si by ultrafast spatial atomic layer deposition of aluminum oxide. Appl. Phys. Lett. **97**, 162103 (2010)
151. F. Benner, P.M. Jordan, C. Richter, D.K. Simon, I. Dirnstorfer, M. Knaut, J.W. Bartha, T. Mikolajick, Atomic layer deposited high-κ nanolaminates for silicon surface passivation. J. Vac. Sci. Technol. B **32**, 03D110 (2014)
152. S. Bordihn, V. Mertens, J.W. Müller, W.M.M. Kessels, Deposition temperature dependence of material and Si surface passivation properties of O_3-based atomic layer deposited Al_2O_3-based films and stacks. J. Vac. Sci. Technol. A **32**, 01A128 (2014)
153. D.K. Simon, P.M. Jordan, I. Dirnstorfer, F. Benner, C. Richter, T. Mikolajick, Symmetrical Al_2O_3-based passivation layers for *p*- and *n*-type silicon. Sol. Energy Mater. Sol. Cells **131**, 72–76 (2014)
154. G. Dingemans, W.M.M. Kessels, Aluminum oxide and other ALD materials for Si surface passivation. ECS Trans. **41**, 293–301 (2011)
155. B. Liao, B. Hoex, A.G. Aberle, D. Chi, C.S. Bhatia, Excellent c-Si surface passivation by low-temperature atomic layer deposited titanium oxide. Appl. Phys. Lett. **104**, 253903 (2014)
156. D. Suh, D.-Y. Choi, K.J. Weber, Al_2O_3/TiO_2 stack layers for effective surface passivation of crystalline silicon. J. Appl. Phys. **114**, 154107 (2013)
157. Y. Wan, J. Bullock, A. Cuevas, Passivation of c-Si surfaces by ALD tantalum oxide capped with PECVD silicon nitride. Sol. Energy Mater. Sol. Cells **142**, 42–46 (2015)
158. T.G. Allen, M. Ernst, C. Samundsett, A. Cuevas, Demonstration of c-Si solar cells with gallium oxide surface passivation and laser-doped gallium p^+ regions. IEEE J. Photovolt. **5**, 1586–1590 (2015)
159. P. Poodt, D.C. Cameron, E. Dickey, S.M. George, V. Kuznetsov, G.N. Parsons, F. Roozeboom, G. Sundaram, A. Vermeer, Spatial atomic layer deposition: a route towards further industrialization of atomic layer deposition. J. Vac. Sci. Technol. A **30**, 010802 (2012)
160. V.I. Kuznetsov, M.A. Ernst, E.H.A. Granneman, Al_2O_3 surface passivation of silicon solar cells by low cost ald technology, in *Photovoltaic Specialists Conference (PVSC), IEEE 40th* (2014), pp. 0608–0611
161. E. Cornagliotti, L. Tous, A. Uruena de Castro, A. Rothschild, R. Russell, V. Lu, S. Radosavjlevic, J. John, J. Toman, M. Aleman, F. Duerinckx, J. Poortmans, J. Szlufcik, B. Dielissen, F. Souren, X. Gay, R. Görtzen, B. Hallam, Integration of spatial ALD aluminum oxide for rear side passivation of *p*-type PERC/PERL solar cells, in *Proc 28th PVSEC* (2013), pp. 976–981
162. L. Black, T. Allen, K.R. McIntosh, Safe and inexpensive Al_2O_3 deposited by APCVD with single-source precursor. *Proc 28th PVSEC* (2013), pp. 1068–1072
163. L.E. Black, K.R. McIntosh, Surface passivation of c-Si by atmospheric pressure chemical vapor deposition of Al_2O_3. Appl. Phys. Lett. **100**, 202107 (2012)
164. L.E. Black, T. Allen, A. Cuevas, K.R. McIntosh, B. Veith, J. Schmidt, Thermal stability of silicon surface passivation by APCVD Al_2O_3. Sol. Energy Mater. Sol. Cells **120**, 339–345 (2014)
165. A.F. Thomson, K.R. McIntosh, Light-enhanced surface passivation of TiO_2-coated silicon: light-enhanced surface passivation of TiO_2-coated silicon. Prog. Photovolt. Res. Appl. **20**, 343–349 (2012)
166. B.S. Richards, J.E. Cotter, C.B. Honsberg, Enhancing the surface passivation of TiO_2 coated silicon wafers. Appl. Phys. Lett. **80**, 1123 (2002)
167. M. Vetter, Surface passivation of silicon by rf magnetron-sputtered silicon nitride films. Thin Solid Films **337**, 118–122 (1999)

168. X. Zhang, A. Cuevas, Plasma hydrogenated, reactively sputtered aluminium oxide for silicon surface passivation. Phys. Status Solidi RRL—Rapid Res. Lett. **7**, 619–622 (2013)
169. T.-T.A. Li, A. Cuevas, Role of hydrogen in the surface passivation of crystalline silicon by sputtered aluminum oxide. Prog. Photovolt. Res. Appl. **19**, 320–325 (2011)
170. X. Zhang, A. Cuevas, A. Thomson, Process control of reactive sputter deposition of AlO_x and improved surface passivation of crystalline silicon. IEEE J. Photovolt. **3**, 183–188 (2013)
171. J. Schmidt, F. Werner, B. Veith, D. Zielke, S. Steingrube, P.P. Altermatt, S. Gatz, T. Dullweber, R. Brendel, Advances in the surface passivation of silicon solar cells. Energy Procedia **15**, 30–39 (2012)
172. S. Vandana, N. Batra, J. Gope, R. Singh, J. Panigrahi, S. Tyagi, P. Pathi, S.K. Srivastava, C. M.S. Rauthan, P.K. Singh, Effect of low thermal budget annealing on surface passivation of silicon by ALD based aluminum oxide films. Phys. Chem. Chem. Phys. **16**, 21804–21811 (2014)
173. G. Seguini, E. Cianci, C. Wiemer, D. Saynova, J.A.M. van Roosmalen, M. Perego, Si surface passivation by Al_2O_3 thin films deposited using a low thermal budget atomic layer deposition process. Appl. Phys. Lett. **102**, 131603 (2013)
174. T. Gebel, M. Voelskow, W. Skorupa, G. Mannino, V. Privitera, F. Priolo, E. Napolitani, A. Carnera, Flash lamp annealing with millisecond pulses for ultra-shallow boron profiles in silicon, Nucl. Instrum. Methods Phys. Res. Sect. B **186**, 287–291 (2002)
175. J. Lehmann, R. Hübner, J.V. Borany, W. Skorupa, T. Mikolajick, A. Schäfer, J. Schubert, S. Mantl, Millisecond flash lamp annealing for $LaLuO_3$ and $LaScO_3$ high-k dielectrics. Microelectron. Eng. **109**, 381–384 (2013)
176. T. Henke, M. Knaut, C. Hossbach, M. Geidel, L. Rebohle, M. Albert, W. Skorupa, J.W. Bartha, Flash-enhanced atomic layer deposition: basics, opportunities, review, and principal studies on the flash-enhanced growth of thin films. ECS J. Solid State Sci. Technol. **4**, P277–P287 (2015)
177. T. Gebel, L. Rebohle, R. Fendler, W. Hentsch, W. Skorupa, M. Voelskow, W. Anwand, R.A. Yankov, Millisecond annealing with flashlamps: Tool and process challenges, in RTP '06. 14th IEEE International Conference on Advanced Thermal Processing of Semiconductors (2006), pp. 47–55
178. R.A. McMahon, M.P. Smith, K.A. Seffen, M. Voelskow, W. Anwand, W. Skorupa, Flash-lamp annealing of semiconductor materials—applications and process models. Vacuum **81**, 1301–1305 (2007)
179. G. Dingemans, M.C.M. van de Sanden, W.M.M. Kessels, Excellent Si surface passivation by low temperature SiO_2 using an ultrathin Al_2O_3 capping film. Phys. Status Solidi RRL—Rapid Res. Lett. **5**, 22–24 (2011)
180. O. Sharia, A.A. Demkov, G. Bersuker, B.H. Lee, Effects of aluminum incorporation on band alignment at the SiO_2/HfO_2 interface. Phys. Rev. B **77**, 085326 (2008)
181. F. Feldmann, M. Bivour, C. Reichel, M. Hermle, S.W. Glunz, Passivated rear contacts for high-efficiency *n*-type Si solar cells providing high interface passivation quality and excellent transport characteristics. Sol. Energy Mater. Sol. Cells **120**, 270–274 (2014)
182. B. Nemeth, D.L. Young, H.-C. Yuan, V. LaSalvia, A.G. Norman, M. Page, B.G. Lee, P. Stradins, Low temperature Si/SiO_x/pc-Si passivated contacts to *n*-type Si solar cells, in *Photovoltaic Specialists Conference (PVSC), IEEE 40th* (2014), pp. 3448–3452
183. U. Römer, R. Peibst, T. Ohrdes, B. Lim, J. Krügener, E. Bugiel, T. Wietler, R. Brendel, Recombination behavior and contact resistance of n^+ and p^+ poly-crystalline Si/mono-crystalline Si junctions. Sol. Energy Mater. Sol. Cells **131**, 85–91 (2014)
184. D. Garcia-Alonso, S. Smit, S. Bordihn, W.M.M. Kessels, Silicon passivation and tunneling contact formation by atomic layer deposited Al_2O_3/ZnO stacks. Semicond. Sci. Technol. **28**, 082002 (2013)
185. D. Zielke, J.H. Petermann, F. Werner, B. Veith, R. Brendel, J. Schmidt, Contact passivation in silicon solar cells using atomic-layer-deposited aluminum oxide layers. Phys. Status Solidi RRL—Rapid Res. Lett. **5**, 298–300 (2011)

186. F. Feldmann, K.-U. Ritzau, M. Bivour, A. Moldovan, S. Modi, J. Temmler, M. Hermle, S.W. Glunz, High and low work function materials for passivated contacts. Energy Procedia **77**, 263–270 (2015)
187. H. Steinkemper, F. Feldmann, M. Bivour, M. Hermle, Numerical simulation of carrier-selective electron contacts featuring tunnel oxides. IEEE J. Photovolt. **5**, 1348–1356 (2015)
188. F. Klow, E.S. Marstein, S.E. Foss, Tunneling contact passivation simulations using Silvaco Atlas. Energy Procedia **77**, 99–105 (2015)
189. J.B. Heng, J. Fu, B. Kong, Y. Chae, W. Wang, Z. Xie, A. Reddy, K. Lam, C. Beitel, C. Liao, C. Erben, Z. Huang, Z. Xu, >23 % High-efficiency tunnel oxide junction bifacial solar cell with electroplated Cu gridlines. IEEE J. Photovolt. **5**, 82–86 (2015)
190. F. Feldmann, M. Bivour, C. Reichel, H. Steinkemper, M. Hermle, S.W. Glunz, Tunnel oxide passivated contacts as an alternative to partial rear contacts. Sol. Energy Mater. Sol. Cells **131**, 46–50 (2014)
191. B.G. Lee, J. Skarp, V. Malinen, S. Li, S. Choi, and H.M. Branz, Excellent passivation and low reflectivity Al_2O_3/TiO_2 bilayer coatings for *n*-wafer silicon solar cells, in *Photovoltaic Specialists Conference (PVSC), 38th IEEE* (2012), pp. 001066–001068
192. W.-C. Wang, M.-C. Tsai, J. Yang, C. Hsu, M.-J. Chen, Efficiency enhancement of nanotextured black silicon solar cells using Al_2O_3/TiO_2 dual-layer passivation stack prepared by atomic layer deposition. ACS Appl. Mater. Interfaces **7**, 10228–10237 (2015)
193. W. Jeon, H.-S. Chung, D. Joo, S.-W. Kang, $TiO_2/Al_2O_3/TiO_2$ nanolaminated thin films for DRAM capacitor deposited by Plasma-enhanced atomic layer deposition. Electrochem. Solid-State Lett. **11**, H19 (2008)
194. L. Aarik, T. Arroval, R. Rammula, H. Mändar, V. Sammelselg, B. Hudec, K. Hušeková, K. Fröhlich, J. Aarik, Atomic layer deposition of high-quality Al_2O_3 and Al-doped TiO_2 thin films from hydrogen-free precursors. Thin Solid Films **565**, 19–24 (2014)
195. W.D. Kim, G.W. Hwang, O.S. Kwon, S.K. Kim, M. Cho, D.S. Jeong, S.W. Lee, M.H. Seo, C.S. Hwang, Y.-S. Min, Y.J. Cho, Growth characteristics of atomic layer deposited TiO_2 thin films on Ru and Si electrodes for memory capacitor applications. J. Electrochem. Soc. **152**, C552 (2005)
196. I. Dirnstorfer, H. Mähne, T. Mikolajick, M. Knaut, M. Albert, K. Dubnack, Atomic layer deposition of anatase TiO_2 on porous electrodes for dye-sensitized solar cells. J. Vac. Sci. Technol. A **31**, 01A116 (2013)
197. R. Peibst, U. Romer, K.R. Hofmann, B. Lim, T.F. Wietler, J. Krugener, N.-P. Harder, R. Brendel, Simple model describing the symmetric I-V characteristics of *p* polycrystalline Si/*n* monocrystalline Si, and *n* polycrystalline Si/*p* monocrystalline Si junctions. IEEE J. Photovolt. **4**, 841–850 (2014)
198. E. Hecht, *Optics*, 4th edn (2014)
199. M.H. Kang, K. Ryu, A. Upadhyaya, A. Rohatgi, Optimization of SiN AR coating for Si solar cells and modules through quantitative assessment of optical and efficiency loss mechanism. Prog. Photovolt. Res. Appl. **19**, 983–990 (2011)
200. H. Nagel, A.G. Aberle, R. Hezel, Optimised antireflection coatings for planar silicon solar cells using remote PECVD silicon nitride and porous silicon dioxide. Prog. Photovolt. Res. Appl. **7**, 245–260 (1999)
201. E. Bustarret, M. Bensouda, M.C. Habrard, J.C. Bruyère, S. Poulin, S.C. Gujrathi, Configurational statistics in a-$Si_xN_yH_z$ alloys: a quantitative bonding analysis. Phys. Rev. B **38**, 8171–8184 (1988)
202. A. Thomson, Y. Wan, N. Lal, R.G. Elliman, Graded silicon nitride films: optics and passivation. J. Vac. Sci. Technol. A **33**, 060610 (2015)
203. S.W. Glunz, High-efficiency crystalline silicon solar cells. Adv. Optoelectron. 1–15 (2007)
204. D. Kray, M. Hermle, S.W. Glunz, Theory and experiments on the back side reflectance of silicon wafer solar cells. Prog. Photovolt. Res. Appl. **16**, 1–15 (2008)
205. J. Rentsch, H.-P. Sperlich, N. Kohn, D. Kania, P. Saint-Cast, S. Schetter, M. Hofmann, R. Preu, Industrial deposition of PECVD AlO_x for rear passivation of PERC type mc-Si solar cells, in *Proc 25th PVSEC*, pp. 1715–1718 (2010)

206. E. Yablonovitch, Statistical ray optics. J. Opt. Soc. Am. **72**, 899 (1982)
207. P. Campbell, M.A. Green, Light trapping properties of pyramidally textured surfaces. J. Appl. Phys. **62**, 243 (1987)
208. H. Seidel, Anisotropic etching of crystalline silicon in alkaline solutions. J. Electrochem. Soc. **137**, 3612–3626 (1990)
209. M. Otto, M. Algasinger, H. Branz, B. Gesemann, T. Gimpel, K. Füchsel, T. Käsebier, S. Kontermann, S. Koynov, X. Li, V. Naumann, J. Oh, A.N. Sprafke, J. Ziegler, M. Zilk, R.B. Wehrspohn, Black silicon photovoltaics. Adv. Opt. Mater. **3**, 147–164 (2015)
210. R. Einhaus, E. Vazsonyi, J. Szlufcik, J. Nijs, and R. Mertens, Isotropic texturing of multicrystalline silicon wafers with acidic texturing solutions, in *Photovoltaic Specialists Conference (PVSC), 26th IEEE* (1997), pp. 167–170
211. K. Kim, S.K. Dhungel, S. Jung, D. Mangalaraj, J. Yi, Texturing of large area multi-crystalline silicon wafers through different chemical approaches for solar cell fabrication. Sol. Energy Mater. Sol. Cells **92**, 960–968 (2008)
212. M. Burresi, F. Pratesi, F. Riboli, D.S. Wiersma, Complex photonic structures for light harvesting. Adv. Opt. Mater. **3**, 722–743 (2015)
213. C. Trompoukis, I. Abdo, R. Cariou, I. Cosme, W. Chen, O. Deparis, A. Dmitriev, E. Drouard, M. Foldyna, E.G. Caurel, I. Gordon, B. Heidari, A. Herman, L. Lalouat, K.-D. Lee, J. Liu, K. Lodewijks, F. Mandorlo, I. Massiot, A. Mayer, V. Mijkovic, J. Muller, R. Orobtchouk, G. Poulain, P. Prod'Homme, P.R. i Cabarrocas, C. Seassal, J. Poortmans, R. Mertens, O.E. Daif, V. Depauw, Photonic nanostructures for advanced light trapping in thin crystalline silicon solar cells: advanced light trapping in thin crystalline silicon solar cells. Phys. Status Solidi A **212**, 140–155 (2015)
214. Z. Yu, A. Raman, S. Fan, Fundamental limit of nanophotonic light trapping in solar cells. Proc. Natl. Acad. Sci. **107**, 17491–17496 (2010)
215. A. Mellor, I. Tobías, A. Martí, M.J. Mendes, A. Luque, Upper limits to absorption enhancement in thick solar cells using diffraction gratings. Prog. Photovolt. Res. Appl. **19**, 676–687 (2011)
216. K.X. Wang, Z. Yu, V. Liu, Y. Cui, S. Fan, Absorption enhancement in ultrathin crystalline silicon solar cells with antireflection and light-trapping nanocone gratings. Nano Lett. **12**, 1616–1619 (2012)
217. C. Trompoukis, O. El Daif, V. Depauw, I. Gordon, J. Poortmans, Photonic assisted light trapping integrated in ultrathin crystalline silicon solar cells by nanoimprint lithography. Appl. Phys. Lett. **101**, 103901 (2012)
218. N. Tucher, J. Eisenlohr, H. Hauser, J. Benick, M. Graf, C. Müller, M. Hermle, J.C. Goldschmidt, B. Bläsi, Crystalline silicon solar cells with enhanced light trapping via rear side diffraction grating. Energy Procedia **77**, 253–262 (2015)
219. M. Peters, M. Rüdiger, H. Hauser, M. Hermle, B. Bläsi, Diffractive gratings for crystalline silicon solar cells-optimum parameters and loss mechanisms: diffractive gratings for crystalline silicon solar cells. Prog. Photovolt. Res. Appl. **20**, 862–873 (2012)
220. R. Rothemund, T. Umundum, G. Meinhardt, K. Hingerl, T. Fromherz, W. Jantsch, Light trapping in pyramidally textured crystalline silicon solar cells using back-side diffractive gratings: Diffractive gratings for pyramidally textured silicon solar cells. Prog. Photovolt. Res. Appl. **21**, 747–753 (2012)
221. J. Eisenlohr, B.G. Lee, J. Benick, F. Feldmann, M. Drießen, N. Milenkovic, B. Bläsi, J.C. Goldschmidt, M. Hermle, Rear side sphere gratings for improved light trapping in crystalline silicon single junction and silicon-based tandem solar cells. Sol. Energy Mater. Sol. Cells **142**, 60–65 (2015)
222. J. Eisenlohr, J. Benick, M. Peters, B. Bläsi, J.C. Goldschmidt, M. Hermle, Hexagonal sphere gratings for enhanced light trapping in crystalline silicon solar cells. Opt. Express **22**, A111–A119 (2014)
223. X. Huang, S. Han, W. Huang, X. Liu, Enhancing solar cell efficiency: the search for luminescent materials as spectral converters. Chem. Soc. Rev. **42**, 173–201 (2013)

224. A. Shalav, B.S. Richards, M.A. Green, Luminescent layers for enhanced silicon solar cell performance: up-conversion. Sol. Energy Mater. Sol. Cells **91**, 829–842 (2007)
225. S. Fischer, J.C. Goldschmidt, P. Löper, G.H. Bauer, R. Brüggemann, K. Krämer, D. Biner, M. Hermle, S.W. Glunz, Enhancement of silicon solar cell efficiency by upconversion: optical and electrical characterization. J. Appl. Phys. **108**, 044912 (2010)
226. F. Lahoz, C. Pérez-Rodríguez, S.E. Hernández, I.R. Martín, V. Lavín, U.R. Rodríguez-Mendoza, Upconversion mechanisms in rare-earth doped glasses to improve the efficiency of silicon solar cells. Sol. Energy Mater. Sol. Cells **95**, 1671–1677 (2011)
227. B.S. Richards, A. Shalav, Enhancing the near-infrared spectral response of silicon optoelectronic devices via up-conversion. IEEE Trans. Electron Devices **54**, 2679–2684 (2007)
228. G.E. Arnaoutakis, J. Marques-Hueso, A. Ivaturi, S. Fischer, J.C. Goldschmidt, K.W. Krämer, B.S. Richards, Enhanced energy conversion of up-conversion solar cells by the integration of compound parabolic concentrating optics. Sol. Energy Mater. Sol. Cells **140**, 217–223 (2015)
229. G.H. Dieke, H.M. Crosswhite, The spectra of the doubly and triply ionized rare earths. Appl. Opt. **2**, 675 (1963)
230. J.C. Goldschmidt, S. Fischer, B. Herter, B. Fröhlich, K.W. Krämer, B.S. Richards, A. Ivaturi, S.K. . MacDougall, J. Marques Hueso, E. Favilla, M. Tonelli, Record efficient upconverter solar cell devices, in *Proc 29th PVSEC* (2014), pp. 1–4
231. S. Fischer, B. Fröhlich, H. Steinkemper, K.W. Krämer, J.C. Goldschmidt, Absolute upconversion quantum yield of β-NaYF4 doped with Er^{3+} and external quantum efficiency of upconverter solar cell devices under broad-band excitation considering spectral mismatch corrections. Sol. Energy Mater. Sol. Cells **122**, 197–207 (2014)
232. P.-A. Hansen, H. Fjellvåg, T. Finstad, O. Nilsen, Structural and optical properties of lanthanide oxides grown by atomic layer deposition (Ln = Pr, Nd, Sm, Eu, Tb, Dy, Ho, Er, Tm, Yb). Dalton Trans. **42**, 10778 (2013)
233. G. Dingemans, A. Clark, J.A. van Delft, M.C.M. van de Sanden, W.M.M. Kessels, Er^{3+} and Si luminescence of atomic layer deposited Er-doped Al_2O_3 thin films on Si(100). J. Appl. Phys. **109**, 113107 (2011)
234. A. Polman, Erbium implanted thin film photonic materials. J. Appl. Phys. **82**, 1 (1997)
235. T. Trupke, M.A. Green, P. Würfel, Improving solar cell efficiencies by down-conversion of high-energy photons. J. Appl. Phys. **92**, 1668 (2002)
236. G. Gao, L. Wondraczek, Near-infrared down-conversion in Mn^{2+}–Yb^{3+} co-doped Zn_2GeO_4. J. Mater. Chem. C **2013**, 1 (1952)

Chapter 3
Nanostructured Cathode Buffer Layers for Inverted Polymer Solar Cells

Zhiqiang Liang and Guozhong Cao

Abstract Polymer solar cells (PSCs) based on polymer donor-fullerene derivative acceptor bulk heterojunction (BHJ) active layer have attracted considerable attention as a promising alternative to conventional silicon-based solar cells due to their cost effectiveness, compatible with flexible substrate and solution-based roll-to-roll processing technique. Inverted PSC seems to be the best candidate that meets the requirements for both high efficiency and good stability. To achieve high power conversion efficiency (PCE) in inverted PSCs, the selection of an appropriate nanostructured cathode buffer layer (CBL) so as to optimize the electron collection and transport is particularly important. Nanostructured zinc oxide (ZnO) CBLs have been the most extensively studied CBLs for inverted PSCs in view of its relatively high electron mobility, high optical transparency, ease of being synthesized with low cost solution methods at low temperature, versatile morphologies, and being environmentally stable. This chapter focuses on the design, fabrication and characterization of the nanostructured CBLs constructed with pristine ZnO, doped-ZnO, and ZnO-based composites as well as the surface modified ZnO-based CBLs for the improvement of PCE and long-term device stability of inverted PSCs.

Z. Liang
Institute of Functional Nano and Soft Materials Laboratory (FUNSOM), Jiangsu Key Laboratory for Carbon-Based Functional Materials and Devices, Collaborative Innovation Center of Suzhou Nano Science and Technology, Soochow University, Suzhou 215123, Jiangsu, China

Z. Liang · G. Cao (✉)
Department of Materials Science and Engineering, University of Washington, Seattle, WA 98195, USA
e-mail: gzcao@u.washington.edu

Q. Li (ed.), *Nanomaterials for Sustainable Energy*,
NanoScience and Technology, DOI 10.1007/978-3-319-32023-6_3

3.1 Introduction

Photovoltaic technology converting sunlight directly into electricity is considered as one of the most promising and mature technologies for a CO_2-free and renewable energy supply [1–4]. In recent years, polymer solar cells (PSCs) have attracted extensive investigation as a potential alternative to conventional silicon-based solar cells, mainly due to their inherent advantages of being low-cost, compatible with flexible substrate and solution-based roll-to-roll processing technique [3–8]. In addition, the rapid energy payback time (EPBT) and low carbon emissions are also the major motivation for PSCs in future energy marketplace [9, 10]. Further improvements in the efficiency and stability are two of critical challenges to fulfil the practical application of PSCs. PSCs with inverted configuration seem to be the best candidate that well meets the requirements of high efficiency, good stability, and being compatible with the low-cost roll-to-roll manufacturing [6, 11–13]. In inverted PSCs, a nanostructured thin layer, usually a *n*-type semiconductor metal oxide, is required to be inserted between cathode and active layer and serves as electron-collecting and hole blocking layer, which is essential for achieving high efficiency PSCs.

For inverted PSCs, the device performance heavily relies on the optical and electrical properties of cathode buffer layer (CBL) as well as the condition of the interface between the CBL and active layer. To achieve high power conversion efficiency (PCE) in inverted PSCs, the selection of an appropriate material to form the high quality CBLs so as to optimize the electron collection and transport is particularly important. The nanostructure and surface energy of the CBLs play an important role in forming a desirable active layer morphology and contact quality between active layer and buffer layer. Zinc oxide (ZnO) has been the most extensively studied material for CBL of inverted PSCs in view of its relatively high electron mobility, optical transparency, easy of being synthesized with low cost solution methods at low temperature, versatile morphologies, and being environmentally stable. A variety of nanostructured ZnO thin films, such as one-dimensional nanorod arrays, two-dimensional planar nano-films and three-dimensional nano-ripple films (Fig. 3.1) have been used to serve as the CBLs in inverted PSCs for electron collection and transport. This chapter focuses on the design, fabrication and characterization of the nanostructured CBLs constructed with pristine ZnO, doped-ZnO, and ZnO-based composites as well as their surface chemistry and morphology modification for the improvement of efficiency and long-term device stability of inverted PSCs. The development of ZnO CBLs will make an important contribution to the fabrication of PSCs with high PCE and long-term stability at large scale for their practical applications.

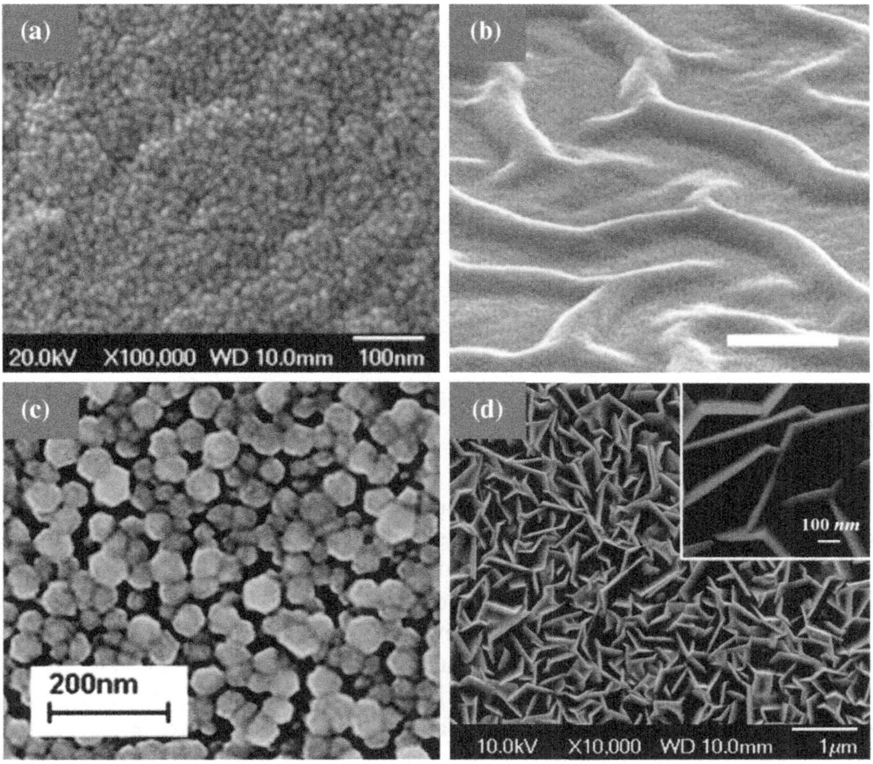

Fig. 3.1 Typical ZnO nanostructured films used in inverted PSCs: **a** planar dense film. Adapted with permission from [14]. Copyright 2012 Wiley-VCH. **b** Nano-ripple film (scale bar: 500 nm). Adapted with permission from [15]. Copyright 2009 Elsevier. **c** Vertical nanorod array. Adapted with permission from [16]. Copyright 2015 Elsevier. **d** ZnO nanowall array. Adapted with permission from [17]. Copyright 2013 Elsevier

3.2 Polymer Solar Cells

According to the charges transport direction, PSCs can generally be divided into conventional and inverted types as simply sketched in Fig. 3.2. For both of the conventional and inverted PSCs, the device architecture consists of a bulk-heterojunction (BHJ) active layer sandwiched between a transparent conducting electrode, such as indium tin oxide (ITO) glass, and a low work-function metal electrode (which usually uses Al material). In addition, the cathode buffer layer and anode buffer layer (ABL) are usually inserted between active layer and electrodes. In literature, the CBL is also named as electron selective layer (ESL), electron collection layer (ECL) or electron transport layer (ETL), and the ABL is also named as hole selective layer (HSL), hole collection layer (HCL) or hole transport layer (HTL). For the sake of clarity and ease in presentation, CBL and ABL are used in this chapter unless specified otherwise.

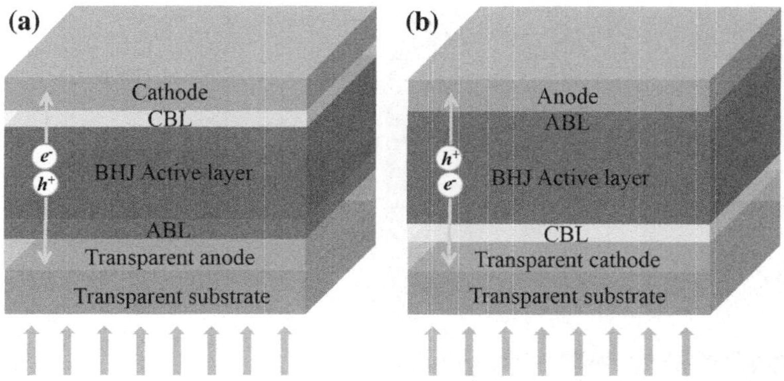

Fig. 3.2 Schematic illustration of device structures of **a** conventional and **b** inverted PSCs with bulk heterojunction active layer

It is worth noting that, in the development process of PSCs, the emergence of the concept of BHJ active layer gives rise to substantially improved device performance of BHJ solar cells compared to the previous bilayer solar cells. In the active layer of PSCs, only excitons formed within a distance of ~10–20 nm from the *p*-type polymer donor/n-type fullerene derivative acceptor interface (D-A interface) can reach the interface and then dissociate into free charge carriers [7, 18–20]. Therefore, the design criterion for BHJ is to maximize the D-A interfacial area to provide possibly more exciton dissociation sites and form two interpenetrated continuous charge transport pathways in each material towards the corresponding electrodes [7, 18, 19]. Figure 3.3a schematically depicts a preferred morphology of BHJ, which provides donor polymer and acceptor domains at around 10–20 nm, which allow excitons to diffuse to the D-A interface and thus achieves efficient charge separation [21]. The concept of BHJ was considered to establish the cornerstones of polymer solar cells [22]. Typically, the BHJ active layer is achieved by a blend of the p-type polymer donor and *n*-type fullerene acceptor materials dissolved in common solvent, and subsequent phase segregation results in the formation of two interpenetrated percolated networks during the annealing process after spin-coating. Many researches have been devoted to develop the BHJ active layer with the preferred morphology with a large D-A interfacial area (Fig. 3.3b–d) [5, 23–27].

3.2.1 *Inverted PSCs*

In recent years, significant progress has been achieved on the improvement of the performance of BHJ PSCs. The state of the art PCEs of single junction PSCs with conventional structure exceeding 9 % [28–30], and higher than 10 % [31] for tandem PSCs have been achieved in small area devices, which promises them a

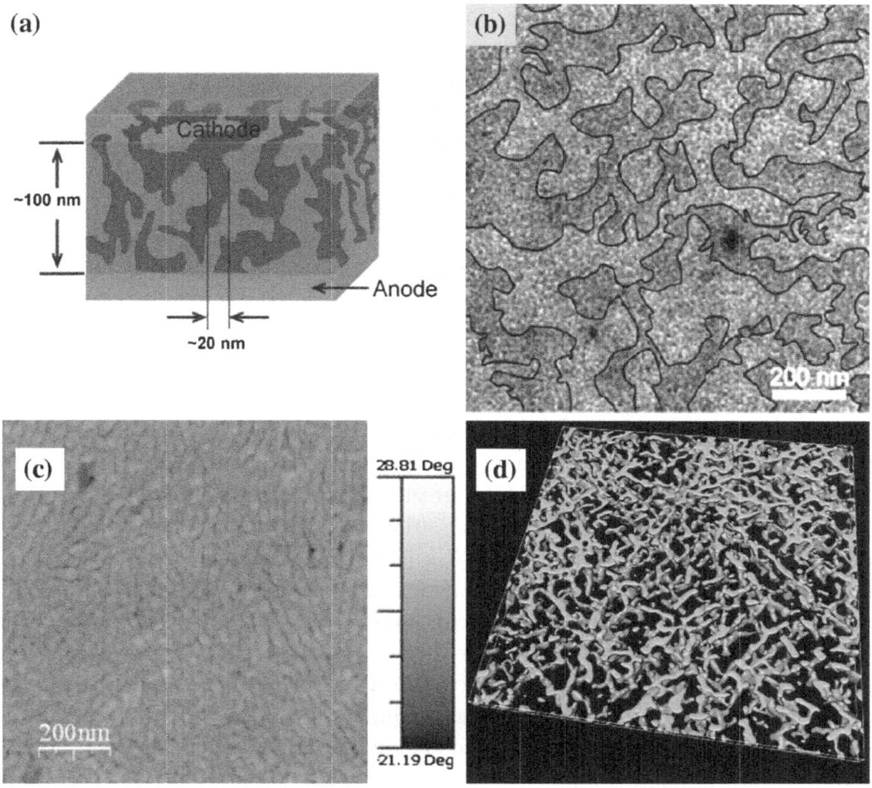

Fig. 3.3 **a** Schematic morphology model of BHJ active layer with interpenetration network of polymer and acceptor. Adapted with permission from [21]. Copyright 2007 American Chemical Society. **b** Bright field TEM image of PTB7/$PC_{61}BM$/DCB+DIO thin film (*black lines* are added to represent the heterojunctions between PTB7-rich and $PC_{61}BM$ rich domains). Adapted with permission from [23]. Copyright 2011 American Chemical Society. **c** Tapping mode AFM phase image of a solvent-annealed high-crystallinity P3HT:PCBM film. Adapted with permission from [26]. Copyright 2007 Wiley-VCH. **d** Three-dimensional electron tomography image of thermally annealed P3HT:PCBM film. Adapted with permission from [25]. Copyright 2009 American Chemical Society

bright future in commercialization. For more details, readers are recommended to consult the abundant literature in the field of PSCs. For instance, there are many papers reviewed the photovoltaic materials of active layers and electrode buffer layers, cell architectures, interfacial layers engineering, and the device operation mechanisms of PSCs [2, 3, 6–8, 11, 12, 20, 31–41].

However, in spite of high PCEs, the PSCs with conventional architecture have suffer from some drawbacks that hinder the solar cells from practical application. One of the major drawbacks for the conventional PSCs is the lack of long-term stability when exposed to air [42–45]. This is because the use of a low-work-function metal (e.g., Al) as cathode, which is sensitive to air and moisture and leads to the oxidization of the cathode quickly. Besides, in the conventional structure PSCs, the acidic poly(3,4-ethylenedioxithiophene):poly(styrene sulfonic acid) (PEDOT:PSS) interfacial layer directly contacts with the ITO glass and can etch the ITO resulting in a degradation of the device performance [38, 42–44]. This has inspired the emergence of inverted PSCs, in which the nature of charge collection is reversed [6, 46–48]. Figure 3.2b is a schematic showing of an inverted structure PSC with a stacked structure of a transparent electrode (e.g., ITO, FTO and AZO), an interfacial layer named as cathode buffer layer (CBL), a BHJ active layer, an anode buffer layer (ABL) usually made of PEDOT:PSS or molybdenum oxide (MoO_3), and a metal electrode with high-work-function such as Ag and Au.

In the inverted device architecture, the contact between the ITO and PEDOT: PSS is avoided, and meanwhile Al for top electrode adopted in conventional PSCs is replaced with an air stable high-work-function metal, such as Au or Ag [13, 46]. As a result, the inverted PSCs exhibit greatly improved ambient stability as compared to the PSCs with a conventional structure, overcoming one big hurdle for possible widespread application of PSCs [36, 47–51]. Moreover, in the case of inverted solar cells, the high-work-function metal anode, such as Ag, can be prepared with either coating or printing technology. These technologies are compatible with all solution processing and can therefore greatly simplify the fabrication process and lower the manufacturing cost of solar cells [52–54]. Another advantage of the inverted geometry is that it possesses higher flexibility in term of the design of multi-junction or tandem solar cells compared to the conventional one [7, 31]. In addition, vertical phase separation mechanism, which describes the P3HT prone to accumulate onto the electrode top and the fullerene derivative onto the bottom, also makes the inverted device configuration more advantageous compared to the normal configuration [13, 22, 34, 55, 56]. In recent years, it has been demonstrated that inverted PSCs can possess both long-term stability and high PCEs [31, 57–62]. Through device engineering and a use of new polymers, the state of the art PCEs of single junction inverted PSCs exceeding 10 % have been achieved in small area devices, which are comparable to and even exceed the efficiencies of conventional one [59, 61, 63–66]. Therefore, inverted PSCs seem to be the best candidate that meets the requirements for high efficiency, good stability, low cost and being compatible with the existing roll-to-roll process [6, 11–13]. The working mechanism of inverted PSCs can be described as the following. Under illumination, photo-generate excitons diffuse to the donor-acceptor (such as P3HT:PCBM) interface, where the excitons dissociate due to a driving force in view of the energy level difference between the donor and acceptor. The separated electrons will diffuse through the acceptor layer and collected by CBL, and ultimately reach the cathode electrode [36].

3.2.2 *Nanostructured ZnO Cathode Buffer Layers for Inverted PSCs*

For inverted PSCs, the device performance critically relies on the type of the CBL material and the condition of the interface between the CBL and the BHJ active layer. In detail, the CBLs in inverted devices are to (1) form an electron selective and transport interlayer, (2) block the reverse holes flowing from the donor polymer to cathode, (3) adjust the energetic barrier between the active layer and the cathode, and (4) prevent the chemical or physical reaction between active layer and cathode. To achieve high power conversion efficiency in inverted PSCs, the selection of an appropriate material to form the high quality CBL so as to optimize the electron collection and transport is particularly important. The key requirements for an ideal CBL include being good at electron extraction and transport, having suitable energy level that facilitates electron transport, and having high transparency, good compactness, and high stability.

Nano-films constructed with semiconducting metal oxides (MOs) including zinc oxide (ZnO) [36, 45, 46, 48, 67, 68], zinc tin oxide (ZTO) [69], titanium sub-oxide (TiO_x) [49, 50, 70–73], aluminum oxide (Al_2O_3) [74], and niobium pentoxide (Nb_2O_5) [75] have been applied to fabricate the CBLs for inverted PSCs. Several polymers, such as hydrophilic conjugated 2,7-carba-zole-1,4-phenylene alternating copolymers PCP-NOH or PCP-EP [76], cationic biopolymer poly(2-(dimethylamino)ethyl methacrylate) (PDMAEMA) [77], poly(amido amine) (PAMAM, generation 2) [77], poly(ethyleneimine) (PEI) [78] and conjugate polyelectrolyte poly [(9,9-bis(3′-(N,N-dimethylamino)propyl)-2,7-fluorene)-alt-2,7-(9,9-ioctylfluorene)] (PFN) [57, 63, 79], and PFPA1 [80] have also been studied for inverted PSCs to serve as efficient CBL materials. In addition, some other materials such as ZnS [81], CdS [82, 83], cesium carbonate (Cs_2CO_3) [84–87], $SrTiO_3$ [88], $SrTiO_3$:ZnO composite [89], MoO_3-Al composites [90], In_2S_3 [91, 92], ionic liquid-functionalized carbon nanoparticles (ILCNs) [93], sodium hydroxide (NaOH) [94], zwitterions [95], amino acid (serine) [96], amino-functionalized fluorene oligomers [97], ionic liquids (ILs) [65], and the low work function metals (such as: Al, Mg and Ca) [98] have also been investigated as CBLs in inverted PSCs. Graphene as a promising two-dimensional nanomaterial with outstanding electronic, optical, thermal, and mechanical properties is also a suitable material for CBLs [99–101].

Among the materials mentioned above, ZnO has attracted the most extensively investigated one for CBL in the inverted PSCs, mainly due to its suitable energy levels, high electron mobility, good transparency, environment stability and low cost [7, 11, 36, 45–47, 67, 102, 103]. Firstly, the good fit of ZnO as a CBL in inverted PSCs first of all originates from its suitable energy levels. Figure 3.4 shows a schematic diagrams of the energy levels and the transport directions of electrons and holes in an inverted structure PSC consisting of ZnO CBL and P3HT:PCBM active layer. The conduction band bottom of ZnO at −4.4 eV is lower than the lowest unoccupied molecular orbital (LUMO) of fullerene derivative acceptor (for example, −3.8 and −3.74 eV for PCBM and ICBA, respectively), meaning that

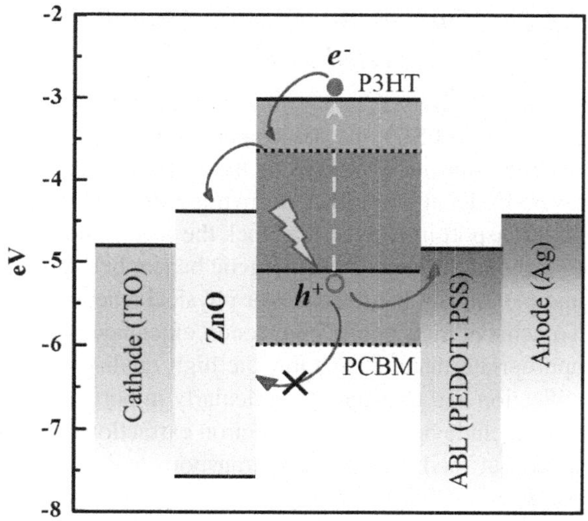

Fig. 3.4 Schematic illustration of the energy level and the main charge transportation of in inverted PSC with ZnO CBL and P3HT:PCBM active layer

ZnO CBL can help extract and collect electrons in the fullerene derivative acceptor [36, 104]. In addition, the valence band top of ZnO at ~−7.8 eV is lower than the highest occupied molecular orbital (HOMO) of the polymer donor (for example, −5.0 eV for P3HT). As a result, the reverse flow of holes from the polymer donor to the ITO cathode is blocked by the great energy barrier at the P3HT/ZnO interface. Therefore, ZnO thin layers as the CBLs can prevent the generation of leakage current at the polymer/ITO interface. That is why the inverted PSCs with ZnO CBLs have been found dramatically improved photovoltaic performance compared to those without a ZnO CBL [48, 56, 68, 105, 106].

In addition, the relatively high electron mobility of ZnO makes it a suitable material for cathode buffer layer to reduce the charge recombination. The good transparency in the whole visible spectrum benefits lowering optical loss and the band edge cut-off of ZnO at around 375 nm can block UV light and accordingly protect the organic materials from the photo-degradation under UV light irradiation [48, 56]. For single junction inverted PSCs, the state of the art PCEs of ~10 % have been achieved by using ZnO or ZnO-based CBLs [58, 59, 61, 64, 65, 107]. Another merit of ZnO as the CBL material is that it can be easily processed via a solution method with subsequent thermal treatment at relatively low temperatures. This makes the ZnO fully compatible to all solution roll-to-roll fabrication on flexible plastic substrates which is the predominant advantage of polymer solar cells [11, 45, 108–111]. Almost all of the reported inverted PSCs fabricated by roll-to-roll processing are using ZnO thin layers as CBL [12, 53, 54, 109, 112, 113]. Shirakawa et al. were the first group to have demonstrated the application of ZnO film as cathode interfacial layer in an organic photovoltaic cell with a stacked structure of ITO/ZnO/C_{60}/PAT6/Au [18]. The first inverted PSCs based on a sol-gel processed ZnO thin film on ITO as the CBL and a P3HT:PCBM BHJ active layer was pioneered by White and co-workers in 2006 [46]. After that, significant progress has been made in the past a few years in this area, with research focused on the preparation, surface modification and characterization of nanostructured ZnO and ZnO based buffer layers.

3.3 Fabrication of Nanostructured ZnO Films for Inverted PSCs

Deposition of ZnO thin films has been a subject of intensive study in the last two decades, and many methods have been developed and improved for the growth of ZnO nano-films. In this section, we will only introduce the well-established deposition methods which have been widely used to fabricate nanostructured ZnO films to serve as the CBLs in inverted PSCs. For inverted PSCs, the device performance can be significantly affected by the method used for the preparation of ZnO CBL. The deposition method can influence the morphology, thickness, crystallinity, and the optical and electrical properties of the resulting ZnO nano-films. The fabrication methods for the nanostructured ZnO CBLs can generally be divided into two groups: solution processing and vapor phase deposition. The former includes, for example, sol-gel method [14, 15, 45, 46, 48, 67, 114–122], solution method derived from pre-fabricated ZnO NPs suspensions or aqueous ZnO hydrate [47, 68, 112, 123–126], chemical bath deposition (CBD) [127], and electro-deposition (ED) [56]. Examples of the latter are (1) atomic layer deposition (ALD) [105, 128–132], (2) spray pyrolysis [56, 133, 134], (3) metal organic chemical vapor deposition (MOCVD) [135], (4) mist pyrolysis chemical vapor deposition (MPCVD) [136], (5) radio-frequency sputtering (RF) [137], (6) magnetron sputtering [138, 139], and (7) fine-channel mist-spray coating [140]. Among these methods, the solution-based methods have been the most popular ways to make ZnO CBLs for inverted PCSs, due to its high-quality, ease of preparation at relatively low temperature, without requiring any vacuum instrument and low cost.

3.3.1 Sol-Gel Processing

Sol-gel processing is a widely used approach for the fabrication of nanostructured ZnO films CBL for inverted PSCs. ZnO thin films created with sol-gel processing present good surface quality, excellent optical transmittances and electrical properties. In addition, sol-gel processing is a low-cost process that doesn't involve any expensive and/or vacuum instruments [14, 45, 141]. Moreover, sol-gel method offers important processing advantages, such as low processing temperature and compatible to roll-to-roll (R2R) manufacturing for the fabrication of inverted PSCs on temperature sensitive flexible plastic substrates [11, 45, 108].

Sol-gel processing through the use of a mixture of zinc acetate [ZnAc] and monoethanolamine (MEA) in 2-methoxy ethanol as precursors is a popular approach for the fabrication of nanostructured ZnO CBLs for inverted PSCs. The typical processes of the preparation of ZnO sol-gel using these precursors as following: (i) equimolar mixture of zinc acetate dehydrate and MEA are first dissolved in 2-methoxy ethanol at room temperature, (ii) then the resulting solution is required to be stirred at a low temperature (typically, around 60 °C) to yield a homogeneous, clear, and transparent sol, (iii) the as-prepared solution should be aged at room temperature for one day in order to make it more glutinous. The sol-gel processed ZnO nano-films employed in inverted PSCs are mostly coated by using the spin-coating method for rigid glass/ITO substrate and the slot-die coating method for flexible plastic substrates in roll-to-roll process, respectively. Typical sol-gel processing of ZnO films includes three main steps: (i) sol-gel preparation, (ii) coating and (iii) post heat-treatment, as sketched in Fig. 3.5. In these three steps, the nature and concentration of precursor, solvent and additive, solution processing temperature and aging time, and the parameters of coating and heat treatment can strongly affect the quality of ZnO nano-films. For more detailed information, readers are recommended to consult the abundant literature in the field of ZnO films derived from solution processed sol-gel method. For instance, *sol–gel-deposited ZnO thin films: A review* published by Znaidi [142] provides a comprehensive coverage on sol-gel processed ZnO thin films. In this section, we mainly focus on the effects of the quality of sol-gel processed ZnO CBLs on the photovoltaic performance of the inverted PSCs.

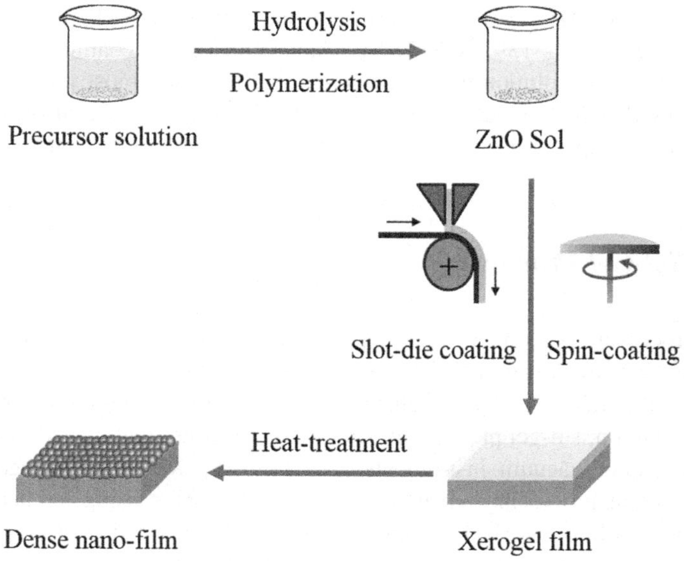

Fig. 3.5 Schematic showing ZnO nano-films derived from sol-gel processing: film from slot-die coating or spin-coating methods

In the sol-gel process, a heat treatment process is adopted to convert the spin-coated gel film to crystalline ZnO and is a factor which can significantly affect the optical and electrical properties of ZnO films. Firstly, the heat treatment temperature is required to be higher than the boiling points of the solvents (2-methoxyethanol and MEA at 125 and 170 °C, respectively) to get the solvents completely evaporated [141, 142]. Secondly, it has been demonstrated that, when heated in air, the ZnAc begin to thermal decompose at 190 °C, its thermal decomposition completed at around 310 °C, and the formation of crystalline ZnO happened at the temperature between 300 and 400 °C. A higher annealing temperature ranging from 400–500 °C is identified as the most efficient temperature for the thermal decomposition of ZnAc to form crystalline ZnO [143].

The electron mobility of sol-gel processed ZnO films as well as the device performance of inverted PSCs based on these ZnO films as the CBL would be promoted by improving the crystallinity of ZnO. However, there have been conflicting reports of the optimized annealing temperature of sol-gel processed ZnO CBLs for inverted PSCs in the literature [45, 48, 51, 118, 122, 144, 145]. The optimal annealing temperature reported in the literature ranged from 100 to 450 °C [26, 30, 35, 134, 140, 161, 162]. For instance, Zhang et al. reported that the inverted PSCs using sol-gel processed ZnO CBLs exhibited a clear improvement in the device performance with increasing the ZnO sintering temperature from 100 to 350 °C [118]. It was observed that the R_S of the inverted devices decreased with an increased sintering temperature and it reached a minimal value when the sintering temperature increased from 100 to 350 °C, as a result an improved device performance (PCE = 2.58 %) was achieved. Such a performance improvement was ascribed to the improved ZnO quality in view of a relatively high sintering temperature, which effectively promoted the electron mobility of sol-gel processed ZnO films by improving the crystallization of ZnO. However, the need of high annealing temperature above 300 °C is incompatible with typical printing processes of PSCs that use flexible substrates, such as polyethylene terephthalate (PET), which usually cannot withstand a high temperature treatment process. Additionally, the high annealing temperature processes will also lead to the degradation of the conductivity of cathode (i.e., ITO) and thus degrades the device performance [118, 145]. For instance, the sheet resistance of the ITO cathode increased linearly with the increasing of annealing temperatures and times, which, in turn, contributes to the reduction of the FF and the PCE of the corresponding solar cells [145].

In order to overcome these constraints, the sol-gel derived ZnO CBLs annealed at relatively low temperature have been studied [45, 48, 51, 118, 145]. For instance, the sol-gel derived ZnO films obtained by annealing at relative low temperatures of 130, 150 and 200 °C have been investigated to serve as the CBLs in inverted PSCs [45]. The ZnO films annealed at 200 °C are able to function as an efficient electron transporting layer in inverted solar cells, leading to a PCE as high as of 6.33 % for the inverted devices constructed with PCDTBT:$PC_{70}BM$ active layer. In addition, the inverted devices based on those low temperatures annealed ZnO buffer layers also exhibit a promising long-term stability [45]. It should be noted that the ZnO film annealed at 200 °C is amorphous instead of crystalline, and its electron transporting property ($\sim 1.3 \times 10^{-4}$–2.6×10^{-5} $cm^2\ V^{-1}\ s^{-1}$) is not as good as that of the crystalline ZnO (~ 1 $cm^2\ V^{-1}\ s^{-1}$) [48, 51, 122]. However, it seems because the electron mobility of this amorphous ZnO is comparable to which of most organic materials for PSCs, the device performance is not apparently affected by the relatively poor conduction of amorphous ZnO [48]. Through using the low temperature annealed amorphous ZnO CBL and PTB7-F20/$PC_{71}BM$ active layer, the inverted PSCs reached a J_{sc} of 13.93 mA cm^{-2}, a V_{oc} of 0.71 V, a FF of 0.65, and a PCE of 6.42 %, plus long term stability [122]. Furthermore, to reduce the degradation of flexible plastic substrates, the use of sol-gel driven amorphous ZnO CBL annealed at a lower temperature (~ 100 °C) for inverted PSCs has also been studied [120]. Encouragingly, these sol-gel driven amorphous ZnO layers annealed at such a low temperature delivered the solar cell performance comparable to the devices based on ZnO films prepared at substantially higher temperature. The inverted PSCs based on this amorphous ZnO CBLs and constructed with low band-gap polymer donors on glass/flexible PET substrates show performances of: PTB7:$PC_{71}BM$ (PCE: 6.5 % (glass)/5.6 % (PET)) and PBDTTPD:$PC_{71}BM$ (PCE: 6.7 % (glass)/ 5.9 % (PET)) [120]. Therefore, more sol-gel processed ZnO CBLs adopted heat treatment at a relative low temperature (<200 °C) due to the comparable performance and compatible with the solution processing technique of inverted PSCs fabricated on flexible substrates.

During the post annealing process, the annealing environment, such as humidity and vacuum, can also affect the electrical property of ZnO CBL to some extent [146]. The vacuum environment during the thermal treatment of solution proceed ZnO films could help further reduce the defects and increase the electron mobility in the low temperature (150 °C) processed ZnO film (up to 0.36 $cm^2\ V^{-1}\ S^{-1}$), and thus improve the electron extraction in inverted PSCs [146]. The concentration of ZnO precursor is another key factor largely affecting the quality on ZnO CBLs derived from a sol-gel route. In general, a relatively low sol concentration (~ 0.1–0.3 M) is more prone to lead to a high quality ZnO films with denser surface, lower surface roughness, and higher transmittance [14, 51, 118, 146]. The low ZnO sol concentration benefit the form small nano-sized ZnO colloids and small grain sizes. The inverted devices based on the dense ZnO CBLs exhibited low reverse current and R_S between the ZnO layer and active layer owing to the full coverage of ITO surface and intimate contact between ZnO CBL and active layer, giving rise to enhanced photovoltaic performance [14, 118]. In contrast, a relatively high sol concentration (~ 0.5–0.75 M) is usually prone to result in a ZnO film with larger surface roughness and even voids. The rough surface and the existence of voids in ZnO buffer layers would result in an increase in the R_S, due to the inferior contact between the ZnO buffer layer and photoactive layer [37, 132, 137]. However, ZnO buffer layers with both dense and smooth surface and large surface area could be obtained by using a slow ramp annealing process to heat treat the wet gel film derived from ZnO sol with relatively high concentrations (i.e., 0.75 M) [138, 139]. This is because the low heating rate in the ramp annealing process provided sufficient time for the gel particles to structurally relax and pile up, resulting in a dense and undulated film [15, 119]. The choice of ZnO precursors could also affect the property of the sol-gel processed ZnO nano-films, and thus impact on the photovoltaic performance and stability of the inverted PSCs [147]. It has been reported that the inverted PSCs with ZnO layers derived from diethylzinc are more stable than the devices with ZnO layers prepared by decomposing zinc acetate under long term illumination [147]. The difference in device stability was attributed to the different chemical nature of ZnO surfaces, specifically the presence of interstitial zinc (Zn_i) defects at ZnO surface and their subsequent interactions with the adjacent active layer. It was explained that the ZnO layers derived from diethylzinc (deZn) possessed less Zni defects than the ZnO layers deposited from zinc acetate [147]. Overall, to achieve a high quality ZnO CBL with the solution processed sol-gel method, the type of ZnO precursor as well as the concentration, and the post heat treatment temperature and environment should be considered comprehensively.

3.3.2 ZnO CBLs Derived from Pre-fabricated ZnO Nanoparticles

The deposition of ZnO nano-films by spin-coating or slot-die coating the pre-fabricated the ZnO nanoparticle (NP) suspensions or colloids is also a widely used solution processing [6, 47, 68, 112, 123–126, 148]. The ZnO NPs used in this solution processing are usually synthesized using a hydrolysis and condensation of zinc acetate dihydrate under low temperature. The resulting ZnO NPs demonstrated a good electron mobility ($\sim$0.066 $cm^2\ V^{-1}\ s^{-1}$) even through no additional post-thermal treatment applied [149]. It has been found that ZnO nano-films with dense and homogenous surface (Fig. 3.6) and controllable thickness can be achieved by spin-coating the pre-fabricated ZnO NP colloids [6, 47]. The inverted PSCs based on the ZnO NPs films as the CBL exhibited performance comparable to that of devices based on sol-gel processed dense ZnO CBLs treated at high temperature (400 °C). However, it has been found that there were a large number of

Fig. 3.6 **a** AFM and **b** SEM images of ZnO NPs thin films. Adapted with permission from [148]. Copyright 2010 American Chemical Society

hydroxyl (−OH) groups formed on the ZnO NPs synthesized by hydrolysis and condensation of zinc acetate dehydrate [126]. These surface hydroxyl groups have potential to depredate the photoactive layer by elevating photocatalytic activity of ZnO, and hence deteriorate the device performance. Fortunately, the hydroxyl groups could be effectively detached from the ZnO NPs by annealing at a low temperature, ~150 °C [126]. The low temperature treatment and the scalable processing capability make the colloidal ZnO NPs well compatible with the solution processing technique for inverted PSCs fabricated on flexible substrates. Krebs et al. has successfully employed the solution processed ZnO NPs layers as CBLs for full roll-to-roll process to manufacture inverted PSCs with large scale [54, 109]. Recently, Spyropoulos and co-workers reported the fabrication of flexible organic tandem solar modules with PCE as high as of ~6 % by using ZnO NP CBL combined high throughput roll-to-roll compatible processing [113].

One of the problems of ZnO NPs suspensions or colloids is that the ZnO NPs in suspensions or colloids are not very stable and are sensitive to the moisture. Usually, an appropriate ligand is required to attach onto the NPs to stabilize them and so as to avoid the NPs from aggregating. The prevention of ZnO NPs from aggregating is important for the fabrication of ZnO CBLs with a smooth surface, and so as to achieve high efficiency inverted PSCs [108, 132–134, 150]. Therefore, a major challenge of this method is to achieve spatial distribution of individual NPs and gain high electron mobility without high temperature annealing process [151, 152]. The improvement of spatial distribution of ZnO NPs can be achieved by using polymeric or organic molecules as a matrix or surfactant [124, 151–153]. It should be noted that the removal of surfactant material from ZnO film is required to optimize the property of ZnO CBLs as well as the device performance, due to the fact that surfactant materials may reduce the electron transport by introducing additional resistance [151, 153]. Besides the surfactant, the solvent of ZnO NP solution also plays an important role in the quality of ZnO NP dispersion and the morphology of ZnO NPs film. For instance, ZnO NPs aggregations in solution can be controlled by adjusting the solvent ratio of chloroform and methanol [150]. When the ratio of chloroform: methanol is 1:1 (V/V), the superior dispersity of ZnO NPs in solution and a consequent ZnO film with compact and homogeneous surface could be obtained. Besides the use of colloidal ZnO NPs, spin coating an aqueous zinc oxide hydrate ($ZnO–xH_2O$) solution is also an approach to fabricate ZnO nano-films [154]. For the ZnO films derived from this method, the post annealing temperature can be as low as 80 °C due to the low energy metal-ammine dissociation and hydroxide condensation/dehydration chemistry. The inverted devices based on this ZnO films as the CBLs and with $PCDTBT:PC_{71}BM$ photoactive layers show a high PCE, up to 6.48 %, which is comparable to those devices based on ZnO CBLs deposited by sol-gel process using zinc acetate and annealed at 200 °C [154].

3.3.3 Atomic Layer Deposition

Atomic layer deposition (ALD) is a precise thin film deposition method, which allows for truly nanometre or even angstrom-level thickness control with conformal films of a variety of materials. Due to the self-limiting growth nature, each time only one atomic or molecular layer can grow. Therefore, conformal and pinhole-free films can be deposited layer by layer at atomic scale. ALD can be considered as a special modification of chemical vapor deposition (CVD) with the precursor gas reactants being introduced into the reaction zone separately and sequentially [155]. Excellent reviews on ALD technique, Atomic layer deposition (ALD): from precursors to thin film structures by Leskelä and Ritala [156] and A Short History of Atomic Layer Deposition: Tuomo Suntola's Atomic Layer Epitaxy by Puurunen [155] have been published.

For the fabrication of ZnO thin films, the ALD technique is regarded as a very promising approach to fabricate ZnO nano-films in view of the capability of ALD in producing high surface quality film at relatively low temperature, in large areas and controlling the film thickness precisely [129]. Moreover, ALD allows for super-conformal coverage of ZnO nano-films, which is benefit to separate the cathode from the active layer, and so as to efficiently block the reverse flow of holes from active layer to cathode. The choice of precursors is a key issue for achieving high quality ZnO nano-films by ALD processing. The common precursors used in ALD include zinc acetate (ZnAc), diethyl zinc (DEZ) and dimethyl zinc (DMZ). As the first precursor used for ALD ZnO processing [157], ZnAc requires a relatively high temperature to react with H_2O (the reported deposition temperature using ZnAc precursor is no less than 280 °C) [158]. For DEZ-based ALD processes, ZnO thin films can be deposited at temperatures much lower than that needed for the ZnAc-based processes. Generally, the common deposition temperatures for DEZ-based ZnO ALD processing are range from room temperature to around 300 °C. So far, DEZ is the most common Zn precursor used in ALD processing [158], and most of the ALD processed ZnO CBLs for inverted PSCs used DEZ as Zn precursor [104, 123–127]. The conversion of the DEZ and water into ZnO according to the following equations [158]:

$$Zn(C_2H_5)_2^* + H_2O\ (g) \rightarrow ZnO^* + 2C_2H_6(g) \tag{3.1}$$

where * denotes a surface species. Besides the Zn precursor, the crystalline, quality of the surface, and optical properties of ALD processed ZnO films are also depend on the deposition parameters, such as deposition temperature, growth chamber pressure, precursor pulse time and purging time. Thus, one can obtain ZnO nano-films with the required quality by adjusting growth conditions. Figure 3.7 shows the AFM images of ALD proceed ZnO films grown at low-temperature (110–170 °C) and using DEZ precursor [159]. All of the films exhibit dense and homogenous surface with roughness vary from 0.9 to 3.9 nm for the growth temperatures of films decrease from 170 to 110 °C. In addition, ALD proceed ZnO

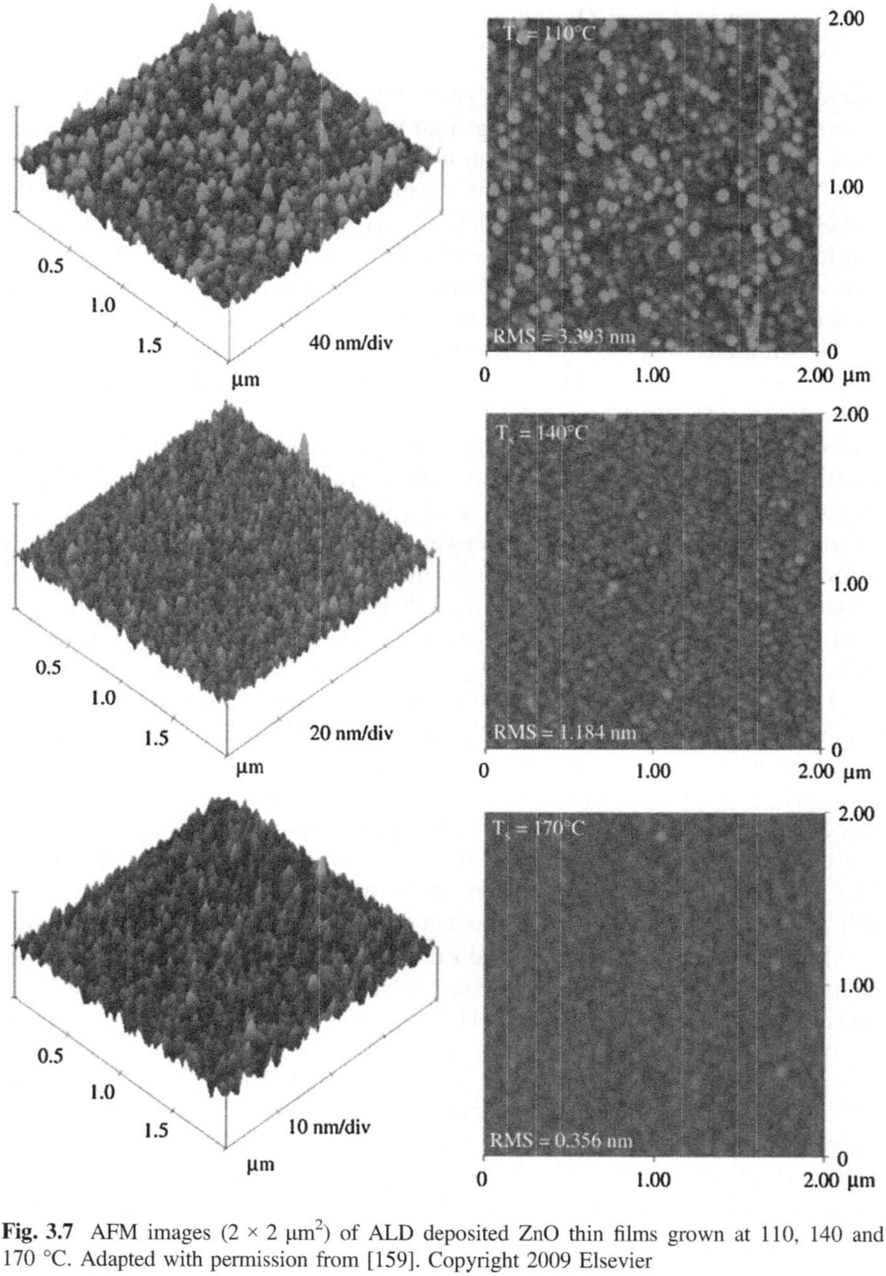

Fig. 3.7 AFM images ($2 \times 2\ \mu m^2$) of ALD deposited ZnO thin films grown at 110, 140 and 170 °C. Adapted with permission from [159]. Copyright 2009 Elsevier

films grown at low-temperature (110–170 °C) are found to be polycrystalline and the crystallographic orientation depend on ALD deposition temperature. For more details about ALD ZnO processing, readers may wish to consult the abundant literature in this field. For instance, *Atomic layer deposition of ZnO: a review* published by Tynell and Karppinen [158] provide a comprehensive coverage on ALD processing and materials of ZnO thin films. In this section, we briefly focus on the effects of optical and electrical properties of ALD processed ZnO CBLs on the photovoltaic performance of the inverted PSCs.

To fabricate flexible inverted PSCs, ZnO CBLs can be deposited on plastic substrate, such as ITO/PEN, by ALD method at low temperatures (45 and 80 °C) [105]. The inverted devices based on ALD ZnO CBL grown at 80 °C presented batter photovoltaic performance than the devices based on ZnO grown at 45 °C, for the reason that the former achieved a smoother and more hydrophobic surface than the latter. The smoother and more hydrophobic surface of ZnO CBL could beneficial to the contact quality between ZnO and active layer, and thus leading to reduced series resistance and improved electron transport to the ITO electrode. It is worthy pointing out that the ALD temperature can also affects the electrical properties of the grown ZnO films [159]. For example, it has been found that both the electron mobility and electron concentration of the ZnO CBL increased with increasing deposition temperature for the reason of increased crystallinity [130]. Although ALD offers the most precise control of thickness and the high quality of the grown ZnO thin films, this method suffer from the complicated deposition instrumentation, vacuum-based process and slow growth rate. To avoid the slow and vacuum-based process in conventional ALD, an atmospheric atomic layer deposition (AALD) method was also developed to grown ZnO thin layer in atmospheric ambient [131]. It has been demonstrated that the AALD ZnO layers possessed compact surface, high electron mobility (3.4 ± 0.1 cm^2 Vs^{-1}) and good transmittance to visible light. The inverted BHJ P3HT:PCBM PSCs with AALD ZnO CBLs exhibited a PCE of 3.6 % and good device stability in air. Compared with the conventional ALD route, such an AALD method allows the ZnO layer to be deposited more rapidly in air, and thus has higher potential for implementing in a throughput roll-to-roll process.

3.4 The Impacts of ZnO CBLs on the Performance of Inverted PSCs

3.4.1 The Impacts of the Morphology of ZnO CBLs on the Performance of Inverted PSCs

For the inverted PSCs with nanostructured ZnO CBLs, the BHJ active layer is deposited on the top of ZnO CBLs, and thus the morphology of ZnO CBLs strongly affect the condition of ZnO/active layer interface. Besides, the photovoltaic performance of the inverted PSCs depends critically on the electrical property of the interface between ZnO CBL and active layer [7, 14, 36, 160]. Therefore, control of

Fig. 3.8 The SEM images of ZnO layers derived from sol concentrations of **a** 0.02, **b** 0.05, **c** 0.1, **d** 0.3, **e** 0.6, and **f** 1 M, respectively. Adapted with permission from [14]. Copyright 2012 Wiley-VCH

the morphology of the nanostructured ZnO CBL is an important aspect to tune the contact quality so as to achieve the full potential of inverted PSCs.

The morphology of ZnO nano-films prepared using a sol-gel method can be easily controlled through adjusting the concentration of precursor sol [14]. Figure 3.8a–f exemplifies the morphology change when the concentration of precursor sol changes from 0.05 to 1 M. The inverted PSC with P3HT:PCBM active layer and using a dense and homogenous ZnO film driven from 0.1 M sol (Fig. 3.8c) exhibited PCE of 3.3 %; such an efficiency is 32 % higher than that for the device made with a rough ZnO film derive from a 1 M sol. The PCE enhancement arises from the higher FF while J_{sc} only increases slightly and the V_{oc} remains almost unchanged. This is because the dense ZnO CBL with homogenous surface allowed the formation of an intimate contact between the ZnO CBL and the BHJ photoactive layer, however a rough surface including voids led to inferior contact and an increase in the contact resistance [14]. These results demonstrated that the ZnO film with dense and homogenous surface benefited to establishing a better contact between the ZnO and the active layer, resulting in the low R_S and high FF of the solar cell. In contrast, the R_S was found to increase with increasing the surface roughness and the content of voids in ZnO film. In addition, the uniform coverage of ZnO CBL derive from 0.1 M sol is crucial to separating the ITO cathode from contacting with the active layer, and blocking the reverse flow of holes from active layer to ITO electrode.

Besides the compactness, the surface area of ZnO CBL, which is mainly determined by the morphology, is another important issue that impacts the device performance by affecting the interfacial area between the CBL and active layer and therefore affecting the electron collection efficiency [15, 56, 119, 132]. Sekine et al.

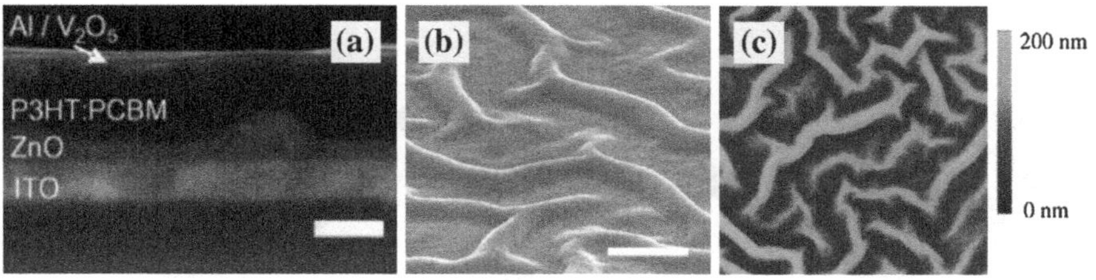

Fig. 3.9 **a** Cross sectional SEM image of the inverted PSC, scale bar: 200 nm and **b** SEM image of ZnO nano-ridge film, scale bar: 500 nm, and **c** AFM image of ZnO nano-ridge film showing a 5 μm × 5 μm surface area. Adapted with permission from [15]. Copyright 2009 Elsevier

firstly reported the use of a unique nano-ridge (also named as nano-ripple in some literature) ZnO films having both large surface area and dense surface in inverted PSCs to serve as a CBL (Fig. 3.9a) [15]. By adopting sol-gel processing and using a ramp annealing method at a rate of 50 °C/min from room temperature to 275 °C, the nano-ridge of ZnO films with peak height around 120 nm and valley to valley distance of about 500 nm can be fabricated, as shown in Fig. 3.9b and c [15]. It was suggested that the low heating rate of the ramp annealing method provided sufficient time for the gel particles to structurally relax and pile up, resulting in a dense and undulated film [15]. The inverted PSCs fabricated based on nano-ridge films showed an average PCE of 3.87 %, while the devices with planar films only showed an efficiency of 3.06 %. The enhancement was concluded to low leakage current due to the improvement in the electron collection efficiency and hole blocking capability of the nano-ridge film, resulting in higher FF, while the V_{oc} and J_{sc} were found to remain almost unchanged. The better charge extraction was thought to result from the large surface area of the ZnO nano-ridge film, which makes the device possess large ZnO/BHJ active layer interfacial area for electron collection and shortens the travelling distance of charges. Meanwhile, the ZnO nano-ridge film is also very dense, making it effective in blocking the hole from transporting from active layer to ITO cathode, and thus reducing the charge recombination and leading to a low leakage current.

To meet the full potential of ZnO nano-ripple CBLs for high performance inverted PSCs, the ripple size and density should be controlled. By adding ZnO nanoparticles in zinc acetate solutions and using a ramp annealing method, ZnO nano-ripple films with controlled ripple size and density can be achieved (Fig. 3.10) [119]. The inverted PSCs based on the ZnO nanoripples with surface structure of ~70 nm high nanoripples gives an enhanced PCE of ~3.2 %. Both the homogenous surface and the higher contact area between ZnO and active layer contributed to the enhanced photovoltaic performance of the inverted PSC. Besides adding ZnO NPs, the ripple morphology also can be tuned via controlling the concentration of the ZnAc sol-gel solution, annealing ramp rate, and the spin-coating parameters [120, 122, 161–163]. It has been found that the surface morphology of the sol-gel derived ZnO thin films depends largely on the solvent drying kinetics [122].

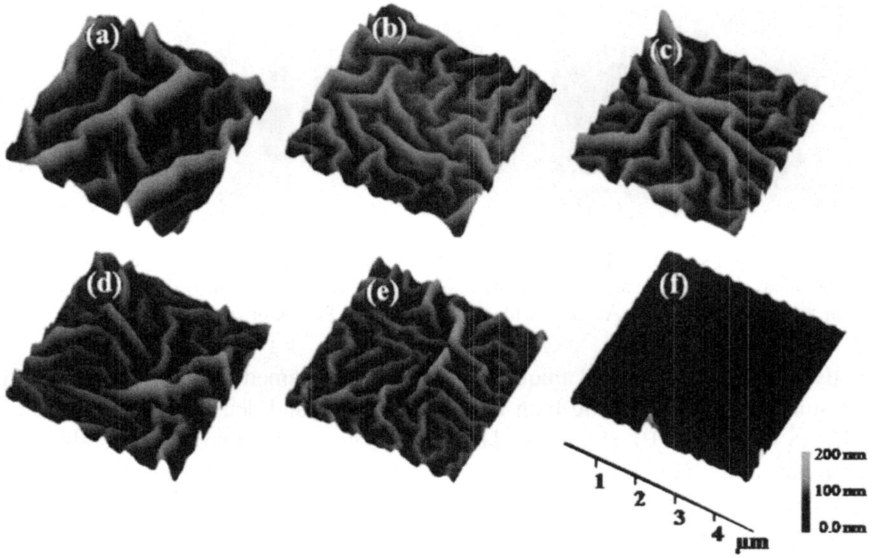

Fig. 3.10 AFM images of **a** ZnO layer was spin-coated using bare zinc sol. **b–f** ZnO layers were spin-coated using mixtures of ZnO nanoparticles and ZnO sol. Amount of ZnO nanoparticles added to 21 ml of ZnO sol was 0.25, 1, 3, 5 and 10 mg ZnO NPs for **b–f**, respectively. All of the ZnO layers were heated to 350 °C with a heating rate of 23 °C/min. Adapted with permission from [119]. Copyright 2011 Elsevier

Although the ripple morphology of ZnO CBLs could result in a high contact area between the ZnO and active layer and thus enhanced the solar cell performance, it has also been found that the ripple ZnO CBLs with a large surface area however harm the device performance compared to the smooth ZnO layers [120, 162]. For instance, it has been reported that nano-ripple ZnO CBLs with a high surface roughness could lead to high surface energy and a small donor/acceptor (D/A) interfacial area in the active layer deposited on the top of ZnO CBLs, and thus lower the J_{sc} [162]. In addition, the large surface area of ZnO CBLs could also result in more trap-assisted recombination at the active layer/ZnO interface compared to the case of smooth ZnO CBL, and accordingly reduces the FF of solar cell [162]. The different or even contrary effects of the large surface area of nano-ripple ZnO CBLs on the device performance may because the dominant factors influencing on the device performance, such as surface area, surface energy and surface defect, are different in different researches. Therefore, to fully achieve the potential of the nano-ripple ZnO CBLs possessing a large surface area, the surface defects of the nano-ripple ZnO should be passivated and the surface energy should be optimized.

Besides the abovementioned ZnO nano-ripples, there are several other ZnO nano-films with large surface area have been investigated to serve as the CBL in inverted PSCs. For instance, the ZnO nano-films with different surface morphology prepared using spry-pyrolysis (SP) and electro-deposition (ED) methods [56], and the ZnO films with large surface area prepared by utilizing a mist pyrolysis chemical vapor deposition (MPCVD) method [136]. It should be noted that the gaps in the ZnO films with rough surface morphology should be wide enough so as to allow the blend of polymer/fullerene to infiltrate easily and thus obtain an intimated contact. In such cases, the ZnO/BHJ active layer interfacial area can be adjusted by controlling the surface roughness of ZnO CBL. It has been demonstrated that the J_{sc} increased with increasing the ZnO/BHJ active layer contact area which provided larger interface for electron extraction and transport. In sum, the appropriate morphology of ZnO CBLs for inverted PSCs should meet the following requirements: (i) having dense and homogenous surface, (ii) being able to provide large surface area, and (iii) allowing the polymers to well contact with ZnO layer.

3.4.2 The Effects of the Thickness of ZnO CBLs on the Performance of Inverted PSCs

For the inverted PSCs, the thickness of nanostructured ZnO CBL can affect the optical transmittance, electrical conduction, and even work function of ZnO CBL, and thus impose a profound impact on the photovoltaic performance of the inverted PSCs. For instance, it has been found that increasing the thickness of ZnO CBLs from 30 to 126 nm could result in a drop of solar cell PCE from 2.5 to 1.5 % [164]. The PCE decrease arose from the low conduction of thick ZnO CBL, which resulted in an increase of the series resistance (R_S) from 1 to 3.2 Ω cm^2. Cheun et al. studied the effects of the thickness of ZnO CBL by varying the thicknesses from 0.1 to 100 nm using an atomic layer deposition (ALD) technique [129]. They demonstrated that the device performance strongly depended on the thickness of ZnO CBL and there was a critical value for the thickness, which was around 10 nm. For ZnO CBL with thickness below this critical value the device demonstrated poor performance and the current–voltage (J-V) curves presented an "S-shaped" kink. For devices with ZnO layers from 10 nm and up to 100 nm in thickness, the S-shaped kink completely disappeared, and the thickness difference of ZnO CBL only caused a little variation in the PCE and FF. Such a S-shaped J-V curve was suggested to be a result of the low conductivity of such an ultra-thin ZnO layers, which had large surface-to-bulk ratio and might cause considerable defects at grain boundaries; these defects act as electron traps reducing the electrical conduction of the CBL [54, 129]. Usually, the low conductivity of ZnO films was ascribed to the presence of oxygen at the grain boundaries or surface [54, 104, 165, 166]. Similar phenomenon of S-shaped kink in J-V curve in inverted devices with ultrathin ZnO CBL (several nanometers) has also been found in Liang's [14] and Sharma's [180] works. In addition, it should be noted that the film thickness and the preparation method are two main factors affecting the coverage of ZnO buffer layer. A full and conformal coverage of ZnO layer is crucial to separate the cathode from the active layer, and so as to block the reverse flow of holes from active layer to cathode [14, 120, 167, 168]. For the ZnO buffer layer derived from sol-gel method, it has been found that a very thin ZnO layer (a few nanometers in thickness) is difficult to achieve a full coverage on the ITO surface, and thus leaving a portion of uncovered ITO in direct contact with the active layer forming current leakage pathways [14, 120, 168]. Therefore, a certain thickness of sol-gel derived ZnO film (exceed 10 nm) is necessary to obtain a full coverage on cathode surface so as to completely separate the cathode from contacting with active layer and thus blocking the reverse hole flow, while not harm its electrical properties, transmittance, and mechanical robustness.

Transmittance of CBL is another important factor that affects the device performance by influencing the light harvest of the BHJ photoactive layer [14, 67, 116, 136]. In order to obtain a high device performance, the light absorption of active layer should be maximized. For inverted PSCs, incident light comes from the transparent cathode side (i.e., ITO or FTO) where the CBL was prepared on. Therefore, CBL with high transmittance in visible region is important in order to make the active layer gain high optical absorption of active layer. The transmittance of ZnO films not only depends on the film thickness, but is also affected by the crystal size, defects and crystallinity, which are related to the preparation methods and post-treatment parameters [169–171]. High quality ZnO thin films possess an excellent transparency in the whole visible range and give rise to minimize optical losses. Through an optimization of the heat treatment process, Liang et al. fabricated high transparent ZnO layers on ITO cathode with the thickness ranging from several nm to ~55 nm using a sol-gel process [14]. The high transmittances of the ITO/ZnO layers will certainly benefits to the light absorption of active layer in inverted PSCs. In addition, the sharp absorption edge of the transmittance spectra of ZnO coated ITO exhibits a gradual red shift toward ~380 nm (corresponding to energy gap, ~3.2 eV of ZnO) with increasing the ZnO film thickness (Fig. 3.11)

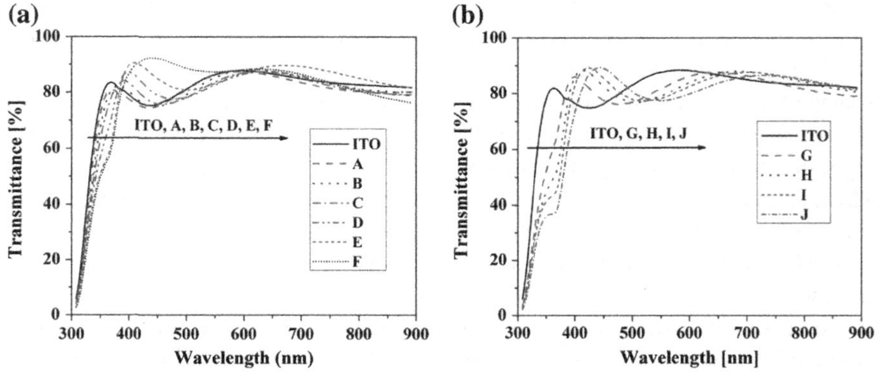

Fig. 3.11 The transmittance spectra of bare ITO and ITO coated with ZnO layers. **a** ZnO layers derived from *A* 0.02, *B* 0.05, *C* 0.1, *D* 0.3, *E* 0.6 and *F* 1 M sol, respectively; **b** ZnO CBLs derived from 0.1 M sol and by spinning (*G*) 3, *H* 5, *I* 7 and *J* 10 times, respectively. Adapted with permission from [14]. Copyright 2012 Wiley-VCH

[14, 121]. The band edge cut-off of ZnO at ~380 nm can help to block the UV light and thus protect the organic materials from the photo-degradation induced by UV light [48, 56]. However, for the high transparent ZnO CBLs, it has been demonstrated the minor difference of the transmittance of ZnO CBL just makes a negligible impact on the light harvest of photo-active layer and device performance [14]. In other words, the light harvest and the device performance are only affected when there is an obvious change in the transmittance of ZnO CBLs.

3.5 Doping of ZnO CBLs in Inverted Polymer Solar Cells

In inverted PSCs, the thickness of ZnO CBLs is typically only a few tens of nanometers in order to ensure good conductivity so as to reduce the serial resistance and improve device performance [51, 164]. However, such a thin CBL makes the devices suffer from low mechanical robustness and reduced protective properties against chemical or physical reaction between the active layer and the electrode [164, 172]. Therefore, such a thin ZnO CBL imposes a challenge for the manufacturing of PSCs with scalable processes, such as the roll-to-roll manufacturing [59, 164, 172]. Doping is an efficient way to enhance the conductivity of ZnO so as to overcome the thickness limitation. Through doping, an increase in the electrical conductivity of ZnO CBL and photovoltaic performance of inverted PSCs has been observed [164, 172–177]. The group-III elements, such as boron (B) [135], aluminum (Al) [164, 172, 175, 177–179], gallium (Ga) [173, 180] and indium (In) [174] have been intensively investigated for the doping of ZnO CBLs. It was explained that the group-III elements could work as an n-type dopant for ZnO to replace Zn sites and generate free electrons. Besides aluminium and the group III elements, the group-II elements (alkaline earth), such as magnesium (Mg), strontium (Sr) and barium (Ba) doped ZnO (named as MZO, ZnSrO and ZnBaO, respectively) layers also have been introduced into inverted PSCs [60, 181]. Besides the metals, the fullerene derivatives have been used to dope ZnO for forming an efficient CBL in inverted PSCs [61, 107].

3.5.1 Metal Doped ZnO Nano-films

Among the various doping methods, aluminium doping is a well-known approach to enhance the conductivity of ZnO layer for inverted PSCs. This seems to be because the doping of ZnO with aluminium can be easily processed and the aluminium-doped ZnO (AZO) possesses better electrical properties than that of intrinsic ZnO (i-ZnO) reference [164, 172]. For instance, the inflection (S-shaped) I-V curve of the inverted PSCs with pristine ZnO CBLs could be removed by introducing an AZO CBL with enhanced electrical properties [175–177]. In addition, it has been found that an increase of the thickness of AZO CBL from ~30 nm to more than 100 nm did not hamper the solar cell performance, while the devices based on 100 nm i-ZnO films suffered from severely decreased FF, V_{oc} and PCE due to an increase in the series resistance [164, 172]. Considering that a thick CBL has relatively good mechanical strength robust which is more suitable to the manufacturing of PSCs by scalable processes, such as the roll-to-roll manufacturing, thus the doping of ZnO with aluminium opens the possibility to incorporate a thick CBL into the inverted devices without sacrificing their efficiency [58, 164, 173]. By using a facial and low temperature (125 °C) solution processed Al doped ZnO nano-films as the CBLs, high efficiency inverted PSCs achieved a remarkable PCE in excess of 10 % on glass substrates and 8 % on plastic substrates [59]. Moreover, the average PCE of inverted devices based on these AZO CBLs and using PTB7: $PC_{70}BM$ active layer decreased marginally from 8.8 to 8.1 %, while the thickness of AZO CBLs increased from 22 to 75 nm. The very slight decrease of the device performance with increasing the AZO thickness indicates again that relative thick AZO can serve as an efficient CBL material for inverted PSCs.

To fabricate an efficient AZO CBL for inverted PSCs, it is worth noting that the device performances are strongly affected by the crystallinity and stoichiometry of AZO [172, 182]. The electronic properties of AZO CBLs, such as work function, conductivity, mobility as well as surface states, are massively determined by the stoichiometry of AZO [154, 164]. Tsai and co-workers have investigated the effects of Al content (from 0 to 12 mol%) on the electronic properties of AZO layers and device performances of inverted PSCs with structure of ITO/AZO/P3HT: PCBM/PEDOT:PSS/Ag [182]. They found that the carrier concentration, mobility and resistivity of AZO thin films strongly depended on Al molar ratio (Fig. 3.12), which furtherly impacted on the performance of inverted devices with the AZO CBLs. In their results, to achieve the best device performance, the optimum Al content of AZO CBLs is 6 mol%. Compared the devices based on pure ZnO layers, the device with optimum AZO CBL exhibits highest PCE of 3.78 %, enhanced by ~13.9 % via improving J_{sc} form 10.73 to 11.12 mA cm^{-2} and FF from 53.7 to 60.8 % [182]. At the optimum Al doping concentration, the authors demonstrated that the band structure of ZnO was modified with an upward energy shift in the Fermi level, which facilitates the electron collection and can also serves as an efficient hole blocking layer. The tunable band-gap combine with the increased carrier concentration and conductivity of AZO films via Al doping, and enhanced optical absorption of P3HT are the reasons of enhanced device performance [182].

Several group-III elements, such as boron (B), gallium (Ga), and indium (In), can also be used to doping ZnO CBLs for high performance inverted PSCs [135, 173, 174, 180]. For instance, the correlation between the device performance and the indium contents of the sol-gel derived indium-doped ZnO (IZO) CBLs has been investigated [174]. Compared to the device with i-ZnO CBL, the photovoltaic performance of the devices using appropriately doped IZO CBL, especially in term of the J_{sc}, were enhanced, due to a combined effect of improvement in charge collection and higher optical transmittance of the electrode/IZO buffer layer stack. The device with 1 at% IZO CBL showed maximum PCE of 3.3 % compared to the PCE of 2.94 % for the device with i-ZnO CBL. Gallium (Ga) has also been used as

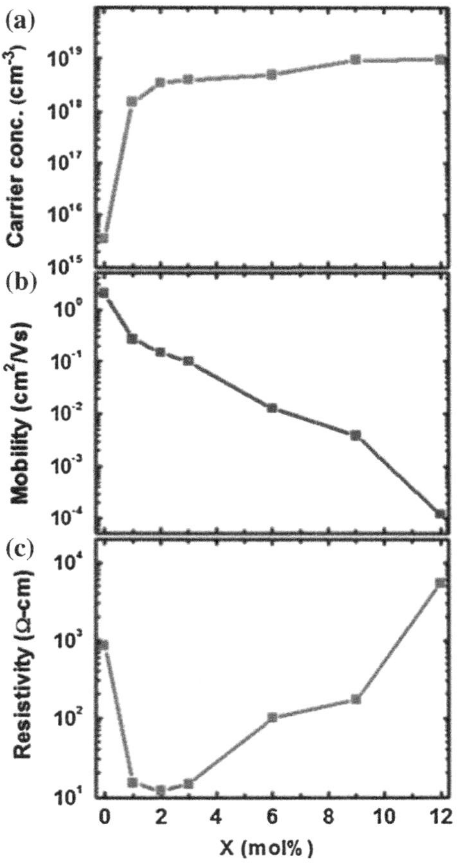

Fig. 3.12 **a** Carrier concentration, **b** mobility, and **c** resistivity of $A_xZn_{1-x}O$ (AZO) thin layers as a function of Al content (X, molar ratio). Adapted with permission from [182]. Copyright 2013 AIP Publishing LLC

a dopant to enhance the electrical conductivity of ZnO CBL. Shin et al. investigated the Ga-doped ZnO (GZO) CBL for the inverted PSCs [173]. By employing a GZO CBL, the PCE of the fabricated devices was improved by about 110 % compared to the devices with i-ZnO CBL, which was ascribed to the higher electron conductivity and smooth surface morphology of the ZnO film with Ga doping. In addition to group-III elements, the group-II elements, such as magnesium (Mg), strontium (Sr) and barium (Ba), doped ZnO (named as MZO, ZnSrO and ZnBaO, respectively) films also have potential for serving as CBLs in inverted PSCs [60, 181]. For instance, by employing the sol-gel processed amorphous Sr and Ba doped ZnO CBLs, the inverted PSCs exhibited better photovoltaic performance and device stability than the devices with pristine ZnO CBLs [181]. In addition, the commonly requirement of light soaking for reaching a maximum photovoltaic performance of inverted device based on i-ZnO CBLs was successfully circumvented. The improvements of device performance were attributed to the doping of ZnO with Sr or Ba reducing the electron trapping on the surface associated with oxygen absorption. It was suggested that the Sr or Ba doping could effectively suppression/reduction of the oxygen adsorption at mobile oxygen vacancy sites on the ZnO surface [181].

For doped ZnO CBLs, the band gap structure is a key parameter that could affect the device performance. The band gap and energy level dependence on the doping content enables researchers to adjust the optical and electrical properties of the doped ZnO CBLs for achieving better device performance [60, 172, 178]. For example, it is well established that the substitutional doping of Mg^{2+} into the Zn site leads to an increase in the bandgap of ZnO [60, 183, 184]. By using Mg doped ZnO (ZMO) CBLs Yin et al. fabricated inverted PSCs with high performance. [60] They developed a solution processed ZMO CBL with tunable band gap and energy levels (Fig. 3.13a), good optical transmittance and electron transporting abilities. With an increase of the Mg content ($x = 0.1 - 0.6$), the CB edge value of the ZMO CBLs are finely adjusted in the range of ~ -4.3 to -3.9 eV (Inset in Fig. 3.13b) to approach the LUMO of $PC_{71}BM$. The corresponding band gap of ZMO films increased from 3.3 to 3.7 eV, which was confirmed by the gradual blue shift of the fundamental absorption edge as the Mg content increases (Fig. 3.13b). A favorable CB level of

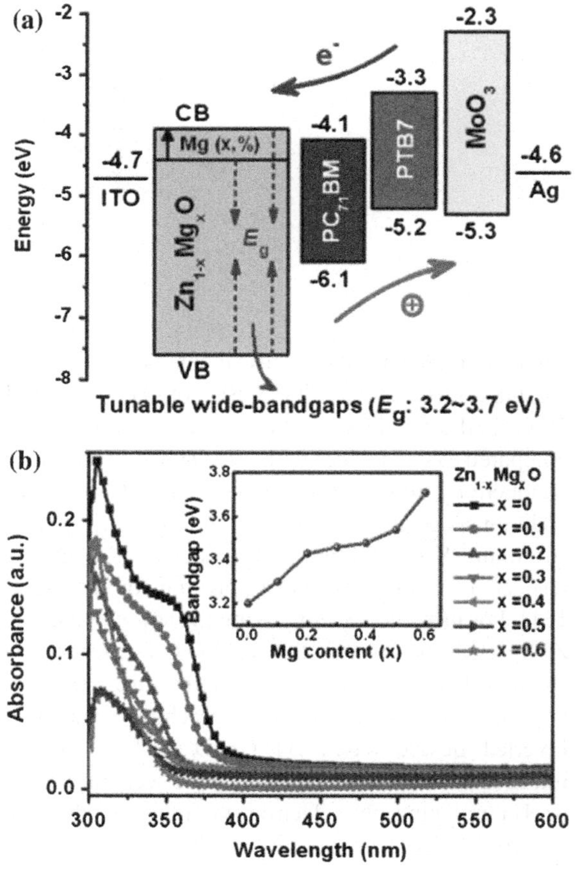

Fig. 3.13 **a** Energy levels of the components in the inverted PSCs with various ZMO CBLs. **b** Optical absorption spectra of ZMO films. The *inset* shows an increase in the bandgap of ZMO films with the increasing of Mg content (*x*). Adapted with permission from [60]. Copyright 2014 Wiley-VCH

ZnO matching well with the $PC_{71}BM$ acceptor would improve the electron extraction/transportation in inverted devices. The electron mobility of the ZMO films was found close to that of pure ZnO film, which is in favor of electron extraction/transportation in the inverted PSCs. Moreover, ZMO layers also exhibit a more hydrophobic surface than ZnO, benefiting to forming high quality contact between the organic active layer and the inorganic metal oxide CBLs. By incorporating the optimized ZMO layers derived form a 0.2 M precursor solution, the inverted PSCs with structure ITO/ZMO/PTB7:$PC_{71}BM/MoO_3$/Ag exhibited a preferable PCEs exceeding 8 % (up to 8.31–8.35 %), which are much higher than those of devices with an i-ZnO CBL (PCE = 7.1 %) under the same conditions [60]. Macleod and co-workers have also reported an improved PCE in inverted PSCs when a sol-gel processed Mg-ZnO is used in place of ZnO as CBLs [184]. It has been found that, as the Mg content increased, the FF and V_{oc} increased substantially for both the devices based on P3HT:PCBM and P3HT:ICBA systems. The maximum PCEs were achieved when the Mg content in the precursor was 10 mol%, and subsequently decreased for higher Mg content due to decay in the J_{sc}. They have also found that the increase in Mg content led to a decrease in the work function and an increase in the bandgap. The increased bandgap in Mg-ZnO was ascribed to a downshift in the valence band maximum, rather than an upshift in the conduction band minimum. The improved device performances were partially ascribed to the reduced recombination caused by substituting of the Zn with Mg. However, it has been suggested that the mechanisms for the improvement of inverted devices with Mg doped ZnO CBLs remain an open question [184]. More systematic investigation needs to be done to elucidate the dominating factors with regard to the improvement in the performance of inverted PSCs when using Mg-ZnO CBL.

3.5.2 *Fullerene Derivatives Doped ZnO Nano-films*

In addition to the metal dopants, fullerene derivatives can also be used to dope ZnO for forming high quality CBL in inverted PSCs [61, 107]. Liao et al. reported a fullerene derivative doped ZnO (ZnO-C_{60}) nanofilm as the CBL of inverted PSCs [107]. For the inverted device with ZnO-C_{60} CBL, the PCE was remarkably improved from 7.64 to 9.35 % for PTB7-th:$PC_{71}BM$ active layer, from 6.65 to 8.21 % for PTB7:$PC_{71}BM$, and from 5.26 to 6.60 % for P3HT:ICBA, compared to the devices with pristine ZnO CBL. It has been demonstrated that the ZnO-C_{60} CBL can enhance the electron collection by producing a fullerene derivative rich cathode surface and promote electron conduction at the interface and in the bulk of ZnO-C_{60}. The high content of fullerene at the surface of ZnO-C_{60} CBL offers a better contact to the fullerene derivative in the active layer, and thus facilitates electron collection. In another work also done by Liao et al. a dual-doped ZnO with Indium and fullerene derivative (BisNPC_{60}-OH) were fabricated by sol-gel processing and applied as CBL in inverted PSCs [61]. Figure 3.14 shows the energy levels diagram of ZnO, InZnO, ZnO-BisC_{60}, InZnO-BisC_{60}, and all of the other

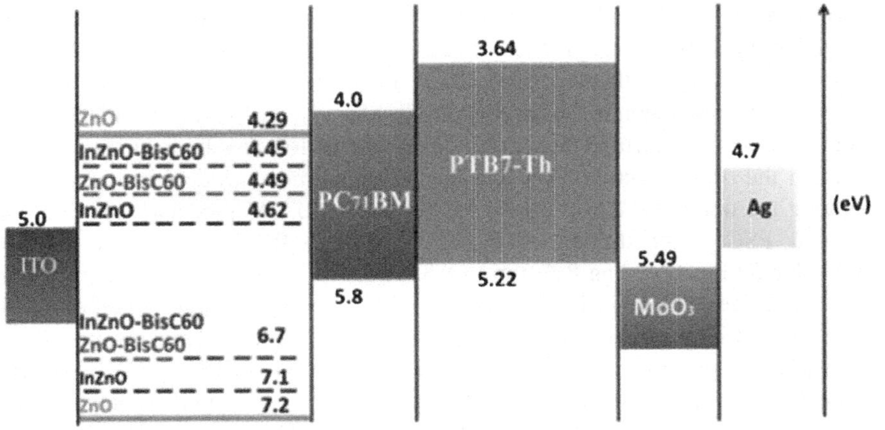

Fig. 3.14 Energy levels diagram for ZnO, InZnO, ZnO-BisC_{60} and InZnO-BisC_{60} determined from ultraviolet photoelectron spectroscopy (UPS) and UV–Vis results and for all the components in the inverted PSCs. Adapted with permission from [61]. Copyright 2014 Nature Publishing Group

components in the inverted PSCs. It has been demonstrated that this dual doped ZnO, InZnO-BisC_{60}, CBL showed dual and opposite gradient dopant concentration profiles, being rich in fullerene derivative at the cathode surface in contact with active layer and rich in Indium at the cathode surface in contact with the ITO surface. For the InZnO-BisC_{60} film, the surface conductivity was improved from 0.015 to 4.06 S cm^{-1}, and the electron mobility was enhanced from 8.25×10^{-5} to 1.09×10^{-2} cm^2 V^{-1} s^{-1}. As a result, with this InZnO-BisC_{60} film as CBL, the PTB7-Th-$PC_{71}BM$ inverted device exhibited a preferable PCE, 10.31 %, significantly higher than the 8.25 % for pristine ZnO.

3.6 One-Dimensional ZnO Nanostructures for Inverted PSCs

Although the BHJ active layer is powerful as a solution for addressing the exciton dissociation sites and the charge transport channels [7, 19, 22], there are still some shortcomings of BHJ active layer that limit the photovoltaic performance of PSCs. One of the limitations is that the very thin thickness of the BHJ active layer leads to low optical absorption and thus non-ideal device performance [185–187]. The use of very thin active layer is, on one hand, because the photo-generated charges in active layer need to transport in the percolation network of the polymer donor and the accepter, and on the other hand the carrier mobility in polymers is relatively low [185, 188]. The charge recombination in a thick active layer will increase because of an increase in the diffusion length of carriers, leading to the device performance decrease. Thus, the optimal thickness of active layer is determined by a trade-off between maximize the optical absorption and meanwhile allow the charges transport efficiently from BHJ active layer to electrode. It has reported that an optimal active layer thickness is typically about only ~100 to 200 nm [185–187, 189].

Research effort on the methods of light absorption enhancement has been driven by these limitations, including using photonic crystals [190], plasmonic enhancement [191–194], and textures [195, 196]. The use of vertical aligned one-dimensional (1 D) ZnO nanostructures is also a potential approach to enhance the light harvest of photoactive layer. This is because the vertical aligned 1 D ZnO nanostructures can permeate into active layer and serve as a direct electron transport pathway so as to enhance charge carrier collection and transport [214, 215]. Subsequently, the enhancement of charge transport avoids the increase of charge recombination in the case of a thick active layer, which benefit to the light harvest of photoactive layer [197, 198]. Figure 3.15 shows a schematic diagram of inverted PSCs with 1 D ZnO nanostructure. In the case of inverted PSC with ZnO nanorods (NRs), after the charge separation at the polymer and fullerene interface, the photo-generated electrons at LUMO level of fullerene transfer to the conduction band of ZnO. Then, the electrons quickly move to cathode electrode along the ZnO NRs, in view of high electron mobility of ZnO NRs (~1 to 5 $cm^2\ s^{-1}\ V^{-1}$ along c-axis) [199] which is several orders of magnitude higher than what are typically found in most of organic semiconducting materials [2, 197, 200]. For the devices with ZnO NRs, the average distance from the generation sites of charge carriers to the surface of ZnO electron collection layer would be shorter than that in the solar cells without ZnO NRs, and as a result the charges recombination can be effectively reduced [197, 201]. To date, the reported vertical ZnO nanostructures, named as nanorods (NRs) [197, 200, 202–210], nanowires (NWs) [201, 211, 212], nanofibers (NFs) [213], and nanopillars [214, 215], and 2 D vertical aligned nanowall arrays [17, 210], combined with dense ZnO films have been employed to fabricate the inverted PSCs.

Preparation of vertical aligned 1 D ZnO nanostructures has been a subject of intensive study, and many methods have been developed and improved for the growth of a variety of 1 D ZnO nanostructures due to their unique properties for electronic and optoelectronic applications. The main fabrication techniques of 1 D

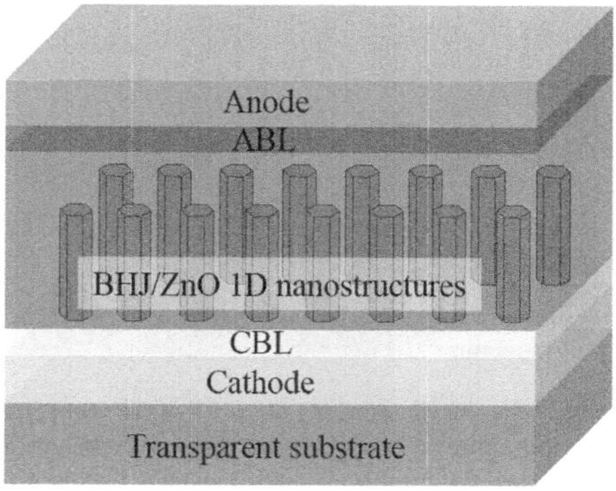

Fig. 3.15 Schematic diagram of inverted PSC structure with 1 D ZnO nanorods

ZnO nanostructures include physical vapor deposition, chemical vapor deposition, pulsed laser deposition, sputtering and several solution chemical methods. Among these methods, hydrothermal growth methods have been a popular way to grow 1 D ZnO nanostructures for inverted PCSs, due to its produce good nanostructure, ease of preparation at relatively low temperature (lower than 100 °C), without using sophisticated and/or vacuum equipment and low cost, and good potential for large-scale production. For more detailed information on the solution growth of 1D ZnO nanostructures, readers are recommended to a comprehensive and excellent review, *One-Dimensional ZnO Nanostructures: Solution Growth and Functional Properties,* published by Xu and Wang [216]. In this section, we mainly focus on the effects of the 1 D ZnO nanostructures on the photovoltaic performance of the inverted PSCs. Figure 3.16 presents the SEM images of the typically vertical aligned 1 D ZnO nanowires used in inverted PSCs.

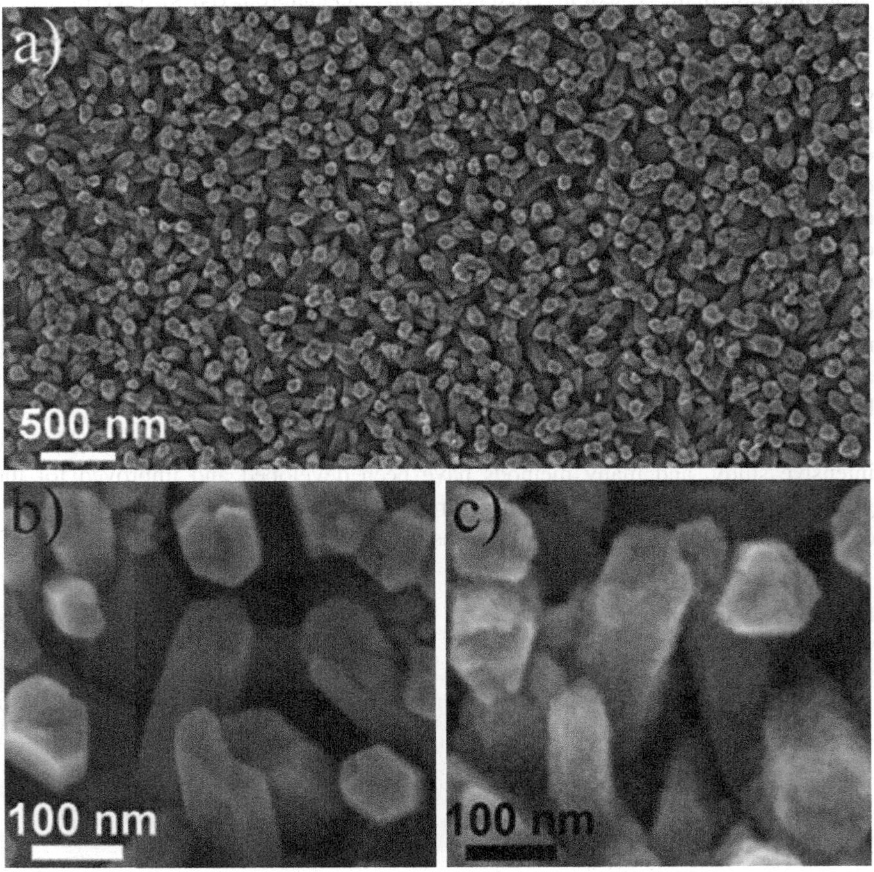

Fig. 3.16 SEM images of 1 D ZnO nanostructures employed in inverted PSCs. FE-SEM micrographs of **a** ZnO nanowires and magnified views of **b** as-electrodeposited and **c** ZnO nanoparticles modified ZnO nanowires. Adapted with permission from [211]. Copyright 2011 Royal Society of Chemistry

Olson et al. demonstrated for the first time that the inverted PSCs fabricated with the blend of a P3HT:PCBM and 1 D ZnO nanostructure prepared with a solution method, and a PCE of 2.03 % was obtained [213]. It has been demonstrated that the ZnO NR array in ZnO NR/P3HT:PCBM system mainly works as electron collector/transport pathway, rather than an accepter material to form heterojunctions with the donor polymer in view of much larger interface between the P3HT and PCBM than the interface between the P3HT and ZnO NR [68, 197, 217]. Although ZnO nanorods may not improve the hole transport, the extraction and transport away of electrons would reduce the probability of electron-hole recombination.

In order to fully achieve the potential of 1 D ZnO nanostructures in inverted PSCs, an electrical coherence at the organic photoactive layer and ZnO interface needs to be provided. Therefore, the infiltration of polymer into the space between ZnO NRs is a critical issue that greatly affects the devices performance. Several methods have been developed to improve the contact between the active layer and ZnO nanostructures, including surface modification of ZnO nanostructures and optimizing the post-treatment processes [198, 206, 210, 211]. For example, coating a solution-processed PCBM interlayer on the surface of ZnO NRs can efficiently enhanced the infiltration of active layer into the intervals between ZnO NRs [206]. Compared to the devices made with bare ZnO NRs, the PCE of inverted PSCs with PCBM modified ZnO NRs were enhanced from 2.35 to 3.2 %. In addition, the functionalization of the ZnO nanowires (NWs) surface with ZnO NPs with the consideration to improve the contact between ZnO and P3HT: PCBM active layer has also been investigated [211]. Figure 3.17 shows the SEM images of the cross-section of inverted devices with the ZnO nanorods with or without surface

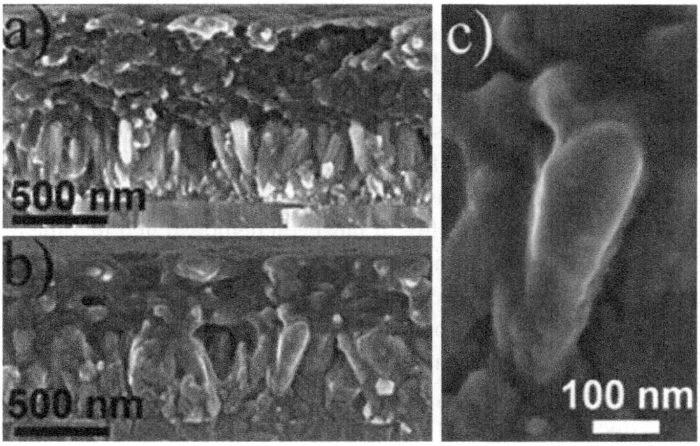

Fig. 3.17 Cross-sectional SEM images of inverted devices **a** with and **b** without fullerene interlayer [206]. Cross-sectional FE-SEM micrographs of inverted devices with **c** unmodified ZnO nanowires, **d** ZnO NPs modified ZnO nanowires, and **e** the higher magnification view of the ZnO nanowires well covered by the P3HT:PCBM blend. Adapted with permission from [211]. Copyright 2011 The Royal Society of Chemistry

modification. It was found that, by modifying the ZnO NWs with a thin layer of ZnO NPs by immersing in a ZnO colloidal suspension, improved P3HT:PCBM blend penetration was achieved (Fig. 3.17b and c). The ZnO NPs modification was found to contribute to enhancing the exciton dissociation and increasing the charge lifetime by reducing charge recombination. Moreover, the density and length of ZnO NRs can also affect the infiltration of organic active layer. For instance, It has been found that the morphology of two-growth ZnO NRs helps to achieve deeper and superior infiltration of organic active layer [16].

Increasing the polymer solidification time is also an approach to improve the infiltration of active layer into ZnO NPs space, and so as to enhance the device performance of inverted PSCs with ZnO NRs [198]. It has been found that the increase of polymer solidification time made the polymer chains have enough time to self-organize, leading to the enhanced crystallinity of polymer and meanwhile improved infiltration of photoactive layer into ZnO NPs space. Overall, for the 1 D ZnO nanostructures applied in inverted PSCs, their length, density, morphology, and surface modification are key parameters that will significantly affect the penetration of polymer and fullerene blend and thereby affect the device performance.

3.7 Surface Modification of ZnO CBLs

The performance of polymer solar cells is critically dependent on the properties of each layer and the interfacial contacts. For ZnO CBLs, without interface engineering, the inorganic ZnO CBL and surface defects may lead to a poor interfacial contact with the polymer based active layer and, thus, result in a poor electron extraction, which is one of the main reasons leading to relatively high charge recombination and series resistance (R_S) [218, 219]. Especially, in the case of ZnO CBLs prepared with widely used sol-gel processing or colloidal nanocrystals, the low-temperature and solution process often lead to a high densities of surface defects, such as dangling bonds, surface groups and charged oxygen molecules [54, 104, 175, 220]. Such surface defects can act as recombination centres for photogenerated charge carriers, decreasing the photocurrent and power conversion efficiency as well as the device stability. It has been demonstrated that the deteriorated electrical properties at the interface between ZnO CBLs and organic active layers are caused by incompatible chemical interfaces and the formation of electron trapping states on the surface of ZnO layers [221]. A phenomenon commonly observed on the J-V curve of inverted PSCs when employing ZnO CBLs is the appearance of an inflection point, i.e. the so-called S-shaped J-V curve feature. It is believed to be due to the accumulation of charges at the ZnO/active layer interface [124, 129, 177, 181]. Therefore, appropriate electrical contacts at interfaces between ZnO CBLs and BHJ active layers are highly desirable and essential to achieve high power conversion efficiency and good stability of inverted PSCs. Engineering the surface of ZnO CBLs through improving the interfacial electrical properties, better aligning energy-level and controlling the surface energy is an effective way to reduce the R_S of solar cells. There have been several efficient surface engineering methods to modify the surface of ZnO CBLs, such as ultraviolet (UV) [54, 114, 124, 129, 166, 177, 181] or UV-ozone treatment [124, 153], thermal and vacuum treatment [146], and incorporation of surface modification interlayer, such as the self-assembled monolayers (SAMs) [148, 222–225].

3.7.1 UV Illumination Treatment of ZnO CBLs

UV or UV-containing light illumination treatment, also known as "light-soaking", is a common way to improve the electrical characteristics and optimize optimum photovoltaic performance of inverted devices with ZnO CBLs [54, 114, 124, 129, 166, 177, 181]. UV illumination can passivate the defects states, such as the negatively charged oxygen molecules trapped at the grain boundaries and surface, and thus remove electron extraction barrier and reduce carrier recombination at ZnO/photoactive layer interface [54, 104, 175]. The metal-oxide/organic active layer interface is identified as the reason of the S-shaped J-V curve due to an electron extraction barrier [165, 166, 177]. Under the dark condition, electrons in the conduction band of ZnO layer are trapped by the chemisorption of O_2 molecules (Fig. 3.18a, step (1)), and therefore forming a depletion layer and band bending next to the surface (Fig. 3.18a, step (2)) [165, 166]. It has been suggested that both the depletion layer and the band bending can result in the formation of a barrier for the electron injection and extraction at the ZnO/BHJ active layer interface and are more pronounced in the case of small ZnO nanocrystals with a high surface-area-to volume ratio. Upon UV illumination, the electron-hole pairs are generated in ZnO, leading to neutralization of O_2^- species and desorption of oxygen from ZnO surface, as shown in Fig. 3.18b, step (3), (4) and (5) [165]. The desorption of O_2 from ZnO surface reduces the electron traps, and thus increases the concentration of mobile electrons [166]. As a result, the inflection point problem (S-shaped), on J-V curve of the inverted devices incorporating ZnO CBLs could be solved by exposing the cells to UV illumination, leading to the device performances improvement (Fig. 3.18b) [54].

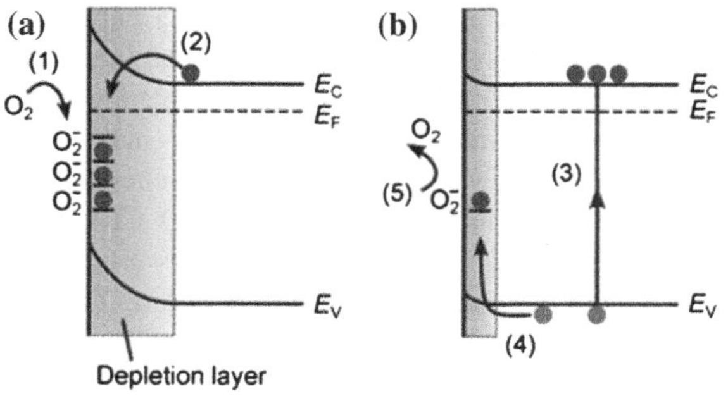

Fig. 3.18 Schematic view of the oxygen absorption and desorption processes in *n*-type ZnO layer in the *dark* (**a**) and UV illumination (**b**). Adapted with permission from [165]. Copyright 2014 American Chemical Society

The UV-ozone treatment (UVO) has also been investigated to modify ZnO CBLs [124, 153]. It has been found that the UVO treatment of ZnO NPs films could effectively passivate the defect states leading to longer carrier life time of devices compared to the devices treated with UV light soaking [124]. By reducing the interfacial recombination via UVO treatment and employing high performance PDTG-TPD:$PC_{71}BM$ active layer, the inverted PSCs with PCEs exceeding 8 % were achieved. According to the PL measurement, it was proposed that the passivation mechanism under UVO treatment of ZnO films is correlated with the reduction of oxygen vacancies (V_O) in both the surface and bulk due to the penetration of oxygen in the porosity NPs films. The defect passivation in ZnO NPs layers could help to reduce the carrier recombination at the ZnO/active layer interface, and thus enhance the device performances [124]. However, the UV exposure approach is limited in the case when the devices are based on substrates with low UV transmittance, UV blocking filters or UV to VIS down-conversion concepts [177]. Many active organic donor polymers must be protected from being exposed to UV light so as to achieve a reasonable device stability [11, 177, 226]. In order to alleviate the requirement for UV illumination, surface modification methods and the use of doped-ZnO CBLs are developed to improve the electronic property of ZnO CBLs and the contact property of ZnO/organic active layer interface.

3.7.2 Fullerene-Based Interlayer Modification of ZnO CBLs

Modifying ZnO CBLs with a functional interlayer derived from fullerene derivatives [62, 148, 222, 224, 225, 227, 228], conjugated polyelectrolyte (CPE) [229, 230], poly(ethyleneimine) (PEI) [78] or some other interfacial modification materials are very important in enhancing charge extraction as well as the overall performance of inverted PSCs. The functional interlayer can achieve different proposes, including to help passivate surface charge traps, tune the energy level offset between CBL and active layer, and affect the morphology of the upper BHJ active layer, depending on the nature of the interfacial modification materials and the processing method. Among these interfacial modification materials, fullerene-based interlayers (C_{60}-SAMs), which are effective in acting as interlayer for the modification of ZnO CBLs, are a popular class of interfacial modification materials for high performance solar cells. In addition, due to similar energy levels and structures with PCBMs in BHJ active layer, the development of the fullerene derivatives as interfacial modification materials has attached wide attention. Figure 3.19 shows the cross-section schematics of the inverted devices with fullerene-based interlayer modified ZnO CBLs. Figure 3.20 shows the chemical structures of several fullerene derivatives used to modify ZnO CBLs. When compared with inverted PSCs based on a bare ZnO CBL, the PCEs of devices with fullerene-based interlayer modified ZnO CBLs can be increased by approximately

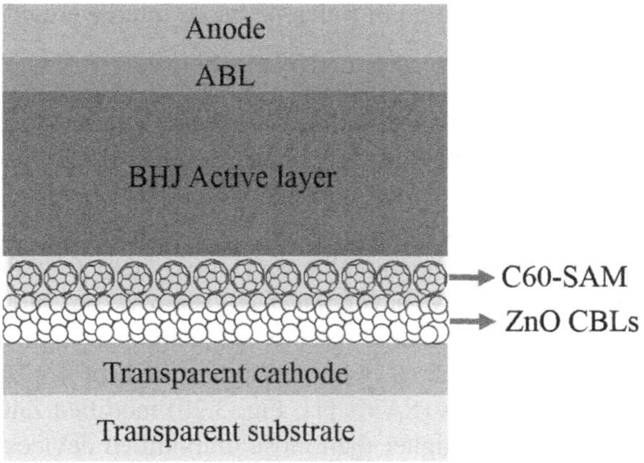

Fig. 3.19 Schematic illustrations of device structure of inverted PSCs with fullerene-based interlayer modified ZnO CBLs

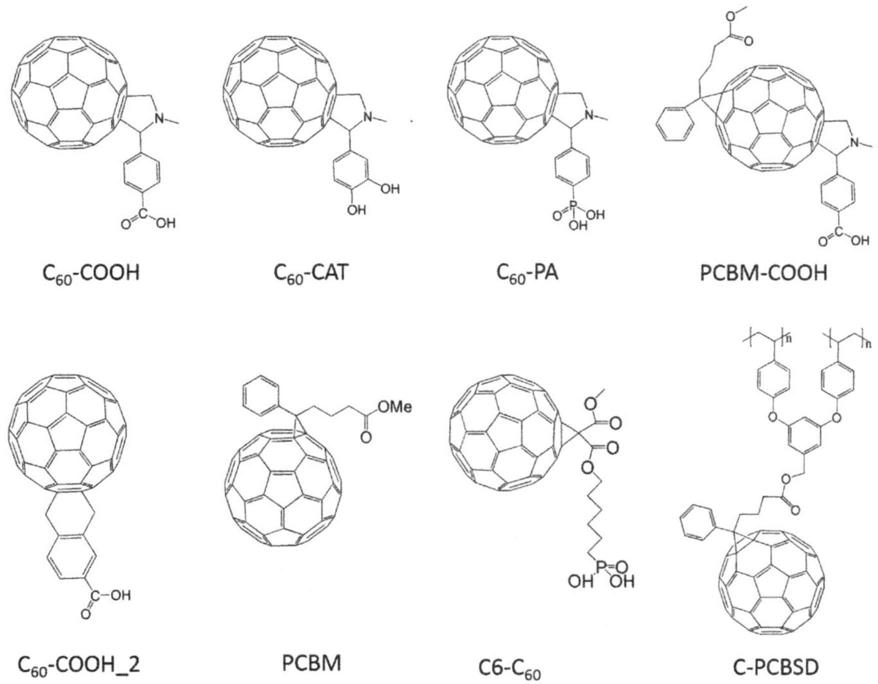

Fig. 3.20 Chemical structures of the fullerene derivatives used to modify ZnO CBLs

10–30 %, depending on the nature of fullerene based interlayer and the BHJ active layer used in device.

The C_{60}-SAMs can be easily processed to form a modification layer on the surface of ZnO CBLs through either a solution immersion technique or solution-based spin-coating method. Jen group developed a series of C_{60}-SAMs containing different anchoring groups (catechol, carboxylic acid, and phosphonic acid), linkage location, and functionalization (Fig. 3.20) [148, 222]. It has been found that these C_{60}-SAMs functional layers could help to reduce the charge recombination at the ZnO/active layer interface, improve the morphology of the active layer, and enhance the charge selectivity, leading to high FF and J_{sc}. For instance, an average PCE of 4.5 % (with the highest PCE of 4.9 %) was achieved for the devices with C_{60}-SAMs (SAMs [1], Fig. 3.20) modified ZnO CBL surface; such efficiency is over 20 % higher than those unmodified devices.

A PCBM-based and cross-linked *n*-type interlayer (C-PCBSD) (Fig. 3.20) has also been synthesized for the modification of ZnO CBL [224, 227]. This C-PCBSD interlayer was considered to have several positive impacts on the interfaces of the active layer, including improved exciton dissociation efficiency, reduced charge recombination, decreased interface contact resistance, and introduced vertical phase separation to reduce the bulk resistance of the active layer as well as passivating the local shunts of the ZnO CBL interface [224, 227]. It has been suggested that, firstly, the C-PCBSD interlayer provided an extra P3HT/C-PCBSD interface area for ultrafast excitons dissociation; secondly, the LUMO energy level of C-PCBSD (3.8 eV) was located between the LUMO of P3HT (3.3 eV) and the conduction band of ZnO (4.4 eV), which makes the electrons can be efficiently transported from active layer to ZnO through C-PCBSD as an energetically favorable downhill pathway [224]. The inverted devices based on C-PCBSD modified ZnO CBL with P3HT:PCBM blend active layer gained an improvement in J_{sc} from 11.6 to 12.8 mA cm^{-2}, and exhibited PCE as high as 4.4 %, which is much higher than 3.5 % for a reference device with unmodified ZnO CBL [224]. An impressive high PCE of 6.2 % has been achieved through introducing C-PCBSD modified ZnO CBL into inverted device based on ICBA: P3HT blend active layer, leading to a 29 % enhancement compared to the device with un-modified ZnO CBL [227]. The improved device photovoltaic performance in terms of J_{sc} and FF of the devices with C-PCBSD modified ZnO CBL was attributed to high shunt resistance (R_{SH}) and low series resistance (R_S) due to the decreased contact resistance and increased electron collection efficiency at the ZnO/C-PCBSD interface. Besides high PCE, the devices with C-PCBSD modified ZnO CBL also demonstrated a good device stability in air without encapsulation. In their results, the PCE of the unencapsulated devices with P3HT:PCBM active layer and based on bare ZnO CBL retained 80 % of its original value after being stored in ambient condition for 35 days. In contrast, the device based on C-PCBSD modified ZnO CBL exhibited a better long-term stability, retained above 85 % of its original value under the same conditions [224]. In sharp contrast, the device with conventional structure (ITO/PEDOT:PSS/PCBM:P3HT/Ca/Al) and without encapsulation suffer a rapid decay in just two days. It was suggested that the coverage of three-dimensional C-PCBSD interlayer on ZnO buffer layer might have passivated the hot spots generated in ZnO CBLs to suppress the leakage current, and thus improved the device lifetime [247].

For the fullerence-based interlayer, the functionality of the C_{60}-SAMs molecule determines the electronic and electrical properties of the CBL. Stubhan et al. investigated the modification of AZO buffer layer with phosphonic acid anchored C_{60}-SAMs (Fig. 3.20) [228]. It has been demonstrated that the phosphonic acid anchored C_{60}-SAMs decorated AZO could increase R_{SH} and decrease the R_S of the inverted PSCs duo to the improved charge transfer from the PCBM to AZO. The FF of inverted P3HT:PCBM solar cell can be increased through surface modification of AZO CBL via phosphonic acid anchored C_{60}-SAMs. More recently, the PCBM has also been applied as an interfacial modification material (Fig. 3.20) [225]. It has been found that the PCBM layer between ZnO and BHJ active layer can effective quench electron-hole recombination by reducing trapped charges at ZnO surface, which resulted in a 16 % increase in the PCE (up to 7.7 %) compared to devices without an additional PCBM layer. The fullerene-based modifiers with weaker acids have also been explored to modifying ZnO CBL [62, 148, 222]. It has been found that the fullerene-based modifiers with ester units exhibit much weaker acidity and better affinity to ZnO surface [148]. A C_{60} pyrrolidine tris-acid ethylester (PyC_{60}) with three ester groups has also been developed to modify ZnO CBL [62]. It has been demonstrated that PyC_{60} can improves the morphology quality of ZnO/PyC_{60} film by filling in the voids of ZnO layer and partly passivating the defects of ZnO NPs. By using this PyC_{60} upper layer modified ZnO NPs film as the CBL, high efficiency (average PCE of 8.62 %) inverted PCSs based on photoactive layer of PTB7:PC_{71}BM blend were achieved. It was indicated that the PyC_{60} could also improve the interfacial contact between the ZnO and active layer, lowering the electron injection barrier. As a result, the ZnO/PyC_{60} CBL effectively decreased R_S and interfacial charge recombination, and increased electron injection and collection efficiency in the inverted PSCs [62].

3.7.3 Non-fullerene Based Interlayer Modification of ZnO CBLs

Non-fullerene based interlayer, such as: mixed SAM [219], poly(ethylene oxide) (PEO) [152], poly(ethyleneimine) (PEI) [78], ruthenium dye N719 [179], conjugated polyelectrolyte (CPE) [229, 230], HfO_2 [130], Ionic liquids (Ils) [65] and ultrathin metal oxides layers [117, 132, 231, 232] have also been studied to modify ZnO CBLs for further improving the performance of inverted PSCs. When compared with inverted PSCs based on bare a ZnO thin layer, the PCEs of devices using non-fullerene based interlayer modified ZnO CBLs can be increased by approximately 5–38 %, depending on the nature of the non-fullerene based interlayer and BHJ active layer used in devices.

Conjugated polyelectrolyte (CPE) is a popular class of materials for high performance PSCs, which are advantages for their flexibility, adjustable properties and low-temperature processing. On one hand, CPE can be used to modify the ITO as a CBL between the cathodic electrode and active layer in inverted PSCs, thereby improving the device performance by changing the electronic and orbital interactions at the interface [57, 63, 79, 233]. On the other hand, CPEs are also an efficient interfacial modification material to modify the ZnO CBLs for high PCE inverted PSCs [229, 230, 234, 235]. In addition, the CPEs possess both a conjugated backbone and highly polar side chains, which ensures their semiconductivity and good water/alcohol solubility, respectively. The low-temperature and solution processed ability of CEPs in polar solvents are compatible with the low-cost printing manufacturing of PSCs. Commonly used CPEs to modify ZnO CBL include poly [(9,9-bis (3′-(*N*,*N*-dimethylamino)propyl)-2,7-fluorene)-alt-2,7-(9,9dioctylfluorene)] dibromide (PFN-Br) [230], poly(9,9′-bis(6″-*N*,*N*,*N* trimethylammoniumhexyl) fluorene-co-alt-phenylene) with bromide counterions (FPQ-Br) [229], poly{3-[2-[4,8-bis(2-ethyl-hexyloxy)-6-methyl-1,5-dithia-*s*-indacen-2-yl]-9-(3-dimethylamino-propyl)-7-methyl-9H-fluoren-9-yl]-propyl}-dimethyl-amine (PBN) [235] and poly[9,9-bis(6′-(*N*,*N*-diethylamino)propyl)-fluorene-*alt*-9,9-bis-(3-ethyl(oxetane-3-ethyloxy)-hexyl) fluorene] (PFN-OX) [234], as summarized in Fig. 3.21. For instance, a thin layer PFN-Br (Fig. 3.21) has been deposited on ZnO CBL to reengineer the interface between ZnO and PBDT-DTNT:$PC_{71}BM$ [230]. The inverted PSCs with device architectures of ITO/ZnO/PFN-Br/PBDT-DTNT: $PC_{71}BM/MoO_3/Ag$ showed a J_{sc} of 17.4 mA cm^{-2}, a V_{oc} of 0.75 V, a FF of 61 % and an average PCE up to 8.4 % was attained. The conduction band edge of ZnO was tuned from −4.46 to –4.08 eV by using a PFN-Br thin layer, leading to an enhancement of the V_{oc} of the device. The enhanced J_{sc} and FF in the inverted PSCs with the ZnO/PFN-Br CBLs were ascribed to the improved contact between PFN-Br interfacial layer and active layer, which was better than the interfacial layer formed between naked ZnO layer and active layer. As a result, over 37 % enhancement in the PCE was obtained. In addition to PFN-Br, another PFN-based cross-linkable water/alcohol soluble CPE, PFN-OX (Fig. 3.21), has also been designed and synthesised for inverted PSCs applications [234]. The PTB7: $PC_{71}BM$ inverted devices based on PFN-OX modified ZnO CBL gained an improvement PCE form 8.45 % (ZnO device) to 9.28 % (ZnO/PFN-OX device) with a FF high up to 74.4 % due to doping of PCBM by PFN-OX. A novel CPE polymer, PBN (Fig. 3.21) having benzodithiophene (BDT)-based molecular structure can also be used to modify ZnO CBO for high efficiency inverted PSCs [235]. The PTB7:$P_{70}BM$ device with ZnO/PBN CBL yield a PCE of 8.6 %, representing a 21.1 % improvement over the control device with pristine ZnO CBL. The introduction of PBN on the ZnO surface can create interfacial dipoles at the ZnO/BHJ active layer interface and reduce the interfacial resistance and improve the charge transport and collection through the PBN layer. The improved contact quality between BHJ active layer and ZnO CBL is due to hydrophobic properties, and displayed a reduced surface energy. An ultrathin (<10 nm) CPE, FPQ-Br (Fig. 3.21), has also been successfully employed as an interlayer between ZnO and

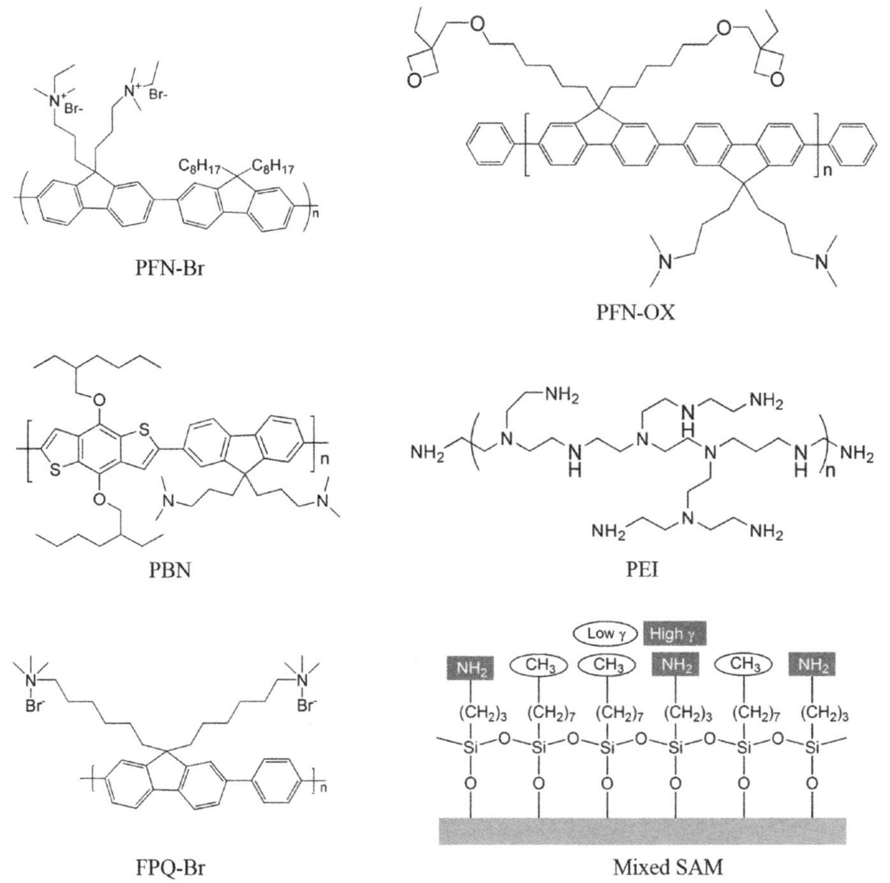

Fig. 3.21 Chemical structures of the non-fullerene based materials used to modify ZnO CBLs

active layer to improve the electron transport and interfacial contact [229]. Compared with the device based on bare ZnO layer, the J_{sc} increased by 11 % and the PCE increased from 6.37 to 7.04 % for the device with ZnO/FPQ-Br CBL.

A mixed SAM can also been used to control the surface energy of the ZnO buffer layer [219]. Through changing the surface composition of the two different SAMs (Fig. 3.21), the surface energies of the ZnO CBL could be tuned over a wide range of values (between 40 and 70 mN m^{-1}) with negligible changes in its work function [219]. It has been demonstrated that a surface treatment to obtain an intermediate surface energy above the neutral range favors the competition between lateral and vertical phase separation, resulting in morphology with desirable domain size. By tuning the surface energy of ZnO CBLs in a given range, the PCE of inverted solar cell increased from 3.27 to 3.70 %. Several other SAMs derived from benzoic acid or benzoyl chloride have also been developed to tune the work function and surface properties of ZnO CBLs [236, 237]. It was demonstrated that the benzoic acid interlayer could form a favorable interface dipole at the interface between the ZnO and active layer, thus to passivate the ZnO surface traps and accordingly reduce the work function and facilitate electron transport.

Poly(ethyleneimine) (PEI) as an inexpensive neutral polymers dissolved in water can also be used to prepare CBLs for inverted PSCs to decrease the ITO work function due to the formation of dipole layer at the ITO-PEI interface (Fig. 3.22) [78, 238]. To combine the advantages of both ZnO and PEI, PEI-coated ZnO layers have been fabricated and applied as the CBLs for inverted PSCs. An appreciable PCE enhancement in inverted PSCs has been attained by using the electron-rich polymer nano-layer PEI modified ZnO CBL [78]. It is worth noting that the thickness of the PEI nano-layer was controlled to be only 2 nm in view of its high insulting character. The enhanced performance was assigned to the lowered conduction band energy of the ZnO layer via the formation of an interfacial dipole layer at the interface between ZnO layer and PEI nano-layer. In addition, the PEI nano-layer could increase the surface roughness of ZnO, which noticeably decreases the series resistance of device. The inverted PSCs with structure of ITO/ZnO-PEI/PTB7:$PC_{71}BM/MoO_3$/Ag resulted in a PCE up to 8.9 %, while the devices using only ZnO or PEI buffer layer show relatively low PCEs, 6.99 and 7.49 %, respectively. The enhanced PCE mainly resulting from the increased J_{sc} and FF for devices with ZnO/PEI CBLs was attributed to the fast electron transport.

Ionic liquids (ILs) with entirely cation-anion combinations exhibits ionic conductivity, low volatility, high electrochemical and thermal stability that make them important candidates for interfacial layer in solar cells. For instance, the IL of 1-butyl-3-methylimidazolium tetrafluoroborate ([BMIM]BF_4) not only can serves as a CBL material but also can be used as a surface modification layer to modify ZnO buffer layer for inverted PSCs. The single-junction PTB7-Th:$PC_{71}BM$

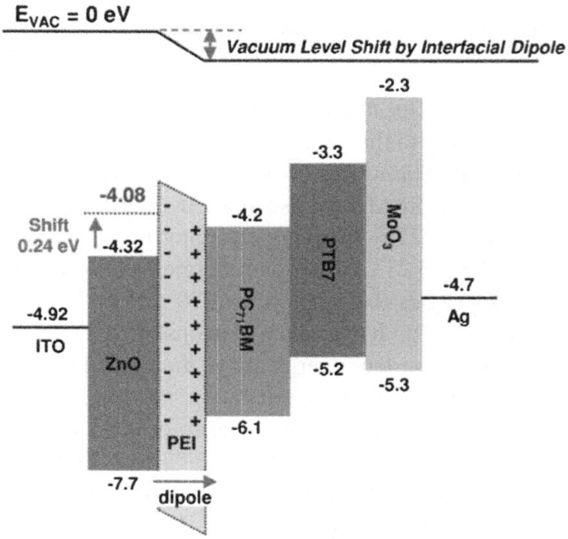

Fig. 3.22 Flat energy band diagram of the device with PEI-coated ZnO CBL where the conduction band energy of the ZnO CBL was lowered by 0.24 eV due to the dipole formation by the PEI nano-layer. Adapted with permission from [78]. Copyright 2014 Wiley-VCH

inverted PSCs using the solution processed ZnO/[BMIM]BF4 combination CBLs have achieved a remarkable PCE exceeding 10 % (10.15 %) for the single-junction PTB7-Th:$PC_{71}BM$ inverted PSCs [65]. The PTB7:$PC_{71}BM$ inverted PSCs with the ZnO/[BMIM]BF4 CBL achieved an average PCE pf 8.71 % (champion PCE of 9.12 %), which is 15.2 % (champion 21.4 %) increase compared to the control devices based on pure ZnO (PCE of 7.51 %). The simultaneously enhanced J_{sc}, FF and PCEs were attributed to the decreased interfacial energy barrier due to the decreased work function of the cathode by forming spontaneous dipolar polarization at the interface. They demonstrated that the IL layer and the ZnO/IL combination layers with low work function, good optical transmittance, improved electron extraction and reduced resistance at the cathode interface, make an excellent and general interfacial layers for inverted PSCs [65]. In addition, the [BzMIM]Cl or ZnO/[BzMIM]Cl combined CBL showed a similar improvement of the device performance to the inverted devices based on [BMIM]BF4 or ZnO/[BMIM]BF4 combined CBL [65].

A facial and general ethanedithiol (EDT) treatment method to passivate the surface defects and modulate the intragap states of ZnO CBLs fabricated with solution-processing at low-temperature has also been developed [220]. It has demonstrated that the covalent bonding of EDT molecules onto ZnO nanocrystals could effectively remove various surface defects of ZnO nanocrystal layers by forming zinc ethanedithiolates during the EDT treatment. Due to the chemical changes through EDT-passivation, the intragap states in pristine ZnO nanocrystals were modulated and a new intragap band was introduced (Fig. 3.23a and b), resulting in enhanced electron selectivity and improved electron transport properties to EDT-treated ZnO CBL. The inverted devices with EDT-treated ZnO CBL exhibited reduced charge recombination and enhanced charge extraction property. The EDT treatment was also found to result in significantly improved of the ambient stability of the inverted PSCs. For the unencapsulated P3HT:$PC_{61}BM$ devices with EDT-treated ZnO CBL, the PCE remained 90 % of its original value after being exposed to dark and ambient conditions for 30 days. In contrast, the PCE of control device with pure ZnO CBL degraded to around 60 % of the original value under the same conditions. The improved device stability was ascribed to the well-passivated EDT-treated ZnO films, which are less susceptible to oxygen and water molecules (as shown in Fig. 3.23c) [220].

Several kinds of metal oxides and salt, such as TiO_2 [117], Cs_2CO_3 [231], hafnium dioxide [130] and even ZnO [132, 232] itself also have potential to modify ZnO CBLs. For instance, an ultrathin TiO_2 layer with a mean thickness less than 3 nm on the surface of ZnO CBL could slightly enhance the photovoltaic performance of inverted PSCs, while a thicker TiO_2 film resulted in reduced photovoltaic performance [117]. Based on the photoelectron spectroscopy study, it was suggested that the ultrathin TiO_2 layer could quench the recombination of electrons and holes on the surface of ZnO CBL [117]. Cs_2CO_3 thin layer can also be used to modify the ZnO CBLs for high performance inverted PSCs [231]. The ZnO CBL coated with a solution-processed Cs_2CO_3 thin layer (ZnO/Cs_2CO_3 bilayer) presented enhanced electron extraction and leakage-current-suppression abilities due to

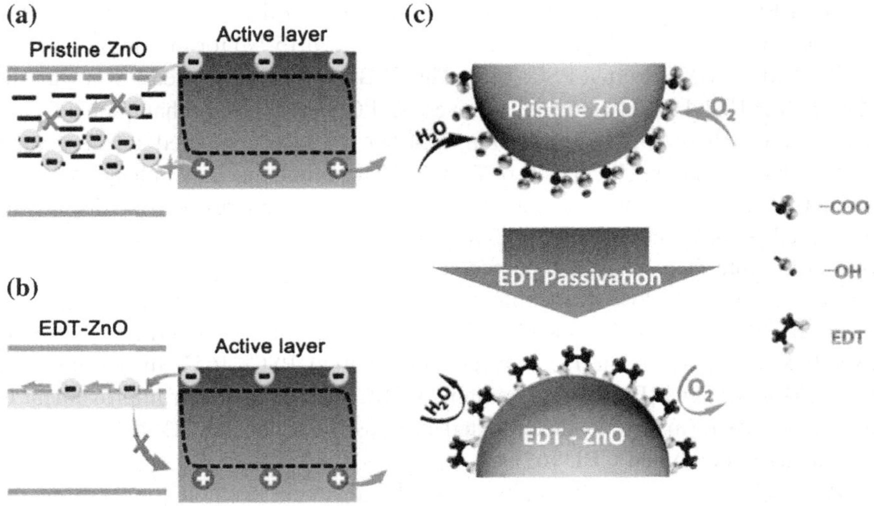

Fig. 3.23 **a** Various intragap states of pristine ZnO layers, which act as recombination centres for photogenerated charges. **b** The intragap states are modified to a new intergap band, which facilitates electron transport in CBLs, thereby suppressing the interfacial bimolecular recombination and enhancing the charge extraction properties of the devices. **c** Schematic view showing that the various surface groups are removed and EDT molecules are covalently bound onto ZnO. Adapted with permission from [220]. Copyright 2015 Wiley-VCH

its modified energy level and improved surface morphology. As a result, about 40 % enhancement in the PCE was obtained. An ultrathin ZnO layer can also be employed to modify ZnO CBLs for inverted PSCs. It has been reported that the wet-chemically prepared ripple-structured ZnO CBLs could be modified by depositing a very thin ZnO layer by ALD method. [132] A 1∼2 nm ALD-ZnO layer can effectively quench the electron-hole recombination caused by surface defects of ZnO nanoripples, and thus enhance the J_{sc} and PCE of inverted devices. Modifying the surface of ZnO CBL surface with a 0.6 nm ALD HfO_2 layer has also been used to enhance the electron injection and hole-blocking, leading to the PCE increasing from 4.1 to 4.5 % [130]. The R_{SH} of the devices increased significantly with the addition of the HfO_2 layer, while the R_S decreased. It was explained that the enhancement in the electron transport arose from the HfO_2 layer which passivated the ZnO CBL and reduced the defect states on ZnO surface, thereby suppressed the loss of electrons caused by charge recombination at the defect sites. It should be noted that the deteriorated performance of device with HfO_2 layer thicker than 0.6 nm indicates that the enhancement in electron transport is reduced due to the insulating effect of the HfO_2 layer, which became dominant in the case of thick HfO_2. Besides the modification of the interface between ZnO and active layer, the modification of the interface between ZnO and ITO electrode has also been investigated to achieve more efficient collection and transport of electrons from ZnO layer to ITO [239]. For instance, inserting a nanoscopic copper hexadecafluorophthalocyanine (F16CuPc) nanowire film between ZnO layer and ITO electrode of inverted PSCs can increase the performance of inverted PSCs by enhancing interfacial electron transport [239].

3.8 ZnO-Based Nanocomposites CBLs

Nanocomposites refer to materials consisting of no less than two phases and with one dispersed in another that is called matrix and forms a three-dimensional network. Recently, metal oxide and organic nanocomposites have attracted more and more research interests for their applications as electrode materials in energy conversion and storage system. In this section, a brief summary on the preparation and characterization of ZnO-based nanocomposites and their applications in inverted PSCs will be presented. For low-temperature solution-processed ZnO CBLs, the general presence of high density electron traps, such as absorbed oxygen, dangling bonds and surface groups, and poor spatial distribution of nanoparticles over a large area restrict their contribution to enhancing the performance of inverted PSCs [151, 153]. The development of uniform ZnO NPs films with low density defects is essential to achieve high performance of inverted PSCs. Polymer modified ZnO films, or known as the so-called ZnO/polymer nanocomposites or hybrids, have been studied to meet such challenges [151, 152, 240–243]. For example, the ZnO/polymer nanocomposites, such as ZnO/poly(ethylene oxide) (PEO) [152], ZnO/poly (ethylene glycol) PEG [241], ZnO/fullerene-end-capped poly(ethylene glycol) (C_{60}-PEG) [244], ZnO/Polyethylenimine (PEI) [243], ZnO/polyacrylic acid (PAA) [245], ZnO/poly(zinc diacrylate) (pZA) [240], ZnO/poly (diallyldimethylammonium chloride) (PDADMAC) [242] and ZnO/Polyvinylpyrrolidone (PVP) [151, 242] have been studied to serve as the efficient CBL in inverted PSCs to further improve the device performance. Figure 3.24 shows the cross-section schematic of the inverted device with ZnO-based nanocomposites CBLs. When compared with inverted PSCs based on a

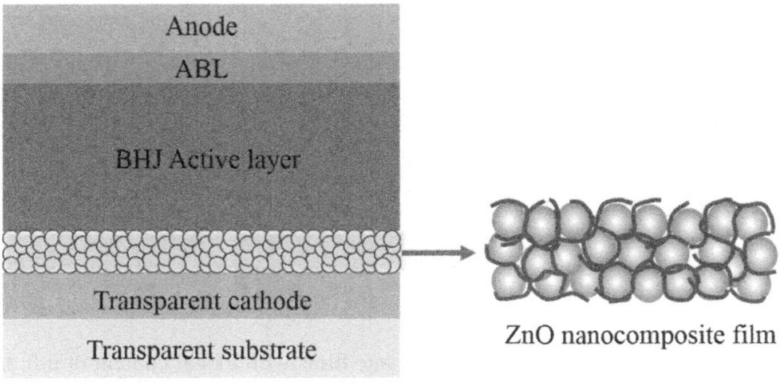

Fig. 3.24 Schematic illustration of the inverted PSC with a ZnO/polymer nanocomposite CBL

pristine ZnO thin layer, the PCEs of inverted devices using ZnO/polymer nanocomposite films as the CBLs can be increased by approximately 10–43 %, depending on both the nature and contents of the incorporated polymer, and the nature of BHJ active layer [151, 152, 240–243].

Poly(ethylene oxide) (PEO) modification to the ZnO NPs surface could effectively passivate the surface traps, suppress the recombination loss of carriers, reduce the R_S as well as improve the electrical coupling of ZnO/active layer in inverted PSCs [152]. For the inverted devices composed of TQ1: PCBM and PCDTBT: $PC_{71}BM$ active layers, the PCE of the devices using 0.05 % (PEO/ZnO wt%) PEO-modified ZnO NPs layer increased to 5.4 and 6.59 % from 4.5 % and 5.39 % for the reference devices based on unmodified ZnO. However, when the PEO content was increased to 1 wt%, the device performance decreased due to the very rough surface of ZnO-PEO films. Compared to pristine ZnO nano-films, the solution processed ZnO/PEI nanocomposite film can also facilitates electron extraction in inverted PSCs [243]. The energy bands of ZnO/PEI composite could be tuned considerably by varying the content of PEI, and thus forming a good electron transport layer. In addition, the content of PEI can also affect the morphology of ZnO/PEI nanocomposite films due to the change in the surface properties of ZnO NPs after blending with PEI. Figure 3.25 shows AFM surface topography images of the ZnO/PEI nanocomposite films containing amounts of PEI vary from 0 to 9 wt%. When the contents of PEI vary from 0 to 7 %, the presence of PEI could decrease the surface roughness with increasing PEI amounts due to the

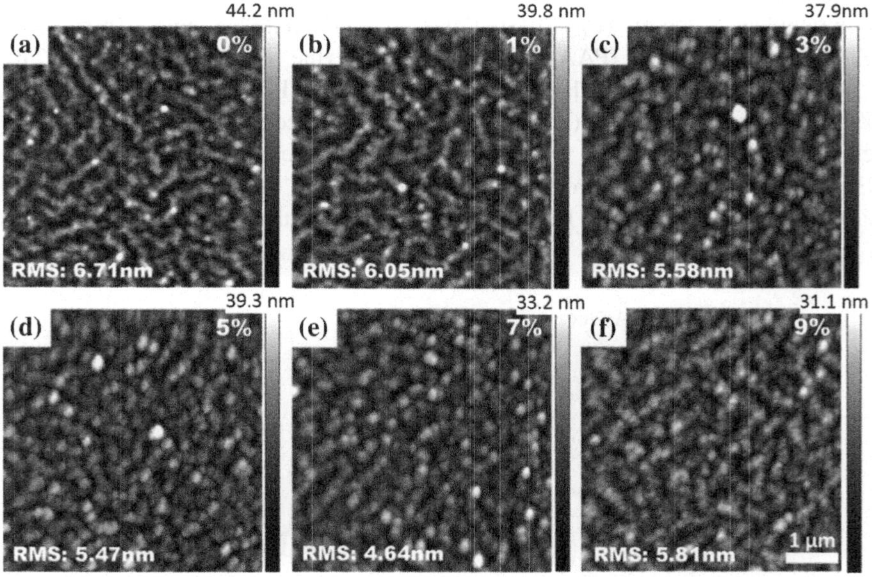

Fig. 3.25 AFM images of ZnO/PEI nanocomposite films with PEI at content of **a** 0, **b** 1, **c** 3, **d** 5, **e** 7, and **f** 9 wt%. Adapted with permission from [243]. Copyright 2015 American Chemical Society

change in the surface properties of the ZnO NPs after blending with PEI. However, furtherly increase the PEI content up to 9 % will result in an increase in the surface roughness of the corresponding ZnO/PEI film. A smoother ZnO/PEI nanocomposite film means fewer surface traps, thereby decreasing the possibility of trap-assisted recombination of electron, leading to a high FF of solar cells. Furthermore, by adjusting the content of PEI, the structural order of ZnO in the ZnO/PEI films was adjusted to align perpendicular to the ITO electrode, facilitating the electron transport vertically in ZnO/PEI nanocomposite film [243].

For ZnO/polymer nanocomposite CBL, besides the polymer content, the molecular weight of polymer also should be tailored to achieve desired interface and electrical properties of ZnO/polymer nanocomposites CBLs as well as the photovoltaic performance of the inverted PSCs [152, 241]. In ZnO/PEG hybrid CBLs, the PEG can passivate the surface traps of ZnO NPs and thus suppress the interfacial charge recombination, decrease the work function and improve the energy level alignment between the $PC_{61}BM$ and ZnO [241]. It has been suggested that the PEG chains wrapped around the ZnO surface could share their lone electron pairs of oxygen in the backbone with ZnO NPs, and thus passivate the shallow surface traps of ZnO. Compared to the devices with ZnO/PEG400 CBL and ZnO/PEG20000 CBL (with average molecular weight of 400 and 20000, respectively), the devices using ZnO/PEG6000 (with averaged molecular weight of 6000) hybrids exhibited the best performance (PCE = 3.3 %). On one hand, for ZnO/PEG400 CBL, the short backbone of PEG (e.g., MW = 400) containing less oxygen which could not effectively passivate ZnO surface traps. One the other hand, for ZnO/PEG20000 CBL, the PEG with too long backbone could lead to the formation of a charge transport barrier because of the insulating nature of PEG with long backbone [241]. In addition, PEG can also affect the morphology of ZnO/PEG hybrid CBLs. As shown in Fig. 3.26, the pristine ZnO NPs are aggregated together forming large aggregates, while the PEG stabilizes the dispersion of large ZnO nanocluster domains with fewer boundaries [241]. Subsequently, the ZnO/PEG nano-films contain less domain boundaries and smoother surface (RMS 8.17 nm) than the pristine ZnO films (RMS 10.47). The low surface roughness of ZnO CBL benefits to reduce the possibility of trap-assisted recombination of electron, leading to a high FF of solar cells. In addition, the ZnO nanocluster containing fewer boundaries benefit to enhancing electron transport and decreasing carrier trapping in the trap sites at boundaries [241, 246, 247]. It should be noted that the morphology of ZnO/PEG hybrid films is suspected to be also depend on the molecular weight of PEG. An appropriate length of PEG backbone, such as $M_w = 6000$, produce a uniform clusters and small domains, while a short PEG backbone, such as $M_w = 400$, is not long enough to wrap around the ZnO NPs surface to form uniform nanoclusters with suitable size. For a too long PEG, such as $M_w = 20000$, the crimp long insulating backbone of PEG result in a barrier of the charge transport.

An amphiphilic fullerene/ZnO hybrids (ZnO@C_{60}-PEG) obtained by in situ grown ZnO from amphiphilic fullerene-end-capped poly(ethylene glycol) (C_{60}-PEG) at relatively low temperatures also can be employed to fabricate CBLs and to improve charge selectivity of inverted PSCs [244]. The C_{60}-PEG could act as an *n*-

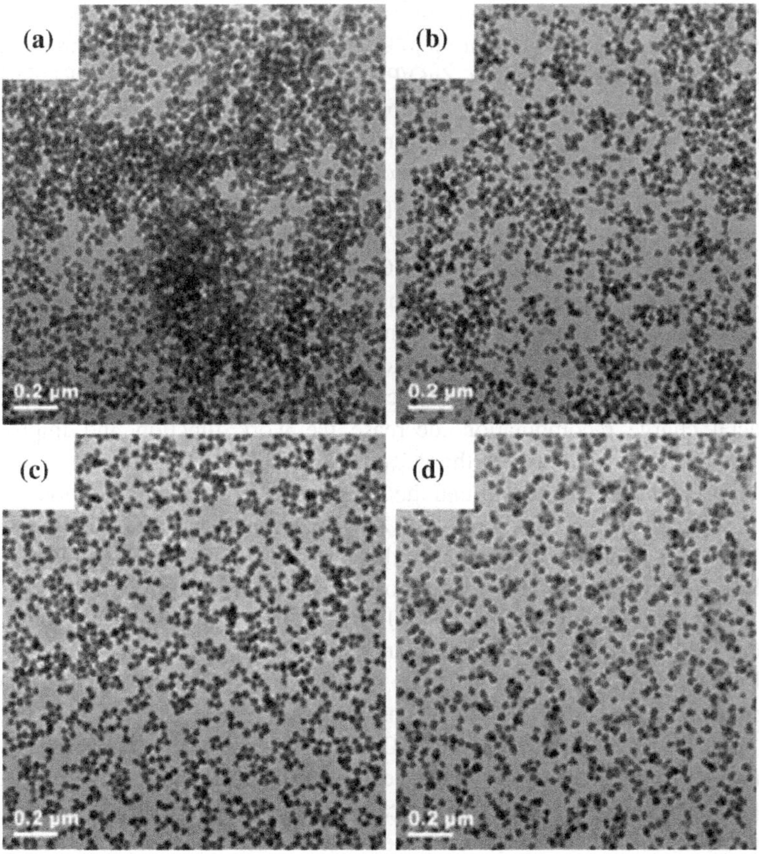

Fig. 3.26 TEM images of **a** ZnO, **b** ZnO/PEG400, **c** ZnO/PEG6000, and **d** ZnO/PEG20000. Adapted with permission from [241]. Copyright 2013 American Chemical Society

dopant to the ZnO while the oxygen-rich PEG side chain of C_{60}-PEG could passivate the defects of the ZnO. In addition, the amphiphilic C_{60}-PEG could facilitate the compatibility of the inorganic ZnO CBL and organic active layer [244]. Compared to the pure ZnO, the ZnO@C_{60}-PEG possessed a decreased work function, less defects, higher electron mobility, which could effectively reduce the recombination of carriers and enhance electron extraction. Consequently, the inverted PSC based on ZnO@C_{60}-PEG CBL and with PTB7:$PC_{71}BM$ active layer reached a high PCE of 8.0 %. Due to the C_{60}-PEG modified ZnO CBL with fewer defects, the inverted PSCs based on ZnO@C_{60}-PEG hybrid CBLs exhibited a better stability than the devices with pristine ZnO CBLs.

The poly(zinc diacrylate) (pZA) can also employee to passivate the defects in ZnO CBL [240]. By in situ cross-linking pZA in ZnO films with a 3D network, a pZA cross-linked ZnO composite films can be prepared. It has been demonstrated that the pZA in pZA cross-linked ZnO composite CBL can decrease the trap-assisted recombination rather than giving better energy alignment between the active layer and the ZnO interlayer [240]. Based on the pZA cross-linked ZnO CBL derived from ZnO NPs solution containing 0.5 wt% zinc diacrylate (ZDA), the inverted PSCs composed of respective P3HT:PCBM and PTB7:$PC_{71}BM$ active layer achieved PCE of 3.26 and 7.37 %, respectively, while the PCEs for the reference devices without pZA were 2.58 and 6.67 % for P3HT:PCBM and PTB7:$PC_{71}BM$ active layers, respectively. The improvement in the device performance was attributed to the enhanced charge generation and transport. In addition, the cross-linked pZA network probable acts as a template during the ZnO NPs distributing process. After the ZnO/pZA network is annealed at 150 °C for 30 min in a nitrogen-filled glovebox, the obtained ZnO/pZA CBL exhibits an ideally dense and homogenous surface morphology, especially for the 0.5 % case [240]. A dense and homogeneous morphology of CBL is preferred to enhance the contact between the buffer layer and the BHJ active layer as well as beneficial for hole blocking and electron collecting.

Several other polyelectrolytes, such as poly (diallyldimethylammonium chloride) (PDADMAC), Poly (acrylic acid sodium salt) (PAS), poly (4-styrenesulfonic acid) (PSS), and Polyvinylpyrrolidone (PVP) have potential for fabrication of ZnO/polymer nanocomposite CBLs for inverted PSCs [242]. Among these polyelectrolytes, the combination of ZnO with PVP and PDADMAC gave the highest improvement in the PCEs. Compared to the pristine ZnO films, the ZnO-PVP nanocomposite films exhibited more uniform distribution of the ZnO nano-clusters in polymer matrix in view of suppressing ZnO aggregation [151]. It has been demonstrated that the PVP can acted as an organic capping molecule and polymeric matrix, promoting the formation of ZnO electron-transporting nanocomposite films with uniform morphology. The inverted devices with UV-ozone treated (10 min) ZnO-PVP nanocomposite CBL and PDTG–TPD:$PC_{71}BM$ active layer showed a PCE of 8.1 % with a J_{sc} of 14.0 mA cm^{-2}, a V_{oc} of 0.86 V, and a FF of 67.3 %. Besides serving as a matrix for suppressing ZnO aggregation and passivating the surface traps in ZnO films, the ZnO/polymer composite exhibit adjustable energy bands [243].

It is worth noting that, a crucial issue for ZnO-polymer composite or organic-stabilized ZnO NP CBLs is the removal of polymer or organic surfactant so as to improve the electrical properties of ZnO CBL as well as the device performance. This is because the insulating polymer or organic surfactant will aggravate the charge transport by introducing additional resistance [151, 153]. Ultraviolet-ozone (UVO) treatment is regarded as a promising approach to remove polymer from ZnO surface and to ensure a good electrical coupling between ZnO and active layer in view of the act that UVO treatment did not alter the size, shape or distribution of ZnO NPs. Through an UV induced decomposition mechanism, UVO treatment could effectively remove the residual organic stabilizer molecules on the surface of ZnO NPs films [153]. Small et al. investigated the UVO treatment duration on the performance of inverted solar cells with ZnO-PVP composite film as CBL and found that ZnO-PVP nanocomposite films with UVO treated for 10 min led to the highest enhancement in the device performance compared to the

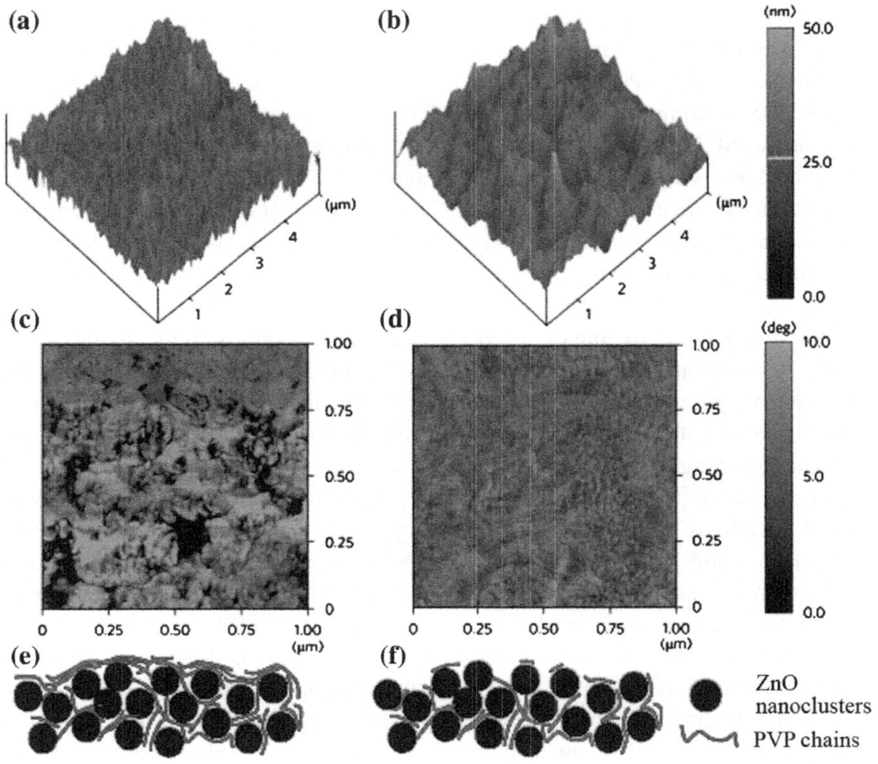

Fig. 3.27 Three-dimensional topography AFM images of **a** as-prepared and **b** 10 min UVO treated ZnO-PVP nanocomposite films. Phase images of **c** as-prepared and **d** 10 min UVO treated ZnO-PVP nanocomposite films. Schematic illustrations of **e** as-prepared and **f** 10 min UVO treated ZnO-PVP nanocomposite films. Adapted with permission from [151]. Copyright 2011 Nature Publishing Group

devices without UVO treatment [151]. It has been found that the UVO treatment removed the top PVP enabling the ZnO NPs to expose at the film surface, which was proven by both AFM (Fig. 3.27a–d) and X-ray photoelectron spectroscopy (XPS) characterizations. The enhanced device performance with UVO treated ZNO-PVP nanocomposite films was ascribed to the modified surface composition promoting charge collection. The UVO treatment time is a crucial factor that determines the morphology and electrical properties of ZNO-PVP nanocomposite films. It has been demonstrated that a too short time treatment results in an incomplete removal of the PVP from the top surface of ZnO-PVP nanocomposite films, while a too long time treatment may induces excess oxygen onto the ZnO surface, which will reduce the electron extraction efficiency [151]. Overall, polymer modified ZnO NPs/polymer hybrid or composite films are of more uniform and with low defects than the unmodified ZnO NPs films. By choosing suitable polymer or organic surfactant material, adjusting the amount and molecular weight of polymer, and employing an optimized UVO treatment process, one can expect to enhance the performance of ZnO/polymer composite CBL as well as the device performance.

The composites of ZnO with other oxides or titanates can also be used to fabricate the CBLs for inverted PSCs. Ta_2O_5-ZnO composite films with varied composition have been fabricated by sol–gel processing and applied as the CBLs in inverted PSCs, and demonstrated enhanced PCE with excellent stability in air [248]. The CBLs incorporated with Ta_2O_5 exert two competing impacts on the photovoltaic performance of solar cells. On one hand, the presence of Ta_2O_5 is likely to induce more positive charges around the Zn atom and form Ta-O-Zn bonding; it can reduce the interface charge recombination between the active layer and the Ta_2O_5-ZnO composite CBL, and result in high PCE; on the other hand, an excessive amount of Ta_2O_5 would block the pathways of charge transport and lead to a drastic reduction in PCE. The integration of 12 and 18 % Ta_2O_5 into ZnO resulted in the efficiency enhanced maximally for P3HT:$PC_{61}BM$ and PSEHTT:$PC_{71}BM$ active layers, respectively; the addition of excessive Ta_2O_5, such as exceeds 24 %, would however lead to a dramatic decrease in the device performance. A dual phase $SrTiO_3$:ZnO nanocomposite films with varied composition ratios fabricated by sol-gel processing have also been applied as the CBLs in inverted PSCs, and demonstrated enhanced devices performance [89]. The localized polar molecules $SrTiO_3$ in the $SrTiO_3$:ZnO nanocomposite CBL are in random orientations in the dark condition or open circuit voltage condition (Fig. 3.28a), while a self-built electric field on the interface between the active layer and CBL was generated by electron and hole transfer to the opposite direction and there is a net interface dipole directed away from the CBL under illumination (Fig. 3.28b). Such a spontaneous polarization is likely to induce a self-built electric field to prevent electron recombination on the interface of the active layer and CBL, which shows an enhancement of FF of device performance. An enhancement in PCE from

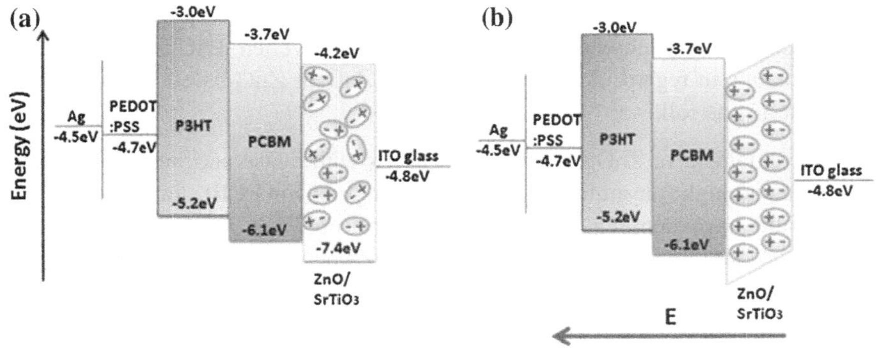

Fig. 3.28 Schematic energy level diagrams of inverted PSCs with $SrTiO_3$: ZnO composite CBLs. **a** The inverted device is under open circuit condition, and there is no net interface dipole. **b** The inverted device is under illumination, electron field was generated by electron and hole transfer to the opposite direction and there is a net interface dipole directed away from the CBL. Adapted with permission from [89]. Copyright 2014 Elsevier

3.58 to 4.1 % was achieved for the device with $SrTiO_3$:ZnO of 90:10 CBL and P3HT:PCBM active layer [89].

3.9 Conclusion and Outlook

The BHJ PSCs is a complex device, so far high PCE can only be achieved when all the following requirements are met: (1) good *p*-type polymer donor and n-type fullerene derivative acceptor with desired electronic properties, (2) BHJ active layer with desired micro and nanostructures, and (3) appropriate interfaces between different layers for efficient charge collection and transport. The PSCs with inverted structure seem to be the best candidate that meets the requirements for both high efficiency and good stability. To achieve high PCE in inverted PSCs, the selection of an appropriate nanostructured CBL so as to optimize the electron collection and transport is particularly important. The key requirements for an ideal CBL include being good at electron extraction and transport, having suitable energy level that facilitates electron transport, and having high transparency, good compactness, and high stability. Nanostructured ZnO CBLs have been the most extensively studied CBLs for inverted PSCs in view of its relatively high electron mobility, optical transparency, ease of being synthesized with low cost solution methods at low temperature, versatile morphologies, and being environmentally stable. The contact quality between ZnO-based CBLs and the BHJ active layer can significantly affect the performance of inverted PSCs. The contact quality between ZnO-based CBLs and the BHJ active layer relies on both the preparation method and the post-treatment of nanostructured ZnO CBLs, which can influence the properties of ZnO CBL in terms of morphology, thickness and transmittance, and surface state. Doping and surface modification are effective ways to tailor the electronic, optical and chemical properties of the ZnO CBL and thus to improve the photovoltaic performance of the inverted PSCs.

To achieve high photovoltaic performance in inverted PSCs, some specific considerations in regard of the nanostructured ZnO or ZnO-based nanocomposite CBLs are list as follows:

(1) Development of ZnO or ZnO-based CBLs with dense packing, large surface area and high transmittance. The morphology of ZnO CBL for inverted PSCs should have dense and homogenous surface, being able to provide large surface area, and allowing the polymers to well contact with ZnO layer. The dense and homogenous surface of ZnO CBLs favors the formation of an intimate contact between ZnO CBL and BHJ photoactive layer, and thus lowers the contact resistance and decrease the series resistance of the device. In addition, the dense surface also makes the ZnO layer having better electrical properties and higher transmittance. The large surface area of ZnO CBLs can enlarge ZnO/BHJ active layer contact area, thus providing larger electron

extraction interface and enabling more electrode to penetrate into the BHJ active layer so as to reduce charge transport distance.

(2) Doping ZnO for higher conductivity and appropriate energy level. An appropriate doping can enhance the electrical conductivity of ZnO and thus improve the photovoltaic performance of inverted PSCs. In addition, the tunable energy level of doped-ZnO CBLs can also facilitates the electron transport form active layer to CBL and can also serves as an efficient hole blocking layer.

(3) Development of interfacial materials for surface modification of ZnO-based CBLs with appropriate surface energy and improved interfacial properties. The surface modification of ZnO-based CBL helps to reduce the charge recombination and improve the charge transport at the interface by passivating the surface defects, tuning the surface energy and work function, and may thus lead to improved photovoltaic performance of the inverted PSCs.

(4) Development of ZnO-based nanocomposite CBLs with low density charge traps, dense and uniform surface, appropriate surface energy, and favourable electron transport way.

References

1. M.A. Green, Third generation photovoltaics: solar cells for 2020 and beyond. Physica E **14**, 65–70 (2002)
2. J. Huang, Z. Yin, Q. Zheng, Applications of ZnO in organic and hybrid solar cells. Energy Environ. Sci. **4**, 3861–3877 (2011)
3. F. He, L. Yu, How far can polymer solar cells go? In need of a synergistic approach. J. Phys. Chem. Lett. **2**, 3102–3113 (2011)
4. Y.J. Cheng, S.H. Yang, C.S. Hsu, Synthesis of conjugated polymers for organic solar cell applications. Chem. Rev. **109**, 5868–5923 (2009)
5. G. Li, V. Shrotriya, J. Huang, Y. Yao, T. Moriarty, K. Emery, Y. Yang, High-efficiency solution processable polymer photovoltaic cells by self-organization of polymer blends. Nat. Mater. **4**, 864–868 (2005)
6. S.K. Hau, H.-L. Yip, A.K.Y. Jen, A review on the development of the inverted polymer solar cell architecture. Polymer Rev. **50**, 474–510 (2010)
7. R. Po, C. Carbonera, A. Bernardi, N. Camaioni, The role of buffer layers in polymer solar cells. Energy Environ. Sci. **4**, 285–310 (2011)
8. G. Li, R. Zhu, Y. Yang, Polymer solar cells. Nat. Photon. **6**, 153–161 (2012)
9. S.B. Darling, F. You, The case for organic photovoltaics. RSC Adv. **3**, 17633–17648 (2013)
10. S. Lizin, S. Van Passel, E. De Schepper, W. Maes, L. Lutsen, J. Manca, D. Vanderzande, Life cycle analyses of organic photovoltaics: a review. Energy Environ. Sci. **6**, 3136–3149 (2013)
11. M. Jørgensen, K. Norrman, S.A. Gevorgyan, T. Tromholt, B. Andreasen, F.C. Krebs, Stability of polymer solar cells. Adv. Mater. **24**, 580–612 (2012)
12. R.R. Søndergaard, M. Hösel, F.C. Krebs, Roll-to-roll fabrication of large area functional organic materials. J. Polym. Sci. Part B: Polym. Phys. **51**, 16–34 (2013)
13. F. Zhang, X. Xu, W. Tang, J. Zhang, Z. Zhuo, J. Wang, J. Wang, Z. Xu, Y. Wang, Recent development of the inverted configuration organic solar cells. Solar Energy Mater. Solar Cells **95**, 1785–1799 (2011)

14. Z. Liang, Q. Zhang, O. Wiranwetchayan, J. Xi, Z. Yang, K. Park, C. Li, G. Cao, Effects of the morphology of a ZnO buffer layer on the photovoltaic performance of inverted polymer solar cells. Adv. Funct. Mater. **22**, 2194–2201 (2012)
15. N. Sekine, C.-H. Chou, W.L. Kwan, Y. Yang, ZnO nano-ridge structure and its application in inverted polymer solar cell. Org. Electron. **10**, 1473–1477 (2009)
16. Y.-C. Ho, P.-Y. Ho, H.-C. Lee, S.-K. Chang, Y.-R. Hong, C.-F. Lin, Enhancing performance of inverted polymer solar cells using two-growth ZnO nanorods. Solar Energy Mater. Solar Cells **132**, 570–577 (2015)
17. Z. Liang, R. Gao, J.-L. Lan, O. Wiranwetchayan, Q. Zhang, C. Li, G. Cao, Growth of vertically aligned ZnO nanowalls for inverted polymer solar cells. Solar Energy Mater. Solar Cells **117**, 34–40 (2013)
18. S. Tomoki, U. Tokiyoshi, H. Yuuki, F. Akihiko, Y. Katsumi, Effect of ZnO layer on characteristics of conducting polymer/C_{60} photovoltaic cell. J. Phys. D Appl. Phys. **37**, 847–850 (2004)
19. G. Yu, J. Gao, J.C. Hummelen, F. Wudl, A.J. Heeger, Polymer photovoltaic cells—Enhanced efficiencies via a network of internal donor-acceptor heterojunctions. Science **270**, 1789–1791 (1995)
20. H. Zeng, X. Zhu, Y. Liang, X. Guo, Interfacial layer engineering for performance enhancement in polymer solar cells. Polymers **7**, 333–372 (2015)
21. X. Yang, J. Loos, Toward high-performance polymer solar cells: The importance of morphology control. Macromolecules **40**, 1353–1362 (2007)
22. L.-M. Chen, Z. Hong, G. Li, Y. Yang, Recent progress in polymer solar cells: Manipulation of polymer: fullerene morphology and the formation of efficient inverted polymer solar cells. Adv. Mater. **21**, 1434–1449 (2009)
23. W. Chen, T. Xu, F. He, W. Wang, C. Wang, J. Strzalka, Y. Liu, J. Wen, D.J. Miller, J. Chen, K. Hong, L. Yu, S.B. Darling, Hierarchical nanomorphologies promote exciton dissociation in polymer/fullerene bulk heterojunction solar cells. Nano Lett. **11**, 3707–3713 (2011)
24. M. Campoy-Quiles, T. Ferenczi, T. Agostinelli, P.G. Etchegoin, Y. Kim, T.D. Anthopoulos, P.N. Stavrinou, D.D.C. Bradley, J. Nelson, Morphology evolution via self-organization and lateral and vertical diffusion in polymer: fullerene solar cell blends. Nat. Mater. **7**, 158–164 (2008)
25. S.S.V. Bavel, E. Sourty, G.D. With, J. Loos, Three-dimensional nanoscale organization of bulk heterojunction polymer solar cells. Nano Lett. **9**, 507–513 (2009)
26. G. Li, Y. Yao, H. Yang, V. Shrotriya, G. Yang, Y. Yang, "Solvent annealing" effect in polymer solar cells based on poly(3-hexylthiophene) and methanofullerenes. Adv. Funct. Mater. **17**, 1636–1644 (2007)
27. X. Yang, J. Loos, S.C. Veenstra, W.J.H. Verhees, M.M. Wienk, J.M. Kroon, M.A.J. Michels, R.A.J. Janssen, Nanoscale morphology of high-performance polymer solar cells. Nano Lett. **5**, 579–583 (2005)
28. S. Zhang, L. Ye, W. Zhao, D. Liu, H. Yao, J. Hou, Side chain selection for designing highly efficient photovoltaic polymers with 2D-conjugated structure. Macromolecules **47**, 4653–4659 (2014)
29. L. Ye, S. Zhang, W. Zhao, H. Yao, J. Hou, Highly efficient 2D-conjugated benzodithiophene-based photovoltaic polymer with linear alkylthio side chain. Chem. Mater. **26**, 3603–3605 (2014)
30. T.L. Nguyen, H. Choi, S.J. Ko, M.A. Uddin, B. Walker, S. Yum, J.E. Jeong, M.H. Yun, T. J. Shin, S. Hwang, J.Y. Kim, H.Y. Woo, Semi-crystalline photovoltaic polymers with efficiency exceeding 9 % in a [similar] 300 nm thick conventional single-cell device. Energy Environ. Sci. **7**, 3040–3051 (2014)
31. J. You, L. Dou, K. Yoshimura, T. Kato, K. Ohya, T. Moriarty, K. Emery, C.-C. Chen, J. Gao, G. Li, Y. Yang, A polymer tandem solar cell with 10.6 % power conversion efficiency. Nat. Commun. **4**, 1–10 (2013)
32. R. Steim, F.R. Kogler, C.J. Brabec, Interface materials for organic solar cells. J. Mater. Chem. **20**, 2499–2512 (2010)

33. R. Po, M. Maggini, N. Camaioni, Polymer solar cells: recent approaches and achievements. J. Phys. Chem. C **114**, 695–706 (2010)
34. L.-M. Chen, Z. Xu, Z. Hong, Y. Yang, Interface investigation and engineering—achieving high performance polymer photovoltaic devices. J. Mater. Chem. **20**, 2575–2598 (2010)
35. W. Cai, X. Gong, Y. Cao, Polymer solar cells: Recent development and possible routes for improvement in the performance. Solar Energy Mater. Solar Cells **94**, 114–127 (2010)
36. H.-L. Yip, A.K.Y. Jen, Recent advances in solution-processed interfacial materials for efficient and stable polymer solar cells. Energy Environ. Sci. **5**, 5994–6011 (2012)
37. Y.-Y. Lai, Y.-J. Cheng, C.-S. Hsu, Applications of functional fullerene materials in polymer solar cells. Energy Environ. Sci. **7**, 1866–1883 (2014)
38. H. Cao, W. He, Y. Mao, X. Lin, K. Ishikawa, J.H. Dickerson, W.P. Hess, Recent progress in degradation and stabilization of organic solar cells. J. Power Sour. **264**, 168–183 (2014)
39. F. Wang, Z.A. Tan, Y. Li, Solution-processable metal oxides/chelates as electrode buffer layers for efficient and stable polymer solar cells. Energy Environ. Sci. **8**, 1059–1091 (2015)
40. I. Etxebarria, J. Ajuria, R. Pacios, Solution-processable polymeric solar cells: a review on materials, strategies and cell architectures to overcome 10 %. Org. Electron. **19**, 34–60 (2015)
41. C.-C. Chueh, C.-Z. Li, A.K.Y. Jen, Recent progress and perspective in solution-processed interfacial materials for efficient and stable polymer and organometal perovskite solar cells. Energy Environ. Sci. **8**, 1160–1189 (2015)
42. K. Kawano, R. Pacios, D. Poplavskyy, J. Nelson, D.D.C. Bradley, J.R. Durrant, Degradation of organic solar cells due to air exposure. Solar Energy Mater. Solar Cells **90**, 3520–3530 (2006)
43. M.P. de Jong, L.J. van Ijzendoorn, M.J.A. de Voigt, Stability of the interface between indium-tin-oxide and poly(3,4-ethylenedioxythiophene)/poly(styrenesulfonate) in polymer light-emitting diodes. Appl. Phys. Lett. **77**, 2255–2257 (2000)
44. K.W. Wong, H.L. Yip, Y. Luo, K.Y. Wong, W.M. Lau, K.H. Low, H.F. Chow, Z.Q. Gao, W.L. Yeung, C.C. Chang, Blocking reactions between indium-tin oxide and poly (3,4-ethylene dioxythiophene):poly(styrene sulphonate) with a self-assembly monolayer. Appl. Phys. Lett. **80**, 2788–2790 (2002)
45. Y. Sun, J.H. Seo, C.J. Takacs, J. Seifter, A.J. Heeger, Inverted polymer solar cells integrated with a low-temperature-annealed sol-gel-derived ZnO film as an electron transport layer. Adv. Mater. **23**, 1679–1683 (2011)
46. M.S. White, D.C. Olson, S.E. Shaheen, N. Kopidakis, D.S. Ginley, Inverted bulk-heterojunction organic photovoltaic device using a solution-derived ZnO underlayer. Appl. Phys. Lett. **89**, 143517–143513 (2006)
47. S.K. Hau, H.-L. Yip, N.S. Baek, J. Zou, K. O'Malley, A.K.Y. Jen, Air-stable inverted flexible polymer solar cells using zinc oxide nanoparticles as an electron selective layer. Appl. Phys. Lett. **92**, 253301–253303 (2008)
48. T. Yang, W. Cai, D. Qin, E. Wang, L. Lan, X. Gong, J. Peng, Y. Cao, Solution-processed zinc oxide thin film as a buffer layer for polymer solar cells with an inverted device structure. J. Phys. Chem. C **114**, 6849–6853 (2010)
49. C.-Y. Li, T.-C. Wen, T.-H. Lee, T.-F. Guo, J.-C.-A. Huang, Y.-C. Lin, Y.-J. Hsu, An inverted polymer photovoltaic cell with increased air stability obtained by employing novel hole/electron collecting layers. J. Mater. Chem. **19**, 1643–1647 (2009)
50. J.-H. Huang, H.-Y. Wei, K.-C. Huang, C.-L. Chen, R.-R. Wang, F.-C. Chen, K.-C. Ho, C.-W. Chu, Using a low temperature crystallization process to prepare anatase TiO_2 buffer layers for air-stable inverted polymer solar cells. Energy Environ. Sci. **3**, 654–658 (2010)
51. Z. Yin, Q. Zheng, S.-C. Chen, D. Cai, Interface control of semiconducting metal oxide layers for efficient and stable inverted polymer solar cells with open-circuit voltages over 1.0 volt. ACS Appl. Mater. Interfaces **5**, 9015–9025 (2013)
52. C. Girotto, B.P. Rand, S. Steudel, J. Genoe, P. Heremans, Nanoparticle-based, spray-coated silver top contacts for efficient polymer solar cells. Org. Electron. **10**, 735–740 (2009)

53. F.C. Krebs, All solution roll-to-roll processed polymer solar cells free from indium-tin-oxide and vacuum coating steps. Org. Electron. **10**, 761–768 (2009)
54. M.R. Lilliedal, A.J. Medford, M.V. Madsen, K. Norrman, F.C. Krebs, The effect of post-processing treatments on inflection points in current–voltage curves of roll-to-roll processed polymer photovoltaics. Solar Energy Mater. Solar Cells **94**, 2018–2031 (2010)
55. Z. Xu, L.-M. Chen, G. Yang, C.-H. Huang, J. Hou, Y. Wu, G. Li, C.-S. Hsu, Y. Yang, Vertical phase separation in poly(3-hexylthiophene): Fullerene derivative blends and its advantage for inverted structure solar cells. Adv. Funct. Mater. **19**, 1227–1234 (2009)
56. S. Schumann, R. Da Campo, B. Illy, A.C. Cruickshank, M.A. McLachlan, M.P. Ryan, D. J. Riley, D.W. McComb, T.S. Jones, Inverted organic photovoltaic devices with high efficiency and stability based on metal oxide charge extraction layers. J. Mater. Chem. **21**, 2381–2386 (2011)
57. Z. He, C. Zhong, S. Su, M. Xu, H. Wu, Y. Cao, Enhanced power-conversion efficiency in polymer solar cells using an inverted device structure. Nat. Photonics **6**, 591–595 (2012)
58. J. Kong, I.W. Hwang, K. Lee, Top-down approach for nanophase reconstruction in bulk heterojunction solar cells. Adv. Mater. **26**, 6275–6283 (2014)
59. L.K. Jagadamma, M. Al-Senani, A. El-Labban, I. Gereige, G.O. Ngongang Ndjawa, J.C.D. Faria, T. Kim, K. Zhao, F. Cruciani, D.H. Anjum, M.A. McLachlan, P.M. Beaujuge, A. Amassian, Polymer solar cells with efficiency >10 % enabled via a facile solution-processed Al-doped ZnO electron transporting layer. Adv. Energy Mater. **5**, 1500204 (2015)
60. Z. Yin, Q. Zheng, S.-C. Chen, D. Cai, L. Zhou, J. Zhang, Bandgap tunable Zn1-xMgxO thin films as highly transparent cathode buffer layers for high-performance inverted polymer solar cells. Adv. Energy Mater. **4**, 1301404 (2014)
61. S.H. Liao, H.J. Jhuo, P.N. Yeh, Y.S. Cheng, Y.L. Li, Y.H. Lee, S. Sharma, S.A. Chen, Single junction inverted polymer solar cell reaching power conversion efficiency 10.31 % by employing dual-doped zinc oxide nano-film as cathode interlayer. Sci. Rep. **4**, 6813 (2014)
62. P. Li, X. Li, C. Sun, G. Wang, J. Li, T. Jiu, J. Fang, Performance enhancement of inverted polymer solar cells with fullerene ester derivant-modified ZnO film as cathode buffer layer. Solar Energy Mater. Solar Cells **126**, 36–41 (2014)
63. Z. He, B. Xiao, F. Liu, H. Wu, Y. Yang, S. Xiao, C. Wang, T.P. Russell, Y. Cao, Single-junction polymer solar cells with high efficiency and photovoltage. Nat. Photonics **9**, 174–179 (2015)
64. Y. Liu, J. Zhao, Z. Li, C. Mu, W. Ma, H. Hu, K. Jiang, H. Lin, H. Ade, H. Yan, Aggregation and morphology control enables multiple cases of high-efficiency polymer solar cells. Nat. Commun. **5**, 5293 (2014)
65. W. Yu, L. Huang, D. Yang, P. Fu, L. Zhou, J. Zhang, C. Li, Efficiency exceeding 10 % for inverted polymer solar cells with a ZnO/ionic liquid combined cathode interfacial layer. J. Mater. Chem. A **3**, 10660–10665 (2015)
66. X. Ouyang, R. Peng, L. Ai, X. Zhang, Z. Ge, Efficient polymer solar cells employing a non-conjugated small-molecule electrolyte. Nat. Photonics **9**, 520–524 (2015)
67. A.K.K. Kyaw, X.W. Sun, C.Y. Jiang, G.Q. Lo, D.W. Zhao, D.L. Kwong, An inverted organic solar cell employing a sol-gel derived ZnO electron selective layer and thermal evaporated MoO[sub 3] hole selective layer. Appl. Phys. Lett. **93**, 221107–221103 (2008)
68. J.-P. Liu, K.-L. Choy, X.-H. Hou, Charge transport in flexible solar cells based on conjugated polymer and ZnO nanoparticulate thin films. J. Mater. Chem. **21**, 1966–1969 (2011)
69. T.Z. Oo, R.D. Chandra, N. Yantara, R.R. Prabhakar, L.H. Wong, N. Mathews, S.G. Mhaisalkar, Zinc tin oxide (ZTO) electron transporting buffer layer in inverted organic solar cell. Org. Electron. **13**, 870–874 (2012)
70. S.K. Hau, H.-L. Yip, O. Acton, N.S. Baek, H. Ma, A.K.Y. Jen, Interfacial modification to improve inverted polymer solar cells. J. Mater. Chem. **18**, 5113–5119 (2008)
71. T. Kuwabara, H. Sugiyama, T. Yamaguchi, K. Takahashi, Inverted type bulk-heterojunction organic solar cell using electrodeposited titanium oxide thin films as electron collector electrode. Thin Solid Films **517**, 3766–3769 (2009)

72. I. Sasajima, S. Uesaka, T. Kuwabara, T. Yamaguchi, K. Takahashi, Flexible inverted polymer solar cells containing an amorphous titanium oxide electron collection electrode. Org. Electron. **12**, 113–118 (2011)
73. Y.-J. Kang, C.S. Kim, D.S. You, S.H. Jung, K. Lim, D.-G. Kim, J.-K. Kim, S.H. Kim, Y.-R. Shin, S.-H. Kwon, J.-W. Kang, Effect of electron transport layer crystallinity on the transient characteristics of inverted organic solar cells. Appl. Phys. Lett. **99**, 073308–073303 (2011)
74. Y. Zhou, H. Cheun, J.W.J. Potscavage, C. Fuentes-Hernandez, S.-J. Kim, B. Kippelen, Inverted organic solar cells with ITO electrodes modified with an ultrathin Al_2O_3 buffer layer deposited by atomic layer deposition. J. Mater. Chem. **20**, 6189–6194 (2010)
75. O. Wiranwetchayan, Z. Liang, Q. Zhang, G. Cao, P. Singjai, The role of oxide thin layer in inverted structure polymer solar cells. Mater. Sci. Applications **2**, 1697–1701 (2011)
76. Y. Zhu, X. Xu, L. Zhang, J. Chen, Y. Cao, High efficiency inverted polymeric bulk-heterojunction solar cells with hydrophilic conjugated polymers as cathode interlayer on ITO. Solar Energy Mater. Solar Cells **97**, 83–88 (2012)
77. M. Jin Tan, S. Zhong, R. Wang, Z. Zhang, V. Chellappan, W. Chen, Biopolymer as an electron selective layer for inverted polymer solar cells. Appl. Phys. Lett. **103**, 063303 (2013)
78. S. Woo, W. Hyun Kim, H. Kim, Y. Yi, H.-K. Lyu, Y. Kim, 8.9 % Single-stack inverted polymer solar cells with electron-rich polymer nanolayer-modified inorganic electron-collecting buffer layers. Adv. Energy Mater. **4**, 1301692 (2014)
79. R. Xia, D.-S. Leem, T. Kirchartz, S. Spencer, C. Murphy, Z. He, H. Wu, S. Su, Y. Cao, J.S. Kim, J.C. deMello, D.D.C. Bradley, J. Nelson, Investigation of a conjugated polyelectrolyte interlayer for inverted polymer:fullerene solar cells. Adv. Energy Mater. **3**, 718–723 (2013)
80. Z. Tang, W. Tress, Q. Bao, M.J. Jafari, J. Bergqvist, T. Ederth, M.R. Andersson, O. Inganäs, Improving cathodes with a polymer interlayer in reversed organic solar cells. Adv. Energy Mater. **4**, 1400643 (2014)
81. T. Kuwabara, M. Nakamoto, Y. Kawahara, T. Yamaguchi, K. Takahashi, Characterization of ZnS-layer-inserted bulk-heterojunction organic solar cells by ac impedance spectroscopy. J. Appl. Phys. **105**, 124513–124516 (2009)
82. J.-J. Zhu, Z.-Q. Xu, G.-Q. Fan, S.-T. Lee, Y.-Q. Li, J.-X. Tang, Inverted polymer solar cells with atomic layer deposited CdS film as an electron collection layer. Org. Electron. **12**, 2151–2158 (2011)
83. J.-J. Zhu, G.-Q. Fan, H.-X. Wei, Y.-Q. Li, S.-T. Lee, J.-X. Tang, Solution-processed inverted polymer solar cells using chemical bath deposited CdS films as electron collecting layer. CrystEngComm **14**, 8090–8096 (2012)
84. G. Li, C.W. Chu, V. Shrotriya, J. Huang, Y. Yang, Efficient inverted polymer solar cells. Appl. Phys. Lett. **88**, 253503–253503 (2006)
85. H.-H. Liao, L.-M. Chen, Z. Xu, G. Li, Y. Yang, Highly efficient inverted polymer solar cell by low temperature annealing of Cs_2CO_3 interlayer. Appl. Phys. Lett. **92**, 173303 (2008)
86. Y.-I. Lee, J.-H. Youn, M.-S. Ryu, J. Kim, H.-T. Moon, J. Jang, Electrical properties of inverted poly(3-hexylthiophene): Methano-fullerene [6,6]-phenyl C_{71}-butyric acid methyl ester bulk hetero-junction solar cell with Cs_2CO_3 and MoO_3 layers. Solar Energy Mater. Solar Cells **95**, 3276–3280 (2011)
87. H. Bin Yang, Y. Qian Dong, X. Wang, S. Yun Khoo, B. Liu, C. Ming Li, Graphene quantum dots-incorporated cathode buffer for improvement of inverted polymer solar cells. Solar Energy Mater. Solar Cells **117**, 214–218 (2013)
88. J.W. Kim, Y.H. Suh, C.L. Lee, Y.S. Kim, W.B. Kim, A nano-grid structure made of perovskite $SrTiO_3$ nanowires for efficient electron transport layers in inverted polymer solar cells. Nanoscale **7**, 4367–4371 (2015)
89. J.-L. Lan, Z. Liang, Y.-H. Yang, F.S. Ohuchi, S.A. Jenekhe, G. Cao, The effect of $SrTiO_3$: ZnO as cathodic buffer layer for inverted polymer solar cells. Nano Energy **4**, 140–149 (2014)
90. J. Liu, S. Shao, G. Fang, B. Meng, Z. Xie, L. Wang, High-efficiency inverted polymer solar cells with transparent and work-function tunable MoO_3-Al composite film as cathode buffer layer. Adv. Mater. **24**, 2774–2779 (2012)

91. M.R. Rajesh Menon, M.V. Maheshkumar, K. Sreekumar, C. Sudha Kartha, K. P. Vijayakumar, Inverted polymer solar cells with indium sulfide electron selective layer. Solar Energy Mater. Solar Cells **94**, 2212–2217 (2010)
92. M.R.R. Menon, M.V. Maheshkumar, K. Sreekumar, C.S. Kartha, K.P. Vijayakumar, Spray pyrolysed In_2S_3 thin films: a potential electron selective layer for large area inverted bulk-heterojunction polymer solar cells. Physica Status Solidi (a). **209**, 199–203 (2012)
93. X. Chen, J. Yang, L.Y.X.C. Haley, J. Lu, F. Zhu, K.P. Loh, Towards high efficiency solution processable inverted bulk heterojunction polymer solar cells using modified indium tin oxide cathode. Org. Electron. **11**, 1942–1946 (2010)
94. K. Sun, H. Zhang, J. Ouyang, Indium tin oxide modified with sodium compounds as cathode of inverted polymer solar cells. J. Mater. Chem. **21**, 18339–18346 (2011)
95. K. Sun, B. Zhao, A. Kumar, K. Zeng, J. Ouyang, Highly efficient, inverted polymer solar cells with indium tin oxide modified with solution-processed zwitterions as the transparent cathode. ACS Appl. Mater. Interfaces **4**, 2009–2017 (2012)
96. A. Li, R. Nie, X. Deng, H. Wei, S. Zheng, Y. Li, J. Tang, K.-Y. Wong, Highly efficient inverted organic solar cells using amino acid modified indium tin oxide as cathode. Appl. Phys. Lett. **104**, 123303 (2014)
97. Q. Zhang, D. Zhang, X. Li, X. Liu, W. Zhang, L. Han, J. Fang, Neutral amine based alcohol-soluble interface materials for inverted polymer solar cells: realizing high performance and overcoming solvent erosion. Chem. Commun. **51**, 10182–10185 (2015)
98. C.Y. Jiang, X.W. Sun, D.W. Zhao, A.K.K. Kyaw, Y.N. Li, Low work function metal modified ITO as cathode for inverted polymer solar cells. Solar Energy Mater. Solar Cells **94**, 1618–1621 (2010)
99. X. Wan, G. Long, L. Huang, Y. Chen, Graphene—a promising material for organic photovoltaic cells. Adv. Mater. **23**, 5342–5358 (2011)
100. J. Liu, Y. Xue, Y. Gao, D. Yu, M. Durstock, L. Dai, Hole and electron extraction layers based on graphene oxide derivatives for high-performance bulk heterojunction solar cells. Adv. Mater. **24**, 2228–2233 (2012)
101. J. Liu, M. Durstock, L. Dai, Graphene oxide derivatives as hole- and electron-extraction layers for high-performance polymer solar cells. Energy Environ. Sci. **7**, 1297–1306 (2014)
102. Y.W. Heo, D.P. Norton, L.C. Tien, Y. Kwon, B.S. Kang, F. Ren, S.J. Pearton, J.R. LaRoche, ZnO nanowire growth and devices. Mater. Sci. Eng. R: Rep. **47**, 1–47 (2004)
103. K. Ellmer, Resistivity of polycrystalline zinc oxide films: current status and physical limit. J. Phys. D Appl. Phys. **34**, 3097–3108 (2001)
104. A. Manor, E.A. Katz, T. Tromholt, F.C. Krebs, Enhancing functionality of ZnO hole blocking layer in organic photovoltaics. Sol. Energy Mater. Sol. Cells **98**, 491–493 (2012)
105. J.-C. Wang, W.-T. Weng, M.-Y. Tsai, M.-K. Lee, S.-F. Horng, T.-P. Perng, C.-C. Kei, C.-C. Yu, H.-F. Meng, Highly efficient flexible inverted organic solar cells using atomic layer deposited ZnO as electron selective layer. J. Mater. Chem. **20**, 862–866 (2010)
106. J. Sun, Y. Zhu, X. Xu, L. Lan, L. Zhang, P. Cai, J. Chen, J. Peng, Y. Cao, High efficiency and high Voc inverted polymer solar cells based-on a low-lying HOMO polycarbazole donor and a hydrophilic polycarbazole interlayer on ITO cathode. J. Phys. Chem. C **116**, 14188–14198 (2012)
107. S.H. Liao, H.J. Jhuo, Y.S. Cheng, S.A. Chen, Fullerene derivative-doped zinc oxide nanofilm as the cathode of inverted polymer solar cells with low-bandgap polymer (PTB7-Th) for high performance. Adv. Mater. **25**, 4766–4771 (2013)
108. F.C. Krebs, J. Fyenbo, D.M. Tanenbaum, S.A. Gevorgyan, R. Andriessen, B. van Remoortere, Y. Galagan, M. Jorgensen, The OE-A OPV demonstrator anno domini 2011. Energy Environ. Sci. **4**, 4116–4123 (2011)
109. F.C. Krebs, T. Tromholt, M. Jorgensen, Upscaling of polymer solar cell fabrication using full roll-to-roll processing. Nanoscale **2**, 873–886 (2010)
110. F.C. Krebs, J. Fyenbo, M. Jorgensen, Product integration of compact roll-to-roll processed polymer solar cell modules: methods and manufacture using flexographic printing, slot-die coating and rotary screen printing. J. Mater. Chem. **20**, 8994–9001 (2010)

111. F.C. Krebs, T.D. Nielsen, J. Fyenbo, M. Wadstrom, M.S. Pedersen, Manufacture, integration and demonstration of polymer solar cells in a lamp for the "Lighting Africa" initiative. Energy Environ. Sci. **3**, 512–525 (2010)
112. F.C. Krebs, S.A. Gevorgyan, J. Alstrup, A roll-to-roll process to flexible polymer solar cells: model studies, manufacture and operational stability studies. J. Mater. Chem. **19**, 5442–5451 (2009)
113. G.D. Spyropoulos, P. Kubis, N. Li, D. Baran, L. Lucera, M. Salvador, T. Ameri, M.M. Voigt, F.C. Krebs, C.J. Brabec, Flexible organic tandem solar modules with 6 % efficiency: combining roll-to-roll compatible processing with high geometric fill factors. Energy Environ. Sci. **7**, 3284–3290 (2014)
114. T. Kuwabara, Y. Kawahara, T. Yamaguchi, K. Takahashi, Characterization of inverted-type organic solar cells with a ZnO layer as the electron collection electrode by ac impedance spectroscopy. ACS Appl. Mater. Interfaces **1**, 2107–2110 (2009)
115. M.-Y. Liu, C.-H. Chang, C.-H. Chang, K.-H. Tsai, J.-S. Huang, C.-Y. Chou, I.-J. Wang, P.-S. Wang, C.-Y. Lee, C.-H. Chao, C.-L. Yeh, C.-I. Wu, C.-F. Lin, Morphological evolution of the poly(3-hexylthiophene)/[6,6]-phenyl-C61-butyric acid methyl ester, oxidation of the silver electrode, and their influences on the performance of inverted polymer solar cells with a sol–gel derived zinc oxide electron selective layer. Thin Solid Films **518**, 4964–4969 (2010)
116. H.-J. Park, K.-H. Lee, B. Kumar, K.-S. Shin, S.-W. Jeong, S.-W. Kim, Inverted organic solar cells with ZnO thin films prepared by sol-gel method. J. Nanoelectron. Optoelectron. **5**, 135–138 (2010)
117. H.O. Seo, S.-Y. Park, W.H. Shim, K.-D. Kim, K.H. Lee, M.Y. Jo, J.H. Kim, E. Lee, D.-W. Kim, Y.D. Kim, D.C. Lim, Ultrathin TiO_2 films on ZnO electron-collecting layers of inverted organic solar cell. J. Phys. Chem. C **115**, 21517–21520 (2011)
118. C. Zhang, H. You, Z. Lin, Y. Hao, Inverted organic photovoltaic cells with solution-processed zinc oxide as electron collecting layer. Jpn. J. Appl. Phys. **50**, 082302 (2011)
119. D.C. Lim, W.H. Shim, K.-D. Kim, H.O. Seo, J.-H. Lim, Y. Jeong, Y.D. Kim, K.H. Lee, Spontaneous formation of nanoripples on the surface of ZnO thin films as hole-blocking layer of inverted organic solar cells. Sol. Energy Mater. Sol. Cells **95**, 3036–3040 (2011)
120. L.K. Jagadamma, M. Abdelsamie, A. El Labban, E. Aresu, G.O. Ngongang Ndjawa, D.H. Anjum, D. Cha, P.M. Beaujuge, A. Amassian, Efficient inverted bulk-heterojunction solar cells from low-temperature processing of amorphous ZnO buffer layers. J. Mater. Chem. A. **2**, 13321–13331 (2014)
121. X. Bao, A. Yang, Y. Yang, T. Wang, L. Sun, N. Wang, L. Han, Low-temperature solution-processed ZnO nanocrystalline interfacial layer with antireflective effect for efficient inverted polymer solar cells. Physica B **432**, 1–4 (2014)
122. H.-Y. Park, D. Lim, K.-D. Kim, S.-Y. Jang, Performance optimization of low-temperature-annealed solution-processable ZnO buffer layers for inverted polymer solar cells. J. Mater. Chem. A **1**, 6327–6334 (2013)
123. S.K. Hau, H.-L. Yip, J. Zou, A.K.Y. Jen, Indium tin oxide-free semi-transparent inverted polymer solar cells using conducting polymer as both bottom and top electrodes. Org. Electron. **10**, 1401–1407 (2009)
124. S. Chen, C.E. Small, C.M. Amb, J. Subbiah, T.-H. Lai, S.-W. Tsang, J.R. Manders, J.R. Reynolds, F. So, Inverted polymer solar cells with reduced interface recombination. Adv. Energy Mater. **2**, 1333–1337 (2012)
125. M.J. Tan, S. Zhong, J. Li, Z. Chen, W. Chen, Air-stable efficient inverted polymer solar cells using solution-processed nanocrystalline ZnO interfacial layer. ACS Appl. Mater. Interfaces **5**, 4696–4701 (2013)
126. W. Qin, X. Xu, D. Liu, C. Ma, L. Yang, S. Yin, F. Zhang, J. Wei, Surface states of ZnO nanoparticles effect on the performance of inverted-organic solar cells. J. Renew. Sustain. Energy **5**, 053106 (2013)

127. Y. Lare, A. Godoy, L. Cattin, K. Jondo, T. Abachi, F.R. Diaz, M. Morsli, K. Napo, M.A. del Valle, J.C. Bernède, ZnO thin films fabricated by chemical bath deposition, used as buffer layer in organic solar cells. Appl. Surface Sci. **255**, 6615–6619 (2009)
128. S. Park, S.J. Tark, J.S. Lee, H. Lim, D. Kim, Effects of intrinsic ZnO buffer layer based on P3HT/PCBM organic solar cells with Al-doped ZnO electrode. Solar Energy Mater. Solar Cells **93**, 1020–1023 (2009)
129. H. Cheun, C. Fuentes-Hernandez, Y. Zhou, W.J. Potscavage, S.-J. Kim, J. Shim, A. Dindar, B. Kippelen, Electrical and optical properties of ZnO processed by atomic layer deposition in inverted polymer solar cells. J. Phys. Chem. C **114**, 20713–20718 (2010)
130. C.-Y. Chang, F.-Y. Tsai, Efficient and air-stable plastics-based polymer solar cells enabled by atomic layer deposition. J. Mater. Chem. **21**, 5710–5715 (2011)
131. R.L.Z. Hoye, D. Muñoz-Rojas, D.C. Iza, K.P. Musselman, J.L. MacManus-Driscoll, High performance inverted bulk heterojunction solar cells by incorporation of dense, thin ZnO layers made using atmospheric atomic layer deposition. Solar Energy Mater. Solar Cells **116**, 197–202 (2013)
132. K.D. Kim, D.C. Lim, J. Hu, J.D. Kwon, M.G. Jeong, H.O. Seo, J.Y. Lee, K.Y. Jang, J.H. Lim, K.H. Lee, Y. Jeong, Y.D. Kim, S. Cho, Surface modification of a ZnO electron-collecting layer using atomic layer deposition to fabricate high-performing inverted organic photovoltaics. ACS Appl. Mater. Interfaces **5**, 8718–8723 (2013)
133. N.O.V. Plank, M.E. Welland, J.L. MacManus-Driscoll, L. Schmidt-Mende, The backing layer dependence of open circuit voltage in ZnO/polymer composite solar cells. Thin Solid Films **516**, 7218–7222 (2008)
134. Y.-J. Noh, S.-I. Na, S.-S. Kim, Inverted polymer solar cells including ZnO electron transport layer fabricated by facile spray pyrolysis. Solar Energy Mater. Solar Cells **117**, 139–144 (2013)
135. Z. Hu, J. Zhang, Y. Zhu, Inverted polymer solar cells with a boron-doped zinc oxide layer deposited by metal organic chemical vapor deposition. Solar Energy Mater. Solar Cells **117**, 610–616 (2013)
136. K.-S. Shin, H.-J. Park, B. Kumar, K.-K. Kim, S.-G. Ihn, S.-W. Kim, Low-temperature growth and characterization of ZnO thin films for flexible inverted organic solar cells. J. Mater. Chem. **21**, 12274–12279 (2011)
137. S.-G. Ihn, K.-S. Shin, M.-J. Jin, X. Bulliard, S. Yun, Y. Suk Choi, Y. Kim, J.-H. Park, M. Sim, M. Kim, K. Cho, T. Sang Kim, D. Choi, J.-Y. Choi, W. Choi, S.-W. Kim, ITO-free inverted polymer solar cells using a GZO cathode modified by ZnO. Solar Energy Mater. Solar Cells **95**, 1610–1614 (2011)
138. F. Zhu, X. Chen, L. Zhou, J. Zhou, J. Yang, S. Huang, Z. Sun, Dependence of the performance of inverted polymer solar cells on thickness of an electron selective ZnO layer deposited by magnetron sputtering. Thin Solid Films **551**, 131–135 (2014)
139. F. Zhu, X. Chen, J. Zhou, Z. Lu, Y. Chen, S. Huang, Z. Sun, Enhanced efficiency of inverted polymer solar cells using two-step sputtered ZnO as cathode interfacial layer. Mater. Res. Exp. **1**, 025020 (2014)
140. J.-H. Lee, S. Yoshikawa, T. Sagawa, Fabrication of efficient organic and hybrid solar cells by fine channel mist spray coating. Solar Energy Mater. Solar Cells **127**, 111–121 (2014)
141. M. Ohyama, H. Kouzuka, T. Yoko, Sol-gel preparation of ZnO films with extremely preferred orientation along (002) plane from zinc acetate solution. Thin Solid Films **306**, 78–85 (1997)
142. L. Znaidi, Sol-gel-deposited ZnO thin films: a review. Mater. Sci. Eng. B **174**, 18–30 (2010)
143. B.S. Ong, C. Li, Y. Li, Y. Wu, R. Loutfy, Stable, Solution-processed, high-mobility ZnO thin-film transistors. J. Am. Chem. Soc. **129**, 2750–2751 (2007)
144. C. Jiang, R.R. Lunt, P.M. Duxbury, P.P. Zhang, High-performance inverted solar cells with a controlled ZnO buffer layer. RSC Adv. **4**, 3604–3610 (2014)
145. P. Morvillo, R. Diana, A. Mucci, E. Bobeico, R. Ricciardi, C. Minarini, Influence of annealing treatments on solution-processed ZnO film deposited on ITO substrate as electron

transport layer for inverted polymer solar cells. Solar Energy Mater. Solar Cells **141**, 210–217 (2015)
146. Z. Lin, J. Chang, C. Jiang, J. Zhang, J. Wu, C. Zhu, Enhanced inverted organic solar cell performance by post-treatments of solution-processed ZnO buffer layers. RSC Adv. **4**, 6646–6651 (2014)
147. B.A. MacLeod, B.J. Tremolet de Villers, P. Schulz, P.F. Ndione, H. Kim, A.J. Giordano, K. Zhu, S.R. Marder, S. Graham, J.J. Berry, A. Kahn, D.C. Olson, Stability of inverted organic solar cells with ZnO contact layers deposited from precursor solutions. Energy Environ. Sci. **8**, 592–601 (2015)
148. S.K. Hau, Y.-J. Cheng, H.-L. Yip, Y. Zhang, H. Ma, A.K.Y. Jen, Effect of chemical modification of fullerene-based self-assembled monolayers on the performance of inverted polymer solar cells. ACS Appl. Mater. Interfaces **2**, 1892–1902 (2010)
149. A.L. Roest, J.J. Kelly, D. Vanmaekelbergh, E.A. Meulenkamp, Staircase in the electron mobility of a ZnO quantum dot assembly due to shell filling. Phys. Rev. Lett. **89**, 036801 (2002)
150. P. Li, T. Jiu, G. Tang, G. Wang, J. Li, X. Li, J. Fang, Solvents induced ZnO nanoparticles aggregation associated with their interfacial effect on organic solar cells. ACS Appl. Mater. Interfaces **6**, 18172–18179 (2014)
151. C.E. Small, S. Chen, J. Subbiah, C.M. Amb, S.-W. Tsang, T.-H. Lai, J.R. Reynolds, F. So, High-efficiency inverted dithienogermole-thienopyrrolodione-based polymer solar cells. Nat. Photonics **6**, 115–120 (2012)
152. S. Shao, K. Zheng, T. Pullerits, F. Zhang, Enhanced performance of inverted polymer solar cells by using poly(ethylene oxide)-modified ZnO as an electron transport layer. ACS Appl. Mater. Interfaces **5**, 380–385 (2012)
153. J.M. Cho, S.-W. Kwak, H. Aqoma, J.W. Kim, W.S. Shin, S.-J. Moon, S.-Y. Jang, J. Jo, Effects of ultraviolet–ozone treatment on organic-stabilized ZnO nanoparticle-based electron transporting layers in inverted polymer solar cells. Org. Electron. **15**, 1942–1950 (2014)
154. Y. Chen, Z. Hu, Z. Zhong, W. Shi, J. Peng, J. Wang, Y. Cao, Aqueous solution processed, ultrathin ZnO film with low conversion temperature as the electron transport layer in the inverted polymer solar cells. J. Phys. Chem. C **118**, 21819–21825 (2014)
155. R.L. Puurunen, A short history of atomic layer deposition: Tuomo Suntola's atomic layer epitaxy. Chem. Vapor Deposition **20**, 332–344 (2014)
156. M. Leskelä, M. Ritala, Atomic layer deposition (ALD): from precursors to thin film structures. Thin Solid Films **409**, 138–146 (2002)
157. M. Tammenmaa, T. Koskinen, L. Hiltunen, L. Niinistö, M. Leskelä, Zinc chalcogenide thin films grown by the atomic layer epitaxy technique using zinc acetate as source material. Thin Solid Films **124**, 125–128 (1985)
158. T. Tynell, M. Karppinen, Atomic layer deposition of ZnO: a review. Semiconductor Sci. Technol. **29**, 043001 (2014)
159. I.A. Kowalik, E. Guziewicz, K. Kopalko, S. Yatsunenko, A. Wójcik-Głodowska, M. Godlewski, P. Dłużewski, E. Łusakowska, W. Paszkowicz, Structural and optical properties of low-temperature ZnO films grown by atomic layer deposition with diethylzinc and water precursors. J. Cryst. Growth **311**, 1096–1101 (2009)
160. S. Oh, I. Jang, S.-G. Oh, S.S. Im, Effect of ZnO nanoparticle morphology and post-treatment with zinc acetate on buffer layer in inverted organic photovoltaic cells. Sol. Energy **114**, 32–38 (2015)
161. X. Yu, X. Yu, Z. Hu, J. Zhang, G. Zhao, Y. Zhao, Effect of sol–gel derived ZnO annealing rate on light-trapping in inverted polymer solar cells. Mater. Lett. **108**, 50–53 (2013)
162. Z. Ma, Z. Tang, E. Wang, M.R. Andersson, O. Inganäs, F. Zhang, Influences of surface roughness of ZnO electron transport layer on the photovoltaic performance of organic inverted solar cells. J. Phys. Chem. C **116**, 24462–24468 (2012)
163. R.B. Ambade, S.B. Ambade, R.S. Mane, S.H. Lee, Interfacial Engineering Importance of Bilayered ZnO Cathode Buffer on the Photovoltaic Performance of Inverted Organic Solar Cells. ACS Appl. Mater. Interfaces **7**, 7951–7960 (2015)

164. T. Stubhan, H. Oh, L. Pinna, J. Krantz, I. Litzov, C.J. Brabec, Inverted organic solar cells using a solution processed aluminum-doped zinc oxide buffer layer. Org. Electron. **12**, 1539–1543 (2011)
165. S. Wilken, J. Parisi, H. Borchert, Role of oxygen adsorption in nanocrystalline ZnO interfacial layers for polymer–fullerene bulk heterojunction solar cells. J. Phys. Chem. C **118**, 19672–19682 (2014)
166. F. Verbakel, S.C.J. Meskers, R.A.J. Janssen, Electronic memory effects in diodes of zinc oxide nanoparticles in a matrix of polystyrene or poly(3-hexylthiophene). J. Appl. Phys. **102**, 083701–083709 (2007)
167. A. Sharma, M. Ionescu, G.G. Andersson, D.A. Lewis, Role of zinc oxide thickness on the photovoltaic performance of laminated organic bulk-heterojunction solar cells. Solar Energy Mater. Solar Cells. **115**, 64–70 (2013)
168. A. Sharma, J.B. Franklin, B. Singh, G.G. Andersson, D.A. Lewis, Electronic and chemical properties of ZnO in inverted organic photovoltaic devices. Org. Electron. **24**, 131–136 (2015)
169. S. O'Brien, L.H.K. Koh, G.M. Crean, ZnO thin films prepared by a single step sol–gel process. Thin Solid Films **516**, 1391–1395 (2008)
170. S. Kamaruddin, K.-Y. Chan, H.-K. Yow, M. Zainizan Sahdan, H. Saim, D. Knipp. Zinc oxide films prepared by sol–gel spin coating technique. Appl. Phys. A: Mater. Sci. Process. **104**, 263–268 (2011)
171. Y. Li, L. Xu, X. Li, X. Shen, A. Wang, Effect of aging time of ZnO sol on the structural and optical properties of ZnO thin films prepared by sol–gel method. Appl. Surface Sci. **256**, 4543–4547 (2010)
172. H. Oh, J. Krantz, I. Litzov, T. Stubhan, L. Pinna, C.J. Brabec, Comparison of various sol–gel derived metal oxide layers for inverted organic solar cells. Solar Energy Mater. Solar Cells **95**, 2194–2199 (2011)
173. K.-S. Shin, K.-H. Lee, H.H. Lee, D. Choi, S.-W. Kim, Enhanced power conversion efficiency of inverted organic solar cells with a Ga-doped ZnO nanostructured thin film prepared using aqueous solution. J. Phys. Chem. C **114**, 15782–15785 (2010)
174. K. Aung Ko Ko, S. Xiaowei, Z. De Wei, T. Swee Tiam, Y. Divayana, H.V. Demir, Improved inverted organic solar cells with a sol–gel derived indium-doped zinc oxide buffer layer. IEEE J. Sel. Top. Quantum Electron. **16**, 1700–1706 (2010)
175. J. Alstrup, M. Jørgensen, A.J. Medford, F.C. Krebs, Ultra fast and parsimonious materials screening for polymer solar cells using differentially pumped Slot-Die coating. ACS Appl. Mater. Interfaces **2**, 2819–2827 (2010)
176. R. Søndergaard, M. Helgesen, M. Jørgensen, F.C. Krebs, Fabrication of polymer solar cells using aqueous processing for all layers including the metal back electrode. Adv. Energy Mater. **1**, 68–71 (2011)
177. S. Trost, K. Zilberberg, A. Behrendt, A. Polywka, P. Görrn, P. Reckers, J. Maibach, T. Mayer, T. Riedl, Overcoming the "light-soaking" issue in inverted organic solar cells by the use of Al:ZnO electron extraction layers. Adv. Energy Mater. **3**, 1437–1444 (2013)
178. S.-W. Cho, Y.T. Kim, W.H. Shim, S.-Y. Park, K.-D. Kim, H.O. Seo, N.K. Dey, J.-H. Lim, Y. Jeong, K.H. Lee, Y.D. Kim, D.C. Lim, Influence of surface roughness of aluminum-doped zinc oxide buffer layers on the performance of inverted organic solar cells. Appl. Phys. Lett. **98**, 023102–023103 (2011)
179. A. Gadisa, Y. Liu, E.T. Samulski, R. Lopez, Minimizing interfacial losses in inverted organic solar cells comprising Al-doped ZnO. Appl. Phys. Lett. **100**, 253903 (2012)
180. M. Thambidurai, J.Y. Kim, J. Song, Y. Ko, H.-J. Song, C.-M. Kang, N. Muthukumarasamy, D. Velauthapillai, C. Lee, High performance inverted organic solar cells with solution processed Ga-doped ZnO as an interfacial electron transport layer. J. Mater. Chem. C. **1**, 8161 (2013)
181. O. Pachoumi, C. Li, Y. Vaynzof, K.K. Banger, H. Sirringhaus, Improved performance and stability of inverted organic solar cells with sol-gel processed, amorphous mixed metal oxide

electron extraction layers comprising alkaline earth metals. Adv. Energy Mater. **3**, 1428–1436 (2013)
182. S.H. Tsai, S.T. Ho, H.J. Jhuo, C.R. Ho, S.A. Chen, J.-H. He, Toward high efficiency of inverted organic solar cells: concurrent improvement in optical and electrical properties of electron transport layers. Appl. Phys. Lett. **102**, 253111 (2013)
183. J. Piris, N. Kopidakis, D.C. Olson, S.E. Shaheen, D.S. Ginley, G. Rumbles, The Locus of free charge-carrier generation in solution-cast Zn1−xMgxO/poly(3-hexylthiophene) bilayers for photovoltaic applications. Adv. Funct. Mater. **17**, 3849–3857 (2007)
184. B.A. MacLeod, P. Schulz, S.R. Cowan, A. Garcia, D.S. Ginley, A. Kahn, D.C. Olson, Improved performance in bulk heterojunction organic solar cells with a sol-gel MgZnO electron-collecting layer. Adv. Energy Mater. **4**, 1400073 (2014)
185. M. Lenes, L.J.A. Koster, V.D. Mihailetchi, P.W.M. Blom, Thickness dependence of the efficiency of polymer:fullerene bulk heterojunction solar cells. Appl. Phys. Lett. **88**, 243502 (2006)
186. D.W. Sievers, V. Shrotriya, Y. Yang, Modeling optical effects and thickness dependent current in polymer bulk-heterojunction solar cells. J. Appl. Phys. **100**, 114509 (2006)
187. T. Kirchartz, T. Agostinelli, M. Campoy-Quiles, W. Gong, J. Nelson, Understanding the thickness-dependent performance of organic bulk heterojunction solar cells: the influence of mobility, lifetime, and space charge. J. Phys. Chem. Lett. **3**, 3470–3475 (2012)
188. W.L. Ma, C.Y. Yang, X. Gong, K. Lee, A.J. Heeger, Thermally stable, efficient polymer solar cells with nanoscale control of the interpenetrating network morphology. Adv. Funct. Mater. **15**, 1617–1622 (2005)
189. S. Loser, B. Valle, K.A. Luck, C.K. Song, G. Ogien, M.C. Hersam, K.D. Singer, T.J. Marks, High-efficiency inverted polymer photovoltaics via spectrally tuned absorption enhancement. Adv. Energy Mater. **4**, 1301938 (2014)
190. D.-H. Ko, J.R. Tumbleston, W. Schenck, R. Lopez, E.T. Samulski, Photonic crystal geometry for organic polymer:fullerene standard and inverted solar cells. J. Phys. Chem. C **115**, 4247–4254 (2011)
191. J.M. Lee, J. Lim, N. Lee, H.I. Park, K.E. Lee, T. Jeon, S.A. Nam, J. Kim, J. Shin, S.O. Kim, Synergistic concurrent enhancement of charge generation, dissociation, and transport in organic solar cells with plasmonic metal-carbon nanotube hybrids. Adv. Mater. **27**, 1519–1525 (2014)
192. S.-H. Jeong, H. Choi, J.Y. Kim, T.-W. Lee, Silver-based nanoparticles for surface Plasmon resonance in organic optoelectronics. Part. Part. Syst. Charact. **32**, 164–175 (2014)
193. C.H. Chou, F.C. Chen, Plasmonic nanostructures for light trapping in organic photovoltaic devices. Nanoscale **6**, 8444–8458 (2014)
194. S.-J. Ko, H. Choi, W. Lee, T. Kim, B.R. Lee, J.-W. Jung, J.-R. Jeong, M.H. Song, J.C. Lee, H.Y. Woo, J.Y. Kim, Highly efficient plasmonic organic optoelectronic devices based on a conducting polymer electrode incorporated with silver nanoparticles. Energy Environ. Sci. **6**, 1949–1955 (2013)
195. J. Li, L. Zuo, H. Pan, H. Jiang, T. Liang, Y. Shi, H. Chen, M. Xu, Texture design of electrodes for efficiency enhancement of organic solar cells. J. Mater. Chem. A. **1**, 2379–2386 (2013)
196. L. Müller-Meskamp, Y.H. Kim, T. Roch, S. Hofmann, R. Scholz, S. Eckardt, K. Leo, A.F. Lasagni, Efficiency enhancement of organic solar cells by fabricating periodic surface textures using direct laser interference patterning. Adv. Mater. **24**, 906–910 (2012)
197. K. Takanezawa, K. Hirota, Q.-S. Wei, K. Tajima, K. Hashimoto, Efficient charge collection with ZnO nanorod array in hybrid photovoltaic devices. J. Phys. Chem. C **111**, 7218–7223 (2007)
198. C.-Y. Chou, J.-S. Huang, C.-H. Wu, C.-Y. Lee, C.-F. Lin, Lengthening the polymer solidification time to improve the performance of polymer/ZnO nanorod hybrid solar cells. Solar Energy Mater. Solar Cells **93**, 1608–1612 (2009)
199. M. Law, L.E. Greene, J.C. Johnson, R. Saykally, P. Yang, Nanowire dye-sensitized solar cells. Nat. Mater. **4**, 455–459 (2005)

200. M. Wang, Y. Li, H. Huang, E.D. Peterson, W. Nie, W. Zhou, W. Zeng, W. Huang, G. Fang, N. Sun, X. Zhao, D.L. Carroll, Thickness dependence of the MoO[sub 3] blocking layers on ZnO nanorod-inverted organic photovoltaic devices. Appl. Phys. Lett. **98**, 103305 (2011)
201. N.K. Elumalai, T.M. Jin, V. Chellappan, R. Jose, S.K. Palaniswamy, S. Jayaraman, H.K. Raut, S. Ramakrishna, Electrospun ZnO nanowire plantations in the electron transport layer for high-efficiency inverted organic solar cells. ACS Appl. Mater. Interfaces **5**, 9396–9404 (2013)
202. K. Takanezawa, K. Tajima, K. Hashimoto, Efficiency enhancement of polymer photovoltaic devices hybridized with ZnO nanorod arrays by the introduction of a vanadium oxide buffer layer. Appl. Phys. Lett. **93**, 063308 (2008)
203. R. Thitima, C. Patcharee, S. Takashi, Y. Susumu, Efficient electron transfers in ZnO nanorod arrays with N719 dye for hybrid solar cells. Solid-State Electron. **53**, 176–180 (2009)
204. Y. Hames, Z. Alpaslan, A. Kösemen, S.E. San, Y. Yerli, Electrochemically grown ZnO nanorods for hybrid solar cell applications. Sol. Energy **84**, 426–431 (2010)
205. S. Yodyingyong, X. Zhou, Q. Zhang, D. Triampo, J. Xi, K. Park, B. Limketkai, G. Cao, Enhanced photovoltaic performance of nanostructured hybrid solar cell using highly oriented TiO_2 nanotubes. J. Phys. Chem. C **114**, 21851–21855 (2010)
206. J.-S. Huang, C.-Y. Chou, C.-F. Lin, Enhancing performance of organic–inorganic hybrid solar cells using a fullerene interlayer from all-solution processing. Solar Energy Mater. Solar Cells **94**, 182–186 (2010)
207. K.H. Lee, B. Kumar, H.-J. Park, S.-W. Kim, Optimization of an electron transport layer to enhance the power conversion efficiency of flexible inverted organic solar cells. Nanoscale Res. Lett. **5**, 1908–1912 (2010)
208. Z. Yuan, J. Yu, N. Wang, Y. Jiang, Well-aligned ZnO nanorod arrays from diameter-controlled growth and their application in inverted polymer solar cell. J. Mater. Sci.: Mater. Electron. **22**, 1730–1735 (2011)
209. Y.-M. Sung, F.-C. Hsu, C.-T. Chen, W.-F. Su, Y.-F. Chen, Enhanced photocurrent and stability of inverted polymer/ZnO-nanorod solar cells by 3-hydroxyflavone additive. Solar Energy Mater. Solar Cells **98**, 103–109 (2012)
210. M. Ahmadi, K. Mirabbaszadeh, M. Ketabchi, Density-controlled ZnO nanorod arrays in polymer solar cells based on poly(3-hexylthiophene) and indene-C_{60} bis-adduct. Electron. Mater. Lett. **9**, 729–734 (2013)
211. J. Ajuria, I. Etxebarria, E. Azaceta, R. Tena-Zaera, N. Fernandez-Montcada, E. Palomares, R. Pacios, Novel ZnO nanostructured electrodes for higher power conversion efficiencies in polymeric solar cells. Phys. Chem. Chem. Phys. **13**, 20871–20876 (2011)
212. H. Woo Choi, K.-S. Lee, N. David Theodore, T.L. Alford, Improved performance of ZnO nanostructured bulk heterojunction organic solar cells with nanowire-density modified by yttrium chloride introduction into solution. Solar Energy Mater. Solar Cells **117**, 273–278 (2013)
213. D.C. Olson, J. Piris, R.T. Collins, S.E. Shaheen, D.S. Ginley, Hybrid photovoltaic devices of polymer and ZnO nanofiber composites. Thin Solid Films **496**, 26–29 (2006)
214. Z. Hu, J. Zhang, Y. Liu, Y. Li, X. Zhang, Y. Zhao, Efficiency enhancement of inverted organic photovoltaic devices with ZnO nanopillars fabricated on FTO glass substrates. Synth. Met. **161**, 2174–2178 (2011)
215. Z. Hu, J. Zhang, Y. Liu, Z. Hao, X. Zhang, Y. Zhao, Influence of ZnO interlayer on the performance of inverted organic photovoltaic device. Solar Energy Mater. Solar Cells **95**, 2126–2130 (2011)
216. S. Xu, Z. Wang, One-dimensional ZnO nanostructures: solution growth and functional properties. Nano Res. **4**, 1013–1098 (2011)
217. P.P. Boix, J. Ajuria, I. Etxebarria, R. Pacios, G. Garcia-Belmonte, J. Bisquert, Role of ZnO electron-selective layers in regular and inverted bulk heterojunction solar cells. J. Phys. Chem. Lett. **2**, 407–411 (2011)
218. A. Manor, E.A. Katz, T. Tromholt, F.C. Krebs, Electrical and photo-induced degradation of ZnO layers in organic photovoltaics. Adv. Energy Mater. **1**, 836–843 (2011)

219. X. Bulliard, S.-G. Ihn, S. Yun, Y. Kim, D. Choi, J.-Y. Choi, M. Kim, M. Sim, J.-H. Park, W. Choi, K. Cho, Enhanced performance in polymer solar cells by surface energy control. Adv. Funct. Mater. **20**, 4381–4387 (2010)
220. S. Bai, Y. Jin, X. Liang, Z. Ye, Z. Wu, B. Sun, Z. Ma, Z. Tang, J. Wang, U. Würfel, F. Gao, F. Zhang, Ethanedithiol treatment of solution-processed ZnO thin films: Controlling the intragap states of electron transporting interlayers for efficient and stable inverted organic photovoltaics. Adv. Energy Mater. **5**, 1401606 (2015)
221. J.J. Intemann, K. Yao, Y.-X. Li, H.-L. Yip, Y.-X. Xu, P.-W. Liang, C.-C. Chueh, F.-Z. Ding, X. Yang, X. Li, Y. Chen, A.K.Y. Jen, Highly efficient inverted organic solar cells through material and interfacial engineering of indacenodithieno[3,2-b]thiophene-based polymers and devices. Adv. Funct. Mater. **24**, 1465–1473 (2013)
222. S.K. Hau, H.-L. Yip, H. Ma, A.K.Y. Jen, High performance ambient processed inverted polymer solar cells through interfacial modification with a fullerene self-assembled monolayer. Appl. Phys. Lett. **93**, 233304–233303 (2008)
223. H.-L. Yip, S.K. Hau, N.S. Baek, H. Ma, A.K.Y. Jen, Polymer solar cells that use self-assembled-monolayer-modified ZnO/metals as cathodes. Adv. Mater. **20**, 2376–2382 (2008)
224. C.-H. Hsieh, Y.-J. Cheng, P.-J. Li, C.-H. Chen, M. Dubosc, R.-M. Liang, C.-S. Hsu, Highly efficient and stable inverted polymer solar cells integrated with a cross-linked fullerene material as an interlayer. J. Am. Chem. Soc. **132**, 4887–4893 (2010)
225. S. Cho, K.D. Kim, J. Heo, J.Y. Lee, G. Cha, B.Y. Seo, Y.D. Kim, Y.S. Kim, S.Y. Choi, D.C. Lim, Role of additional PCBM layer between ZnO and photoactive layers in inverted bulk-heterojunction solar cells. Sci. Rep. **4**, 4306 (2014)
226. T. Tromholt, M.V. Madsen, J.E. Carlé, M. Helgesen, F.C. Krebs, Photochemical stability of conjugated polymers, electron acceptors and blends for polymer solar cells resolved in terms of film thickness and absorbance. J. Mater. Chem. **22**, 7592 (2012)
227. Y.-J. Cheng, C.-H. Hsieh, Y. He, C.-S. Hsu, Y. Li, Combination of indene-C_{60} bis-adduct and cross-linked fullerene interlayer leading to highly efficient inverted polymer solar cells. J. Am. Chem. Soc. **132**, 17381–17383 (2010)
228. T. Stubhan, M. Salinas, A. Ebel, F.C. Krebs, A. Hirsch, M. Halik, C.J. Brabec, Increasing the fill factor of inverted P3HT:PCBM solar cells through surface modification of Al-doped ZnO via phosphonic acid-anchored C_{60} SAMs. Adv. Energy Mater. **2**, 532–535 (2012)
229. J. Jo, J.R. Pouliot, D. Wynands, S.D. Collins, J.Y. Kim, T.L. Nguyen, H.Y. Woo, Y. Sun, M. Leclerc, A.J. Heeger, Enhanced efficiency of single and tandem organic solar cells incorporating a diketopyrrolopyrrole-based low-bandgap polymer by utilizing combined ZnO/polyelectrolyte electron-transport layers. Adv. Mater. **25**, 4783–4788 (2013)
230. T. Yang, M. Wang, C. Duan, X. Hu, L. Huang, J. Peng, F. Huang, X. Gong, Inverted polymer solar cells with 8.4 % efficiency by conjugated polyelectrolyte. Energy Environ. Sci. **5**, 8208–8214 (2012)
231. G. Cheng, W.-Y. Tong, K.-H. Low, C.-M. Che, Thermal-annealing-free inverted polymer solar cells using ZnO/Cs_2CO_3 bilayer as electron-selective layer. Solar Energy Mater. Solar Cells **103**, 164–170 (2012)
232. K.-D. Kim, D.C. Lim, M.-G. Jeong, H.O. Seo, B.Y. Seo, J.Y. Lee, Y. Song, S. Cho, J.-H. Lim, Y.D. Kim, Enhanced stability of organic photovoltaics by additional ZnO layers on rippled ZnO electron-collecting layer using atomic layer deposition. Bull. Korean Chem. Soc. **35**, 353–356 (2014)
233. Z. Hu, K. Zhang, F. Huang, Y. Cao, Water/alcohol soluble conjugated polymers for the interface engineering of highly efficient polymer light-emitting diodes and polymer solar cells. Chem. Commun. **51**, 5572–5585 (2015)
234. K. Zhang, C. Zhong, S. Liu, C. Mu, Z. Li, H. Yan, F. Huang, Y. Cao, Highly efficient inverted polymer solar cells based on a cross-linkable water-/alcohol-soluble conjugated polymer interlayer. ACS Appl. Mater. Interfaces **6**, 10429–10435 (2014)

235. H.I. Kim, T.T.T. Bui, G.-W. Kim, G. Kang, W.S. Shin, T. Park, A Benzodithiophene-based novel electron transport layer for a highly efficient polymer solar cell. ACS Appl. Mater. Interfaces **6**, 15875–15880 (2014)
236. S.I. Yoo, T.T. Do, Y.E. Ha, M.Y. Jo, J. Park, Y.-C. Kang, J.H. Kim, Effect of self-assembled monolayer treated ZnO on the photovoltaic properties of inverted polymer solar cells. Bull. Korean Chem. Soc. **35**, 569–574 (2014)
237. Y.E. Ha, M.Y. Jo, J. Park, Y.-C. Kang, S.-J. Moon, J.H. Kim, Effect of self-assembled monolayer treated ZnO as an electron transporting layer on the photovoltaic properties of inverted type polymer solar cells. Synth. Met. **187**, 113–117 (2014)
238. H. Kang, S. Hong, J. Lee, K. Lee, Electrostatically self-sssembled nonconjugated polyelectrolytes as an ideal interfacial layer for inverted polymer solar cells. Adv. Mater. **24**, 3005–3009 (2012)
239. S.M. Yoon, S.J. Lou, S. Loser, J. Smith, L.X. Chen, A. Facchetti, T. Marks, Fluorinated copper phthalocyanine nanowires for enhancing interfacial electron transport in organic solar cells. Nano Lett. **12**, 6315–6321 (2012)
240. Z. Wu, T. Song, Z. Xia, H. Wei, B. Sun, Enhanced performance of polymer solar cell with ZnO nanoparticle electron transporting layer passivated by in situ cross-linked three-dimensional polymer network. Nanotechnol. **24**, 484012 (2013)
241. T. Hu, F. Li, K. Yuan, Y. Chen, Efficiency and air-stability improvement of flexible inverted polymer solar cells using ZnO/poly(ethylene glycol) hybrids as cathode buffer layers. ACS Appl. Mater. Interfaces **5**, 5763–5770 (2013)
242. J.P. Tiwari, S. Pillai, S. Parakh, F. Ali, A. Sharma, S. Chand, A futuristic approach towards interface layer modifications for improved efficiency in inverted organic solar cells. Appl. Phys. Lett. **104**, 041114 (2014)
243. H.C. Chen, S.W. Lin, J.M. Jiang, Y.W. Su, K.H. Wei, Solution-processed zinc oxide/polyethylenimine nanocomposites as tunable electron transport layers for highly efficient bulk heterojunction polymer solar cells. ACS Appl. Mater. Interfaces **7**, 6273–6281 (2015)
244. T. Hu, L. Chen, K. Yuan, Y. Chen, Amphiphilic fullerene/ZnO hybrids as cathode buffer layers to improve charge selectivity of inverted polymer solar cells. Nanoscale **7**, 9194–9203 (2015)
245. M. Eita, A.E. Labban, F. Cruciani, A. Usman, P.M. Beaujuge, O.F. Mohammed, Ambient layer-by-layer ZnO assembly for highly efficient polymer bulk heterojunction solar cells. Adv. Funct. Mater. **25**, 1558–1564 (2015)
246. B. Sun, H. Sirringhaus, Solution-processed zinc oxide field-effect transistors based on self-assembly of colloidal nanorods. Nano Lett. **5**, 2408–2413 (2005)
247. Y. Natsume, H. Sakata, Zinc oxide films prepared by sol-gel spin-coating. Thin Solid Films **372**, 30–36 (2000)
248. J.-L. Lan, S.-J. Cherng, Y.-H. Yang, Q. Zhang, S. Subramaniyan, F.S. Ohuchi, S.A. Jenekhe, G. Cao, The effects of Ta_2O_5–ZnO films as cathodic buffer layers in inverted polymer solar cells. J. Mater. Chem. A. **2**, 9361–9370 (2014)

Chapter 4
Nanomaterials for Stretchable Energy Storage and Conversion Devices

Keyu Xie and Bingqing Wei

Abstract With the continuous progress of the energy use and demand, functionalized energy storage and conversion devices (ESCDs) are urgently needed. In the meantime, stretchable ESCDs are attracting intensive attention due to their great potential for specific applications, such as wearable electronics, an electronic skin, implant electronics, and other collapsible gadgets. Design and synthesis of nanomaterials are at the core in the development of highly stretchable supercapacitors, batteries, and solar cells for such applications. This chapter first provides a brief summary of research development on the stretchable ESCDs in the past decade through various strategies of device design and manufacturing approach. After that, the focuses are on the advanced engineering of nanomaterials as active materials to achieve the stretchability of these ESCDs while maintaining a stable and functional performance. Finally, some of the challenges and the important directions in the areas of design and synthesis of nanomaterial facing the stretchable ESCDs are discussed concisely.

4.1 Introduction

With the rapidly increasing population and economic development, the world's energy consumption is escalating dramatically, resulting in one of the most serious global challenges of our time-energy scarcity. Hence, a crowd of research in the areas of advanced technologies for energy storage (e.g., batteries and supercapacitors) and conversion (e.g., solar cells) during the past decade has been driven by the

K. Xie · B. Wei
Center for Nano Energy Materials, State Key Laboratory of Solidification Processing and School of Materials Science and Engineering, Northwestern Polytechnical University, Xi'an 710072, China

B. Wei (✉)
Department of Mechanical Engineering, University of Delaware, Newark, DE 19716, USA
e-mail: weib@udel.edu

Q. Li (ed.), *Nanomaterials for Sustainable Energy*,
NanoScience and Technology, DOI 10.1007/978-3-319-32023-6_4

demands of a wide range of applications, from micro- or nano-electromechanical systems to portable electronic devices, and to hybrid vehicles and renewable energy systems [1–6]. Specially, the emerging of stretchable electronics, which show extensive application possibilities in various fields ranging from "epidermal" health/wellness monitors to sensitive robotic skins, soft surgical tools to electronic "eyeball" imaging devices with arbitrarily shaped surfaces [7, 8], is starved for the corresponding stretchable energy storage and conversion devices (ESCDs) [9–11]. As a result, stretchable batteries [12–14], supercapacitors [15, 16], and organic/inorganic solar cells [17–21], have all sprung up and became the research frontier of ESCDs in the past several years.

Realization of the stretchability of these ESCDs usually brings out a more challenging type of mechanics, in which these devices must accommodate a large strain deformation, typically of arbitrary form. In contrast, from the materials points of view, conventional active materials, such as inorganic materials (e.g., Si, MnO_2) and conductive polymers (e.g., PEDOT:PSS), used in the ESCDs during stretching usually remain well below the typical levels required to induce fracture (e.g., <1 % for metal/inorganic materials and <5 % for conductive polymers) [7]. Even for the carbon nanotube (CNT), only a small strain value below 10 % can be sustained during the stretching. Obviously, almost all the active materials are not stretchable enough to accommodate a significant deformation for stretchable ESCD applications.

Thus, strategies for device design and manufacturing approach, to achieve the stretchability of these ESCDs while maintaining a stable and functional performance, are highly desired. The following strategies for achieving stretchable ESCDs were proposed and demonstrated: (1) Wave structural configuration (Fig. 4.1a): Wave structural configuration, which can accommodate large tensile and compressive strains by changing the wave amplitudes and wavelengths to avoid potentially destroying the materials themselves, has become one of the most well-developed strategies to achieve the mechanical stretchability of electronic devices; (2) Interconnect-island mesh configuration (Fig. 4.1b): In this configuration, rigid material islands are interconnected by metal interconnects with in-plane/out-plane designs, which offer the stretchability; (3) Textile structural configuration (Fig. 4.1c): Many textile items have exceptional characteristics of stretchability [22]. For example, the textile based on Spandex can be stretched up to 200 % of its original length and recovered completely when released [23]. Because most textile items are hierarchical structures consisting of up to 4 levels, i.e. fiber, yarn, fabric, and the final product, and within each level, there are a great variety of fibres or fabric structures available, a full range of options for optimizing the geometries of these layouts to achieve a high stretchability could be provided; (4) Origami/Kirigami patterned configuration (Fig. 4.1d): The principles of origami/kirigami are originated from the folding/cutting of paper to achieve the desired shape with unprecedented deformation capabilities, including folding, bending, twisting, and stretching. The newly emerged origami/kirigami-based methodology shows great potential to expand to other deformable electronics [24–26].

Fig. 4.1 **a** Wave structural configuration: a buckled SWNT macrofilm on an elastomeric PDMS substrate. Reproduced with permission [16] Copyright 2009, Wiley-VCH. **b** Interconnect-island mesh configuration: a stretchable GaAs photovoltaic module; Reproduced with permission [22] Copyright 2011, Wiley-VCH. **c** Textile structural configuration: a dye-sensitized solar cell textile integrated into the fabric. Reproduced with permission [22] Copyright 2014, Wiley-VCH. **d** Origami patterned configuration: an origami battery in the completely unfolded state using 45 °C Miura folding. Reproduced with permission [24] Copyright 2014, Nature Publishing Group

In spite of the fast updating of these major strategies mentioned above and their derivatives, there is still one of the most challenging problems need to be resolved: what kind of materials can be used to realize the functionalities of these stretchable ESCDs while endowing them with excellent mechanical properties? Notice that the stretchable ESCDs usually consist of nearly all conceivable classes of materials: liquid, gel, solid, organic and inorganic, insulator, semiconductor, and metal, which span a range of more than 12 orders of magnitude in the elastic modulus [27]. Fortunately, the emergence of nanomaterials and nanotechnology may help us to find the possible answers.

In this chapter, we intend to introduce recent work dedicated to advancing the engineering of nanomaterials as the active materials for the stretchable ESCDs. This review will also present various methods that have been developed for the fabrication of active nanostructured materials for the stretchable ESCDs with excellent mechanic performance and electronic/electrochemical performance.

4.2 Carbon Materials

Carbon-based materials are one of the common materials in constructing different ESCDs, such as the anode electrodes in Lithium-ion batteries (LIBs), the electrodes in supercapacitors, the electrodes or the catalyst supporters in polymer electrolyte membrane fuel cells (PEMFCs), and the counter electrodes in dye-sensitized solar cells (DSSCs). The promising applications of carbon-based materials are based on their diversity features in structures, such as activated carbons, templated carbons, carbide-derived carbons, activated carbon fabrics, carbon fibers, carbon aerogels, carbon nanotubes, carbon nanohorns, carbon onions, and the emerging graphene or graphene-based composite. They have been widely used and extensively studied as electrode materials, owing to their high conductivity, high surface area (up to 2000 $m^2\ g^{-1}$), excellent anti-corrosion property, high-temperature stability, controllable porous structure, compatibility in composite materials, and relatively low cost. There is no exception that carbon-based materials will play an important role in building the stretchable ESCDs by considering their unique and superior mechanical robustness, stable and flexible integrity features. Followings are just a few examples under scrutiny in possibly stretchable ESCD constructions [28].

4.2.1 Carbon Nanotubes

CNTs, with ideal nanoscale dimensions, high mechanical performance, and excellent electrical conductivity, have been widely utilized in exploiting stretchable electronic devices [29–31].

One of the most widespread methods to fabricate stretchable electrodes is buckling CNTs on an elastomeric substrate. In our previous work, we had pioneered a new direction to fabricating stretchable supercapacitors with two sinusoidal single-walled carbon nanotube (SWNT) macrofilms as the stretchable electrodes, an organic electrolyte, and a polymeric separator [16]. The key to achieving the stretchability of the supercapacitor is that the sinusoidal SWNT macrofilm electrodes were prepared via the buckling strategy. The detail fabrication process includes the following three steps: first, fabrication of thin SWNT macrofilms by a floating chemical vapor deposition (CVD) method; second, attachment of the SWNT macrofilms on pre-strained elastomeric substrates (poly (dimethylsiloxane), PDMS) through a chemical bond; and last, release of the pre-strain applied to the PDMS substrates. This relaxation leads to the spontaneous formation of microscale, periodic, "wavy"-like structures in the SWNT macrofilm, which enable these systems not only to bend but also to stretch and compress (Fig. 4.2a, c). Figure 4.2b shows the cyclic voltammograms (CVs) of the stretchable supercapacitor measured at a high scan rate of 1000 mV s^{-1}. The CVs of the stretchable supercapacitors with 0 and 30 % applied strains retained a similarly rectangular shape, indicating an ideal capacitive behavior. Meanwhile, the specific

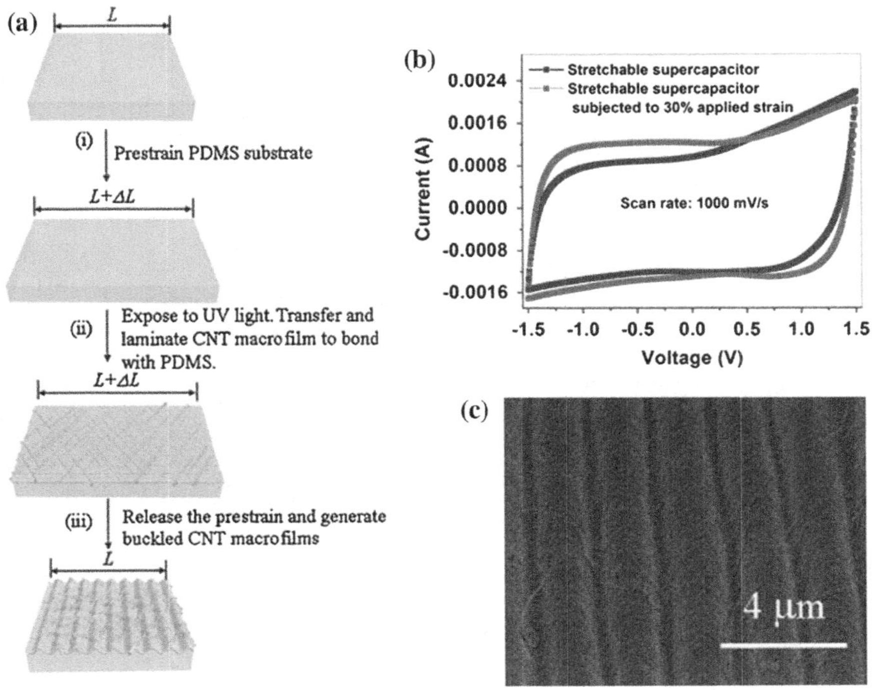

Fig. 4.2 **a** Illustration of the fabrication procedures of the buckled SWNT macrofilm on a PDMS substrate. **b** Cyclic voltammograms of the stretchable supercapacitors measured at the scan rate of 1000 mV s^{-1}. **c** SEM image of a buckled SWNT macrofilm. The buckling wavelength is 2 μm. Reproduced with permission [16] Copyright 2009, Wiley-VCH

capacitance of the stretchable supercapacitors with or without applied strain does not alter up to 1000 charge-discharge cycles, which concludes an excellent electrochemical stability [28]. To further enhance the stretchability while retaining high electric conductivity, the architectures of this wave structural configuration are carefully tailored by finely designing the CNTs with various morphologies and spatial distributions, such as CNT fibers [32], CNT ribbons [33], aligned CNTs [34], and hierarchically reticulated CNTs [35]. The similar buckling strategy can also be applied to other energy storage devices. Weng et al. fabricated an ultra-highly stretchable lithium-ion battery (LIB) based on two arched CNT composite electrodes via an improved buckling strategy [36]. The stretchable LIB consisted of an arched anode, a gel electrolyte, and an arched cathode, and finally packaged by an elastic polymer. Both the arched anode and cathode were fabricated according to the following steps: (1) Aligned CNT sheets, which had been dry-drawn from the spinnable CNT arrays, were stacked on a Cu foil with a thickness of *ca.* 100 nm (Fig. 4.3a); (2) Lithium titanium oxide (LTO, anode) or lithium manganese oxide (LMO, cathode) nanoparticles blended with CNT powders were then coated on the CNT sheets, followed by stacking another CNT sheet

Fig. 4.3 **a** CNT sheet on Cu foil. **b** An arched structure of the CNT/LTO-CNT composites with a pre-strain of 450 %. A gum-like LIB without stretching (**c**) and under stretching with increasing strain of 400 % (**d**). Reproduced with permission [36] Copyright 2015, Wiley-VCH

of ca.100 nm to form a sandwich structure; (3) an un-curing PDMS fluid was introduced into the sandwich structure, which was then placed onto a pre-stretched PDMS substrate with the pre-strain of 450 %; (4) after curing, removing the Cu foil and releasing the pre-strain, the arched sandwich-like composite electrodes were achieved (Fig. 4.3b). As a result, the stretchability of the designed arched electrode and LIB achieved 400 % (Fig. 4.3c, d), and the active components can be strongly anchored in the arched structure for a stable performance under stretching.

Inspired by previous fundamental and applied research into electronic textiles (e-textile), the concept of textile structural configuration is a current subject of the ESCD research. By using a simple "dipping and drying" process, where the fabric textiles were immersed in an SWNT ink, a highly stretchable supercapacitor has been presented by Hu and co-workers (Fig. 4.4a). The stretchable SWNT/fiber electrodes with high conductivity (125 S cm^{-1}) and low sheet resistance (<1 Ω sq^{-1}) were synthesized and used for the stretchable supercapacitors [37]. No obvious change in the specific capacitance of the stretchable supercapacitors was observed before and after being stretched up to 120 % strain for 100 times (Fig. 4.4b). As we mentioned above, fiber is the lowest structural level for most textile items. Begin with the lowest structural level to the highest structural level, Peng's group has developed a new and general method to produce stretchable and wearable photovoltaic textiles based on elastic and electrically conducting fibers. The fiber electrodes, with high and stable electronic properties under stretch, were firstly prepared by winding aligned multi-walled CNT (MWNT) sheets on stretched rubber fibers (Fig. 4.4c) [38]. A modified Ti wire as a working electrode was then

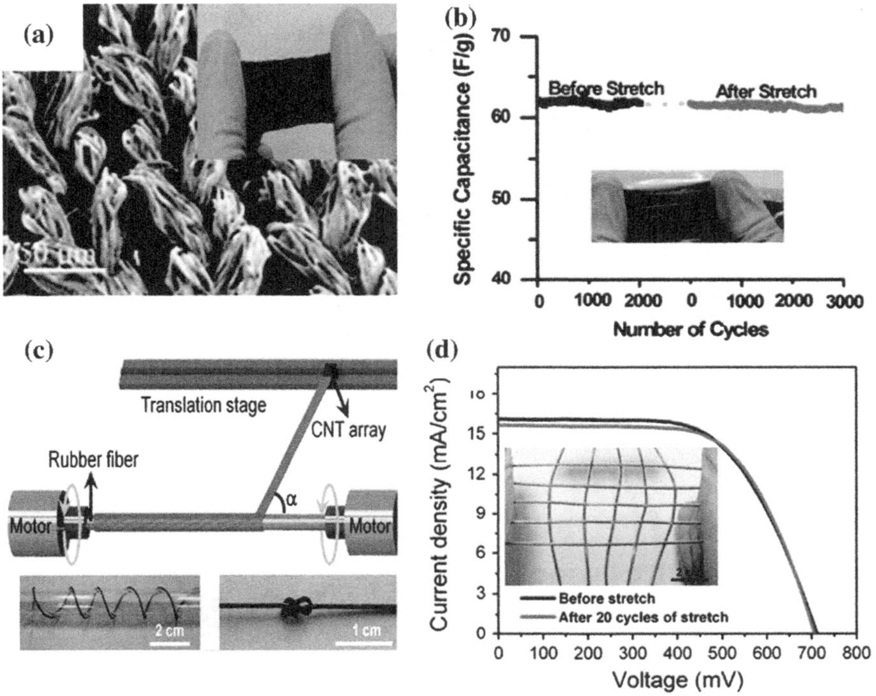

Fig. 4.4 **a** SEM image of a fabric sheet coated with SWNTs on the fabric fiber surface. The inserted optical image shows the excellent stretchability of the conductive textile. Reproduced with permission [37] Copyright 2010, American Chemical Society. **b** The cycle performance of a stretchable supercapacitor before and after stretching to 120 % strain. The charge-discharge current density is 1 mA cm^{-2}. The *inset* is a photograph of the supercapacitor under 120 % strain. Reproduced with permission [37] Copyright 2010, American Chemical Society. **c** Schematic illustration of fabrication of elastic conducting fibers. *Bottom* Photographs of the elastic conducting fibers in different forms. Reproduced with permission [38] Copyright 2014, Wiley-VCH. **d** I-V curves of a stretchable DSSC before and after a stretch for 20 cycles at a strain of 30 %. *Inset* photographs of a flexible and stretchable photovoltaic textile after stretch. Reproduced with permission [38] Copyright 2014, Wiley-VCH

twisted onto the elastic MWNT fiber, followed by coating photoactive materials to make a wire-shaped dye-sensitized solar cell (DSSC). The DSSCs were finally woven into the desired stretchable, wearable photovoltaic textile, and a high energy conversion efficiencies of 7.13 % was maintained under stretch (Fig. 4.4d) [38].

Origami/Kirigami-based approaches represent a potential alternative to enable high stretchability over existing methods that use elastomeric materials and mechanically designed structures such as buckling and serpentine shapes. For origami-based strategies, the ancient art of paper folding, compact deformable three-dimensional structures can be created from two-dimensional sheets through high degrees of folding along predefined creases. Following this idea, Jiang's group first reported an origami LIB, with excellent mechanical deformability, including

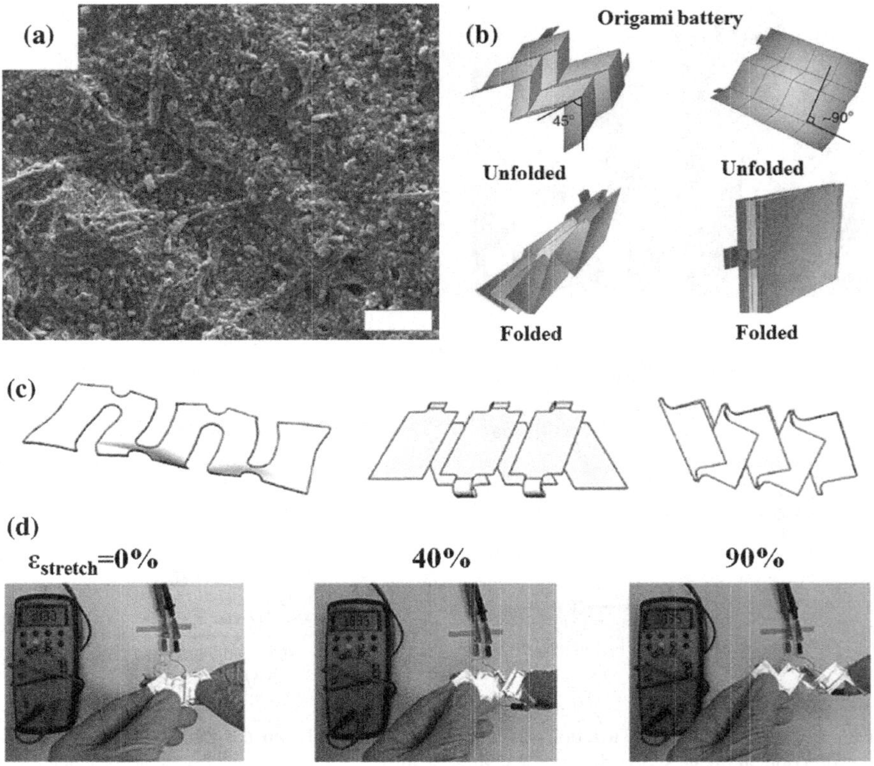

Fig. 4.5 **a** SEM image of $LiCoO_2$-CNT/paper cathode. Reproduced with permission [24] Copyright 2014, Nature Publishing Group. **b** Two examples of origami LIBs using Miura folding. Reproduced with permission [24] Copyright 2014, Nature Publishing Group. **c** Illustrations of three kirigami patterns. Reproduced with permission [25] Copyright 2015, Nature Publishing Group. **d** Photograph of stretching a kirigami LIB while it was connected to a voltmeter. Reproduced with permission [25] Copyright 2015, Nature Publishing Group

significant system-level linear and areal deformability, large twistability, and bendability, and up to 74 % areal coverage [24]. In order to achieve good deformability and electrical conductivity in the origami LIB, CNT-coated paper was used as the current collector, on which the cathode materials ($LiCoO_2$) and anode materials ($Li_4Ti_5O_{12}$) were mixed with the conductive agent and binder to form the slurries, and coated on the CNT-coated paper, respectively (Fig. 4.5a). The origami LIBs were realized by folding these layers based on specific origami patterns of Miura-ori. Two examples of origami LIBs using Miura folding are shown in Fig. 4.5b. However, the origami LIBs have two shortcomings [25]. First, their foldability is limited from the folded state to the planar state. Although it can be tuned to different folding patterns, the same constraint is still prescribed by the planar state. Second, the folded state involves uneven surfaces, which introduces inconvenience when integrating with planar systems, though this issue can be somewhat circumvented. In order to overcome these two shortcomings, the authors further explored the kirigami approach, by combining folding and cutting at the same time, to define patterns that form an even surface after stretching and the stretchability is not limited by the planar state [25]. Three kirigami patterns, cut-N-twist, cut-N-shear, and zigzag-cut patterns, were introduced in stretchable LIBs (Fig. 4.5c). Specifically, the LIBs using cut-N-twist patterns show a high level of stretchability up to 150 % (Fig. 4.5d). An additional benefit is the commercially standard manufacturing technologies that are used in the fabrication of origami/kirigami-based LIBs, which when combined with other deformable electronic devices may lead to direct practical applications [24, 25].

4.2.2 Graphene

Graphene has been attracting considerable attention for its use in stretchable ESCDs owing to its fascinating properties, such as extremely high electron mobility, high surface area, and high elasticity [39, 40]. Various synthesis methods for graphene have been developed, including micromechanical cleavage, epitaxial growth, CVD, and chemical solution methods. Due to the low cost, high efficiency, reliability, and relatively easy to realize large-scale production, the latter two strategies are broadly used to prepare high-quality graphene for stretchable ESCDs.

Chou et al. explored the applicability of the widely used pre-straining-then-buckling method to realize the stretchability of graphene electrodes for transparent and stretchable supercapacitor [41]. The CVD graphene film on a copper foil, which was a monolayer and of high quality, was first coated with poly (methyl methacrylate) (PMMA). After the Cu foil had been etched away in solution, the floating PMMA/graphene film was scooped up by another single-layer graphene on copper foil. After this cycle was repeated four times, the laminated multilayer graphene was attached to a pre-stretched PDMS film, and then the pre-stretching was released. Figure 4.6a shows the morphology of the buckled graphene layer with nanoscale, ripple-like features, which endow the electrode with high stretchability. Another transparent and stretchable supercapacitor based on wrinkled graphene electrode was constructed by Dai and his colleagues [42]. The fabrication process is different with Chou's work. First, a wrinkled copper foil was produced by sliding a tweezer with an appropriately wrinkled structure over it. The resultant wrinkled copper foil was then used as the substrate for the graphene growth by CVD. Followed by coating a thin layer of PDMS onto the top surface of the as-prepared graphene sheet and removing the copper substrate, a piece of a stretchable wrinkled graphene sheet on PDMS was obtained (Fig. 4.6b). Lastly, the transparent and stretchable all-solid-state supercapacitors were assembled by pressing two of the PDMS supported graphene electrodes together with a transparent layer of the polymer electrolyte. The transparent and stretchable supercapacitors shown a high transparency (57 % at 550 nm) and could be stretched up to 40 % strain without obvious performance change over hundreds of stretching

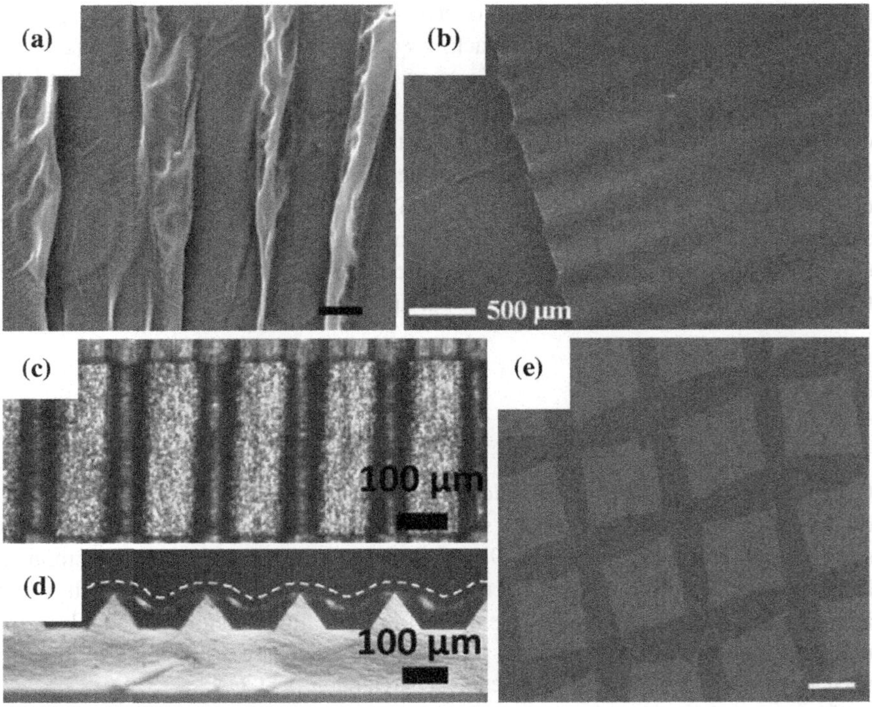

Fig. 4.6 CVD graphene for stretchable ESCDs: **a** SEM image of a four-layer buckled graphene film on PDMS substrate. Reproduced with permission [41] Copyright 2014, American Chemical Society. **b** Typical SEM image of the wrinkled graphene sheet on the PDMS substrate from the top view. Reproduced with permission [42] Copyright 2014, American Chemical Society. **c** Top- and **d** cross-section images of stretchable microsupercapacitors with suspended wavy structures. Reproduced with permission [43] Copyright 2015, Wiley-VCH. **e** SEM image of graphene woven-fabric. Reproduced with permission [44] Copyright 2014, Wiley-VCH

cycles. In 2015, graphene microribbons with a novel suspended wavy structure have been developed by Qi et al. for highly stretchable microsupercapacitors [43]. The fabrication process of the graphene microribbons includes photolithography, synchronous reduction, assembly strategy, and thermal evaporation processes. The graphene microribbon electrodes were then transferred onto a tripod-structured PDMS substrate to form the graphene microribbons with a novel suspended wavy structure (Fig. 4.6c). The electrochemical performance of the stretchable microsupercapacitors remained nearly unchanged when stretched up to 100 %. After being stretched-released for 5000 cycles, no significant decrease of the capacitance was observed. The excellent electrochemical, as well as mechanical performance, may be attributed to the following two design features [43]: (1) the suspended wavy structured electrode arrays decreased the stain concentration on the electrode fingers in the stretching/relaxing processes so that the detaching and cracking of the electrode materials could be prevented; and (2) it ensured the electrode fingers keeping relatively constant distance in the stretching/releasing process, so the stability of the microsupercapacitors could be further enhanced. In addition to the wavy structural CVD graphene, a graphene woven-fabric (GWF) film was also synthesized by using CVD on a copper mesh, which can be effectively removed with a chemical solution (Fig. 4.6d) [44]. Compared to its graphene film counterparts, GWF possesses superior flexibility and strength, presumably because they retain the network configuration of the copper mesh. After transferring it to a pre-strained PDMS, a stretchable GWF electrode was fabricated and used for stretchable supercapacitor application [45].

Apart from the CVD graphene, the chemically/electrochemically reduced graphene oxides (rGO) were also introduced for stretchable supercapacitor. Zang et al. fabricated a chemically reduced rGO paper in the form of hydrogel by vacuum filtration of a rGO solution [46]. The as-prepared rGO paper was then bonded onto a pre-strained elastomer film via a dry-transfer method. Followed by using the prestraining-then-buckling method along two directions, a crumpled rGO paper was fabricated (Fig. 4.7a). The crumpled rGO paper has demonstrated not only

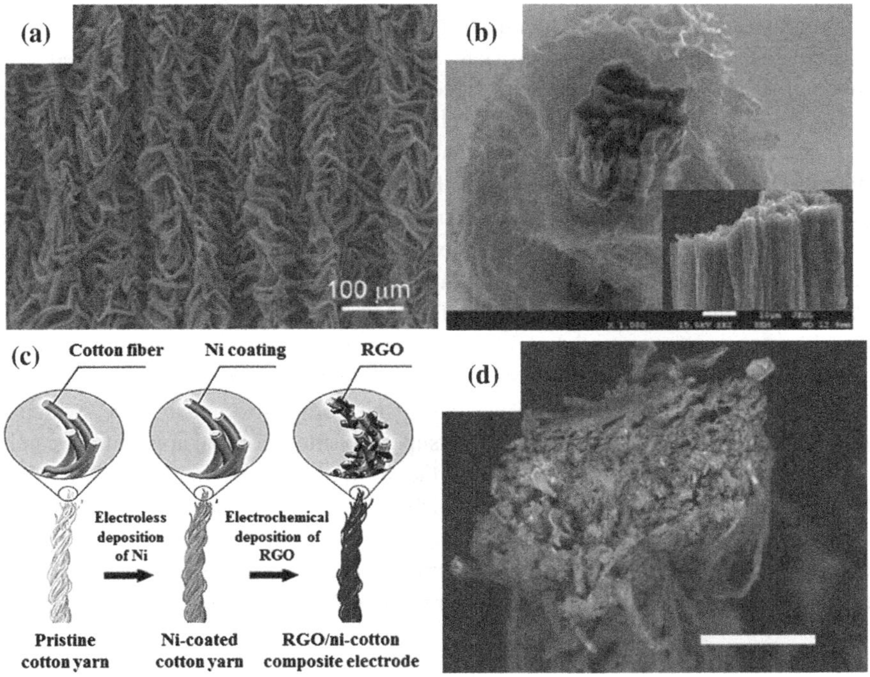

Fig. 4.7 Chemical/electrochemical reduced graphene for stretchable ESCDs: **a** SEM image of crumpled graphene paper. Reproduced with permission [46] Copyright 2014, Nature Publishing Group. **b** SEM image of a graphene fiber@3D-graphene (*inset* graphene fiber). Reproduced with permission [47, 48] Copyright 2012, 2013, Wiley-VCH. **c** Schematic illustration of the fabrication of RGO/Ni cotton yarn composite electrodes. Reproduced with permission [49] Copyright 2015, Nature Publishing Group. **d** SEM image of typical RGO/Ni cotton composite electrode. Reproduced with permission [49] Copyright 2015, Nature Publishing Group. Scale bar: 250 μm

outstanding capacitive performance with the specific capacitance in the range of 166–196 F g^{-1} at a discharge rate of 1 A g^{-1}, but also extremely high stretchability of 300 % linear strain and 800 % areal strain. By using aqueous graphite oxide (GO) suspensions as precursors and a glass pipeline as reaction vessel, Qu et al. developed a facile one-step dimensionally confined hydrothermal strategy to fabricate neat rGO fibers (rGF, Fig. 4.7b Inset) [47]. By electrochemically electrolyzing GO aqueous suspension on the rGF, a unique all graphene core-sheath fiber (rGF@3D-G) was prepared, in which a core of rGF is covered with a sheath of 3D porous network-like graphene framework (Fig. 4.7b) [48]. When two rGF@3D-G fibers were interwound together with gel electrolyte, an all-solid-state fiber supercapacitor, which can be managed to construct a spring supercapacitor with highly compressible and stretchable properties, was assembled. Moreover, the spring supercapacitor can also be conveniently woven into a textile for wearable electronics. Another novel and wearable supercapacitor yarns, consisted of hierarchical graphene-metallic textile composite electrodes, were constructed by Zheng and co-workers [49]. The hierarchical composite electrodes consist of rGO sheets anchored on the surface of Ni-coated cotton yarns, which are fabricated by highly scalable electroless deposition of Ni and electrochemical deposition of graphene on commercial cotton yarns, sequentially (Fig. 4.7c, d). A high energy density and power density of the all solid-state supercapacitor yarn was achieved (6.1 mWh cm^{-3} and 1400 mW cm^{-3}), respectively. The newly-developed supercapacitor yarn with fantastic mechanical characteristics is integratable into various wearable electronic devices as stretchable ESCDs. In consideration of ESCDs' practical application, two industrial methods: machine coating and pressure spray printing were introduced by Li et al. to fabricate a rGO-based stretchable supercapacitor with 3D interdigital electrodes (Fig. 4.8a) [50]. In this supercapacitor, each interdigital electrode contains a layer of solid electrolyte film by means of machine coating and a sandwich layer of rGO thin film, which consists of two rGO thin layers as active materials with a layer of Ag nanowires as current collector (rGO/Ag-NWs/rGO), through pressure spray printing (Fig. 4.8b). Based on the novel concept of 3D interdigital electrode, the volumetric capacitance, which is usually unsatisfactory to graphene-based supercapacitors for real application, can be adjusted to a higher level. Encouraging, the volumetric capacitance can reach 77.9 F cm^{-3} with the 5 μm rGO layer, which is one of the highest values of the carbon-based supercapacitors. Meanwhile, the variation of the discharge capacitance of the supercapacitor with 30 % strain after 4000 charge-discharge cycles is negligible ($\sim$3 %), indicating a stable mechanical-electrochemical performance.

4.2.3 CNT/Graphene Hybrid

In order to take full advantage of potential merits of both CNT and graphene, a hierarchically structured hybrid microfiber, consisting of SWNTs and nitrogen-doped rGO sheets (Fig. 4.9a, b), was proposed for fiber-shape supercapacitors [51].

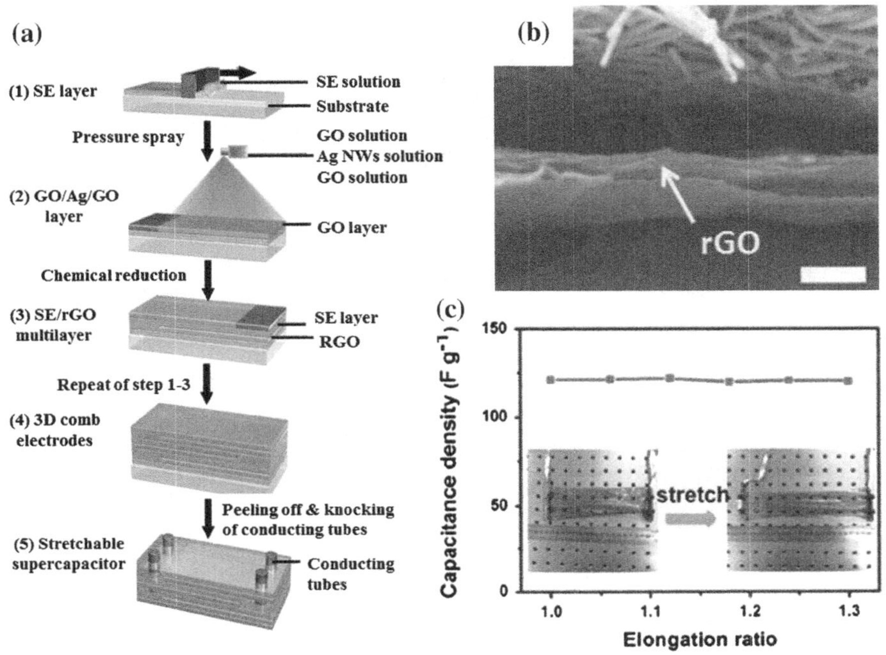

Fig. 4.8 **a** Schematic illustration of the in situ fabrication of solid electrolyte/rGO/Ag NWs/rGO (SRARS)-based stretchable supercapacitor. **b** Magnifying view of the cross-section of solid electrolyte/rGO/Ag NWs/rGO layer. Scale bar: 500 nm. **c** Changes of capacitance density with different ε at current density of 0.5 A g^{-1}. *Inset* optical images of as-fabricated two-layer SRARS film with ε at 1.0 and 1.3, respectively. Reproduced with permission [50] Copyright 2015, Wiley-VCH

To fabricate the hybrid microfiber, a scalable hydrothermal method was developed to continuously produce the fibers using a silica capillary column as a microreactor. During the fabrication process, nitric-acid-treated SWNTs and GO were first mixed in the presence of ethylenediamine (EDA) to form a homogeneous aqueous dispersion, which was then injected into a flexible fused-silica capillary column served as the hydrothermal reactor by a peristaltic pump. After heated at 220 °C for 6 h and cooling to room temperature, the resultant wet fiber was pushed out by nitrogen flow into a water reservoir. The large specific surface area (396 m^2 g^{-1}) and high electrical conductivity (102 S cm^{-1}) endows the CNT/graphene hybrid with high packing density and a large ion-accessible surface area. Hence, the fiber-shaped supercapacitors fabricated with two fibers show ultrahigh volumetric energy densities comparable to those of thin-film lithium batteries, without compromising cyclability and rate capability. Almost simultaneously, a coaxial wet-spinning assembly strategy is reported by Gao and his colleagues to readily synthesize polyelectrolyte-wrapped CNT/graphene core-sheath fibers that are applied directly as contactable and interweaving yarn supercapacitor electrodes for stretchable electronics [52]. Figure 4.9c shows a continuous coaxial fiber with the core of mixture of rGO and CNTs and the

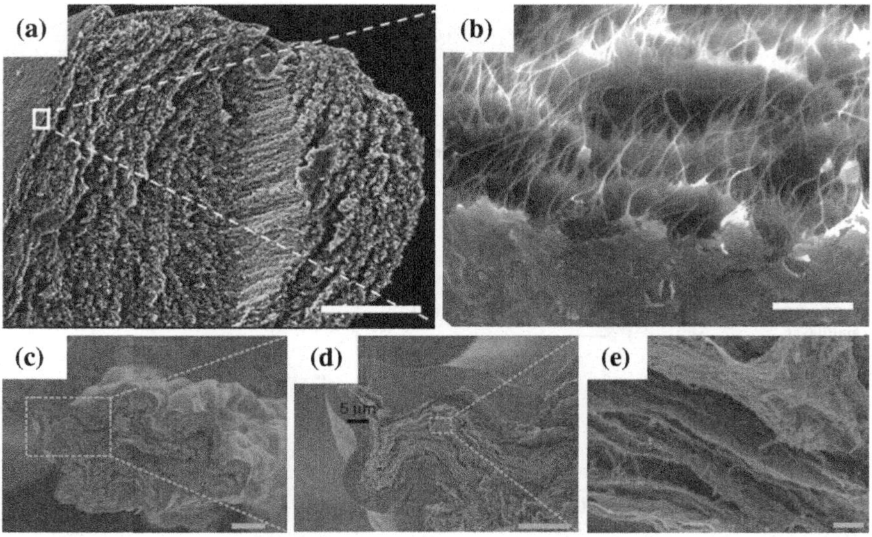

Fig. 4.9 **a**, **b** SEM images of the cross-section of CNT/graphene hybrid fiber. Scale bar: 15 μm for (**a**) and 0.5 μm for (**b**). Reproduced with permission [51] Copyright 2014, Nature Publishing Group. **c–e** rGO + CNT@CMC spun with GO and CNT mixture and CMC. Scale bar: 20, 20 and 0.2 μm for (**c**), (**d**) and (**e**), respectively. Reproduced with permission [52] Copyright 2014, Nature Publishing Group

shell of sodium carboxymethyl cellulose (CMC). It can be seen that the CNTs were densely coated on the surfaces of rGO and separated the adjacent rGO sheets aligned with layered structures (Fig. 4.9d, e). As a result, the rGO/CNT@CMC fibers are used directly as safe electrodes to assembly two-ply yarn supercapacitors with ultra-high capacitance. The combination of scalable coaxial wet-spinning technology and excellent performance of yarn supercapacitors makes the CNT/graphene hybrid a promising alternative material for wearable and stretchable electronics.

4.2.4 Carbon Fiber

Along with its good electrical conductivity and interwoven network structure, carbon fiber is regarded as a replacement to inactive polymer fiber backbone in a conventional textile structural configuration for further increasing the energy density of stretchable supercapacitors. Hence, highly conductive carbon fibers are knitted to construct a fabric as the conductive current collector (Fig. 4.10a) [53]. After painting activated carbon by the screen printing technique and employing a polymer electrolyte (Fig. 4.10b–e), the stretchable supercapacitors, with the highest reported capacitance per area for an all-carbon textile-supercapacitor (0.5 F cm^{-2}), have been constructed. The textile-supercapacitor can also be bent at 90°, 135°, and 180° when simultaneously stretched to 50 %, though accompanied with different degrees of capacitance decreases.

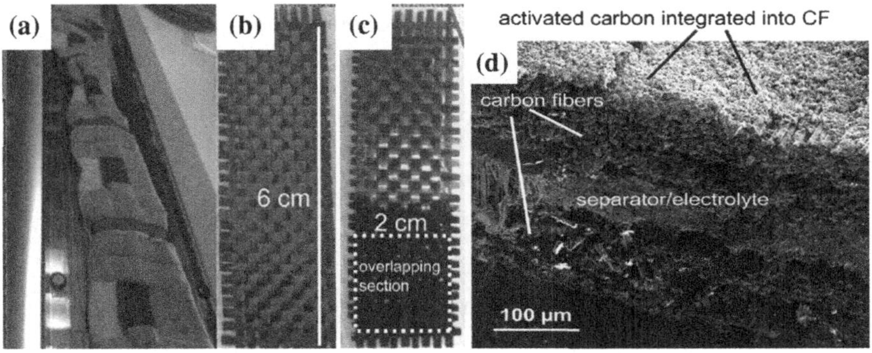

Fig. 4.10 **a** Four current collectors knitted at once, demonstration of how lengths of electrodes can be knitted quickly and efficiently, **b** carbon fiber woven fabric before printing. **c** Carbon fiber woven fabric after printing. Only the overlapping section is coated in the electrolyte. **d** SEM micrograph of the cross-section of a YP17 woven carbon fiber device, ~350 μm thick. Reproduced with permission [53] Copyright 2013, Royal Society of Chemistry

4.2.5 Carbon Grease

Carbon grease, which normally consists of carbon powders dispersed into a viscous matrix, is one of the most common materials used as electrodes for stretchable electronic devices at the early stage [54]. By integrating carbon black/silicon oil paste into a stretchable acrylic elastomer, Kaltenbrunner et al. have utilized the carbon grease as current collectors for ultracompliant electrochemical dry gel cells (MnO_2–Zn primary cell) [12]. These dry gel cells can withstand stretch ratios up to 100 % before failure. Though carbon grease is easy to handle and shows good stretchability, it also has its own shortcomings [12, 54, 55]. First, the long term stability of the carbon grease is often impaired by the viscous matrix drying or diffusing into the elastomeric substrate. Second, the fluidity of the carbon grease negatively impacts the large-scale homogeneity of the electrode properties. More importantly, when stretched, significant decreases in electrical conductivity of the carbon grease current collectors and a loss in electrical contact between the active materials and the current collectors have appeared, hurting the performance of the stretchable dry gel cells to a large extent. Therefore, significant efforts are still urgently needed to solve these problems for the easy-fabricated carbon grease.

4.3 Conjugated Polymer

Owing to their excellent electrical and optical properties, conjugated polymers, such as poly(3,4-ethylenedioxythiophene):poly(styrenesulfonate) (PEDOT:PSS), polypyrrole (PPy) and polyaniline (PANI), are promising materials in stretchable ESCDs. For instance, organic electronic devices often show high compliance [56].

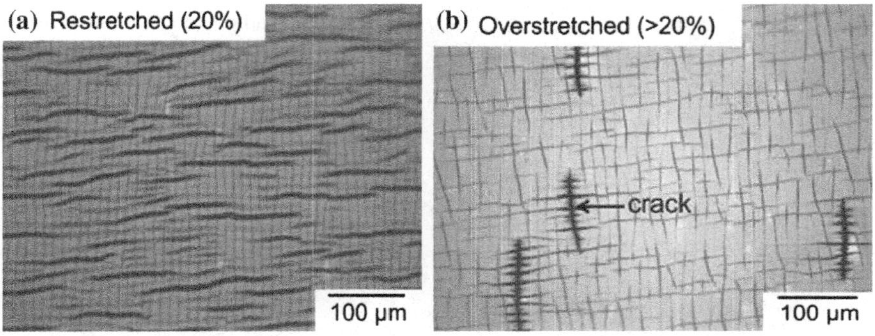

Fig. 4.11 Optical images of PEDOT:PSS/P3HT:PCBM films on a pre-strained PDMS substrate. Reproduced with permission [20] Copyright 2011, Wiley-VCH

In a recent demonstration, Lipomi and co-workers have fabricated the first stretchable organic solar cell based on the buckling strategy [20]. The stretchable organic solar cell was produced by depositing the spin-coated PEDOT:PSS transparent conductor, the P3HT:PCBM active layer, and a conformal eutectic gallium-indium top electrode on a pre-strained PDMS membrane. After removing the clips and separating the device from the glass, the device contracted spontaneously, and buckled structures in the PEDOT:PSS/P3HT:PCBM film were formed. If a tensile strain is not beyond the pre-strain (20 %), no degradation can be found in electronic performance of the PEDOT:PSS/P3HT:PCBM film even up to 1000 stretching cycles (Fig. 4.11a). When the film was stretched over the value of the pre-strain, cracks appeared in the film (Fig. 4.11b). As for the device, the short circuit current (J_{SC}) and the open circuit voltage (V_{OC}) increased with the strain. When the organic solar cells were stretched to 18.5 % after eleven full cycles, the device exhibited the following figures of merit under a flux of 95 mW cm^{-2}: J_{SC} is 7.4 mA cm^{-2}; V_{OC} is 415 mV; fill factor (FF) is 0.38; and the power conversion efficiency (PCE) is 1.2 %. Stretching beyond 22.2 % produced cracks in the active layer, and concomitant reductions in J_{SC}, FF, and PCE. However, the stretchable organic solar cell presented there, is far from the state of the art in organic solar cells in terms of efficiency.

A creative work of Kaltenbrunner et al. has demonstrated an ultrathin and light-weight organic solar cell on plastic foil less than 2 μm thick (the entire device), with equal PCE to their thick counterparts [21]. In the fabrication procedure, the P3HT:PCBM is firstly deposited on a PEDOT:PSS-coated polyethylene terephthalate (PET) substrate foil (1.4 μm), and the Ca/Ag (115 nm) is then evaporated onto it as the top electrode. Thus, the whole organic solar cell is just 1.9 μm thick, and only one-quarter of the thickness is composed of the active solar cell (Fig. 4.12). By adhering the flexible organic solar cell on a 100-μm-thick pre-stretched 3 M VHB 4905 elastomer, the devices can survive quasi-linear compression to below 70 % of their original area. The optimized device performs with V_{OC} = 580 mV, J_{SC} = 11.9 mA cm^{-2}, FF = 61 %, and PCE = 4.2 % under one sun illumination and

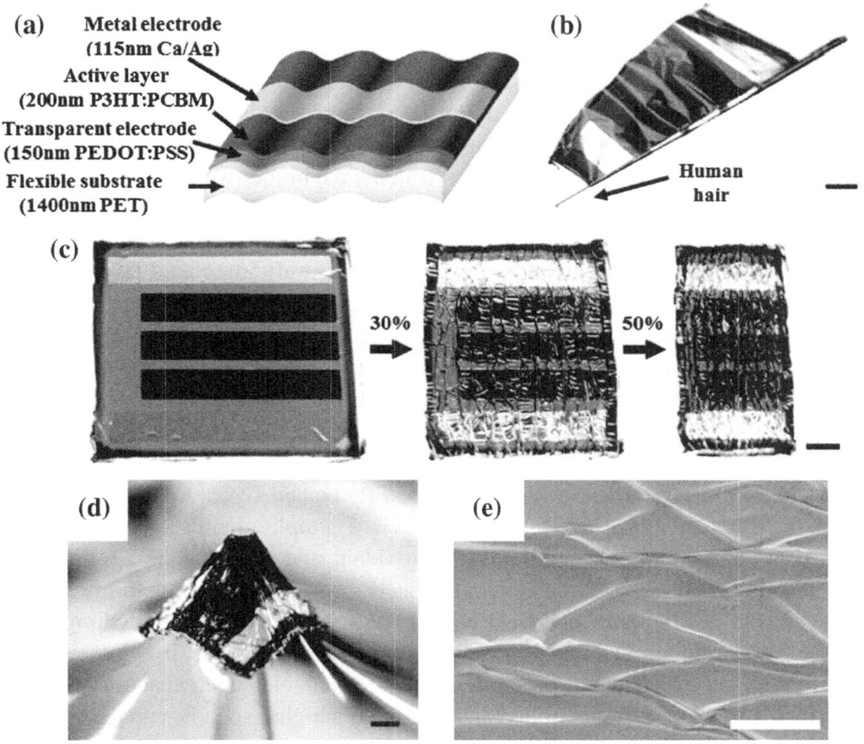

Fig. 4.12 Sub-2-μm-thick organic solar cells. **a** Scheme of the ultra-light and flexible organic solar cell. The layer thicknesses are drawn to scale. **b** Extreme bending flexibility demonstrated by wrapping a solar cell around a 35-μm-radius human hair. Scale bar (also in **c**, **d**) 2 mm. **c** Stretchable solar cells made by attaching the ultrathin solar cell to a pre-stretched elastomer. They are shown flat (*left*) and at 30 % (*middle*) and 50 % (*right*) quasi-linear compression. **d** The device attached to the elastomeric support, under three-dimensional deformation by pressure from a 1.5 mm-diameter plastic tube. **e** SEM image of the PET surface of the solar cell in a compressed state. The radius of curvature for the shown wrinkles is estimated to be on the order of 10 μm. Scale bar 500 μm. Reproduced with permission [21] Copyright 2012, Nature Publishing Group

remains fully functional down to 70 % compression. Biaxial compression/tension is a more important property in practical applications of the stretchable ESCDs. The device performances under reversible biaxial compression of 0, ~10, ~20, and 50 % are evaluated. According to the accepted definitions for stretchable conducting films, the solar cell can afford 400 % tensile strain during the re-stretching process [21]. Only a slight decrease in the photovoltaic performances of the devices can be seen under the biaxial compression of 20 %. Even under a very high biaxial compression of more than 50 %, the device can remain functional. Nevertheless, during the 22 consecutive compression-stretch cycles to 50 % compression, the J_{SC}, FF, and the output power decreased gradually, resulting in a 27 % decrease in power after 22 cycles. Even so, this work does provide new insights into the design and fabrication of the stretchable organic solar cells for practical applications.

In another systematic work, stretchable e-textile fabricated by simply soaking various fabrics, including 100 % cotton, 60 % cotton/40 % polyester, 95 % cotton/5 % Lycra, 60 % polyester/40 % rayon, 100 % polyester, 50 % nylon/50 % polyurethane Spandex, and 80 % nylon/20 % Spandex, in a conductive polymer aqueous dispersion (PEDOT:PSS), have also been studied by Ding and co-workers [23]. It has been found that those fabrics which have higher water uptake capability usually resulted in higher conductivities upon soaking in PEDOT:PSS. Because of the simplicity of the production process, this work is valuable for academic and the industrial community to develop real, wearable, and truly integrated fabric ESCDs.

The first application of inherently conducting polymers as stretchable battery electrodes has been reported by Wallace et al. [13]. Buckled PPy with a wave structural configuration can endure 2000 stretching cycles with 30 % tensile strain applied whilst retaining its electrical conductivity and electrochemical properties for Mg batteries. A cell voltage of 1.12 V was obtained after 2320 in situ stretching cycles at 30 % strain.

In the textile structural configuration, the PPy coated Lycra fabric electrode, in which PPy was synthesized on the fabric via a simple chemical polymerization process with ammonium persulfate as an oxidant and naphthalene-2,6-disulfonic acid disodium salt as a dopant, demonstrated an excellent stretchability as well (Fig. 4.13) [57]. The fabricated electrode can be subjected to 1000 stretching cycles with 100 % strain applied and still delivers a stable capacitance. It is interesting to note that the electrochemical properties of the stretchable supercapacitors could be improved with in situ strain applied and exhibited a higher specific capacitance. In a two-electrode system, the specific capacitance increased from 108.5 F g^{-1} with 0 % strain to 117.6, 119.6, and 125.1 F g^{-1} with 20, 40, and 60 % elongations, respectively, at a current density of 1.0 A g^{-1}. Another strategy to fabricate the PPy-based stretchable electrode with textile structures is electrochemical deposition of PPy on smartly-tailored stainless steel meshes (Fig. 4.13) [58]. By cutting along a particular direction, a knitted stainless steel mesh was found to become very stretchable. The fabricated solid-state supercapacitors, based on the PPy coated stainless steel mesh electrodes, possess a specific capacitance of 170 F g^{-1} at a current density of 0.5 A g^{-1} and it can be effectively enhanced to 214 F g^{-1} with a 20 % strain. Moreover, the supercapacitors achieve remarkable capacitance retentions of 98 % and 87 % at a very high current density of 10 A g^{-1} after 10,000 cycles under 0 and 20 % strains, respectively, indicating an excellent cycle stability.

Similar to fabricate the PPy-based stretchable electrode, PANI coated carbon fiber was prepared by directly electrochemical deposition of PANI on the carbon fiber as the positive electrode (Fig. 4.14a–d) [59]. By coupling a functionalized carbon fiber as a negative electrode and a gel electrolyte, a fiber-shaped supercapacitor was constructed (Fig. 4.14a). The as-prepared supercapacitor shows a volumetric energy density up to 2 mWh cm^{-3} and a maximum power density of 11 W cm^{-3}. Apart from good rate capability, long cycle life, and high volumetric capacitance, the proposed supercapacitor also owns an excellent stretchability. Figure 4.14e shows a striking example of the fiber-shaped supercapacitor with high stretchability. The capacitance of the stretchable supercapacitor remains almost

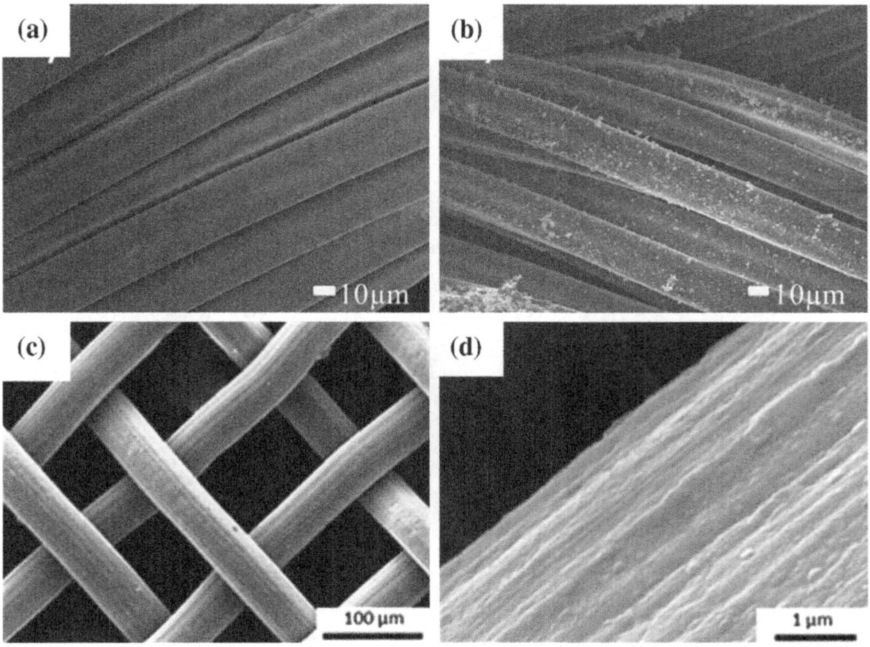

Fig. 4.13 SEM images of **a** blank and **b** PPy coated nylon lycra fabric. Reproduced with permission [57] Copyright 2012, Elsevier. **c** An overall view of the uniformly coated 2D hierarchical PPy-mesh electrode, and **d** High magnification of a PPy film showing a rough surface. Reproduced with permission [58] Copyright 2015, Elsevier

unchanged up to 100 % strain (Fig. 4.14f). The CV curves of the stretchable supercapacitor under different applied tensile strains (0, 10, 20, 40, 60, 80, and 100 %) at a scan rate of 100 mV s^{-1} are also shown in the inset of Fig. 4.14f. Under different stretch stains, there is negligible difference in the CV curves, indicating the feasibility of the stretchable supercapacitor for stretchable electronics. Moreover, Zhong et al., for the first time, fabricated conductive and stretchable CNT reinforced shape memory polymer (SMP) fibers (polyvinyl alcohol, PVA) via wet-spinning for supercapacitor applications. Followed by using the layer by layer (LBL) method, a densely packed MWCNT layer serving as a conductive current collector were prepared on the SMP substrate. PANI was then continuously deposited on the surface of the MWCNT as an active material to further promote capacitance of the fibers. After the infiltration of a PVA-based gel electrolyte into the porous MWCNT and PANI layer, the conductivity of ~50 S cm and stretchability of more than 400 % are achieved (Fig. 4.15) [60]. The resultant shape memory polymer fibrous supercapacitor (SMPFSC) shows an extremely high capacitance (>427 F cm^3).

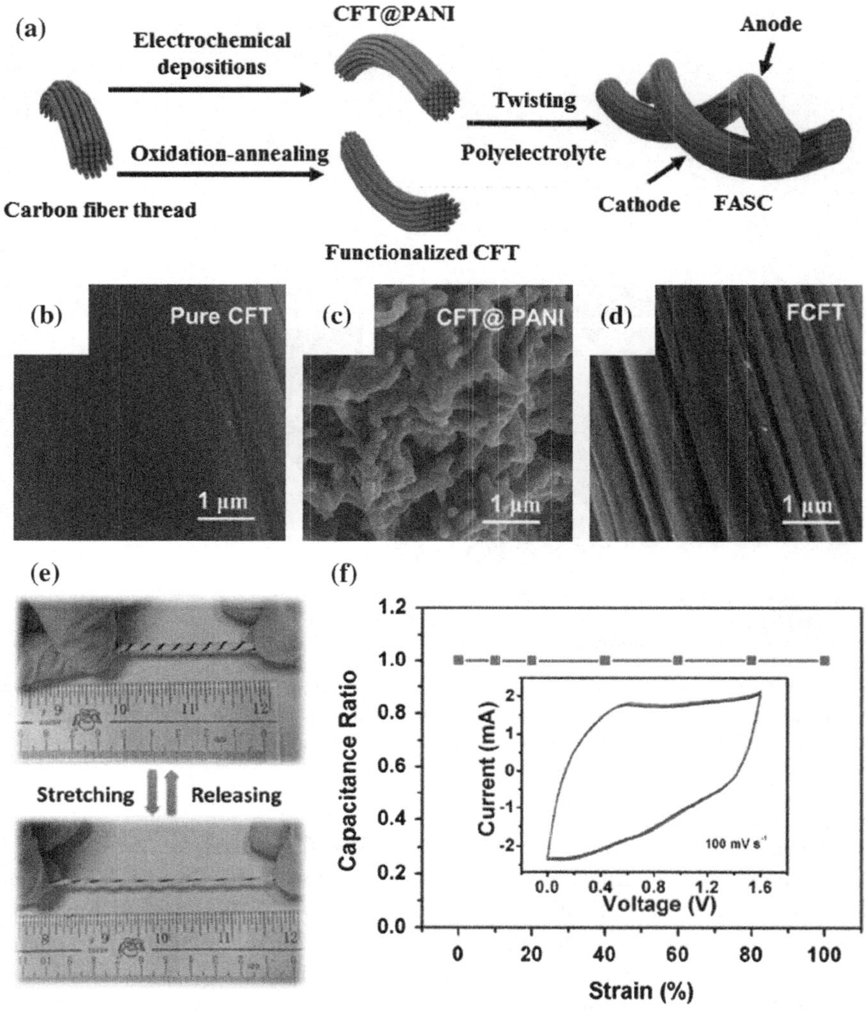

Fig. 4.14 **a** Schematic diagram of the fabrication procedure of a fiber-shaped asymmetric supercapacitor (FASC). **b** SEM image of a carbon fiber thread (CFT). **c** SEM image of CFT@PANI electrode. **d** SEM image of functionalized carbon fiber thread (FCFT) electrode. **e** Digital images of the stretchable FASC before and after stretching. **f** Capacitance ratio of the FASC under different applied strains (0, 10, 20, 40, 60, 80, and 100 %) at a scan rate of 100 mV s^{-1}. *Inset* shows the CV curves of the FASC measured at different applied tensile strains. Reproduced with permission [59] Copyright 2015, Elsevier

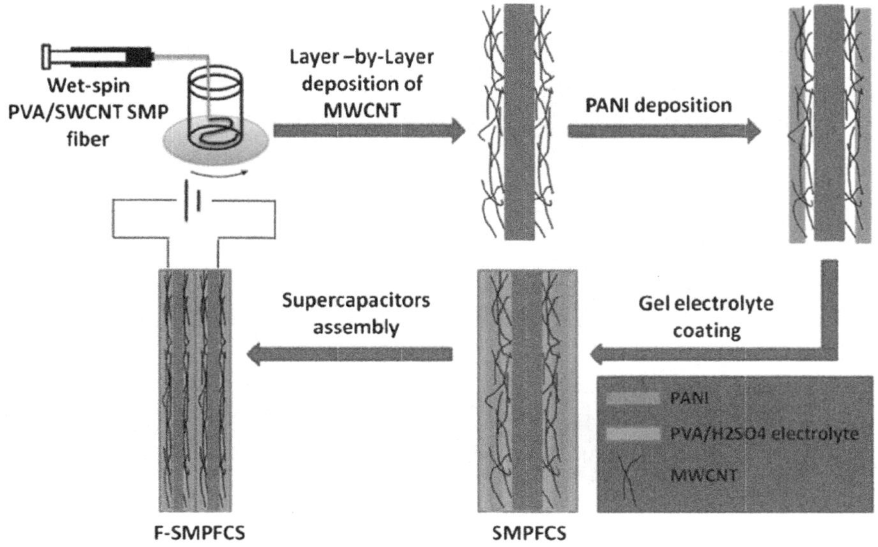

Fig. 4.15 Schematic of the SMPFCS and F-SMPFCS fabrication process. Reproduced with permission [60] Copyright 2015, Elsevier

4.4 Metal Oxides

Metal oxides have been of particular interest for ESCDs during the past several decade. With low-cost, abundant and environmentally friendly nature, metal oxides, specially MnO_2 [61], Fe_3O_4 [6], and TiO_2 [62, 63], are widely investigated for supercapacitor, battery, and solar cell applications. However, the hard and brittle characteristics of these metal oxides have greatly limited their applications in stretchable ESCDs. Hence, various strategies, as mentioned in the introduction part, have been employed to construct the metal oxide based stretchable ESCDs.

Take MnO_2 as an example for stretchable and high-performance supercapacitor application, Xu et al. proposed an asymmetric stretchable wire-shaped supercapacitor with a MnO_2/CNT hybrid fiber as positive electrode (Fig. 4.16b), which was prepared by precipitating nanostructured MnO_2 particles from the potassium permanganate precursor on the aerogel CNT fiber, a pristine CNT fiber as negative electrode (Fig. 4.16a) and PVA-KOH gel as electrolyte [61]. The asymmetric wire-shaped supercapacitor can work at a potential window from 0.8 to 1.5 V, achieving tripled energy density and doubled power density compared to its symmetric counterpart while accomplishing stretchability of up to 100 % through the prestraining-then-buckling approach. These excellent electrochemical and mechanical characteristics made the supercapacitor a promising energy storage device for smart textiles. Choi et al. chose a commercial nylon fiber as the coiled polymer substrate [64]. When wrapped with CNT sheet and electrodeposited MnO_2 consequent consecutively (Fig. 4.16c–e), two coiled MnO_2/CNT/nylon fiber were

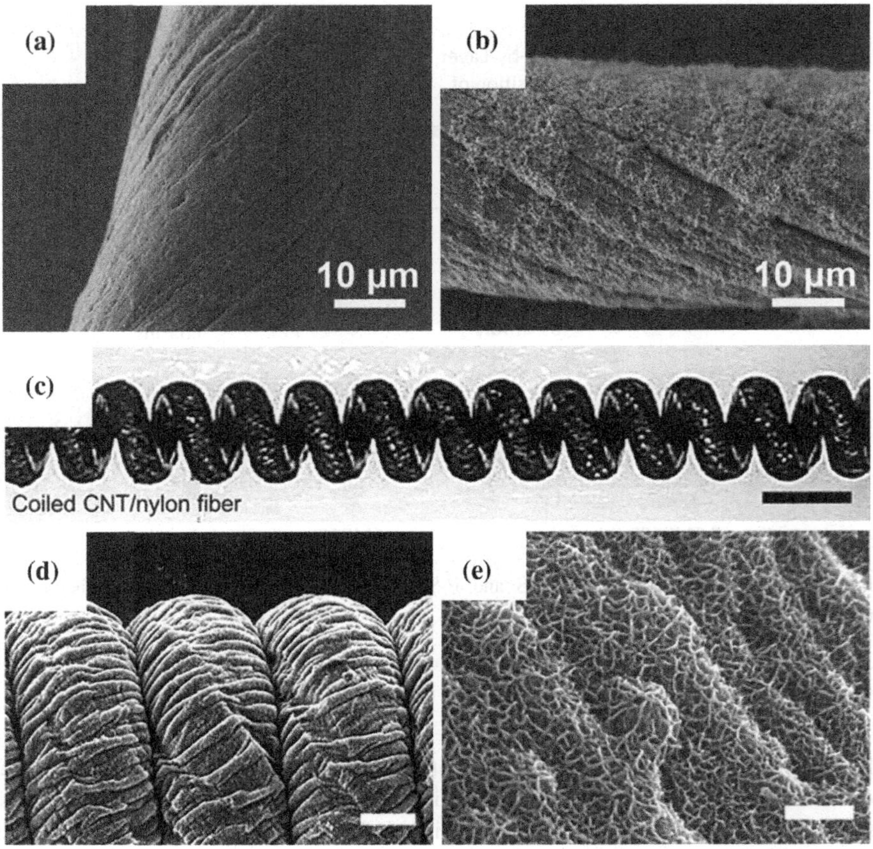

Fig. 4.16 SEM morphologies of **a** pristine CNT fiber electrode and **b** MnO_2/CNT hybrid fiber electrode. Reproduced with permission [61] Copyright 2015, American Chemical Society. **c** Optical microscope image of a stretched, coiled CNT/nylon electrode fiber (scale bar 300 μm). Reproduced with permission [64] Copyright 2015, Nature Publishing Group. **d** SEM image of a coiled CNT/nylon fiber electrode showing the wrinkles in the CNT layer that form during pre-coiling twist insertion, which increases the external surface layer that is available for electrolyte infiltration (scale bar 50 μm). Reproduced with permission [64] Copyright 2015, Nature Publishing Group. **e** SEM image of the surfaces of CNT/nylon electrodes after mesoporous MnO_2 deposition (scale bar 150 nm). Reproduced with permission [64] Copyright 2015, Nature Publishing Group

placed parallel and coated with an aqueous PVA gel containing LiCl to complete the assembly of the highly stretchable supercapacitor. The maximum linear, areal capacitances, as well as areal energy and power densities, are 5.4 mF cm, 40.9 mF cm^2, 2.6 μWh cm^2 and 66.9 μW cm^2, respectively. In addition, retention of supercapacitor performance during a large strain (50 %) elastic deformation is also demonstrated for these supercapacitors incorporated into the wristband of a glove.

In 2015, a strongly bonded MnO_2/CNT hybrid electrode was synthesized by our group for dynamically stretchable pseudocapacitors for the first time [65].

The MnO_2/CNT hybrid film electrodes were prepared by a modified precipitation method. Briefly, a purified CNT film was immersed in ethanol, while a $KMnO_4$ aqueous solution was dropwise added. By adjusting the precipitation time, thin layers of nanostructured MnO_2 were deposited on the surface of the CNT bundles. After being carefully washed and dried, the stretchable MnO_2/CNT electrode was fabricated by using the prestraining-then-buckling strategy. When two MnO_2/CNT electrodes and one elastomeric polyurethane separator were assembled with 1 M $TEABF_4$/PC electrolyte, a stretchable pseudocapacitor was constructed. Moreover, the dynamically stretchable pseudocapacitor displays an excellent pseudo-capacitive stability under both electrochemical charge-discharge cycles and mechanically stretching and releasing cycles.

4.5 Lithium Metal Oxides

Nowadays, significant efforts have been devoted to the search for lithium metal oxides (such as $LiCoO_2$, $LiMn_2O_4$, $LiFePO_4$, and $Li_4Ti_5O_{12}$) as the electrode materials for high-performance LIBs.

Stretchable film batteries with a $LiFePO_4$ cathode, a $Li_4Ti_5O_{12}$ anode, and polyethylene oxide (PEO) solid electrolyte for smart textile applications have been demonstrated by Liu and co-workers [66] (Fig. 4.17). The flexible anode and cathode film fabrication started by mixing appropriate amounts of $LiFePO_4$ or $Li_4Ti_5O_{12}$ (37.5 wt%), PEO (50 wt% as binders), and conductive carbon black (12.5 wt%). After being assembled with PEO solid electrolyte, the flexible film batteries were first cut into stripes and then integrated into a textile during weaving with a manual Dobby loom to achieve the stretchability of the LIBs. However, it has been found that the registered voltage (not exceeding 1.0 V) is always significantly lower than the theoretical value of the $LiFePO_4$–$Li_4Ti_5O_{12}$ cells ($\sim$1.8 V). One of the possible reasons is that the ionic conductivity of the solid polymer electrolyte is relatively low in this work, resulting in a big drop in the open circuit voltage. Though the operating voltage of a single battery is relatively low, when the conductive threads were used to weave textile electrodes and to connect the individual battery stripes in series, the output voltage can be increased enough to power up a 3 V light-emitting diode for practical applications.

A great advance in design and manufacturing of highly stretchable LIBs has been achieved by Xu and co-workers [14]. A set of design and manufacturing concepts for a highly stretchable lithium-ion battery technology that exploits thin, low modulus silicone elastomers as substrates, with a segmented design in the active materials, and unusual 'self-similar' interconnect structures between them, has been proposed (Fig. 4.18). For accommodating the two competing goals to simultaneously achieving a high areal capacity and a high stretchability, which require a large coverage of active areas and long distances between these areas, respectively, the devices consisted of pouch cells in which arrays of small-scale storage components are connected by 'self-similar' structures of wires in serpentine

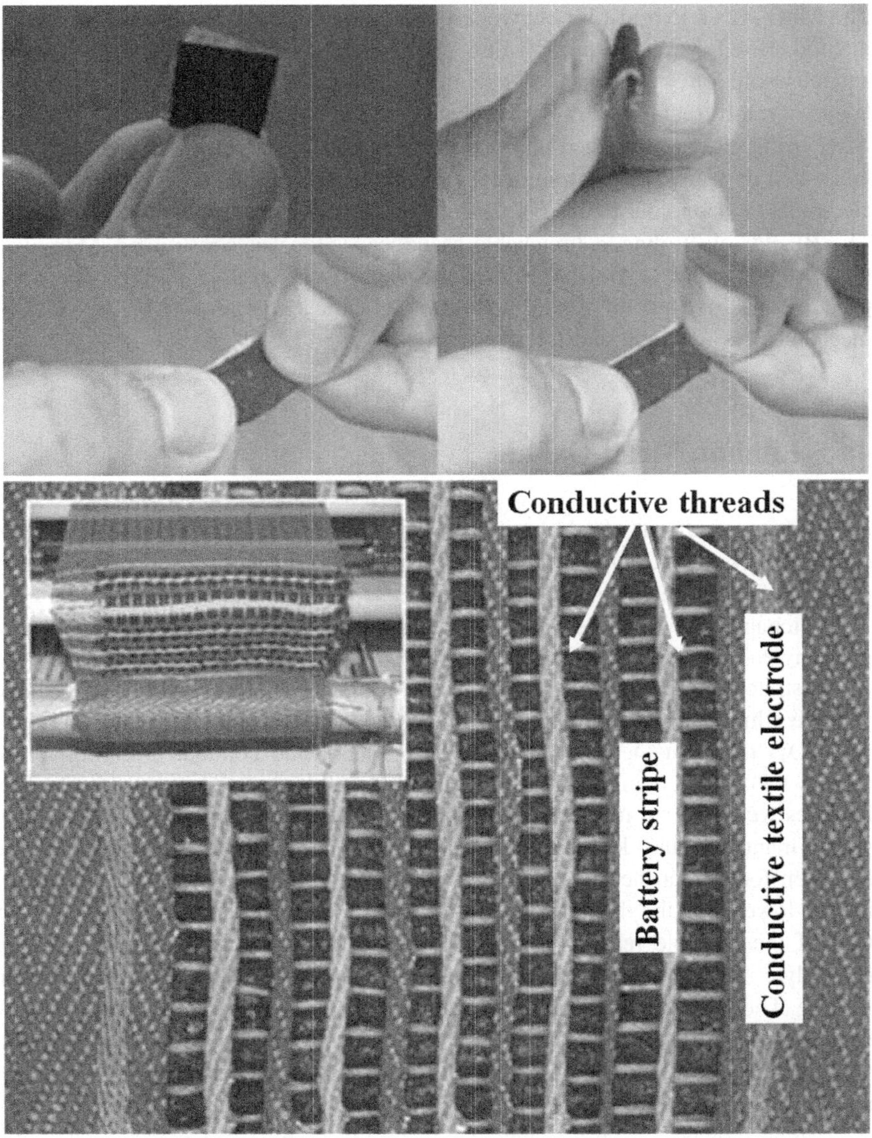

Fig. 4.17 *Top row* photographs of a flexible battery made with the binding individual cathode, anode, and polymer electrolyte films. *Middle row* resulting battery is highly stretchable. *Bottom row* battery stripes woven into a textile using Dobby loom. The stripes are connected in series with conductive threads. Two textile electrodes are formed by the conductive threads at the textile extremities. Reproduced with permission [66] Copyright 2012, The Electrochemical Society

configurations as interconnects to offer a high system-level stretchability and a low interconnect resistance. Remarkably, the imaginative design enables the lithium-ion battery with a reversible level of stretchability up to 300 %, which significantly

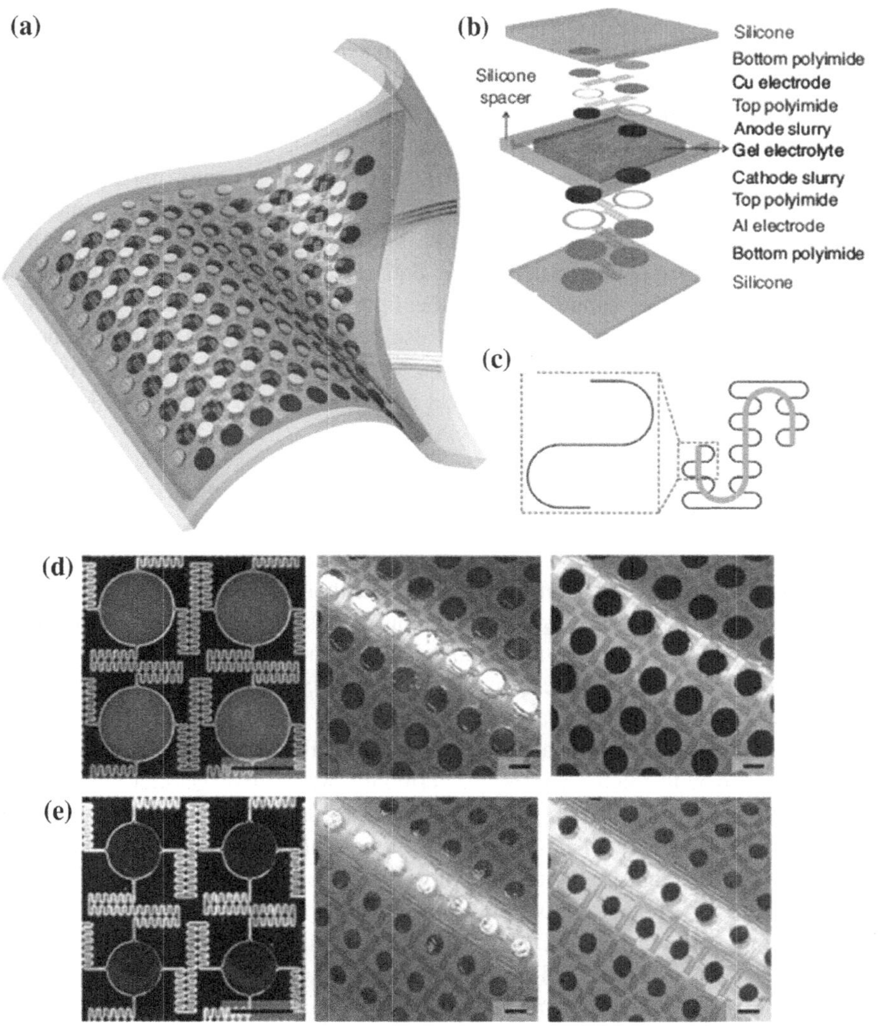

Fig. 4.18 Aspects in battery layout and design. **a** Schematic illustration of a completed device, in a state of stretching and bending. **b** Exploded view layout of the various layers in the battery structure. **c** Illustration of 'self-similar' serpentine geometries used for the interconnects (*black* 1st level serpentine; *yellow* 2nd level serpentine). **d** Optical images of the Al electrode pads and self-similar interconnects on a Si wafer (*left panel*; top down view; ~4 unit cells), after transfer printing on a sheet of silicone (*middle panel*; top down view, in a bent geometry), and with moulded slurries of $LiCoO_2$ (*right panel*; top down view, in a bent geometry). **e** Optical images of the Cu electrode pads and self-similar interconnects on a Si wafer (*left panel*; top down view; ~4 unit cells), after transfer printing on a sheet of silicone (*middle panel*; top down view, in a bent geometry), and with moulded slurries of $Li_4Ti_5O_{12}$ (*right panel*; top down view, in a bent geometry). Scale bars in (**d**) and (**e**) are 2 mm. Reproduced with permission [14] Copyright 2013, Nature Publishing Group

exceeds those of previous reports in stretchable batteries and/or battery electrodes (Fig. 4.18), while the coverage of the active material is as high as 50 %, enabling a high areal capacity of ~1.1 mAh cm^2. Finite element analysis indicates that the effective modulus (66.8 kPa) of the whole lithium-ion battery is only slightly higher than that of the silicone substrate materials (60.0 kPa), while obviously lower than that of the human epidermis (140–600 kPa), thereby providing the potential for integration onto skin or other biological tissues [14]. As a consequence, the lithium-ion battery is not only stretchable but also extremely soft and compliant. Furthermore, under a biaxial stretching and releasing up to 300 %, the loss in output power of the battery can be negligible. As an additional advance, a stretchable wireless charging system has also been introduced to charge the battery without establishing physical connections to external power supplies, which is a valuable characteristic for many practical applications such as implanted devices.

4.6 Elemental and Compound Semiconductors

Inorganic solar cells, based on elemental and compound semiconductors (such as Si, CdTe, GaAs, and $CuInSe_2$), are commonly consisted of hard and brittle inorganic materials [67–70]. Research in exploring new materials demonstrates a tremendous feasibility to overcome the intrinsic constraints of these inorganic materials for stretchable inorganic solar cells.

The pioneering work to achieve the stretchability of single-crystal Si in a thin-film form with micrometer-scale, periodic, "wavy"-like geometries and integrate it into an intrinsically stretchable photovoltaic device have been accomplished by Khang and co-workers [71]. The details of the fabrication process to produce the Si wave buckles includes the following three steps: First, fabrication of thin single-crystal Si ribbons supported on a Si wafer by conventional lithographic processing; second, a pre-strained elastomeric substrate (PDMS) is introduced to bond with the Si ribbons; and last, peeling back the PDMS, with the Si ribbons bonded on its surface, and then releasing the pre-strain. This relaxation leads to the spontaneous formation of microscale, periodic, "wavy"-like geometries in the Si ribbons. The photoelectrical responses of the stretchable photovoltaic device based on a p-n-junction photodiode were also investigated. No systematic variation in the electrical properties of the devices has been observed when stretched or compressed to a certain extent. The photocurrent density was ~35 mA cm^{-2}, and unsurprisingly, did not change substantially, even after 100 cycles of compressing (to ~5 % applied strain), stretching (to ~15 % applied strain), and releasing. In this wave structural configuration, however, the mechanical effects dramatically change the photoelectronic properties of the device when the strain has exceeded a certain degree. If the compressive strain is too large, the PDMS tends to bend in ways that made the device operation difficult. When the tensile strain is too large, the Si ribbons may fracture or slip, depending on the Si thickness, the ribbon length, and the strength of bonding between the Si ribbons and PDMS substrates.

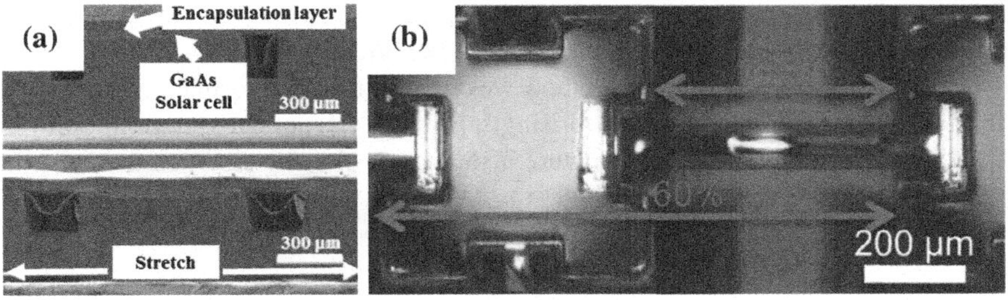

Fig. 4.19 **a** SEM cross-sectional image of an encapsulated cell module (*top*). The interconnects with arc-shaped structures are clearly seen. SEM image of a similar module, stretched uniaxially by 15 % (*bottom*). The interconnects move freely to accommodate the applied strain. Reproduced with permission [17] Copyright 2011, Wiley-VCH. **b** Top-view of a single GaInP/GaAs photovoltaic module in a state of 60 % overall biaxial strain. Reproduced with permission [18] Copyright 2012, Wiley-VCH

As a consequence, stretchable interconnects, combined with a high electrical conductance, were widely used as sophisticated designs to arrive at the high stretchability by localizing strain at these interconnects [19, 72]. Nevertheless, to balance the high areal coverage of the active material regions and the length of interconnections in the interconnect-island mesh configuration, Lee et al. reported a new design for stretchable, ultrathin, single junction GaAs solar cells that exploit elastomeric substrate with surface relief that confines strains at the locations of the interconnections, and away from the devices (Fig. 4.19a) [17]. Compared to similar configurations with conventional, flat substrates, the maximum shear and normal stresses for the surface-relieved substrate are about 8 and 7 times lower than those for the flat substrate, respectively. And the efficiency ($\sim$12.5 %) of the newly-designed GaAs solar microcells is constant over more than 500 cycles by biaxial stretching to 20 % followed by relaxation, while the conventionally-designed solar microcells suffer high interfacial stresses that can cause adverse effects on the module under the same cycle mechanisms. With optimizing in geometry, a representative module with a $\sim$70 % areal coverage of islands has been finally fabricated. Meanwhile, the open circuit voltage of seven microcells connected in series was almost the same in relaxed and stretched states. When these devices are biaxially stretched up to 20 %, these characteristics do not change. Lately, Lee et al. further optimized the geometrical structures in the interconnect-island mesh configuration to not only enhance the stretchability and the areal coverage but also provide a natural form of strain-limiting behavior, which is helpful to avoid destructive effects of excessive deformations for high-efficiency dual-junction GaInP/GaAs solar cells [18]. In this configuration, raised islands, with rectangular notches in the central regions of the four edges, provide a surface for mounting interconnected arrays of microscale solar cells. When the lateral dimensions of the islands are 500 μm × 500 μm, separated by trenches with widths of $\sim$76 μm, the notches on the islands have widths of $\sim$120 μm and depths of $\sim$60 μm, the height of the islands and the thickness of the underlying base layer are 350 and 160 μm, respectively, and the areal coverage of the islands corresponds to 67 % of the overall area. Meanwhile, the solar cell module can be stretchable up to 60 % biaxially as shown in Fig. 4.19b.

In addition, Kirigami-based method was also introduced to realize the stretchability of GaAs solar cell [26]. Surprisingly, with an elegant cut pattern in thin-film GaAs solar cells, the kirigami structure also endows the stretchable solar cell with excellent solar tracking property.

4.7 Metals

In terms of high conductivity and long-term stability, metals (Au, Ag, Cu) though usually present a high Young's modulus and a low limit of elasticity, remain to be promising candidates for the stretchable ESCDs. Meanwhile, the rapid development of the microelectronics and microelectromechanical system industries has helped create a broad range of micro-fabrication technologies to fabricate stretchable metal materials for electronic devices [54, 73]. Yan et al. have presented the construction of a stretchable silver-zinc battery based on embedded Ag nanowire elastic conductors, which were fabricated using a lithographic filtration method (Fig. 4.20) [74]. These Ag nanowire electrodes possess bifunctionality by simultaneously providing electroactive materials and current collectors. Moreover, these Ag electrodes can maintain their functions even when stretched up to 80 % and are stable over 1000 cycles. Meanwhile, fully stretchable batteries are also demonstrated with both Ag and Zn electrodes being stretchable, although the cyclic stability of the stretchable batteries still needs to be improved.

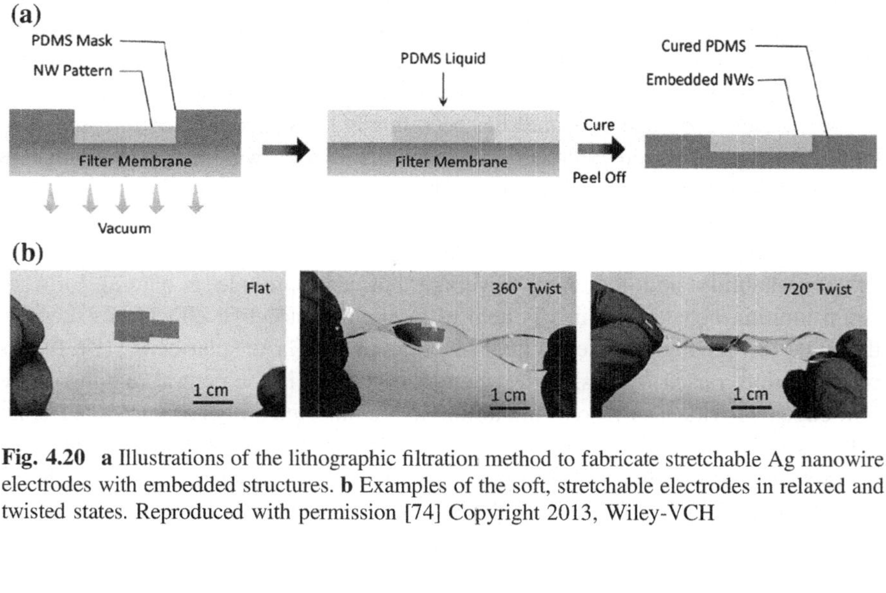

Fig. 4.20 **a** Illustrations of the lithographic filtration method to fabricate stretchable Ag nanowire electrodes with embedded structures. **b** Examples of the soft, stretchable electrodes in relaxed and twisted states. Reproduced with permission [74] Copyright 2013, Wiley-VCH

4.8 Summary and Outlook

Although significant advancements have been achieved in a relatively short period, there remain many challenges to the development of stretchable ESCDs. For most stretchable ESCDs, the stretchability of these devices (usually <30 %) is much lower than the demands of fully stretchable integrated devices, which require over 100 % stretchability [12, 16, 20, 75]. To this end, new nanomaterial design and synthesis are the keys to the development of highly stretchable ESCDs for practical applications.

Till now, significant efforts are devoted to synthesizing new materials with terrific mechanical properties. For example, Sun et al. have synthesized a novel and highly stretchable hydrogel with ionically and covalently crosslinked networks, which can be stretched beyond 20 times its initial length [76]. However, synthesis of new materials is difficult. Meanwhile, it is also a significant challenge for new materials to meet all specific requirements for a variety of applications. For instance, PDMS is widely used as an elastic substrate for stretchable electronics due to its excellent mechanical properties and chemical stability. Yet, this material can be slightly dissolved in very dilute HF, which makes it not suitable as a substrate for application in stretchable LIBs, in which commercial electrolyte, lithium salt ($LiPF_6$) will decompose and generate HF during the charge and discharge process of the LIBs [77]. To sum up, many challenges in new materials design and synthesis is still need to be conquered.

Undoubtedly, stretchable ESCDs, such as supercapacitors, batteries, and solar cells, are very promising and hold potential for practical applications. In the past few years, research in this area has witnessed fast developments and substantial achievements, which has benefitted directly from the progress on the synthesis of new materials in these devices. However, there are still many challenges that need to be overcome on the road to a large-scale application of stretchable ESCDs. The research in this area is exciting and relatively open-ended which provides researchers from various research fields (e.g., chemistry, materials, mechanics, nanoscience, nanotechnology, and so on) excellent opportunities to explore and develop novel and new concepts. Some interesting and important directions in the areas of material synthesis and design for the stretchable ESCDs are exampled: (1) experimental study and theoretical prediction in new elastic materials; (2) enabling existing materials with improved mechanical properties via the development of nanoscience and nanotechnology; (3) investigation of fundamental insights of the stretchable ESCDs with the aim to further develop more specific stretchable ESCDs for different applications.

In conclusion, similar to the development paths of stretchable electronics, the development of better material synthesis and design will usher in the next-generation of stretchable ESCDs technologies. Meanwhile, there is no doubt that the nanoscience and nanotechnology will keep on playing the key roles in developing stretchable ESCDs with high-performance.

Acknowledgments This work was supported by the financial assistance by the National Natural Science Foundation of China (No. 51302219 and 51472204), the Natural Science Foundation of Shannxi Province (No. 2015JM2045), the Research Fund of the State Key Laboratory of Solidification Processing (NWPU), China (Grant No. 06-QP-2014), the Fundamental Research Funds for the Central Universities (No. 3102014JCQ01019), and the Specialized Research Fund for the Doctoral Program of Higher Education of China (No. 20136102120024 and 20136102140001).

References

1. P.G. Bruce, S.A. Freunberger, L.J. Hardwick, J.M. Tarascon, Li-O_2 and Li-S batteries with high energy storage. Nat. Mater. **11**, 19–29 (2012)
2. M.F. El-Kady, V. Strong, S. Dubin, R.B. Kaner, Laser scribing of high-performance and flexible graphene-based electrochemical capacitors. Science **335**, 1326–1330 (2012)
3. H.G. Zhang, X.D. Yu, P.V. Braun, Three-dimensional bicontinuous ultrafast-charge and-discharge bulk battery electrodes. Nat. Nanotechnol. **6**, 277–281 (2011)
4. J. Burschka, N. Pellet, S.J. Moon, R. Humphry-Baker, P. Gao, M.K. Nazeeruddin, M. Gratzel, Sequential deposition as a route to high-performance perovskite-sensitized solar cells. Nature **499**, 316–319 (2013)
5. W. Gao, N. Singh, L. Song, Z. Liu, A.L. Reddy, L. Ci, R. Vajtai, Q. Zhang, B.Q. Wei, P.M. Ajayan, Direct laser writing of micro-supercapacitors on hydrated graphite oxide films. Nat. Nanotechnol. **6**, 496–500 (2011)
6. K.Y. Xie, Z. Lu, H. Huang, W. Lu, Y. Lai, J. Li, L. Zhou, Y. Liu, Iron supported C@Fe_3O_4 nanotube array: a new type of 3d anode with low-cost for high performance lithium-ion batteries. J. Mater. Chem. **22**, 5560–5567 (2012)
7. J.-H. Ahn, J.H. Je, Stretchable electronics: materials, architectures and integrations. J. Phys. D Appl. Phys. **45**, 103001 (2012)
8. D.H. Kim, N. Lu, R. Ma, Y.S. Kim, R.H. Kim, S. Wang, J. Wu, S.M. Won, H. Tao, A. Islam, K.J. Yu, T.I. Kim, R. Chowdhury, M. Ying, L. Xu, M. Li, H.J. Chung, H. Keum, M. McCormick, P. Liu, Y.W. Zhang, F.G. Omenetto, Y. Huang, T. Coleman, J.A. Rogers, Epidermal electronics. Science **333**, 838–843 (2011)
9. B.G. Choi, J. Hong, W.H. Hong, P.T. Hammond, H. Park, Facilitated ion transport in all-solid-state flexible supercapacitors. ACS Nano **5**, 7205–7213 (2011)
10. A. Chirila, S. Buecheler, F. Pianezzi, P. Bloesch, C. Gretener, A.R. Uhl, C. Fella, L. Kranz, J. Perrenoud, S. Seyrling, R. Verma, S. Nishiwaki, Y.E. Romanyuk, G. Bilger, A.N. Tiwari, Highly efficient Cu(In, Ga)Se_2 solar cells grown on flexible polymer films. Nat. Mater. **10**, 857–861 (2011)
11. N. Singh, C. Galande, A. Miranda, A. Mathkar, W. Gao, A.L. Reddy, A. Vlad, P.M. Ajayan, Paintable battery. Sci. Rep. **2**, 481 (2012)
12. M. Kaltenbrunner, G. Kettlgruber, C. Siket, R. Schwodiauer, S. Bauer, Arrays of ultracompliant electrochemical dry gel cells for stretchable electronics. Adv. Mater. **22**, 2065–2067 (2010)
13. C. Wang, W. Zheng, Z. Yue, C.O. Too, G.G. Wallace, Buckled, stretchable polypyrrole electrodes for battery applications. Adv. Mater. **23**, 3580–3584 (2011)
14. S. Xu, Y. Zhang, J. Cho, J. Lee, X. Huang, L. Jia, J.A. Fan, Y. Su, J. Su, H. Zhang, H. Cheng, B. Lu, C. Yu, C. Chuang, T.I. Kim, T. Song, K. Shigeta, S. Kang, C. Dagdeviren, I. Petrov, P. V. Braun, Y. Huang, U. Paik, J.A. Rogers, Stretchable batteries with self-similar serpentine interconnects and integrated wireless recharging systems. Nat. Commun. **4**, 1543 (2013)
15. X. Li, T. Gu, B.Q. Wei, Dynamic and galvanic stability of stretchable supercapacitors. Nano Lett. **12**, 6366–6371 (2012)

16. C. Yu, C. Masarapu, J. Rong, B.Q. Wei, H. Jiang, Stretchable supercapacitors based on buckled single-walled carbon-nanotube macrofilms. Adv. Mater. **21**, 4793–4797 (2009)
17. J. Lee, J. Wu, M. Shi, J. Yoon, S.I. Park, M. Li, Z. Liu, Y. Huang, J.A. Rogers, Stretchable GaAs photovoltaics with designs that enable high areal coverage. Adv. Mater. **23**, 986–991 (2011)
18. J. Lee, J. Wu, J.H. Ryu, Z. Liu, M. Meitl, Y.-W. Zhang, Y. Huang, J.A. Rogers, Stretchable semiconductor technologies with high areal coverages and strain-limiting behavior: demonstration in high-efficiency dual-junction GaInP/GaAs photovoltaics. Small **8**, 1851–1856 (2012)
19. J. Yoon, A.J. Baca, S.I. Park, P. Elvikis, J.B. Geddes, L. Li, R.H. Kim, J. Xiao, S. Wang, T.H. Kim, M.J. Motala, B.Y. Ahn, E.B. Duoss, J.A. Lewis, R.G. Nuzzo, P.M. Ferreira, Y. Huang, A. Rockett, J.A. Rogers, Ultrathin silicon solar microcells for semitransparent, mechanically flexible and microconcentrator module designs. Nat. Mater. **7**, 907–915 (2008)
20. D.J. Lipomi, B.C. Tee, M. Vosgueritchian, Z. Bao, Stretchable organic solar cells. Adv. Mater. **23**, 1771–1775 (2011)
21. M. Kaltenbrunner, M.S. White, E.D. Glowacki, T. Sekitani, T. Someya, N.S. Sariciftci, S. Bauer, Ultrathin and lightweight organic solar cells with high flexibility. Nat. Commun. **3**, 770 (2012)
22. S. Pan, Z. Yang, P. Chen, J. Deng, H. Li, H. Peng, Wearable solar cells by stacking textile electrodes. Angew. Chem. Int. Ed. **53**, 6110–6114 (2014)
23. Y. Ding, M.A. Invernale, G.A. Sotzing, Conductivity trends of PEDOT-PSS impregnated fabric and the effect of conductivity on electrochromic textile. ACS Appl. Mater. Interfaces **2**, 1588–1593 (2010)
24. Z. Song, T. Ma, R. Tang, Q. Cheng, X. Wang, D. Krishnaraju, R. Panat, C.K. Chan, H. Yu, H. Jiang, Origami lithium-ion batteries. Nat. Commun. **5**, 3140 (2014)
25. Z. Song, X. Wang, C. Lv, Y. An, M. Liang, T. Ma, D. He, Y.-J. Zheng, S.-Q. Huang, H. Yu, H. Jiang, Kirigami-based stretchable lithium-ion batteries. Sci. Rep. **5**, 10988 (2015)
26. A. Lamoureux, K. Lee, M. Shlian, S.R. Forrest, M. Shtein, Dynamic kirigami structures for integrated solar tracking. Nat. Commun. **6**, 8092 (2015)
27. S. Wagner, S. Bauer, Materials for stretchable electronics. MRS Bull. **37**, 207–213 (2012)
28. K.Y. Xie, B.Q. Wei, Materials and structures for stretchable energy storage and conversion devices. Adv. Mater. **26**, 3592–3617 (2014)
29. Y. Kim, J. Zhu, B. Yeom, M. Di Prima, X. Su, J.G. Kim, S.J. Yoo, C. Uher, N.A. Kotov, Stretchable nanoparticle conductors with self-organized conductive pathways. Nature **500**, 59–63 (2013)
30. T. Yamada, Y. Hayamizu, Y. Yamamoto, Y. Yomogida, A. Izadi-Najafabadi, D.N. Futaba, K. Hata, A stretchable carbon nanotube strain sensor for human-motion detection. Nat. Nanotechnol. **6**, 296–301 (2011)
31. T. Sekitani, H. Nakajima, H. Maeda, T. Fukushima, T. Aida, K. Hata, T. Someya, Stretchable active-matrix organic light-emitting diode display using printable elastic conductors. Nat. Mater. **8**, 494–499 (2009)
32. M. Zu, Q. Li, G. Wang, J.-H. Byun, T.-W. Chou, Carbon nanotube fiber based stretchable conductor. Adv. Funct. Mater. **23**, 789–793 (2013)
33. F. Xu, X. Wang, Y. Zhu, Y. Zhu, Wavy ribbons of carbon nanotubes for stretchable conductors. Adv. Funct. Mater. **22**, 1279–1283 (2012)
34. Y. Zhu, F. Xu, Buckling of aligned carbon nanotubes as stretchable conductors: a new manufacturing strategy. Adv. Mater. **24**, 1073–1077 (2012)
35. L. Cai, J. Li, P. Luan, H. Dong, D. Zhao, Q. Zhang, X. Zhang, M. Tu, Q. Zeng, W. Zhou, S. Xie, Highly transparent and conductive stretchable conductors based on hierarchical reticulate single-walled carbon nanotube architecture. Adv. Funct. Mater. **22**, 5238–5244 (2012)
36. W. Weng, Q. Sun, Y. Zhang, S. He, Q. Wu, J. Deng, X. Fang, G. Guan, J. Ren, H. Peng, A gum-like lithium-ion battery based on a novel arched structure. Adv. Mater. **27**, 1363–1369 (2015)

37. L. Hu, M. Pasta, F.L. Mantia, L. Cui, S. Jeong, H.D. Deshazer, J.W. Choi, S.M. Han, Y. Cui, Stretchable, porous, and conductive energy textiles. Nano Lett. **10**, 708–714 (2010)
38. Z. Yang, J. Deng, X. Sun, H. Li, H. Peng, Stretchable, wearable dye-sensitized solar cells. Adv. Mater. **26**, 2643–2647 (2014)
39. C. Lee, X. Wei, J.W. Kysar, J. Hone, Measurement of the elastic properties and intrinsic strength of monolayer graphene. Science **321**, 385–388 (2008)
40. K.I. Bolotin, K.J. Sikes, Z. Jiang, M. Klima, G. Fudenberg, J. Hone, P. Kim, H.L. Stormer, Ultrahigh electron mobility in suspended graphene. Solid State Commun. **146**, 351–355 (2008)
41. P. Xu, J. Kang, J.-B. Choi, J. Suhr, J. Yu, F. Li, J.-H. Byun, B.-S. Kim, T.-W. Chou, Laminated ultrathin chemical vapor deposition graphene films based stretchable and transparent high-rate supercapacitor. ACS Nano **8**, 9437–9445 (2014)
42. T. Chen, Y. Xue, A.K. Roy, L. Dai, Transparent and stretchable high-performance supercapacitors based on wrinkled graphene electrodes. ACS Nano **8**, 1039–1046 (2014)
43. D. Qi, Z. Liu, Y. Liu, W.R. Leow, B. Zhu, H. Yang, J. Yu, W. Wang, H. Wang, S. Yin, X. Chen, Suspended wavy graphene microribbons for highly stretchable microsupercapacitors. Adv. Mater **27**, 5559–5566 (2015)
44. X. Zang, Q. Chen, P. Li, Y. He, X. Li, M. Zhu, X. Li, K. Wang, M. Zhong, D. Wu, H. Zhu, Highly flexible and adaptable, all-solid-state supercapacitors based on graphene woven-fabric film electrodes. Small **10**, 2583–2588 (2014)
45. X. Zang, M. Zhu, X. Li, X. Li, Z. Zhen, J. Lao, K. Wang, F. Kang, B.Q. Wei, H. Zhu, Dynamically stretchable supercapacitors based on graphene woven fabric electrodes. Nano Energy **15**, 83–91 (2015)
46. J. Zang, C. Cao, Y. Feng, J. Liu, X. Zhao, Stretchable and high-performance supercapacitors with crumpled graphene papers. Sci. Rep. **4**, 6492 (2014)
47. Z. Dong, C. Jiang, H. Cheng, Y. Zhao, G. Shi, L. Jiang, L. Qu, Facile fabrication of light, flexible and multifunctional graphene fibers. Adv. Mater. **24**, 1856–1861 (2012)
48. Y. Meng, Y. Zhao, C. Hu, H. Cheng, Y. Hu, Z. Zhang, G. Shi, L. Qu, All-graphene core-sheath microfibers for all-solid-state, stretchable fibriform supercapacitors and wearable electronic textiles. Adv. Mater. **25**, 2326–2331 (2013)
49. L. Liu, Y. Yu, C. Yan, K. Li, Z. Zheng, Wearable energy-dense and power-dense supercapacitor yarns enabled by scalable graphene-metallic textile composite electrodes. Nat. Commun. **6**, 7260 (2015)
50. F. Li, J. Chen, X. Wang, M. Xue, G.F. Chen, Stretchable supercapacitor with adjustable volumetric capacitance based on 3D interdigital electrodes. Adv. Funct. Mater. **25**, 4601–4606 (2015)
51. D. Yu, K. Goh, H. Wang, L. Wei, W. Jiang, Q. Zhang, L. Dai, Y. Chen, Scalable synthesis of hierarchically structured carbon nanotube-graphene fibres for capacitive energy storage. Nat. Nanotechnol. **9**, 555–562 (2014)
52. L. Kou, T. Huang, B. Zheng, Y. Han, X. Zhao, K. Gopalsamy, H. Sun, C. Gao, Coaxial wet-spun yarn supercapacitors for high-energy density and safe wearable electronics. Nat. Commun. **5**, 3754 (2014)
53. K. Jost, D. Stenger, C.R. Perez, J.K. McDonough, K. Lian, Y. Gogotsi, G. Dion, Knitted and screen printed carbon-fiber supercapacitors for applications in wearable electronics. Energ. Environ. Sci. **6**, 2698–2705 (2013)
54. S. Rosset, H.R. Shea, Flexible and stretchable electrodes for dielectric elastomer actuators. Appl. Phys. A **110**, 281–307 (2012)
55. A.M. Gaikwad, A.M. Zamarayeva, J. Rousseau, H. Chu, I. Derin, D.A. Steingart, Highly stretchable alkaline batteries based on an embedded conductive fabric. Adv. Mater. **24**, 5071–5076 (2012)
56. T. Sekitani, T. Someya, Stretchable, large-area organic electronics. Adv. Mater. **22**, 2228–2246 (2010)
57. B. Yue, C. Wang, X. Ding, G.G. Wallace, Polypyrrole coated nylon lycra fabric as stretchable electrode for supercapacitor applications. Electrochim. Acta **68**, 18–24 (2012)

58. Y. Huang, J. Tao, W. Meng, M. Zhu, Y. Huang, Y. Fu, Y. Gao, C. Zhi, Super-high rate stretchable polypyrrole-based supercapacitors with excellent cycling stability. Nano Energy **11**, 518–525 (2015)
59. H. Jin, L. Zhou, C.L. Mak, H. Huang, W.M. Tang, H.L.W. Chan, High-performance fiber-shaped supercapacitors using carbon fiber thread (CFT)@polyanilne and functionalized CFT electrodes for wearable/stretchable electronics. Nano Energy **11**, 662–670 (2015)
60. J. Zhong, J. Meng, Z. Yang, P. Poulin, N. Koratkar, Shape memory fiber supercapacitors. Nano Energy **17**, 330–338 (2015)
61. P. Xu, B.Q. Wei, Z. Cao, J. Zheng, K. Gong, F. Li, J. Yu, Q. Li, W. Lu, J.-H. Byun, B.-S. Kim, Y. Yan, T.-W. Chou, Stretchable wire-shaped asymmetric supercapacitors based on pristine and MnO_2 coated carbon nanotube fibers. ACS Nano **9**, 6088–6096 (2015)
62. K.Y. Xie, M. Guo, X. Liu, H. Huang, Enhanced efficiencies in thin and semi-transparent dye-sensitized solar cells under low photon flux conditions Using TiO_2 nanotube photonic crystal. J. Power Sources **293**, 170–177 (2015)
63. K.Y. Xie, M. Guo, H. Huang, Photonic crystals for sensitized solar cells: fabrication, properties, and applications. J. Mater. Chem. C **3**, 10665–10686 (2015)
64. C. Choi, S.H. Kim, H.J. Sim, J.A. Lee, A.Y. Choi, Y.T. Kim, X. Lepro, G.M. Spinks, R.H. Baughman, S.J. Kim, Stretchable, weavable coiled carbon nanotube/MnO_2/polymer fiber solid-state supercapacitors. Sci. Rep. **5**, 9387 (2015)
65. T. Gu, B.Q. Wei, Fast and stable redox reactions of MnO_2/CNT hybrid electrodes for dynamically stretchable pseudocapacitors. Nanoscale **7**, 11626–11632 (2015)
66. Y. Liu, S. Gorgutsa, C. Santato, M. Skorobogatiy, Flexible, solid electrolyte-based lithium battery composed of $LiFePO_4$ cathode and $Li_4Ti_5O_{12}$ anode for applications in smart textiles. J. Electrochem. Soc. **159**, A349–A356 (2012)
67. A. Ruland, C. Schulz-Drost, V. Sgobba, D.M. Guldi, Enhancing photocurrent efficiencies by resonance energy transfer in CdTe quantum dot multilayers: towards rainbow solar cells. Adv. Mater. **23**, 4573–4577 (2011)
68. G. Mariani, P.S. Wong, A.M. Katzenmeyer, F. Leonard, J. Shapiro, D.L. Huffaker, Patterned radial GaAs nanopillar solar cells. Nano Lett. **11**, 2490–2494 (2011)
69. R.J. Knuesel, H.O. Jacobs, Self-tiling monocrystalline silicon; a process to produce electrically connected domains of Si and microconcentrator solar cell modules on plastic supports. Adv. Mater. **23**, 2727–2733 (2011)
70. Q. Guo, S.J. Kim, M. Kar, W.N. Shafarman, R.W. Birkmire, E.A. Stach, R. Agrawal, H.W. Hillhouse, Development of $CuInSe_2$ nanocrystal and nanoring inks for low-cost solar cells. Nano Lett. **8**, 2982–2987 (2008)
71. D.Y. Khang, H. Jiang, Y. Huang, J.A. Rogers, A stretchable form of single-crystal silicon for high-performance electronics on rubber substrates. Science **311**, 208–212 (2006)
72. B.Y. Ahn, E.B. Duoss, M.J. Motala, X. Guo, S.I. Park, Y. Xiong, J. Yoon, R.G. Nuzzo, J.A. Rogers, J.A. Lewis, Omnidirectional printing of flexible, stretchable, and spanning silver microelectrodes. Science **323**, 1590–1593 (2009)
73. J. Ge, H.B. Yao, X. Wang, Y.D. Ye, J.L. Wang, Z.Y. Wu, J.W. Liu, F.J. Fan, H.L. Gao, C.L. Zhang, S.H. Yu, Stretchable conductors based on silver nanowires: improved performance through a binary network design. Angew. Chem. Int. Ed. **52**, 1654–1659 (2013)
74. C. Yan, X. Wang, M. Cui, J. Wang, W. Kang, C.Y. Foo, P.S. Lee, Stretchable silver-zinc batteries based on embedded nanowire elastic conductors. Adv. Energy Mater. **4**, 1301396 (2013)
75. Z. Niu, H. Dong, B. Zhu, J. Li, H.H. Hng, W. Zhou, X. Chen, S. Xie, Highly stretchable, integrated supercapacitors based on single-walled carbon nanotube films with continuous reticulate architecture. Adv. Mater. **25**, 1058–1064 (2013)
76. J.Y. Sun, X. Zhao, W.R. Illeperuma, O. Chaudhuri, K.H. Oh, D.J. Mooney, J.J. Vlassak, Z. Suo, Highly stretchable and tough hydrogels. Nature **489**, 133–136 (2012)
77. J. Li, K.Y. Xie, Y.Q. Lai, Z.A. Zhang, F.Q. Li, X. Hao, X.J. Chen, Y.X. Liu, Lithium oxalyldifluoroborate/carbonate electrolytes for $LiFePO_4$/artificial graphite lithium-ion cells. J. Power Sources **195**, 5344–5350 (2010)

Chapter 5
Piezoelectric Nanomaterials for Energy Harvesting

Kory Jenkins and Rusen Yang

Abstract Piezoelectric nanomaterials have many important and unique properties which make them suitable for a wide range of applications including consumer goods, motor vehicles, medical devices and scientific instrumentation. Vibrations, biomechanical motions, and waves are examples of ever-present mechanical energy in the environment. The small size and unique electrical and mechanical properties inherent in piezoelectric nanomaterials make them ideal for harvesting these renewable energy sources. ZnO nanostructures have been widely studied for energy harvesting due to their enhanced piezoelectricity and scalable, reproducible synthesis techniques. Single atomic layer MoS_2 is an emerging material for energy harvesting due to the existence of piezoelectric properties not found in the bulk material. Bio-inspired piezoelectric materials such as peptide nanotubes have excellent mechanical properties and are promising candidates for nanogenerator devices. Many different nanogenerator architectures have been developed and steadily improving performance has led to their incorporation into self-powered nanodevice systems.

5.1 Introduction to Piezoelectric Nanomaterials

Piezoelectricity has historically found uses in many applications in sensing and actuation. The recent discovery of piezotronics, which combines the piezoelectric and semiconducting properties of materials, has opened new avenues for piezoelectricity research [1]. With the advent of micro and nanoscale devices, new opportunities and challenges have arisen for small scale power generation. Energy harvesting at the nanoscale has applications for wearable and implantable medical devices, sensors for structural health monitoring, and many other areas. The energy harvesting performance of nanogenerators has been steadily improved and many

K. Jenkins · R. Yang (✉)
Department of Mechanical Engineering, University of Minnesota,
Minneapolis, MN 55455, USA
e-mail: yangr@umn.edu

Q. Li (ed.), *Nanomaterials for Sustainable Energy*,
NanoScience and Technology, DOI 10.1007/978-3-319-32023-6_5

opportunities still exist for improving efficiency and designing devices that are robust enough for commercial applications. There are also many exciting opportunities to explore nanogenerators based on emerging materials such as single atomic layer MoS_2. This chapter will explore some of the important and emerging materials for nanoscale energy harvesting and will highlight significant milestones and demonstrations of energy harvesting with these materials.

5.1.1 Piezoelectricity

Piezoelectricity is a property of crystals with non-central symmetry wherein charge polarization is induced by applied strain. Piezoelectric materials also exhibit the converse piezoelectric effect whereby an applied bias may induce strain in the piezoelectric material. There are 21 noncentrosymmetric crystal classes. Among these 21 classes, the cubic class 432 has piezoelectric charges which cancel along the <111> axes and therefore does not display piezoelectricity. Piezoelectric materials such as zinc oxide (ZnO) nanowires have seen much recent success in nanoscale energy harvesting applications as will be illustrated later in this chapter. More recently there has been significant interest in monolayer MoS_2 which has the potential for single atomic layer energy harvesting. Biologically inspired materials including diphenylalanine (FF) peptide nanotubes are also being investigated for their impressive structural and piezoelectric effects. The electromechanical coupling in piezoelectric materials makes them well suited for mechanical energy harvesting at the nanoscale. Piezoelectric materials can also have semiconducting properties which can play an integral part in energy harvesting performance. This coupling of piezoelectric and semiconducting properties is known as the piezotronic effect [1]. Semiconducting properties of nanomaterials such as ZnO have been instrumental in the development of nanogenerators [2]. There are many previously neglected sources of mechanical energy that nanomaterials are uniquely suited to collect including biomechanical energy and low intensity vibrations from the environment [3].

5.2 Properties and Synthesis of Piezoelectric Nanomaterials

Nanomaterials are especially suited for energy harvesting applications due to their favorable scaling laws [4]. The small size of nanomaterials makes them amenable to applications where space is limited such as implantable medical devices. Less material is required to fabricate nanostructures and the synthesis processes may be highly scalable [5]. Nanomaterials may also have favorable electrical properties not inherent in the bulk materials such as increased piezoelectric coefficient. Some materials such as MoS_2 may even switch from an indirect to a direct bandgap [6, 7].

ZnO nanowires have improved elastic modulus over the bulk material and typical synthesis routes can produce high quality, single crystal materials which are resistant to fatigue failure [8, 9].

5.2.1 Zinc Oxide

Zinc oxide (ZnO) is a piezoelectric and semiconducting material that has been central to recent developments in nanoscale energy harvesting research. ZnO exists as 3 distinct crystal structures, namely wurtzite, zinc blende (sphalerite), and rock salt [10]. While all 3 crystal structures lack inversion symmetry and are therefore piezoelectric, it is the hexagonal wurtzite form that has most prominently been explored for energy harvesting applications [11, 12]. ZnO can be synthesized into nanowires, a quasi-1D, high aspect ratio structure capable of generating large piezopotentials along the length of the wire [13]. Zinc oxide is a direct, wide bandgap semiconductor with a bandgap of 3.37 eV [14]. The electron affinity of ZnO is 4.2 eV and it can form Schottky contacts with high work function metals such as gold and platinum. The formation of Schottky and Ohmic contacts with ZnO has been extensively reviewed by Brillson and Lu [15].

The mechanical properties of ZnO nanowires illustrate the often favorable scaling properties of nanostructures. Agrawal et al. measured the elastic modulus of ZnO nanowires using transmission electron microscopy (TEM) with a MEMS based material testing system. The authors tested nanowire samples with diameters ranging from 20 to 80 nm [9]. As the diameter decreased, the modulus was observed to increase from 140 to 160 GPa. Gao et al. investigated the piezoelectric potential in ZnO nanowires using a finite element model of a 1200 nm long nanowire as shown in Fig. 5.1 [13]. Under a tensile load of 85 nN they calculated a potential difference of about 0.4 V along the c-axis of the wire. The potential was reversed under compressive strain. They also investigated a torsional load, which did not create a net potential along the length of the wire, and only generated very small localized potential gradients. However, tension or compression combined with torsion still produced similar results to the pure tensile/compressive cases, confirming that the c-axis strain generates the dominating piezopotential along the length of the wire. The high potential values generated in ZnO nanowires are promising for nanogenerator applications. The potential in a laterally bent nanowire was similarly investigated by Gao and Wang using analytical and finite element methods [16]. This lateral deflection will form the basis for some of the first working nanogenerators which will be discussed shortly. Using a nanowire 600 nm in length and 50 nm in diameter, the authors were able to calculate at potential of almost ± 0.3 V across the diameter of a laterally deflected nanowire under 80 nN transverse load as shown in Fig. 5.2.

One of the most popular methods for synthesizing ZnO nanowires is the hydrothermal method. The hydrothermal method is a low temperature, aqueous solution method that can be used to produce vertical, well-aligned arrays of

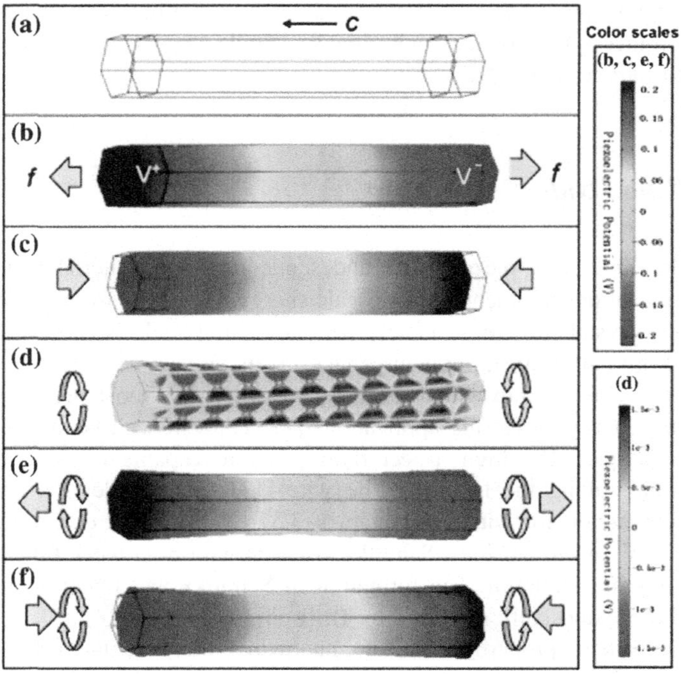

Fig. 5.1 Finite element model of **a** 1200 nm long ZnO nanowire. Piezoelectric potential distribution in nanowire under 85 nN tension (**b**) and compression (**c**). Piezoelectric potential under twist due to a 60 nN force pair (**d**). Piezoelectric potential under simultaneous twisting and tensile force (**e**) and twisting and compressive force (**f**). Reprinted with permission from [13]. Copyright 2009, AIP Publishing LLC

nanowires [5]. A popularly used variation on this method is to prepare precursor solution using a zinc salt such as zinc chloride or zinc nitrate [17]. Hexamethylenetetramine (HMTA) decomposes to formaldehyde and ammonia and is added to the solution usually in a 1:1 ratio to act as a pH buffer [18–20]. Additionally, ammonia can complex with zinc ions, thereby reducing unwanted homogenous nucleation, which can occur in the solution away from the substrate [21]. Growth typically takes place below 100 °C [17]. The low-temperature nature of the process is compatible with many substrate materials such as PET and silicon, making it ideal for device fabrication. Typically the substrate is coated with a zinc oxide seed layer, for example by sputtering [12]. The substrate is floated upside down on the surface of the solution and heated. Zinc hydroxyl species are formed by the reaction which then dehydrate to form zinc oxide nanowires on the seed layer [17]. The hydrothermal method usually results in nanowires of lengths from few to 10's of micrometers. Many methods have been explored to increase growth rate of hydrothermally grown nanowires, including microwave and laser assisted growth [22, 23]. The hydrothermal method is highly scalable, low cost, and repeatable. It is compatible with a wide range of substrates and materials making it

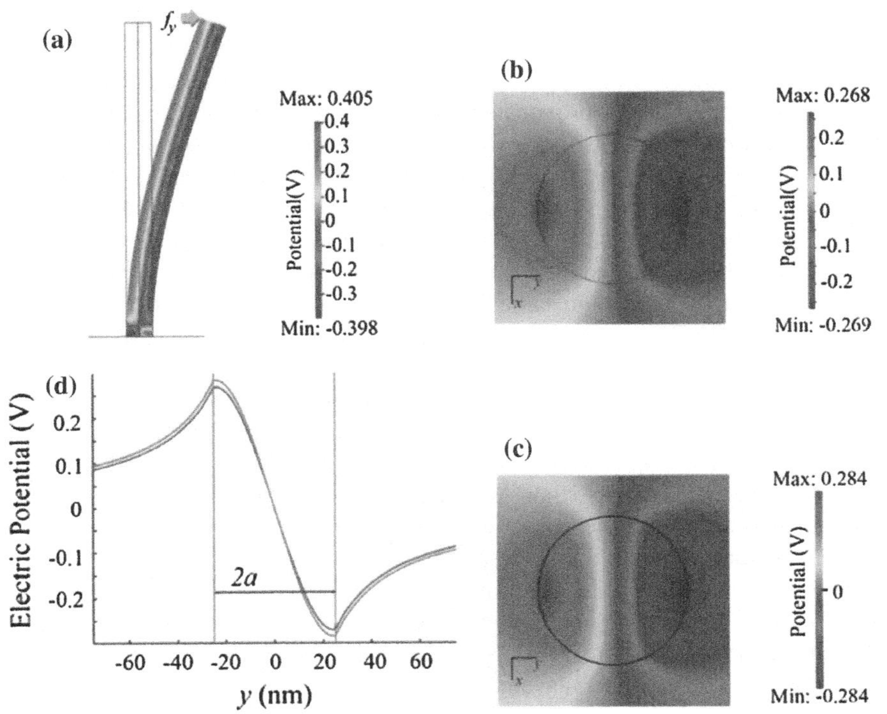

Fig. 5.2 **a** Finite element simulation results for piezoelectric potential in a laterally deformed nanowire under 80 nN transverse force. The nanowire length and diameter are 600 and 50 nm, respectively. **b** Cross-sectional view of the potential of the wire in (**a**). **c** Analytical results for piezoelectric potential in the cross section. **d** Comparison of piezoelectric potential from the FEA result (**b**) and the analytical result (**c**). Reprinted with permission from [16]. Copyright 2007 American Chemical Society

suitable for device fabrication. Zinc oxide nanowires have favorable piezoelectric, mechanical, and semiconducting properties which have been used to great advantage in the development of nanoscale energy generators. An overview of the development of these nanogenerators will be presented later in this chapter.

5.2.2 Molybdenum Disulfide

While 1D materials such as ZnO remain popular, there has been much recent interest in single atomic layer, or so called '2D' materials. One such material is molybdenum disulfide (MoS_2), a transition metal dichalcogenide (TMDC). MoS_2 is an inorganic compound with a hexagonal crystal structure [24]. For single atomic layer MoS_2, the two orthogonal crystal directions are often referred to as "armchair" and "zigzag" due to their respective appearances. The lattice constants are a = b = 3.16 Å and c = 12.29 Å [25]. These values agree closely with recent experimental work using XRD [26].

Bulk MoS_2 has an *indirect* bandgap of about 1.2 eV. However monolayer MoS_2 has a *direct* bandgap of approximately 1.8 eV [27, 28]. The direct bandgap makes monolayer MoS_2 amenable to optoelectronics applications. MoS_2 can tolerate large strains, making it possible to further tune the bandgap by means of strain engineering [29]. Scalise et al. calculated a Young's modulus of 435 GPa for single layer MoS_2 based on first principals using density functional theory. Using this model the authors predicted a transition to indirect band gap semiconductor at 2 % strain and a transition to metallic behavior at 8 % strain [30]. The prediction of monolayer MoS_2 being a direct bandgap semiconductor was experimentally verified using several different optical spectroscopy techniques [7].

The experimental study of piezoelectricity in single and few-layer MoS_2 flakes was reported by Wu et al. for energy harvesting applications [6]. In this work the authors exfoliated single layer MoS_2 flakes onto a PET substrate and deposited metal electrodes. Periodic strain was applied in the 'armchair' direction. The authors investigated the piezoelectric effect in even and odd-number layered samples up to 6 layers. Piezoelectric response was found only in odd-number layered samples and decreased with increasing number of layers. The piezoelectric effect was not observed in even-number layered samples or in the bulk material. This is due to the crystal structure of MoS_2 and is consistent with experimental results showing that odd-number layered MoS_2 exhibits noncentrosymmetry while even-number layered samples are centrosymmetric [31]. Zhu et al. reported a piezoelectric coefficient e_{11} = 2.9 E − 10 C/m using atomic force microscopy [32].

Recent work in the synthesis of monolayer MoS_2 has sought to overcome challenges to synthesizing large-area, single crystal MoS_2. The formation of single layer MoS_2 typically results in equilateral triangular structures which are single crystal in nature. For MoS_2, a single atomic layer has a thickness of about 0.7 nm [33]. A characteristic example of triangular MoS_2 crystals is shown in Fig. 5.3 [34].

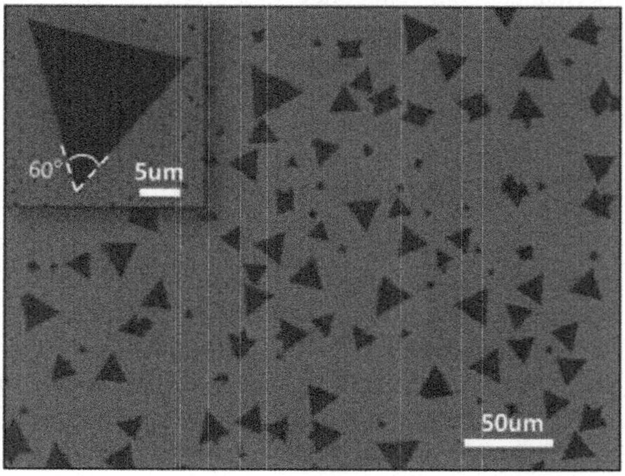

Fig. 5.3 Typical MoS_2 *equilateral triangle* monolayer structure imaged by scanning electron microscopy. The crystals were grown on a Si/SiO_2 substrate. Reprinted with permission from [34]. Copyright 2013 American Chemical Society

Like graphene, monolayer MoS_2 can be synthesized by mechanical exfoliation [7]. In light of challenges associated with achieving large area growth, chemical vapor deposition (CVD) has emerged as a promising method. Two main synthesis routes from CVD include direct growth from MoS_2 powder and from sulfurization of molybdenum trioxide (MoO_3) [33, 35]. Growth is typically done on Si/SiO_2 substrates and commonly results in the previously mentioned single crystal, equilateral triangular morphologies as seen in Fig. 5.3. Piranha solution is often used because of the importance of substrate cleanliness for growth [26]. Because these crystals form preferentially near cracks and imperfections on the substrate, nucleation can be controlled by various surface modification techniques [33]. Najamei and colleagues investigated the importance of grain boundary formation on these triangular domains to better understand the development of large area MoS_2 growth. The investigation revealed two mechanisms for grain boundary formation. These are the traditional chemical bonding between in-plane grains and the overlapping of adjacent grains [33].

Van der Waals (vdW) epitaxy is a promising approach being explored to permit growth on substrates with a high degree of lattice mismatch with MoS_2. The advantage of vdW epitaxy is that it does not require a chemical bond between the substrate and material as in the case of conventional lattice-matching epitaxy. Rather, vdW epitaxy relies on the van der Waals interactions between the material and the substrate [36]. Graphene has been successfully used for vdW epitaxy growth of monolayer MoS_2 by Shi et al. and by Ago and colleagues [37, 38]. Shi et al. used copper foil with a layer of graphene deposited by CVD [37]. Despite the lattice mismatch with MoS_2, graphene is cited as being an ideal surface for vdW epitaxy due to its inherent lack of dangling bonds which precludes the formation of strong chemical bonds between the substrate and growth material. The authors used $(NH_4)_2MoS_4$ as the single precursor for the growth and observed the formation of hexagonal MoS_2 flakes up to several micrometers in lateral dimension. During the growth process the flakes merged to form single to few-layer films as verified by TEM [37]. MoS_2 was found to have closely followed the orientation of the graphene template. This finding was confirmed by Ago et al. who reported two preferred orientations for triangular MoS_2 crystals grown on graphene templates [38]. Figure 5.4 shows the SEM and AFM images for single crystal, monolayer MoS_2 flakes grown by vdW epitaxy on graphene.

The prediction and discovery of piezoelectricity in single layer MoS_2 has led to many works on the synthesis and characterization of monolayer MoS_2 and its material and electrical properties. MoS_2 holds great potential for energy harvesting applications. Initial explorations of MoS_2 for energy harvesting applications will be presented later in the chapter.

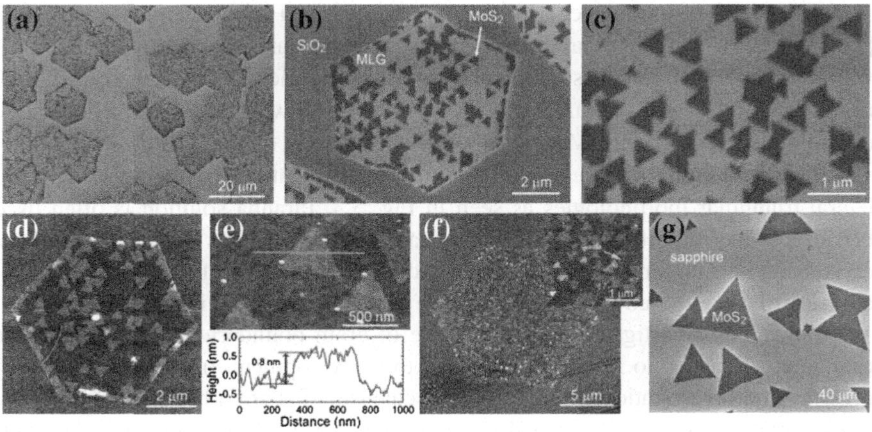

Fig. 5.4 MoS_2 crystals grown by vdW epitaxy on hexagonal graphene monolayers as imaged by SEM (**a–c**). Corresponding AFM images (**d–f**). *Inset* of (**e**) shows topography profile of MoS_2 crystal. Growth of MoS_2 on Sapphire substrate (**g**). Reprinted with permission from [38]. Copyright 2015 American Chemical Society

5.2.3 *Diphenylalanine (FF) Peptide*

Diphenylalanine (FF) peptide is a unique bioinspired material with good piezoelectric and mechanical properties. Since the observed self-assembly of diphenylalanine into nanotubes, there has been much interest in studying the unique properties of this material [39]. Kol et al. investigated the mechanical properties of diphenylalanine peptide nanotubes by atomic force microscopy [40]. By repeatedly measuring the point stiffness of individual peptide nanotubes, the authors were able to calculate that the stiffness for a peptide nanotube was about 160 N/m on average. Using the stiffness and topography measurements from the AFM scans, a finite element model of a peptide nanotube was created with assumed linear elasticity. Based on this model a Young's modulus of 19 GPa was calculated.

The piezoelectric effect in FF peptide nanotubes was investigated by Kholkin and colleagues [41]. Piezoresponse Force Microscopy (PFM) was used to scan nanotubes of various sizes along the long axis of the nanotube. The authors used the PFM response data and calibrated the measurements to a sample of $LiNbO_3$ with known piezoelectric properties. For the peptide nanotubes, a d_{15} piezoelectric coefficient of 60 pm/V was calculated based on these results. Heredia et al. investigated the temperature dependence of the piezoelectric effect in FF nanotubes. Piezoelectricity was found to decrease with increasing temperature and was lost completely around 140 °C, corresponding to a crystalline phase transition [42].

Diphenylalanine peptide nanostructures are typically synthesized by solution-based self-assembly methods [43–46]. A recent work by Nguyen et al. explored the epitaxial growth of vertically aligned FF peptide microrods [47]. Figure 5.5a–c shows schematically the growth process. First, a seed layer was

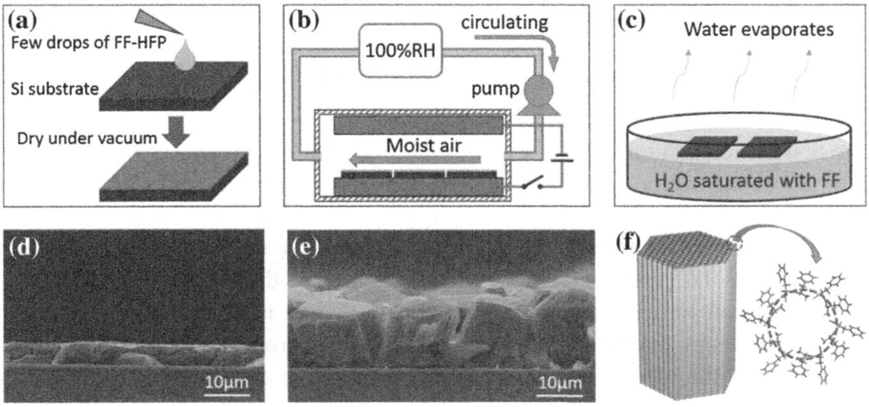

Fig. 5.5 Synthesis process for FF peptide microrods starting with deposition of FF peptide layer on substrate (**a**) and crystallization to form seed layer (**b**) followed by self-assembly of peptide microrods (**c**). SEM images show side profile of seed layer (**d**) and microrods (**e**). A schematic of the FF peptide microrods shows the hexagonal nanochannels (**f**). Reprinted from Nano Energy, vol 17, Nguyen et al., Epitaxial growth of vertically aligned piezoelectric diphenylalanine peptide microrods with uniform polarization, 323–329, Copyright 2015, with permission from Elsevier [47]

prepared on a substrate of ITO-coated silicon. A precursor solution was prepared by dissolving diphenylalanine lyophilized powder (FF) in hexafluoro-2-propanol (HFP). A small quantity of the solution was placed onto the substrate which was then placed in a vacuum desiccator. The pressure was quickly reduced and the solution evaporated, leaving behind a thin film. Crystallization of the film into the seed layer was achieved by placing the substrate in a chamber with circulating air at ~ 100 % relative humidity. To grow the FF peptide microrod structure, a second precursor solution was prepared by dissolving FF powder into DI water. The solution was heated before the substrate was floated face down in this solution for 6 h. After the solution had cooled and evaporated, the substrate was removed revealing the desired microrod structure. The growth process is compatible with many substrates including silicon, ITO, and gold. The morphology was verified by SEM in Fig. 5.5e, f which showed ~ 10 μm long microrods with uniform vertical growth. PFM measurements revealed d_{33} piezoelectric coefficients of 9.4 and 9.9 pm/V for the seed layer and microrods respectively. As grown, the microrod domains exhibit random polarization resulting in zero net piezoelectric effect. By applying an external electric field during growth, the authors were able achieve uniform polarization and a net piezoelectric effect. The favorable structural and electrical properties of diphenylalanine peptide nanostructures make them ideal candidates for future work in energy harvesting applications.

5.3 Energy Harvesting with Piezoelectric Nanomaterials

5.3.1 Energy Harvesting with Zinc Oxide

The first demonstration of energy harvesting from ZnO nanowires was published in 2006 by Wang's group [2]. In this work the authors used an array of zinc oxide nanowires to convert mechanical energy from an atomic force microscope (AFM) into electrical energy. The wires were grown on an Al_2O_3 substrate using a vapor-liquid-solid process. Gold was used to catalyze the reaction, resulting in a small gold nanoparticle on top of each wire. Although most of the gold nanoparticles did not stay on the tips of the wires, those that remained either partially or fully covered the top of the wires. By controlling the density and length of the wires, the authors were able to space the wires so that AFM could easily probe individual wires. The individual wires were scanned in contact mode with a force of 5 nN. Silver paste was used for the bottom electrode, which formed an Ohmic contact with the nanowires. The platinum coated tip of the AFM formed a Schottky contact with the zinc oxide resulting in a metal-semiconductor-metal junction. As the tip was scanned the output voltage was recorded across an external resistive load. Scans of the wires showed output voltages on the order of 6–9 mV with an estimated lifetime of 0.6 s based on the scan rate. This experiment was similarly repeated with non-piezoelectric materials to verify the source of the voltage was in fact from the piezoelectric effect.

The deflection of the individual wires could be estimated from the AFM topography map. From this the authors estimated the energy required to elastically deform the wires and compared this to the energy of the measured piezoelectric discharge. A comparison between these energies resulted in an estimated efficiency of 17–30 %. In this experiment the piezoelectric potential arises from the lateral bending of the nanowire, and the potential distribution is similar to that seen in Fig. 5.2. For the majority of nanowires without a capping gold particle, the Pt-coated AFM tip formed a Schottky contact with ZnO. On the tensile side, the laterally deformed wire has a positive potential and forms a reverse-biased Schottky contact with the AFM tip. As the AFM tip scans across the wire it contacts the negative, or compressed side, forming a forward-biased Schottky contact. Therefore, the flow of current is dependent on whether the tip is contacting the tensile or compressive side of the wire. Nanowires that remained completely capped with a gold nanoparticle did not show any voltage output because the displacement potential in the wire was neutralized by free carriers. The effectiveness of the nanowire based energy generator is therefore dependent on both the piezoelectric and semiconducting properties of zinc oxide. In conclusion the authors estimated an output power density of 10 pW/μm^2, demonstrating the energy harvesting potential of zinc oxide nanowires.

The use of vertically aligned arrays of ZnO nanowires for energy harvesting in a standalone device was presented by Wang et al. [11]. The authors grew ZnO nanowires approximately 1 μm in length on GaN and sapphire substrates. To mimic the effect of the AFM tip, a silicon wafer was patterned with an array of triangular teeth and coated with platinum. This top electrode was placed atop the nanowire array and held in place by polymer strips which supported and connected it to the bottom substrate. This can be thought of as being equivalent to an array of single nanowires, each with their own AFM tip. The device was approximately 2 mm^2 and was subjected to 41 kHz ultrasonic waves. As the row of teeth move up and down from the vibration, some of the wires are deflected laterally. Similar to the AFM study, the platinum coated teeth make a Schottky contact with the tips of the nanowires. As the individual nanowire is deflected, the tooth contacts the tensile side creating a reversed Schottky contact and a buildup of charge occurs. As the relative motion of the tooth and wire continues, the tooth moves to the negative side where it forms a forward-biased contact and discharges. The net effect of this charging and discharging due to the multiple, dynamic Schottky contacts is a mostly constant DC output. The magnitude of the output voltage of the parallel nanowires was up to 7 mV with nanoamp scale output currents. The authors estimate that such a device could achieve a power output of 10 $\mu W/cm^2$.

Xu et al. further expands on the ZnO nanowire array generator by stacking multiple layers of nanowire/nanotip layers together [12]. Xu and colleagues prepared silicon substrates with an insulating layer of Al_2O_3 on each side. Chromium and ZnO layers were sputtered on each side to serve as a bottom electrode and seed layer, respectively. Using the previously discussed hydrothermal method, unique ZnO nanostructures were grown on each side of the substrate. Long nanowires were grown on one side of the substrate and short, conical nanotips were grown on the other side. The nanotips were sputtered with a thin layer of gold to serve as the top electrode, again analogous to an AFM tip. Like platinum, gold also forms a Schottky contact with ZnO. The silicon substrates were stacked so that the nanowire side of the lower layer was in contact with the nanotip side of the upper layer. Figure 5.6 shows the fabrication of the individual nanogenerator layers and their assembly into a multilayer nanogenerator. The generator was then subject to ultrasonic vibrations as previously discussed. As with the previous generator design, the working principal here is the switching between reverse and forward-biased Schottky contacts as the nanotips move from the stretched side of the nanowire to the compressed side. Despite variations in individual wire response, the net effect of the entire population of wires is a relatively steady DC output current. To test the benefits of the stacked nanogenerator architecture, the output currents were measured for two individual nanogenerators to be 13 and 10 nA separately, and 22 nA when connected in parallel. Connecting the nanogenerators in antiparallel subsequently reduced the combined output current to 3 nA. Two nanogenerators which separately gave output voltages of 2.6 and 0.8 mV yielded an output voltage of 3.5 mV when connected in series. Lastly a total of 4 nanogenerators were connected to give an output voltage of 62 mV and an output current of 105 nA. Voltage and current output for the 4-layer nanogenerator are shown in Fig. 5.7. The authors estimate an output power density of 0.11 $\mu W/cm^2$ from this configuration.

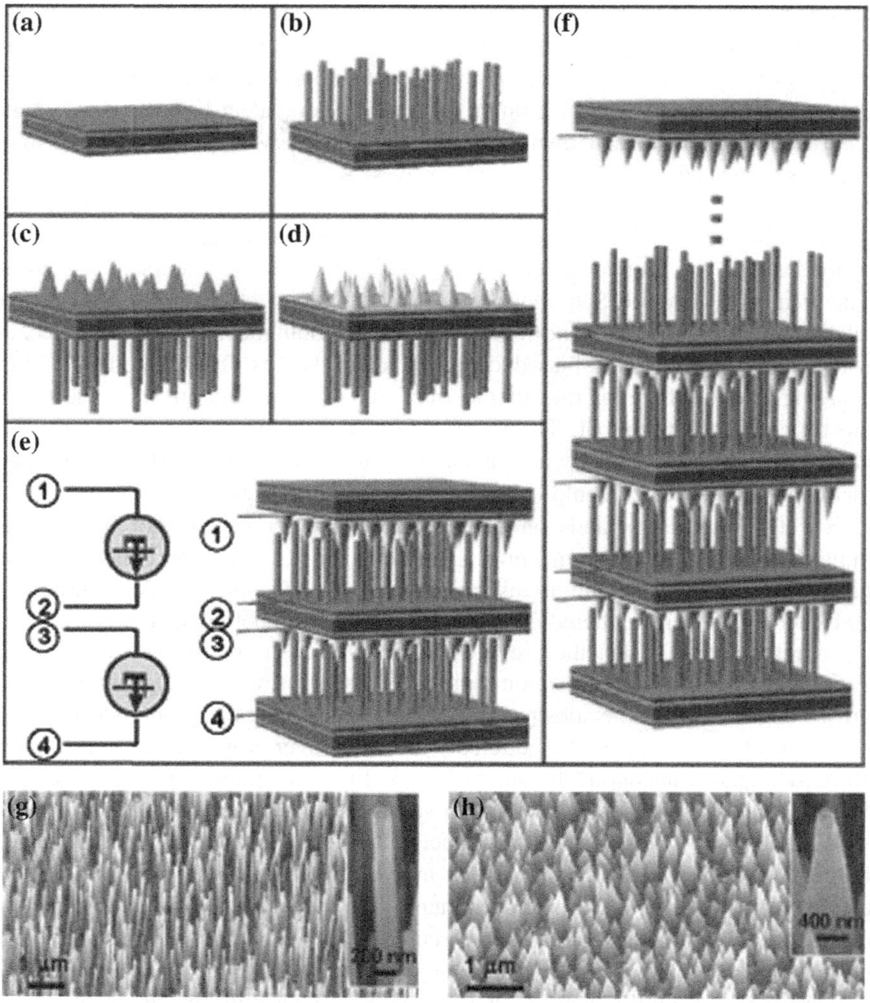

Fig. 5.6 **a–d** show the fabrication steps of an individual nanogenerator layer beginning with **a** preparation of substrate with ZnO seed layer and **b** growth of vertical nanowires on the substrate. ZnO nanotips are grown on the opposite side (**c**) and coated with Au (**d**). An assembled two-layer generator is shown in (**e**) with corresponding schematic. Multiple nanogenerator elements are stacked for enhanced energy generation performance (**f**). Scanning electron microscope images **g** and **h** show the vertically aligned nanowire and nanotip arrays, respectively. Reprinted with permission from [12]. Copyright 2008 American Chemical Society

Yang et al. takes a different approach and explores nano energy harvesting with a single laterally oriented ZnO microwire [48]. The single wire generator (SWG) was constructed by placing a single ~200–300 μm long nanowire on a 50 μm thick Kapton substrate and applying silver paste to form electrodes on both ends before packaging with a flexible polymer film. When the polymer substrate is bent, the top

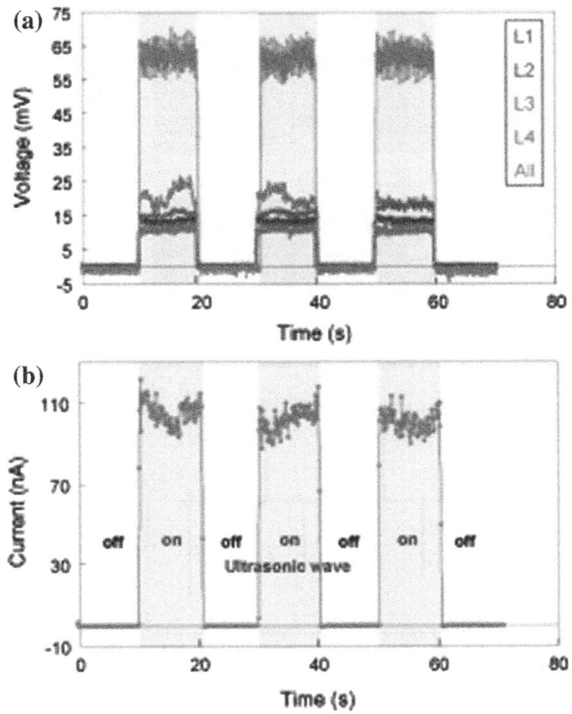

Fig. 5.7 **a** The output voltage from four nanogenerator layers connected in series with ultrasonic vibration switched on and off. The total voltage and individual layer voltages are shown. **b** Combined output current measured from the four-layer nanogenerator. Reprinted with permission from [12]. Copyright 2008 American Chemical Society

surface where the nanowire is located experiences tension. When the ZnO microwire is placed in tension the piezoelectric potential generated in the wire causes the flow of electrons. Once the strain is released, the flow of electrons is reversed. The result of periodically applied strain is therefore an alternating current. Although silver paste was used on both sides of the wire, experimental results showed that one of the ZnO–Ag junctions formed a Schottky contact while the other contact was Ohmic. This is determined by whether the end of the wire terminates with zinc or oxygen.

Testing of the device was accomplished by applying a periodic strain of ~0.05–0.1 %. The device was shown to be stable over 2 h of testing. The output voltage and current were found to be dependent on the strain rate, with higher strain rates corresponding to higher outputs. This single wire generator was able to output up to 50 mV and 750 pA with an efficiency of 6.8 %. The design is mechanically robust and may be able to overcome challenges associated with wear faced by the previous vertical nanowire generators. To further illustrate the applications for such a nanogenerator, Yang and colleagues used this generator for extracting biomechanical energy [3]. The device was connected to a human finger as seen in Fig. 5.8a which was flexed periodically to generate output voltages up to 25 mV and currents of about 150 pA. The current-voltage characteristics of the nanogenerator are shown in Fig. 5.8b. Output voltage and current from the finger flexing

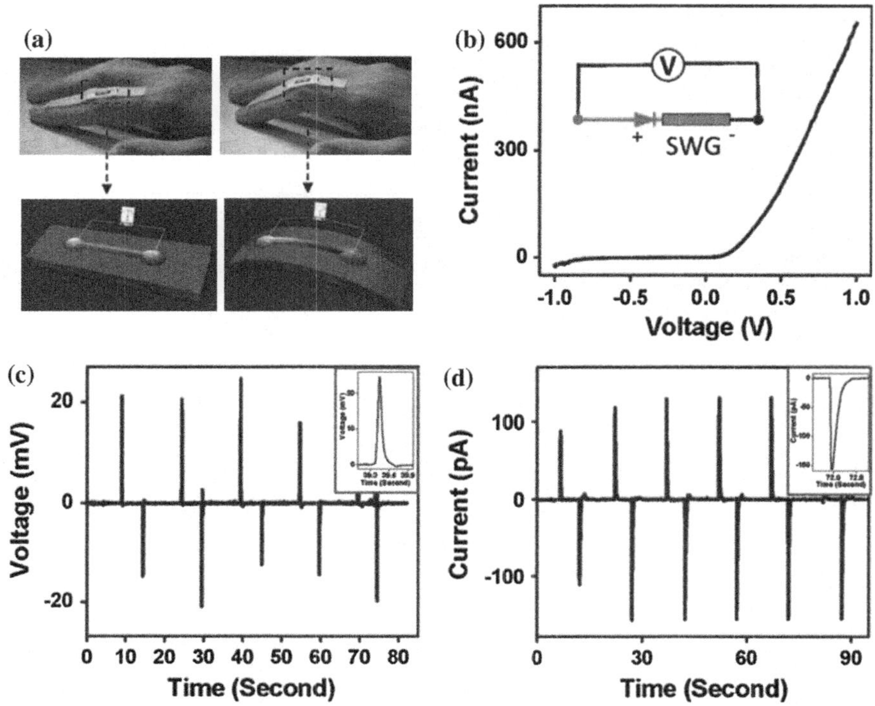

Fig. 5.8 **a** The lateral wire nanogenerator is attached to the human finger and used to collect energy from the flexing motion. **b** The rectifying current-voltage (I-V) characteristics of the nanogenerator are shown along with an electrical schematic. The measured voltage (**c**) and current (**d**) peaks corresponding to finger flexing. Reprinted with permission from [3]. Copyright 2009 American Chemical Society

motion are presented in Fig. 5.8c, d, respectively. Next the generator was attached to a jacket which could be worn by a hamster. Mechanical energy was harvested from the motion of the hamster running on a wheel which created 50–100 mV and 0.5 nA output. This work demonstrated the potential for nanoscale energy harvesters to extract energy from biomechanical motions.

The concept of a lateral nanowire generator was further expanded to take advantage of rows of many lateral, aligned nanowires as shown in Fig. 5.9 [49]. Using the low-temperature hydrothermal method, the authors patterned rows of ZnO seed layers and grew rows of aligned nanowires parallel to the flexible substrate. The authors patterned 700 rows in total, estimating about 20,000 nanowires per row. A gold electrode was deposited over one end of each of the nanowire rows, forming a Schottky contact with all of the nanowires in that row. This nanogenerator design achieved an output voltage of 1.26 V and 28.8 nA. Results for the output voltage and current are shown in Fig. 5.10. The authors demonstrated the application of this nanogenerator by using it to power a nanoscale pH sensor and a UV sensor.

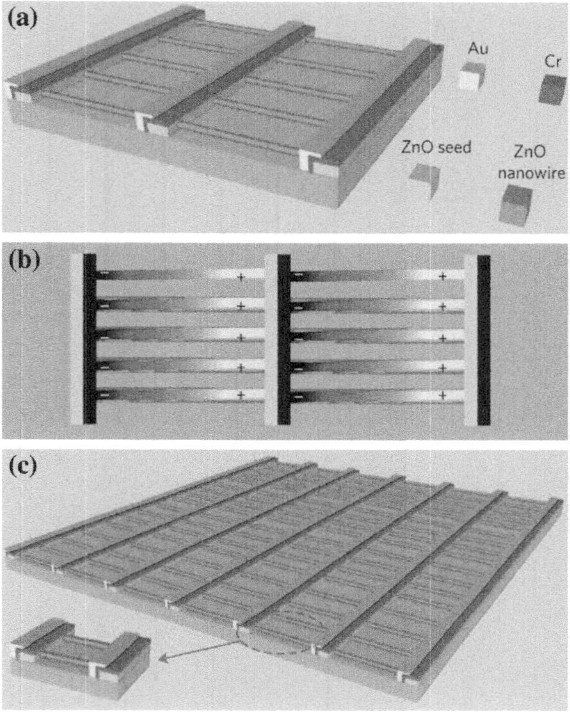

Fig. 5.9 **a** Concept image of a nanogenerator based on rows of lateral nanowires. The main components are shown including substrate, seed layer, ZnO nanowires, and common gold and chromium electrodes. **b** Polarization generated in the nanowire rows under applied strain. **c** Scalability of the lateral nanowire-based nanogenerator is illustrated. Reprinted by permission from Macmillan Publishers Ltd: Nature Nanotechnology, [49], Copyright 2010

5.3.2 Energy Harvesting with Molybdenum Disulfide

Wu and colleagues demonstrated the possibility of energy harvesting based on 2D materials with the investigation of piezoelectricity in single layer MoS_2 [6]. Single layer triangular MoS_2 flakes were prepared by exfoliation with a typical flake size of about 5 μm × 10 μm. Once the atomic layer thickness was verified, the authors used single harmonic generation (SHG) to identify the crystal structure orientation of a single flake. This information was superimposed onto an optical image of the flake so as to identify the "armchair" and "zigzag" directions of the lattice. Figure 5.11a shows the superposition of the MoS_2 flake with the lattice orientation derived from the SH data in Fig. 5.11b. The flake was placed on a PET substrate as seen in Fig. 5.11c. The SH data was then used to orient and deposit metal electrodes (Cr/Pd/Au) so that the flake would be strained in the armchair direction when the substrate was flexed. By applying a periodic strain, the authors were able to measure the output voltage and current across a load resistor as shown in Fig. 5.11d. Up to 18 mV open-circuit voltage was recorded at 0.64 % strain with a maximum output current of 27 pA. Next the authors explored the power output by varying the external load resistance on the nanogenerator operating at 0.53 % strain. Using an external load of 220 MΩ, the authors calculated power density of 2 mW/m^2 with about 5 % efficiency. Figure 5.12 shows the response data for the

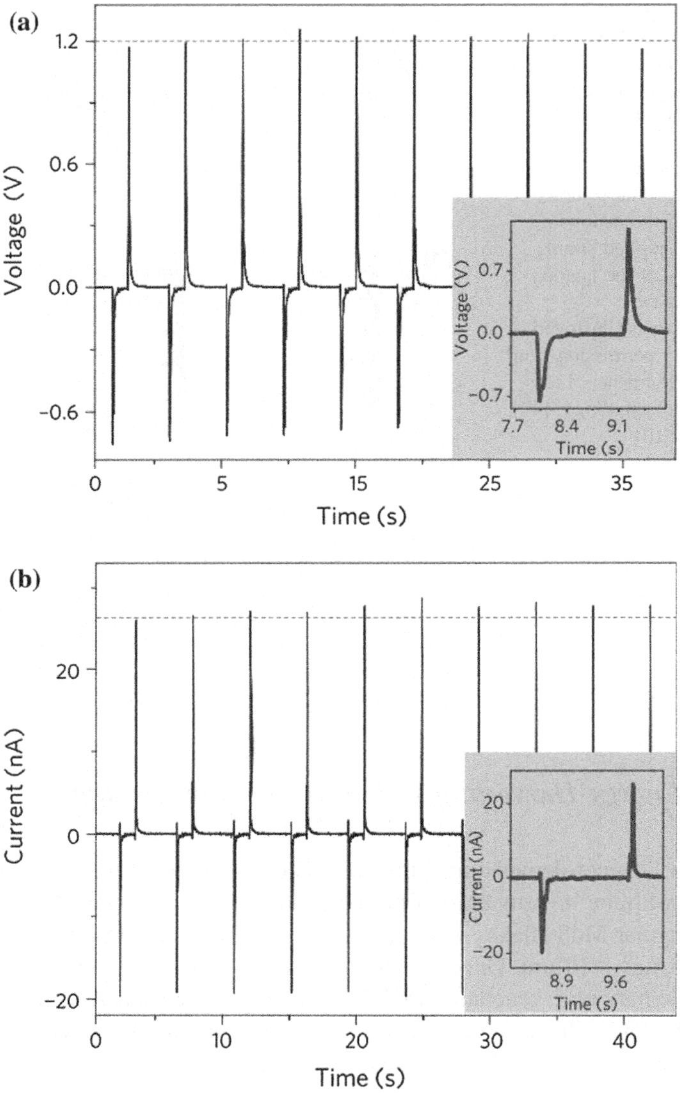

Fig. 5.10 Measurements from a lateral nanowire generator with many rows. **a** Output voltage and **b** current are shown for periodic applied strain. Reprinted by permission from Macmillan Publishers Ltd: Nature Nanotechnology, [49], Copyright 2010

nanogenerator. In similar fashion to the previous ZnO NW based generators, the authors then moved from a single MoS_2 flake to an array of connected flakes. The authors grew new MoS_2 monolayer triangles, this time using chemical vapor deposition (CVD). These flakes were similarly transferred to a flexible PET substrate. By noting differences in edge morphologies, the authors were easily able to identify the armchair direction on the flakes [50]. Up to four flakes were connected

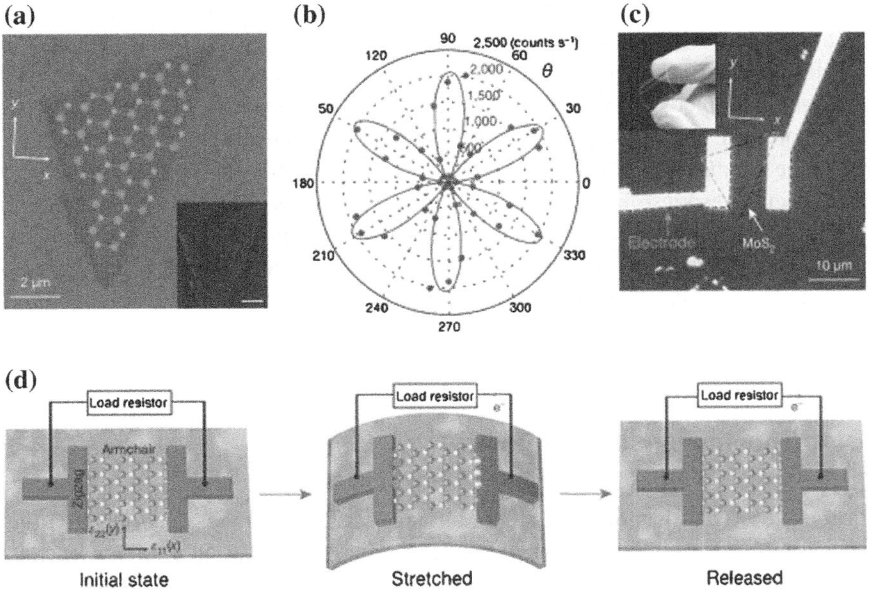

Fig. 5.11 **a** Using SHG the lattice orientation was found and combined with optical microscopy images to identify the armchair and zigzag directions on a single MoS_2 flake. **b** Single harmonic data plotted in polar coordinates. **c** Fabricated energy harvesting device with the flake and electrodes positioned such that strain occurs in the armchair direction. **d** Stretching and relaxation of the device generates polarization in the material. The output can be measured across an external load. Reprinted by permission from Macmillan Publishers Ltd: Nature, [6], Copyright 2014

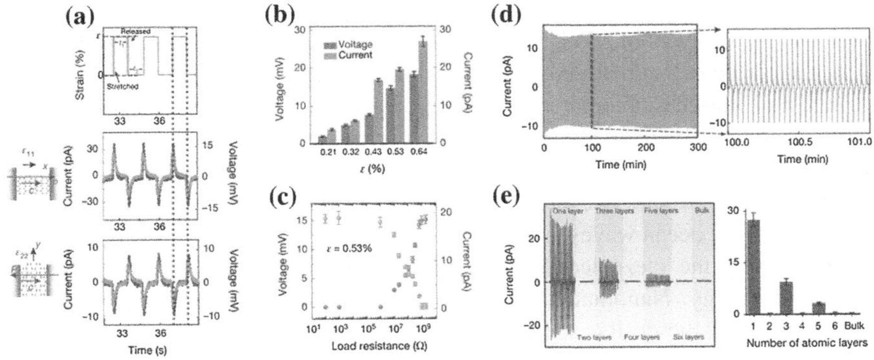

Fig. 5.12 **a** Periodic strain (*top*) applied to a MoS_2 device with current and voltage outputs for strain applied in the armchair (*middle*) and zigzag (*bottom*) directions. **b** Output voltage and current as a function of strain. **c** For a strain of 0.53 %, the output current and voltage vary with the external resistive load value. **d** Long-term stability of the device under period strain. **e** Measured current output decreases as the number of MoS_2 layers increases. Note lack of response in even-number layered configurations. Reprinted by permission from Macmillan Publishers Ltd: Nature, [6], Copyright 2014

with the output current and voltage increasing with the number of flakes, demonstrating the potential for nanogenerators based on arrays of MoS_2 monolayer structures [6]. As shown in Fig. 5.12e, no piezoelectric potential was observed in even-number layered MoS_2 flakes. A recent work by Zhou et al. further investigates the theoretical performance of nanogenerators based on multiple layers of MoS_2 [51]. The results of this simulation study were consistent with the observed lack of piezoelectric response in even-number layered nanogenerators and the decrease in piezoelectric response with increasing odd-number of layers. Interestingly, the results suggest that the capacitance of MoS_2 nanogenerators will be an order of magnitude less than that of nanowire-based generators. This suggests that MoS_2 nanogenerators may be especially well suited for high frequency energy harvesting.

5.4 Conclusions and Outlook

Piezoelectric nanomaterials are important smart and functional materials and they have found wide application including consumer products, automobiles, medical diagnostics, and sophisticated scientific instruments. Piezoelectric nanomaterials have attracted significant research efforts due to their small size, low power consumption, superior mechanical robustness and better performance, and they are expected to have great potential in renewable energy and physical sensor applications. ZnO nanostructures are the most widely studied piezoelectric nanomaterial because of their strong enhanced piezoelectricity which is further improved at the nanoscale. Well-established synthesis approaches can grow ZnO nanowires on almost any substrate. In addition to ZnO nanomaterials, many other materials have also been discovered for developing piezoelectric energy harvesters. For instance, atomically thin layers of MoS_2 showed the piezoelectric effect although bulk MoS_2 is not a piezoelectric material. Peptide supramolecular structures are among the strongest piezoelectric biomaterials. New piezoelectric nanomaterials and structures continue to be developed and various piezoelectric nanodevices have been demonstrated.

Ubiquitous mechanical energy can be found in the environment in the forms of vibration, wind, ocean waves, and human movements. Harvesting these mechanical energies using the piezoelectric effect provides a novel approach for generating renewable energy. Nanomaterials offer unique structural and electrical properties which allow for the design of many interesting nanogenerator architectures. There are two major working modes for energy generation. One is the bending mode, in which the nanowire is bent and the piezoelectric potential on the side face contributes to the power output. Nanogenerators working in this mode often use nanowires perpendicular to a substrate. The second mode is the stretch-compression model, in which the nanowire is stretched and/or compressed. Nanogenerators working in this model rely on the potential across the length of the nanowire for power output, and the nanowire can be either along, or normal to, the surface of the substrate. These nanogenerators have been successfully used to harvest mechanical energy from muscle movement, heart beats, and rain drops. The performance of nanogenerators has rapidly improved to the point where they can be used to power nanodevices. Further development of nanogenerators and their successful integration with other sensors, telemetry, and control units are expected to create autonomous and self-powered nanosystems which will have a broad impact on our society.

References

1. Y. Zhang, Y. Liu, Z.L. Wang, Fundamental theory of piezotronics. Adv. Mater. **23**, 3004–3013 (2011)
2. Z.L. Wang, J. Song, Piezoelectric nanogenerators based on zinc oxide nanowire arrays. Science **312**, 242–246 (2006)
3. R. Yang, Y. Qin, C. Li, G. Zhu, Z.L. Wang, Converting biomechanical energy into electricity by a muscle-movement-driven nanogenerator. Nano Lett. **9**, 1201–1205 (2009)
4. M.J. Madou. *Fundamentals of Microfabrication: the Science of Miniaturization* (CRC press, 2002)
5. L. Vayssieres, Growth of arrayed nanorods and nanowires of ZnO from aqueous solutions. Adv. Mater. **15**, 464–466 (2003)
6. W. Wu, L. Wang, Y. Li, F. Zhang, L. Lin, S. Niu, D. Chenet, X. Zhang, Y. Hao, T.F. Heinz, J. Hone, Z.L. Wang, Piezoelectricity of single-atomic-layer MoS_2 for energy conversion and piezotronics. Nature **514**, 470–474 (2014)
7. K.F. Mak, C. Lee, J. Hone, J. Shan, T.F. Heinz, Atomically thin MoS(2): a new direct-gap semiconductor. Phys. Rev. Lett. **105**, 136805 (2010)
8. Z.L. Wang, ZnO nanowire and nanobelt platform for nanotechnology. Mater. Sci. Eng. R. **64**, 33–71 (2009)
9. R. Agrawal, B. Peng, E.E. Gdoutos, H.D. Espinosa, Elasticity size effects in ZnO nanowires—a combined experimental-computational approach. Nano Lett. **8**, 3668–3674 (2008)
10. H. Morkoç, Ü. Özgür, *Zinc Oxide : fundamentals, Materials and Device Technology* (Wiley-VCH, Weinheim, 2009)
11. X. Wang, J. Song, J. Liu, Z.L. Wang, Direct-current nanogenerator driven by ultrasonic waves. Science **316**, 102–105 (2007)
12. S. Xu, Y. Wei, J. Liu, R. Yang, Z.L. Wang, Integrated multilayer nanogenerator fabricated using paired nanotip-to-nanowire brushes. Nano Lett. **8**, 4027–4032 (2008)
13. Z. Gao, J. Zhou, Y. Gu, P. Fei, Y. Hao, G. Bao, Z.L. Wang, Effects of piezoelectric potential on the transport characteristics of metal-ZnO nanowire-metal field effect transistor. J. Appl. Phys. **105**, 113707 (2009)
14. U. Özgür, Y.I. Alivov, C. Liu, A. Teke, M.A. Reshchikov, S. Doğan, V. Avrutin, S.J. Cho, H. Morkoç, A comprehensive review of ZnO materials and devices. J. Appl. Phys. **98**, 041301 (2005)
15. L.J. Brillson, Y. Lu, ZnO Schottky barriers and ohmic contacts. J. Appl. Phys. **109**, 121301 (2011)
16. Y. Gao, Z.L. Wang, Electrostatic potential in a bent piezoelectric nanowire. The fundamental theory of nanogenerator and nanopiezotronics. Nano Lett. **7**, 2499–2505 (2007)
17. S. Xu, Z.L. Wang, One-dimensional ZnO nanostructures: solution growth and functional properties. Nano Res. **4**, 1013–1098 (2011)
18. K.M. McPeak, T.P. Le, N.G. Britton, Z.S. Nickolov, Y.A. Elabd, J.B. Baxter, Chemical bath deposition of ZnO nanowires at near-neutral ph conditions without hexamethylenetetramine (Hmta): understanding the role of Hmta in ZnO nanowire growth. Langmuir **27**, 3672–3677 (2011)

19. S. Xu, Z.L. Wang, Oxide nanowire arrays for light-emitting diodes and piezoelectric energy harvesters. Pure Appl. Chem. **83**, 2171–2198 (2011)
20. S. Xu, N. Adiga, S. Ba, T. Dasgupta, C.F. Wu, Z.L. Wang, Optimizing and improving the growth quality of ZnO nanowire arrays guided by statistical design of experiments. ACS Nano **3**, 1803–1812 (2009)
21. J.H. Tian, J. Hu, S.S. Li, F. Zhang, J. Liu, J. Shi, X. Li, Z.Q. Tian, Y. Chen, Improved seedless hydrothermal synthesis of dense and ultralong ZnO nanowires. Nanotechnology **22**, 245601 (2011)
22. H.E. Unalan, P. Hiralal, N. Rupesinghe, S. Dalal, W.I. Milne, G.A.J. Amaratunga, Rapid synthesis of aligned zinc oxide nanowires. Nanotechnology **19**, 255608 (2008)
23. J. Yeo, S. Hong, M. Wanit, H.W. Kang, D. Lee, C.P. Grigoropoulos, H.J. Sung, S.H. Ko, Rapid, one-step, digital selective growth of ZnO nanowires on 3d structures using laser induced hydrothermal growth. Adv. Funct. Mater. **23**, 3316–3323 (2013)
24. J. Brivio, D.T. Alexander, A. Kis, Ripples and layers in ultrathin MoS_2 membranes. Nano Lett. **11**, 5148–5153 (2011)
25. P. Young, Lattice parameter measurements on molybdenum disulphide. J. Phys. D Appl. Phys. **1**, 936 (1968)
26. K.K. Liu, W. Zhang, Y.H. Lee, Y.C. Lin, M.T. Chang, C.Y. Su, C.S. Chang, H. Li, Y. Shi, H. Zhang, C.S. Lai, L.J. Li, Growth of large-area and highly crystalline MoS_2 thin layers on insulating substrates. Nano Lett. **12**, 1538–1544 (2012)
27. K. Kam, B. Parkinson, Detailed photocurrent spectroscopy of the semiconducting group Vib transition metal dichalcogenides. J. Phys. Chem. **86**, 463–467 (1982)
28. S. Lebegue, O. Eriksson, Electronic structure of two-dimensional crystals from ab initio theory. Phys. Rev. B. **79**, 115409 (2009)
29. S. Manzeli, A. Allain, A. Ghadimi, A. Kis, Piezoresistivity and strain-induced band gap tuning in atomically thin MoS_2. Nano Lett. **15**, 5330–5335 (2015)
30. E. Scalise, M. Houssa, G. Pourtois, V. Afanas'ev, A. Stesmans, Strain-induced semiconductor to metal transition in the two-dimensional honeycomb structure of MoS_2. *Nano Res.* **5**, 43–48 (2011)
31. Y. Li, Y. Rao, K.F. Mak, Y. You, S. Wang, C.R. Dean, T.F. Heinz, Probing symmetry properties of few-layer MoS_2 and H-Bn by optical second-harmonic generation. Nano Lett. **13**, 3329–3333 (2013)
32. H. Zhu, Y. Wang, J. Xiao, M. Liu, S. Xiong, Z. J. Wong, Z. Ye, X. Yin, X. Zhang, Observation of piezoelectricity in monolayer molybdenum disulfide, arXiv preprint arXiv: 1408.7074 (2014)
33. S. Najmaei, Z. Liu, W. Zhou, X. Zou, G. Shi, S. Lei, B.I. Yakobson, J.C. Idrobo, P.M. Ajayan, J. Lou, Vapour phase growth and grain boundary structure of molybdenum disulphide atomic layers. Nat. Mater. **12**, 754–759 (2013)
34. S. Wu, C. Huang, G. Aivazian, J.S. Ross, D.H. Cobden, X. Xu, Vapor-solid growth of high optical quality MoS_2 monolayers with near-unity valley polarization. ACS Nano **7**, 2768–2772 (2013)
35. Y.H. Lee, X.Q. Zhang, W. Zhang, M.T. Chang, C.T. Lin, K.D. Chang, Y.C. Yu, J.T. Wang, C.S. Chang, L.J. Li, T.W. Lin, Synthesis of large-area MoS_2 atomic layers with chemical vapor deposition. Adv. Mater. **24**, 2320–2325 (2012)
36. M.I. Utama, Q. Zhang, J. Zhang, Y. Yuan, F.J. Belarre, J. Arbiol, Q. Xiong, Recent developments and future directions in the growth of nanostructures by Van Der Waals epitaxy. Nanoscale **5**, 3570–3588 (2013)
37. Y. Shi, W. Zhou, A.Y. Lu, W. Fang, Y.H. Lee, A.L. Hsu, S.M. Kim, K.K. Kim, H.Y. Yang, L. J. Li, J.C. Idrobo, J. Kong, Van Der Waals epitaxy of MoS(2) layers using graphene as growth templates. Nano Lett. **12**, 2784–2791 (2012)
38. H. Ago, H. Endo, P. Solis-Fernandez, R. Takizawa, Y. Ohta, Y. Fujita, K. Yamamoto, M. Tsuji, Controlled Van Der Waals epitaxy of monolayer MoS_2 triangular domains on graphene. ACS Appl. Mater. Interfaces. **7**, 5265–5273 (2015)

39. M. Reches, E. Gazit, Casting metal nanowires within discrete self-assembled peptide nanotubes. Science **300**, 625–627 (2003)
40. N. Kol, L. Adler-Abramovich, D. Barlam, R.Z. Shneck, E. Gazit, I. Rousso, Self-assembled peptide nanotubes are uniquely rigid bioinspired supramolecular structures. Nano Lett. **5**, 1343–1346 (2005)
41. A. Kholkin, N. Amdursky, I. Bdikin, E. Gazit, G. Rosenman, Strong piezoelectricity in bioinspired peptide nanotubes. ACS Nano **4**, 610–614 (2010)
42. A. Heredia, I. Bdikin, S. Kopyl, E. Mishina, S. Semin, A. Sigov, K. German, V. Bystrov, J. Gracio, A.L. Kholkin, Temperature-driven phase transformation in self-assembled diphenylalanine peptide nanotubes. J. Phys. D Appl. Phys. **43**, 462001 (2010)
43. J. Zhang, X. Wu, Z. Gan, X. Zhu, Y. Jin, Unidirectionally aligned diphenylalanine nanotube/microtube arrays with excellent supercapacitive performance. Nano Res. **7**, 929–937 (2014)
44. M. Sakurai, P. Koley, M. Aono, A new approach to molecular self-assembly through formation of dipeptide-based unique architectures by artificial supersaturation. Chem. Commun. **50**, 12556–12559 (2014)
45. S. Scanlon, A. Aggeli, Self-assembling peptide nanotubes. Nano Today **3**, 22–30 (2008)
46. M. Reches, E. Gazit, Controlled patterning of aligned self-assembled peptide nanotubes. Nat. Nanotechnol. **1**, 195–200 (2006)
47. V. Nguyen, K. Jenkins, R. Yang, Epitaxial growth of vertically aligned piezoelectric diphenylalanine peptide microrods with uniform polarization. Nano Energy **17**, 323–329 (2015)
48. R. Yang, Y. Qin, L. Dai, Z.L. Wang, Power generation with laterally packaged piezoelectric fine wires. Nat. Nanotechnol. **4**, 34–39 (2009)
49. S. Xu, Y. Qin, C. Xu, Y. Wei, R. Yang, Z.L. Wang, Self-powered nanowire devices. Nat. Nanotechnol. **5**, 366–373 (2010)
50. A.M. van der Zande, P.Y. Huang, D.A. Chenet, T.C. Berkelbach, Y. You, G.-H. Lee, T.F. Heinz, D.R. Reichman, D.A. Muller, J.C. Hone, Grains and grain boundaries in highly crystalline monolayer molybdenum disulphide. Nat. Mater. **12**, 554–561 (2013)
51. Y. Zhou, W. Liu, X. Huang, A. Zhang, Y. Zhang, Z. L. Wang, Theoretical study on two-dimensional MoS_2 piezoelectric nanogenerators. *Nano Res.*1–8 (2016)

Chapter 6
Discotic Liquid Crystals for Self-organizing Photovoltaics

Hari Krishna Bisoyi and Quan Li

Abstract Development of photovoltaic solar cells using organic materials as active components is an emerging area of contemporary research endeavor toward sustainable energy harvesting. Among the organic optoelectronic materials, disc-shaped polycyclic aromatic compounds capable of exhibiting columnar liquid crystal (LC) properties are unique. Liquid crystalline materials simultaneously display the anisotropic properties of crystalline solids and flow properties of isotropic liquids. The order and dynamics of the molecules of liquid crystalline materials make them stimuli responsive smart soft materials. LCs are ubiquitous in our daily life as the active switching materials in flat panel display such as the screens of laptops, computers, cameras and mobile phones. Interestingly, the nematic LC phase of disc-like molecules has been used to fabricate optical compensation films which enable enlarging the viewing angle of LC display devices. However the columnar LC phase of disc-like compounds has been found to possess remarkable semiconducting properties. Like other LCs, they show promising characteristics such as controllable alignment and self-healing of structural defects which render them appealing candidates for use in organic semiconducting devices. In this chapter, we present the state-of-the-art of organic photovoltaics employing discotic LCs as active components as well as facilitating additives in the active layer of photovoltaic devices. Various discotic cores such as phthalocyanine, porphyrin, hexabenzocoronene, triphenylene, perylene diimide and others with different electronic properties have been used in organic photovoltaics. The discotic LCs act as electron and/or hole transporting materials in the active layer, which have been tested both in bilayer and bulk heterojunction device configurations. Though this research enterprise is in early stage, with adequate attention these fascinating self-organizing materials could contribute toward the sustainable energy harvesting from solar radiation in future.

H.K. Bisoyi · Q. Li (✉)
Liquid Crystal Institute and Chemical Physics Interdisciplinary Program,
Kent State University, Kent, OH 44242, USA
e-mail: qli1@kent.edu

Q. Li (ed.), *Nanomaterials for Sustainable Energy*,
NanoScience and Technology, DOI 10.1007/978-3-319-32023-6_6

6.1 Introduction

6.1.1 Organic Photovoltaic Solar Cells

Human civilization's increasing energy needs, with ever growing industrialization in our world, have threatened the nonrenewable fossil fuel reservoirs of nature. The unwavering usage of fossil fuels for driving the economic prosperity of nations not only depleting these blessed natural resources but also causing challenging environmental problems which imparts negative consequences on the climate change and earth's ecosystem. However, of late, it has been globally realized that depending solely on the limited natural resources for energy demands is no longer sustainable. It is high time to look for sustainable and inexhaustible energy sources. The sun provides the largest energy source known to man, with more energy from sunlight striking the earth in 1 h than all of the energy consumed on the planet in an entire year. Therefore, the only fully renewable source that has the capability to meet the world's large and growing energy demand is solar energy. In this context, harvesting solar energy which the earth receives as sun light has been recognized as one of the sustainable methods to meet present and future energy requirements. Consequently, photovoltaic (PV) solar cells (also mentioned in previous chapters) have been developed to harvest solar energy on earth. PV solar cells use photovoltaic effect to capture sun light and convert it into electrical energy. Thus produced electrical energy can be utilized directly in the place of production or it can be stored and transported to other places for satisfying the energy requirements. The harvesting and use of solar energy are a sustainable technology for carbon free energy future. Therefore different types of PV solar cells such as silicon solar cells, dye sentitized solar cells (DSSCs), organic photovoltaic solar cells, quantum dot solar cells, pervoskite solar cells are in constant development [1–14]. Silicon solar cells are an established technology, their solar energy conversion efficiency, reliance and robustness are unparalleled. Though silicon solar cells have been commercialized and deployed in public domains, their production and maintenance are expensive. As a viable alternative enterprise, organic photovoltaic (OPV) solar cells are being explored. They hold promise for fabricating light-weight and flexible devices via the low-cost and high-throughput production process. Due to these advantages, extensive research efforts around the world have been devoted to understanding and improving the performance of OPV solar cells during the last two decades. OPV solar cells use organic semiconducting materials as the active components to harvest solar energy. Conjugated polymers, oligomers and small molecules can be employed as organic semiconductive active elements in OPV solar cells [15–20]. It has been established that organic materials can act as either *p*-type (hole transporting) or *n*-type (electron transporting) semiconductors in solar cells.

In OPV solar cells, materials with delocalized π-electrons absorb sunlight, create photo-generated charge carriers, and transport these charge carriers to the electrodes. For efficient photovoltaic devices, the photo-generated charges need to be

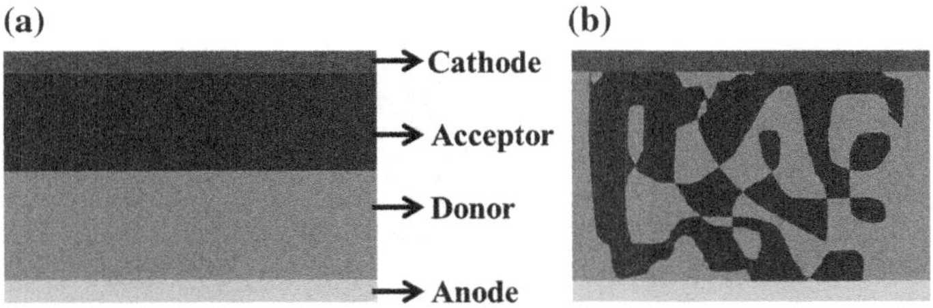

Fig. 6.1 Device configuration of bilayer (**a**) and bulk heterojunction (**b**) organic solar cells

transported to the appropriate electrodes within their lifetime. Therefore these devices usually contain one hole transporting materials and another electron transporting material. Depending on the donor and acceptor materials configurations in the active layer of the devices they can be either bilayer devices or bulk heterojunction (BHJ) devices (Fig. 6.1). The bilayer devices contain a thin layer of donor material and a thin layer of acceptor material separated by a heterojunction interface. In the BHJ device configuration, both the donor and acceptor materials are mixed to form the BHJ. In the BHJ, the donor and acceptor form a nanoscale interpenetrating network which provides separate side by side percolation pathways for hole and electron transport. The bulk heterojunction concept has dramatically increased (orders of magnitude) the interfacial area between the donor and acceptor phases and resulted in improved efficiency solar cells. Upon light illumination, the donor material absorbs light and creates excitons (bound electron-hole pairs), these excitons diffuse to the donor-acceptor interface and get separated as free charge carriers which are then transported to the electrodes. The process is schematically depicted in Fig. 6.2a. It is apparent that various materials parameters play decisive role in determining the operation and efficiency of solar cells. First and foremost is the absorption profile of the donor material. The donor should be able to absorb over a broad range of wavelengths covering the sunlight and it should possess high absorption coefficient for efficient absorption. Since the excitons need to be diffused to the interface so that they can dissociate into free charges, the excitons diffusion length is another key parameter for determining the efficiency of the solar cells. Following the dissociation of excitons into free charge carriers at the donor-acceptor interfaces, they need to be fast transported to the appropriate electrodes for charge collection before they recombine. In this respect, the materials ought to possess high charge carrier mobility values. The charge carriers need a driving force to reach the electrodes. A gradient in the chemical potentials of electrons and holes is built up in a donor-acceptor junction that is determined by the difference between the highest occupied molecular orbital (HOMO) level of the donor and the lowest unoccupied molecular orbital (LUMO) level of the acceptor. Therefore proper alignment of the frontier energy levels of the donor and acceptor materials is another necessary aspect that is often taken into consideration. A schematic of the HOMO and LUMO levels of the electron donor and acceptor materials is shown in Fig. 6.2b. Power conversion efficiency (PCE), external

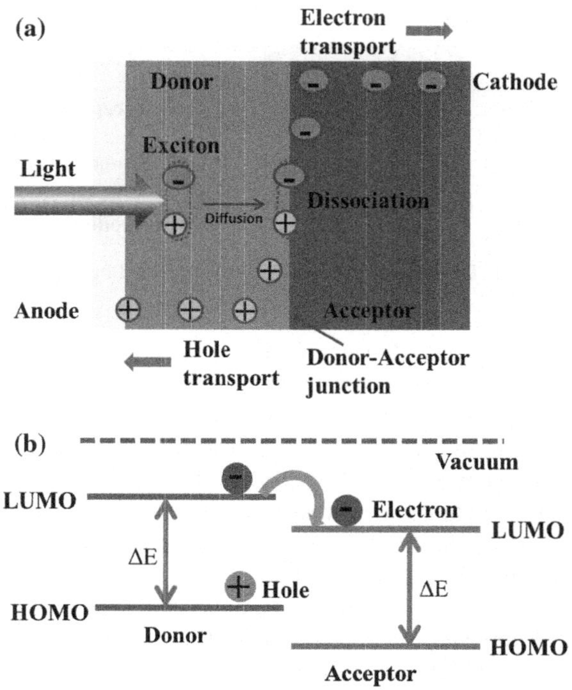

Fig. 6.2 Schematic presentation of the process of exciton formation, diffusion, dissociation and charge transport in the active layer of organic photovoltaic solar cell (**a**); HOMO-LUMO levels of donor and acceptor materials (**b**)

quantum efficiency, open circuit voltage, short circuit current and fill factor are the critical parameters which determine the worthiness and efficiency of solar cells.

6.1.2 Discotic Liquid Crystals

Liquid crystals (LCs) are regarded as the fourth state of matter after solid, liquid and gaseous states. In this intriguing state of matter, the molecules simultaneously exhibit order and mobility. This unique combination of contrasting properties renders LCs as one of the most appealing functional soft materials with stimuli responsive attributes. Though they are well known as the active switching components in flat panel display devices, diverse applications of LCs have been demonstrated, [21–57]. LCs has been broadly classified into two major classes: thermotropic and lyotropic. When the LC phase formation is driven by the effect of temperature, they are referred to as thermotropic whereas when the LC phase formation is driven by concentration of a solute in a suitable solvent, they are referred to as lyotropic LCs. Depending on the molecular organization in the LC phase they have been classified as nematic, smectic, columnar etc. LC phases are exhibited by both molecular and macromolecular materials of different shapes. Recently, the LCs formed by disc-shaped molecules are considered as a new generation of organic semiconductors as well as mesophase semiconductors owing to their remarkable electronic and optoelectronic device performances [58–84]. The supramolecular columnar phases furnished by disc-shaped molecules not only act as model system for the study of charge and energy migration in self-organized systems but also as functional materials in organic electronic and optoelectronic devices such as organic light emitting diodes (OLED), organic field effect transistors (OFET), and organic photovoltaic (OPV) solar cells. Furthermore, the columnar phases are well-defined nanostructures with fascinating one-dimensional (1D) semiconducting characteristics. This is due to the fact that charge carrier mobility values in the columnar phase are several orders of magnitude higher along the column axes than across them. The peripheral chains (usually flexible alkyls) linked to the rigid cores form an insulating hydrocarbon mantle prohibiting charge carriers from transporting between columns.

Molecules which exhibit discotic liquid crystalline phases, i.e. discotic LCs (DLCs), are typically made of central disc-shaped cores (Scheme 6.1) peripherally substituted by varying number of flexible alkyl chains. The liquid crystallinity in these materials is a consequence of the antagonistic relationship between the central aromatic cores and radial aliphatic chains which leads to nanosegregation of the two constituents. Hence by tailoring the shape, size, and nature of the central core as well as the position, number, length and nature of attached side chains, DLC materials with different ability to self-organize into different LC morphologies have been realized. The disc-like molecules spontaneously self-assemble into infinite 1D stacks, which in turn self-organize to various 2D liquid crystalline lattices where the third dimension has no translational order. Mesophases formed by disc-shaped molecules are primarily two types: nematic (N_D) and columnar (Col) as shown in Fig. 6.3. However, some discotic compounds are known to exhibit lamellar and

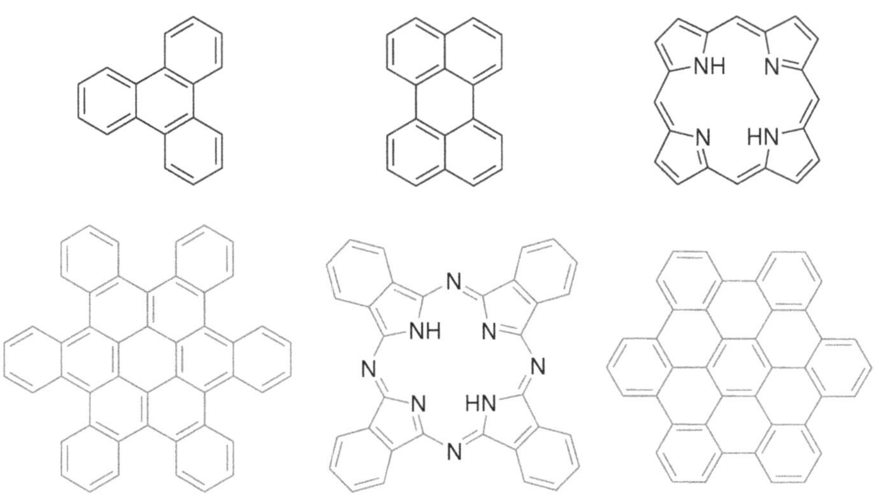

Scheme 6.1 Some common aromatic cores of discotic liquid crystals (DLCs)

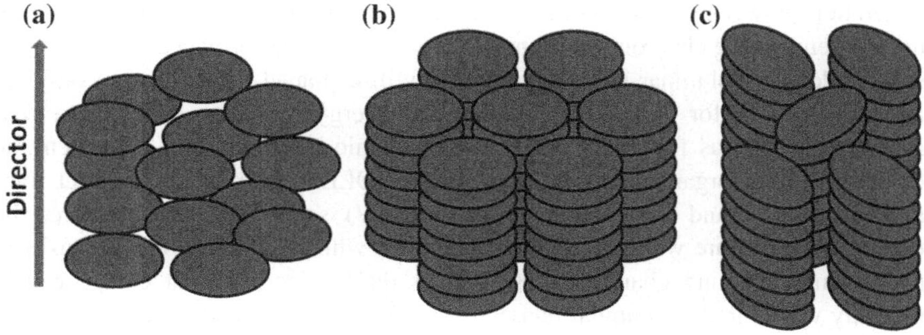

Fig. 6.3 Schematic representations of commonly observed discotic liquid crystalline phases. **a** Discotic nematic, **b** columnar hexagonal, **c** columnar rectangular

cubic phases of different symmetries. In DLCs, the columnar phases are ubiquitous probably due to strong π-π intermolecular interactions among the aromatic cores. The thermotropic (temperature dependent) phase behavior of DLCs, like any other LCs, is usually studied by using differential scanning calorimetry (DSC), polarizing optical microscopy (POM), small and wide angle X-ray diffraction (XRD). DSC is used to determine the temperature of phase transitions and enthalpy changes related to each transition. The fluid character of the mesophases and their characteristic textures are observed under a POM. The supramolecular organizations and corresponding packing parameters in each DLC phase can be established in detail by XRD studies.

DLCs possess certain advantageous traits when compared to other organic semiconducting materials such as conducting polymers and single crystals for device applications. They are self-assembling in nature, possess one-dimensional conductivity, and their charge carrier mobility values are comparable to conjugated polymers. Unlike single crystals, DLCs can be facilely and economically processed from solutions and isotropic melts. Their optical and electronic properties can be tuned at will by appropriately editing the primary molecular structures. Interestingly, they exhibit self-healing of structural defects owing to their dynamic characteristics. Generally, conjugated polymers are polydisperse in nature and their thin films are often polycrystalline with grain boundaries. These grain boundaries act as charge trapping sites and thus greatly undermine their device performances. On the other hand, organic single crystals demonstrate the best performances in semiconducting devices, however their processing cost is prohibitively high and growth of single crystals of adequate size takes very long time and is tedious. Therefore, their scalable use in devices is not economically viable. In this context, columnar DLCs could be viable alternatives since their unique self-organizing properties promise low cost solution techniques for the fabrication of semiconducting devices in general and organic photovoltaic solar cells in particular.

As mentioned earlier, DLCs are quasi-one-dimensional semiconductors, i.e. charge and excitons migrate much faster along the columnar stacks than between the columns. This implies that the columns must be appropriately aligned in a

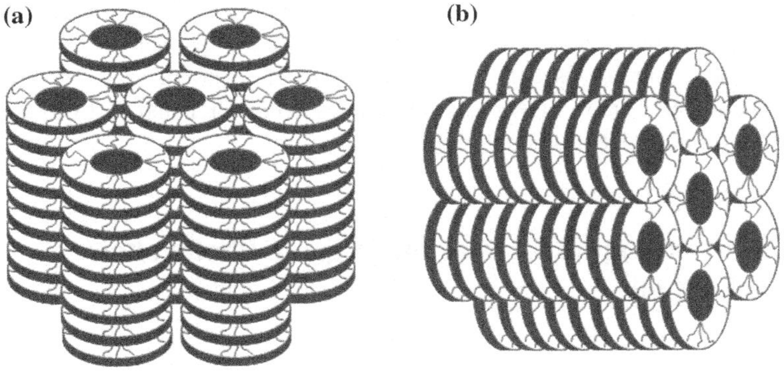

Fig. 6.4 Schematic representations of homeotropic alignment (**a**) and homogeneous alignment (**b**) of the columnar phase of discotic liquid crystals

direction that the current is desired to flow in a device architecture. The columns of DLCs can align either perpendicular to the substrate surface, i.e. homeotropic alignment or parallel to the substrate surface, i.e. planar alignment. This is illustrated in Fig. 6.4. In the homeotropic alignment, the molecular cores/planes of the discotic molecules lie flat or 'face-on' with respect to the substrate surface, therefore the columnar axes or the director is perpendicular to the surface. In the planar alignment, the discotic cores are oriented orthogonal or 'edge-on' to the substrate plane, hence the column axes lie parallel to the substrate surface. Planar alignment of columns on a substrate is required for OFET applications, while a homeotropic alignment is preferred for solar cell and OLED applications. DLCs upon cooling from their isotropic phases directly transform to columnar phase and are difficult to align with control over their molecular orientations. Though some DLCs have been observed to spontaneously develop homeotropic alignment of acceptable quality such incidences are rare. Therefore different techniques have been developed for controlled homeotropic alignment of DLCs. These techniques include thermal annealing, chemical structure modification of the DLCs, surface modification of substrates, electric field, infrared irradiation etc. [23, 24]. Each technique has its pros and cons and yield films with varying degrees of order. Where a single technique is not effective, multiple techniques are adopted in combination to achieve the desired homeotropic alignment and to improve long range ordering of DLCs.

As noted earlier, charge carrier mobility is a key parameter in determining the performance of a semiconducting material and DLCs are no exception. Therefore, various techniques have been adopted to determine the charge carrier mobility values of DLCs such as pulse radiolysis time resolved microwave conductivity (PRTRMC), time of flight (TOF), space charge limited current (SCLC) and field effect transistor (FET). When comparing the charge carrier mobility values by above methods, it is found that they present different values for an identical sample. The charge mobility determined by the TOF method is often found to be lower than the PR-TRMC value. This is because TOF mobility is measured over relatively thick samples (usually tens of microns) on a millisecond or microsecond timescale and therefore is more susceptible to defects and grain boundaries. Nevertheless, TOF experiments lead to higher mobility values over OFET and SCLC techniques which yield one order of magnitude lower.

6.2 Discotic Liquid Crystals in Organic Photovoltaic Solar Cells

Self-organizing DLCs have been employed as active components in organic photovoltaic solar cells. DLCs with relatively large aromatic cores absorb in the visible region of the solar spectrum and some compound show large absorption coefficients. Photoconductivity measurements have yielded charge carrier mobility values ranging from 10^{-3} to 1.0 $cm^2V^{-1}s^{-1}$. These carrier mobility values are large enough to yield highly efficient devices. Such high values are a consequence of large π-orbital overlap between adjacent molecules in supramolecular columns of DLCs. Moreover, the exciton diffusion length in some columnar phases has been observed to reach about 70 nm which is very large as compared to conjugated polymers. Depending on the ionization potential or electron affinity, DLCs can act as hole transporting materials as well as electron transporting materials depending upon their detailed molecular structure. Interestingly by changing the substitution pattern of the central aromatic cores, they can be made electron rich or electron deficient as necessary [85, 86]. As discussed, semiconducting discotic LCs bring two advantageous characteristics, i.e. high charge carrier mobility of single crystals and processability of conjugated polymers in one class of materials. Among the most promising columnar DLC materials that have been investigated so far for solar cell applications are such as porphyrin, phthalocyanine, hexabenzocoronene, triphenylene, and perylene derivatives. Figure 6.5 shows the structure of a donor/acceptor all-columnar photovoltaic device based on DLCs.

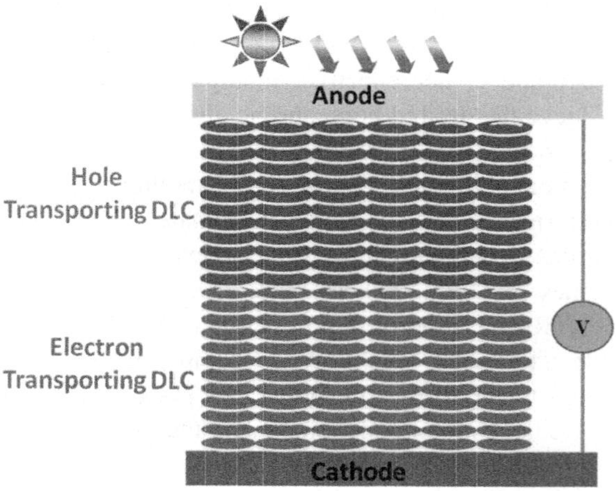

Fig. 6.5 Schematic presentation of a bilayer solar cell constructed from donor and acceptor discotic liquid crystals

6.2.1 Liquid Crystalline Porphyrins in OPV

Porphyrin derivatives are one of the early systems in the area of DLCs [87–90]. Interestingly they are the first DLC materials where photovoltaic effect was demonstrated by Gregg et al. [87–89]. The use of porphyrin as the building block of discotic materials has been inspired by nature since this core forms the integral part of the best photoreceptor in nature, chlorophyll. Moreover, porphyrin and its derivatives have many desirable features as semiconducting materials, such as highly conjugated disc plane, high stability, efficient absorption of sunlight and a small gap between the HOMO and LUMO energy levels. The high charge carrier mobility value of the porphyrin derivative **1** (Scheme 6.2) has led Bard and co-researchers to evaluate its potential in photovoltaic devices. Gregg et al. [87] reported photovoltaic effect in symmetrical cells of the liquid crystalline porphyrin derivative. The photovoltaic cells were made of two indium tin oxide (ITO) conducting glass electrodes and were filled with LC porphyrin by capillary action. They described the changes in photoconversion properties observed in thin films of the discotic liquid crystalline Zn porphyrin as the long range film order is varied through the crystal-mesophase transition. Though a photovoltaic effect in such symmetrical cells (both electrodes have identical work function) seems unusual, it has been explained as resulting from two simple processes: first, upon photoillumination the concentration of excitons is much higher at the illuminated electrode interface than at the dark electrode; and second, excitons dissociation at the porphyrin/ITO interface is inherently asymmetric and electron injection into the ITO is more facile than hole injection. It has been possible to measure the change in photovoltaic effect as the semiconductor is heated through a reversible phase transition. The photocurrents in the ordered films are estimated to be approximately four orders of magnitude higher than in the disordered films. Remarkably, the photovoltaic cell is entirely controlled by interfacial kinetics rather than by a gradient of electrochemical potential.

Recently, we have designed and synthesized new discotic porphyrin derivatives **2** and **3** (Scheme 6.2) which exhibit hexagonal columnar phases [91–94]. The high propensity of these DLCs toward spontaneous formation of large area homeotropic alignment (Fig. 6.6) encouraged us to investigate them in photovoltaic devices. Polarized optical microscopy textures of **2** in homeotropically aligned cells at room temperature with 2 and 8 μm samples both with ITO–ITO and ITO–gold electrode pairs are shown in Fig. 6.6. It can be seen that the alignment of the material in all four of the cells tends to be homeotropic. Thinner films with 2 μm thickness in Fig. 6.6 have almost perfect homeotropic alignment. Interestingly, these LC materials are the first light-harvesting liquid crystalline porphyrins in which homeotropic alignment has been achieved. In order to understand the photovoltaic properties of the materials, the photovoltaic performance of material **2** was investigated based on bilayer and bulk-heterojunction solar cells, which were fabricated by sandwiching a blend film of porphyrin **2** and [6,6]-phenyl-C_{61}-butyric acid methyl ester (PCBM) (1:1 w/w) between the ITO anode coated with a conductive

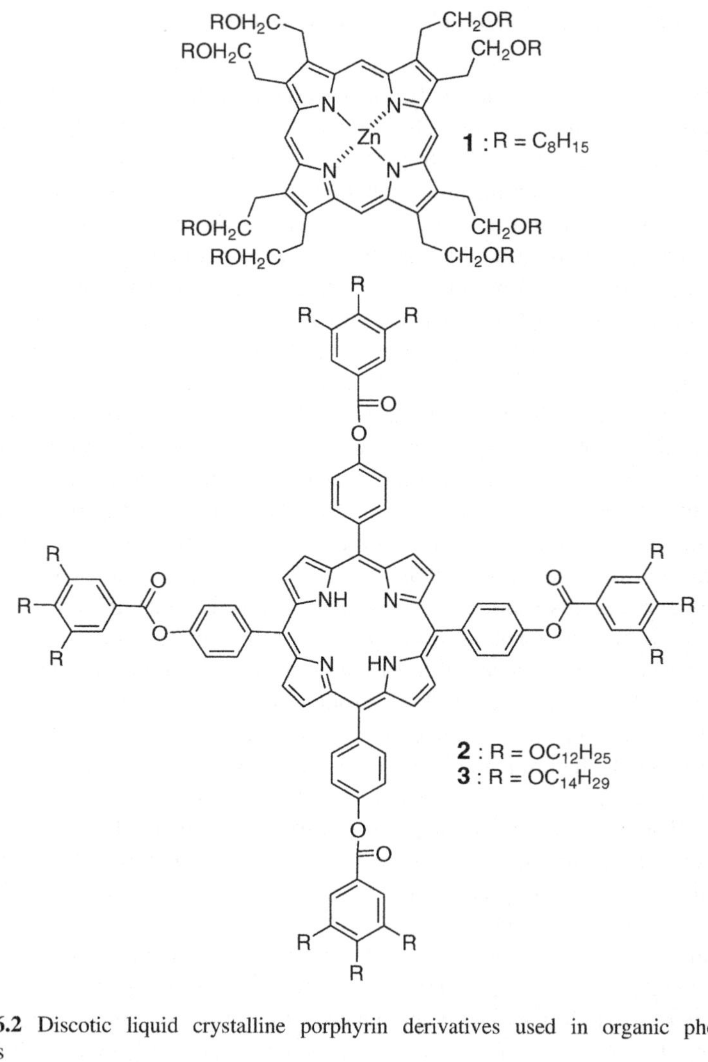

Scheme 6.2 Discotic liquid crystalline porphyrin derivatives used in organic photovoltaic solar cells

polymer poly(3,4-ethylenedioxythiophene) (PEDOT) and Ca/Al cathode. In such solar cells, porphyrin acts as an electron donor, whereas PCBM acts as an electron acceptor. In the as-prepared bilayer solar cell containing **2** the power conversion efficiency was 0.07 %. The relatively poor performance of the bilayer device was attributed to the limited interfacial area between the donor and acceptor layers. Upon thermal annealing, the power conversion efficiency was improved to 0.14 %. This improvement could be due to thermally induced alignment of porphyrin layer. The as-prepared bulk-heterojunction device exhibited power conversion efficiency of 0.22 %. However, upon annealing the device performance was improved by about 300–400 %. The post-annealed device power conversion efficiency was

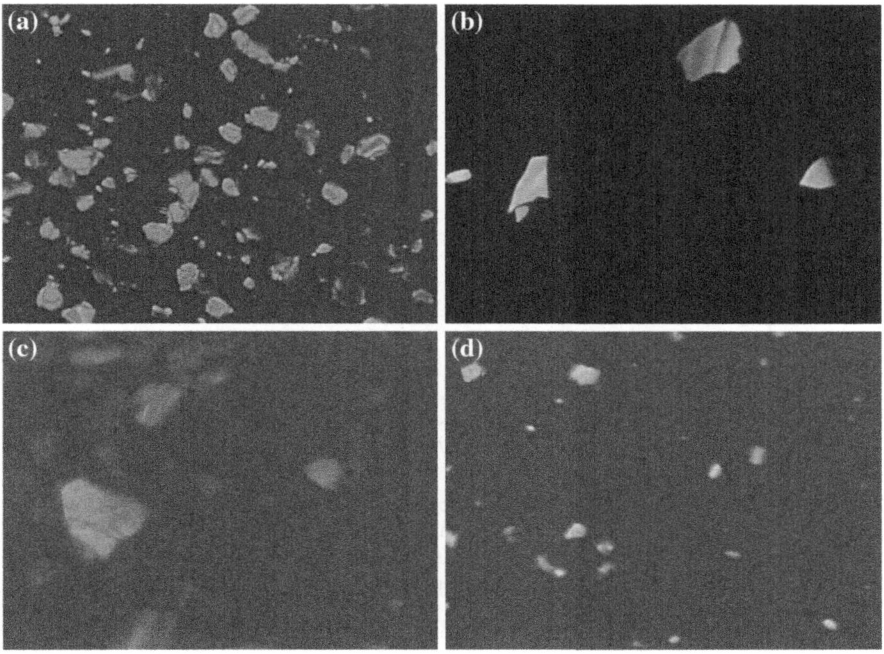

Fig. 6.6 Crossed-polarized optical textures of **2** in homeotropically aligned cells (the *dark areas* represent homeotropic alignment where the planes of the *disk-shaped* molecules are parallel to the substrates). *Bright* domains represent areas where the molecular planes are tilted with respect to the substrates. **a** 8 μm ITO–ITO cell; **b** 2 μm ITO–ITO cell; **c** 8 μm ITO–gold cell; **d** 2 μm ITO–gold cell

found to 0.71 %. This improvement in performance has been attributed to the enhanced columnar ordering of the porphyrin LC material upon annealing. It is clear that compared to the bilayer devices, the bulk-heterojunction solar cells showed much improved performance because of the maximized interfacial area for the charge separation and collection. The current density-voltage (J-V) characteristics of the bilayer photovoltaic devices fabricated using **2** and **3** are shown in Fig. 6.7.

6.2.2 Liquid Crystalline Phthalocyanines in OPV

Phthalocyanine (Pc) derivatives are a promising class of DLCs which exhibit remarkably high charge carrier mobility values in their columnar phases. The versatility, architectural flexibility, non-toxicity, and ease of processing make them eligible candidates for use in electronics. Interestingly, Pc derivatives can act both as electron acceptor and electron donor with respect to other moieties in

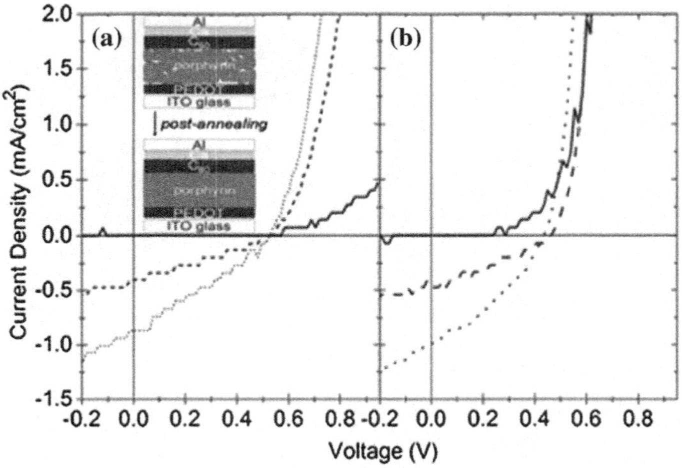

Fig. 6.7 *J-V* characteristics of the bilayer solar cells based on **a 2** and **b 3** under dark (*solid line*), and before (*dash line*) and after (*dot line*) postannealing under the illumination of AM 1.5 G and 100 mW/cm^2. The *inset* of (**a**) schematically shows possible ordering in the porphyin-based bilayer-heterojunction solar cell by postannealing. Reproduced with permission from [91]. Copy right 2007 American Institute of Physics

donor-acceptor ensembles. The chemical flexibility of Pcs allows the preparation of a large variety of related structures and consequently, the tailoring of the physical, electronic, and optical properties as well as the improvement of processability. Since the discovery of liquid crystallinity in Pcs, many compounds have been prepared with variations in the nature, number, length and positions of the flexible side chains. In addition, the linking groups, central metal ions all have substantial influence on the structure and thermal stability of the resultant mesophases. They have been investigated as potential small molecule organic semiconductors for electronic and optoelectronics applications. Particularly, the high intrinsic charge-carrier mobilities reported for several liquid crystalline Pc dervatives have highlighted their potentials for solar cell applications [95–101]. Petritsch et al. [95] investigated a double-layer photovoltaic device comprising a discotic phthalocyanaine derivative **4** (Scheme 6.3) as electron-donor (hole transporting) and a perylene derivative **5** as electron-acceptor (electron transporting) material. They compared the performances of two different devices one containing a pristine phthalocyanine layer whereas the other device was thermally treated to obtain a higher ordered mesophase. Instead of expected improved efficiency due to better transport properties of the higher ordered mesophase they observed a smaller efficiency for the heated device compared to the pristine device when both were measured in vacuum. The factor affecting the performance of the heated device was found to be dewetting of the Pc upon heating to isotropic phase. When the Pc thin film was heated to isotropic temperature it forms isolated droplets instead of a continuous liquid film. This study suggests that it is important to have better

Scheme 6.3 Chemical structures of discotic liquid crystalline phthalocyanine and perylene diimide derivatives used in solar cells

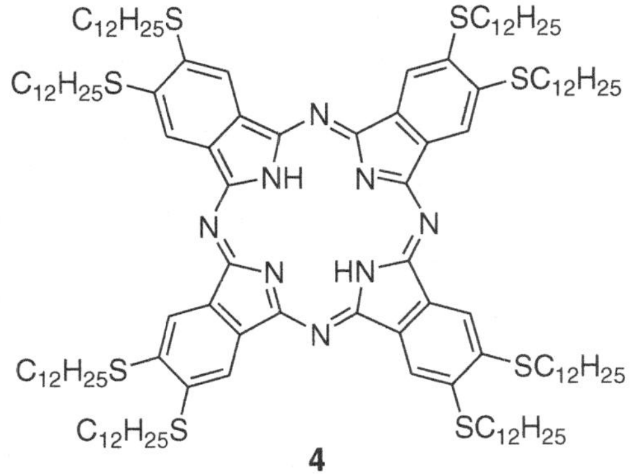

4

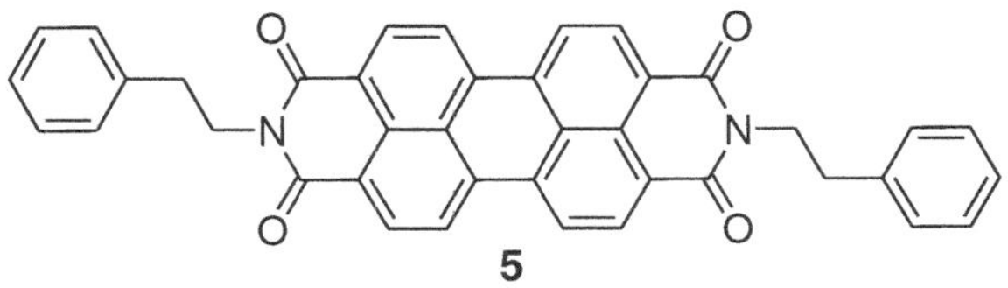

5

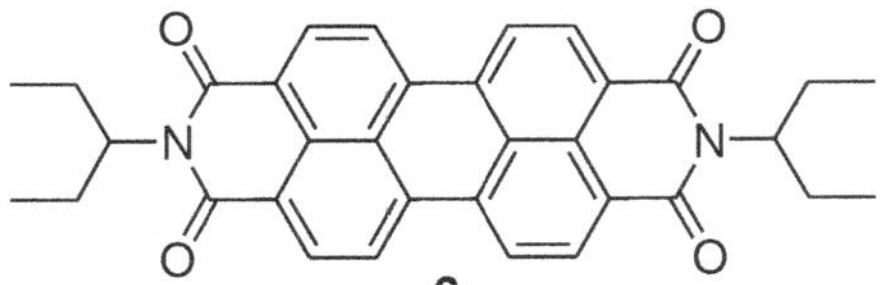

6

interaction of the discotic molecules with substrate surfaces for better device performance. Subsequently they have tested the performance of donor/acceptor solar cells with three different architectures. The architectures comprise a blend and a double layer structure as well as laminated device structure [96]. In the blend and laminated device structures, the liquid crystalline Pc **4** was used as the donor material. The blend device contains 1:1 mixture of Pc and perylene diimide **6** and its structure is shown in Fig. 6.8. This cell utilizes light from the entire visible spectrum. This device was observed to exhibit external quantum efficiency of 1 %, open circuit voltage of 110 mV and a fill factor of 25 %. The laminated device showed lower external quantum efficiency than the blend structure indicating that there is no clear efficiency gain due to the lamination method.

Hori et al. [97] have fabricated bulk heterojunction solar cells by utilizing Pc **7** (Scheme 6.4) as the donor and [6,6]-phenyl-C_{61}-butyric acid methyl ester (PCBM) as the acceptor component. This solution processed solar cell was found to display strong dependence of active layer thickness on its photovoltaic properties. The optimized active layer thickness was determined to be 120 nm. Moreover, by inserting MoO_3 hole transporting buffer layer between the positive electrode and active layer the energy conversion efficiency was enhanced to 3.2 %. Subsequently

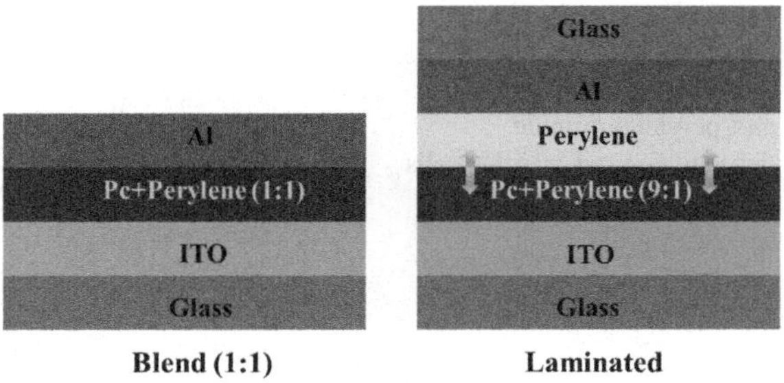

Fig. 6.8 Two types of D/A solar cell architectures using molecular phthalocyanine and perylene diimide. Adapted with permission from [96]. Copy right 2000 Elsevier Science B.V

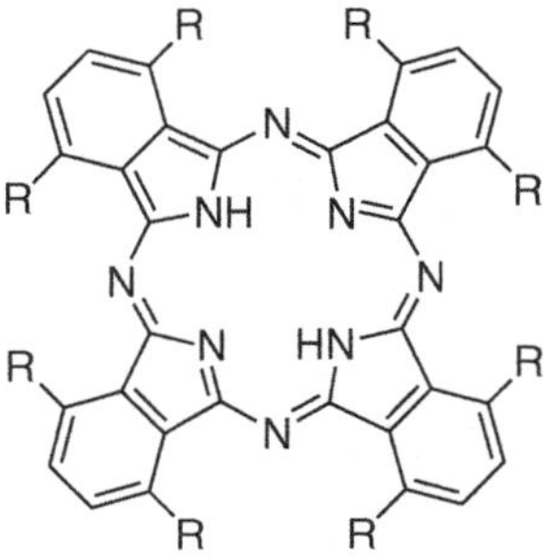

7 : R = C_6H_{13} **8** : R = C_7H_{15}

Scheme 6.4 Chemical structures of phthalocyanine derivatives **7** and **8**

Dao et al. [100] significantly enhanced the efficiency of the bulk heterojunction solar cell by adding processing additive 1,8-diiodooctane (DIO) to the active layer. They could achieve power conversion efficiency more than 4.1 %. The marked improvement in efficiency has been attributed to the alteration of active layer surface morphology and phase separation between the components promoted by the processing additive and solvent. The active layer containing 1,8-diiodooctane represents a distinct morphology as shown in Fig. 6.9. Hori et al. studied the effect of Pc **7** doping on the photovoltaic properties of a bulk heterojunction solar cell based on poly(3-hexylthiophene) (P3HT) and PCBM. Since compound **7** possesses near-infrared absorption, by doping the active layer with **7** the photosensitivity of the solar cell in the long-wavelength region was improved. This led to a power conversion efficiency of 3.0 % from 2.3 % at a composition ratio of P3HT:**7**: PCBM = 10:3:10. However, on increasing the amount of Pc **7** in the active layer the device performance declined as depicted in Fig. 6.10 [99]. Photoconversion in the bulk heterojunction solar cell active layer has been discussed by taking into account the existence of microphase separated and highly ordered P3HT domains and hexagonal columnar structures of the Pc **7** in the active layer. The different pathways leading to improved cell performance is schematically shown in Fig. 6.11. It has been noticed that the efficiency of the device does not improve by doping the active layer with structurally similar compound **8**.

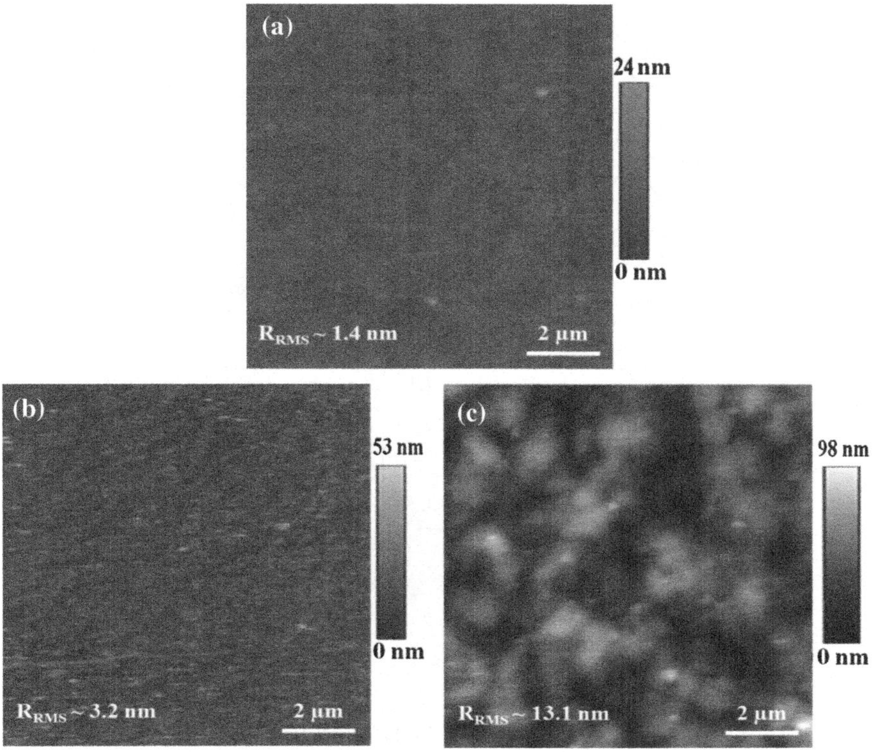

Fig. 6.9 AFM images of surface of **7**:PCBM composite thin films: **a** without DIO, **b** with 0.2 % v/v of DIO, and **c** with 0.8 % v/v of DIO. Reproduced with permission from [100]. Copy right 2013 Elsevier B.V

6.2.3 Liquid Crystalline Hexabenzocoronenes in OPV

Hexa-peri-hexabenzocoronene (HBC) derivatives have emerged as enabling discotic liquid crystalline materials. Owing to the large size of the central polycyclic aromatic core, these compounds possess very strong intermolecular π-π interaction. As a consequence they display some of the highest charge carrier mobility values among DLCs. This unique quality has rendered them as promising small molecule organic semiconducting materials. They have been tested in organic photovoltaic solar cells as self-organizing hole transporting materials [102–110]. Schmidt-Mende reported a bulk heterojunction photovoltaic solar cell based on HBC derivative **9** (Scheme 6.5) as a donor and perylenediimide **6** as an acceptor [102]. The solution processed device exhibited a maximum external quantum efficiency of 34 % and a power conversion efficiency of 2 %. The active layer morphology analysis in this well performing device revealed that both the donor and acceptor form vertically segregated domains with large interfacial area (Fig. 6.12). The high efficiency has been thought to result from efficient

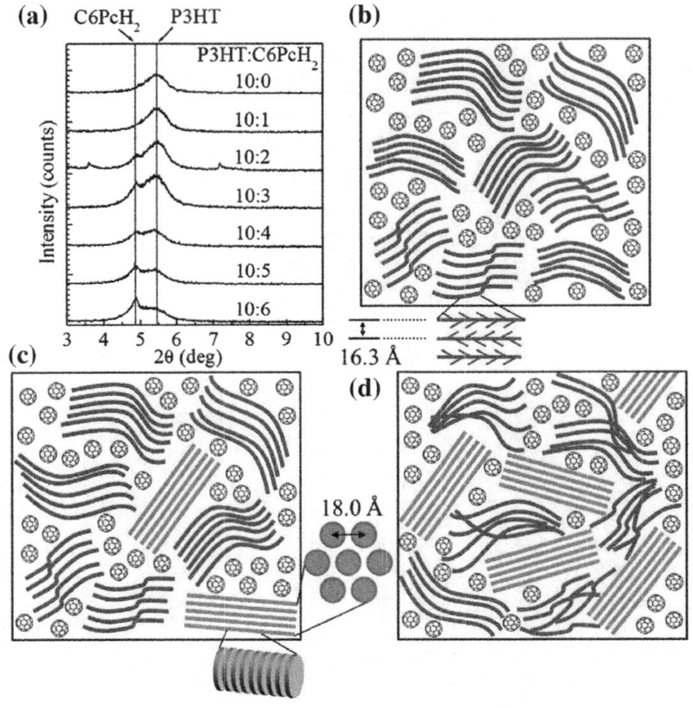

Fig. 6.10 **a** XRD patterns of the P3HT:**7**:PCBM bulk heterojunction films and schematic diagrams of micro phase separation in the P3HT:**7**:PCBM active layers, the composition ratios of which are **b** 10:0:10, **c** 10:3:10, and **d** 10:6:10. Reproduced with permission from [99]. Copy right 2011 Elsevier B.V

photo-induced charge transfer between the HBC and perylene diimide, as well as from effective transport of charges through vertically segregated perylene diimide and HBC π systems. Later it was learned that the occurrence of the columnar LC phase in HBC **9** at room temperature also plays a role in the enhanced efficiency of the devices. This fact was found by comparing the efficiency of devices containing HBC **10** without room temperature columnar LC phase with HBC **9**.

Photovoltaic devices with HBC derivative **11** as hole conductor show lower efficiencies. The lower device performance of these devices has been attributed to the different film morphology occurring from spin coating of mixture of **11** and perylene diimides. To learn what happens when two discotic liquid crystalline compounds (one acting as a donor and the other as an acceptor) are used as the active layer in a solar cell, devices were fabricated based on HBC **9** and **10** and the perylene tetraester **12** (Scheme 6.5). The current-voltage-characteristics of the all-columnar two-layer devices indicated a photovoltaic effect with an open circuit voltage of 0.8 V, a short circuit current of 3.9 mA/cm^2, and a filling factor of 31 %. Interestingly, the spectral distribution of the photocurrent of different samples indicated that excitations of the perylene derivative or of the HBC derivative both

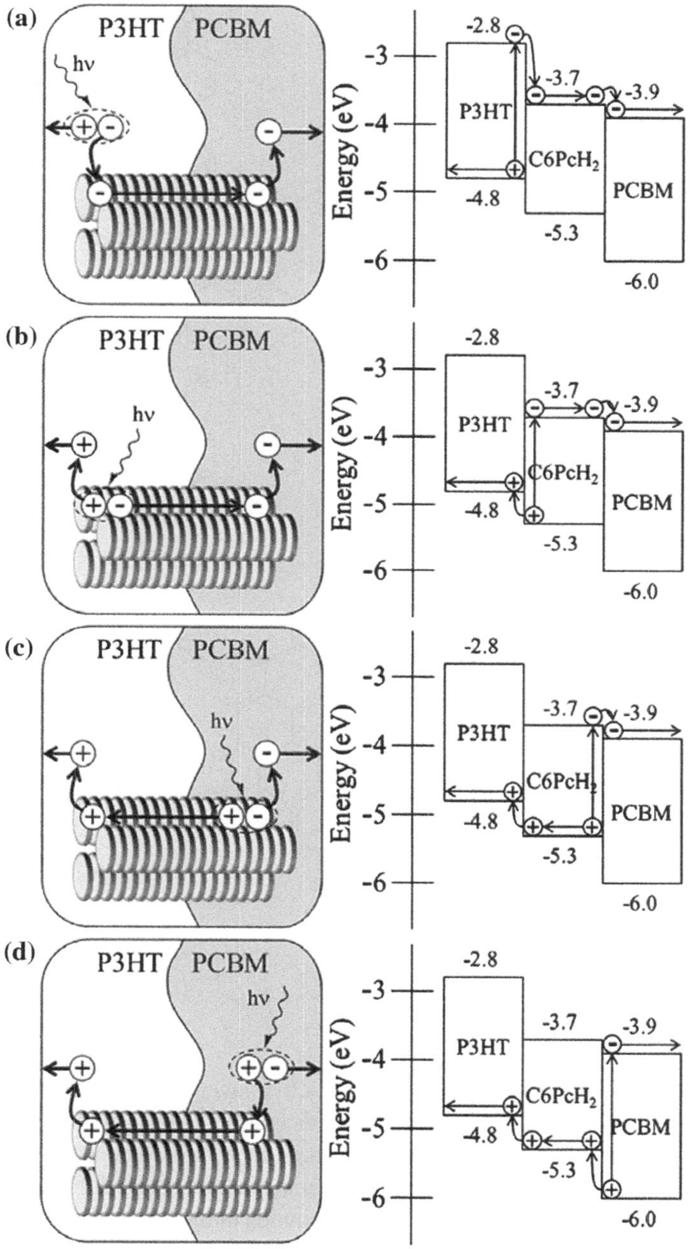

Fig. 6.11 Schematic diagrams of charge generation and transfer in the solar cells with P3HT:**7**: PCBM bulk heterojunction active layer (*left side*) and energy diagrams of P3HT, **7** and PCBM in the solar cells (*right side*). **a**, **b** indicate the charge separation between P3HT and C6PcH2 in the case of excitation at P3HT and **7**, respectively. **c**, **d** indicate the charge separation between **7** and PCBM in the case of excitation at **7** and PCBM, respectively. Reproduced with permission from [99]. Copy right 2011 Elsevier B.V

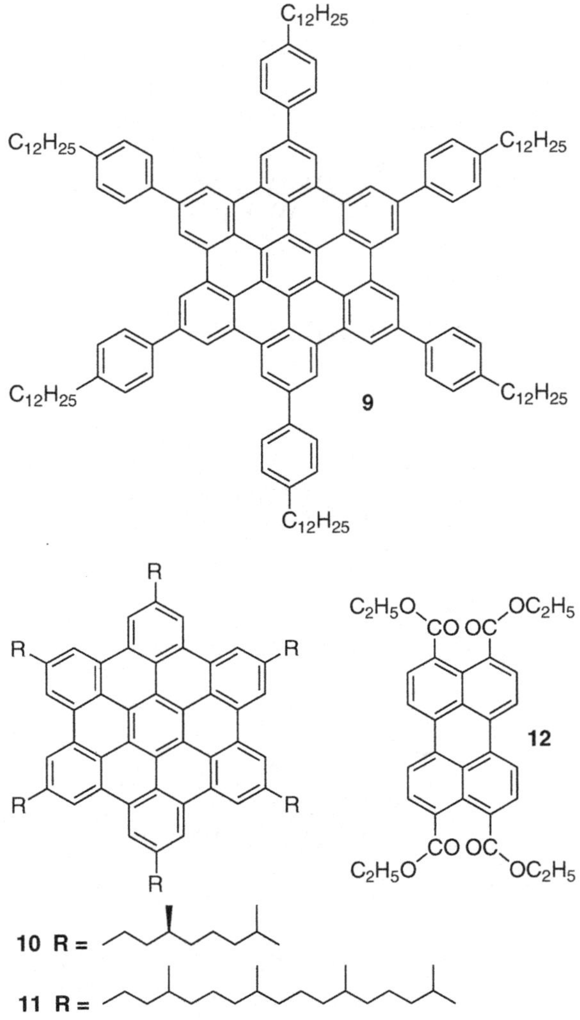

Scheme 6.5 Discotic hexabenzocoronene derivatives used in bulk heterojunction solar cells

can lead to a photovoltaic effect. Since the active layer thin film morphology plays critical role for better performance of devices, the detail structure, the factors affecting it and control of morphology has been investigated by using HBC derivatives **13-17** (Scheme 6.6) with peripheral substituents of different nature. It is observed that better crystallinity of the compounds in the active layer leads to better device performances. Moreover, short alkyl chains on the HBC core allows intimate mixing with the acceptor component.

Li et al. [106] investigated the photovoltaic behavior of three HBC derivatives **18-20** (Scheme 6.7) in bulk heterojunction configuration and determined the influence of the branched alkyl side chains on the device performance. In these devices perylene diimide **6** was employed as the acceptor material. The HBC

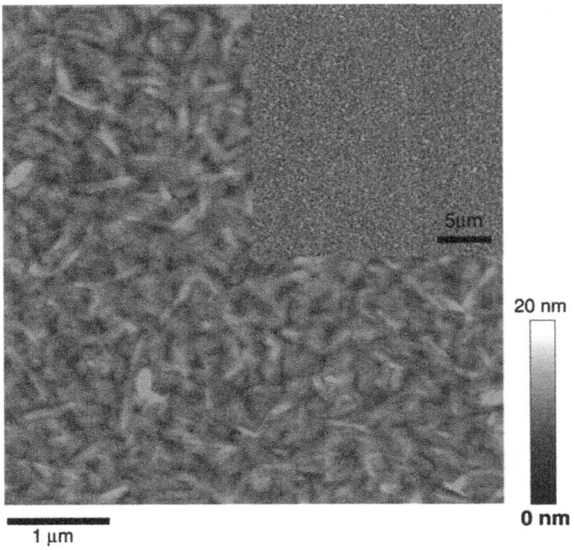

Fig. 6.12 Tapping-mode AFM image of a film spin-coated from a 40:60 blend solution of HBC derivative **9** and perylene diimide **6**. Reproduced with permission from [102]. Copy right 2001 American Association for Advancement of Science

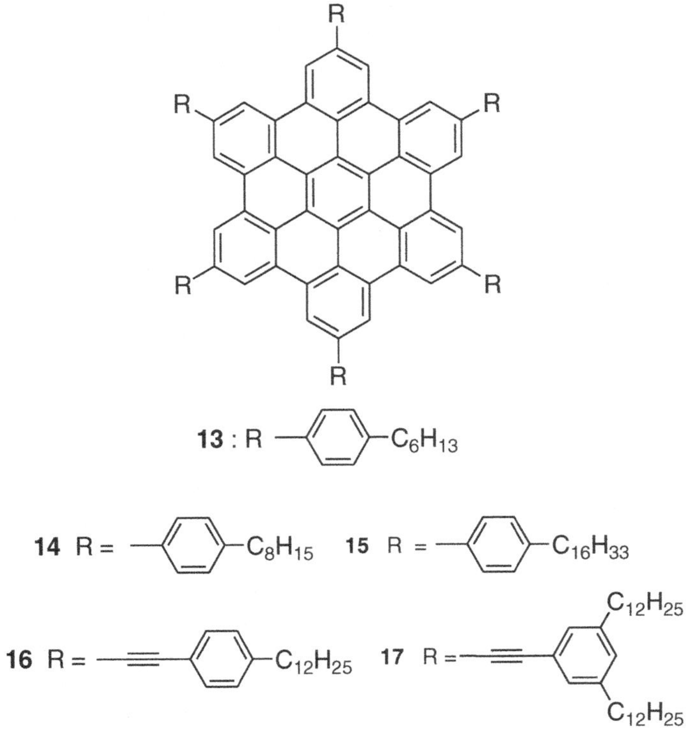

Scheme 6.6 Chemical structures of HBC derivatives **13-17**

Scheme 6.7 Chemical structures of HBC derivatives **18-20**

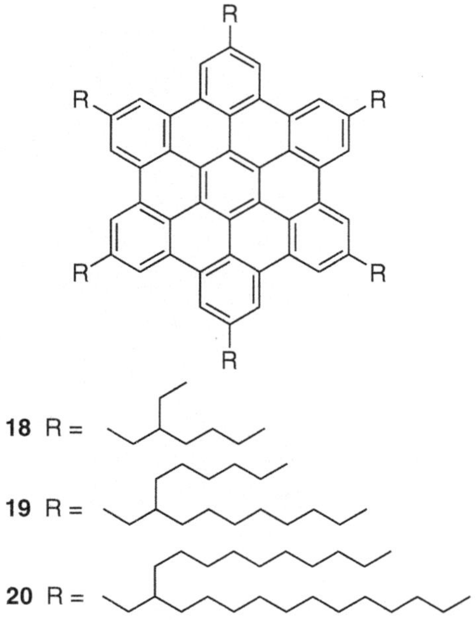

derivative **18** with shortest side chain was found to be the best donor for the perylene diimide. The device containing **18** exhibited highest external quantum efficiency of 12 % corresponding to a power efficiency of 1.5 %. Scanning electron microscopy imaging suggested the important role of the morphology of the active film in determining the performance of the devices. As shown in Fig. 6.13, although similar columnar crystals were observed on the surface of all three blends, the bulk morphology appeared to be quite different in their cross-section images. Thin films of **19/6** and **20/6** were "homogeneous" across the thickness with cylindrical crystals evenly dispersed from bottom to top. However, in the film of **18/6**, the bottom layer is rather featureless with some crystals protruding from the surface. The difference in film morphologies have been ascribed to the different crystallinity of the HBC derivatives, the HBC with shortest alkyl chains is more crystalline than others. Similar study on the influence of normal alkyl side chains of different chain length on the active layer morphology has been undertaken to understand the structure-morphology relationships.

Kang et al. [111] have fabricated a heterojunction solar cell active layer by solution processing using the contorted hexabenzocoronene LC compound **21** (Scheme 4.8) as the donor material and phenyl-C_{70}-butyric acid methyl ester ($PC_{70}PM$) as an acceptor. It was observed that the compound **21** forms nested structure around the $PC_{70}BM$ which results a self-assembled molecular scale *p*-*n* junction in the active layer of the solar cell. The power conversion efficiency is dramatically enhanced by the mode of donor and acceptor assembly within the active layer film. Interestingly, the stoichiometry between donor and acceptor controls a hierarchy of self-assembly that is responsible for the active layer film

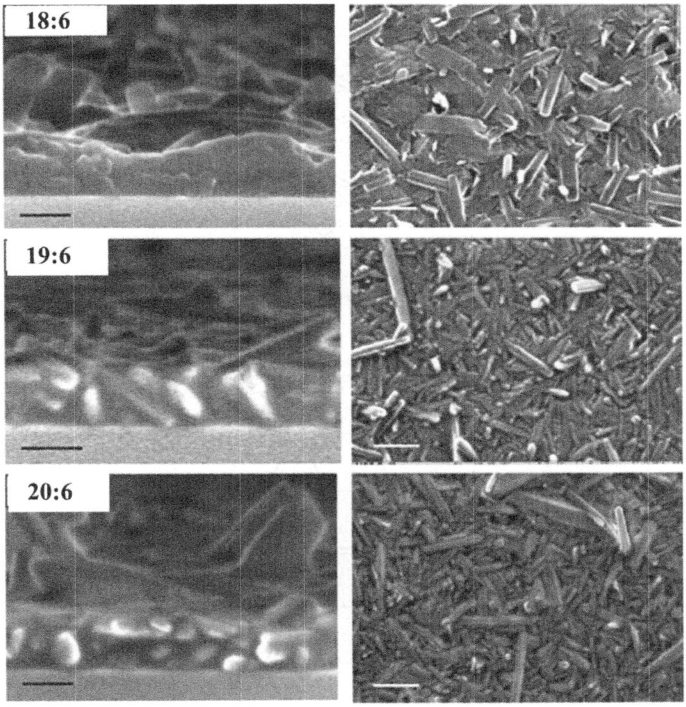

Fig. 6.13 SEM images of the cross-section view of the blend sample (scale bar: 100 nm) shown on the *left*. The top view images are shown on the *right* (scale bar: 500 nm). All the films were annealed at 120 °C for 1 h in a nitrogen atmosphere. Reproduced with permission from [106]. Copy right 2007 WILEY-VCH Verlag GmbH & Co. KGaA, Weinheim

morphology and thus influences the device performance. The device with 10:90 wt % of **21**:$PC_{70}BM$ exhibited 2.41 % efficiency. However the device efficiency decreased with increase in the amount of the DLC compound **21**. Figure 6.14 shows the active layer structure. This study demonstrated that good device efficiency can be achieved by minimal use of donor material.

Feng et al. [112] have designed, synthesized and investigated the supramolecular organization of triangle-shaped discotic liquid crystalline compounds **22** and **23** (Scheme 6.8) with swallow-tailed alkyl substituents. Subsequently they fabricated solution processed bulk heterojunction solar cells by using **22** and **23** as the donor component and perylenediimide **6** as the acceptor material. The highest device efficiency was observed for blends with the ratio of **22**:**6** = 4:6. The blend active layers morphology in the devices were investigated by scanning electron microscopy. Homogeneous bicontinuous phase was apparent from the SEM image across the film thickness with cylindrical perylenediimide crystals evenly dispersed within the LC matrix from bottom to top of the film (Fig. 6.15). The morphology displayed is highly favorable for the transport of holes and electrons in the corresponding separated continuous percolation pathways.

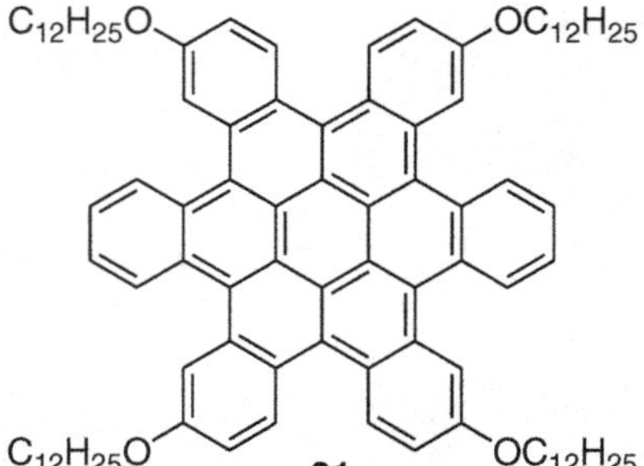

21

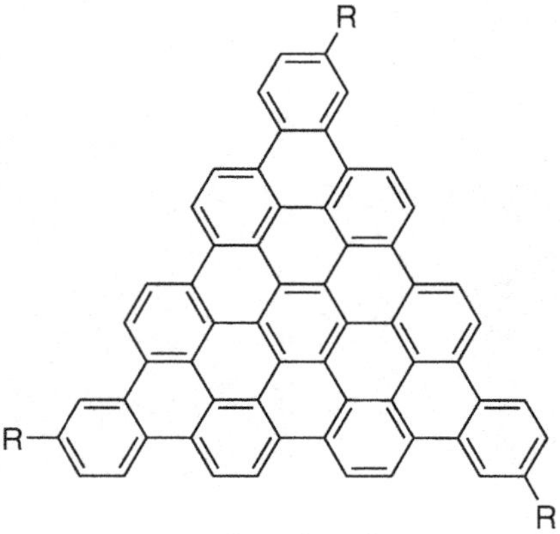

22 R =

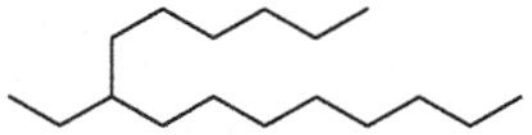

23 R =

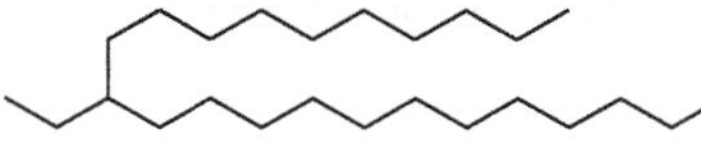

Scheme 6.8 Chemical structures of discotic liquid crystal compounds used in photovoltaic solar cells

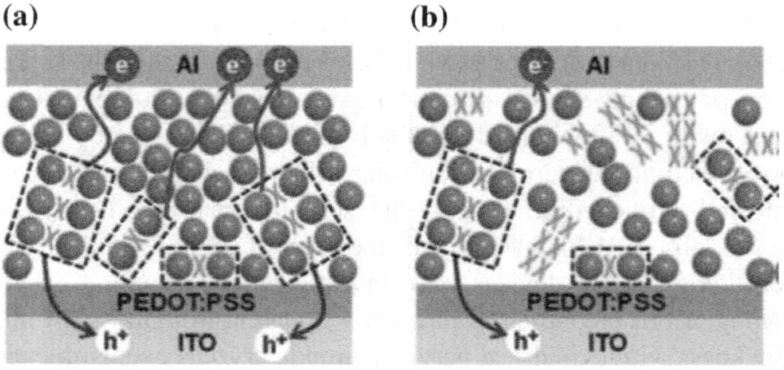

Fig. 6.14 Schematic model of the intermolecular *p–n* junction assembly in the PC70BM medium at **a** 10/90 and **b** 30/70 wt%. Reproduced with permission from [111]. Copy right 2012 American Chemical Society

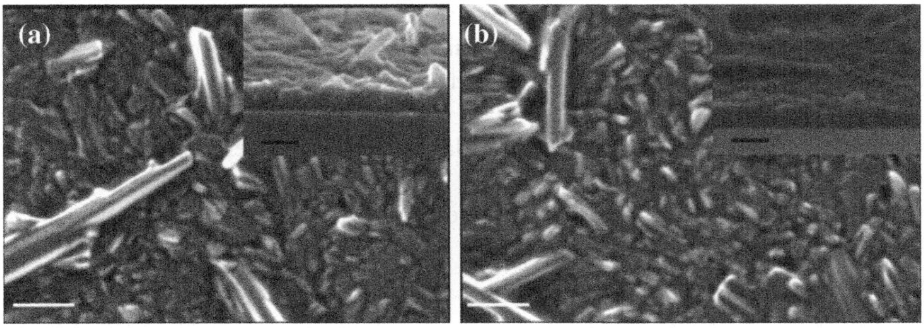

Fig. 6.15 SEM images: *top view images* (scale bar for 100 nm) and *cross-section view* (*inset*, scale bar for 500 nm) of **a** **23**:PDI and **b** **22**:PDI (both in a ratio of 4:6). All films were annealed at 120 °C for 1 h. Reproduced with permission from [112]. Copy right 2008 WILEY-VCH Verlag GmbH & Co. KGaA, Weinheim

6.2.4 Liquid Crystalline Perylenebisimides in OPV

Perylene diimide derivatives are well known for their electron transporting characteristics in organic semiconducting devices. Excellent electronic absorption and fluorescence emission properties of this material have received attention of scientists working in various areas of materials science. These dye molecules upon appropriate substitution with peripheral flexible alkyl chains are able to exhibit columnar LC phases. Electron mobility in such columnar assemblies of perylene diimide has also been measured. Kim et al. [113] fabricated organic donor/acceptor heterojunction photovoltaic devices using a liquid crystalline perylene diimide **24** (Scheme 6.9) as the electron transporting layer and a zinc phthalocyanine **25** as the hole transporting layer. The solar cell performances were examined under

Scheme 6.9 Chemical structure of liquid crystalline perylene diimide derivative **24**

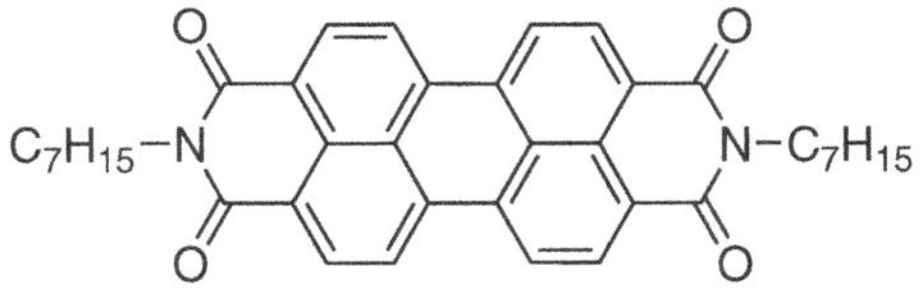

24

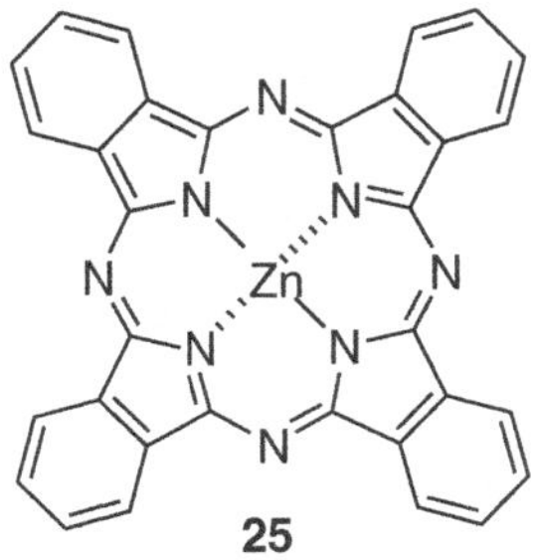

25

illumination with different organic layer thicknesses. A cell incorporating 25 nm thick organic layer showed a short circuit photocurrent of 1.6 mA/cm^2, an open circuit photovoltage of 0.6 V, and a fill factor of 0.4 with white Xe light illumination. Cells with 100 and 50 nm layer thicknesses exhibited poor performance compared to the cell with 25 nm organic layers. The better performance of the later device has been attributed to the efficient separation of electron-hole pairs and the ready transport of free charges to each electrode contact in thin layers. This study indicates that liquid crystalline perylene diimides can be used as self-organizing materials for solar energy conversion.

6.2.5 Liquid Crystalline Triphenylenes in OPV

Triphenylene derivatives are the most widely studied DLCs. Triphenylene derivatives are thermally and chemically stable, their chemistry is fairly accessible, they show a variety of mesophases, and their one-dimensional charge and energy migration properties offer tremendous potential applications. Upon suitable chemical structure modification, triphenylene derivatives can form columnar and nematic phases. However, the columnar phase of triphenyle derivatives have been tested as hole transporting materials in semiconducting devices. Oukachmih et al. [114] fabricated two-layer organic photovoltaic cells using the triphenylene ether **26** (Scheme 6.10) as a hole transporting material, and perylene derivatives **12**, **27** and **28** as electron transporting materials. Notably the electron affinities of the perylene derivatives used in the study are different. The photovoltaic device configuration is shown in Fig. 6.16. Three devices were studied and showed external quantum efficiencies of around 3 %. High exciton diffusion lengths and absorption coefficients were found to contribute to the high observed photocurrents in these devices. Though the organic/organic interface was found to be the main origin of the

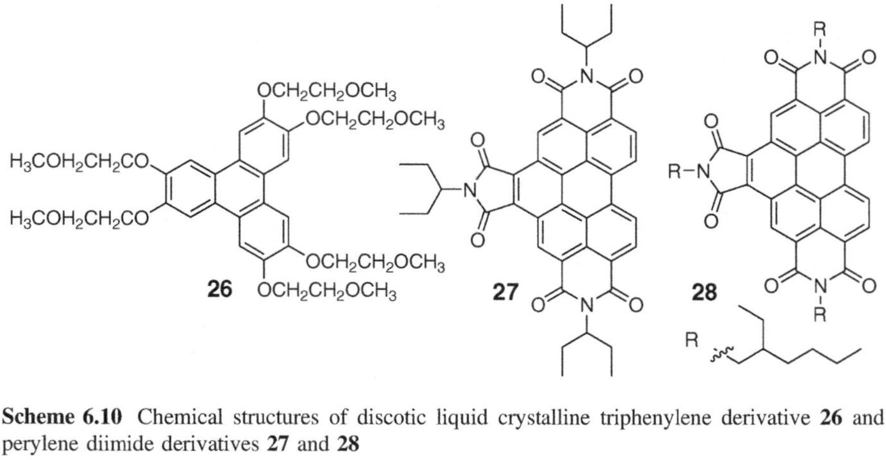

Scheme 6.10 Chemical structures of discotic liquid crystalline triphenylene derivative **26** and perylene diimide derivatives **27** and **28**

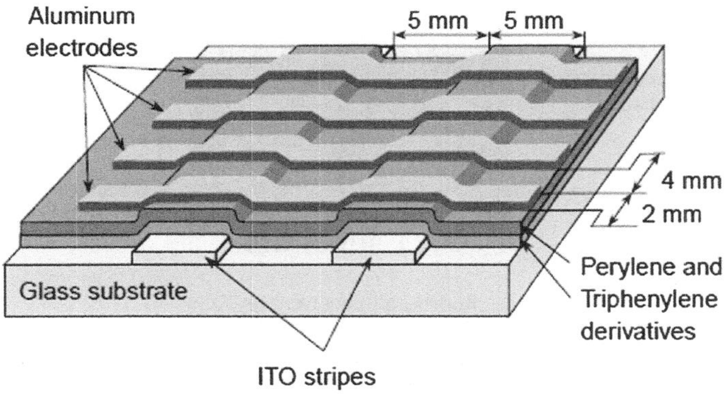

Fig. 6.16 Photovoltaic cell configuration which uses triphenylene derivative **26**. Reproduced with permission from [114]. Copy right 2004 Elsevier B.V

photocurrent generation, nevertheless the photovoltaic parameters were found to be dependent on the ITO/organic interface as well. In particular, it was demonstrated that ITO treatments with argon plasma and UV–ozone modify the open-circuit voltage of the photovoltaic devices.

Zheng et al. [115] demonstrated that it is possible to increase the efficiency of a bulk-heterojunction solar cell based on poly(3-hexylthiophene) and [6,6]-phenyl-C_{61}-butyric acid methyl ester active layer by inserting LC triphenylene derivative **29** (Scheme 6.11). They inserted the compound **29** at the interface between the active layer and a hole transporting layer. It was observed that the discotic LC molecules distribute in the active layer of the device to form more

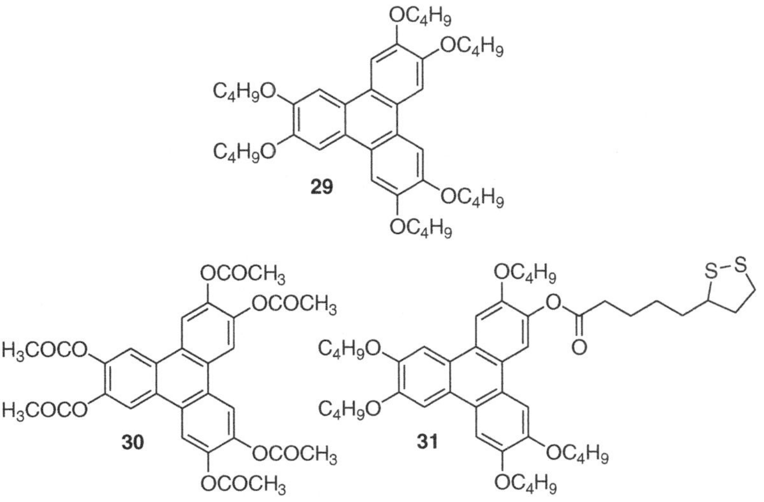

Scheme 6.11 Chemical structures of triphenylene derivatives used as additives in solar cells

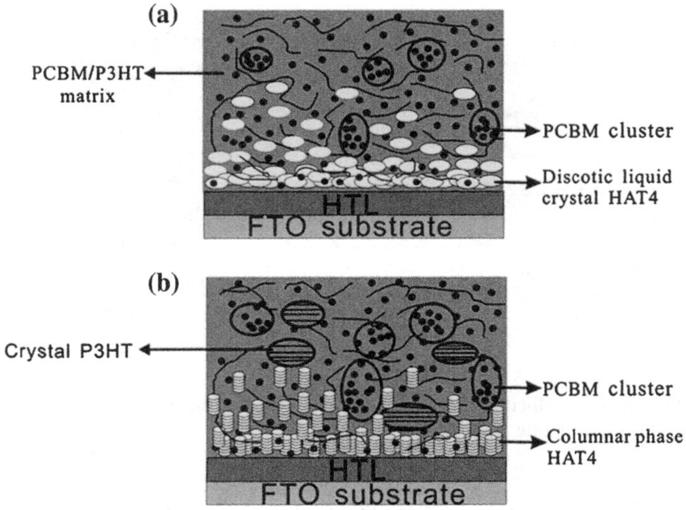

Fig. 6.17 Schematic pictures showing the **29** (HAT4)/P3HT:PCBM microscopic process during annealing: **a** before and **b** after thermal annealing at 150 °C for 10 min. Reproduced with permission from [115]. Copy right 2011 Elsevier B.V

efficient carrier transport channels leading to improved charge mobility. About 43 % increase in the power conversion efficiency was obtained in the devices compared to the device without the DLC compound **29** after thermal annealing. Similar results were obtained in the devices when different hole transporting layers were tested. Furthermore the influence of the thickness of the discotic layer was also determined in different devices. Figure 6.17 schematically shows the probable way of increasing the charge carrier transport in the device facilitated by self-assembled discotic columns.

Jeong et al. [116] have shown that by adding the LC discotic compound **30** (Scheme 6.11) to a bulk-heterojunction organic solar cell based on P3HT:PCBM, the efficiency can be enhanced to 3.97 % from 3.03 %. The efficiency improvements, indicates that the discotic additives can remain embedded in the active layer and facilitate facile charge transport. Recently, Chen et al. [117] reported hybrid solar cells based on a blend of P3HT and triphenylene **31** functionalized zinc oxide (ZnO) nanoparticles. The discotic modified ZnO nanoparticles possess well dispersability and the DLC ligands improve the compatibility between P3HT polymer and ZnO nanoparticles. The compatibility of the components is found to be beneficial for charge separation and transfer efficiency. The device efficiencies containing ZnO nanoparticle and triphenylene modified ZnO nanoparticles were compared and it was observed that the power conversion efficiency of the hybrid solar cell increased from 0.46 to 0.95 % upon surface functionalization of ZnO nanoparticles. It is believed that the discotic surface layer not only promotes the compatibility but also increases the interface area between polymers and nanocrystals. This study suggests that power conversion efficiencies of hybrid solar

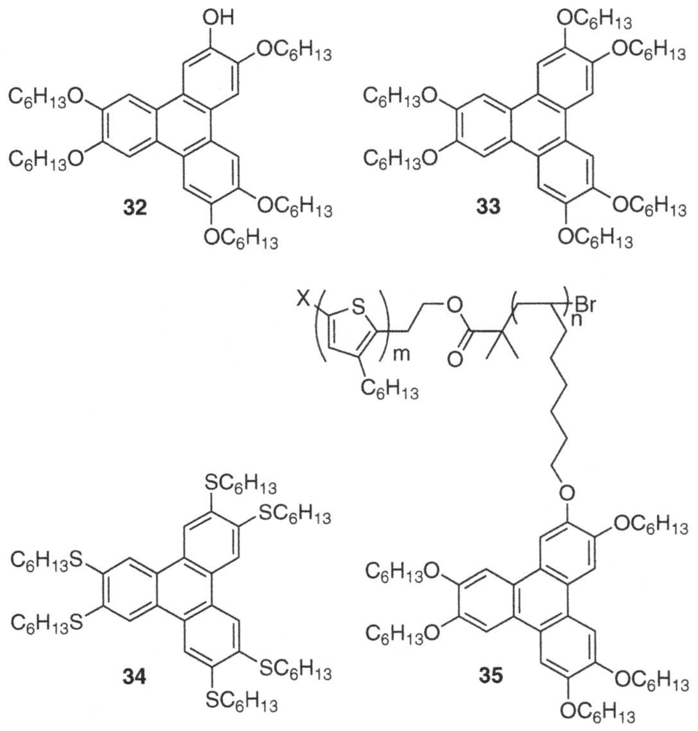

Scheme 6.12 Chemical structures of triphenylene derivatives used in solar cells

cells can be improved by combining low band gap polymers with inorganic nanocrystals as the active layer.

Shi et al. [118] fabricated electron transporting nanoarrays by cladding ZnO nanorods with a dye and with a mixture of dye and discotic liquid crystal **32** (Scheme 6.12). These nanoarrays were integrated into inverted bulk-heterojunction polymer solar cells (Fig. 6.18). Owing to these treatments of the ZnO nanorods, the interfacial contact between cathode and the bulk-heterojunction active layer increases which benefits charge generation, transport and collection in the devices. The power conversion efficiency of the polymer solar cells with only dye coating and with a mixture of dye and discotic LC coating were found to be 7.3 and 8.0 % respectively. This work provides a novel way of enhancing the efficiency of polymer solar cells by using DLC additives.

Bajpai et al. [119] investigated the effect of addition of disc-shaped liquid crystalline material **29** (Scheme 4.11) in P3HT:PCBM bulk heterojunction solar cells. A schematic representation of the device structure is shown in Fig. 6.19. These disc-shaped molecules organise into ordered columnar hexagonal structures through intermolecular π–π interactions as monitored by polarised light optical microscopy. Current–voltage characteristics of the device prepared with LC layer

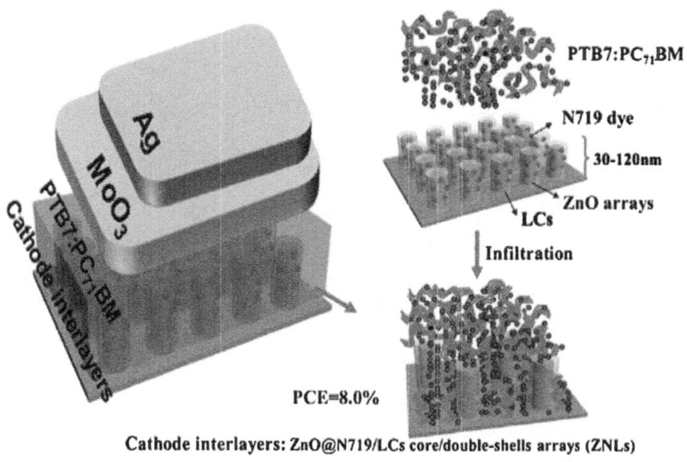

Fig. 6.18 Device structure of a polymer solar cell using ZnO nanoarrays containing discotic liquid crystal compound **32**. Reproduced with permission from [118]. Copy right 2014 American Chemical Society

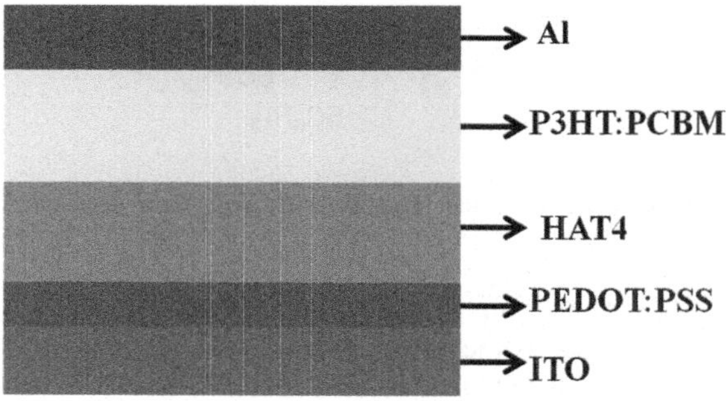

Fig. 6.19 Device architecture of a bulk heterojunction solar cell containing **29** (HAT4) as a hole transporting layer

exhibited a short-circuit current of 10.5 mA cm^{-2} and a fill factor of 35 %. The resultant power conversion efficiency of the device was 1.54 %. The influence of varying the thickness of LC layer and annealing on these solar cells was also studied. Maximum value of PCE was achieved when the thickness of the LC layer was 10 nm. Increasing the thickness beyond 10 nm led to the decrease in efficiency. It was reasoned that the inferior efficiency of thick layer devices is due to the creation of traps by increasing LC layer thickness. Since annealing process is known to enhance the efficiency of such cells, post annealing studies on the solar cells were performed. Indeed it was found that the power conversion efficiency of the solar cells containing self-organized DLCs in the active layer increased to 2.3 %.

In a novel approach, Selvaraj et al. [120] have employed triphenylene derivatives **33** and **34** (Scheme 6.12), for the first time, as iodine-free redox electrolyte in

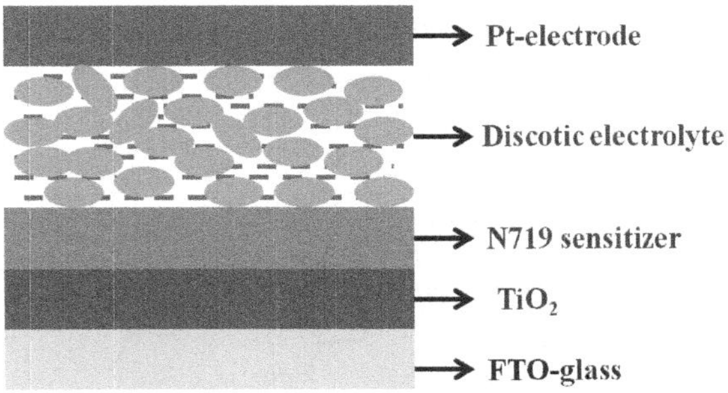

Fig. 6.20 Device architecture of a DSSC with discotic liquid crystal electrolytes

dye sensitized solar cells. Power conversion efficiency of about 0.45 % has been observed in the cells. The surface profile of the electrolytes was investigated by scanning electron microscopy. The geometry of DSSC with discotic electrolytes is depicted in Fig. 6.20. These preliminary results demonstrate the potential of DLC molecules as charge transporter and mediator in iodine-free DSSCs.

Very recently, Yuan et al. [121] reported performance enhancement of P3HT: PCBM solar cells by incorporation of discotic liquid crystalline block copolymer **35** (Scheme 6.12). The solar cells based on the self-assembled liquid crystalline block copolymers blended with PCBM show poor photovoltaic performance, even after thermal treatment at liquid crystalline temperature, due to the low conductivity of the non-conjugated liquid crystalline block. However, when the liquid crystalline block copolymer was utilized as compatibilizers in P3HT:PCBM blends, the photovoltaic performance of P3HT:PCBM solar cells could be significantly improved after liquid crystalline state annealing. The morphology and device performance was improved after thermal annealing at the liquid crystalline temperature originating from the self-orientation of the liquid crystalline block copolymer **35** in P3HT: PCBM blends. Moreover, it was observed that the self-assembly of LC block at the donor and acceptor interface can enhance the crystallization and ordering of P3HT chains and guarantee the formation of interpenetrating networks, subsequently resulting in the improvement of efficient exciton separation of the active layer. Due to the greater compatibility with fullerene acceptors and the more efficient charge transport caused by the self-assembled columnar phase from discotic LC block, the power conversion efficiency of the device was improved to 4.03 %.

6.2.6 Liquid Crystalline Decacyclene in OPV

Tri-substituted decacyclene derivatives are one example of DLC materials with interesting physicochemical properties. 1,7,13-Alkanoyldecacyclene is a C3 symmetrical compound with a decacyclene core and three peripheral alkyl chains. Recently discotic 1,7,13-heptanoyldecacyclene **36** (Scheme 6.13) was synthesized and it was found to show columnar LC phase. Hirota et al. [122] investigated the

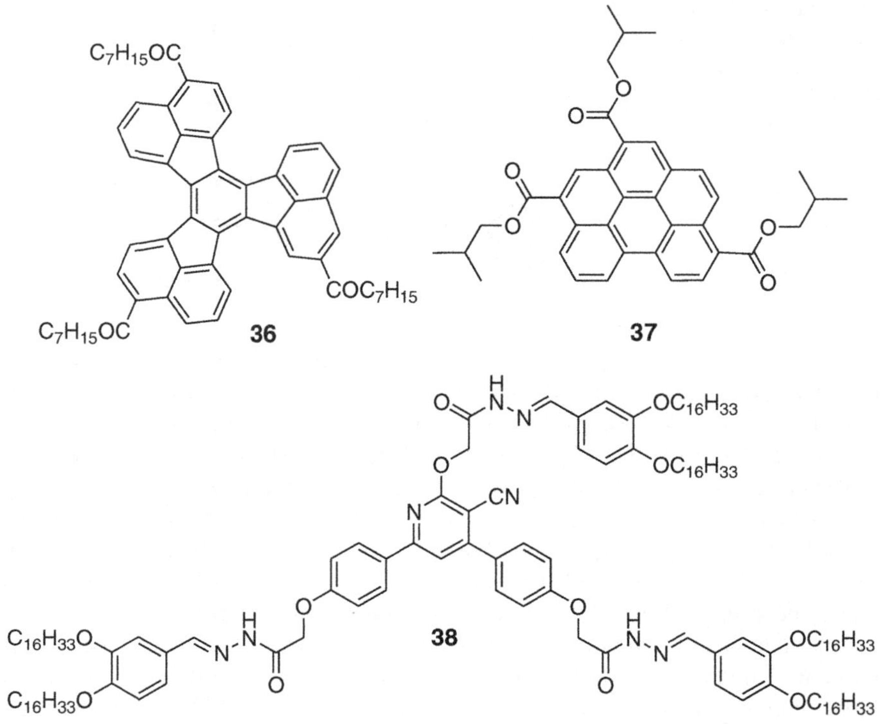

Scheme 6.13 Chemical structures of some discotic liquid crystalline compounds used in solar cells

electrochemical and optoelectronic characteristics of this LC compound. Compound **36** was found to be electron deficient so it was expected to function as an acceptor material. Interestingly when this compound was combined with poly [2-methoxy-5-(2′-ethylhexyloxy)-1,4-phenylenevinylene] (MEH-PPV), photoluminescence quenching was observed. Spin-coated photovoltaic devices containing MEH-PPV and decacyclene derivative **36** exhibited a relatively high open circuit voltage. When the device was annealed at a temperature where the LC phase transition of the decacyclene occurs, the photovoltaic performance of the device was significantly improved. Morphological studies conducted with atomic force microscope of the active layer revealed large crystal-like assemblies of **36** imbedded in the matrix of MEH-PPV which was not observed before annealing. Such morphological changes upon annealing have been thought to improve electron transport through compound **36**. The external quantum efficiency of the photovoltaic device was found to increase from 1.3 to 2.5 % upon annealing. Similarly power conversion efficiency was improved to 0.30 % from 0.14 %. By playing with the peripheral alkyl chains, different decacyclene derivatives with better miscibility with the conducting polymer can be prepared which might improve the power conversion efficiency.

6.2.7 Other Liquid Crystalline Discotic Compounds in OPV

Bock et al. [123] have synthesized discotic liquid crystalline benzoperylene triester **37** (Scheme 6.13) and benzoperylene triimide **27** derivatives. Upon combining **27** as an acceptor and **37** as a donor in a photovoltaic device, they obtained a high open circuit voltage of 1.16 V. Even though these simple solar cells based on a planar donor-acceptor heterojunction were fabricated by conventional thermal vacuum evaporation yet without any attempt to obtain uniformly aligned films of the two materials, they already yielded a short circuit current of 0.24 mA/cm^2 and a reasonable fill factor of 40 %, and thus an external quantum efficiency of 13 % at 400 nm monochromatic irradiation, corresponding to a power efficiency of 2.0 %. Considerably higher currents, better fill factors and thus better efficiencies should be attainable once uniformly aligned layers can be realized in which all column axes and thus the paths of good electron and exciton mobility are perpendicular to the substrate.

The well known BHJ solar cell comprised of P3HT as an electron donor and PCBM as an electron acceptor exhibits quite promising power conversion efficiency. However, inherent space charge effects and the unfavorable morphology of such BHJ structure are the major drawbacks for achieving high energy conversion efficiency. In addition, obtaining an ordered bicontinuity of the microphase, which ensures optimum charge-carrier photogeneration, extraction and transfer to the electrodes are the most critical challenges in organic polymer solar cells. To avoid these challenges, various approaches such as thermal annealing, post-fabrication annealing at high temperature, and the use of processing additives have been successfully demonstrated to develop an optimum morphology with minimum phase separation as well as the formation of nano-crystalline domains in the blends to achieve an improved PCE. Towards this end, Ahipa et al. [124] have prepared a trihydrazone functionalized cyanopyridine liquid crystal material **38** (Scheme 6.13) capable of exhibiting ambient temperature hexagonal columnar phase. This compound has been introduced into the P3HT:PCBM photoactive layer as a processing additive for developing nanostructured crystallite and an improved bicontinous phase separation in the prepared film with enhanced photophysical and conducting properties in the construction of an efficient solar cell. It is observed that the addition of the liquid crystal compound **38** drastically increases the current passing through the P3HT:PCBM film. Moreover, the structural anisotropic nature of the material **38** helps to obtain well-ordered morphology with nanostructured crystallite formation as well as enhanced current in the P3HT: PCBM film. To determine the interpenetrating network in the photoactive layer, epi-fluoroscence microscopy images were recorded under UV light. It is interesting to note that in the case of ternary P3HT:**38**:PCBM film an enhanced interpenetrating network was perceived (Fig. 6.21).

Similar results were observed in the surface topography images recorded by atomic force microscope. Thus, in the ternary film, an excellent interpenetrating network with desired channel thickness, which contributes significantly to an

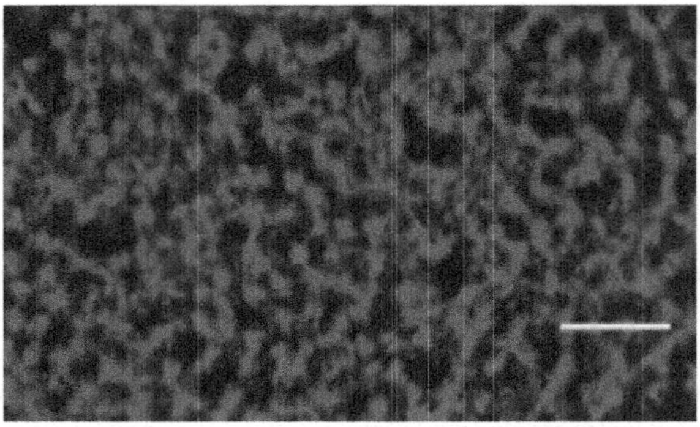

Fig. 6.21 Epi-fluorescence microscopy (scale bar: 100 mm) images under UV light of P3HT:**38**: PCBM (1:0.5:0.5) active layer film morphology. Reproduced with permission from [124]. Copy right The Royal Society of Chemistry and the Centre National de la Recherche Scientifique 2015

efficient exciton dissociation and migration in the photovoltaic devices, was observed. Additionally current measurements over the ternary film exhibit a profound increase in the mean current value as determined in the device.

6.3 Conclusions and Outlook

In this chapter, the use of DLCs as active components, i.e., donors or acceptors in organic photovoltaic solar cells, has been discussed. Various DLCs with distinct optoelectronic properties based on porphyrin, phthalocyanine, hexabenzocoronene, perylene diimide, triphenylene, decacyclene etc. have been used in photovoltaic solar cells. Perylene diimide and fullerene derivatives have been used as electron transporting components in the active layer of self-organizing discotic photovoltaic solar cells. Power conversion efficiency exceeding 4 % has been achieved in discotic photovoltaic devices. In some cases, columnar phase forming DLCs have been deployed as facilitating additives in the active layer of conventional polymer solar cells to dramatically enhance their efficiencies. The effect of active layer morphology on the device performances have been investigated to a large extent and clarified. Similarly, the effect of molecular structure of the DLCs on device characteristics has been studied. Overall, the proof-of-principle studies discussed here clearly show that the columnar nanostructures of DLCs can be appealing contenders for organic photovoltaic solar cells in the quest for sustainable energy harvesting. However, lots of challenges remain to be met before the opportunities can be capitalized on DLCs. It is necessary to design and synthesize new DLC materials which can furnish about 10 % efficiency which has been demonstrated with polymer solar cells. Unfortunately, though DLCs are the first small molecules that have been tested in organic solar cells, they have not been able to catch up with other recent systems which display >8 % efficiency [125]. New stable electron transporting room temperature DLCs should be realized in order to construct all-columnar bilayer and bulk heterojunction solar cells with high efficiencies. The introduction of new polycyclic aromatic discotic cores with appropriate frontier energy levels and broad absorption with high absorption coefficients would provide additional thrust to this already active endeavor. Since the ground work done so far is encouraging, one can expect some breakthroughs in the use of DLCs in the active layer of organic photovoltaic solar cells in the future.

Acknowledgements The preparation of this chapter benefited from the support to Quan Li by Ohio Third Frontier, US Department of Energy (DOE), US Department of Defense Multidisciplinary University Research Initiative (DoD MURI), US Air Force Office of Scientific Research (AFOSR), US Army Research Office (ARO), US National Aeronautics and Space Administration (NASA), and US National Science Foundation (NSF).

References

1. S. Gunes, H. Neugebauer, N.S. Sariciftci, Conjugated polymer-based organic solar cells. Chem. Rev. **107**, 1324–1338 (2007)
2. M. Gratzel, Dye-sensitized solar cells. J. Photochem. Photobio. C: Photochem. Rev. **4**, 145–153 (2003)
3. P.V. Kamat, Quantum dot solar cells. Semiconductor nanocrystals as light harvesters. J. Phys. Chem. C **112**, 18737–18753 (2008)
4. M. Liu, M.B. Johnston, H.J. Snaith, Efficient planar heterojunction pervoskite solar cells by vapour deposition. Nature **501**, 395–398 (2013)
5. F.C. Krebs, Fabrication and processing of polymer solar cells: a review of printing and coating techniques. Sol. Enrgy Mater. Sol. Cells **93**, 394–412 (2009)
6. C. Liu, K. Wang, X. Gong, A.J. Heeger, Low bandgap semiconducting polymers for polymeric photovoltaics. Chem. Soc. Rev. (2015)
7. Y. Sun, G.C. Welch, W.L. Leong, C.J. Takacs, G.C. Bazan, A.J. Heeger, Solution-processed small-molecule solar cells with 6.7 % efficiency. Nat. Mater. **11**, 44–48 (2012)
8. P. Kumar, S. Chand, Recent progress and future aspects of organic solar cells. Prog. Photovolt: Res. Appl. **20**, 377–415 (2012)
9. G. Li, R. Zhu, Y. Yang, Polymer solar cells. Nat. Photon. **6**, 153–161 (2012)
10. Y. Liu, C.-C. Chen, Z. Hong, J. Gao, Y. Yang, H. Zhou, L. Dou, G. Li, Y. Yang, Solution-processed small-molecule solar cells: breaking the 10 % power conversion efficiency. Sci. Report **3**, 3356 (2013)
11. Y.-W. Su, S.-C. Lan, K.-H. Wei, Organic photovoltaics. Mater. Today **15**, 554–562 (2012)
12. J.W. Rumer, I. McCulloch, Organic photovoltaics: crosslinking for optimal morphology and stability. Mater. Today **18**, 425–435 (2015)
13. R.A. Street, Electronic structure and properties of organic bulk-heterojunction interfaces. Adv. Mater. (2015). doi: 10.1002/adma.201503162
14. C.B. Nielsen, S. Holliday, H.-Y. Chen, S.J. Cryer, I. McCulloch, Non-fullerene electron acceptors for use in organic solar cells. Acc. Chem. Res. **48**, 2803–2812 (2015)
15. Y. Lin, X. Zhan, Oligomer molecules for efficient organic photovoltaics. Acc. Chem. Res. (2015)
16. B. Walker, A.B. Tamayo, X.-D. Dang, P. Zalar, J.H. Seo, A. Garcia, M. Tantiwiwat, T.-Q. Nguyen, Nanoscale phase separation and high photovoltaic efficiency in solution-processed, small-molecule bulk heterojunction solar cells. Adv. Funct. Mater. **19**, 3063–3069 (2009)

17. B. Walker, C. Kim, T.-Q. Nguyen, Small molecule solution-processed bulk heterojunction solar cells. Chem. Mater. **23**, 470–482 (2011)
18. J. Roncali, Molecular bulk heterojunctions: an emerging approach to organic solar cells. Acc. Chem. Res. **42**, 1719–1730 (2009)
19. J. Roncali, P. Leriche, P. Blanchard, Molecular materials for organic photovoltaics: small is beautiful. Adv. Mater. **26**, 3821–3838 (2014)
20. M.T. Lloyd, J.E. Anthony, G.G. Malliaras, Photovoltaics from soluble small molecules. Mater. Today **10**, 34–41 (2007)
21. Q. Li (ed.), *Liquid Crystals Beyond Displays: Chemistry, Physics, and Applications* (Wiley, New Jersey, 2012)
22. Q. Li (ed.), *Self-organized Organic Semiconductors: From Materials to Device Applications* (Wiley, New Jersey, 2011)
23. Q. Li (ed.), *Intelligent Stimuli Responsive Materials: From Well-Defined Nanostructures to Applications* (Wiley, New Jersey, 2013)
24. Q. Li (ed.), *Nanoscience with Liquid Crystals: From Self-Organized Nanostructures to Applications* (Springer, Heidelberg, 2014)
25. Q. Li (ed.), *Anisotropic Nanomaterials: Preparation, Properties, and Applications* (Springer, Heidelberg, 2015)
26. B. Bahadur (ed.), *Liquid Crystals: Applications and Uses*, vol. 1–3 (World Scientific, Singapore, 1990)
27. P.G. de Gennes, J. Prost, *The Physics of Liquid Crystals* (Oxford University Press, Oxford, 1993)
28. S. Chandrasekhar, *Liquid Crystals* (Cambridge University Press, Cambridge, U.K., 1992)
29. P.J. Collings, *Liquid Crystals: Natures Delicate Phase of Matter* (Princeton University Press, Princeton, 2002)
30. P.J. Collings, M. Hird, *Introduction to Liquid Crystals: Chemistry and Physics* (Taylor & Francis, London, U. K., 1997)
31. P. Oswald, P. Pieranski, *Smectic and Columnar Liquid Crystals: Concepts and Physical Properties Illustrated by Experiments* (Taylor & Francis, CRC Press, Boca Raton, Florida, 2005)
32. J.W. Goodby, P.J. Collings, T. Kato, C. Tschierske, H. Glesson, P. Raynes (eds.), *Handbook of Liquid Crystals*, 2nd edn. (Wiley-VCH, Weinheim, 2014)
33. P. Oswald, P. Pieranski, *Nematic and Cholesteric Liquid crystals: Concepts and Physical Properties Illustrated by Experiments* (Taylor & Francis, CRC Press, Boca Raton, Florida, 2005)
34. G.W. Gray, G.R. Luckhurst (eds.), *The Molecular Physics of Liquid Crystals* (Academic Press, London, U.K., 1997)
35. S. Kumar (ed.), *Liquid Crystals: Experimental Study of Physical Properties and Phase Transitions* (Cambridge University Press, Cambridge, U.K., 2001)
36. A. Jakli, A. Saupe, *One- and Two-dimensional Fluids: Properties of Smectic, Lamellar and Columnar Liquid Crystals* (CRC PressFlorida, Boca Raton, 2006)
37. G.W. Gray, P.A. Windsor, *Liquid Crystals and Plastic Crystals*, vol. 1–2 (Ellis Horwood Ltd, Chichester, 1974
38. M.R. Fisch, *Liquid Crystals, Laptops and Life* (World Scientific, Singapore, 2004)
39. R.H. Chen, *Liquid Crystal Displays: Fundamental Physics and Technology* (Wiley, New Jersey, 2011)
40. G.H. Brown, J.J. Wolken, *Liquid Crystals and Biological Structures* (Academic Press, New York, 1979)
41. S.J. Woltman, G.D. Jay, G.P. Crawford (eds.), *Liquid Crystals: Frontiers in Biomedical Applications* (World Scientific, New Jersey, 2007)
42. T.J. Sluckin, D.A. Dunmur, H. Stegemeyer (eds.), *Crystals That Flow: Classic Papers from the History of Liquid Crystals* (CRC Press, Boca Raton, Florida, 2004)
43. D. Dunmur, T. Sluckin, *Soap, Science and Flat-Screen TVs—A History of Liquid Crystals* (Oxford University Press, Oxford, 2011)

44. A.G. Petrov, *The Lyotropic state of Matter: Molecular Physics and Living Matter Physics* (Gordon & Breach Science Pub, Amsterdam, The Netherlands, 1999)
45. N. Garti, P. Somasundaran, R. Mezzenga (eds.), *Self-Assembled Supramolecular Architectures: Lyotropic Liquid Crystals* (Wiley, New Jersey, 2012)
46. C. Tschierske (ed.), *Liquid Crystals: Materials Design and Self-Assembly* (Springer, Berlin, 2012)
47. J.L. Serrano (ed.), *Metallomesogens: Synthesis, Properties and Applications* (Wiley-VCH, Weinheim, Germany, 1996)
48. H.-S. Kitzerow, C. Bahr (eds.), *Chirality in Liquid Crystals* (Springer, New York, 2001)
49. J.W. Goodby, *Ferroelectric Liquid Crystals: Principles, Properties and Applications* (Gordon and Breach, Philadelphia, PA, 1991)
50. S.T. Lagerwall, *Ferroelectric and Antiferroelectric Liquid Crystals* (Wiley-VCH, Weinheim, Germany, 1999)
51. I. Musevic, R. Blinc, B. Zeks, *The Physics of Ferroelectric and Antiferroelectric Liquid Crystals* (World Scientific, Singapore, 2000)
52. A.M. Donald, A.H. Windle, S. Hanna, *Liquid Crystalline Polymers* (Cambridge University Press, Cambridge, 2006)
53. M. Warner, E.M. Terentjev, *Liquid Crystal Elastomers* (Oxford University Press, New York, 2003)
54. H.K. Bisoyi, Q. Li, Light-directing chiral liquid crystal nanostructures: from 1D to 3D. Acc. Chem. Res. **47**, 3184–3195 (2014)
55. Y. Wang, Q. Li, Light-driven chiral molecular switches or motors in liquid crystals. Adv. Mater. **24**, 1926–1945 (2012)
56. L. Wang, Q. Li, Stimuli-directing self-organized 3D liquid crystalline nanostructures: from materials design to photonic applications. Adv. Funct. Mater. **26**, 10–28 (2016)
57. H.K. Bisoyi, Q. Li, Light-directed dynamic chirality inversion in functional self-organized helical superstructures. Angew. Chem. Int. Ed. **55**, 2994–3010 (2016)
58. S. Kumar, *Chemistry of Discotic Liquid Crystals: From Monomers to Polymers* (CRC Press, Boca Raton, Florida, 2011)
59. R.J. Bushby, S.M. Kelly, M. O'Neill (eds.), *Liquid Crystalline Semiconductors: Materials, Properties and Applications* (Springer, Dordrecht, The Netherlands, 2013)
60. D. Adam, P. Schumacher, J. Simmerer, L. Haussling, K. Siemensmeyer, K.H. Etzbachi, H. Ringsdorf, D. Haarer, Fast photoconduction in the highly ordered columnar phase of a discotic liquid crystal. Nature **371**, 141–143 (1994)
61. X. Feng, V. Marcon, W. Pisula, M.R. Hansen, J. Kirkpatrick, F. Grozema, D. Andrienko, K. Kremer, K. Mullen, Towards high charge-carrier mobilities by rational design of the shape and periphery of discotics. Nat. Mater. **8**, 421–426 (2009)
62. S. Laschat, A. Baro, N. Steinke, F. Giesselmann, C. Hagele, G. Scalia, R. Judele, E. Kapatsina, S. Sauer, A. Schreivogel, M. Tosoni, Discotic liquid crystals: from tailor-made synthesis to plastic electronics. Angew. Chem. Int. Ed. **46**, 4832–4887 (2007)
63. S. Kumar, Self-organization of disc-like molecules: chemical aspects. Chem. Soc. Rev. **35**, 83–109 (2006)
64. R.J. Bushby, O.R. Lozman, Discotic liquid crystals 25 years on. Curr. Opin. Colloid Interface Sci. **7**, 343–354 (2002)
65. S. Kumar, Recent developments in the chemistry of triphenylene-based discotic liquid crystals. Liq. Cryst. **31**, 1037–1059 (2004)
66. R.J. Bushby, K. Kawata, Liquid crystals that affected the world: discotic liquid crystals. Liq. Cryst. **38**, 1415–1426 (2011)
67. B.R. Kaafarani, Discotic liquid crystals for opto-electronic applications. Chem. Mater. **23**, 378–396 (2011)
68. H.K. Bisoyi, S. Kumar, Discotic nematic liquid crystals: science and technology. Chem. Soc. Rev. **39**, 264–285 (2010)
69. J. Wu, W. Pisula, K. Mullen, Graphene as potential material for electronics. Chem. Rev. **107**, 718–747 (2007)

70. H.K. Bisoyi, S. Kumar, Liquid-crystal nanoscience: an emerging avenue of soft self-assembly. Chem. Soc. Rev. **40**, 306–319 (2011)
71. S. Chandrasekhar, S.K. Prasad, Recent developments in discotic liquid crystals. Contemporary Phys. **40**, 237–245 (1999)
72. S. Sergeyev, W. Pisula, Y.H. Geerts, Discotic liquid crystals: a new generation of organic semiconductors. Chem. Soc. Rev. **36**, 1902–1929 (2007)
73. D. Janietz, Structure formation control of disc-shaped molecules. Mol. Cryst. Liq. Cryst. **396**, 251–264 (2003)
74. C.D. Simpson, J. Wu, M.D. Watson, K. Mullen, From graphite molecules to columnar superstructures-an exercise in nanoscience. J. Mater. Chem. **14**, 494–504 (2004)
75. D. Janietz, Structure formation of functional sheet-shaped mesogens. J. Mater. Chem. **8**, 265–274 (1998)
76. W. Pisula, X. Feng, K. Mullen, Tuning the columnar organization of discotic polycyclic aromatic hydrocarbons. Adv. Mater. **22**, 3634–3649 (2010)
77. K. Kawata, Orientation control and fixation of discotic liquid crystal. Chem. Rec. **2**, 59–80 (2002)
78. H. Mori, The wide view (WV) film for enhancing the field of view of LCDs. J. Display Technol. **1**, 179–186 (2005)
79. K. Ohta, K. Hatsusaka, M. Sugibayashi, M. Ariyoshi, K. Ban, F. Maeda, R. Naito, K. Nishizawa, A.M. Van de Craats, J.M. Warman, Discotic liquid crystalline semiconductors. Mol. Cryst. Liq. Cryst. **397**, 25–45 (2003)
80. M. Funahasi, Development of liquid-crystalline semiconductors with high carrier mobilities and their application to thin-film transistors. Polym J. **41**, 459–469 (2009)
81. B.R. Kaafarani, Discotic liquid crystals for opto-electronic applications. Chem. Mater. **23**, 378–396 (2011)
82. M. O'Neill, S.M. Kelly, Ordered materials for organic electronics and photonics. Adv. Mater. **23**, 566–584 (2011)
83. J. Wu, W. Pisula, K. Mullen, Graphene as potential material for electronics. Chem. Rev. **107**, 718–747 (2007)
84. W. Pisula, M. Zorn, J.Y. Chang, K. Mullen, R. Zentel, Liquid crystalline ordering and charge transport in semiconducting materials. Macromol. Rapid Commun. **30**, 1179–1202 (2009)
85. E. Grelet, H. Bock, T. Brunet, J. Kelber, O. Thiebaut, P. Jolinat, S. Mirzaei, P. Destruel, Toward organic photovoltaic cells based on the self-assembly of discotic columnar liquid crystals. Mol. Cryst. Liq. Cryst. **542**, 182–189 (2011)
86. H. Bock, N. Buffet, E. Grelet, I. Seguy, J. Navarro, P. Destruel, Novel columnar LCs for a new generation of organic solar cells. Proc. SPIE **6911**(69110N), 1–8 (2008)
87. B.A. Gregg, M.A. Fox, A.J. Bard, Photovoltaic effect in symmetrical cells of a liquid crystal porphyrin. J. Phys. Chem. **94**, 1586–1598 (1990)
88. B.A. Gregg, M.A. Fox, A.J. Bard, 2,3,7,8,12,13,17,18,-Octakis(β-hydroxyethyl)porphyrin and its liquid crystalline derivatives: synthesis and characterization. J. Am. Chem. Soc. **111**, 3024–3029 (1989)
89. B.A. Gregg, Effect of long range order, temperature and phase on the photoconversion properties of liquid crystal porphyrin films. Mol. Cryst. Liq. Cryst. **257**, 219–227 (1994)
90. P.G. Schouten, J.M. Warman, M.P. de Haas, M.A. Fox, H.-L. Pan, Charge migration in supramolecular stacks of peripherally substituted porphyrins. Nature **353**, 736–737 (1991)
91. Q. Sun, L. Dai, X. Zhou, L. Li, Q. Li, Bilayer- and bulk-heterojunction solar cells using liquid crystalline porphyrins as donors by solution processing. Appl. Phys. Lett. **91**, 253505 (2007)
92. L. Li, S.-W. Kang, J. Harden, Q. Sun, X. Zhou, L. Dai, A. Jakli, S. Kumar, Q. Li, Nature-inspired light-harvesting liquid crystalline porphyrins for organic photovoltaics. Liq. Cryst. **35**, 233–239 (2008)
93. X. Zhou, S.-W. Kang, S. Kumar, R.R. Kulkarni, S.Z.D. Cheng, Q. Li, Self-assembly of porphyrin and fullerene supramolecular complex into highly ordered nanostructure by simple thermal annealing. Chem. Mater. **20**, 3551–3553 (2008)

94. S.-W. Kang, Q. Li, B.D. Chapman, R. Pindak, J.O. Cross, L. Li, M. Nakata, S. Kumar, Microfocus X-ray diffraction study of the columnar phase of porphyrin-based mesogens. Chem. Mater. **19**, 5657–5663 (2007)
95. K. Petritsch, R.H. Friend, A. Lux, G. Rozenberg, S.C. Moratti, A.B. Holmes, Liquid crystalline phthalocyanines in organic solar cells. Synth. Metals **102**, 1776–1777 (1999)
96. K. Petritsch, J.J. Dittmer, E.A. Marseglia, R.H. Friend, A. Lux, G.G. Rozenberg, S.C. Moratti, A.B. Holmes, Dye-based donor/acceptor solar cells. Solar Energy Mater. Solar Cells **61**, 63–72 (2000)
97. T. Hori, N. Fukuoka, T. Masuda, Y. Miyake, H. Yoshida, A. Fujii, Y. Shimizu, M. Ozaki, Bulk heterojunction organic solar cells utilizing 1,4,8,11,15,18,22,25-octahexyl-phthalocyanine. Sol. Energy Mater. Sol. Cells **95**, 3087–3092 (2011)
98. Q.D. Dao, T. Hori, K. Fukumura, T. Masuda, T. Kamikado, A. Fujii, Y. Shimizu, M. Ozaki, Efficiency enhancement in mesogenic-phthalocyanine-based solar cells with processing additives. Appl. Phys. Lett. **101**, 263301 (2012)
99. T. Hori, T. Masuda, N. Fukuoka, T. Hayashi, Y. Miyake, T. Kamikado, H. Yoshida, A. Fujii, Y. Shimizu, M. Ozaki, Non-peripheral octahexylphthalocyanine doping effects in bulk heterojunction polymer solar cells. Org. Electron. **13**, 335–340 (2012)
100. Q.-D. Dao, T. Hori, K. Fukumura, T. Masuda, T. Kamikado, A. Fujii, Y. Shimizu, M. Ozaki, Effect of processing additives on nanoscale phase separation, crystallization and photovoltaic performance of solar cells based on mesogenic phthalocyanine. Org. Electron. **14**, 2628–2634 (2013)
101. T. Masuda, T. Hori, K. Fukumura, Y. Miyake, D.Q. Duy, T. Hayashi, T. Kamokado, H. Yoshida, A. Fujii, Y. Shimizu, M. Ozaki, Photovoltaic properties of 1,4,8,11,15,18,22,25-octaalkylphthalocyanine doped polymer bulk heterojunction solar cells. Jpn. J. Appl. Phys. **51**, 02BK15 1–4 (2012)
102. L. Schmidt-Mende, A. Fechtenkotter, K. Mullen, E. Moons, R.H. Friend, J.D. MacKenzie, Self-organized discotic liquid crystals for high-efficiency organic photovoltaics. Science **293**, 1119–1122 (2001)
103. L. Schmidt-Mende, A. Fechtenkotter, K. Mullen, R.H. Friend, J.D. MacKenzie, Efficient organic photovoltaics from soluble discotic liquid crystalline materials. Physica E **14**, 263–267 (2002)
104. L. Schmidt-Mende, M. Watson, K. Mullen, R.H. Friend, Organic thin film photovoltaic devices from discotic materials. Mol. Cryst. Liq. Cryst. **396**, 73–90 (2003)
105. T. Hasheider, S.A. Benning, M.W. Lauhof, H.-S. Kitzerow, H. Bock, M.D. Watson, K. Mullen, Organic heterojunction photovoltaic cells made of discotic, mesogenic materials. Mol. Cryst. Liq. Cryst. **413**, 461–472 (2004)
106. J. Li, M. Kastler, W. Pisula, J.W.F. Robertson, D. Wasserfallen, A.C. Grimsdale, J. Wu, K. Mullen, Organic bulk-heterojunction photovoltaics based on alkyl substituted discotics. Adv. Funct. Mater. **17**, 2528–2533 (2007)
107. H.C. Hesse, J. Weickert, M. Al-Hussein, L. Dossel, X. Feng, K. Mullen, L. Schmidt-Mende, Discotic materials for organic solar cells: effect of chemical structure on assembly and performance. Sol. Energy Mater. Sol. Cells **94**, 560–567 (2010)
108. C. Liu, A. Fechtenkotter, M.D. Watson, K. Mullen, A.J. Bard, Room temperature discotic liquid crystalline thin films of hexa-peri-hexabenzocoronene: synthesis and optoelectronic properties. Chem. Mater. **15**, 124–130 (2003)
109. M. Al-Hussein, H.C. Hesse, J. Weickert, L. Dossel, X. Feng, K. Mullen, L. Schmidt-Mende, Structural properties of the active layer of discotic hexabenzocoronene/perylene diimide bulk heterojunction photovoltaic devices: the role of alkyl side chain length. Thin Solid Films **520**, 307–313 (2011)
110. J.P. Schmidtke, R.H. Friend, M. Kastler, K. Mullen, Control of morphology in efficient photovoltaic diodes from discotic liquid crystals. J. Chem. Phys. **124**(174704), 1–6 (2006)
111. S.J. Kang, S. Ahn, J.B. Kim, C. Schenck, A.M. Hiszpanski, S. Oh, T. Schiros, Y.-L. Loo, C. Nuckolls, Using self-organization to control morphology in molecular photovoltaics. J. Am. Chem. Soc. **135**, 2207–2212 (2012)

112. X. Feng, M. Liu, W. Pisula, M. Takase, J. Li, K. Mullen, Supramolecular organization and photovoltaics of triangle-shaped discotic graphenes with swallow-tailed alkyl substituents. Adv. Mater. **20**, 2684–2689 (2008)
113. J.Y. Kim, A.J. Bard, Organic donor/acceptor heterojunction photovoltaic devices based on zinc phthalocyanine and a liquid crystalline perylene diimide. Chem. Phys. Lett. **383**, 11–15 (2004)
114. M. Oukachmih, P. Destruel, I. Seguy, G. Ablart, P. Jolinat, S. Archambeau, M. Mabiala, S. Fouet, H. Bock, New organic discotic materials for photovoltaic conversion. Sol. Energy Mater. Sol. Cells **85**, 535–543 (2005)
115. Q. Zheng, G. Fang, W. Bai, N. Sun, P. Qin, X. Fan, F. Cheng, L. Yuan, X. Zhao, Efficiency improvement in organic solar cells by inserting a discotic liquid crystal. Sol. Energy Mater. Sol. Cells **95**, 2200–2205 (2011)
116. S. Jeong, Y. Kwon, B.-D. Choi, H. Ade, Y.S. Han, Improved efficiency of bulk heterojunction poly(3-hexylthiophene):[6, 6]-phenyl-C_{61}-butyric acid methyl ester photovoltaic devices using discotic liquid crystal additives. Appl. Phys. Lett. **96**, 183305 (2010)
117. X. Chen, L. Chen, Y. Chen, Self-assembly of dscotic liquid crystal decorated ZnO nanoparticles for efficient hybrid solar cells. RSC Adv. **4**, 3627–3632 (2014)
118. Y. Shi, L. Tan, Y. Chen, Dye-sensitized nanoarrays with discotic liquid crystals as interlayer for high-efficiency inverted polymer solar cells. ACS Appl. Mater. Interfaces **6**, 17848–17856 (2014)
119. M. Bajpai, N. Yadav, S. Kumar, R. Srivastava, R. Dhar, Bulk heterojunction solar cells based on self-assembling disc-shaped liquid crystalline material. Liq. Cryst. **43**, 305–313 (2016)
120. A.R.K. Selvaraj, V. Lakshminarayanan, R. Dhar, S. Kumar, Dye-sensitized solar cells with iodine-free discotic electrolytes. Liq. Cryst. **42**, 1815–1822 (2015)
121. K. Yuan, L. Chen, Y. Chen, Photovoltaic performance enhancement of P3HT/PCBM solar cells driven by incorporation of conjugated liquid crystalline rod-coil block copolymers. J. Mater. Chem. C **2**, 3835–3845 (2014)
122. K. Hirota, K. Tajima, K. Hashimoto, Physicochemical study of discotic liquid crystal decacyclene derivative and utilization in polymer photovoltaic devices. Synth. Met. **157**, 290–296 (2007)
123. S. Archambeau, H. Bock, I. Seguy, P. Jolinat, P. Destruel, Organic solar cells with an ultra thin organized hole transport layer. J. Mater. Sci.: Mater. Electron. **18**, 919–923 (2007)
124. T.N. Ahipa, K.M. Anoop, R.K. Pai, Hexagonal columnar liquid crystals as a processing additive to a P3HT:PCBM photoactive layer. New J. Chem. **39**, 8439–8445 (2015)
125. D. Meng, D. Sun, C. Zhong, T. Liu, B. Fan, L. Huo, Y. Li, W. Jiang, H. Choi, T. Kim, J.Y. Kim, Y. Sun, Z. Wang, A.J. Heeger, High-performance solution-processed non-fullerene organic solar cells based on selenophene-containing perylene bisimide acceptor. J. Am. Chem. Soc. **137**, 11156–11162 (2015)

Chapter 7
Vertically-Aligned Carbon Nanotubes for Electrochemical Energy Conversion and Storage

Feng Du, Quanbin Dai, Liming Dai, Qiuhong Zhang, Thomas Reitz and Levi Elston

Abstract Vertically-aligned carbon nanotubes (VA-CNTs) have a large surface area, high electronic conductivity and electrochemical accessibility, and mechanical/chemical/electrochemical stability. These unique properties make VA-CNTs promising electrode materials for energy conversion and storage devices, including fuel cells, lithium batteries, and supercapacitors. This chapter provides an overview on recent development of VA-CNT electrodes with and without heteroatom-doping for efficient energy conversion and storage by summarizing our work on the discovery of nitrogen-doped VA-CNTs as a highly active cathode for ORR in fuel cells, vertically aligned nitrogen doped coral-like carbon fiber arrays (VA-NCCFs) as a high-performance air cathode in Li-air batteries, as well as VA-CNTs and their 3D derivatives as porous electrodes in high-performance Li-ion batteries and supercapacitors.

7.1 Introduction

The flow of energy impacts the lives of everyone around the globe, where continuously increasing world population, rapid economic expansion worldwide, and ever-increasing human reliance on energy-based appliances are driving up the world's energy demand, which is predicted to double by 2050 [1]. The most popular energy form in use today is electricity produced from fossil fuels with concomitant greenhouse gas (CO_2) emissions that are detrimental to the environment. Therefore, it is necessary to generate green and renewable energy that can be

F. Du · Q. Dai · L. Dai (✉)
Center of Advanced Science and Engineering for Carbon (Case4Carbon), Department of Macromolecular Science and Engineering, Case Western Reserve University, Cleveland, OH 44106, USA
e-mail: liming.dai@case.edu

Q. Zhang · T. Reitz · L. Elston (✉)
Air Force Research Laboratory (AFRL), WPAFB, Fairborn, OH 45433, USA
e-mail: levi.elston@us.af.mil

Q. Li (ed.), *Nanomaterials for Sustainable Energy*,
NanoScience and Technology, DOI 10.1007/978-3-319-32023-6_7

easily transported or stored. In this context, it is highly desirable to develop various energy-related devices, such as fuel cells, supercapacitors, and batteries with an emphasis on efficient energy conversion and storage.

Although the efficiency of energy conversion and storage devices depends on a variety of factors, their overall performance strongly relys on the structure and property of the electrode materials used. For instance, fuel cell are one of the most efficient and environmentally favorable energy conversion technologies. A typical fuel cell usually consists of an anode, a cathode, and an electrolyte, which converts chemical energy directly into electricity by reducing oxygen gas at the cathode and oxidizing fuel (e.g., hydrogen gas) at the anode to produce water as the only by-product. The oxygen reduction reaction (ORR) at the cathode is slow, which requires an active catalyst to speed it up. To improve the performance of fuel cells, electrocatalysts are essential for an efficient ORR reaction at the cathode. To date, platinum (Pt) nanoparticles have been known to be the best catalyst for ORR [2–5]. However, the high cost and scarcity of Pt-based catalysts have limited commercialization of fuel cell technology [6, 7]. Consequently, ORR catalysts based on non-Pt metals [8–10] and metal-free carbon materials [11–14] have attracted a great deal of interest.

Having high electrical conductivity, large surface area , and good mechanical/thermal/chemical stability, carbon nanomaterials have been made into various nanostructured electrodes not only for highly efficient fuel cells but also for high-performance lithium batteries [15, 16], supercapacitors [17–19], and even solar cells [20, 21]. Of particular interest, carbon nanotubes have been fabricated into electrodes with various porous 3-D nanostructures with a large surface area and good mechanical stability as well as excellent transport properties for both electrons and electrolytes [17, 22, 23].

7.2 VA-CNTs for Efficient Energy Conversion and Storage

Since the discovery of carbon nanotubes (CNTs) by Iijima in 1991 [24], CNTs have attracted considerable attention as novel electronic and electrochemical materials due to their unique structure and properties [25–30]. Carbon nanotubes (CNTs) can be categorized into single-walled carbon nanotubes (SWNTs) and multi-walled carbon nanotubes (MWNTs). As seen in Fig. 7.1, a SWNT can be conceptually viewed as a single layer graphene sheet that has been rolled up into a tiny tube form around an (n, m) lattice vector [31], which determines the diameter and chirality of the carbon nanotube. Depending on their diameter and the helicity of the arrangement of graphite rings along the tube length, carbon nanotubes can exhibit different properties, making them potentially useful in diverse applications ranging from optoelectronic to energy-related devices. Based on the (n, m) indices, SWNTs can be further classified into "zigzag" (n = 0), "armchair" (n = m), and "chiral"

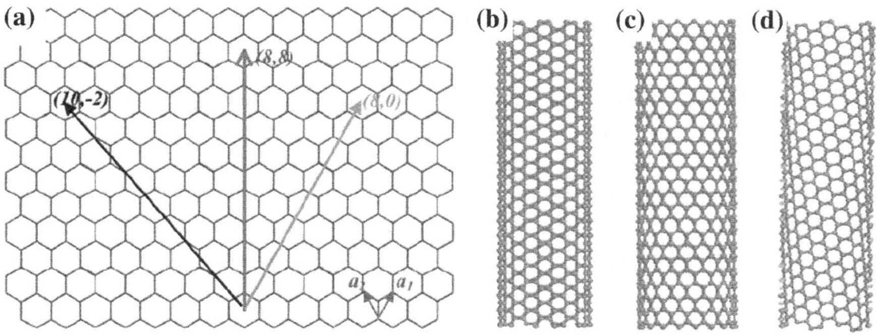

Fig. 7.1 **a** Schematic honeycomb structure of a graphene sheet. Single-wall carbon nanotubes can be formed by folding the sheet along lattice vectors. The two basis vectors a_1 and a_2 are shown. Folding of the graphene sheet along the (*8, 8*), (*8, 0*), and (*10, −2*) vectors into armchair (**b**), zigzag (**c**), and chiral (**d**) tubes, respectively. Reproduced with permission from [31]. Copyright 2002 American Chemical Society

(others) nanotubes. A carbon nanotube is "metallic" (electrically conducting) if 2n + m = 3q with q being an integer [32]. A MWNT consists of a SWNT core wrapped with concentric multilayered graphitic tubes with an interlayer distance around 3.4 Å. MWNTs usually are considered to be metallic.

Macroscopically, CNTs can be produced in either a randomly entangled non-aligned form [33] or a well-aligned (both horizontally and perpendicularly) form [34–36]. Compared with non-aligned CNTs, vertically-aligned CNTs (VA-CNTs) have multiple advantages as electrode materials for electrochemical energy conversion and storage, including well-defined inter-tube spacing and large surface area for improved charge and/or ion transport/storage, strong contact of the ends of VA-CNT arrays with common electrodes to reduce the contact resistance, and very good directional thermal conductivity along the nanotube length. These advantages, coupled with good chemical/thermal stabilities intrinsically associated with CNTs, make VA-CNTs attractive as electrode materials for high-performance energy devices. Therefore, various synthetic technologies have been developed to produce VA-CNTs for energy-related and many other applications [37]. Among them, thermal chemical vapor deposition (CVD) is the most popular method for producing VA-CNTs: examples include pyrolysis of acetylene on cobalt or iron catalyst within an alumina membrane template [38], and template-free pyrolysis of iron (II) phthalocyanine (FePc) [34], ferrocene [39, 40] or hydrocarbons [23, 37, 41]. As shown later in this chapter, VA-CNTs produced with or without functionalization have been widely used in fuel cells [13], lithium batteries [16, 42, 43], and supercapacitors [44, 45] for energy conversion and storage.

7.3 VA-CNTs for Energy Conversion

Fuel Cells. Fuel cells directly generate electricity by electrochemically reducing oxygen and oxidizing fuel into water. This energy conversion technology currently receives intensive research and development focus because of its high energy conversion efficiency, virtually no pollution, and potential large-scale applications [46]. However, fuel cells require a catalyst for oxygen reduction reaction (ORR) [3, 4]. Traditionally, platinum has been regarded as the state-of-the-art ORR catalyst. Although alkaline fuel cells (AFCs) with platinum-loaded carbon as an electrocatalyst for ORR have been developed for Space Shuttle Programs (e.g., Apollo) [5], the limited resources and high cost of platinum have also made these catalysts the primary barrier to mass market fuel cells for commercial applications [5]. Apart from its high cost, Pt-based ORR catalysts have also suffered from their susceptibility to time-dependent drift [6] and CO deactivation [7]. Thus, efforts are needed to identify an alternative material: one that is readily available, cost effective, and shows comparable or better catalytic effects than Pt for cathodic ORR in fuel cells. The search for efficient alternatives to replace/reduce Pt for electrochemical reduction of oxygen in fuel cells began in the 1960s and has grown extensively. Recent research efforts in reducing/replacing the expensive platinum-based electrode in fuel cells have led to the development of certain transition metal chalcogenide [46], carbon-nanotube-supported metal particles [9, 10], enzymatic electrocatalytic systems [12], and even conducting polymer coated membranes [11]. Although tremendous progress has been made, most non-noble catalysts still remain either too expensive or too low in efficiency when compared to Pt itself [47, 48].

Along with the more than a half century of research and development in nonprecious metal ORR catalysts, we have recently discovered that N-doped VA-CNTs (VA-NCNTs) show high electrocatalytic activities for ORR [13]. In particular, we prepared multiwalled VA-CNTs by pyrolysis of iron (II) phthalocyanine (FePc) [49] in the presence and absence of additional NH_3 vapor [50]. After the complete removal of residual Fe, if any, by acid washing and electrochemical oxidation, the resultant N-doped VA-CNT (Fig. 7.2a, b) demonstrated a much higher electrocatalytic activity, lower overpotential, better long-term operational stability, and no methanol-crossover or CO-poisoning effects [13, 51] in alkaline electrolytes [3, 4], with respect to the commercially available platinum electrode (e.g., 20 wt% platinum on Vulcan XC-72R; E-TEK). Quantum mechanics calculations using B3LYP hybrid DFT (Gaussian 03) indicate that carbon atoms adjacent to nitrogen dopant in the VA-NCNT structure possess a high positive charge density to balance the strong electron affinity of the nitrogen atom (Fig. 7.2c). The nitrogen-induced charge delocalization could change not only the electrochemical potential for ORR but also the chemisorption mode of oxygen molecules into side-on adsorption (Yeager model) at the N-doped CNT surface (Fig. 7.2d) [13, 52]. Compared to the usual end-on adsorption (Pauling model) at the conventional CNT surface, the parallel diatomic adsorption on the NCNT could effectively weaken the O-O bonding to facilitate ORR at the VA-NCNT electrodes. Thus, the VA-NCNTs exhibited

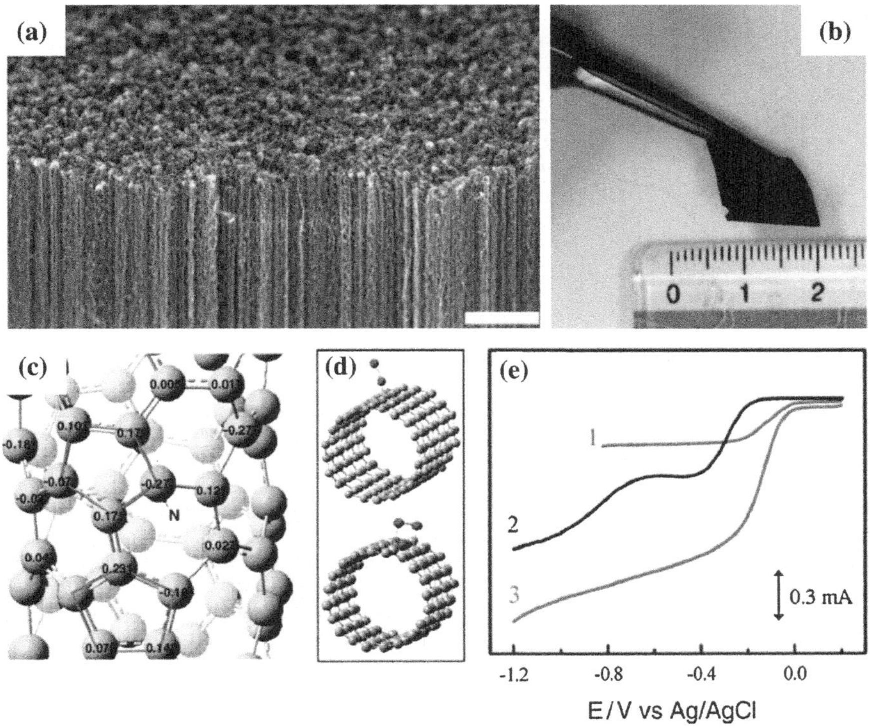

Fig. 7.2 **a** SEM image of the *as-synthesized* VA–NCNTs prepared by pyrolysis of iron phthalocyannine (FePc) on a quartz substrate (*Scale bars*, 2 μm). **b** Digital photograph of the VA-NCNT array after having been transferred onto a polystyrene-nonaligned CNT conductive nanocomposite film. **c** Calculated charge density distribution for the NCNTs. **d** Schematic representations of possible adsorption modes of an oxygen molecule at the nitrogen-free carbon nanotubes (CCNTs) (*top*) and NCNTs (*bottom*). **e** The steady-state voltammograms for oxygen reduction in air saturated 0.1 M KOH at the Pt-C (*curve 1*), vertically-aligned commercially available carbon nanotubes (Nitrogen-free) (VA-CCNT) (*curve 2*), and VA-NCNT (*curve 3*) electrodes. Reproduced with permission from [13]. Copyright 2009 American Association for the Advancement of Science

superior ORR performance to a Pt/C electrode (Fig. 7.2e). Later, these VA-NCNTs were found to be active and stable even in an acidic medium [14, 53, 54].

By extension, we have successfully prepared vertically-aligned carbon nanotubes containing both nitrogen and boron heteroatoms (VA-BCN). Due to a synergetic effect arising from co-doping of carbon nanotubes with boron and nitrogen, the resultant VA-BCN nanotubes have shown a higher electrocatalytic activity for ORR in an alkaline medium than VA-CNTs doped with either boron or nitrogen only (Fig. 7.3a) [55]. Good tolerance to methanol and carbon monoxide and excellent durability were also observed for the VA-BCN nanotube electrode (Fig. 7.3b–d).

A similar synergetic effect was also found from the co-doping of VA-CNTs with both phosphorus (P) and nitrogen (N) [56]. The P,N-codoped VA-CNTs were synthesized by injection-assisted chemical vapor deposition using a mixture of

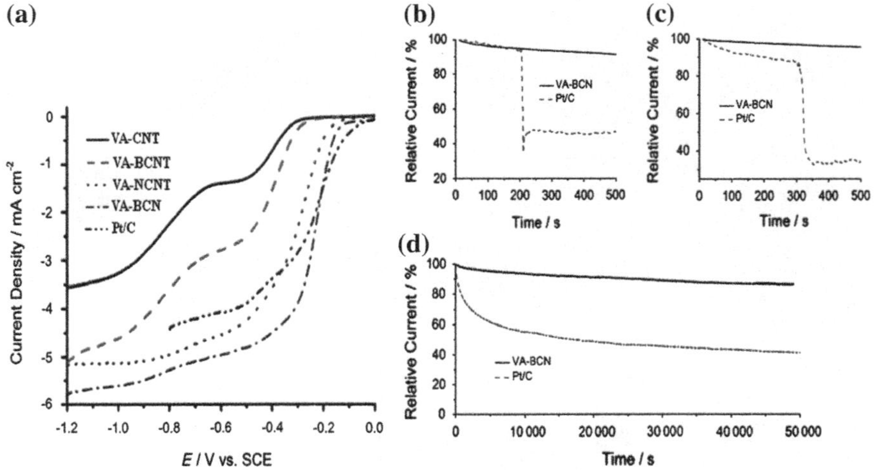

Fig. 7.3 **a** Linear-sweep voltammetry curves of various electrodes in oxygen-saturated 0.1 M KOH electrolyte at a scan rate of 10 mVs^{-1} and a rotation rate of 1000 rpm. Chronoamperometric response for ORR at VA-BCN and Pt/C electrodes: **b** on addition of 3M methanol after about 200 s and **c** on introduction CO after about 300 s at -0.3 V. **d** Durability evaluation of Pt/C and VA-BCN nanotube electrodes for 50000 s at -0.2 V and a rotation rate of 1000 rpm. Reproduced with permission from [55]. Copyright 2011 WILEY-VCH Verlag GmbH & Co

ferrocene, pyridine, and triphenylphosphine as precursors. A high ORR electrocatalytic activity, superb long-term durability, and good tolerance to methanol and CO were observed which significantly outperformed the P- or N-doped VA-CNTs.

The above results clearly indicate that heteroatom-doped VA-CNTs could act as very efficient and active metal-free electrocatalysts for ORR in fuel cells. The discovery of metal-free carbon-based ORR catalysts has a large impact on not only the commercialization of fuel cell technology but also many other catalytic processes beyond fuel cells, including the oxygen evolution reaction in rechargeable metal-air batteries discussed below.

7.4 VA-CNTs for Energy Storage

Batteries. Along with the above-mentioned progress in the development of heteroatom-doped VA-CNTs as effective metal-free catalysts for ORR in fuel cells, VA-CNTs with and without heteroatom-doping have also been shown to exhibit similar or even greater capacities than that of the graphite in lithium-ion batteries [57, 58]. Compared with the conventional graphite electrode, it was believed that the VA-CNT electrode could allow for significantly improved electron/ion transports and rate capabilities [59–62]. However, relatively low capacities were observed for a lithium-ion battery (LIB) using a VA-CNT electrode at low discharge rates due to poor contact between the CNT array and the current collector [63–65]. In order to solve this contact problem, Welna et al. [42] investigated the electrochemical property of a FePc-generated VA-CNT cathode in a LIB (Fig. 7.4).

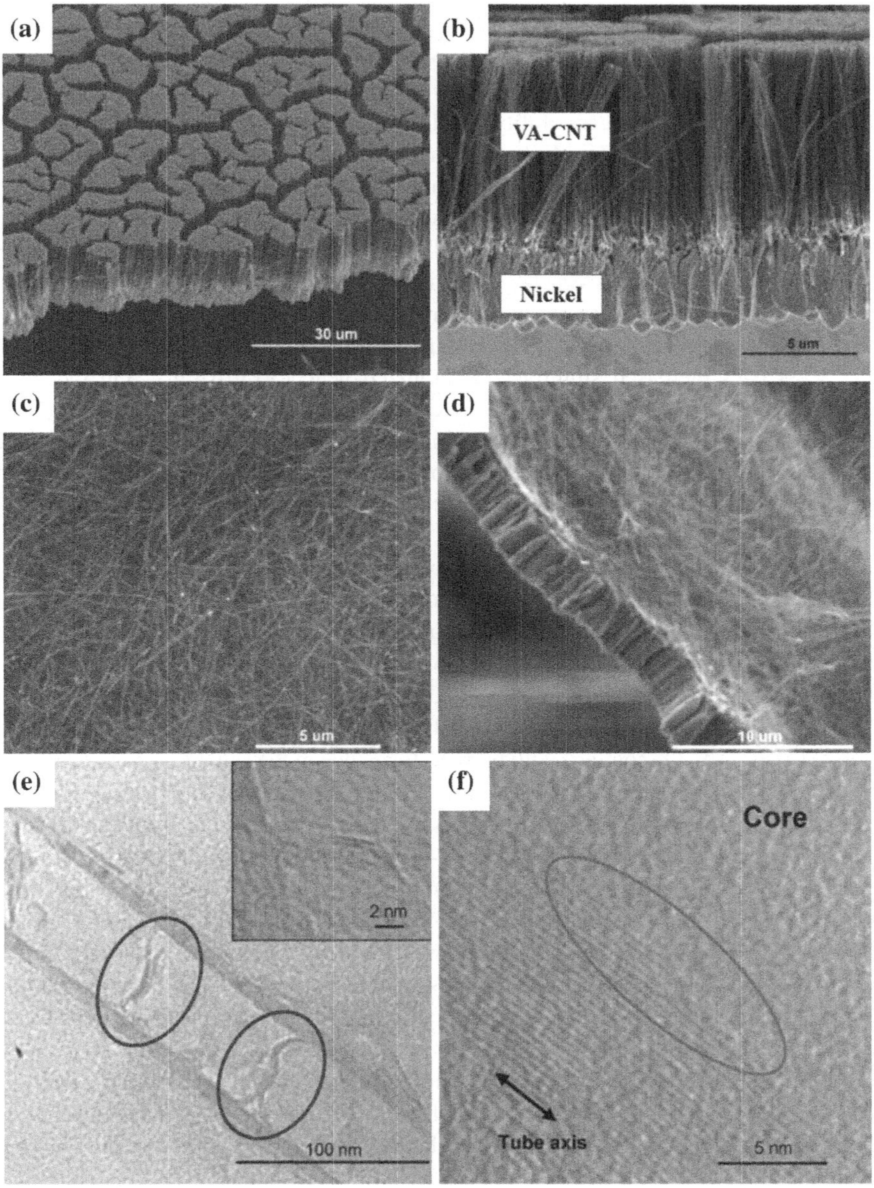

Fig. 7.4 SEM images of a VA-CNT (**a**, **b**) and an nonaligned CNT (**c**, **d**) electrode. TEM images of a VA-CNT electrode, showing the loose graphene layers (**e**) and the graphene nanotube edges (**f**) within the internal nanotube structure. The distance between the graphene nanotube layers is 0.335 nm. Reproduced with permission from [42], Copyright 2011 Elsevier

As shown in Fig. 7.4e, f, the internal core of the nanotubes contains some irregular structures, which can significantly increase the effective surface area of the nanotubes and provide more defect sites at which lithium-ions could be adsorbed [66–68]. The intimate contact between each of individual VA-CNTs and the nickel film current collector provides a low contact resistance for effective and efficient connectivity throughout the electrode.

It was shown that the VA-CNT electrodes can lead to higher lithium-ion storage capacities and significantly higher rate capabilities than those of the graphite electrode, and that the accessibility of the electrolyte into the electrodes was significantly increased by the alignment structure, as opposed to their non-aligned counterparts, as shown in Fig. 7.5. Figure 7.5a shows the specific capacity as a function of cycle for VA-CNT and nonaligned CNT electrodes while the voltage profile of cycles 2–34 for both VA-CNT and nonaligned CNT electrodes were shown in Fig. 7.5b, c. As can be seen in Fig. 7.5b, after the irreversible capacity loss of the first cycle (Q_{irr}), the reversible capacity (Q_{rev}) of the VA-CNTs decreased from 980 mAh g^{-1} to a minimum near the tenth cycle. Thereafter, Q_{rev} of the VA-CNTs increased slightly and stabilized near 750 mAh g^{-1}. On the other hand, Q_{rev} for the nonaligned CNTs continuously decreased from 158 to 58 mAh g^{-1} by cycling 34 times (Fig. 7.5b). Compared to the nonaligned CNTs, the alignment of the nanotubes in the direction of ion diffusion may allow lithium ions to more readily access the open nanotube ends and significantly increased the Q_{rev} of the VA-CNT electrodes.

VA-CNTs have also been used in Li-ion batteries not only as electrode materials [42, 69], but also as conductive substrates for the deposition of electroactive materials to develop high-capacity and high-rate electrode materials. For example, a variety of materials (e.g., conjugated polymers, [70] manganese oxide, [71] and vanadium oxide [72]) have been deposited onto VA-CNTs for cathode applications. Figure 7.6A illustrates the use of VA-CNTs as both cathode (with V_2O_5 coating) and anode (without V_2O_5 coating) with high performance in the same Li-ion battery [16].

It was found that the VA-CNT anode and the V_2O_5-VA-CNT cathode showed high capacity (600 and 368 mAh g^{-1}, respectively) high rate capability in an ionic liquid electrolyte (i.e., N–Ethyl–N, N-dimethyl-2-methoxyethylammonium bis (trifluormethylsulfonyl)imide, [EDMMEA]-[TFSI]), and the resultant battery cell showed an excellent energy density (297 Wh kg^{-1}, maximum 847 Wh kg^{-1}) and power density (12 kW kg^{-1}, maximum 35 kW kg^{-1}) (estimated from active material) (Fig. 7.6B). Furthermore, the use of environmentally benign ionic liquid electrolytes ensured a high safety margin and long lifetime for these batteries based on the VA–CNT electrodess.

More recently, we have developed vertically aligned nitrogen doped coral-like carbon fiber (VA-NCCF) arrays by a modified CVD method [43]. When used as an oxygen electrode in a nonaqueous Li–O_2 battery, this unique coral-like carbon fiber structure (Fig. 7.7a–c) provided a large free space for efficient Li_2O_2 deposition and enhanced electron/electrolyte/reactant transport. Meanwhile, the N-doping-induced catalytic activity reduced the charging over-potential for minimizing the electrolyte decomposition and facilitated Li_2O_2 deposition along the VA-NCCF surfaces.

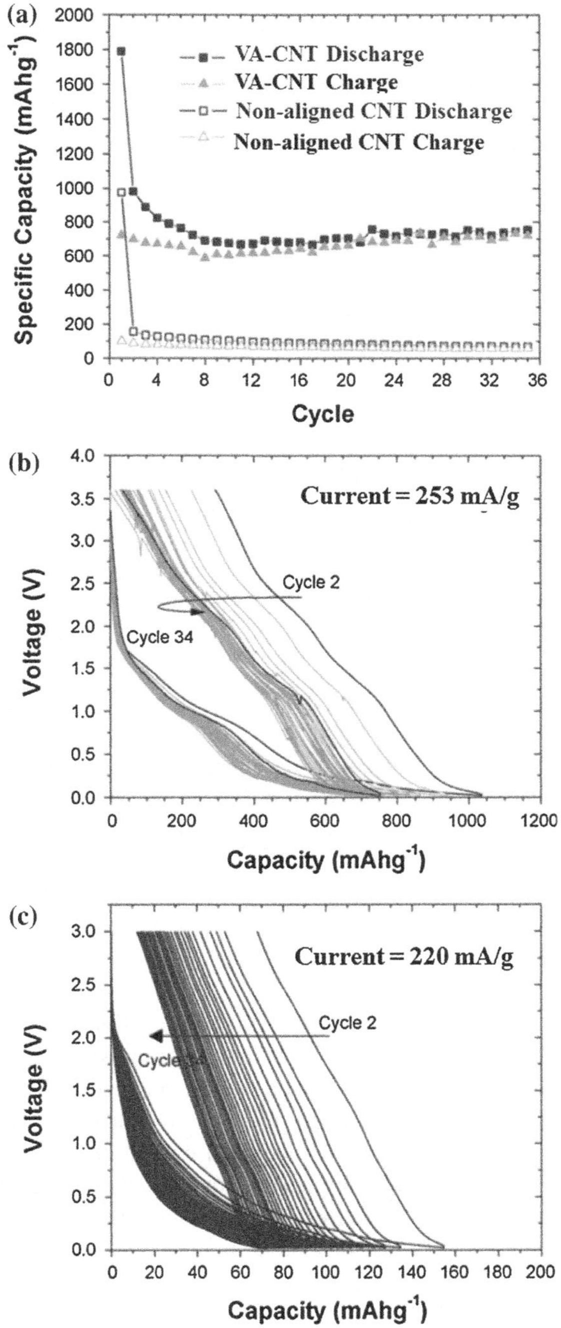

Fig. 7.5 Specific capacity as a function of cycle for VA-CNT and nonaligned CNT electrodes at similar discharge rates (**a**). Voltage profile of cycles 2–34 for VA-CNT (**b**) and nonaligned CNT (**c**) electrodes. Reproduced with permission from [42], Copyright 2011 Elsevier

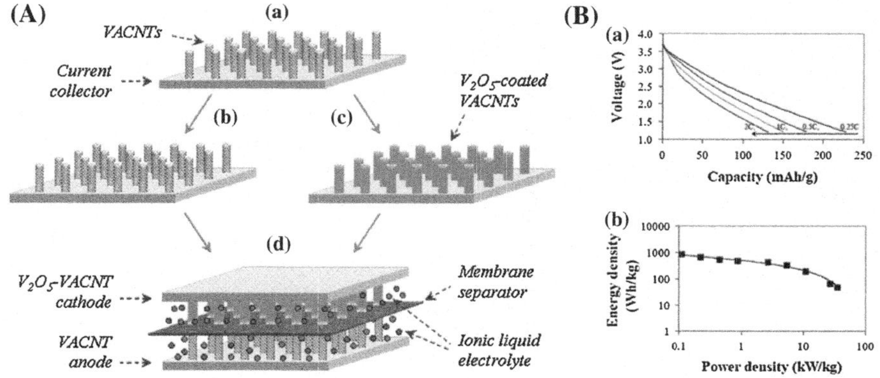

Fig. 7.6 **A** Schematic representation of a Li-ion battery by using VA-CNT on both electrodes: **a** Growth of VA-CNTs on a nickel foil (as current collector). **b** Direct use of VA-CNTs as the anode. **c** Electrochemical deposition of V_2O_5 on VA-CNTs to prepare the V_2O_5-VA-CNT cathode. **d** Assembly of the VA-CNT anode, the V_2O_5 -VA-CNT cathode, a membrane separator, and an ionic liquid electrolyte to fabricate the battery. **B** **a** Discharge curves of a VA-CNT/V_2O_5-VA-CNT battery recorded at the rates increasing from 0.25C to 2C. The capacity is defined by the total active-material-mass of the anode and the cathode. Cut-off voltage: 1.2–3.7 V. **b** Ragone plot of the VA-CNT/V_2O_5-VA-CNT battery. Reproduced with permission from [16], Copyright 2012 PCCP Owner Societies

Consequently, these rationally designed VA-NCCF electrode exhibited an energy efficiency as high as 90 % in a full discharge/charge cycle with a low over-potential of only 0.3 V between the charge/discharge plateaus (Fig. 7.7d). Due to the low overpotential introduced by N-doping, the electrolyte decomposition has been minimized, and the full battery ran for more than 150 cycles with good reversibility even at high specific capacity (up to 1000 mAh g^{-1}) per cycle (Fig. 7.7e). The above studies clearly demonstrated that the use of rationally designed VA-CNTs and VA-NCCFs with well-defined hierarchical structures and/or heteroatom-doping could dramatically improve the energy storage performance for Li-ion and Li-air batteries.

Supercapacitors. VA-CNTs with rationally designed architectures of a high charge transport capability, high mesoporosity, and high electrolyte accessibility are also one of the most attractive electrode materials for developing high performance supercapacitors [73, 74]. Compared with only tens of F g^{-1} capacitance for the commercially available activated carbon electrode, VA-CNT arrays (Fig. 7.8a, b) with the vertically-aligned structure and the well-defined tube-spacing can provide a more electrolyte-accessible surface and improved charge storage/delivery properties since each of the constituent aligned tubes can be connected directly onto a common electrode to allow them to effectively participate in the charging/discharging process. Using vertically aligned and plasma etched CNTs with unique properties, Lu et al. [44] have developed CNT electrodes to show high charge storage/delivery capability in ionic liquid electrolytes. The charge storage can be further improved by opening top end-caps of the VA-CNTs through plasma etching [34, 75]

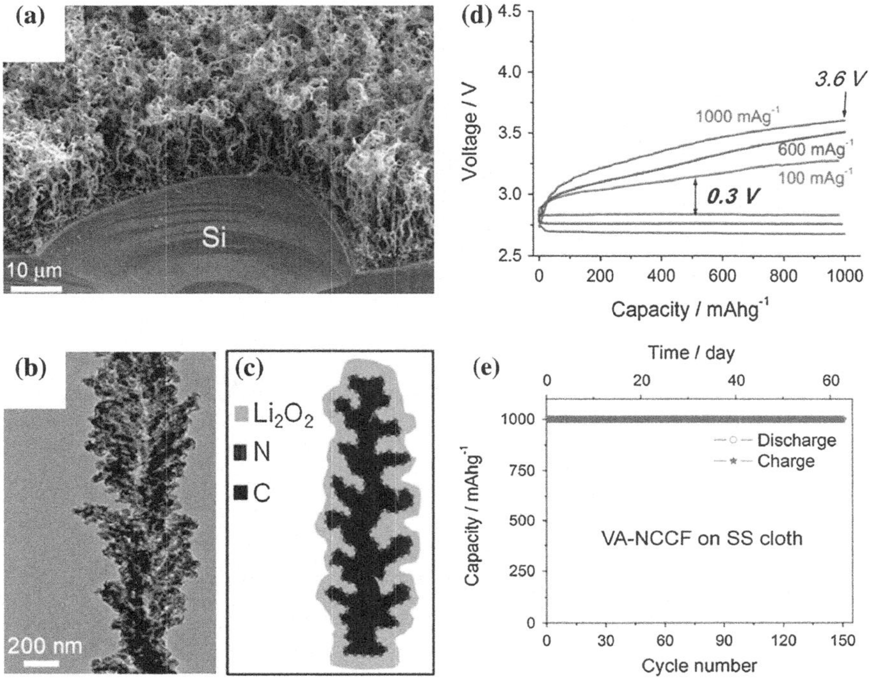

Fig. 7.7 **a** SEM image of a VA–NCCF array grown on a Si wafer by CVD. **b** TEM image of an individual VA-NCCF. **c** The sketch of Li_2O_2 grown on a coral-like carbon nanofiber. **d** Rate performance of the VA-NCCF electrode under current densities of 100, 600 and 1000 mA g^{-1}. **e** Cycling performance of the VA-NCCF array on the stainless steel (SS) cloth substrate. Reproduced with permission from [43]. Copyright 2014 American Chemical Society

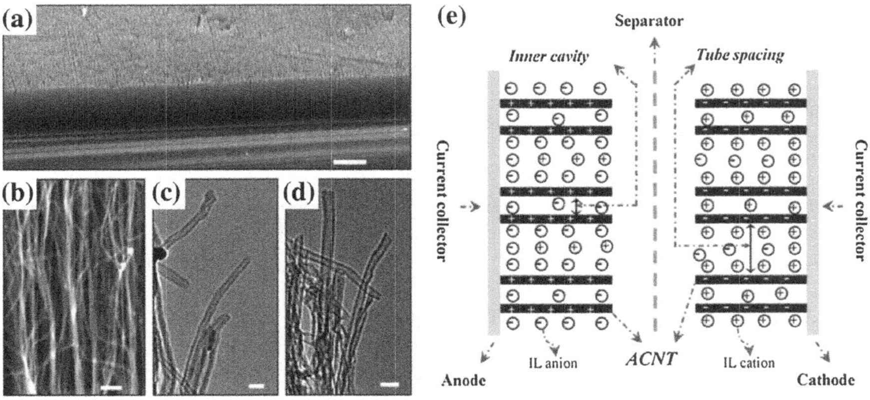

Fig. 7.8 **a** SEM image of a plasma etched VA-CNT electrode (scale bar: 100 μm). **b** Higher magnification view of the electrode (scale bar: 100 nm). TEM images of the CNTs before (**c**) and after (**d**) plasma etching (scale bar: 20 nm). **e** Schematic of an electrochemical capacitor based on plasma-etched VA-CNT electrodes and ionic liquid electrolyte. Reprinted with permission from [44]. Copyright 2009 Elsevier

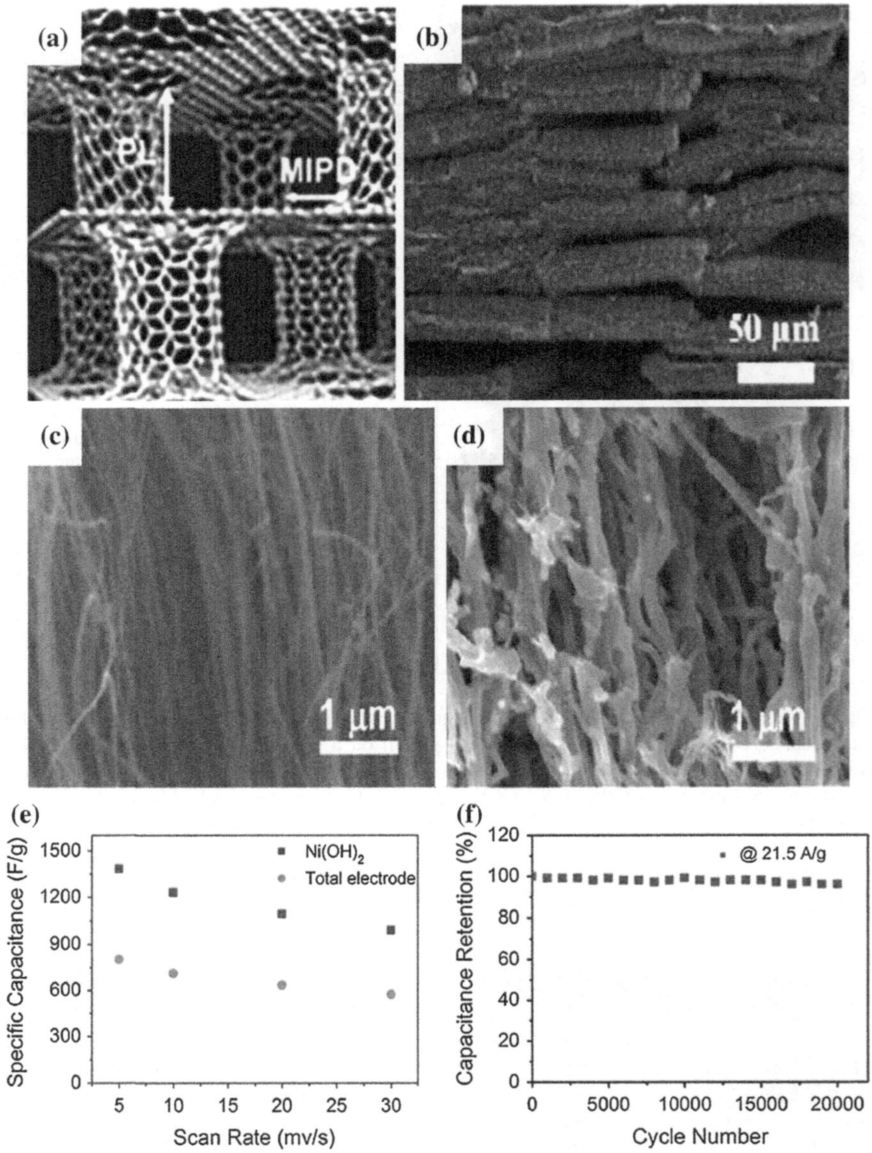

Fig. 7.9 **a** Schematic picture of a 3D pillared VA-CNT-graphene nanostructure. **b** A typical SEM image of 3D pillared VA-CNT-graphene architecture. SEM images of **c** before and **d** after electrochemical deposition of $Ni(OH)_2$ on the 3D VA-CNT-graphene. **e** Specific capacitance (calculated from both the $Ni(OH)_2$ mass and the total electrode mass. **f** Average specific capacitance versus cycle number for the $Ni(OH)_2$-coated VA-CNT-graphene electrode at a galvanostatic charge and discharge current density of 21.5 A/g. Reproduced with permission from [45]. Copyright 2011 American Chemical Society

(Fig. 7.8c, d) to allow the electrolyte access to the otherwise inaccessible inner cavity of the VA-CNTs (Fig. 7.8e). Combining the enhanced double-layer capacitance (due to the vertical alignment and plasma-etched end-cap-opening and defect formation) and the additional pseudocapacitance (due to the plasma-etched oxygenated functionalization), the VA-CNT electrode showed a remarkably high capacitance of 440 Fg^{-1} in an ionic liquid electrolyte [EMIM][Tf_2N] [44].

To further improve the performance of VA-CNT supercapacitors, metal oxide (e.g., MnO_2 [76], TiO_2 [77] and RuO_2 [78]) can be deposited onto the surface of CNTs to introduce pseudocapacitance. In particular, sputter coating technology has been used for depositing metal (e.g., Ru) [79] and metal oxide (e.g., RuO_2, MnO_2) [80] on aligned carbon nanotube array with and without N-doping, respectively, to achieve capacities as high as 1380 F/g [80]. More interestingly, we have developed a rational strategy for creating the 3D pillared vertically-aligned carbon nanotube (VA-CNT)-graphene architectures by intercalated growth of VA-CNTs into thermally-expanded highly ordered pyrolytic graphite (HOPG) (Fig. 7.9a, b) [45]. In conjunction with electrodeposition of nickel hydroxide to introduce the pseudo-capacitance, these 3D pillared VA-CNT-graphene architectures with a controllable nanotube length were demonstrated to show a high specific capacitance (1384 F/g at a scan rate of 5 mV/s and 970 F/g at 30 mV/s, Fig. 7.9e), remarkable rate capability, and long-term stability ($\sim$96 % retention rate after 20000 consecutive cycles). Additionally, the 3D structure significantly outperformed many electrode materials currently used in the state-of-the-art supercapacitors, including RuO_2 [78], MnO_2 [76], NiO [81], Mn_3O_4 [82], and $Ni(OH)_2$ as well as metal oxide or conducting polymer composites with CNTs or graphene [83–85].

7.5 Conclusions

As apparent from the above discussions, the rising global energy demand and the environmental impact of traditional energy resources pose an ever-increasing urgency to develop clean and renewable energy conversion and storage systems, including fuel cells, supercapacitors, and batteries. Generally speaking, the performance of energy conversion and storage devices depends strongly on the structure and property of their electrode materials. With the rapid development in nanoscience and nanotechnology, graphitic carbon nanomaterials (e.g., CNTs) have been playing an increasingly important role in the development of high-performance electrode materials for efficient energy conversion and storage. Considerable efforts have recently been made to utilize graphitic carbon nanomaterials for energy applications, and tremendous progress has been achieved in developing high-performance energy conversion and storage devices based on graphitic carbon electrodes. Of particular interest, VA-CNTs have been demonstrated to show additional advantages for efficient energy conversion and storage devices, including fuel cells, batteries, and supercapacitors. In this chapter, we have provided an overview on recent development of VA-CNT electrodes with well-defined architectures and doping structures for efficient energy conversion and storage by summarizing our work on the discovery of nitrogen-doped VA-CNTs as a highly active cathode for ORR in fuel cells, N-doped VA-NCCFs as a high-performance air cathode in Li-air batteries, as well as VA-CNTs and their 3D derivatives with or without N-doping as porous electrodes in high-performance Li-ion batteries and supercapacitors. The combination of the vertical alignment and rationally-designed architecture with the heteroatom-doping has created synergetic effects, which provides the basis for the development of high-performance energy devices. Vast opportunities remain for developing VA-CNT electrode materials with novel hierarchical structures for highly efficient energy conversion and storage, which is an area where further research and development is needed.

Acknowledgments The authors thank our colleagues for their contributions to the work cited. We are also grateful for financial support from AFRL/DAGSI, AFOSR, and NSF.

References

1. D.B. Botkin, *Powering the Future: A Scientist's Guide to Energy Independence* (FT Press, 2010)
2. B.C.H. Steele, A. Heinzel, Materials for fuel-cell technologies. Nature **414**, 345–352 (2001)
3. S. Basu (ed.), *Recent Trends in Fuel Cell Science and Technology* (Springer, 2007)
4. A.J. Appleby, Electrocatalysis of aqueous dioxygen reduction. J. Electroanal. Chem. **357**, 117–179 (1993)
5. K. Kordesch, J. Gsellmann, M. Cifrain, S. Voss, V. Hacker, R.R. Aronson, C. Fabjan, T. Hejze, J. Daniel-Ivad, Intermittent use of a low-cost alkaline fuel cell-hybrid system for electric vehicles. J. Power Sources **80**, 190–197 (1999)
6. X. Yu, S. Ye, Recent advances in activity and durability enhancement of Pt/C catalytic cathode in PEMFC. J. Power Sources **172**, 145–154 (2007)
7. M. Winter, R.J. Brodd, What are batteries, fuel cells, and supercapacitors? Chem. Rev. **104**, 4245–4269 (2004)
8. K. Gong, P. Yu, L. Su, S. Xiong, L. Mao, Polymer-assisted synthesis of manganese dioxide/carbon nanotube nanocomposite with excellent electrocatalytic activity toward reduction of oxygen. J. Phys. Chem. C **111**, 1882–1887 (2007)
9. G. Che, B.B. Lakshmi, E.R. Fisher, C.R. Martin, Carbon nanotube membranes for electrochemical energy storage and production. Nature **393**, 346–349 (1998)
10. J. Yang, D-J. Liu, N.N. Kariuki, L.X. Chen, Aligned carbon nanotubes with built-in FeN4 active sites for electrocatalytic reduction of oxygen. Chem. Comm. 329–331 (2008)
11. B. Winther-Jensen, O. Winther-Jensen, M. Forsyth, D.R. MacFarlane, High rates of oxygen reduction over a vapor phase-polymerized PEDOT electrode. Science **321**, 671–674 (2008)
12. J.P. Collman, N.K. Devaraj, R.A. Decréau, Y. Yang, Y.-L. Yan, W. Ebina, T.A. Eberspacher, C.E.D. Chidsey, A cytochrome C oxidase model catalyzes oxygen to water reduction under rate-limiting electron flux. Science **315**, 1565–1568 (2007)
13. K. Gong, F. Du, Z. Xia, M. Durstock, L. Dai, Nitrogen-doped carbon nanotube arrays with high electrocatalytic activity for oxygen reduction. Science **323**, 760–764 (2009)
14. L. Dai, Y. Xue, L. Qu, H.-J. Choi, J.-B. Baek, Metal-free catalysts for oxygen reduction reaction. Chem. Rev. **115**, 4823–4892 (2015)
15. M. Endo, C. Kim, K. Nishimura, T. Fujino, K. Miyashita, Recent development of carbon materials for Li ion batteries. Carbon **38**, 183–197 (2000)

16. W. Lu, A. Goering, L. Qu, L. Dai, Lithium-ion batteries based on vertically-aligned carbon nanotube electrodes and ionic liquid electrolytes. Phys. Chem. Chem. Phys. **14**, 12099–12104 (2012)
17. T. Chen, L. Dai, Carbon nanomaterials for high-performance supercapacitors. Mater. Today **16**, 272–280 (2013)
18. E. Frackowiak, K. Metenier, V. Bertagna, F. Beguin, Supercapacitor electrodes from multiwalled carbon nanotubes. Appl. Phys. Lett. **77**, 2421–2425 (2000)
19. W-C. Fang, Synthesis and electrochemical characterization of vanadium oxide/carbon nanotube composites for supercapacitors. J. Phys. Chem. C **112**, 11552–11555 (2008)
20. H. Ago, K. Petritsch, M.S.P. Shaffer, A.H. Windle, R.H. Friend, Composites of carbon nanotubes and conjugated polymers for photovoltaic devices. Adv. Mater. **11**, 1281–1285 (1999)
21. J. van de Lagemaat, T.M. Barnes, G. Rumbles, S.E. Shaheen, T.J. Coutts, C. Weeks, I. Levitsky, J. Peltola, P. Glatkowski, Organic solar cells with carbon nanotubes replacing In2O3:Sn as the transparent electrode. Appl. Phys. Lett. **88**, 233503–1–3 (2006)
22. L. Dai, D.W. Chang, J.-B. Baek, W. Lu, Carbon nanomaterials for advanced energy conversion and storage. Small **8**, 1130–1166 (2012)
23. L. Dai, *Carbon Nanotechnology: Recent Developments in Chemistry, Physics, Materials Science and Device Applications* (Elsevier Science, Boston, 2006)
24. S. Iijima, Helical microtubules of graphite carbon. Nature **354**, 56–58 (1991)
25. M. Cinke, J. Li, B. Chen, A. Cassell, L. Delzeit, J. Han, M. Meyyappan, Pore structure of raw and purified HiPco single-walled carbon nanotubes. Chem. Phys. Lett. **365**, 69–74 (2002)
26. C. Niu, E.K. Sichel, R. Hoch, D. Moy, H. Tennent, High power electrochemical capacitors based on carbon nanotube electrodes. Appl. Phys. Lett. **70**, 1480–1482 (1997)
27. M.-F. Yu, O. Lourie, M.J. Dyer, K. Moloni, T.F. Kelly, R.S. Ruoff, Strength and breaking mechanism of multiwalled carbon nanotubes under tensile load. Science **287**, 637–640 (2000)
28. B.G. Demczyk, Y.M. Wang, J. Cumings, M. Hetman, W. Han, A. Zettl, R.O. Ritchie, Direct mechanical measurement of the tensile strength and elastic modulus of multiwalled carbon nanotubes. Mater. Sci. Eng. A Struct. **334**, 173–178 (2002)
29. E.T. Thostenson, C. Li, T.-W. Chou, Nanocomposites in context. Compos. Sci. Technol. **65**, 491–516 (2005)
30. E. Pop, D. Mann, Q. Wang, K. Goodson, H. Dai, Thermal conductance of an individual single-wall carbon nanotube above room temperature. Nano Lett. **6**, 96–100 (2006)
31. H. Dai, Carbon nanotubes: synthesis, integration, and properties. Acc. Chem. Res. **35**, 1035–1044 (2002)
32. R. Saito, M. Fujita, G. Dresselhaus, M.S. Dresselhaus, Electronic structure of chiral graphene tubules. Appl. Phys. Lett. **60**, 2204–2206 (1992)
33. X. Lv, F. Du, Y. Ma, Q. Wu, Y. Chen, Synthesis of high quality single-walled carbon nanotubes at large scale by electric arc using metal compounds. Carbon **43**, 2013–2032 (2005)
34. S. Huang, L. Dai, A. Mau, Patterned growth and contact transfer of well-aligned carbon nanotube films. J. Phys. Chem. B **103**, 4223–4227 (1999)
35. Y. Ma, B. Wang, Y. Wu, Y. Huang, Y. Chen, The production of horizontally aligned single-walled carbon nanotubes. Carbon **49**, 4098–4110 (2011)
36. D. Yuan, L. Ding, H. Chu, Y. Feng, T.P. McNicholas, J. Liu, Horizontally aligned single-walled carbon nanotube on quartz from a large variety of metal catalysts. Nano Lett. **8**, 2576–2579 (2008)
37. H. Chen, A. Roy, J.-B. Baek, L. Zhu, J. Qu, L. Dai, Controlled growth and modification of vertically-aligned carbon nanotubes for multifunctional applications. Mater. Sci. Eng. Rep. **70**, 63–91 (2010)
38. J. Li, C. Papadopoulos, J. Xu, Highly-ordered carbon nanotube arrays for electronics applications. Appl. Phys. Lett. **75**, 367–369 (1999)
39. C. Rao, R. Sen, B. Satishkumar, A. Govindaraj, Large aligned-nanotube bundles from ferrocene pyrolysis. Chem. Comm. 1525–1526 (1998)

40. B.Q. Wei, R. Vajtai, Y. Jung, J. Ward, R. Zhang, G. Ramanath, P.M. Ajayan, Microfabrication technology: organized assembly of carbon nanotubes. Nature **416**, 495–496 (2002)
41. P.J.F. Harris, *Carbon Nanotubes and Related Structures—New Materials for the Twenty-First Century* (Cambridge University Express, Cambridge, 2001)
42. D.T. Welna, L. Qu, B.E. Taylor, L. Dai, M.F. Durstock, Vertically aligned carbon nanotube electrodes for lithium-ion batteries. J. Power Sources **196**, 1455–1460 (2011)
43. J. Shui, F. Du, C. Xue, Q. Li, L. Dai, Vertically aligned N-doped coral-like carbon fiber arrays as efficient air electrodes for high-performance nonaqueous Li–O_2 batteries. ACS Nano **8**, 3015–3022 (2014)
44. W. Lu, L. Qu, K. Henry, L. Dai, High performance electrochemical capacitors from aligned carbon nanotube electrodes and ionic liquid electrolytes. J. Power Sources **189**, 1270–1277 (2009)
45. F. Du, D. Yu, L. Dai, S. Ganguli, V. Varshney, A.K. Roy, Preparation of tunable 3D pillared carbon nanotube-graphene networks for high-performance capacitance. Chem. Mater. **23**, 4810–4816 (2011)
46. K. Gong, P. Yu, L. Su, S. Xiong, L. Mao, Polymer-assisted synthesis of manganese dioxide/carbon nanotube nanocomposite with excellent electrocatalytic activity toward reduction of oxygen. J. Phys. Chem. C **111**, 1882–1887 (2007)
47. Y. Bing, H. Liu, L. Zhang, D. Ghosh, J. Zhang, Nanostructured Pt-alloy electrocatalysts for PEM fuel cell oxygen reduction reaction. Chem. Soc. Rev. **39**, 2184–2202 (2010)
48. D. Zhao, G. Yin, J. Wei, Non-platinum cathode electrocatalysts in polymer electrolyte membrane fuel cells. Prog. Chem. **21**, 2753–2759 (2009)
49. L. Dai, A. Patil, X. Gong, Z. Guo, L. Liu, Y. Li, D. Zhu, Aligned nanotubes. ChemPhysChem **4**, 1150–1169 (2003)
50. L.S. Panchakarla, A. Govindaraj, C.N.R. Rao, Nitrogen- and boron-doped double-walled carbon nanotubes. ACS Nano **1**, 494–500 (2007)
51. S. Maldonado, K.J. Stevenson, Influence of nitrogen doping on oxygen reduction electrocatalysis at carbon nanofiber electrodes. J. Phys. Chem. B **109**, 4707–4716 (2005)
52. Z. Shi, J. Zhang, Z.-S. Liu, H. Wang, D.P. Wilkinson, Current status of Ab Initio quantum chemistry study for oxygen electroreduction on fuel cell catalysts. Electrochim. Acta **51**, 1905–1916 (2006)
53. W. Xiong, F. Du, Y. Liu, A. Perez, M. Supp, T.S. Ramakrishnan, L. Dai, L. Jiang, 3-D carbon nanotube structures used as high performance catalyst for oxygen reduction reaction. J. Am. Chem. Soc. **132**, 15839–15841 (2010)
54. J. Shui, M. Wang, F. Du, L. Dai, N-doped carbon nanomaterials are durable catalysts for oxygen reduction reaction in acidic fuel cells. Sci. Adv. **1**, e1400129 (2015)
55. S. Wang, E. Iyyamperumal, A. Roy, Y. Xue, D. Yu, L. Dai, Vertically aligned BCN nanotubes as efficient metal-free electrocatalyst for the oxygen reduction reaction: a synergetic effect by co-doping with boron and nitrogen. Angew. Chem. Int. Ed. **50**, 11756–11760 (2011)
56. D. Yu, Y. Xue, L. Dai, Vertically aligned carbon nanotube arrays co-doped with phosphorus and nitrogen as efficient metal-free electrocatalysts for oxygen reduction. J. Phys. Chem. Lett. **3**, 2863–2870 (2012)
57. E. Frackowiak, F. Béguin, Electrochemical storage of energy in carbon nanotubes and nanostructured carbon. Carbon **40**, 1775–1787 (2002)
58. P. Poizot, S. Laruelle, S. Grugeon, L. Dupont, J.M. Tarascon, Nano-sized transition-metal oxides as negative electrode materials for lithium-ion batteries. Nature **407**, 496–499 (2000)
59. D.N. Futaba, K. Hata, T. Yamada, T. Hiraoka, Y. hayamizu, Y. Kakudate, O. Tanaike, H. Hatori, M. Yumura, S. Iijima, Shape-engineerable and highly densely packed single-walled carbon nanotubes and their application as super-capacitor electrodes. Nat. Mater. **5**, 987–994 (2006)
60. Y. Honda, T. Haramoto, M. Takeshige, H. Shiozaki, T. Kitamura, M. Ishikawa, Aligned MWCNT sheet electrodes prepared by transfer methodology providing high-power capacitor performance. Electrochem. Solid-State Lett. **10**, A106–A110 (2007)

61. H. Zhang, G. Cao, Y. Yang, Z. Gu, Comparison between electrochemical properties of aligned carbon nanotube array and entangled carbon nanotube electrodes. Electrochem. Soc. **155**, K19–K22 (2008)
62. J. Zhao, Q.Y. Gao, C. Gu, Y. Yang, Preparation of multi-walled carbon nanotube array electrodes and its electrochemical intercalation behavior of Li ions. Chem. Phys. Lett. **358**, 77–82 (2002)
63. J. Chen, Y. Liu, A.I. Minett, C. Lynam, J. Wang, G. Wallace, Flexible, aligned carbon nanotube/conducting polymer electrodes for a lithium-ion battery. Chem. Mater. **19**, 3595–3597 (2007)
64. V.L. Pushparaj, M.M. Shaijumon, A. Kumar, S. Murugesan, L. Ci, R. Vajtai, R.J. Linhardt, O. Nalamasu, P.M. Ajayan, Flexible energy storage devices based on nanocomposite paper. Proc. Natl. Acad. Sci. **104**, 13574–13577 (2007)
65. C. Masarapu, S. Subramanian, H. Zhu, B. Wei, Long-cycle electrochemical behavior of multiwall carbon nanotubes synthesized on stainless steel in Li ion batteries. Adv. Funct. Mater. **19**, 1008–1014 (2009)
66. B. Cao, A. Kleinhammes, X.P. Tang, C. Bower, L. Fleming, Y. Wu, O. Zhou, Electrochemical intercalation of single-walled carbon nanotubes with lithium. Chem. Phys. Lett. **307**, 153–157 (1999)
67. J. Dahn, T. Zheng, Y. Liu, J. Xue, Mechanisms for lithium insertion in carbonaceous materials. Science **270**, 590–593 (1995)
68. Z.-H. Yang, H.-Q. Wu, Electrochemical intercalation of lithium into raw carbon nanotubes. Mater. Chem. Phys. **71**, 7–11 (2001)
69. G.X. Wang, J. Yao, H.K. Liu, S.X. Dou, J.H. Ahn, Growth and lithium storage properties of vertically aligned carbon nanotubes. Met. Mater. Int. **12**, 413–416 (2006)
70. M. Hughes, G.Z. Chen, M.S.P. Shaffer, D.J. Fray, A.H. Windle, Electrochemical capacitance of a nanoporous composite of carbon nanotubes and polypyrrole. Chem. Mater. **14**, 1610–1613 (2002)
71. H. Zhang, G. Cao, Z. Wang, Y. Yang, Z. Shi, Z. Gu, Growth of manganese oxide nanoflowers on vertically-aligned carbon nanotube arrays for high-rate electrochemical capacitive energy storage. Nano Lett. **8**, 2664–2668 (2008)
72. W-C. Fang, Synthesis and electrochemical characterization of vanadium oxide/carbon nanotube composites for supercapacitors. J. Phys. Chem. C **112**, 11552–11555 (2008)
73. R.H. Baughman, A.A. Zakhidov, W.A. de Heer, Carbon nanotubes–the route toward applications. Science **297**, 787–792 (2002)
74. L. Dai, A.W.H. Mau, Controlled synthesis and modification of carbon nanotubes and C-60: carbon nanostructures for advanced polymer composite materials. Adv. Mater. **13**, 899–913 (2001)
75. S. Huang, L. Dai, Plasam etching for purification and controlled opening of aligned carbon nanotubes. J. Phys. Chem. B. **106**, 3543–3545 (2002)
76. S. Chen, J. Zhu, Q. Han, Z. Zheng, Y. Yang, X. Wang, Shape-controlled synthesis of one-dimensional MnO_2 via a facile quick-precipitation procedure and its electrochemical properties. Cryst. Growth Des. **9**, 4356–4261 (2009)
77. Q. Wang, Z.H. Wen, J.H. Li, A hybrid supercapacitor fabricated with a carbon nanotube cathode and a TiO_2-B nanowire anode. Adv. Funct. Mater. **16**, 2141–2146 (2006)
78. J.H. Jang, A. Kato, K. Machida, K. Naoi, Supercapacitor performance of hydrous ruthenium oxide electrodes prepared by electrophoretic deposition. J. Electrochem. Soc. **153**, A321–A328 (2006)
79. J.-S. Ye, H.F. Cui, X. Liu, T.M. Lim, W.-D. Zhang, F.-S. Sheu, Preparation and characterization of well-aligned carbon nanotubes-ruthenium oxide nano composites for supercapacitors. Small **1**, 560–565 (2005)
80. W.-C. Fang, O. Chyan, C.-L. Sun, C.-T. Wu, C.-P. Chen, K.-H. Chen, L.-C. Chen, J.-H. Huang, Arrayed CN_xNT-RuO_2 nanocomposites directly grown on Ti-buffered Si substrate for supercapacitor application. Electrochem. Comm. **9**, 239–244 (2007)

81. C. Yuan, X. Zhang, L. Su, B. Gao, L. Shen, Facile synthesis and self-assembly of hierarchical porous NiO nano/micro spherical superstructures for high performance supercapacitors. J. Mater. Chem. **19**, 5772–5777 (2009)
82. N. Nagarajan, H. Humadi, I. Zhitomirsky, Cathodic electrodeposition of MnO_x films for electrochemical supercapacitors. Electrochim. Acta **51**, 3039–3045 (2006)
83. U.M. Patil, K.V. Gurav, V.J. Fulari, C.D. Lokhande, O.S. Joo, Characterization of honeycomb-like "β-$Ni(OH)_2$" thin films synthesized by chemical bath deposition method and their supercapacitor application. J. Power Sources **188**, 338–342 (2009)
84. J.-W. Lang, L.-B. Kong, W.-J. Wu, M. Liu, Y.-C. Luo, L. Kang, A facile approach to the preparation of loose-packed $Ni(OH)_2$ nanoflake materials for electrochemical capacitors. J. Solid State Electrochem. **13**, 333–340 (2009)
85. H. Zhang, G. Cao, W. Wang, K. Yuan, B. Xu, W. Zhang, J. Cheng, Y. Yang, Influence of microstructure on the capacitive performance of polyaniline/carbon nanotube array composite electrodes. Electrochim. Acta **54**, 1153–1159 (2009)

Chapter 8
Graphene-Based Electrochemical Microsupercapacitors for Miniaturized Energy Storage Applications

Hao Yang and Wu Lu

Abstract With the development of miniaturized electronic devices, there grows an urgent demand for micropower sources that can be integrated into those systems. The micropower sources should be able to provide high energy density, high power density and long life cycle to power devices such as microelectromechanical (MEMS) systems, wireless sensors, radio frequency identification (RFID) tags and so on. Among all the candidates, the thin, flexible and durable miniaturized graphene-based supercapacitors exhibit great potentials. In this chapter, the development in graphene-based microsupercapacitors including the history, electrode materials, device configuration, fabrication strategies, and energy storage performance, is reviewed.

8.1 Introduction

Recently, the research on micropower sources has been greatly stimulated by the fast development of micro devices like MEMS, nanorobotics, wireless sensor networks, micro medical devices and so on [1, 2]. For example, the nanorobots have great potential in medical technology to identify or kill cancer cells and in environmental detection of toxic chemicals [3]. As the number of nanorobots increases and they are distributed in discrete locations, it is desired to have self-powered nanorobots with integrated micro power sources. This self-powered feature is also useful in many other applications.

Currently, micropower sources are dominated by microbatteries which are based on redox reactions between electrodes and electrolytes. These microbatteries share the same disadvantages as normal batteries such as limited lifetime, low power density, and safety concerns. The technical and economic challenges prevent microbatteries from applications that require good durability and high power

H. Yang · W. Lu (✉)
Department of Electrical and Computer Engineering, The Ohio State University, Columbus, OH 43210, USA
e-mail: lu.173@osu.edu

Q. Li (ed.), *Nanomaterials for Sustainable Energy*,
NanoScience and Technology, DOI 10.1007/978-3-319-32023-6_8

density. Microsupercapacitors, on the other side, can provide fast charge-discharge rate, outstanding power density several orders higher than batteries, and long lifetime up to millions of cycles [4–8]. Development of microsupercapacitors could lead to many new microelectronic applications that are not possible before. While the development of microsupercapacitors is still at its early stage, some significant advances have been made in recent years. This chapter reviews the latest advances in microsupercapacitors, from the history, electrode materials, device configuration, energy storage performance, fabrication strategies, to future development.

8.2 Supercapacitors

Supercapacitors, also called electrochemical capacitors, due to their high charge-discharge rate, long cycle life, and high power density, are desired in many applications, like heavy-duty cranes, electric vehicles, uninterrupted power supplies and so on [9–11]. Supercapacitors are unique energy storage devices with balanced energy density and power density bridging the gap between batteries and capacitors (Fig. 8.1) [12, 13]. Compared with conventional capacitors, the energy density of supercapacitors is orders of magnitude higher. As a result, supercapacitors can be used in high power applications that require reasonable energy levels. There are two energy storage mechanisms in supercapacitors: the electric double-layer capacitance (EDLC) and pseudocapacitance (Fig. 8.2). The ion transport in EDLC is faster and more reversible than the redox reaction in pseudocapacitors, leading to the higher power density and longer cycling life of EDLCs [14]. But the pseudocapacitors have a larger energy density than EDLCs. Due to the high energy storage requirements in many applications, hybrid supercapacitors that combine the

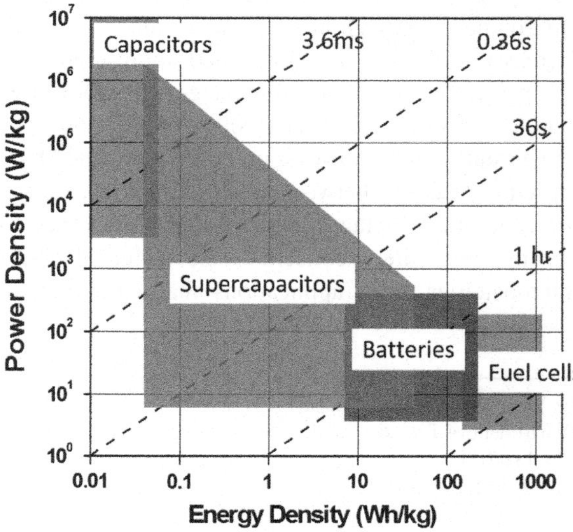

Fig. 8.1 Ragone plot shows the energy performance between capacitors, supercapacitors, batteries, and fuel cells

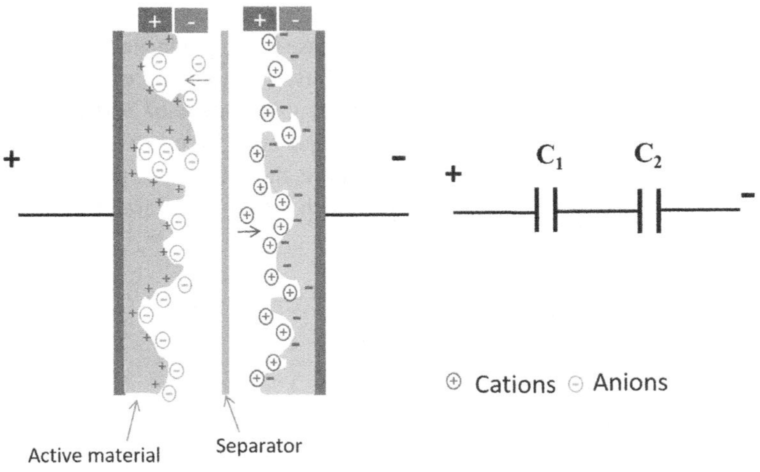

Fig. 8.2 Schematic of the supercapacitor energy storage mechanism

Fig. 8.3 **a** Schematic showing conductive wrapping of graphene/MnO_2 (GM) to introduce an additional electron transport path. **b** Schematic of graphene/MnO_2/CNT and graphene/MnO2/conducting polymer systems formed by wrapping of GM nanostructures with CNTs (GMC) or conducting polymers (GMP), where graphene is *black*, MnO_2 is *rose*, CNTs is *yellow*, and conducting polymer is *blue*. Reprinted with permission from [16]. Copyright 2011 American Chemical Society

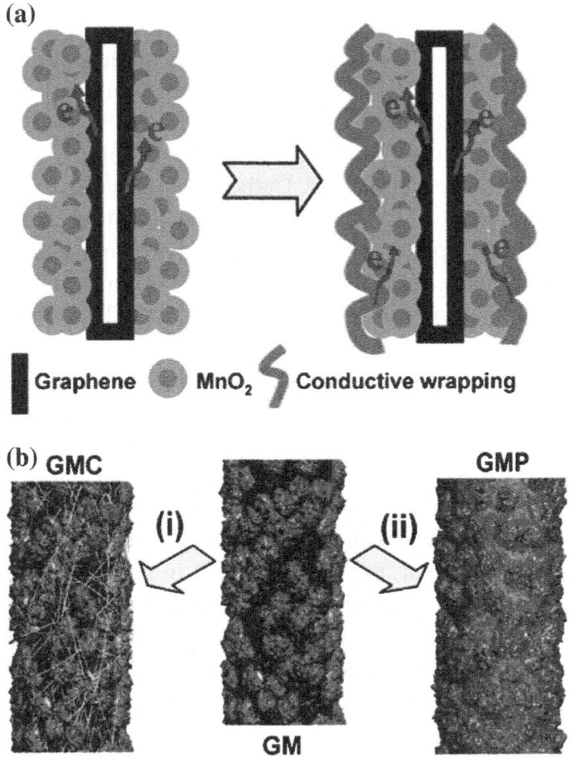

advantages of EDLCs and pseudocapacitors have also been investigated [15, 16]. Yu et al. [16] reported a ternary system based on graphene/MnO_2/CNT and grahphene/MnO_2/polymer composites by a solution coating process (Fig. 8.3). This ternary hybrid material provides a high specific capacitance of 380 F/g. However, these hybrid supercapacitors have the compromise between energy density and long cycling stability. Another approach to improve the capacitive performance is to develop novel porous electrode materials with better pore size uniformity, ion accessibility, and minimal agglomeration.

8.2.1 Electrode Materials

To improve the energy density, a porous electrode material with a high surface area and high ion accessibility is desired. Carbon materials, due to their low cost, various microtexture and processability, are more attractive than other electrode materials. Activated carbons, carbon aerogels, and carbon nanotubes have been widely used as electrode materials [17]. Activated carbons are the most common materials in EDL supercapacitors nowadays. The reported specific surface area of activated carbon ranges from 500 to 2000 m^2/g [18, 19], but only a fraction of which can contribute to energy storage. This is because the activated carbon materials have poor uniformity in pore sizes ranging from micropores ($\sim$0.3 nm) to macropores [20]. In such materials, the micropores are inaccessible to electrolyte solutions [21] and macropores result in a low surface volume ratio, both of which lead to a low specific capacitance [22]. Carbon nanotubes have a moderate surface area as well as a good conductivity. Supercapacitors based on carbon nanotubes can't exhibit good capacitive performance [23], and the manufacturing difficulties plus cost further limit their possibility of implementation in actual energy storage devices.

A more promising candidate for the EDL supercapacitor electrode material to overcome the above mentioned drawbacks is graphene. Graphene, due to its unique lattice structure, has appealing electrical properties, chemical stability, and high surface area. Ideally a monolayer of sp^2 bonded carbon atoms can reach a specific capacitance up to $\sim$550 F/g in $BMIMPF_6$ electrolyte as well as a high surface area of 2675 m^2/g, which basically set the upper limits for all carbon materials. Unlike the CVD-grown graphene, the graphene material needed in energy storage applications requires a highly porous structure instead of a planar thin layer structure. In addition to that, the agglomeration in graphene electrodes is common and acceptable, although undesirable. Therefore the problem becomes how to achieve graphene materials through an efficient and low-cost method that provides minimum agglomeration and a porous structure. With improvements in the exfoliation and reduction techniques of graphite oxide (GO), graphene-based materials can be produced in large-scale at a very low cost. A variety of oxidation-exfoliation methods have been employed to minimize the agglomeration between the graphene sheets, such as chemical reduction [21, 24], microwave irradiation [25], thermal annealing [26, 27], and powerful sonication [28]. Generally chemical reduction methods involve harmful reducing agents and are not suitable for mass production [24]; Thermal annealing can produce high quality graphene sheets but at a cost of very high temperature (~1050 °C) [26]. The irradiation and sonication methods serve as pure exfoliation techniques and always require the help of chemical reduction [25, 28]. The reported graphene-based supercapacitors have achieved 85.6 Wh/kg (154 F/g) under hydrazine reduction [21], 70 Wh/kg (166 F/g) with microwave exfoliation [29], and 1.36 mWh/cm^3 (276 F/g) by laser scribing [30]. Though tremendous progress has been made, these state-of-the-art values of supercapacitors in energy density is still significantly lower than the requirements of many applications [14, 17]. The key challenge to increase the energy storage density of graphene supercapacitors is to develop more effective synthesis approaches offering an ideal pore size, good uniformity, and high ion accessibility.

8.2.2 Electrolytes

Another key to achieve high capacitance is using electrodes with high surface area and electrical conductivity. There are three types of electrolytes used in supercapacitors: aqueous electrolyte, organic electrolyte, and ionic liquid. Aqueous electrolytes such as KOH, NaOH and H_2SO_4 have smaller ion sizes so that ions can access more surface area. The conductivities of acid and alkaline solutions are advantageously high owing to the proton transport resulting in a low equivalent series resistance (ESR) and high power capability. However, the decomposition voltage of aqueous electrolyte is theoretically 1.23 V or practically 1.3–1.4 V in kinetic terms [31]. Another issue of aqueous electrolyte is the corrosion nature of the acid or alkaline solution that may damage the electrode or other components of the device, leaving problems like reliability and self-discharge. Then, non-aqueous electrolytes were employed to break through the potential limit. Organic electrolytes, such as tetraethylammonium tetrafluoroborate ($C_8H_{20}BF_4N$) and triethylme-thylammonium tetrafluoroborate ($C_7H_{18}BF_4N$) in acetonitrile (C_2H_3N), have been studied in supercapacitors with a relatively high potential window around 2–2.5 V. But the large organic solvent molecules set a higher requirement of pore size of the porous electrodes. And the drawbacks like electrolyte depletion upon charge and low safety also limit the use of organic electrolytes [32]. Ionic liquids, on the other hand, have a maximum operating voltage about 4–4.5 V as well as a moderate ion size. AC supercapacitor cells filled with N-butyl-N-methylpyrrolidinium bis(trifluoromethanesulfonyl)imide ($PYR_{14}TFSI$) ionic liquid demonstrated a high cycling stability for 40000 cycles and a high stable specific capacitance of 60 F/g [33]. So the choice of electrolytes largely depends on reliability and performance requirements of the applications of supercapacitors. To choose between aqueous electrolyte, organic electrolyte and ionic liquid, one needs to consider the tradeoffs between operating voltage range, specific capacitance and ESR or power density.

8.2.3 Solid Electrolyte

Conventional supercapacitors with liquid electrolytes consist of a separator as isolation between electrodes. In addition, there is also a trend of using solid electrolytes like poly(vinyl alcohol) (PVA)-H_2SO_4 polymer gelled electrolyte in many applications especially for solid state microsupercapacitors. The use of those solid electrolytes results in a simplified fabrication process with no need for separator and no possible leakage of electrolyte (Fig. 8.4). In the microsupercapacitors and flexible applications, the solid state electrolyte is essential.

A solid polymer electrolyte (SPE) is a solvent free electrolyte by mixing a high molecular weight polymer matrix with a low lattice energy metal salt. The ion conductivity arises from the covalent bonding between the backbones of the polymers and the ions [34]. The electron donors in the polymer form a solvation on the surface of the cations. Thus, the solvation layer offers a separation between the cation and anion, and leads to the ionic hopping generating the ionic conductivity [34]. The rapid segmental movement of the polymer matrix and the Lewis-type interaction between the cations and the donors contribute to the ionic conductivity [35]. The polymer should be flexible enough to allow the ions to migrate through. Therefore SPE serves three

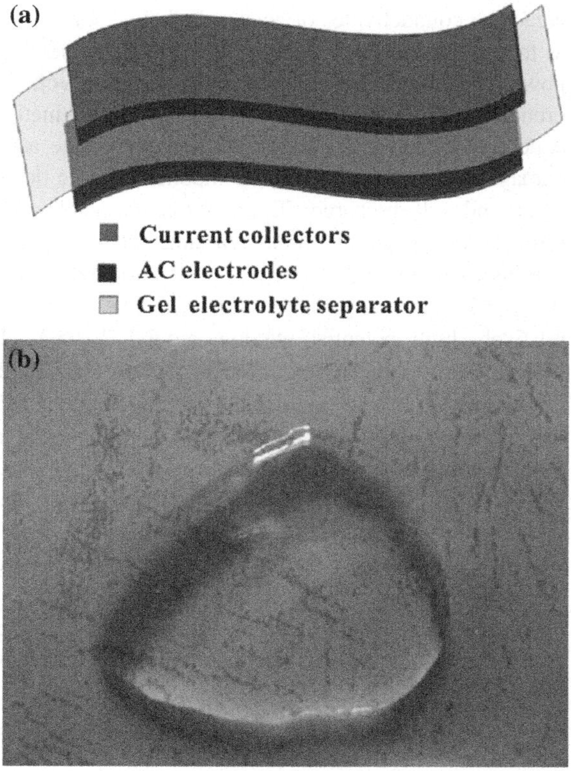

Fig. 8.4 **a** Schematic of a supercapacitor based on gel electrolyte. **b** The appearance of the PVA-H_2SO_4 electrolyte. Adapted with permission from [37]. Copyright 2012 Elsevier

principle roles in a cell: it provides the medium channel which generates the ionic conductivity; it acts as a separator that insulates the anode from cathode; it works as binder to enable good electrical contact with electrodes [34]. PVA with metal salts and immobilized ionic liquid have also been investigated [36–40]. The PVA system has better ionic conductivity, while the immobilized ionic liquid system has a higher potential window. A typical PVA polymer solid electrolyte is PVA-KOH [36]. The PVA-KOH electrolyte can be prepared by a solution casting method. Specifically, the PVA and KOH are dissolved in water and heated with agitation for 2–4 h, after which the solid electrolyte is packed with activated carbon electrodes. The alkaline PVA polymer electrolyte has a high ionic conductivity of the order of 10 mS/cm. The supercapacitor based on PVA-KOH electrolyte exhibited a specific capacitance of 100–112 F/g and good cycling stability for 1000 cycles. Another example of SPE is the immobilized ionic liquid. A copolymer poly(vinylidene fluoride-hexafluoropropylene) (PVDF-HFP) is used as the polymer matrix to support the ionic liquid 1-ethyl-3-methylimidazolium tetrafluoroborate ($EMIMBF_4$) [39]. In addition, graphene oxide is homogeneously added to generate a ion transport network to improve the ionic conductivity. The PVDF-HFP-$EMIMBF_4$ electrolyte exhibited an ionic conductivity of 25 mS/cm and a potential window up to 3.5 V. The specific capacitance of the assembled cells was reported to be in the range of 30–150 F/g.

8.2.4 Performance Evaluation

To evaluate the energy storage performance, the specific capacitance, energy density (energy per mass, volume or area), and power density (power per mass, volume or area) are commonly reported and compared. Depending on the specific application, those parameters can be given in gravimetric, volumetric or areal values. For energy storage devices in electrical vehicles, cranes or other applications caring more about weight, the gravimetrical values play a more important role. For compact or portable electronics, the volumetric values of energy storage devices are often used. For wearable and flexible devices, the performance is usually evaluated in areal parameters.

The system capacitance can be extracted from cyclic voltammetry (CV) and galvanostatic charge-discharge (GCD) measurements. Based on CV curves, the system capacitance C_T can be determined by the expression voltammetric charge/potential window [14, 31, 41, 42].

$$C_T = \frac{\int IdV}{\Delta V(dV/dt)} \tag{8.1}$$

where I is the current through the device, $\int IdV$ is the integral area of CV curve, ΔV is the potential window, dV/dt is the scan rate with a unit of mV/s. Another way to extract the capacitance is by GCD curves [14]:

$$C_T = \frac{I}{(\Delta V/\Delta t)} \tag{8.2}$$

where ΔV is the potential drop, and Δt is the discharging time. Practically, GCD curve is composed of a voltage drop due to the ESR and a linear discharging curve. The $\Delta V/\Delta t$ can be extracted by fitting the linear discharging curve.

In addition, the system capacitance can also be extracted from the impedance spectrum. The capacitance can be found using this equation [29]:

$$C_S = -\frac{1}{2\pi f Z''} \tag{8.3}$$

where f is the frequency and Z'' is the imaginary impedance.

Depending on the assembled structure, supercapacitors are studied in symmetrical two-electrode systems or asymmetrical three-electrode systems. The structure difference will result in the difference in the calculation of specific capacitance of single electrode. The specific capacitance of a three-electrode system can be written as:

$$C_S = \frac{C_T}{m} = \begin{cases} \frac{\int IdV}{m\Delta V(dV/dt)} (from\, CV) \\ \frac{I}{m(\Delta V/\Delta t)} (from\, GCD) \end{cases} \tag{8.4}$$

The m can be the mass, volume or area of the electrode. In the symmetrical two-electrode system, there are two similar specific capacitors connected in series ($1/C_T = 1/mC_s + 1/mC_s$, m can be the mass, volume or area of each electrode) [43]. Thus, the specific capacitance of a two-electrode system is [14, 25, 29]:

$$C_S = \frac{2C_T}{m} = \begin{cases} \frac{2\int IdV}{m\Delta V(dV/dt)} (from\, CV) \\ \frac{2I}{m(\Delta V/\Delta t)} (from\, GCD) \end{cases} \tag{8.5}$$

The energy densities at different current densities are obtained using the following formula [14, 29, 30]:

$$E = \frac{1}{2}\frac{C_T}{2m}V^2 = \frac{1}{8}C_s V^2 \tag{8.6}$$

The corresponding power density can be extracted from the following equation.

$$P = \frac{E}{\Delta t} \tag{8.7}$$

The maximum power density at the matched impedance condition is evaluated with [14, 17, 29, 30]

$$P = \frac{V^2}{8mR_{ESR}} \tag{8.8}$$

where $R_{ESR} = V_{drop}/4I$ is the equivalent series resistance, V_{drop} is the initial voltage drop at the beginning of the discharge.

8.3 Interdigital Microsupercapacitors

When scaling from macrosupercapacitors to microsupercapacitors, most of the principles apply, but the structural design and ionic diffusion behavior may differ [44]. Typically, macrosupercapacitors utilize the sandwiched structure with two electrodes, electrolyte and separator. Due to limited area on integrated circuits, the microsupercapacitors use a planar interdigital design [45–47]. In macrosupercapacitors, the dimensions are large. The power density is predominantly confined by the thickness of electrodes and separator [37]. By increasing the thickness of electrodes, the ion diffusion length increases, thus lowering the charge-discharge rate and power density. In the interdigital microsupercapacitors, a time-dependent mass transport introduces the rapidly varying fields that are spatially inhomogeneous. The capacitance is mainly affected by the width ratio between the electrode and gap, W_E/W_G [48]. A wider electrode width W_E provides more available area for charge storage and improves the capacitance. As the gap width W_G increases, the longer ion migration path results in a higher ESR value. As a result, the power density is lowered. For planar interdigital microsupercapacitors, a wide electrode width and minimal gap spacing can help to improve the specific capacitance, energy density and power density.

The planar interdigital microsupercapacitor was firstly reported by Sung et al. in 2003 with polypyrrole (PPy) electrodes and liquid electrolyte by photolithography and electrochemical polymerization techniques [49]. The gold/platinum "microelectrode" arrays (current collector) were patterned on SiO_2/Si wafer by wet etching. After that, the conducting polymers were electrochemically deposited on the microelectrode arrays. The interdigital fingers' width and spacing were set equal at 50 μm. Later Sung et al. investigated solid state microsupercapacitors utilizing gel polymer electrolyte [50, 51] to avoid the potential leakage of electrolyte (Fig. 8.5).

After the first microsupercapacitor report, more and more attention was gathered on the development of novel electrode materials and their fabrication methods like layer-by-layer assembly [52], inkjet printing [53], and so on. Pseudocapacitive materials like metal oxides [54, 55] and other conducting polymers [56] were studied in the microsupercapacitors. To improve the power density and cycling stability, many alternative electrode materials have been explored. Carbon materials

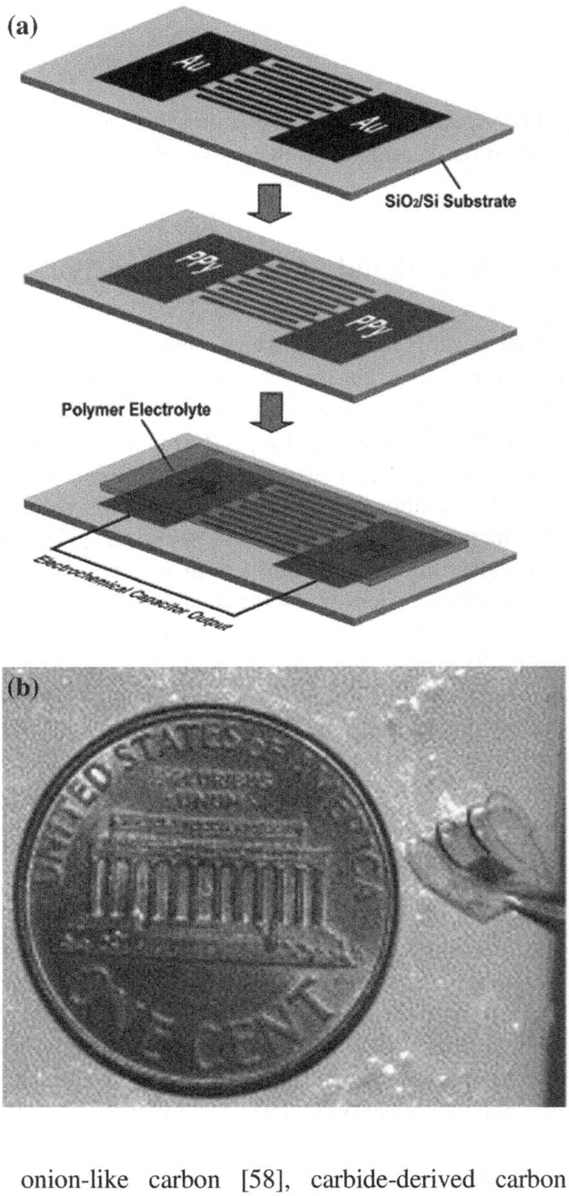

Fig. 8.5 **a** Fabrication procedure of flexible micro-supercapacitors. (**b**) Pictures of the flexible micro-supercapacitor. The all solid state cell can be bended and shows minimal performance change. Reprinted with permission from [51]. Copyright 2004 Elsevier

like activated carbon [57], onion-like carbon [58], carbide-derived carbon (CDC) [59, 60], and carbon nanotubes [61, 62] have been intensively studied. Chmiola and Huang et al. demonstrated the CDC-based microsupercapacitor on a silicon wafer [59, 60]. The CDC film was produced on TiC coated ceramices by chlorination at 450–500 °C for 5 min. During the heating, the Ti was extracted from TiC forming $TiCl_4$ leaving porous CDC coating (Fig. 8.6). Based on the CDC electrodes, the microsupercapacitors show a high volumetric capacitance of 180 F/cm^3 in $TEABF_4$ and 160 F/cm^3 in 1 M H_2SO_4 at a scan rate of 20 mV/s with a

2 μm thick CDC layer [59]. As the CDC layer thickness increases, the volumetric capacitance drops significantly. This is likely due to the microstructural rearrangement in thicker films from surface stress relaxing, which leads to porosity collapse and perturbation of the interconnected structure. The electrolyte starvation with thicker films may also help to decrease the capacitance, as the thicker films require more ions migrating from electrolyte to the porous structure.

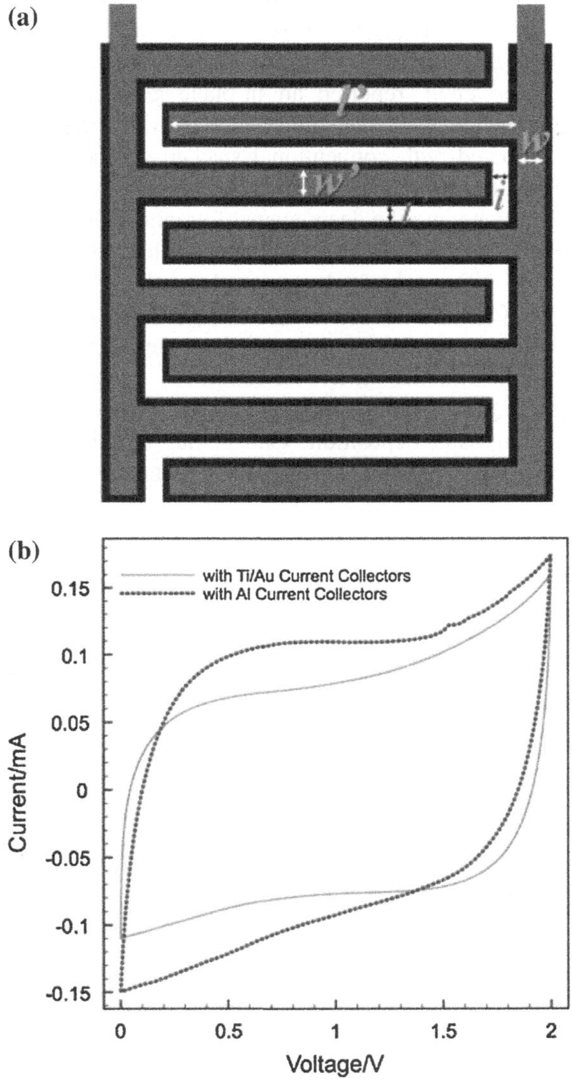

Fig. 8.6 **a** The schematic of the TiC-CDC microsupercapacitor. **b** The CV scans with Ti/Au and Al current collectors in 1 M NEt_4BF_4 in PC. Reprinted with permission from [60]. Copyright 2010 the American Association for the Advancement of Science

8.4 Graphene-Based Microsupercapacitors

The microsupercapacitor electrode materials should be able to provide a high surface area, long cycling stability and good ion accessibility at the same time. Thus the energy storage performance can be improved within a limited area with a well-designed structure. Graphene, due to the low cost, high surface area, processing feasibility, mechanical stability, non-toxicity, has been intensively studied in supercapacitors [14]. With 2D sp^2 bonded carbon atoms, the electrolyte ions can take advantage of the whole surface area of graphene with a short ion diffusion distance [2].

Yoo et al. reported an in-plane graphene supercapacitors with CVD-grown pristine graphene and multilayer reduced graphene oxide (rGO) (Fig. 8.7) [63]. Compared with conventional stacked geometry with incomplete surface accessibility, the in-plane geometry of graphene increases the possibility of electrolyte ions to migrate into the deep layers of graphene and allows for full utilization of surface area. The in-plane supercapacitors achieved a specific capacitance of 80 uF/cm^2 with CVD graphene electrodes. A much higher value of 394 uF/cm^2 was observed in rGO electrodes. The higher specific capacitance per area in rGO electrodes is because more layers are stacked on the rGO electrodes.

Conventional microfabrication of interdigital structures was completed by printing and lithography techniques involving complicate steps like patterning, etching, and deposition. Gao et al. [64] demonstrated this laser reduction and patterning of rGO directly on free-standing hydrated GO films (Fig. 8.8). The

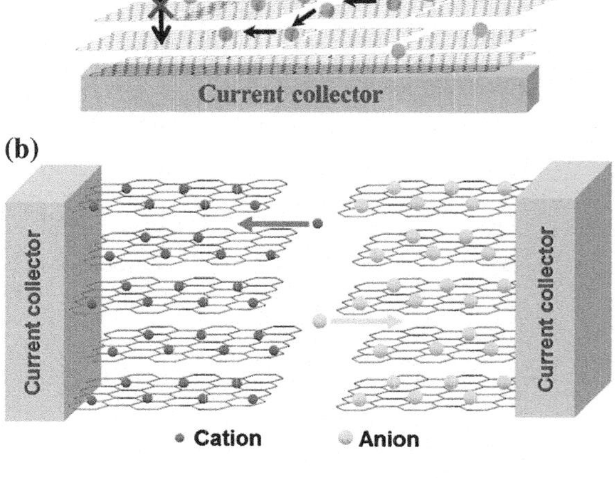

Fig. 8.7 **a** Schematic of the conventional stacked geometry for the fabrication of supercapacitors. **b** Schematic of the in-plane supercapacitor device. Reprinted with permission from [63]. Copyright 2010 American Chemical Society

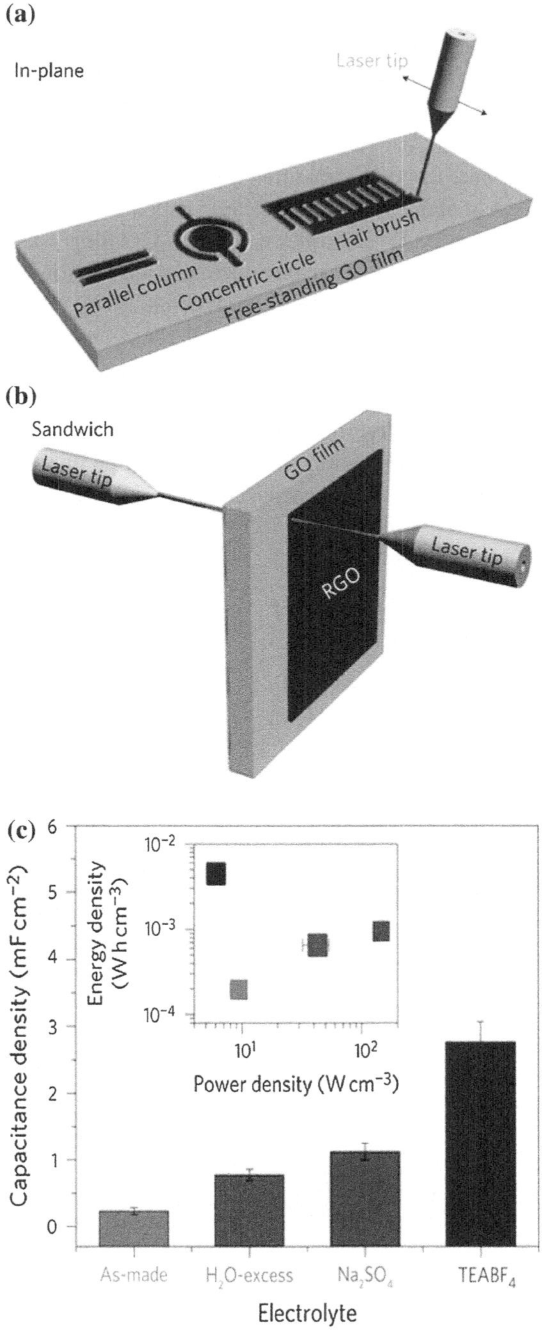

Fig. 8.8 Schematics of laser-patterning of free-standing hydrated GO films to fabricate RGO–GO–RGO devices with **a** in-plane and **b** sandwich geometries. **c** Comparison of specific capacitance of a sandwiched device as prepared, and with excess DI water, aqueous electrolyte and organic electrolyte. Reprinted with permission from [64]. Copyright 2011 Macmillan Publishers Limited

in-plane and conventional sandwiched structure were patterned to study the effect of structure on device performance. The in-plane concentric circular structure achieved a specific capacitance of 0.51 mF/cm^2 (volumetric 3.1 F/cm^3). The energy density of this device was 4.3×10^{-4} Wh/cm^3 at the power density of 1.7 W/cm^3. After 10000 cycles, around 70 % of the initial capacitance was retained in the in-plane circular structure, while 65 % of that in the sandwiched was retained.

A scalable fabrication method called laser scribed graphene (LSG) using a standard LightScribe DVD burner writing on GO films was demonstrated by El-Kady et al. [7] The fabricated LSG microsupercapacitor is shown in Fig. 8.9. After the laser scribing, LSG films showed an electrical conductivity of 2.35×10^3 S/m and a surface area of 1500 m^2/g characterized by methylene blue adsorption. The utilized ionogel electrolyte, fumed silica mixed with 1-butyl-3-methylimidazolium bis(trifluoromethylsulfonyl)imide ([BMIM][Tf_2N]), allowed an wide operation window of 2.5 V, while the conventional hydrogel-polymer electrolyte, poly(vinyl alcohol) (PVA)-H_2SO_4, can only give an operation window of 1 V. The improvement in operation voltage resulted in a high specific capacitance of 2.35 F/cm^3 at a current density of 16.8 mA/cm^3.

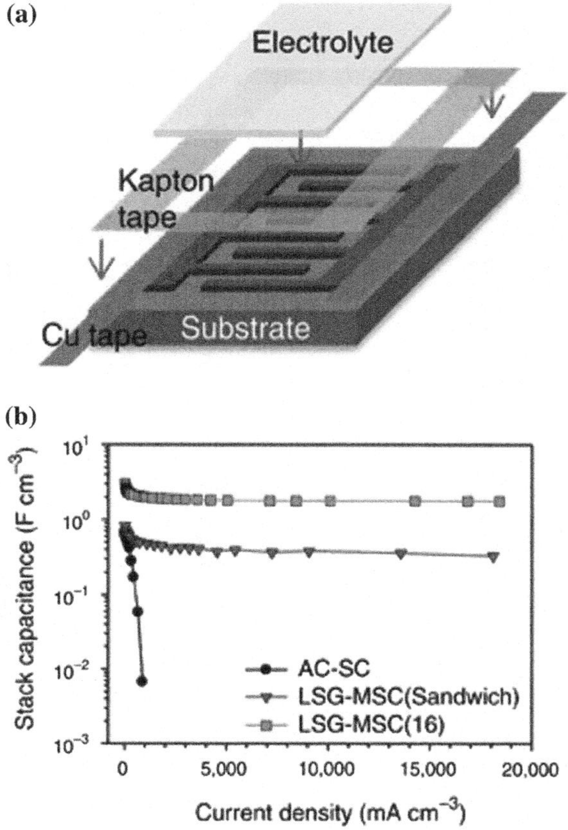

Fig. 8.9 **a** The schematic of a LSG microsupercapacitor. **b** Volumetric stack capacitance of sandwiched and interdigite structure LSG microsupercapacitor. A commercial activated carbon based supercapacitor was used as comparison. Reprinted with permission from [7]. Copyright 2013, Macmillan Publishers Limited

The LSG microsupercapacitor exhibited an ultra-low leakage current less than 150 nA after 12 h, higher power density of 200 W/cm^3, and low self-discharge.

3D graphene/carbon nanotube carpets (G/CNTCs) microsupercapacitors were fabricated in situ on nickel electrodes [65]. The interdigital G/CNTCs microsupercapacitor structure is shown in Fig. 8.10. The G/CNTCs microsupercapacitor exhibited a very good capacitive performance with an impedance angle of -81.5 at a frequency of 120 Hz. In ionic liquid electrolyte, the device provided a high energy density of 2.42 mWh/cm^3 and a maximum power density of 115 W/cm^3. The outstanding performance of this G/CNTCs material was attributed to the seamless junction of graphene/carbon nanotube at the interface of different carbon allotropic forms. Given the good capacitive performance and high energy density, such G/CNTCs microsupercapacitors can be exploited as compact ac filtering units or micro power sources.

To further enhance the capacitance and energy density, a hybrid 2D nanostructure of graphene and metal oxide MnO_2 nanosheets was studied [66]. The δ-MnO_2/graphene material was obtained by chemically intergrating δ-MnO_2 nanosheets on graphene with electrostatic interaction. The schematic of this δ-MnO_2/graphene hybrid structure is given in Fig. 8.11. This planar structure based

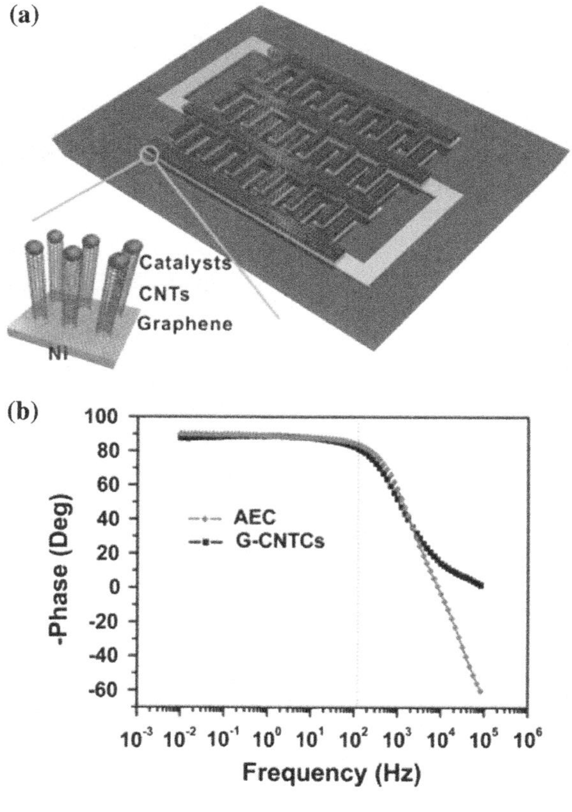

Fig. 8.10 **a** Schematic of the structure of G/CNTCs-MCs. The *inset* shows the enlarged Ni-G-CNTCs pillar structure; **b** Impedance phase angle versus frequency. Reprinted with permission from [65]. Copyright 2012 American Chemical Society

on δ-MnO_2 increased the electrochemically active surface for adsorption of electrolyte ions and introduce pseudocapacitive reactions leading to the enhancement of the specific capacitance. And the δ-MnO_2 layer worked as a spacer between densely stacked graphene sheets, and the extended interlayer spacing allowed more ions to penetrate efficiently into the films. The electrodes were prepared by transferring δ-MnO_2/graphene film on a PET substrate. The hybrid microsupercapacitors using a gel electrolyte PVA/H_3PO_4 exhibited high specific capacitance of 267 F/g at a current density of 0.2 A/g. After 7000 charge-discharge cycles, 92 % of the initial capacitance retained showing excellent cycling stability. In addition, the δ-MnO_2/graphene electrodes showed superior flexibility with over 90 % capacitance retention after 1000 times of folding and unfolding.

In addition to graphene-based microsupercapacitors, the electrochemical performance of graphene quantum dots (GQDs) was also investigated. GQDs have appealing chemical and physical properties like abundant edge defects, good electrical conductivity, chemical stability, and easy functionalization, and become a promising candidate for supercapacitor electrodes [2, 67]. Liu et al. reported this

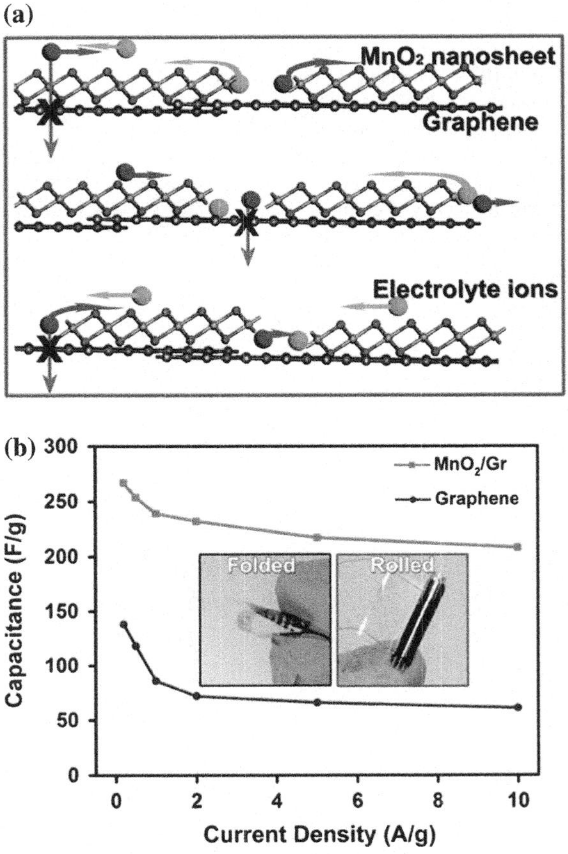

Fig. 8.11 **a** Schematic of the ion transport within the 2D δ-MnO2/graphene hybrid structures. **b** Comparison of specific capacitance values for the supercapacitors based on hybrid δ-MnO2/graphene and graphene. Reprinted with permission from [66]. Copyright 2013 American Chemical Society

symmetrical GQDs microsupercapacitors using simple electro-deposition approach [68]. The GQDs was electrophoretic deposited on the interdigital gold electrodes (Fig. 8.12). The electrochemical properties in aqueous electrolyte (0.5 M Na_2SO_4) and ionic liquid (2 M $EMIMBF_4$/AN) were demonstrated. The prepared GQDs can

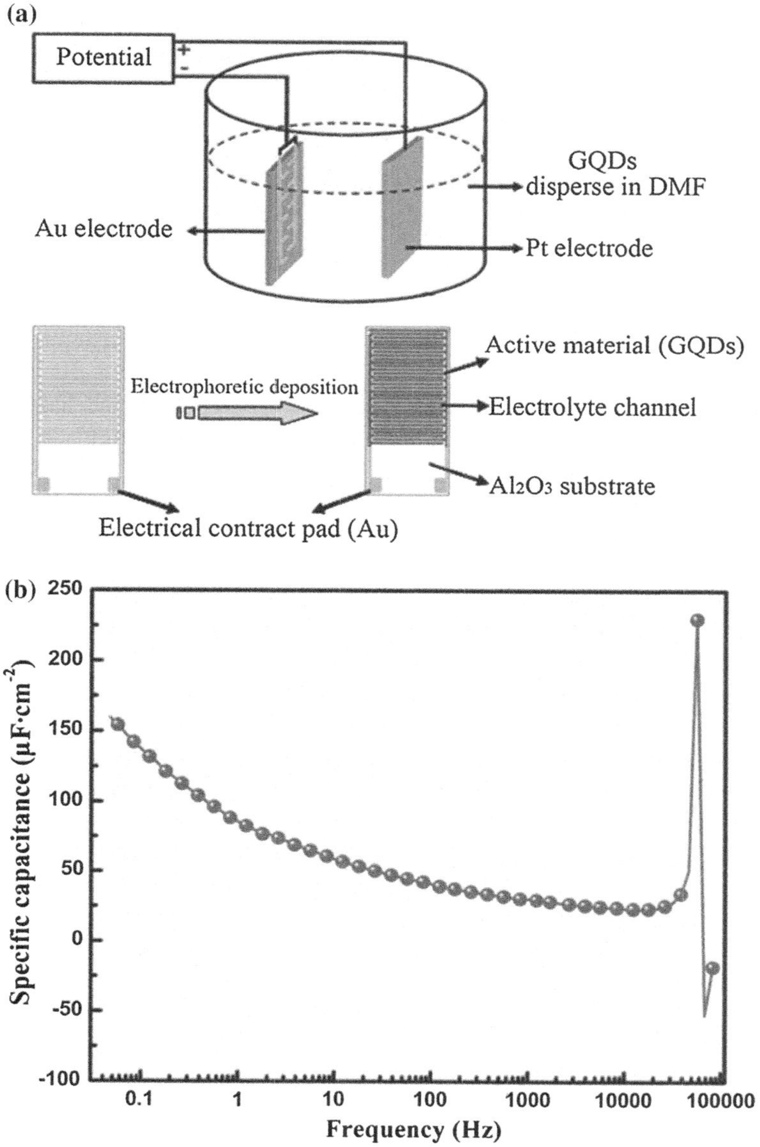

Fig. 8.12 **a** Schematic of the electrophoretic deposition of GQDs on the interdigital finger electrodes. **b** Specific capacitance versus frequency. Reprinted with permission from [67]. Copyright 2013 WILEY-VCH Verlag GmbH & Co. KGaA, Weinheim

work at ultrahigh scan rate of 1000 V/s, and showed short relaxation time constant of 103.6 us in aqueous electrolyte as well as excellent cycling stability with 97.8 % capacitance retention after 5000 cycles. The symmetrical microsupercapacitors achieved a specific capacitance up to 534.7 uF/cm^2 and an energy density of 0.074 uWh/cm^2 (with specific power of 7.5 uW/cm^2) in aqueous electrolyte. The device assembled with ionic liquid electrolyte showed a smaller time constant of 53.8 us, a specific capacitance of 468.1 uF/cm^2 and an energy density of 0.474 uWh/cm^2 (with specific power of 56.7 uW/cm^2). This energy performance improvement in ionic liquid is due to the higher voltage window of 2.7 V. In addition, asymmetrical GQDs/MnO_2 microsupercapacitors were prepared displaying two times higher specific capacitance of 1107.4 uF/cm^2 and energy density of 0.154 uWh/cm^2. The excellent energy storage performance was mainly attributed to the unique properties of GQDs like high surface area, abundant active sites and accessible edges.

8.5 Conclusion and Outlooks

Graphene-based microsupercapacitors is a relatively new area and has attracted great attention in the last few years. Significant progress has been made in development of electrode materials, electrolytes, and design and fabrication of microsupercapacitors. Further improvement in graphene-based microsupercapacitors can be achieved with the integration of graphene with novel nanomaterials enhancing the porous morphology and introducing electroactive sites. By integrating spacers like function groups or carbon nanotubes, the porous network between graphene sheets can provide higher utilization of the accessible area for charge storage. The specific capacitance and energy density can be improved by mixing graphene with pseudocapacitive materials. Another key to obtain high performance microsupercapacitor is to develop novel and low-cost film fabrication techniques that are compatible with industrial device fabrication processes. Methods like photolithography, printing, laser scribing and layer-by-layer assembly are quite effective in improving the energy storage performance. Future development of graphene-based microsupercapacitors will focus on the optimization of electrode material composition, thin film manufacturing techniques, electrode design and interfacial integrity of the components. In the future, graphene-based microsupercapacitors can enable many miniaturized energy storage applications requiring both high power and energy density.

References

1. D.R. Rolison, R.W. Long, J.C. Lytle et al., Multifunctional 3D nanoarchitectures for energy storage and conversion. Chem. Soc. Rev. **38**, 226–252 (2009)

2. H.S. Wu, X. Feng, H.M. Cheng, Recent advances in graphene-based planar micro-supercapacitors for on-chip energy storage. Nat. Sci. Rev. **1**, 277–292 (2014)
3. G.M. Patel, G.C. Patel, R.B. Patel, J.K. Patel, M. Patel, Nanorobot: A versatile tool in nanomedicine. J. Drug Target. **14**, 63 (2006)
4. J. Chmiola, C. Largeot, P.L. Taberna et al., Monolithic carbide-derived carbon films for micro-supercapacitors. Science **328**, 480–483 (2010)
5. D. Pech, M. Brunet, H. Durou et al., Ultrahigh-power micrometre-sized supercapacitors based on onion-like carbon. Nat. Nanotechnol. **5**, 651–654 (2010)
6. W. Gao, N. Singh, L. Song et al., Direct laser writing of micro-supercapacitors on hydrated graphite oxide films. Nat. Nanotechnol. **6**, 496–500 (2011)
7. M.F. El-Kady, R.B. Kaner, Scalable fabrication of high-power graphene micro-supercapacitors for lexible and on-chip energy storage. Nat. Commun. **4**, 1475 (2013)
8. Z.S. Wu, K. Parvez, X.L. Feng et al., Graphene-based in-plane micro-supercapacitors with ultrahigh power and energy densities. Nat. Commun. **4**, 2487 (2013)
9. M. Armand, J.M. Tarascon, Building better batteries. Nature **451**, 652–657 (2008)
10. H. Wu, G. Yu, L. Pan, N. Liu, M.T. McDowell, Z. Bao, Y. Cui, Stable Li-ion battery anodes by in-situ polymerization of conducting hydrogel to conformally coat silicon nanoparticles. Nat. Commun. **2013**, 4 (1943)
11. C.M. Park, J.H. Kim, H. Kim, H. Sohn, Li-alloy based anode materials for Li secondary batteries. J. Chem. Soc. Rev. **39**, 3115–3141 (2010)
12. P. Simon, Y. Gogotsi, Materials for electrochemical capacitors. Nat. Mater. **7**, 845–854 (2008)
13. R. F. Service, New 'supercapacitor' promises to pack more electrical punch. Science **313**, 902–905 (2006)
14. H. Yang, S. Kannappan, A.S. Pandian, J.H. Jang, Y.S. Lee, W. Lu, Nanoporous graphene materials by low-temperature vacuum-assisted thermal process for electrochemical energy storage. J. Power Sources **284**, 146–153 (2015)
15. D. Wang, R. Kou, D. Choi, Z.O Yang, Z. Nie, J. Li, L.V. Saraf, D. Hu, J. Zhang, G. L. Graff, J. Liu, M.A. Pope, and I.A. Aksay, Ternary self-assembly of ordered metal oxide − graphene nanocomposites for electrochemical energy storage. ACS Nano. **4**, 1587–1595 (2010)
16. G. Yu, L. Hu, N. Liu, H. Wang, M. Vosgueritchian, Y. Yang, Y. Cui, Z. Bao, Enhancing the supercapacitor performance of graphene/MnO2 nanostructured electrodes by conductive wrapping. Nano Lett. **11**, 4438–4442 (2011)
17. A. Burke, Ultracapacitors: why, how, and where is the technology. J. Power Sources **91**, 37–50 (2000)
18. D. Qu, Studies of the activated carbons used in double-layer supercapacitors. J. Power Sources **109**, 403–411 (2002)
19. J.A. Fernandez, T. Morishita, M. Toyoda, M. Inagaki, F. Stoeckli, T.A. Centeno, Performance of mesoporous carbons derived from poly(vinyl alcohol) in electrochemical capacitors. J. Power Sources **175**, 675–679 (2008)
20. Y. Wang, Y. Xia, Recent progress in supercapacitors: from materials design to system construction. Adv. Mater. **25**, 5336–5342 (2013)
21. C. Liu, Z. Yu, D. Neff, A. Zhamu, B.Z. Jang, Graphene-based supercapacitor with an ultrahigh energy density. Nano Lett. **10**, 4863–4868 (2010)
22. A. Burke, R&D considerations for the performance and application of electrochemical capacitors. Electrochima Acta **53**, 1083–1091 (2007)
23. H. Pan, J. Li, Y. Feng, Carbon nanotubes for supercapacitor. Nanoscale Res. Lett. **5**, 654–668 (2010)
24. S. Park, R.S. Ruoff, Chemical methods for the production of graphenes. Nat. Nanotech. **4**, 217–224 (2009)
25. Y. Zhu, S. Murali, M.D. Stoller, A. Velamakanni, R.D. Piner, R.S. Ruoff, Microwave assisted exfoliation and reduction of graphite oxide for ultracapacitors. Carbon **48**, 2118–2122 (2010)
26. H.C. Schniepp, J.L. Li, M.J. McAllister, H. Sai, M. Herrera-Alonso, D.H. Adamson, R.K. Prud'homme, R. Car, D.A. Saville, I.A. Aksay. Functionalized single graphene sheets derived from splitting graphite oxide. J. Phys. Chem. B. 110, 8535–8539 (2006)

27. M. Jin, T.H. Kim, S.C. Lim, D.L. Duong, H.J. Shin, Y.W. Jo, H.K. Jeong, J. Chang, S. Xie, Y. H. Lee, Facile physical route to highly crystalline graphene. Adv. Funct. Mater. **21**, 3496–3501 (2011)
28. S. Stankovich, D.A. Dikin, R.D. Piner, K.A. Kohlhaas, A. Kleinhammes, Y. Jia, Y. Wu, S.T. Nguyen, R.S. Ruoff, Synthesis of graphene-based nanosheets via chemical reduction of exfoliated graphite oxide. Carbon **45**, 1558–1565 (2007)
29. Y. Zhu, S. Murali, M.D. Stoller, K.J. Ganesh, W. Cai, P.J. Ferreira, A. Pirkle, R.M. Wallace, K.A. Cychosz, M. Thommes, D. Su, E.A. Stach, R.S. Ruoff, Carbon-based supercapacitors produced by activation of graphene. Science **332**, 1537–1541 (2011)
30. M.F. El-Kady, V. Strong, S. Dubin, R.B. Kaner, Laser scribing of high-performance and flexible graphene-based electrochemical capacitors. Science **335**, 1326–1330 (2012)
31. B.E. Conway, *Electrochemical Supercapacitors: Scientific Fundamentals and Technological Applications* (Plenum Publishers, 1999)
32. Y. Chen, X. Zhang, D. Zhang, P. Yu, Y. Ma, High performance supercapacitors based on reduced graphene oxide in aqueous and ionic liquid electrolytes. Carbon **49**, 573–580 (2011)
33. A. Balducci, R. Dugas, P.L. Taberna, P. Simon, D. Plée, M. Mastragostino, S. Passerini, High temperature carbon–carbon supercapacitor using ionic liquid as electrolyte. J. Power Sources **165**, 922–927 (2007)
34. R.T. Subramaniam, L. Chiam-Wen, L.P. Yee, E. Morris, *Recent Advances in Plasticizers* (InTech, 2012)
35. S. Ganesan, B. Muthuraaman, V. Mathew, J. Madhavan, P. Maruthamuthu, S.A. Suthanthiraraj, Performance of a new polymer electrolyte incorporated with diphenylamine in nanocrystalline dye–sensitized solar cell. Solar Energy Mater. Solar Cells **92**, 1718–1722 (2008)
36. C.C. Yang, S.T. Hsu, W.C. Chien, All solid-state electric double-layer capacitors based on alkaline polyvinyl alcohol polymer electrolytes. J. Power Sources **152**, 303–310 (2005)
37. H. Yu, J. Wu, L. Fan, Y. Lin, K. Xu, Z. Tang, C. Cheng, S. Tang, J. Lin, M. Huang, Z. Lan, A novel redox-mediated gel polymer electrolyte for high-performance supercapacitor. J. Power Sources **198**, 402–407 (2012)
38. G. Merle, S.S. Hosseiny, M. Wessling, K. Nijmeijer, New cross-linked PVA based polymer electrolyte membranes for alkaline fuel cells. J. Membr. Sci. **409–410**, 191–199 (2012)
39. G.P. Pandey, Y. Kumar, S.A. Hashmi, Ionic liquid incorporated polymer electrolytes for supercapacitor application. Indian J. Chem. - Section A **49**, 743–751 (2010)
40. G.P. Pandey, S.A. Hashmi, Performance of solid-state supercapacitors with ionic liquid 1-ethyl-3-methylimidazolium tris(pentafluoroethyl) trifluorophosphate based gel polymer electrolyte and modified MWCNT electrodes. Electrochim. Acta **105**, 333–341 (2013)
41. C.C. Hu, T.W. Tsou, Ideal capacitive behavior of hydrous manganese oxide prepared by anodic deposition. Electrochem. Commun. **4**, 105–109 (2002)
42. C.C. Hu, M.J. Liu, K.H. Chang, Anodic deposition of hydrous ruthenium oxide for supercapacitors. J. Power Sources **163**, 1126–1131 (2007)
43. V. Khomenko, E. Frackowiak, F. Béguin, Determination of the specific capacitance of conducting polymer/nanotubes composite electrodes using different cell configurations. Electrochim. Acta **50**, 2499–2506 (2005)
44. G. Xiong, C. Meng, R.G. Reifenberger, P.P. Irazoqui, T.S. Fisher, Review of graphene-based electrochemical microsupercapacitors. Electroanalysis **26**, 30–51 (2014)
45. W. Sun, X.Y. Chen, Fabrication and tests of a novel three dimensional micro supercapacitor. Microelectron. Eng. **86**, 1307–1310 (2009)
46. J.H. Sung, S. Kim, K.H. Lee, Fabrication of all-solid-state electrochemical microcapacitors. J. Power Sources **133**, 312–319 (2004)
47. J. Chmiola, C. Largeot, P.L. Taberna, P. Simon, Y. Gogotsi, Monolithic carbide-derived carbon films for micro-supercapacitors. Science **328**, 480–483 (2010)
48. C.W. Shen, X.H. Wang, W.F. Zhang, F.Y. Kang, A high-performance three-dimensional micro supercapacitor based on self-supporting composite materials. J. Power Sources **196**, 10465–10471 (2011)

49. J.-H. Sung, S.J. Kim, K.H. Lee, Fabrication of microcapacitors using conducting polymer microelectrodes. J. Power Sources, 124, 343–350 (2003)
50. J.H. Sung, S. Kim, K.H. Lee, Fabrication of all-solid-state electrochemical microcapacitors. J. Power Sources **133**, 312–319 (2004)
51. J.H. Sung, S.J. Kim, S.H. Jeong et al., Flexible micro-supercapacitors. J. Power Sources **162**, 1467–1470 (2006)
52. M. Beidaghi, C.L. Wang, Micro-supercapacitors based on interdigital electrodes of reduced graphene oxide and carbon nanotube composites with ultra high power handling performance. Adv. Funct. Mater. **22**, 4501–4510 (2012)
53. D. Pech, M. Brunet, P.L. Taberna et al., Elaboration of a microstructured inkjet-printed carbon electrochemical capacitor. J. Power Sources **195**, 1266–1269 (2010)
54. M.Q. Xue, Z. Xie, L.S. Zhang et al., Microfluidic etching for fabrication of flexible and all-solid-state micro supercapacitor based on MnO_2 nanoparticles. Nanoscale **3**, 2703–2708 (2011)
55. J. Feng, X. Sun, C.Z. Wu et al., Metallic few-layered VS_2 ultrathin nanosheets: high two-dimensional conductivity for in-plane supercapacitors. J. Am. Chem. Soc. **133**, 17832–17838 (2011)
56. W.W. Liu, X.B. Yan, J.T. Chen et al., Novel and high-performance asymmetric micro-supercapacitors based on graphene quantum dots and polyaniline nanofibers. Nanoscale **5**, 6053–6062 (2013)
57. H. Durou, D. Pech, D. Colin et al., Wafer-level fabrication process for fully encapsulated micro-supercapacitors with high specific energy. Microsyst. Technol. **18**, 467–473 (2012)
58. D. Pech, M. Brunet, H. Durou et al., Ultrahigh-power micrometre-sized supercapacitors based on onion-like carbon. Nat. Nanotechnol. **5**, 651–654 (2010)
59. J. Chmiola, C. Largeot, P.L. Taberna, P. Simon, Y. Gogotsi, Monolithic carbide-derived carbon films for micro-supercapacitors. Science **328**, 480–483 (2010)
60. P. Huang, M. Heon, D. Pech et al., Micro-supercapacitors from carbide derived carbon (CDC) films on silicon chips. J. Power Sources **225**, 240–244 (2013)
61. Y.Q. Jiang, Q. Zhou, L. Lin, Planar MEMS supercapacitor using carbon nan-otube forests, *IEEE 22nd International Conference on Micro Electro Mechanical Systems*, 587–590 (2009)
62. L. Li, C. Chen, J. Xie, Z. Shao, F. Yang, The preparation of carbon nanotube/MnO_2 composite fiber and its application to flexible micro-supercapacitor. J. Nanomater, 821071 (2013)
63. J.J. Yoo, K. Balakrishnan, J.S. Huang et al., Ultrathin planar graphene super-capacitors. Nano Lett. **11**, 1423–1427 (2011)
64. W. Gao, N. Singh, L. Song, Z. Liu, A.L.M. Reddy, L.J. Ci, R. Vajtai, Q. Zhang, B.Q. Wei, P. M. Ajayan, Direct laser writing of micro-supercapacitors on hydrated graphite oxide films. Nat. Nanotechnol. **6**, 496–500 (2011)
65. J. Lin, C.G. Zhang, Z. Yan et al., 3-dimensional graphene carbon nanotube carpet-based microsupercapacitors with high electrochemical performance. Nano Lett. **13**, 72–78 (2013)
66. L.L. Peng, X. Peng, B.R. Liu et al., Ultrathin two-dimensional MnO_2/graphene hybrid nanostructures for high-performance, flexible planar supercapacitors. Nano Lett. **13**, 2151–2157 (2013)
67. W. Liu, X. Yan, J. Chen, Y. Feng, Q. Xue, Novel and high-performance asymmetric micro-supercapacitors based on graphene quantum dots and polyaniline nanofibers. Nanoscale **5**, 6053–6062 (2013)
68. W.W. Liu, Y.Q. Feng, X.B. Yan et al., Superior micro-supercapacitors based on graphene quantum dots. Adv. Funct. Mater. **23**, 4111–4122 (2013)

Chapter 9
Incorporating Graphene into Fuel Cell Design

Edward P. Randviir and Craig E. Banks

Abstract Since the fabrication and subsequent physical characterisation of graphene in 2004 and 2005, a host of potential uses have been suggested and researched. To date, some have had some success, while others are significantly lacking in critical research due to unforeseen problems with the electrochemical properties of graphene. Of the many applications, fuel cells were predicted to gain benefit from graphene; the focus of this book chapter is to assess whether graphene has a place in fuel cells, and disseminate the current state of research across the globe for this purpose. It is evident that the applicability of graphene as a fuel cell anode, cathode, or proton exchange membrane, is dramatically dependent upon the type of graphene used in the fuel cell design. Pristine graphenes lack the necessary active sites for low energy electrochemical reactions required in fuel cells, and therefore their use as anodes or cathodes is seemingly limited. However pristine graphene may permit protons across atomic scale defects in its lattice and prove to be a good material for a proton exchange membrane when coupled with its tensile strength. Laser-induced graphene is a relatively new technique for the fabrication of graphene but offers a potential route towards manufacture of 3D graphene architectures that could be useful for cathodic reactions such as oxygen reduction. Reduced graphene oxides in their many guises could also be useful, provided that a replicable standard can be formulated for fuel cell anode and cathodes. Graphene oxides also have potential for proton exchange membranes but may suffer from longevity issues as a result of the decreased hydrophobicity allowing swelling of the material when in an aqueous environment. Nitrogen-doped graphenes are also potential candidates for the future of fuel cell research for both anodic and cathodic reactions. The future of graphene as a fuel cell material is predicted to be several years away because the research in this area has not solved issues of longevity, reproducibility, and temperature tolerance, while maintaining a good cell voltage and current density.

E.P. Randviir · C.E. Banks (✉)
Faculty of Science and Engineering, Manchester Metropolitan University, Chester Street, Manchester M1 5GD, UK
e-mail: c.banks@mmu.ac.uk

Q. Li (ed.), *Nanomaterials for Sustainable Energy*,
NanoScience and Technology, DOI 10.1007/978-3-319-32023-6_9

9.1 It's All Gone Graphene

If ever the powers-that-be devised a list of material "wonders" of the world akin to that of the Ancient or Modern Wonders, then graphene would almost certainly appear on that list, perhaps as the equivalent of the Panama Canal. This particular Modern Wonder is a relatively long but thin channel that allows the transport of goods between the Atlantic and Pacific oceans without the need to navigate the treacherous Cape Horn route surrounding South America. Similarly, graphene itself is the thinnest material known to man, and exhibits exceptional thermal and electronic conductive properties that allow electronic cargo, or electrons, to flow through it without the need for overcoming huge thermodynamic barriers; the analogy is thus realised.

Graphene is a one atom thick hexagonal carbon lattice, described by some as the "mother of all graphitic materials" [1]. On a physical level, each carbon atom in graphene bonds to three other carbon atoms in a sp^2 hybridized mode, with extra bond strength derived from the overlap of pi bonding and anti-bonding orbitals above and below its planar structure (see Fig. 9.1). Such a unique structure allows graphene to exhibit some of the unparalleled range of properties that it is celebrated for today. Among its many properties (Table 9.1) [2–8], the electronic conductivity of graphene is one of the most exciting properties that may hold the key to many of the technological challenges society is faced with. The electron mobility of graphene (>200,000 $cm^2\ V^{-1}\ s^{-1}$) [4] is significantly higher than common semi-conducting materials, such as silicon (1400 $cm^2\ V^{-1}\ s^{-1}$) [2] and indium

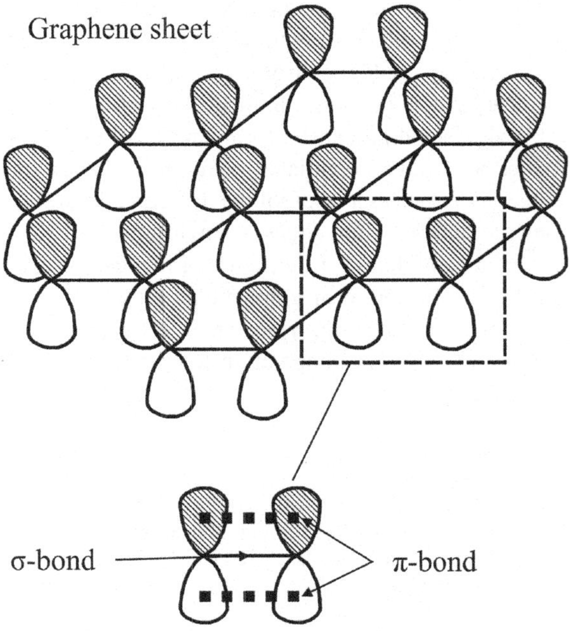

Fig. 9.1 Schematic diagram of a graphene sheet, with its sigma- and pi-bonding modes that incorporate the p bonding and anti-bonding orbitals

Table 9.1 Non-exhaustive list of the properties of graphene

Property	Magnitude	References
High Young's modulus	2.4 TPa	[3]
High electron mobility	>200,000 $cm^2 V^{-1} s^{-1}$	[4]
High tensile strength	130 GPa	[5]
Low resistivity (2D)	100 Ω	[6]
High optical transparency	97.7 % transmission	[7]
High specific surface area	2630 $m^2 g^{-1}$	[8]

antimonide (77,000 $cm^2 V^{-1} s^{-1}$) [9], allowing potential application for conductors and semiconductors. There is significant research towards expanding the band gap between the conduction band and Fermi level of graphene, because it fundamentally exhibits a band gap of zero, meaning that its application for semiconductors in electronic devices is limited as a single sheet. Properties such as the high optical transparency have made graphene a popular material in photovoltaic cell research due to the need for transparent electrodes that exhibit a high conductivity. It is also predicted to be the future for flexible electronics because of its combination of high electron mobility and impressive Young's Modulus and tensile strength. This allows graphene to be flexible without significant distortion of its fundamental properties.

The vast array of applications for graphene born out of its unique versatility could be discussed in detail for seemingly ages. The exciting and wide-ranging properties have been intensively researched throughout the academic world; a simple Web of Knowledge search yields over 21,000 research articles containing the keyword "graphene" in 2015 alone [10]. This itself provides a sufficient challenge insofar as de-convoluting the most current and cutting edge research in what is an almost endless wave of graphene articles. Consequently, this chapter is focussed specifically upon the current research directions for graphene as a material utilized within fuel cells.

9.2 Barriers to Commercialisation of Graphene

It is futile to discuss graphene only in terms of its wonderful benefits and all-encompassing ability (see above). While the picture painted by academia is that of a new age material wonder, in reality there are huge barriers to making graphene into a commercial success [11]; not least the production of it! The breakthrough research papers that are cited to excess (correctly or incorrectly) are celebrated for their relative simplicity in graphene production. It is true that graphene can be relatively simple to make by peeling it from a slab of Highly Ordered Pyrolytic Graphite (HOPG) with a piece of adhesive tape [6, 12]. Yet, if single layer graphene is ever to be a success, it needs to be fabricated on a scale much bigger than this.

Therein lies a huge issue for graphene commercialisation at the moment—there are very few methods that can synthesize long range ordered single layer graphene sheets. Perhaps the best method for this at the moment is Chemical Vapour Deposition (CVD) of graphene upon a metallic substrate such as nickel or copper [13]. In fact the technology has advanced towards roll-to-roll production of CVD graphene with sheet sizes in excess of 30 in [14]. However even these graphenes are suspected to be defective or contain metal impurities [15], and more common CVD methods do not have the ability to be scaled up to a financially feasible and replicable industrial standard.

Resultantly many have had to settle for lesser quality graphenes for meaningful research purposes. Whilst it is true that the fundamental properties of graphene have been established and characterised for single/double graphene layers, such properties have not been able to be put into practise due to the lack of production method for large single layer graphene sheets. Researchers have therefore had to compromise on graphene quality in many applications in order to exploit some potential benefits, and while the properties are not quite the same, there have been some significant breakthroughs in graphene research that have allowed improvements in data storage [16], smart screens [17], optical sensors [18] and electronics [19], and even biosensors [20]. Figure 9.2 [21] provides an interesting digression on the quality of graphene *versus* the application, and it is immediately evident that some areas of research can be accelerated even without large single layer graphene sheets (see above). Fortunately, for fuel cells, the quality required falls within liquid exfoliation and epitaxial growth upon silicon carbide routes, though future application of single-layer graphenes to fuel cell research is not completely off the horizon. Further information on the production methods of graphene can be found in recent reviews [22–24].

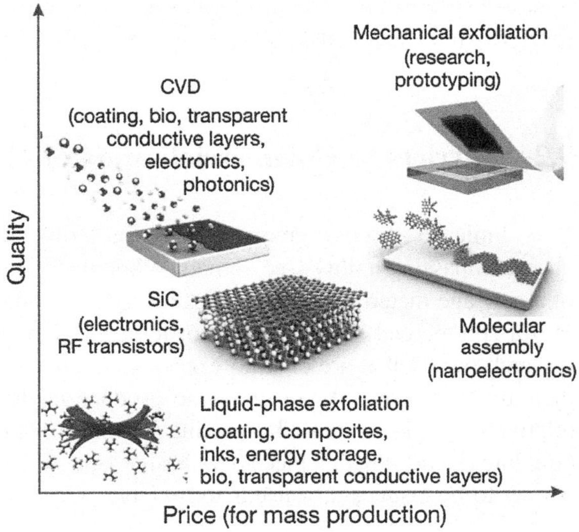

Fig. 9.2 Potential graphene applications correlated against the quality of precursor graphene. Reproduced from [21] with permission from Nature

9.3 Fuel Cells

The improvement of fuel cells is potentially vital towards achieving a more sustainable planet. With the ever-increasing demand for fuel consumption (directly or indirectly), society must turn towards more sustainable fuels and less towards fossil fuels, which are in finite supply. Several fuels have been considered as greener solutions, such as bioethanol, biodiesel, hydrogen, or methanol, but the technology to produce sufficient energy from these without a combustion engine has yet to be perfected. Fuel cells are an attractive proposition because they are theoretically more efficient than a combustion engine and create greener waste products [25]. They have the ability to operate under a range of different circumstances, and a range of different fuels, and thus hold some promise for future applications in energy and transport.

The origin of fuel cells dates back to the mid-1800s; the first reported fuel cell in this era was the sulphuric acid fuel cell that used iron, copper, copper sulphate, and porcelain in its design. The principle of this fuel cell, much like a battery, is the difference in reactivity between the copper and iron drives a load through transport of electrons from the most reactive to the least reactive metal. In this sense, the electrolyte (sulphuric acid/copper sulphate) acts as an electron reservoir to drive a current through the load. Modern fuel cells are similar to this, at least in terms of the driving force originating from electrochemical reactions occurring in tandem at an anode and a cathode. Fuel cells are differentiated from batteries in that they are refuelled chemically, rather than electrochemically [26, 27].

There are several types of fuel cell that are still researched today; the important graphene-related types are listed in Table 9.2 [28–31]. Perhaps of most importance are Proton Exchange Membrane Fuel Cells (PEMFCs), the general design of which is pictured in Fig. 9.3 [32]. Briefly, a PEMFC contains three major sections: an anode, a cathode, and a Proton Exchange Membrane (PEM). A fuel that undergoes oxidation reactions, such as methanol, is fed to the anode that is normally

Table 9.2 Non-exhaustive list of fuel cell types incorporating graphene in modern research

Fuel cell	Example(s)	Advantage(s)	Disadvantage(s)	References
Enzymatic	Glucose	Selective good fuel conversion rate	Can usually only operate at low temperature	[28]
Solid oxide	Hydrocarbon Hydrogen	High temperature Small design	Carbon fouling Peroxide creation	[29]
PEM	Methanol Bioethanol	Potential long lifetime reduced crossover reactions	Cathode fouling	[30]
Microbial	Organic wastes	Environmental clean-up energy from waste	Can usually only operate at low temperature	[31]

constructed from a porous platinum-based material. In this reaction, protons are liberated (see 9.1).

$$CH_3OH + H_2O \rightarrow CO_2 + 6H^+ + 6e^- \quad (9.1)$$

In the oxidation of methanol, six protons are liberated at the electrode surface, while the six resultant electrons are scavenged by the electrode, to be transported to the cathode via the load powered by the fuel cell. A negative electric field is created at the fuel cell cathode, causing migration of charged species in the solution phase towards the PEM, which is specifically tailored to allow protons to be transported from the anode to the cathode and thus be recycled by the system. Normally, the reaction at the cathode is the Oxygen Reduction Reaction (ORR), and the protons are required to create water, or more unfavourably, hydrogen peroxide (H_2O_2). Provided the PEMFC does not deteriorate through electrode/membrane fouling or temperature destabilisation, the only thing that should limit its operation is the fuel input (See Fig. 9.3).

The current of the circuit is dependent upon the surface area of the anode and cathode, as well as the electron transfer efficiency between the fuel and the catalyst. The potential delivered by the fuel cell is the difference between the electrode potentials of the anodic and cathodic processes. One must therefore understand the required current density and cell voltage requirements prior to fuel cell design in order to select appropriate anodes and cathodes. Of course, this is a very basic description. The deliverance of high current to a system can have adverse effects upon the electron transfer at the anode, and therefore the potential of the fuel cell. Many researchers devote their attention towards tailoring fuel cells to exhibit desirable current and potential outputs, which are application-dependent.

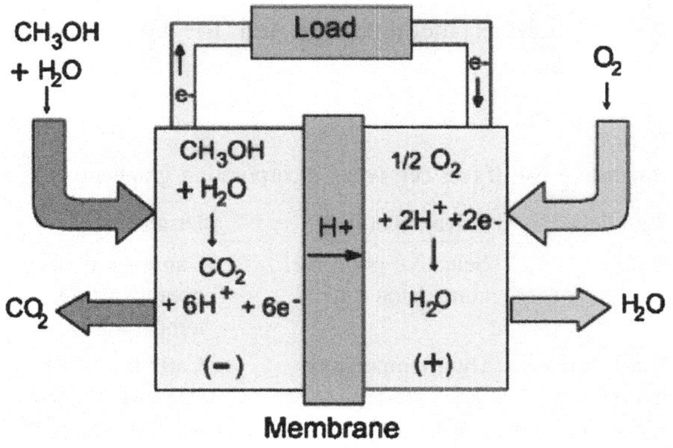

Fig. 9.3 Generic fuel cell design. In this example, methanol is used as the fuel. Methanol is oxidised on a working anode in one side, producing protons and electrons. The electrons are conducted through the circuit, creating a current to power the load, while the protons diffuse through a membrane to the cathode region. The protons reduce oxygen to produce water. Reproduced from [32] with permission from the Brazilian Chemical Society

In terms of graphene fuel cells, there are a number of opportunities for graphene to be successfully incorporated into the fuel cell design, and perhaps even scope to replace Pt as a fuel cell catalyst. Graphene could potentially be used as a fuel cell catalyst; that is, it could facilitate the electron transfer reactions between the anode and the fuel, perhaps improving current densities or voltages for a specific application if the electrode potential is sufficient. Secondly, if graphene has limited use as a fuel cell catalyst, its surface area could be cleverly manipulated as a fuel cell catalyst support. This would allow anodes and/or cathodes to exhibit much greater current densities if necessary, especially if 3D graphene architectures could be developed. These first two applications apply for both oxidation and reduction reactions. Thirdly, graphene may be incorporated into the fuel cell as a material that permits only protons to transfer to the cathode from the anode, which would improve both the efficiency and lifetime of a fuel cell by the reduction of crossover reactions. With these points in mind, we turn towards literature examples, where graphene has been incorporated into fuel cell design for one or more of these purposes; the remainder of the chapter shall focus upon pertinent examples.

9.4 Graphene Fuel Cells

The term "graphene fuel cells" is used here in a very loose sense, because it specifically relates to graphene's use as either a catalyst, catalyst support, or as part of a PEM within the fuel cell. This section has been written to discuss graphene-based applications.

9.4.1 Pristine Graphene

Pristine graphene is defined in this text as a single-layer, defect-free, impurity-free graphene with a high percentage of H terminations and a minimal amount of oxygenated terminating species. Unfortunately, such graphene is practically impossible to produce in any realistic quantity! The adhesive tape method originally reported in 2004 offers a highly pure form [6], but the graphene electrode created from this quantity of graphene is so small that practicality and scalability becomes an issue. Some solution-based graphenes are highly pure; in particular ones that are prepared through sonication of graphite in a solvent such as ethanol [33]. Wet chemically fabricated graphenes tend to contain impurities, surfactants, metal ions, and defects [22]. Perhaps the highest purity method is CVD graphene, where graphene is grown from a gaseous carbon source on a metal support inside a rig. CVD methods have advanced to the point where roll-to-roll graphene can be produced in 30 square inch quantities [14]. Such graphenes are rarely applied (nor are they probably useful) for fuel cell applications. CVD graphene has, however, been reported to be beneficial for Anaerobic Electrochemical Membrane

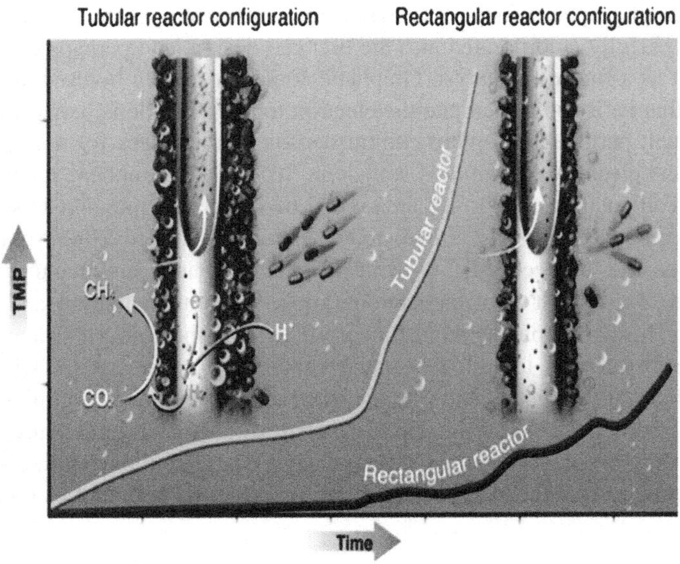

Fig. 9.4 Schematic of the substrate interactions between tubular and rectangular electrode configurations in the AnEMBR. The tubular system allows a bigger surface area for microbially-based reduction reactions, specifically for the genesis of molecular hydrogen from wastewaters. Reprinted with permission from [34] Copyright 2015 American Chemical Society

BioReactors (AnEMBRs) [34]. The general principle of the process is the same as a fuel cell, so the device has been discussed. The purpose of the original research article was to assess how differences in specific cathode surface area and electrode spacing contributed to the overall cell performance, using rectangular- and tubular-shaped electrodes as comparisons (Fig. 9.4). The Schematic provided in Fig. 9.4 depicts a larger biofilm forming around the tubular design due to its higher specific surface area, leading to a higher transmembrane pressure when the device is in operation. The pressure is indicative of hydrogen production initiated by an external electrical load. The paper celebrates the rectangular design as a success due to its apparent longevity compared to the tubular design; the high surface area of the latter increases current density, but also creates a bigger biofilm due to the electric field created at the surface. The biofilm therefore acts as a resistive layer between the cathode and the electrolyte over a period of time, and is the major contributor to device failure. The rectangular system forms a biofilm that is ten times less thick than the tubular system (0.4 μm) [34], meaning that the hydrogen evolving from the reaction can act as a scrubber to replenish the electrode surface. The graphene layer in this case provides the platform for microbes to anchor upon and transduce electrical current into the evolution of hydrogen through microbial interactions. The increase in surface area means a higher percentage of graphitic defects to act as "current suppliers" to the microbes, as the pristine nature of the CVD graphene cannot lend itself to electrochemical reactions without an excessive electrical load. The latter point is perhaps the reason why the device requires an external electrical load, alas differentiating it from a fuel cell!

Pristine graphene has, however, been investigated as an electrode material for the ORR in acidic conditions [35], where the electrochemical reduction potentials of solution-based pristine graphene were found to be extremely high. Consequently, the thermodynamic barrier required to catalyse the ORR is enormous for pristine graphene in acidic media, effectively meaning that pristine graphene electrodes require extremely high temperatures for efficient operation. The reason for this is thought to be due to pristine graphene's largely unreactive basal surface providing a slow rate of electron transfer for almost every electroactive species in solution [36]. This idea is at the centre of much debate, but there is compelling evidence to suggest that electron transfer in graphitic materials comes from the edges of a graphene sheet, rather than the basal plane [22]. Resultantly, pristine graphene itself should not be considered as a useful material for electrocatalysis in fuel cells per se, but may be useful for its large surface area if materials can be grafted upon its surface.

Pristine graphene may also be useful as a PEM in a fuel cell. Research has demonstrated that nothing will penetrate a graphene sheet, including H^+ ions, but the introduction of tiny holes (approximately 1 Å diameter) in a graphene sheet will allow H^+ ions to permeate the graphene sheet only, and further H^+ transport can be achieved by modifying graphene sheets with metal particles [37]. Without atomic level perforations, it is expected that H^+ would take billions of years to penetrate a graphene sheet [37], yet graphene with holes may act as a molecular colander; such is the *modus operandi* for a PEM. Further reports indeed support the notion that pristine graphene is impermeable to even protons, which is attributed to the dense electron cloud surrounding graphene; yet there is evidence to suggest that protons can transfer through graphene at rare point defects, as reported by Achtyl et al. [38]. Their work uses a novel approach to measuring proton transfer across graphene by observing potential differences of the graphene support material, which is a fused silica that undergoes acid/base reactions with protons that are transferred across graphene. The phenomenon is predictably pH-dependent, as demonstrated by means of pH alteration. Intuitively, anionic species present on a surface will trap proton movement through the membrane because of an ion-ion interaction, indicating that graphene-based membranes may benefit from a pH closer to the Point of Zero Charge (PZC) for graphene, which would almost certainly rule out alkaline fuel cells due to the formation of negatively charged surface species. Figure 9.5 depicts computational images for water-mediated proton transfer through atomic defects (proton channels), and shows the extent to which different types of graphene defects will permit protons to transfer through the graphene membrane [38].

Aside from its use as part of a PEM, pristine graphene itself has yet to establish itself as an outright electrode material in fuel cell design. However, many graphene variants have appeared as candidates for electrode improvements in fuel cell applications. Therefore the remainder of the chapter will focus upon graphene's variants, such as reduced graphene oxide and nitrogen-doped graphene.

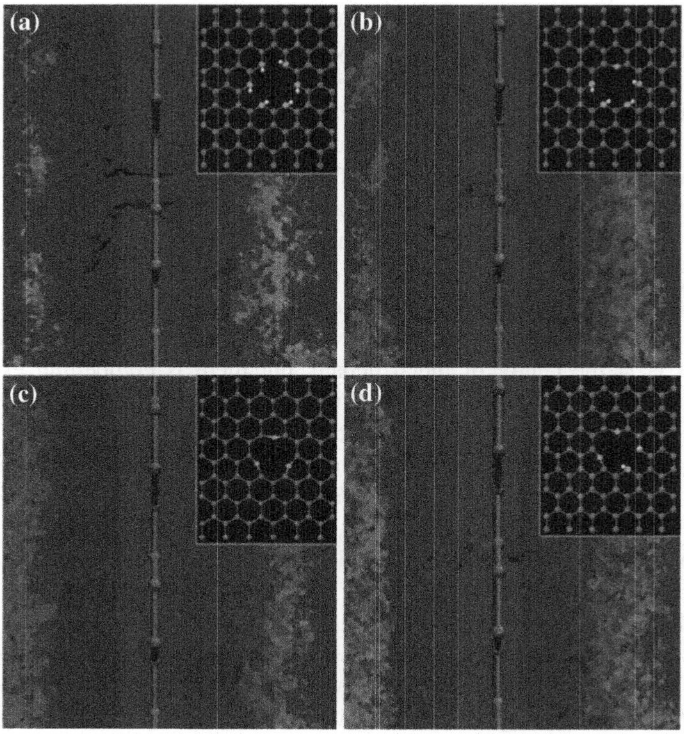

Fig. 9.5 Proton channel formation from computational experiments performed in [38]. Each image represents a different atomic defect: six OH groups (**a**); four OH groups and one oxygen atom (**b**); two OH groups and two oxygen atoms (**c**); and three oxygen atoms (**d**)

9.4.2 Laser-Induced Graphene

One production method for graphene that is used to a far lesser extent than wet synthesis or mechanical exfoliation is the so-called Laser-Induced Graphene (LIG) method. LIG is a method of graphene production that uses the energy from a CO_2 laser to reduce polyaniline into graphene [39]. The resultant graphene is not a 2D graphene sheet, however, but a 3D graphene network that can be patterned by the fabrication process and tailored for the desired application (e.g. Supercapacitors). Of course, any material exhibiting a 3D network may show significant advantages in the case of fuel cells because a porous 3D structure can lead to higher current densities; something which some fuel cells require.

It is well established that graphene exhibits favourable electrochemical responses as part of nanocomposite materials [40–42], however 3D graphene networks of graphene/metal nanoparticle materials are observed less frequently, perhaps due to a lack of production method. The LIG method does offer a route towards such 3D graphene/nanoparticle networks, however, provided the polyaniline substrate is

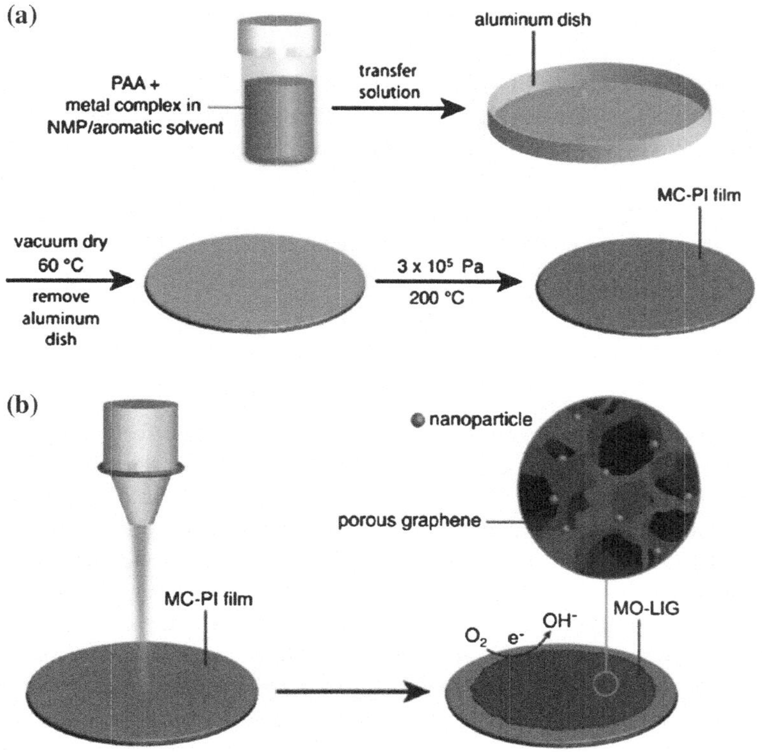

Fig. 9.6 Schematic diagram for the production of a metal nanoparticle/3D graphene network based upon LIG. Reprinted with permission from [43]. Copyright 2015 American Chemical Society

prepared correctly. Figure 9.6 depicts the fabrication process for LIG with metal particles embedded into the network [43]. Briefly, the polyaniline and metal complex is dissolved in an aromatic solvent and cast onto an aluminium dish, where it is vacuum dried. The resultant polyaniline/metal complex film is irradiated with the CO_2 laser, which reduces ambient oxygen on the polyaniline surface, leaving the 3D graphene network as the end product. The LIG without metal intercalation displays a significant overpotential for the ORR when compared to a typical Pt/C cathode in alkaline conditions. The metal-impregnated 3D graphene networks display differing catalytic performances to the ORR, all of which are superior to its LIG parent material. A cobalt metal complex incorporated into the LIG reduces the ORR potential to that of Pt/C [43]. Thus, a potential method for cheaper ORR cathodes may be achievable through this route. One problematic aspect of these electrodes is that they favour the two electron pathway for the ORR, meaning that O_2 reduced to HO_2^-, which can be a problematic intermediate that attacks electrode surfaces and reduces the longevity of the cathode.

The fabrication of LIG is a relatively new concept in graphene research, and for that reason there are few research articles focussing upon its use as a fuel cell anode or cathode. However its 3D nature may be an exciting property that can be exploited for fuel cell anodes and cathodes, particularly when hybridised with metal complexes. We expect many further research articles emerging for LIG in the near future. The formation of the 3D network also bypasses the problem of restacking/annealing of graphene sheets, polyaniline, and metal particles, as has been the case for many other types of graphene [44].

9.4.3 Reduced Graphene Oxide

One aspect of graphene research that should be a candidate for condemnation is researchers naming Reduced Graphene Oxide (RGO) as "graphene" in their research manuscripts. While it is true that reducing graphene oxide by thermal, electrochemical, or chemical routes creates a more "graphene-like" material, the resultant RGO is usually full of defects and oxygenated species that dictate the observed electrochemistry when incorporated into electrode design. Researchers are essentially creating graphitic oxides rather than graphenes in these instances. In fuel cell design, the problems arisen for sensing applications are therefore similar because fuel cells inherently function using electrochemical reactions. Nevertheless, RGO may be a useful prospect for fuel cell electrodes because the introduction of oxygenated species alters the conductive properties of graphene. The oxygenated species in RGO interact with metal particles, and provide nucleation points for catalysis, making RGO a potentially attractive material for fuel cell electrodes, either as standalone electrodes or as catalyst supports for noble metals such as Pt. Indeed, research has shown that the strong oxidising and reducing agents used within the fabrication procedures for wet chemical graphenes may have a detrimental effect upon microbial consortia in microbial fuel cells, thus limiting their use in such applications [44]. Furthermore the production method can be particularly hazardous due to the strong acids and potent oxidising agents used, meaning strict temperature control is required in the experimental design; these problems could prove costly for mass production.

There are several production methods for RGO [22] using an array of reducing agents such as $NaBH_4$ or H_2O_2. The fabrication method itself may be questioned because some methods may leave trace metals within the RGO structure that may give false positive readings in terms of catalytic behaviour; there are copious amounts of literature indicating that contaminants are responsible for the behaviour of many graphenes and RGOs [11, 45–48]. Resultantly, searching academic literature for RGO that is repeatable and metal-free is challenging, however some examples are worthy of consideration for fuel cell design, especially since it may be easier to scale RGO up to a replicable standard than other graphene forms.

The use of RGO as a fuel cell anode may be realised due to RGOs ability to anchor Pt nanoparticles upon its surface. The defects and oxygenated species attributed to RGO make ideal nucleation points for Pt nanoparticles, which effectively increases the electroactive surface area of Pt. The large exposed Pt surface can then reduce oxidation potentials of methanol, meaning that the operating temperature of the fuel cell can theoretically be less because there is a smaller thermodynamic barrier for oxidation due to the favourable interaction between Pt and RGO, when compared to Pt on carbon [49]. The current density for this specific example was also 2.5 times that of Pt on carbon, indicating a fuel cell anode that creates more current and requires less energy for conversion. However, the problem with procedures such as this lies within the use of the Hummers method for graphene oxide production. The method uses strong acids and oxidising agents to exfoliate graphene sheets from bulk graphite and create graphene oxide. However, metal ions such as manganese can impregnate the graphene oxide product, and poor quality acids in the production process can leave residual metal ions in the graphene oxide product, which can influence or dominate the electrochemical responses in fundamental investigations.

There are other very similar procedures reported in the literature that point towards RGO/Pt materials as effective fuel cell anodes. Li and co-workers report RGO/Pt with an electrochemically active surface area of 36.47 $m^2 g^{-1}$, which is estimated from H^+ adsorption on Pt in the −0.2–0.0 V region. The improved surface area can be exploited to create high current densities in fuel cells and indicates that RGOs surface area can be successfully exploited as a catalyst support [50]. Unfortunately the method utilises Hummers-type graphene oxide, and one should exercise caution as has already been discussed previously.

The field has seen a move away from traditional Pt on RGO systems, however, as there is only so much information one can gather from such systems. Today, many researchers look towards creating new composites from RGO and metal ions or compounds (including Pt mixtures) that may rival Pt-based systems in the future. Manganese dioxide and Pt decorated RGO is one such example of a methanol catalyst, with reports suggesting that MnO_2/Pt/RGO electrocatalysts can exhibit a "specific activity" of 1.98 mA cm^{-2}, which is approximately twice as much as Pt/C reported in the same work [51]. The RGO in this case provides a conductive link between the catalytic particles and the underlying electrode material. One has to question the robustness of their study, however, as they appear to rush to conclusions despite the graphene oxide being poor quality and the current density output dropping by 30 % after only 200 voltammetric scans. Ferrous NanoParticles (FeNPs) are also a material of choice, in particular for ORR catalysts because they exhibit catalytic activity towards oxygen. A synthetic route for the preparation of FePt/RGO was reported by Chen et al. and opened a new route to the design of a new ORR catalyst with a high temperature stability and electrocatalytic effect when compared to Pt/C [52]. The FePt nanoparticles are synthesized via a solvothermal method reported in their paper, and grafted upon Hummers-type graphene oxide using ethylene glycol. Figure 9.7 shows the activity of several substrates tested within the work, and highlights how RGO influences the current output for several electrodes. For example, at a voltammetric half-wave potential of 0.90 V, the FeNPs worked up with 2 mg L^{-1} GO suspension show over four times higher

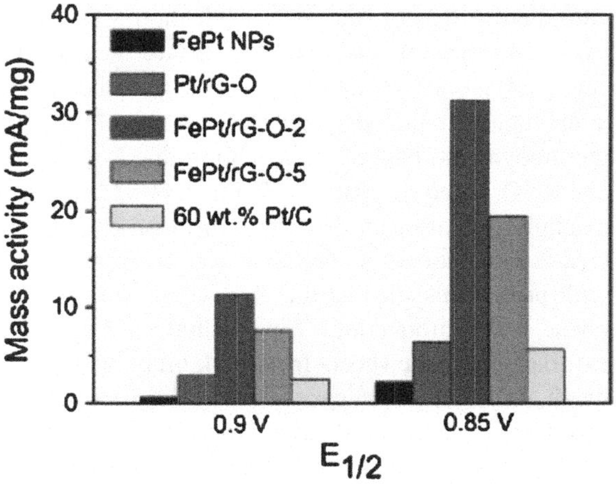

Fig. 9.7 The current response per unit of mass for several cathodes designed within [44] FePt NPs (*black*); Pt/RGO (*red*); FePt/2 mg L^{-1} RGO (*purple*); FePt/5 mg L^{-1} RGO (*green*); Pt/C (*pale green*). Adapted from [52] with permission from Elsevier

current activity per unit of mass compared to Pt on carbon. This effect is further pronounced at a voltammetric half-wave potential of 0.85 V. A 60 % increase in oxidation current is reported when using 2 mg mL^{-1} GO suspension in the fabrication process. The potential cycling stability of the ORR catalyst was demonstrated to be reasonably good, with an approximate 20 % loss in electrochemically active surface area after 750 voltammetric cycles.

The ability of Graphene Oxides (GO) has been explored as an option to scale up graphene as a membrane material, with many examples reporting successful proton transport across the oxygenated species in GO. The oxygenated species in GO may create some advantages over pristine graphene, such as a reduction in hydrophobicity increasing the water retention of GO, a property that may improve proton movement and exhibit an improved operation at higher temperature compared to typical PEMs such as Nafion. However, it may also cause membrane swelling over longer periods and reduce the lifetime of a GO membrane. Indeed, examples of GO as PEMs have demonstrated exactly this, with power densities nearing that of a Nafion membrane fuel cell [53].

Graphene-based foams have also been developed for fuel cell design. Researchers pay tribute to the characteristics of a foam that improves the surface area of a material, whilst incorporating graphene into the foam to allow conductivity. The advantages of using foams is that the pore size is big enough to allow microbial consortia to grow upon the surface; one such example of a foam consists of a GO/agarose gel formation that is freeze-dried to create the graphene-based foam [54]. A typical anode material for an MFC would be a porous conductive material such as a carbon cloth, as it offers a large surface area with a big enough pore size for biofilms to form as part of the anode. The graphene-based foam however produces a reported power density over 4 times that of a fuel cell with a carbon cloth anode.

There are many other examples of original research articles reporting improved electrocatalysis for both anodes and cathodes. The advantage of such methods is their ability to create a graphene sample relatively quickly and in a large amount that may be able to be scaled up towards mass production. However, while many attempts are seemingly successful, there is an obvious issue with technology transfer into industry. The obvious flaw with these electrodes is reproducibility, and the development of a reproducible wet chemical method for the graphene oxide precursor. The quality of wet chemical graphenes is contentious and may not be completely practical for an industrial electrode standard to be used in fuel cell designs. The same could be said for the synthesis of the nanoparticles used in many of these applications. Finally, the true stability of such electrodes is not covered in any detail, but for potential cycling in ideal lab conditions, so more robust testing of electrode stability is desirable to move RGO-based fuel cell electrodes forward.

9.4.4 N-Doped Graphene

It would be forgivable to wonder why one would tinker with graphene by introducing dopants in the design of fuel cells. If graphene exhibits such a unique versatility and applicability, then why would anybody wish to change (or threaten to change) the wide-ranging properties that this material has to offer. Yet graphene has been problematic in some respects, not least because it is a zero band gap material. This essentially means that the conductive properties of graphene in its pristine form are difficult to control. Nitrogen-doped graphene (N-graphene) on the other hand allows researchers to tune the conductive properties of graphene and broaden its applications for supercapacitors, sensors, and batteries [55].

Indeed several research papers have investigated N-graphene as a material for application within fuel cells as a result of its "tunability". One such application within fuel cell design is that of a catalyst support. It is widely accepted that Pt is a preferential catalyst for the ORR cathode in fuel cells, mainly due to its direct 4-electron reaction pathway with oxygen that creates water as a by-product instead of harmful materials such as H_2O_2. Therefore instead of reinventing the wheel, many researchers exploit the large surface area of graphenes to anchor Pt nanoparticles in place, meaning that the effective surface area of the cathode can be much greater than if platinum was used alone. The advantage of N-graphene is its stronger interaction within Pt nanoparticles, and its improved conductivity on its basal plane. The N heteroatom within the ring structure provides a nucleation point for catalytic reactions (much in the same way as RGO) and the combination of Pt and N-graphene has been shown to improve the durability of the catalyst for the ORR, something that is required in abundance for a fuel cell due to the temperatures reached during operation. This property derives from the introduction of a shift in atomic charge distribution due to the introduction of an unpaired electron from the N atom. The delineation of atomic charge across the graphene structure creates nucleation points for anchoring of Pt nanoparticles, and catalytic sites for electron transfer [55].

One problem with N-graphene for fuel cell design is knowing exactly what type of N-dopants one has introduced to the graphene, and the relative effect of these dopants upon the ORR. Research has shown that changing the type of N-dopant significantly affects the catalytic behaviour of N-graphenes; amino-type dopants are generally more inactive than pyridine-type dopants and graphitic-type dopants, according to Zhang et al. Their work demonstrates that N-graphenes have a potential application for supporting Pt for the oxidation of methanol at fuel cell anodes [56]. Increased current densities are demonstrated for N-graphenes that are fabricated at high temperatures and show that this change in production method creates more graphitic- and pyridine-type N dopants into the structure.

Aside from catalyst supports, N-graphenes are also capable of catalysing electrochemical reactions *without* the need for a noble metal catalyst. As discussed previously, the purpose of incorporating N atoms in a graphene structure is to tailor conductive properties, and therefore it makes sense that tweaking the conductivity of graphene may have a significant effect upon its ability to oxidise or reduce materials such as methanol, oxygen, or hydrogen in a fuel cell. Resultantly, N-doped graphene was reported in 2010 to be an effective metal-free electrocatalyst for the ORR [57]. The approach, reported by Qu and co-workers, reported the preparation of CVD N-graphene in the presence of ammonia and its catalytic effect towards the ORR. N-graphene in this context was demonstrated to display significantly improved current densities over a 1.0 V potential range, as is demonstrated in Fig. 9.8, which highlights a current density with a magnitude over three times higher than that of a Pt catalyst upon a carbon substrate at certain voltammetric potentials. Furthermore, Fig. 9.8 demonstrates that graphene-only electrodes display no advantages in terms of current density when compared to composite materials, which is likely due to the electrochemically inactive nature of graphene. A further property of this electrode is its resistance to crossover reactions, due to its high selectivity towards the ORR under the specified conditions. Crucially, it exhibited excellent stability over 200,000 potential cycles, indicating that the ORR cathode could be used repeatedly without a significant loss in current density. This

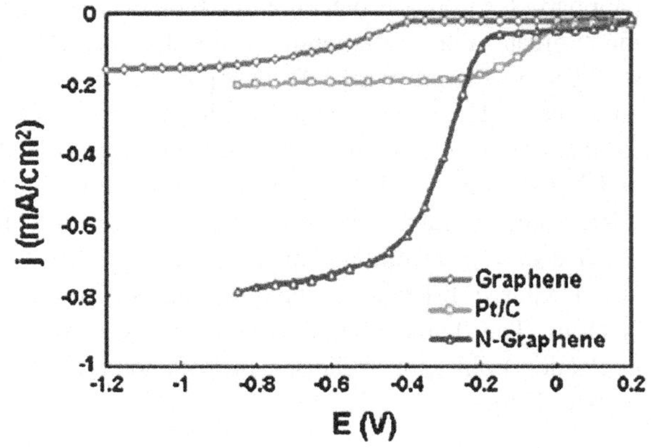

Fig. 9.8 Current-potential curves for (*red*) graphene, (*green*) Pt/C, and (*blue*) N-graphene towards the ORR in alkaline conditions. The N-graphene displays a much larger current density output over the experimental potential range, indicating a possibility for N-graphene as a fuel cell cathode. Reprinted with permission from [57]. Copyright 2015 American Chemical Society

property makes it an attractive proposition for future cathode design. Of course, the disadvantage in this case is the CVD method, as it can only create minimal amounts of N-graphene and therefore scaling up may be difficult until improved fabrication methods are found.

N-graphene coupled with cobalt nanoparticles has also been researched as a potential alternative to cathodes for the ORR in fuel cells. The fabrication procedure for such a material pyrolyses a mixture of $Co(NO_3)_2$, sucrose, and urea together at 800 °C in the presence of argon to directly form N-graphene [58]. The ring structure of urea in this case provides the N hetetroatoms for N-graphene, while the 6 membered ring of sucrose forms the lattice structure required for graphitic materials. This method is yet another metal-free method for fuel cell cathodes that could theoretically drive down the cost of cathodes due to the lack of need for noble metals. Polarisation curves of the N-graphene indicate a reduced reduction potential for the ORR when compared to pristine graphene, and onset potentials for the ORR exhibiting a minimal overpotential of 21 mV compared to Pt/C [58].

N-graphene in its several forms offers very similar properties to RGO in terms of its nucleation points for nanoparticles, and therefore creates an opportunity to exploit the surface area of graphene using electrochemical methods. N-graphene is restricted by reproducibility as it is difficult to control the amount and type of N defect within the graphene structure. N-graphene electrodes require bulk manufacture with a reproducible fabrication method if they are to be incorporated into fuel cell design in the future, but the proof-of-concept approaches disseminated within this chapter look promising in terms of their stability, in particular towards the ORR for fuel cell cathodes.

9.5 Conclusions and Outlook

The exponential rise in graphene research (as alluded to in the introduction) has made it difficult for researchers to properly assess the latest advancements in graphene research, and has likely affected fuel cell research. Whether this effect has been positive or not is yet to be determined. From a pristine graphene perspective, the likelihood of this material taking its place as a fuel cell catalyst to replace Pt seems highly unlikely, due to the barriers to mass production for single layer graphene sheets. Even if the desired defect-free, single layer graphene with large sheet sizes could be made en masse, there would be drawbacks including cost and electrochemical unreactivity. Therefore, it is likely that if graphene is ever to replace Pt as a fuel cell catalyst it will do so in one of its less pure forms, such as RGO or N-graphene. There is a glimmer of hope for graphene as a material for a PEM, but more research is required in this area first before industry is likely to take a leap of faith with graphene.

RGO and N-graphene have indeed shown potential for use as fuel cell anodes and cathodes, for electrochemical oxidation of fuels such as methanol, and of course for the problematic ORR. Some literature examples report that N-graphene could even replace Pt as an electrocatalyst for the ORR, which is certainly an encouraging sign for the field. However, most examples focus upon the ability for such graphenes to bind to Pt nanoparticles, thus improving the electrocatalytic surface area for Pt. If there is one property of graphene that could be exploited for improves current densities, then its high surface area is exactly that; it is therefore no surprise that RGO and N-graphenes have demonstrated a capability to improve current densities through clever manipulation of the reactive sites associated with N- and O-based functional groups.

Despite the positive signs, research for the incorporation of graphene into fuel cell design is relatively slow when compared to other applications of graphene. In fact, it seems that almost every area of graphene research that involves electrochemical reactions is slow in one way or another. It appears likely that many different combinations of RGO and N-graphenes and nanoparticle composites will be tested in the coming years, in order to find an electrocatalyst for fuel cell anodes and cathodes, and perhaps some may make it past the proof-of-concept stage and become more replicable, rigorous, and practical. We believe it is unlikely that there will be a graphene-based fuel cell on the market in the foreseeable future as there is not enough high quality research output for graphene as a fuel cell catalyst in order to sell the idea to a potential investor, in what is an already stagnant research area; however, we hope to be proved wrong.

References

1. A.K. Geim, K.S. Novoselov, The rise of graphene. Nat. Mater. **6**, 183–191 (2007)
2. J.C. Whitaker, *The Electronics Handbook*, 2nd edn. (Taylor & Francis, 2005)
3. J.-U. Lee, D. Yoon, H. Cheong, Estimation of Young's modulus of graphene by Raman spectroscopy. Nano Lett. **12**, 4444–4448 (2012)
4. K.I. Bolotin, K.J. Sikes, Z. Jiang, M. Klima, G. Fudenberg, J. Hone et al., Ultrahigh electron mobility in suspended graphene. Solid State Commun. **146**, 351–355 (2008)
5. C. Lee, X. Wei, J.W. Kysar, J. Hone, Measurement of the elastic properties and intrinsic strength of monolayer graphene. Science **321**, 385–388 (2008)
6. K.S. Novoselov, A.K. Geim, S.V. Morozov, D. Jiang, Y. Zhang, S.V. Dubonos et al., Electric field effect in atomically thin carbon films. Science **306**, 666–669 (2004)
7. R.R. Nair, P. Blake, A.N. Grigorenko, K.S. Novoselov, T.J. Booth, T. Stauber et al., Fine structure constant defines visual transparency of graphene. Science **320**, 1308 (2008)
8. D.A.C. Brownson, D.K. Kampouris, C.E. Banks, An overview of graphene in energy production and storage applications. J. Power Sources **196**, 4873–4885 (2011)
9. C.-H. Kuo, J.-M. Wu, S.-J. Lin, W.-C. Chang, High sensitivity of middle-wavelength infrared photodetectors based on an individual InSb nanowire. Nanoscale Res. Lett. **8**, 327 (2013)
10. Web of Knowledge Literature Search, 2015
11. E.P. Randviir, D.A.C. Brownson, C.E. Banks, A decade of graphene research: production, applications and outlook. Mater. Today **17**, 426–432 (2014)
12. K.S. Novoselov, D. Jiang, F. Schedin, T.J. Booth, V.V. Khotkevich, S.V. Morozov et al., Two-dimensional atomic crystals. Proc. Natl. Acad. Sci. U.S.A. **102**, 10451–10453 (2005)
13. P.Y. Huang, C.S. Ruiz-Vargas, A.M. van der Zande, W.S. Whitney, M.P. Levendorf, J.W. Kevek et al., Grains and grain boundaries in single-layer graphene atomic patchwork quilts. Nature **469**, 389–392 (2011)

14. S. Bae, H. Kim, Y. Lee, X. Xu, J.-S. Park, Y. Zheng et al., Roll-to-roll production of 30-inch graphene films for transparent electrodes. Nat. Nanotechnol. **5**, 574–578 (2010)
15. D.A.C. Brownson, C.E. Banks, The electrochemistry of CVD graphene: progress and prospects. Phys. Chem. Chem. Phys. **14**, 8264–8281 (2012)
16. M.F. El-Kady, R.B. Kaner, Scalable fabrication of high-power graphene micro-supercapacitors for flexible and on-chip energy storage. Nat. Commun. **4**, 1475 (2013)
17. F. Bonaccorso, Z. Sun, T. Hasan, A.C. Ferrari, Graphene photonics and optoelectronics. Nat. Photonics **4**, 611–622 (2010)
18. K.P. Loh, Q. Bao, G. Eda, M. Chhowalla, Graphene oxide as a chemically tunable platform for optical applications. Nat. Chem. **2**, 1015–1024 (2010)
19. Z. Wei, D. Wang, S. Kim, S.-Y. Kim, Y. Hu, M.K. Yakes et al., Nanoscale tunable reduction of graphene oxide for graphene electronics. Science **328**, 1373–1376 (2010)
20. Y. Shao, J. Wang, H. Wu, J. Liu, I.A. Aksay, Y. Lin, Graphene based electrochemical sensors and biosensors: a review. Electroanalysis **22**, 1027–1036 (2010)
21. K.S. Novoselov, V.I. Falko, L. Colombo, P.R. Gellert, M.G. Schwab, K. Kim, A roadmap for graphene. Nature **490**, 192–200 (2012)
22. D.A.C. Brownson, D.K. Kampouris, C.E. Banks, Graphene electrochemistry: fundamental concepts through to prominent applications. Chem. Soc. Rev. **41**, 6944–6976 (2012)
23. S. Park, R.S. Ruoff, Chemical methods for the production of graphenes. Nat. Nanotechnol. **4**, 217–224 (2009)
24. C. Soldano, A. Mahmood, E. Dujardin, Production, properties and potential of graphene. Carbon **48**, 2127–2150 (2010)
25. G. Pearson, M. Leary, A. Subic, J. Wellnitz, in *Sustainable Automotive Technologies*, ed. by S. Hung, A. Subic, J. Wellnitz. Performance Comparison of Hydrogen Fuel Cell and Hydrogen Internal Combustion Engine Racing Cars (Springer, Heidelberg, 2011), pp. 85–91
26. B.C.H. Steele, A. Heinzel, Materials for fuel-cell technologies. Nature **414**, 345–352 (2001)
27. A.J. Appleby, Fuel Cell Handbook,1988
28. I. Willner, Y.M. Yan, B. Willner, R. Tel-Vered, Integrated enzyme-based biofuel cells–a review. Fuel Cells **9**, 7–24 (2009)
29. D. Sarantaridis, A. Atkinson, Redox cycling of Ni-based solid oxide fuel cell anodes: a review. Fuel Cells **7**, 246–258 (2007)
30. V.M. Vishnyakov, Proton exchange membrane fuel cells. Vacuum **80**, 1053–1065 (2006)
31. Z. Du, H. Li, T. Gu, A state of the art review on microbial fuel cells: a promising technology for wastewater treatment and bioenergy. Biotechnol. Advances **25**, 464–482 (2007)
32. T. Iwasita, Fuel cells: spectroscopic studies in the electrocatalysis of alcohol oxidation. J. Brazilian Chem. Soc. **13**, 401–409 (2002)
33. D.A.C. Brownson, L.J. Munro, D.K. Kampouris, C.E. Banks, Electrochemistry of graphene: not such a beneficial electrode material? RSC Adv. **1**, 978–988 (2011)
34. C.M. Werner, K.P. Katuri, A.R. Hari, W. Chen, Z. Lai, B.E. Logan, et al., Graphene-coated hollow fiber membrane as the cathode in anaerobic electrochemical membrane bioreactors—effect of configuration and applied voltage on performance and membrane fouling. Environ. Sci. Technol. (2015)
35. E.P. Randviir, C.E. Banks, The oxygen reduction reaction at graphene modified electrodes. Electroanalysis **26**, 76–83 (2014)
36. D.A.C. Brownson, C.E. Banks, *The Handbook of Graphene Electrochemistry*, 1st edn. (Springer, 2014)
37. S. Hu, M. Lozada-Hidalgo, F.C. Wang, A. Mishchenko, F. Schedin, R.R. Nair et al., Proton transport through one-atom-thick crystals. Nature **516**, 227–230 (2014)
38. J.L. Achtyl, R.R. Unocic, L. Xu, Y. Cai, M. Raju, W. Zhang, et al., Aqueous proton transfer across single-layer graphene. Nat. Commun. **6** (2015)
39. J. Lin, Z. Peng, Y. Liu, F. Ruiz-Zepeda, R. Ye, E.L.G. Samuel, et al., Laser-induced porous graphene films from commercial polymers. Nat. Commun. **5** (2014)

40. X. Chen, G. Wu, J. Chen, X. Chen, Z. Xie, X. Wang, Synthesis of "clean" and well-dispersive Pd nanoparticles with excellent electrocatalytic property on graphene oxide. J. Am. Chem. Soc. **133**, 3693–3695 (2011)
41. Z.-S. Wu, S. Yang, Y. Sun, K. Parvez, X. Feng, K. Müllen, 3D Nitrogen-doped graphene aerogel-supported Fe_3O_4 nanoparticles as efficient electrocatalysts for the oxygen reduction reaction. J. Am. Chem. Soc. **134**, 9082–9085 (2012)
42. E. Yoo, T. Okata, T. Akita, M. Kohyama, J. Nakamura, I. Honma, Enhanced electrocatalytic activity of Pt subnanoclusters on graphene nanosheet surface. Nano Lett. **9**, 2255–2259 (2009)
43. R. Ye, Z. Peng, T. Wang, Y. Xu, J. Zhang, Y. Li et al., In situ formation of metal oxide nanocrystals embedded in laser-induced graphene. ACS Nano **9**, 9244–9251 (2015)
44. M.V. Kannan, G. Gnana kumar, Current status, key challenges and its solutions in the design and development of graphene based ORR catalysts for the microbial fuel cell applications. Biosens. Bioelectron. **77**, 1208–1220 (2016)
45. A. Ambrosi, M. Pumera, The CVD graphene transfer procedure introduces metallic impurities which alter the graphene electrochemical properties. Nanoscale **6**, 472–476 (2014)
46. L. Wang, A. Ambrosi, M. Pumera, "Metal-free" catalytic oxygen reduction reaction on heteroatom-doped graphene is caused by trace metal impurities. Angew. Chem. Int. Ed. **52**, 13818–13821 (2013)
47. C.H.A. Wong, Z. Sofer, M. Kubešová, J. Kučera, S. Matějková, M. Pumera, Synthetic routes contaminate graphene materials with a whole spectrum of unanticipated metallic elements. Proc. Natl. Acad. Sci. U.S.A. **111**, 13774–13779 (2014)
48. A. Ambrosi, C.K. Chua, B. Khezri, Z. Sofer, R.D. Webster, M. Pumera, Chemically reduced graphene contains inherent metallic impurities present in parent natural and synthetic graphite. Proc. Natl. Acad. Sci. U.S.A. **109**, 12899–12904 (2012)
49. Y. Xin, J.-G. Liu, Y. Zhou, W. Liu, J. Gao, Y. Xie et al., Preparation and characterization of Pt supported on graphene with enhanced electrocatalytic activity in fuel cell. J. Power Sources **196**, 1012–1018 (2011)
50. Y. Li, W. Gao, L. Ci, C. Wang, P.M. Ajayan, Catalytic performance of Pt nanoparticles on reduced graphene oxide for methanol electro-oxidation. Carbon **48**, 1124–1130 (2010)
51. A.T.E. Vilian, M. Rajkumar, S.-M. Chen, C.-C. Hu, K.M. Boopathi, C.-W. Chu, High electrocatalytic performance of platinum and manganese dioxide nanoparticle decorated reduced graphene oxide sheets for methanol electro-oxidation. RSC Adv. **4**, 41387–41397 (2014)
52. D. Chen, X. Zhao, S. Chen, H. Li, X. Fu, Q. Wu et al., One-pot fabrication of FePt/reduced graphene oxide composites as highly active and stable electrocatalysts for the oxygen reduction reaction. Carbon **68**, 755–762 (2014)
53. T. Bayer, S.R. Bishop, M. Nishihara, K. Sasaki, S.M. Lyth, Characterization of a graphene oxide membrane fuel cell. J. Power Sources **272**, 239–247 (2014)
54. L. Yang, S. Wang, S. Peng, H. Jiang, Y. Zhang, W. Deng et al., Facile fabrication of graphene-containing foam as a high-performance anode for microbial fuel cells. Chem. Eur. J. **21**, 10634–10638 (2015)
55. H. Wang, T. Maiyalagan, X. Wang, Review on recent progress in nitrogen-doped graphene: synthesis, characterization, and its potential applications. ACS Catal. **2**, 781–794 (2012)
56. L.-S. Zhang, X.-Q. Liang, W.-G. Song, Z.-Y. Wu, Identification of the nitrogen species on N-doped graphene layers and Pt/NG composite catalyst for direct methanol fuel cell. Phys. Chem. Chem. Phys. **12**, 12055–12059 (2010)
57. L. Qu, Y. Liu, J.-B. Baek, L. Dai, Nitrogen-doped graphene as efficient metal-free electrocatalyst for oxygen reduction in fuel cells. ACS Nano **4**, 1321–1326 (2010)
58. G. Zhang, W. Lu, F. Cao, Z. Xiao, X. Zheng, N-doped graphene coupled with Co nanoparticles as an efficient electrocatalyst for oxygen reduction in alkaline media. J. Power Sources **302**, 114–125 (2016)

Chapter 10
Mesoporous Materials for Fuel Cells

Jin Zhang and San Ping Jiang

Abstract Fuel cell is the most efficient and environmentally friendly energy conversion technology to directly convert the chemical energy of fuels such as hydrogen, methane, methanol, ethanol and hydrocarbons into electricity with high efficiency and very low greenhouse gas emission. In fuel cells, porosity is a singular attribute, which controls not only the transport of fuel/oxidant to reactive sites but also the length or area of the three phase boundary or electrode/electrolyte interface where the electrochemical reaction occurs. Mesoporous materials with well-defined and highly ordered pore arrays in the range of 2–50 nm in diameter have attracted increasing attention as effective electrode and electrolyte materials for fuel cells in particular the low temperature proton exchange membrane fuel cells (PEMFCs) due to their unique water retention properties of ordered mesopores. This chapter starts with a brief review of the application of mesoporous materials in high temperature solid oxide fuel cells (SOFCs), followed by a detailed description and discussion of the advances in the synthesis and application of the mesoporous structured materials in PEMFCs, including mesoporous polymer membranes, such as mesoporous sulfonated block copolymers and *meso*-Nafion membrane, mesoporous inorganic/polymer composite membranes, and mesoporous inorganic materials such as mesoporous silica. The development of mesoporous carbon and metal oxide as electrocatalysts and catalyst supports in PEMFCs is also reviewed. The fundamental relationship between space symmetry, pore size, porosity, ordering level of the mesoporous materials and their electrochemical and fuel cell performance has been discussed.

Abbreviation

APTES	3-thiopropyltrimethoxysilane
AST	Accelerated stress test
ATMP	Amino tris(methylene phosphonic acid)
BCPs	Block copolymers

J. Zhang · S.P. Jiang (✉)
Fuels and Energy Technology Institute and Department of Chemical Engineering, Curtin University, Perth, WA6102, Australia
e-mail: S.Jiang@curtin.edu.au

Q. Li (ed.), *Nanomaterials for Sustainable Energy*,
NanoScience and Technology, DOI 10.1007/978-3-319-32023-6_10

BMIm-BF4	1-butyl-3-methylimidazolium tetrafluoroborate
BMIm-TFSI	1-butyl-3-methylimidazolium bis(trifluoromethane sulfone) imide
BTCA	Benzotriazole-5-carboxylic acid
CIM	Conventional impregnation method
CMETCS	2-(carbomethoxy)ethyltrichloro silane
CSPTMS	2-(4-chlorosulfonylphenyl)ethyltrimethoxysilane
DEPTES	Diethylphosphatoethyltriethoxy silane
DFT	Density functional theory
EIm-TfO	*N*-Ethylimidazolium Trifluoromethanesulfonate
EISA	Evaporation-induced self-assembly
FSAS	1,2,2-Trifluoro-2-hydroxy-1-trifluoromethyl-ethanesulfonic acid sultone
HEX	Hexagonally packed cylinders
HMS	Hollow mesoporous silica
HPL	Hexagonally perforated lamellae
HPAs	Heteropolyacids
HPW	Phosphotungstic acid
ICE	Internal combustion engine
IEC	Ion exchange capacity
IPN	Interpenetrating polymer network
KU	Keggin unit
LAM	Lamellae
MEA	Membrane-electrode-assembly
MPTMS	Mercaptopropyltrimethoxy silane
MSA	Methane sulphuric acid
OSPN	Organosiloxane network
PA	Phosphoric acid
PAC	Proton acceptance capacity
PBI	Polybenzenemidazole
PEM	Proton exchange membrane
PEMFC	Proton exchange membrane fuel cell
PESf	Polyethersulfone
PETMS	Phenethytrimethoxy silane
PFS	Poly(4-fluorostyrene)
PFSA	Perfluorosulfonic acid
PILs	Protic ion liquids
PMMA	Poly(methyl methacrylate)
PMO	Periodic mesoporous organic silica
PS-*b*-sPHS	Poly(styrene-block-sulfonated hydrostyrene
PSS-PMB	Poly(styrenesulfonate-methylbutylene)
PTES	Phenyltriethoxysilane
RH	Relative humidity
SAN	Poly(styrene-co-acrylonitrile)
SAXS	Small angle x-ray scattering

SDAs	Structural directing agents
SPEEK	Sulfonated poly(ether ether keton)
sPHS	Sulfonated polyhydroxystyrene
SMBS	Sulfonated cube mesoporous benzene-silica
SPI	Sulfonated polyimide
sPPO	Sulfonated poly(2,6-dimethyl-1,4-phenylene oxide)
SPPSU	Sulfonated poly(phenylsulfone)
sPS-PMMA	Sulfonated polystyrene-poly(methyl methacrylate)
sSEBS	Sulfonated polystyrene-*block*-poly(ethylene-*ran*-butylene)-*block*-polystyrene
TEOS	Tetraethyl orthosilicate
TPS	3-(trihydroxysilyl)-1-propanesulfonic acid
VIM	Impregnation method

10.1 Introduction

Very different from conventional internal combustion engine (ICE), fuel cells directly transfer the chemical energy of the fuels into electricity with high efficiency, low pollution and environmental friendly. Thus, fuel cells are considered as one of the leading technologies for future power generation and are potential and viable candidates to moderate the fast increase in power requirements and to minimize the impact of the increased power consumption on the environment [1]. Fuel cells are also versatile devices ranging from low temperature fuel cells such as proton exchange membrane fuel cells (PEMFCs) to high temperature (500–1000 °C) solid oxide fuel cells (SOFCs). SOFCs allow internal reforming, promote rapid kinetics with non-precious materials and offer high flexibilities in fuel choice. The other advantages of fuel cells are the quiet operation, and flexible modulability.

Among various types of fuel cells, proton exchange membrane fuel cell (PEMFC) shows low operation temperature and fast start-up and shut down cycle, and is considered to be the most promising power source for applications ranging from portable electronic devices, transportation vehicles to power stations. Figure 10.1 shows the operation principle of a PEMFC. The critical components of PEMFC are the membrane electrode assembly (MEA) with proton exchange membrane (PEM) sandwiched between the electrocatalytic anode and cathode. In the case of H_2 fuel, H_2 is oxidized at the anode and the proton is transported to the cathode through the PEM. The electrons produced flow to the cathode through an external load, reducing ambient oxygen to water. PEM has the essential function for proton transport as well as separator to prevent the mixing between fuel and oxidant. Perfluorosulfonic acid (PFSA) membranes such as Nafion® are the state-of-the-art PEMs with high chemical stability and outstanding proton conductivity at hydration condition. The high proton conductivity of Nafion results

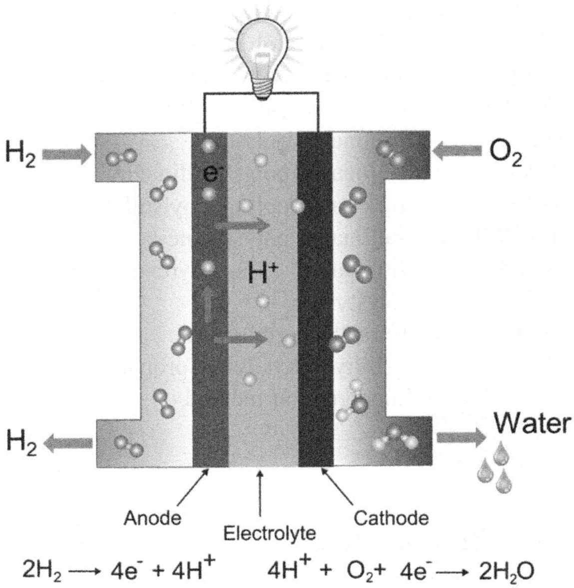

Fig. 10.1 Scheme of a proton exchange membrane fuel cell

from microphase separation derived from unique skeleton of hydrophobic C–F backbone grafted with hydrophilic $-SO_3H$ side chains. However, the proton conductivity of Nafion is extremely dependent on the hydration of the membrane, which requires a high level of relative humidity (RH) for the stable operation of Nafion membrane based fuel cells. It not only complicates water management but also increases the cost. On the other hand, the high cost and scarcity of Pt, the most common catalyst used in fuel cells, results in high cost of fuel cells and hence limits the wide application of fuel cells. Herein, development of PEM with high proton conductivity at elevated high temperature or/and low humidity and catalysts with low cost and high activity are the most critical challenge for PEMFCs.

In fuel cells, porosity is a singular attribute, which controls not only the transport of fuel/oxidant to reactive sites but also the length or area of the three phase boundary or electrode/electrolyte interface where electrochemical reaction occurs. Mesoporous materials, characterized by well-defined and ordered pore arrays in the range of 2–50 nm in diameter, were first reported by the scientists in Mobil on the MCM-41S series of mesoporous silica materials [2, 3]. Mesoporous materials show unique properties of high hydrophilicity of the –OH groups on the highly ordered *meso*-channel and capillary condensation effect of the mesopores [4]. Thus, mesoporous materials can store and release water at elevated temperatures [5]. When the mesoporous structure is introduced in the PEMs, it can facilitate water-assisted proton transportation at high temperature and low relative humidity. Moreover, high surface area and highly ordered mesoporous channels of mesoporous materials not only increase the utilization of catalyst, but also fasten the mass transport in the electrode. Consequently, mesoporous materials as a versatile material have been extensively used in PEMFC to reduce the RH dependence of PEMs, simplifying the water management system and increasing the efficiency of catalyst and reducing the cost of the fuel cells.

Mesoporous materials have also found applications in high temperature SOFCs, however, the biggest challenge in the application of mesoporous materials in SOFC is the collapse of the mesopores during the high processing (up to 1400 °C) and operation (800–1000 °C) temperatures. Thus, in this chapter, the feasibility of application of mesoporous materials in SOFCs is briefly reviewed but the emphasis will be on the development of various types of mesoporous materials including mesoporous organic polymers, mesoporous organic/inorganic materials, and mesoporous inorganic materials for low temperature PEM fuel cells. Mesoporous carbon materials with outstanding performance as catalyst support in the electrode are also reviewed. The structure-activity relationship of the mesoporous materials based PEMs and catalysts and their properties of cell performance are discussed.

10.2 Mesoporous Materials in SOFCs

Ozin's group is probably the first to study the feasibility of mesoporous structured materials in high temperature SOFCs via a soft chemistry strategy in which glycometalates and metal complexes are co-assembled with a supramolecular surfactant template to produce a single phase binary or ternary mesoporous yttria-zirconia or metal-yttria-zirconia [6, 7]. Nickel and platinum are incorporated into the mesoporous structure as nanoclusters with dimensions comparable to the size of pores. The mesostructure showed thermal stability up to 800 °C. Hung et al. [8] prepared mesoporous yttria-stabilized zirconia (YSZ) powders using cetyltrimethylammonium bromide (CTAB) as the surfactant and urea as the hydrolyzing agent and $ZrO(NO_3)\cdot 6H_2O$ and $Y(NO_3)_3\cdot 6H_2O$ as inorganic precursors and showed that the mesoporous structure collapsed at temperatures equal or higher than 900 °C. Hard template route based on mesoporous silica is also used to synthesize highly crystalline and thermally stable gadolinium doped ceria (GDC) powders [9]. The mesoporous structured doped ceria shows high BET surface area (>120 $m^2\ g^{-1}$) and could be used as the composite electrodes for O_2 reduction reaction of SOFCs [10]. However, the use of mesoporous structured YSZ or doped ceria as electrolyte materials in SOFCs is limited due to the collapse of the structure at the processing temperature of the cells, typically in the range of 1300–1400 °C.

Most recently, Su et al. [11] reported a simple and effective method for the synthesis of mesoporous $Ba_{0.5}Sr_{0.5}Co_{0.8}Fe_{0.2}O_{3-\delta}$ (BSCF) oxide with high specific area of 25 $m^2\ g^{-1}$. In this method, original BSCF was prepared by conventional EDTA-citric acid complexing method, followed by treatment in H_2O_2, forming mesoporous BSCF (see Fig. 10.2). The ordered mesopores are in the range of 3.2 nm and high power density of 1.8 W cm^{-2} was demonstrated at 800 °C in hydrogen/air. The stability of the mesoporous structure under the continuous

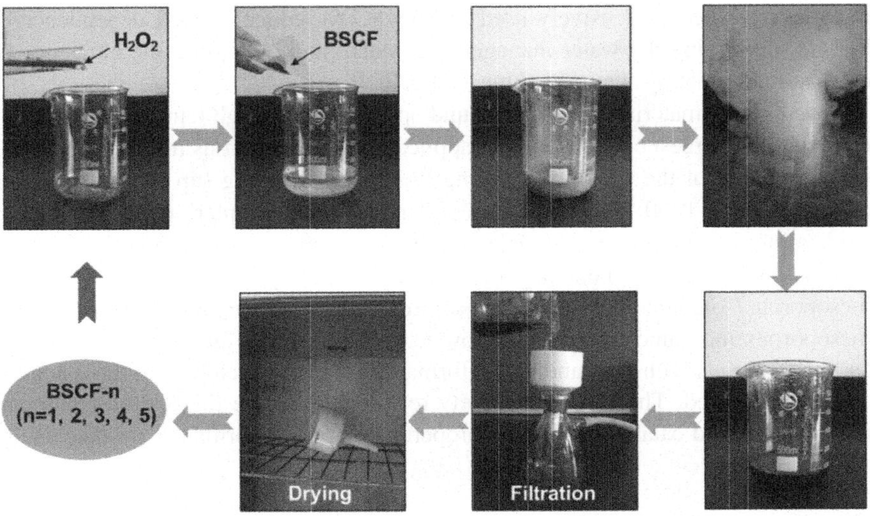

Fig. 10.2 Schematic of the preparation of mesoporous BSCF-n (n = 1, 2, 3, 4, and 5), where n is the cycle number (Reproduced from [11] with permission. Copyright (2015) Elsevier)

operation at such high temperature has not been evaluated. Irregular mesopores could also be formed during the solution thermal decomposition of nitrate precursors, but the performance is much poorer [12].

10.3 Mesoporous Polymer Based PEM

Because the conductivity of Nafion membrane is extremely sensitive for hydration, numerous approaches have been applied to increase the water retention capability of Nafion at low RH and high temperature. Various kinds of hydrophilic fillers including SiO_2 particles [13], heteropolyacids [14], carbon nanotubes [15], TiO_2 particles [16], TiO_2 nanotubes [17, 18] and ZrO_2 particles [19] are introduced into the Nafion membrane in order to enhance the water retention. However, addition of hydrophilic fillers would reduce the mechanical strength as well as the thermal and chemical stability of the composite membrane. On the other hand, mesoporous materials possess periodic and interconnected channels and high surface areas, which show advantages as PEMs in fuel cells, as shown in Fig. 10.3 [20]. The incorporation of mesoporous structures into PEMs such as Nafion could significantly reduce the dependence of proton conductivity upon the relative humidity. Given that mesoporous structure is introduced into the matrix of Nafion membrane, the water retention of Nafion membrane would be increased with minimum impact on the stability and mechanical strength of Nafion membrane. Besides, inspired by the amphiphilic polymer molecules, sulfonated block copolymers (BCPs) are also assembled into mesoporous membranes to retain the water in the mesopores.

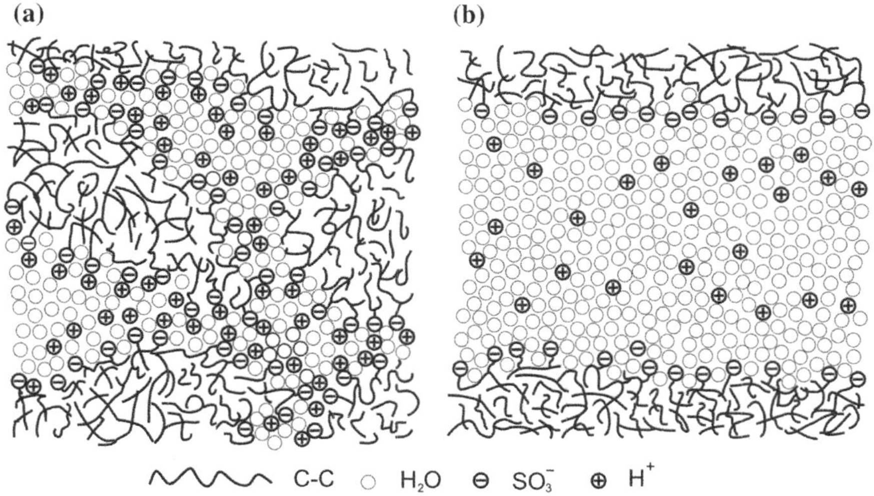

Fig. 10.3 Schematic of the microstructures of **a** random proton channels and **b** highly ordered mesoporous channels in polymer matrix

10.3.1 Mesoporous Nafion Membrane

Generally, in Nafion, hydrophilic ionic domains are randomly embedded in the hydrophobic C-F matrix. In the ionic domains, immobilized water molecules are strongly bonded on the $-SO_3H$ groups by hydrogen bond [21], while mobile water stays in the middle region the ionic channels (see Fig. 10.3a). At low RH, the H_2O molecules on the side-chain region of Nafion molecules lose firstly, and then followed by the H_2O loss in ionic cluster domains. The water loss in the ionic domains leads to a shrinkage of the ionic channels in the Nafion membrane, which in turn reduces the proton conductivity [22]. In order to decrease the water loss in the ionic domain, ordered mesoporous channels is introduced to replace the random ionic domains in Nafion membrane [23, 24]. One approach to fabricate mesoporous Nafion membrane is soft templated method. For soft template method, non-ionic block copolymer surfactant, PEO_{127}-PPO_{148}-PEO_{127} (Pluronic F108) is employed during the formation of Nafion membranes. Then the surfactant embedded in Nafion membrane is removed by reflux with hot water, resulting in the formation of mesoporous structure in Nafion (Fig. 10.4a). The mesoporous Nafion membrane shows higher water retention and proton conductivity in comparison with pristine Nafion membrane particularly under reduced RH, due to the capillary condensation effect within the confined mesoporous channels [25]. However, the low resolution and broad small angle x-ray scattering (SAXS) peak of as-synthesized mesoporous Nafion indicate the low order of mesopores because of the flexible non-ionic block polymer chains [24].

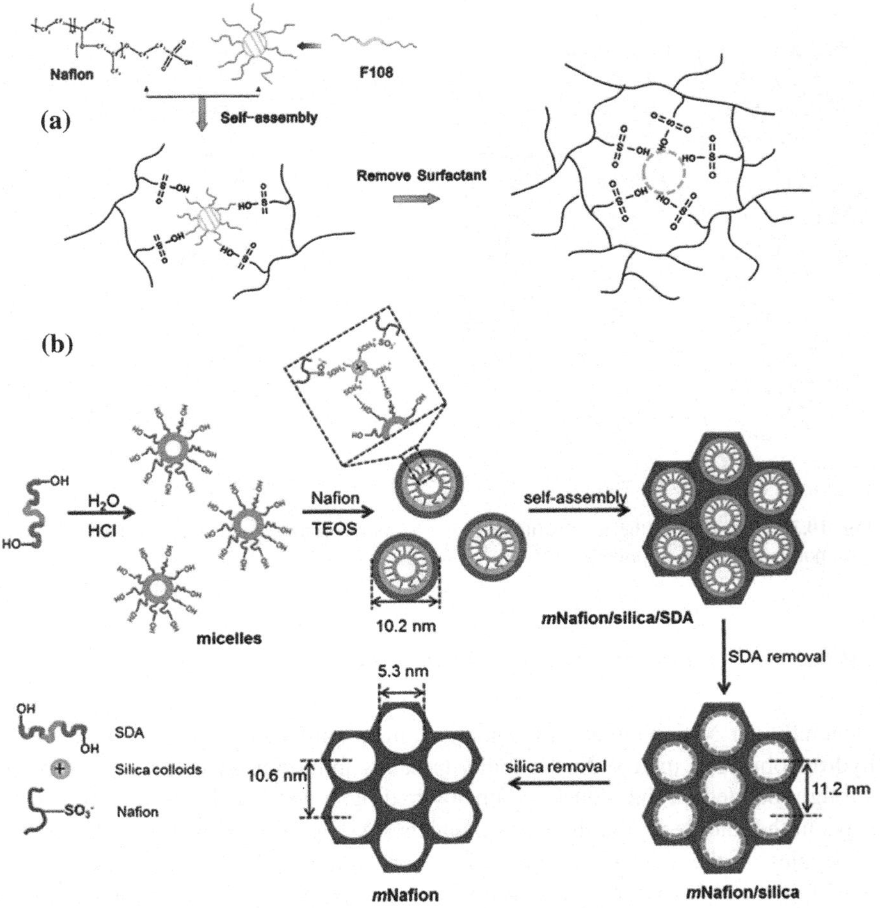

Fig. 10.4 Scheme of mesoporous Nafion membrane formation via **a** soft template method (Reproduced from [23] with permission. Copyright (2012) The Royal Society of Chemistry) and **b** hard template method (Reproduced from [27] with permission. Copyright (2014) The Royal Society of Chemistry)

In order to achieve long-range periodic mesoporous structures, mesoporous silica as hard template is employed to form highly ordered mesoporous Nafion membrane, and the synthesis procedures were shown in Fig. 10.4b [26]. The presence of mesoporous silica hard template significantly increases the degree of order of the mesoporous structure of Nafion membrane. In the initial stage, the presence of acidic HCl solution initiates and promotes the hydrolysis of tetraethyl orthosilicate (TEOS), forming protonated and positively charged silica colloids that play the role as bonding sites between Nafion ionomers and the structural directing agents (SDAs).

Driven by the electrostatic forces between the positively charged silica colloids and the negatively charged sulfonic acid groups of Nafion ionomers, self-assembly of Nafion and silica colloids occurs simultaneously, forming Nafion-silica pairs. Then the Nafion-silica pairs anchor to the ether block of the SDAs through hydrogen bonding. Nafion ionomers attached to the silica colloids stabilize the silica colloids, inhibiting the agglomeration and grain growth of silica colloids and forming an ordered mesoporous structure phase. Highly ordered mesoporous Nafion ionomers are achieved by removing the SDAs via refluxing and silica mediators via HF etching. The resulting mesoporous orientation leads to the increasing exposure of sulfonic groups to the hydrophilic phase and the formation of highly ordered proton transporting channels, facilitating fast proton mobility [27].

The water retention ability of mesoporous Nafion membrane synthesized via hard template at low relative humidity is indicated by the steep change in power output against RH (S curve) [27]. Taking mesoporous Nafion membrane with space symmetry of *P6mm* (2D-H Nafion in abbreviation) as an example, the initial voltage is about 0.68 V at 100 %RH and 80 % RH, while the cell voltage slightly decreases to 0.65 V when the RH decreases to 60 % RH. Furthermore, when the RH drops to 0 % RH, the cell voltage only slightly decreases from 0.56 to 0.47 V after 20 h test with current load of 100 mA cm^{-2} at 60 °C. On the other hand, Nafion 112 membrane based fuel cell can only be operated at high level of hydration (>80 % RH). When RH decreases to 60 %, the cell voltage goes to 0 within very short time under the identical condition as the mesoporous Nafion membrane. In the case of 0 % RH, the Nafion membrane cell ceases to operate with 20 min. The outstanding performance of mesoporous Nafion based cells at low RH is most likely due to the capillary condensation effect of water within the ordered and long-range periodic mesopores.

10.3.2 Mesoporous Block Copolymers

Mesoporous sulfonated block copolymers have been widely investigated as potential PEMs because of their well-ordered nanostructure with precise control in structure and morphology [28–31]. Generally, sulfonated block copolymer molecules have long C–C backbone grafted by $-SO_3H$ terminated side chains. The side chains not only improve movement of the $-SO_3H$ group, but also block the methanol crossover of the membranes. Moreover, the thermodynamic incompatibility between hydrophilic chains and hydrophobic chains maximizes the contact between similar blocks, while minimizes the contact between dissimilar blocks [32, 33]. Thus, mesoporous structure is achieved in BCPs by the microphase separation of hydrophilic and hydrophobic segments. For instance, when hydrophilic sulfonated polyhydroxystyrene (sPHS) is grafted onto the backbone of polystyrene to form poly(styrene-*block*-sulfonated hydrostyrene) (PS-*b*-sPHS), the membrane with ordered proton conductive channels possesses higher proton conductivities than sulfonated polystyrene-*block*-poly(ethylene-*ran*-butylene)-*block*-polystyrene (sSEBS) membranes with random proton channels [34]. Moreover, when the sulfonated poly (2,6-dimethyl-1,4-phenylene oxide) (sPPO) is grafted by hydrophobic poly (4-fluorostyrene) (PFS) side chains, well-ordered mesoporous structure with cylindrical *meso*-channels are formed for the proton transportation [35].

Molecular weight and temperature affect the mesoporous structure of block copolymers. Lamellae (LAM) morphology is discovered in the low molecular weight copolymer of poly(styrenesulfonate-methylbutylene) (PSS-PMB), while mesoporous structures, such as hexagonally perforated lamellae (HPL) and hexagonally packed cylinders (HEX), are found for the high molecular weight copolymer [36]. The molecular weight effect on the morphology of the block copolymers is consistent with the self-consistent field theory for the phase behaviour of copolymers [37]. Moreover, low molecular weight copolymer of PSS-PMB shows phase transitions from HPL to LAM when temperature is changed. On the other hand, high molecular weight copolymers constantly express the mesoporous structure against the temperature [38].

The proton conductivity of the PSS-PMB membranes under low RH depends strongly on the size of the mesopores. The vapour pressure of water confined in *meso*-channels, p_W^{vap}, is given by Kelvin and Young-Laplace equation [36, 39]:

$$\ln\left(\frac{p_W^{vap}}{p_{W,o}^{vap}}\right) = -\frac{2aV_w\gamma\cos\theta}{wRT} \tag{10.1}$$

where $p_{\mathrm{w,o}}^{\mathrm{vap}}$, γ and V_{w} are the normal vapor pressure, the surface tension and molar volume of water, respectively. R and T are the ideal-gas constant and the absolute temperature, respectively. θ is the contact angle between water and the pore surface, while w is the pore size (diameter of cylindrical pores and width of lamellar pores). The value of a is 1 for lamellae, and 2 for cylinders. Given that the pores in PSS-PMB membrane are perfectly wetting, that is, $\theta = 0$, calculation indicates that a rapid decrease of the pressure within the hydrophilic pores occurred when the pore size close to 5 nm. When the width of hydrophilic ionic channels is lower than 5 nm, the membrane prefers to retain water at low RH and elevated temperatures [36].

Besides the pore size, the proton conductivity of block copolymer membrane is also dominated by the morphology of the membranes. For instance, sulfonated polystyrene-poly(methyl methacrylate) (sPS-PMMA) self-assembles into LAM (Fig. 10.5a), HEX (Fig. 10.5b), HPL (Fig. 10.4c) morphologies, while the proton conductivity of sPS-PMMA diblock copolymers follows the trend of ISO < HEX < HPL < LAM, when the proton conductivity of the copolymer is normalized by the volume fraction of the conductive domains, as shown in Fig. 10.5e [40]. That is because lamellar phase possesses larger swelling and greater water uptake than others morphologies, leading to high efficiency in proton dissociation and diffusion from $-SO_3H$.

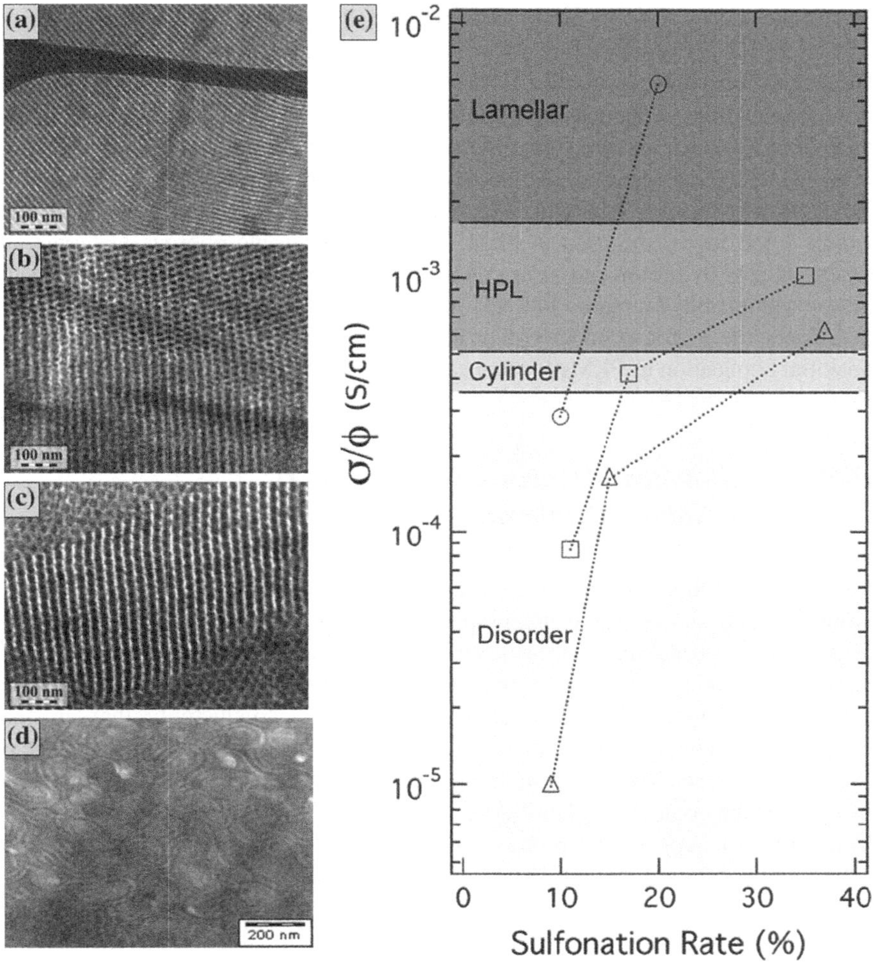

Fig. 10.5 TEM images of vapor annealing samples: **a** sPS-PMMA, SI_20 (LAM); **b** sPS-PMMA, S2_17 (HEX); **c** sPS-PMMA S2_35 (HPL); **d** PS-PMMA, lamellar structure and **e** the normalized proton conductivity σ/Ø as a function of the sulfonation degree for the three series: S1(*circle*), S2 (*square*) and S3(*rectangle*). Ø refers to the volume fraction of the PS, sPS and water domains (Reproduced from [40] with permission. Copyright (2008) American Chemistry Society)

10.4 Sulfonated Mesoporous Silica Base PEMs

Mesoporous inorganic materials possess desirable traits including large surface area, chemical inertness and mechanical stable framework. And the materials have been extensively studied as potential PEM candidates. Mesoporous materials can be embedded in polymeric membranes to form the organic/inorganic composite PEMs. The addition of mesoporous materials can enhance the water uptake of polyelectrolyte membrane and increase its proton conductivity at high temperature and low RH as well as the cell performance [41]. Moreover, the functionalized mesoporous materials are able to form effective proton conducting path by its highly ordered channels to improve the proton conductivity. Mesoporous silica is one of the commonly investigated mesoporous materials with highly ordered mesopores and high thermal stability [42]. Pure mesoporous silica shows negligible proton conductivity [43] and the conductivity of the mesoporous silica film is lower than 10^{-5} S cm^{-1} at 90 % RH, which is too low to be used in PEM. However, when mesoporous materials are functionalized by proton carrier or conductors, the proton conductivity of the materials is significantly increased [44, 45]. Thus, functionalization treatment is essential to enhance the proton conductivity of inorganic mesoporous materials or fillers for potential application as PEMs for fuel cells.

10.4.1 Sulfonated Mesoporous Silica Fillers for Nafion Membrane

Because of the high surface area, mesoporous silica contains large amount of OH groups, which shows promising property for water adsorption. Thus, the introduction of hexagonal mesoporous silica MCM-41 into Nafion membrane improves its thermal stability and enhances its water retention properties [46]. However, the composite membrane generally shows lower proton conductivity than pristine Nafion as the conductivity of mesoporous silica filler is much less than that of Nafion membrane. The decrease in conductivity becomes more severe with the increase of the content of silica fillers added.

In order to improve the proton conductivity of *meso*-silica/Nafion composite membranes while maintain the high water uptake, mesoporous silica can be functionalized by proton conducting groups or proton carriers prior to be introduced into polymer matrix [47]. Functionalized groups in the highly ordered channels of mesoporous silica are able to form effective proton conducting path to improve the proton conductivity. For instance, composite membrane filled with sulfonated mesoporous silica (SMPS) shows 50 times higher of proton conductivity than composite membrane with mesoporous silica [48]. The $-SO_3H$ groups in SMPS create efficient proton diffusion channels by hydrogen bond network in the interior surface of the mesopores of *meso*-silica. This is also confirmed by the observation that proton conductivity is increased by over one order of magnitude at 40 % RH, 80 °C when 1 wt% of sulfonated mesoporous organiosilicate is added into PA/PBI membrane [49].

Sulfonated *meso*-silica/Nafion composite membrane shows higher water uptake capability than recast Nafion membrane, which is due to high surface area and high porosity of sulfonated mesoporous silica. There are two types of proton diffusion mechanism with vehicle [50] and Grotthuss mechanism [51]. For vehicle mechanism, protons are diffused with the aid of water molecules. In Nafion membrane, the proton diffusion in the highly hydrated state is primarily assumed as vehicle mechanism. For Grotthuss mechanism, the protons transfers by hopping from one site to the neighbouring site by the hydrogen bond construction and reorientation of the water molecules. When the RH decreases from 98 to 80 %, the activation energy calculated in the range of 20 to 95 °C doubles from 9.1 to 18.2 kJ mol^{-1}, which is in the range of characteristic of Grotthuss mechanism (14–40 kJ mol^{-1}) [52]. This is due to that fact that when RH decreases, the number of hydrated proton carriers is also reduced. Consequently, Grotthuss mechanism will be a preferential pathway for proton transportation with minimum hydration water molecules. On the other hand, sulfonated mesoporous silica has high propensity to retain water and inhibits the volatility of water in Nafion at high temperature and low RH and then shows low dependence of the proton conductivity on hydration. The proton activation energy slightly increases from 7.1 to 7.6 kJ mol^{-1} when the RH decreases from 98 to 80 %. Therefore, the proton diffusion in the sulfonated mesoporous silica based composite membrane seems to occur by the vehicle mechanism even at crucial condition in terms of high temperature and low RH.

Besides mesoporous silica, hollow silica can also be incorporated into proton exchange membrane as micro-water reservoir [53], leading to the improved water uptake of the composite membrane. Moreover, when combining the water reservoir of hollow materials and periodic proton transport path of mesopores, hollow mesoporous silica (HMS) is developed as superior proton conductive filler under high temperature and low relative humidity [54]. Hollow mesoporous silica is able to be synthesized via two protocols: template method and self-templating etching method. Hard templates including SiO_2 [55], carbon [56], polystyrene (PS) [57] and soft templates like emulsion droplets and vesicle are used as core [58], and following by coverage with mesoporous silica shell. When the cores are etched by HF, NaOH and Na_2CO_3 solutions or calcined at high temperature, HMS is obtained. Self-templating etching contains surface-protected etching [59, 60], and structural difference-based selective etching [61]. Different from the hard template method, the template, SiO_2 for instance, is dissolved in alkaline solution and redeposited into HMS via the assistance of surfactant micelles. Jiang et al. synthesized HMS via hard template method (SiO_2 as template, Fig. 10.6a) and functionalized it by amino acids with sulfonic acid, phosphoric acid and carboxylic acid groups on the surface, as shown in Fig. 10.6d [62]. The acid-base HMS particles were incorporated into Nafion matrix and the composite membranes possessed high proton conductivity as 1.02×10^{-2} S cm^{-1} at 26.1 % RH, 80 °C, which is 11.1 times higher than that of recast Nafion membrane. However, the water uptake of Nafion/HMS (4.0 wt%) increased from 19.9 to 41.2 %, while the swelling ratio slightly raised from 7.42 to 8.11 %, from 25 to 80 °C, respectively. That is due to the water reservoir of HMS microspheres with large volume of lumen, leading to the retention of water under low RH, as shown in Fig. 10.6e [63].

The filler content significantly affects the performance of the composite membranes. When 3 wt% of sulfonated mesoporous silica is embedded into Nafion membrane, the conductivity of the composite membrane is increased from 1.0 to 1.2×10^{-1} S cm^{-1} at 100 % RH, 60 °C; on the contrary, the proton conductivity of pristine Nafion membrane decreases to 8.7×10^{-1} S cm^{-2} at the same condition

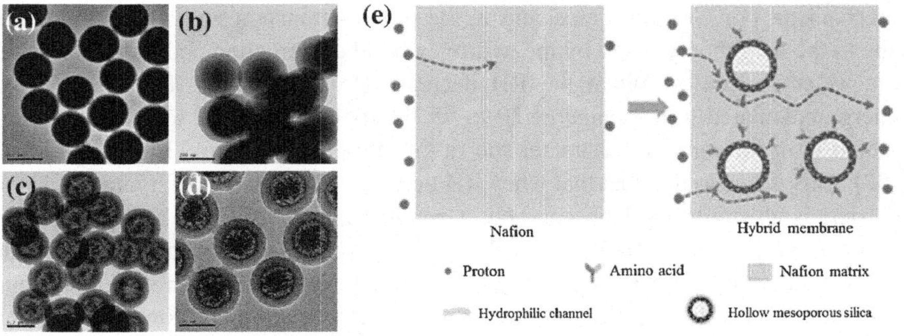

Fig. 10.6 TEM images of **a** solid SiO_2; **b** core-shell SiO_2; **c** hollow mesoporous silica (HMS); **d** sulfonic acid functionalized HMS. **e** Is the schematic for the water reservoir of functionalized HMS in Nafion composite membrane (Reproduced from [62] with permission. Copyright (2015) The Royal Society of Chemistry)

when 3 wt% pristine mesoporous silica is added [64]. However, when large amount of sulfonated mesoporous silica (>10 wt%) are added into Nafion matrix, the excessive fillers form the insulative phase by agglomerates, which prevents the proton transport in the conductive Nafion matrix (Fig. 10.7). Nevertheless, by optimizing the synthesis method, the filler content in Nafion membrane can be further improved with little detrimental effect on the conductivity of Nafion membrane. Pereira et al. [52] synthesized perfluorosulfonic acid/mesoporous silica hybrid membranes using evaporation-induced self-assembly (EISA). Silica phase grows in Nafion matrix in the presence of surfactant, water, solvent, and catalyst. And the surfactant was removed by boiling in 3 v/v% H_2O_2 solution. Mesoporous silica nanoparticles with average size of 50 nm were homogenously dispersed in the Nafion matrix with dense and smooth cross section image even when the filler content increased to 13 wt%. That is presumably due to the in situ growth of the

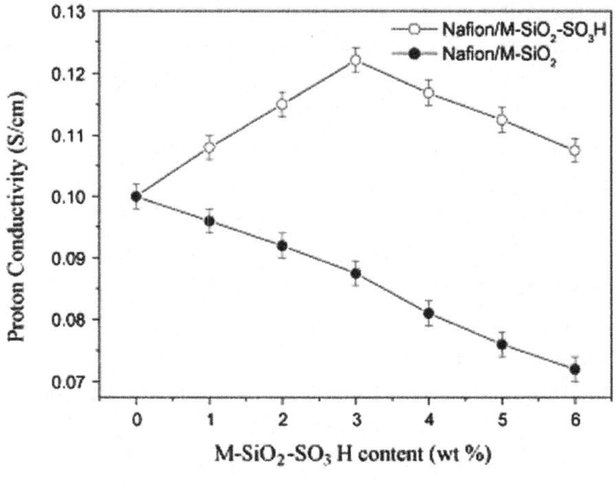

Fig. 10.7 Proton conductivity of native and modified *meso*-silica in Nafion membranes under 100 % RH, 60 °C (Reproduced from [64] with permission. Copyright (2007) Elsevier)

Table 10.1 Proton conductivities of Nafion based hybrid membrane with different fillers, and functional groups

Fillers and content	Function groups	Agents	Conductivity, mS cm^{-1}	References
MCM-41, 3 wt%	N/A	N/A	~100@100 °C, 80 % RH	[46]
Mesoporous silica, 3 wt%	N/A	N/A	87@60 °C, 100 % RH	[64]
Mesoporous silica, 3 wt%	$-SO_3H$	FSAS	108@60 °C, 100 % RH	[64]
Mesoporous silica, 5 wt%	$-SO_3H$	TPS	94@40 °C, 95 % RH	[65]
Mesoporous silica, 13 wt%	$-SO_3H$	CSPTMS	22@RT, 100 % RH	[52]
KIT-6, 2.5 wt%	$-SO_3H$	FSAS	N/A	[66]
MCM-41, MCM-48	$-SO_3H$	PETMS	N/A	[67]
		MPTMS		
	$-PO_3H_2$	DEPTES	N/A	
	–COOH	CMETCS	N/A	
Mesoporous silica, 5 wt%	$-SO_3H$	MPTMS	N/A	[47]
Al-MCM-41,0.5 wt%	$-SO_3H$	H_2SO_4	291@80 °C, 100 % RH	[68]
Pheny-*meso*-silica, 3 wt%	$-SO_3H$	H_2SO_4	12.9@RT, 100 % RH	[93]
HMS, 4 wt%	$-SO_3H$	Cysteine	10.2@80 °C, 21.6 % RH	[62]
	$-PO_3H_2$	Phosphoserine	7.91@80 °C, 21.6 % RH	
	–COOH	Aspartic acid	4.85@80 °C, 21.6 % RH	

inorganic phase entrapped in the polymeric network and well-dispersed at submicronic level creating strong interaction with Nafion. Table 10.1 summarises the proton conductivities of inorganic-organic hybrid membrane with Nafion matrix.

10.4.2 *Sulfonated Mesoporous Silica for Alternative Polymer*

Although Nafion membrane is the state-of-the-art proton exchange membrane in the PEMFC applications, it suffers from high cost, and complex synthesis procedures as well as toxic to the environmental. Alternative PEMs with high performance, low cost and simple synthesis way are urgent to be developed. Sulfonated aromatic polymers such as sulfonated polyimide (SPI), sulfonated polyetheretherketon (SPEEK) and sulfonated poly(phenylsulfone) (SPPSU) are the polymers to be developed as alternative PEMs to replace Nafion in PEMFC.

Sulfonated polyimide (SPI) membrane demonstrates high thermal and chemical stability, high ion exchange capacity (IEC) as well as proton conductivity in hydrated condition. The effects of mesoporous silica for SPI have been comprehensively investigated by Liu's group in terms of filler content and fabrication method [69–71]. Similarly to fillers in Nafion membrane, sulfonated mesoporous silica in SPI indicates same bell shape for the proton conductivity against the filler content and whilst it still shows higher proton conductivity than Nafion membrane in the same condition. The mesoporous structure of sulfonated mesoporous silica is favourable for the water retention and proton diffusion by the periodic proton transportation channels inside the mesopores. On the other hand, when excess filler is embedded into the matrix, filler prefers to aggregate and block the water uptake and decrease IEC of the composite membrane [69]. Liu et al. [70] indicated that 7.0 wt% filler was the optimum for sulfonated mesoporous silica with particle size ~300 nm and pore size of 2.6 nm in SPI matrix for proton conductivity in the temperature range from 25 to 80 °C, 100 % RH.

Sulfonated mesoporous silica has two functions for the incorporation in SPI membranes. First, sulfonated mesoporous silica has high surface area and large amount of Si–OH groups are exposed to absorb water. Second, the embedded sulfonated mesoporous silica limits the mobility of the polymer chains due to the strong interaction between mesoporous silica particle and SPI molecule. With the increase of the content of filler in the composite, microphase with connected structure is formed and hence facilitates proton conductivity. Thus, the conductivity of SPI increases with the increase of filler content until the aggregation of sulfonated mesoporous silica fillers. However, the content of sulfonated mesoporous silica in SPI is limited by simple blending and casting method. The simple blending method is easy to cause agglomeration of the functionalized mesoporous silica particles, leading to the severe proton conductivity and performance loss for the composite membrane. To eliminate the aggregation of the nanoparticles at high content, in situ synthesis of mesoporous silica in SPI was developed by Dai et al. [71]. In comparison with blending method, the amount of mesoporous silica could reach up to 30 wt% in the SPI matrix by the in situ sol-gel and self-assembly approach, while the properties of the composite membrane in terms of water uptake and proton conductivity improved with the increase of filler content in SPI [71].

Similar to SPI, SPEEK and SPPSU and others proton exchange membrane also indicate excellent proton conductivity and stability, simple fabrication as well as low cost [72–74]. The mesoporous structure of sulfonated mesoporous silica maintains the water content at a high level even at low RH because of capillary condensation effect. Moreover, the periodic sulfonated proton diffusion channels facilitate the proton transportation. Herein, blending of sulfonated mesoporous silica into SPEEK and SPPSU improves thermal properties, water uptake, proton conductivity and cell performance of the pure membrane. For instance, when sulfonated SBA-15 particles were added into SPPSU, the composite membrane showed high proton conductivity of 5.9×10^{-3} S cm^{-1} after 130 min elapsed time under dehydrated condition (50 % RH) at 80 °C, higher than 3.6×10^{-3} S cm^{-1} of the pristine SPPSU membrane [75]. When sulfonated cube mesoporous benzene-silica (SMBS) ($Im\bar{3}m$) was added in SPEEK, the proton conductivity of the composite membrane increased up to 70 °C at 40 % RH and whilst Nafion 117 membrane sharply dropped when temperature was higher than 60 °C under the same RH [76].

In order to further increase the conductivity of the composite membrane, high degree of sulfonation for SPEEK is required. However, high level of sulfonation degree results in high water uptake, leading to the decrease of the mechanical strength of SPEEK. And the membrane becomes too brittle in dry state to make the

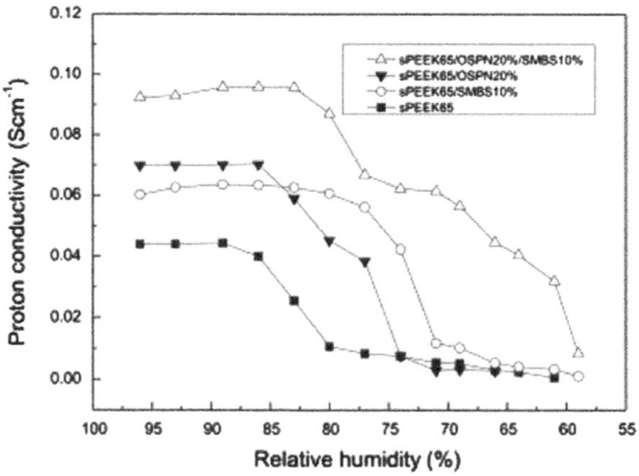

Fig. 10.8 Proton conductivity of sPEEK65/OSPN/SMBS composite membranes at different relative humidity (100 °C) (Reproduced from [79] with permission. Copyright (2013) Elsevier)

MEA. One classic approach to increase the mechanical strength of SPEEK at high level of sulfonation degree is to employ interpenetrating polymer network (IPN) [77, 78]. When 10 wt% SMBS is added into the IPN structure with organosiloxane network (OSPN, 20 wt%) and SPEEK (65 % in sulfonation degree), the ternary composite membrane shows a higher elongation before breaking than SPEEK, whilst it exhibites a higher water uptake than that of the pristine SPEEK membrane [79]. Furthermore, the ternary composite membrane demonstrates higher proton conductivity than that of pure SPEEK membrane as well as the binary composite membrane either with OSPN or SMBS, especially in the low RH (Fig. 10.8). This is due to the high water retention at low RH, derived from the capillary condensation effect of the 2D periodic cylindrical channels of SMBS.

10.4.3 Sulfonation of Mesoporous Silica

The organic functionalization of mesoporous silica generally follows two protocols, either by post-grafting of the functional groups to accessible pore surfaces after the formation of inorganic framework [80, 81] or by co-condensation of the functional species as the inorganic framework [82]. Post-grafting functionalization method bases on the reaction of organic silanes with the framework of mesoporous silica [83]. Sulfonation of *meso*-silica is generally achieved by thiol path [84] such as 3-(trihydroxysilyl)-1-propanesulfonic acid (TPS), mercaptopropyltrimethoxy silane (MPTMS) and sulfonic path [85, 86] such as 1,2,2-trifluoro-2-hydroxy-1-trifluoromethyl-ethanesulfonic acid sultone (FASA) under a relatively severe conditions and longer reaction time. This protocol allows for a wide range of functional groups to be anchored on the surface of mesoporous silica with little deterioration effect on the periodical properties of silica support. Moreover, alternative groups can be anchored on the surface of silica support (i.e. $-NH_2$) as a bridge to anchor the functional organic groups onto the surface of silica support if appropriate silane moieties are not available [62]. However, the challenges for the post-grafting protocol are the control of the acid density and distributions of grafted functional groups. The other one is the co-condensation one-step method that involves the self-assembly of hydrolyzed tetraalkoxysilanes ($(RO)_4Si$) with hydrolyzed trialkoxyorganosilanes($(RO)_3SiR'$), where R and R′ are organic species, via structural-directing agent to functionalize mesoporous silica. However, the loading and diversity of the functionalities are limited by the collective compatibility of various components under the synthesis conditions. Furthermore, high concentration of the silane precursor species are detrimental for the mesoporous structural ordering of materials [87]. In comparison with post grafting method, co-condensation method achieves higher functional sites with higher IEC for mesoporous silica, larger uniform pores, higher surface area and better long-range order. Moreover, it can be conducted at mild and simple synthetic conditions and saves time and material in contrast to post grafting method [88].

In comparison with sulfonation of mesoporous silica, sulfonation of periodic mesoporous organosilica (PMO) is more simple and can be carried out under relatively mild conditions [89]. PMO was discovered by three independent research groups, Ozin et al. [90], Stein et al. [91], and Inagaki et al. [92] in 1999. PMO has both organic and inorganic groups as integral part of the porous framework where organic groups are located within the pore walls as bridges between Si centres. Tsai et al. synthesized the PMO of mesoporous phenyl-silica via both TEOS and phenyltriethoxysilane (PTES) as the precursors and then PMO was sulfonated by concentrated sulfuric acid to form the sulfonated PMO [93]. When 3 wt% of sulfonated mesoporous phenyl-silica were added into Nafion matrix, the composite membrane expressed 2 times higher water uptake and slightly higher ion exchange capacity in comparison with pristine Nafion 212 membrane. Moreover, the composite membrane achieved power density of 414 mW cm^{-2} at 65 °C under dry condition, which is a milestone for the application of Nafion-based composite membrane without external humidification [93].

The introduction of heterogeneities in silica support, such as Al, significantly improves water retention of the mesoporous matrix materials at elevated temperatures (100–200 °C) because of the strong BrØnsted acidity [94]. 2D^{27}Al{^{1}H} HETCPR NMR spectra indicated dominant 2D intensity correlation at 3.0 ppm in the ^{27}Al dimension and 4.0 ppm in the ^{1}H dimension, indicating the interactions between adsorbed water and Al^{IV} aluminosilica species, as shown in Fig. 10.9a [95]. On the contrary, the 2D^{29}Si {^{1}H} HETCPR NMR spectra demonstrates the strong dipole-dipole couplings between the Q^3 ^{29}Si species and hydroxyl groups by the strong correlated signal intensity at 1.6 ppm in the ^{1}H dimension and −102 ppm in the ^{29}Si dimension. The covalently grafted PFSA moieties interact strongly with Q^3 ^{29}Si sites, as the position of Q^3 ^{29}Si shifted from −115 ppm to −102 ppm [96]. By comparison, triflic acid species show the selective interaction with ungrafted Q^3 or Q^4(^{1}Al) ^{29}Si species and interactions with six-coordinate aluminosilica ^{27}Al mesopores surface sites, as shown in Fig. 10.9b. The proton diffusion rate in aluminosilicate is expected to be similar to the aqueous solutions, in which protons move along the hydrogen bonds [68]. However, the replacement of Si atoms by Al atoms in

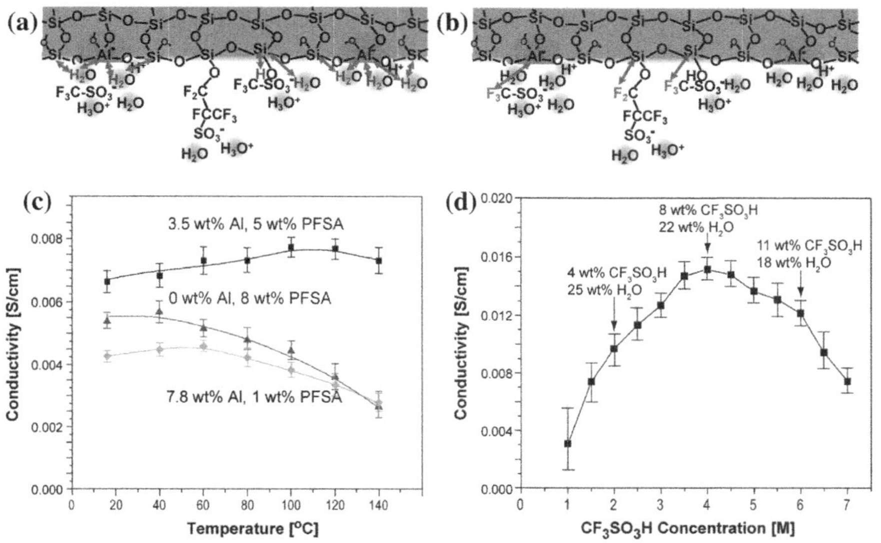

Fig. 10.9 Schematic diagram depicting **a** the interacitons of water and hydroxyl species with surface ^{29}Si and ^{27}Al moieties and **b** interactions of grafted PFSA and triflic acid species with surface ^{29}Si and ^{27}Al moieties in the aluminosilicate framework. **c** Is the proton conductivities as a function of temperature at 50 % RH for identical 65 μm thick cubic mesoporous silica films containing different concentrations of surface-grafted aluminosilicate and PFSA moieties. **d** Is the proton conductivity as a function of mesopores-filling triflic acid concentrations in free-standing 105 μm thick 5 wt% PFSA- and 3.5 wt% Al aluminosilicate grafted cubic mesoporous silica film at ambient condition at 20 °C and ~40 % RH (Reproduced from [95] with permission. Copyright (2011) American Chemistry Society)

the interior mesopores surface is at the cost of surface silanol sites available for proton conductive functional groups, independent of temperature or RH. Nevertheless, the trade-off relationship between the proportion of the functional groups and aluminosilicate content of the silica support can be optimized. Athens et al. [95] optimized the contents for silanol sites replaced by aluminosilicate and PFSA, corresponding to high hydrophilicity and strong acidity (Fig. 10.9c, d), respectively. The optimized content for aluminosilicate and perfluorosulfonic-acid surface grafting densities in multifunctionalized mesoporous silica film with 3.5 wt% Al and 5 wt% PFSA yielded the maximum proton conductivity of 7.7×10^{-3} S cm^{-1} in the temperature range of 20 to 140 °C under 50 % RH, higher than the pure PFAS-grafted *meso*-silica and high aluminosilicate replaced cubic mesoporous films.

10.4.4 Pore Structure and Acidity of Mesoporous Silica

Pore structure in terms of dimensionality, pore size and morphology also influence proton transport under conditions of similar acid loading. Mesoporous MCM-48 with large pore size and three dimensional (3D) interconnected pore structures has demonstrated higher proton conductivity than small pore zeolite beta and one dimensional pore structure MCM-41 with pore size less than 4 nm [67]. For the sake of further improving proton conductivity, sulfonated KIT-6 with pore size of 8 nm and three dimensional interconnected pore structures are incorporated into Nafion membrane with superior properties in terms of proton conductivity and cell performance under 72 % RH at 120 °C [66]. Enhanced water uptake, provision of strong acid sites and high bulk to surface water ratio are the critical factors for high proton conductivity at low RH conditions.

Acidity of the organic groups in mesoporous silica materials is a critical factor that affects the activity of Nafion-based hybrid membranes. Aryl sulfonic acid (S-MP) mesoporous silica exhibits the highest conductivity values followed by propyl sulfonic acid (S-PE) containing samples, the phosphonic acid (P) functionalized mesoporous silica, and finally carboxylic acid (C) functionalized mesoporous silica, when the four types of organic functional groups have the similar acid loading in MCM-41 [67]. Tominaga et al. [65] synthesized three types of silica materials of SiO_2 sphere (p-SiO_2), mesoporous silica (Ne-MPSi) and sulfonated mesoporous silica (Su-MPSi) with acidity of 0.06, 0.38 and 0.46 mmol g^{-1}, respectively. With 5 wt% of the fillers embedded in Nafion matrix, Nafion/Su-MPSi composite showed the highest proton conductivity at 40 °C, 95 % RH via the strong acid sites mainly on the internal surface and hence the composite could achieve high proton conductivity.

Pore size and the acidity control the relationship between the proton conductivity and the water sorption process in the mesopores via Kelvin equation [97]. According to the Kelvin equation, the relative vapour pressure at the capillary condensation becomes large with the increase of contact angel between the pore surface and water molecule and diameter of pores. The increase of acidity in the mesopores decreases the contact angle due to the hydrophilic property of the sulfonic acid group. Thus, the conductivity of the mesoporous silica thin film drops steeply at medium RH as the steep decrease of water adsorption, respectively (Fig. 10.10a) [98]. Moreover, the RH at the steep change in proton conductivity shiftes lower with the decrease in pore size and increase in acid density, which corresponded well with the water vapour adsorption-desorption isotherms. Shannon and the colleges found out that the proton conductivity of sulfonated mesoporous silica film was almost constant down to as low as 20 % RH, while the membrane with larger pores showed the decline, beginning at a higher humidity lever (50–60 %) [97]. The small pore size and high density of sulfonic acid groups can effectively conduct protons in the presence of a small amount of water (Fig. 10.10b), while large pore size and low density of sulfonic acid groups need large amount of water to make proton channels (Fig. 10.10c). Thus, the pore size and acidity of mesoporous silica control the RH position at the steep change by Kelvin equation and the mesoporous electrolytes with small pore size and high acid density achieve high proton conductivity at low RH.

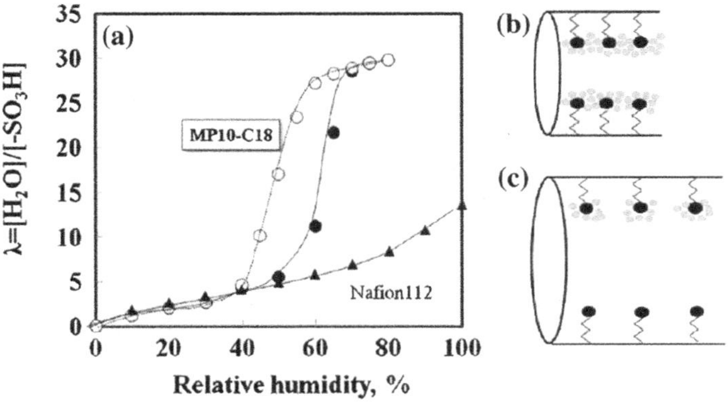

Fig. 10.10 Water uptake (**a**) of sulfonic acid functionalized mesoporous silica films at 25 °C against RH values. *Solid and open circles* show water adsorption and desorption processes, respectively. Water uptake of Nafion 112 and Nafion thin film show the adsorption process. Water sorption at high (**b**) and low (**c**) acid density at low RH and proton channel formation (Reproduced from [98] with permission. Copyright (2013) American Chemistry Society)

Overall, the confinement of $-SO_3H$ group in the mesopores of silica materials reduces the dependence of proton conductivity on hydration. And the water retention of the sulfonated mesoporous silica could be tuned by the pore size, space symmetries of the pore structure, and acidity density of the acid groups. However, the intrinsic proton conduction mechanism of $-SO_3H$ groups requires the incorporation of water molecules. And the boiling point of liquid water limits the proton transportation at elevated temperature. The non-aqueous mediator is able to transfer the protons at elevated temperature and anhydrous conditions. The replacement of water by non-aqueous mediators, such as imidazole, triazole, phosphoric acid and ion liquid for proton transportation has been attracted great attention.

10.5 Non-sulfonated Mesoporous Silica for PEMs

Heterocyclic compounds such as imidazole and triazole play a role of proton transfer mediator similar as water. The proton transfer in the heterocyclic moieties is determined by Grotthuss-type mechanism where proton transfers from one heterocyclic ring to a neighbour one via hydrogen bond construction and deconstruction, and subsequent reorientation of the heterocycle [99]. Low dependence on humidity and high proton conductivity under high temperature and low humidity, therefore, are expected for the heterocyclic moieties [100].

10.5.1 Imidazole

Imidazole groups are known to be involved in proton transferring biological systems and ab initio calculation shows that imidazole groups can act as proton acceptor, donor and proton shuttle species because of its amphoteric and self-dissociation character with acidic proton at –NH– and basic sites at =N– [101, 102]. Accordingly, protonation of imidazole gives symmetrical imidazolium cation and protons are transferred from one imidazole group to the other imidazole group [100]. Moreover, imidazole groups based materials (i.e. polybenzenemidazole, PBI) combined with phosphoric acid have been employed as PEM for fuel cell application at high temperature and anhydrous conditions [103–105].

Three types of imidazole groups have been grafted on the interior mesopores surface of MCM-41, as shown in Fig. 10.11 [106]. Because of the occupation of one N atom in N-imi-MCM-41 for the anchorage of imidazole group, the proton acceptance capacity (PAC) of N-imi-MCM-41 is lower than that of imi-MCM-41 with two N-atoms for proton uptake, resulting in two orders of magnitude lower in conductivity than that of imi-MCM-41. Furthermore, when increase the flexible chain from 3 carbon atoms to five carbon atoms by combining the imidazole group by a peptide bond formation on amino-anchors (pep-imi-MCM-41), the PAC and proton conductivity of MCM-41 are significantly improved from 0.21 mmol g^{-1} to 0.79 mmol g^{-1} and 10^{-5} S cm^{-1} to 10^{-4} S cm^{-1} (100 % RH, 140 °C), respectively. That is due to the high imidazole loadings (PAC values) and long spacing

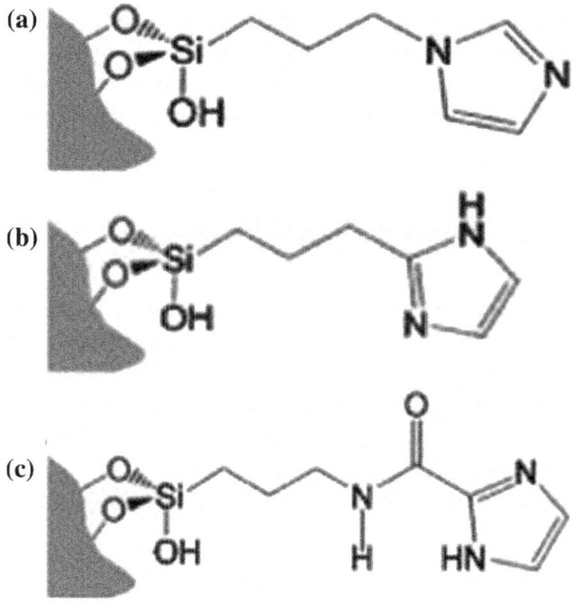

Fig. 10.11 Three types of imidazole groups grafted on mesoporous silica surface **a** N-imi-MCM-41, **b** imi-MCM-41 and **c** pep-imi-MCM-41 (Reproduced from [106] with permission. Copyright (2009) Elsevier)

between the silica host and the protonic groups, which allowed the anchored imidazole group to reach each other easily [107].

Moreover, as the amphoteric character of imidazole diminishes the dependence of proton conductivity on humidity, imidazole functionalized mesoporous silica can be employed as fillers in sulfonated polymer for proton conductivity improvement. Furthermore, ionic path is formed by the hydrogen bonding interaction between the imidazole moieties and sulfonic acid groups, which also enhances proton diffusion [108]. For instance, when imidazole functionalized mesoporous silica was introduced into Nafion membrane, a maximum conductivity of 1.06×10^{-2} S cm^{-1} was achieved at 130 °C under anhydrous condition with 5 wt% filler loading. Moreover, the conductivity of the composite membrane was stable under the severe condition for 76 h [109]. However, the proton conductivity dropped for the further durability test due to the structural reconfiguration process above the glass transition point of Nafion membrane ($\sim$126 °C) and enduing distorted cross-linking geometry of the polymer matrix and discontinuity of proton conducting network path [109].

10.5.2 Triazole

Triazole has similar molecular structure as imidazole, and hence may transfer proton via the similar mechanism as imidazole. 1H-1,2,4-triazole (mp: 120 °C and bp: 256 °C) has a proton conductivity of 1.2×10^{-3} S cm^{-1} at 120 °C under anhydrous state [110], while 1H-1,2,3-triazole (mp: 23 °C and bp: 203 °C) has proton conductivity of 1.5×10^{-4} S cm^{-1} at room temperature in liquid phase [111]. This indicates that self-dissociation of 1H-1,2,4-triazole and 1H-1,2,3-triazole generates protons and the proton diffusion rate is very high, similar to that in imidazole. Moreover, acidity of 1H-1,2,3-triazole ($pK_a1 = 1.17$, $pK_a2 = 9.26$) is higher than imidazole ($pK_a1 = 7.18$, $pK_a2 = 14.52$), indicating higher proton conductivity than imidazole [112]. However, triazole is small molecular and also suffers from the leaching out problem during the fuel cell operation, resulting in the continuous decrease in the conductivity. Generally, triazole molecules are grafted on the interior pore surface of mesoporous silica by chemical bonds via the reaction between the triazole precursor and the hydroxyl groups in mesoporous silica. For instance, benzotriazole-5-carboxylic acid (BTCA) was anchored onto SBA-15 by the groups of carboxylic acid and hydroxyl group [113]. The adding of BTCA/SBA-15 particles into Nafion membrane significantly enhanced the proton conductivity of matrix under 10 % RH, 150 °C and the proton conductivity increased with the increase of the content fillers in the range of 2.3–5.6 wt% [113]. The enhancement effect is due to the guide effect in the aligned mesoporous channels of SBA-15 by the intermolecular proton transportation among adjacent BTCA molecules, and the improved water retention of mesoporous silica.

10.5.3 Protic Ionic Liquids

Protic ionic liquids (PILs) are non-aqueous proton carriers with high ionic conductivity, non-volatility and chemical stability as well as thermal stability up to 300 °C [114, 115]. Ion liquids compose of BrØnsted acids and bases and the interacted hydrogen bond acts as the proton carrier path. They have been extensively introduced into polymer to improve its conductivity at anhydrous conditions [116]. However, ion liquids are in the form of small molecules and thus easy to leach from the composite membrane [117]. On the other hand, mesoporous materials are able to confine the ion liquid in the mesopores, which hinders the release of ionic liquid [118]. For instance, pure PILs/poly(styrene-co- acrylonitrile) (SAN) and SiO_2 sphere composite membrane almost lost all the ion liquid after 4 h extraction by distilled water, while mesoporous PILs/silica/SAN composite membrane retained 20 wt% ion liquid under the same extraction process [119]. The addition of mesoporous materials into the PILs/polymer composite has two competitive effects in terms of proton conductivity [120]. The periodic channels in the mesoporous materials facilitate the proton transportation in the inorganic materials. At the same time, mesoporous inorganic fillers block the movement of polymer molecule chain and make the proton channels in the matrix tortuous, resulting in disturbance for the proton diffusion. That is why the proton conductivity of PILs composite membrane shows volcano shape against the addition of mesoporous silica [120]. And generally, the optimum filler content is in the range of 1–5 wt%, depending on the properties of polymer matrix [121]. Moreover, the pore size of mesoporous silica affects the proton diffusion of ion liquid in the mesopores of inorganic fillers. The higher the ratio of surface area/pore volume, that is, the smaller the pore size, the higher of the proton conductivity is obtained, as the small pores favour the formation of PIL conductive channels. For instance, when the pore size of SBA-15 decreases from 17.4 to 9.5 nm, the proton conductivity of poly(methyl methacrylate) (PMMA)/PIL/SBA-15 composite membrane increases from 2.8 × S cm^{-1} to 9.4 × S cm^{-1} at 160 °C, 0 % RH [122]. Furthermore, the mesopores with small size favour the PIL retention because of strong capillary forces within the mesoporous silica materials. Besides, the morphology of mesoporous silica will also affect the proton conductivity of PILs based composite membranes. Hwang et al. [122] discovered that platelet SBA-15 with short pore channels was more effective in proton conductivity improvement of PMMA/IL/SBA-15 composite membrane than fibre-shape SBA-15 with long channels. That is due to the short distance and improved efficiency for proton diffusion in platelet SBA-15.

10.5.4 Phosphoric Acid

Similar to protonic small molecules including imidazole, triazole and ion liquid, phosphoric acid (PA) also shows high proton conductivity and thermal stability at elevated temperate and low relative humidity [123]. Phosphoric acid has been extensively applied in PEM combined with basic polymers, such as PBI [124]. However, the acid-base composite membrane suffers from acid leaching problem during the prolonged operation in fuel cells [125]. Impregnated PA into the mesopores of mesoporous silica is a good way to solve the leaching problem of PA. Zhao et al. [126] introduced phosphorated mesoporous silica into Nafion membrane, and the proton conductivity of Nafion membrane was substantially increased to 3.39×10^{-1} S cm^{-1} at 100 % RH, 115 °C by addition of 5.0 wt% fillers. Besides mesoporous silica, hollow mesoporous silica also was employed by Jiang et al. as PA reservoir because of the large volume to improve the conductor adsorption. When HMS was impregnated by amino tri(methylene phosphoric acid) (ATMP), the addition of HMS-ATMP with 7.5 wt% significantly increased the proton conductivity of chitosan by two orders of magnitude from 1.1×10^{-4} S cm^{-1} to 1.2×10^{-2} S cm^{-1} [54].

Zeng et al. [127] fabricated novel PA/*meso*-silica composite membrane based on the sintered mesoporous silica disk. They found that sintering treatment for the *meso*-silica disk significantly increased the mechanical strength of mesoporous silica disk and improved the operation temperature by getting rid of the polymeric binder. After PA impregnation to the mesopores of sintered mesoporous silica disk (PA/*meso*-silica) with 27.7 wt% PA loading, superior conductivity was achieved in comparison with PA/PBI membrane as well as PA-porous-silica membrane at elevated temperature range (Fig. 10.12a). Outstanding proton conductivity of 6.0×10^{-2} S cm^{-1} was achieved at 200 °C under anhydrous condition while excellent power output of 689 and 200 mW cm^{-2} was obtained in H_2/O_2 and methanol/O_2 system without humidification at 190 °C, respectively. Although PA/*meso*-silica based fuel cell demonstrates excellent cell performance, it suffers two critical challenges. One is the sintering treatment temperature. When the silica membrane is treated at temperature higher than 650 °C, the mesoporous structure is intact at 750 °C and completely collapses at 850 °C [128]. The other one is the degradation of PA during the high temperature operation. The PA degradation follows the reversible reactions:

$$\text{First stage: } 2H_3PO_4 \leftrightarrow H_4P_2O_7 + H_2O \uparrow \tag{10.2}$$

$$\text{Second stage: } H_3PO_4 + H_4P_2O_7 \leftrightarrow H_5P_3O_{10} + H_2O \uparrow \tag{10.3}$$

According to the reactions, the directions of the reactions are determined by the water content and temperature. Thus, when external water is added into the fuel cell, the reaction goes to the backward and forward of PA degradation is impeded. The hypothesis is confirmed by the superior performance of PA/*meso*-silica by adding 1 or 3 % water content, otherwise the performance decreased along with the fuel cell operation, as shown in Fig. 10.12b.

However, the thermal and chemical stability of PA/*meso*-silica composite particles are not satisfied for long term use at elevated temperature. Yi et al. [130] found out that the structure of mesoporous silica was intact under heat treatment under 80 °C for 24 h, while the structure was completely collapsed and the particle

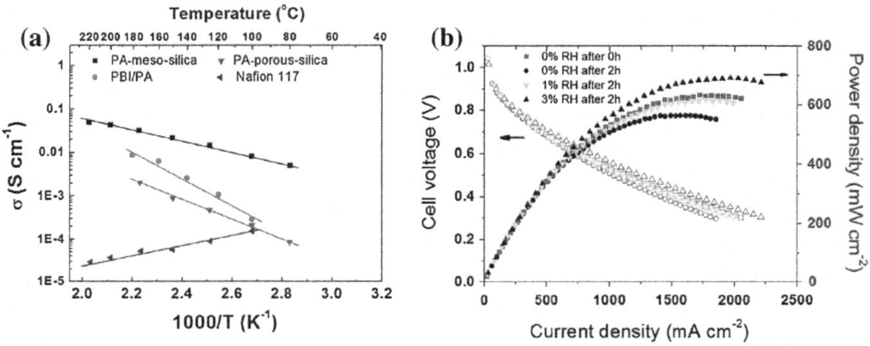

Fig. 10.12 **a** The proton conductivity of composite membranes against with test temperature under anhydrous condition square legend for PA/*meso*-silica membrane, *triangle* for PA-porous-silica, *circle* for PBI/PA membrane from [129], *light grey triangle* for Nafion 117 membrane. **b** Single cell performance of PA/*meso*-silica membrane based fuel cell in H_2/O_2 at 190 °C. The *numbers* in the figure are the RH of hydrogen gas (Reproduced from [128] with permission. Copyright (2013) American Chemistry Society)

size also grew big when the composite particles were annealed above 200 °C for 24 h. FT-IR results shows that the Si–O–Si asymmetric stretching band at 1086 cm^{-1} was blue shifted with increase of the annealing temperature due to the characteristic vibration peak of P–O–P and Si–O–P. Besides peak round 110 ppm of amorphous silica (Si-$(OSi)_4$) by ^{29}Si MAS NMR, a new peak was detected by around 210 ppm, corresponding to the substance of ($Si_3(PO_4)_4$ or $Si(HPO_4)_2$). That is due to the formation of phosphor silicate between PA and mesoporous silica.

10.5.5 Alternative Mesoporous Materials for PEM

Besides highly ordered mesoporous silica, mesoporous structure has also been introduced into other materials including titanium dioxide (TiO_2) [131], zirconium phosphate ($Zr_3(PO_4)_4$) [132], iron phosphate ($FePO_4$) [133], lanthanum phosphate ($LaPO_4$) [134, 135] and titanium phosphate ($Ti_3(PO_4)_4$) [136] as PEMs. Mesoporous anatase TiO_2/Nafion composite membrane (3.0 wt%) indicates 5.7 times higher performance (669 mW cm^{-2}) than that of recast Nafion membrane at 50 % RH, 120 °C, as mesoporous anatase TiO_2 confines high content of water in the mesopores [131]. The surface area of the filler in the composite membrane determines the composite membrane based cell performance at low temperature (<90 °C), while the amount of water in the mesopores of fillers plays critical role for the performance improvement at high temperature (120 °C).

Acid metal phosphates are solid proton conductors, while proton diffusion of metal acid phosphate is determined by the surface functional group –OH in the interlayer with the presence of water [137]. Given that mesoporous structure is introduced into the acid metal phosphate, its surface area would be significantly enlarged. Thus, the proton conductivity of the acid metal phosphate would be substantially increased due to the large amount of exposed –OH group. The mesoporous acid metal phosphate materials could be synthesized by the acid-based pair synthesis way via template, such as Pluronic, P123, F127 and F108, C14-Br, CTAB, and SDS. When worm-like mesoporous structure is introduced into $Zr_3(PO_4)_4$ with pore size in the range of 2–4 nm and surface area between 78 and 174 $m^2\ g^{-1}$, the conductivity of the material reaches up to 4.1×10^{-6} S cm^{-1} at 84 % RH, 22 °C [132].

Moreover, organic groups can also be embedded into the frame of metal phosphate to improve the conductivity of the substrate, similar to the PMO. Sulfonated mesoporous iron oxophenyl phosphate is achieved by oxophenyl sulfonation by fuming sulfuric acid for mesoporous iron oxophenyl phosphate. Proton conductivity of 1.0×10^{-2} S cm^{-1} is obtained for sulfonated mesoporous oxophenyl phosphate at 100 % RH and room temperature [133].

In order to further increase the proton conductivity of the mesoporous metal phosphate materials, proton conductors such as PA have been impregnated into the mesopores. The impregnation of PA significantly increases the proton conductivity of metal phosphate around 1.0×10^{-2} S cm^{-1} at elevated temperature between 120 and 180 °C, 20 % RH at the PA content as low as 5.0 wt%. And conductivity value

Table 10.2 The properties of mesoporous materials for proton exchange membranes

Matrix	Fillers and content	Function groups	Agents	Conductivity, mS cm^{-1}	References
N/A	SBA-15, 5.6 wt%	Triazole	BTCA	0.852@150 °C, 10 % RH	[113]
Nafion	Sulfonated SBA-16	Ion liquid	BMIm-BF4	11.4@160 °C, 0 % RH	[118]
SAN	MCM-41, 6.0 wt%	Ion liquid	Elm-TfD	12@160 °C, 0 % RH	[119]
Nafion	*Meso*-silica, 3.0 wt%	Ion liquid	N/A	54.6@90 °C, 30 % RH	[120]
PMMA	Platelet SBA-15, 5.0 wt%	Ion liquid	BMIm-IFSI	5.3@160 °C, 0 % RH	[122]
N/A	MCM-41, 5.0 wt%	Ion liquid	N/A	67.4@150 °C, 0 % RH	[121]
N/A	Sintered *meso*-silica disk	PA	PA	60@190 °C, 0 % RH	[127]
Nafion	HMS, 7.5 wt%	$-PO_3H_2$	ATMP	12@80°C, 78 % RH	[54]
PMMA	MCM-41, 5.0 wt%	$-PO_3H_2$	$POCl_3$	339@80 °C, 100 % RH	[126]
Nafion	HMS, 4.0 wt%	$-SO_3H$	Cysteine	10.2@80 °C, 21.6 % RH	[62]
		$-PO_3H_2$	Phosphoserine	7.91@80 °C, 21.6 % RH	
		-COOH	Aspartic acid	4.85@80°C, 21.6 % RH	
SPI	*Meso*-silica, 7.0 wt%	$-S0_3H$	MPTMS	42.5@80 °C, 100 % RH	[70]
SPEEK	*Meso*-silica, 10.0 wt%	$-SO_3H$	H_2SO_4	17@70 °C, 40 % RH	[76]
SPPSU	SBA-15, 0.5 wt%	$-SO_3H$	MPTMS	5.9@80 °C, 50 % RH	[75]

is comparable with Nafion membrane at fully hydration state [134]. Properties of selected mesoporous PEM materials are listed in Table 10.2.

10.6 Mesoporous Silica Based Inorganic PEMs

The addition of mesoporous silica filler improves the conductivity and cell performance of the polymer matrix under high temperature and low RH. However, low loading of the protonic functional groups and the proton conductors in the mesopores limits the proton conductivity of mesoporous materials based PEMs. Moreover, the glass transition temperature (T_g) of the polymer matrix restricts the application of the composite membrane at higher temperature than T_g, as the high temperature deteriorates the mechanical strength of the PEMs. In order to improve the loading of the proton conductors in the mesopores of silica and the working temperature of the mesoporous silica based PEMs, mesoporous silica is employed as the skeleton rather than the fillers. In mesoporous silica based inorganic PEMs, proton conductors are confined in the mesopores of the *meso*-silica and the composite particles are made into PEMs. Heteropolyacids (HPAs) and phosphoric acid (PA) are the most common proton conductors used for *meso*-silica based inorganic PEMs.

HPAs are strong Brønsted acid with unique Keggin unit (KU), as shown in Fig. 10.13 [138]. In the primary structure of Keggin unit, the central heteroatom in a tetrahedral coordination is surrounded by 12 metal-oxygen octahedra. There are 12 terminal oxygen atoms (O_t) attached to only one addendum atom, while the left oxygen atoms are bridged by two addenda atoms. Then the KUs are linked by adsorbed water molecules via hydrogen bonds to form a secondary structure. There are four typical compounds including phosphotungstic acid ($H_3PW_{12}O_{40}{\cdot}nH_2O$,

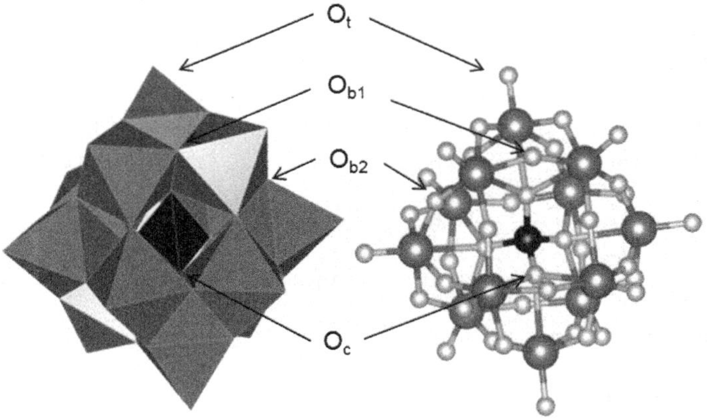

Fig. 10.13 The Keggin structure of anion phosphotungstic acid ($PW_{12}O_{40}^{3-}$): O_t, O_{b1}, O_{b2}, and O_c labelled the four types of oxygen in the sturcute. Oxygen atoms in *light grey*, tungsten in *grey* and phosphor in *black* (Reproduced from [138] with permission. Copyright (2014) Elsevier)

HPW), silicotungstic acid ($H_4SiW_{12}O_{40}{\cdot}nH_2O$, SiWA), phosphomolybdic acid ($H_3PMo_{12}O_{40}{\cdot}nH_2O$, PMoA) and silicomolybdic acid ($H_4SiMo_{12}O_{40}{\cdot}nH_2O$, SiMoA). Among the four HPAs, HPW shows the highest proton conductivity of 0.18 S cm^{-1} in the fully hydrated state at room temperature [139]. Moreover, HPW shows excellent thermal stability up to 450 °C [138]. Although HPW and PMoA pellet are tried to be employed as the PEM in H_2/O_2 system [140], HPAs are easily dissolved in water that is the by-product of the fuel cell operation, resulting in the loss of proton conductivity and cell performance. In order to hinder the leaching of HPAs, HPW molecules are confined in the *meso*-silica to form the composite membrane. And the synthesis and application of HPW/*meso*-silica in PEMFC have recently been systematically reviewed by Jiang [141, 142].

10.6.1 Synthesis of HPW/Meso-Silica

Two main approaches have been employed to incorporate HPW into the mesopores of *meso*-silica matrix: one is impregnation method (post-synthesis) and the other is co-condensation method (one-step self-assembly). Figure 10.14 shows the procedures for the two synthesis methods for HPW/*meso*-silica materials.

10.6.1.1 Post-synthesis Method

The post-synthesis method allows for the incorporation of various proton conductors into mesopores of silica materials, such as imidazole, triazole, ion liquid, PA and HPW. Generally, impregnating HPW into mesoporous silica by post-synthesis

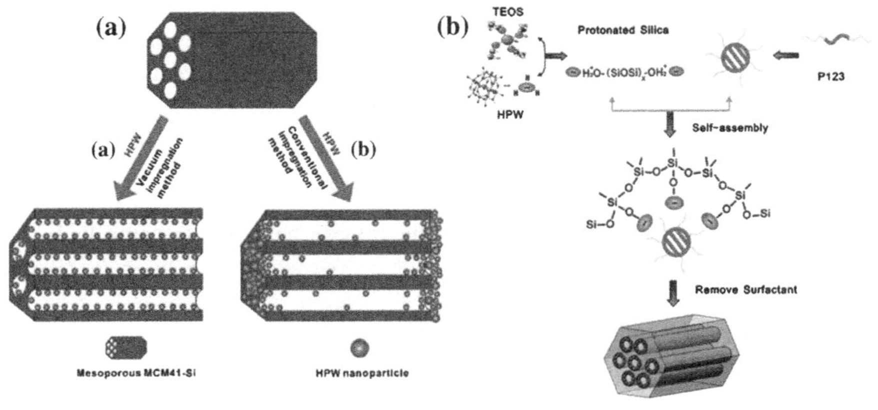

Fig. 10.14 **a** Post-synthesis of HPW/*meso*-silica materials (Reproduced from [143] with permission. Copyright (2010) John & Wiley Sons), **b** Co-condensation synthesis of HPW/*meso*-silica materials (Reproduced from [25] with permission. Copyright (2011) The Royal Society of Chemistry)

method is either by capillary force during the stirring process (conventional impregnation method, CIM) or by vacuum impregnation method (VIM), as shown in Fig. 10.14a. Take MCM-41 for example, it shows highly ordered parallel channels in [110] direction (Fig. 10.15a) and hexagonal channels in [100] direction (Fig. 10.15b), while it shows diameter of 3.0 nm and wall thickness of 2.0 nm [143]. The empty pore space is indicated by the white dot in the view of [100] direction (Fig. 10.15b), while the pore space changes to dark when HPW is impregnated into the channels of MCM-41, as shown in Fig. 10.15d. The distribution of HPW along the channels is indicated by the ordered but not continuous black dots (Fig. 10.15 c). Such phase contrast is also observed in the cesium substituted HPA/*meso*-silica system [144].

In comparison with CIM, VIM shows much more efficient in terms of HPW loading and distribution [145, 146]. HPW molecules are homogenously dispersed in the nanochannels of *meso*-silica by the VIM impregnation (Fig. 10.15c, d), while the HPW molecules are accumulated on the entrance or the edges of *meso*-channels of mesoporous silica in CIM method (Fig. 10.15e). By extending the vacuum treatment time in VIM method, the content of HPW in mesoporous silica improves from 2.5 to 80 wt% while the distance between HPW molecules decrease from 6 to 0 nm [147]. The higher loading of HPW particles inside the *meso*-silica, the lower activation energy barriers will be for the proton transportation in the *meso*-silica matrix. However, the shape of the interconnected channels of HPW/*meso*-silica nanocomposites is not as regular as that of the pristine *meso*-silica host, probably due to the impregnation and insertion of HPW particles.

In addition, the post-synthesis method also shows advantages in the flexibility of easy control of pore size and space symmetries of mesoporous silica for HPW impregnation. The pore size and volume of the mesopores can be tuned by using different types of the SDA [43] and post-synthesis hydrothermal treatment conditions [148]. Several types of mesoporous silica have been impregnated by HPW to form the HPW/*meso*-silica membrane including mesoporous silica in 2D hexagonal (*P6mm*), 3D face-centred cubic ($Fm\bar{3}m$), 3D body-centred ($Im\bar{3}m$), to 3D cubic bicontinuous ($Ia\,\bar{3}d$) symmetries [149]. Among them, HPW/*meso*-silica composite membranes with 3D mesostructures display significantly higher proton conductivity and stability as a function of relative humidity in comparison to 2D *meso*-silica based PEM.

However, the disadvantages of the post-synthesis method are that neither the loading nor the distribution of proton conductors can be precisely controlled.

10.6.1.2 One-Spot Self-assembly Method

Very different from post-synthesis, HPW incorporated into the mesoporous silica matrix can be easily and precisely controlled via one-pot self-assembly synthesis process. The self-assembly between HPW and silica makes the HPW molecules homogeneously anchored and trapped inside the mesopores of *meso*-silica.

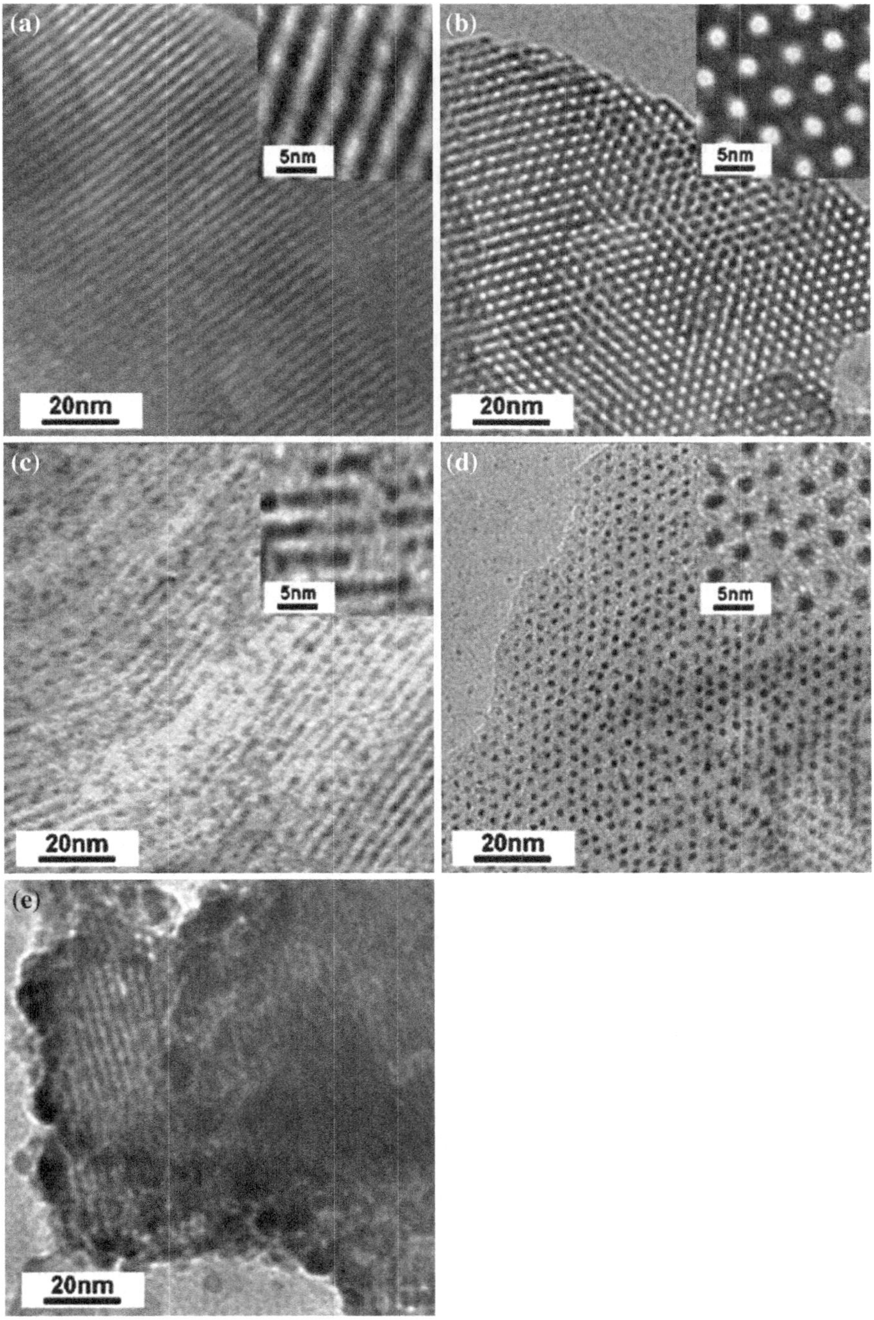

Fig. 10.15 TEM image of MCM-41 (**a**, **b**) and 30 wt% HPW/MCM-41 electrolyte by VIM (**c**, **d**), and CIM (**e**) viewed in [110] direction (**a**, **c**) and [100] direction (**b**, **d**) the pore axis; *inset* high-resolution images (Reproduced from [143] with permission. Copyright (2010) John & Wiley Sons)

HPW dissociates protons and forms negatively charged $PW_{12}O_{40}^{3-}$ ions in water. Meanwhile, the exposed SiOH groups on the surface of *meso*-silica are able to absorb protons in acid solution to form $SiOH_2^+$ [150, 151]. Thus, when HPW molecules are mixed the silica colloids derived from silica precursor, HPW/silica ion pairs are formed in acid solution by the self-assembly between negative charged $PW_{12}O_{40}^{3-}$ ions and positive charged $SiOH_2^+$ colloids via the electrostatic force. When the addition of structure-directing agent, Pluronic 123, HPW/silica ion pairs surround around the micelles of P123 template via the cooperative hydrogen bonds to form the global micelles and then transform into tube-cumulated mesoporous HPW/silica [152]. Finally, the ternary component colloids form ordered mesoporous silica with HPW confined in the mesopores. The surfactant template is then removed by calcination at 400 °C.

The structure and size of mesopores of mesoporous silica by the one-pot self-assembly synthesis method are highly dependent on the content of functional group [153, 154]. That is indicated by the limited HPW content incorporated into the *meso*-silica structure with high density and without any significant detrimental effect on the mesoporous structure [152, 155]. The HPW/*meso*-silica composite particles show uniform and long-range ordered 2D hexagonal mesoporous arrays viewing in [110] direction. The diameter of the nanochannels in HPW/*meso*-silica enlarges from 3.2 to 3.5 nm when the HPW loading in *meso*-silica increases from 10 wt% to 25 wt%, respectively. The uniformly ordered mesoporous arrays are still maintained when the HPW content is less than 25 wt%, while the nanochannels become discontinuous and distorted when the loading of HPW increases to 35 wt%. The shrinkage of the mesoporous structure is also reflected by detrimental of the mesochannels and their macroscopic alignment from top-surface feature [156].

*10.6.2 Conductivity and Cell Performance of HPW/*Meso-*Silica Composite Membrane*

Similar to the sulfonic acid functionalized mesoporous silica materials, the proton conductivity of HPW/*meso*-silica nanocomposites depends strongly on the density of functional sites, that is, HPW content. In the case of HPW/*meso*-silica nanocomposites via the impregnation route, the conductivity of the composite membrane improves with the increase of HPW loading and the threshold of HPW content is ~10 wt% (Fig. 10.16a) [147]. That is due to the fact that the conductivity of the HPW/*meso*-silica composite membrane is too low to be used as PEM in fuel cells when the loading of HPW is lower than 10 wt%. On the other hand, when HPW content is higher than 65 wt%, the conductivity is almost independent of the HPW content with constant value of 7.0×10^{-2} S cm^{-1} at 25 °C and 100 % RH (Fig. 10.16b). A similar trend is also observed on one-pot self-assembled HPW/*meso*-silica nanocomposites with the maximum conductivity on 25 wt% HPW/*meso*-silica [155].

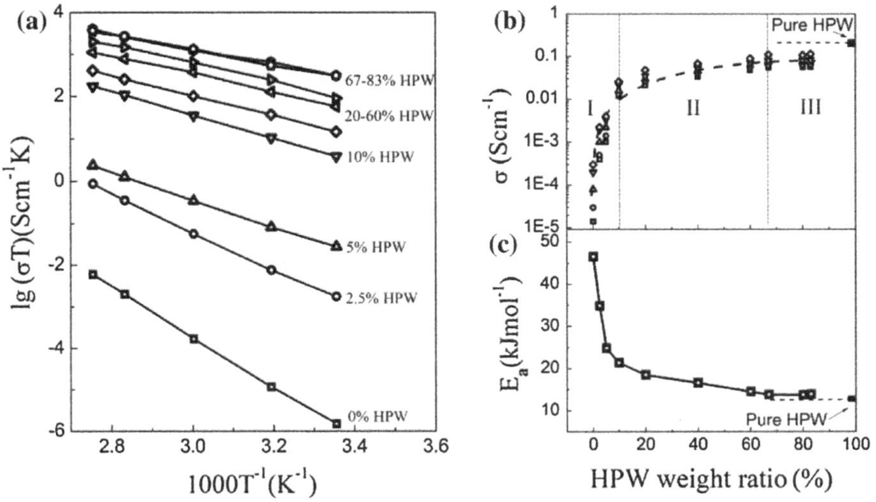

Fig. 10.16 **a** Proton conductivity of HPW/*meso*-silica membrane with different HPW loading under various temperatures, **b** proton conductivity versus HPW loading at 25 °C, and **c** activation energy versus HPW content of HPW/*meso*-silica membrane. Then conductivity and activation energy of pure HPW were from [140] (Reproduced from [147] with permission. Copyright (2011) American Chemistry Society)

Besides the HPW loading, the proton conductivity of HPW/*meso*-silica is also affected by the pore size and structure of the mesoporous silica matrix [149]. When the pore size of mesoporous silica decreases from 12.1 to 5.0 nm, the proton conductivity of HPW/*meso*-silica membrane increases from 4.9×10^{-2} S cm^{-1} to 6.8×10^{-2} S cm^{-1} at 100 % RH and ambient temperature [148]. Similar trend is found in the AMPS/macroporous silica composite membrane with improved proton conductivity by reducing the pore size of silica host [157]. For the structure of mesoporous silica, body-centred cubic ($Im\bar{3}m$) based HPW/*meso*-silica membrane shows the highest proton conductivity of 6.1×10^{-2} S cm^{-1} at 25 °C and 1.4×10^{-2} S cm^{-1} at 150 °C under 100 % RH.

One of the outstanding properties of HPW/*meso*-silica based PEM is that the proton conductivity of HPW/*meso*-silica shows low sensitivity toward the change in RH. The proton conductivity of HPW/*meso*-silica nanocomposite keeps almost steady with RH decrease until steep drop when RH is lower than 20 % RH, 80 °C [148]. The sharp drop in conductivity results from a change in the hydration state of HPW as the conductivity of HPW is strongly dependent on the number of water molecules contained in the HPW KU. Nevertheless, the RH dependence of the conductivity is far less sensitive as compared to that of Nafion membranes. The much less sensitivity of the proton conductivity on the RH indicates the high water retention of the HPW KU-type nanoparticles anchored in the ordered *meso*-silica structure.

The stability of the HPW/*meso*-silica is required for practical application of the materials in fuel cells and it relates to the impregnation method of HPW, pore size and synthesis method of mesoporous silica. Leaching of proton functional groups from PEM decreases proton conductivity and cell performance as well as stability of the PEMs [125]. The stability of the anchored HPW inside the mesopores can be assessed by monitoring the proton conductivity of the PEMs at an accelerated durability test with a constant water flow. Although the conductivity of the different composite HPW/*meso*-silica membranes decreases at the initial stage, the values are stabilized after 2–3 h measurements. The initial decay in the conductivity is most likely due to the leaching of HPW during the flushing of water. The stability of the HPW/*meso*-silica relates to the impregnation method of HPW. In comparison with 30 wt% HPW/*meso*-silica membrane from CIM, the conductivity of VIM 30 wt% HPW/*meso*-silica decreases from 2.3×10^{-2} and stabilizes at 1.4×10^{-2} S cm^{-1} after 5 h test. The loss in proton conductivity is 39 %, lower than 72 wt% of CIM 30 wt% and 100 % of mixed 30 wt% composite membranes [143]. The results indicate that mesoporous silica materials are able to stabilize the proton conductors inside the mesopores because of the confinement effect of mesoporous structure. Furthermore, the stability of the impregnated HPW in *meso*-silica also depends on the pore size of mesoporous materials. The mesoporous silica with a pore size of ~5 nm is beneficial for confining more HPW molecules in the mesopores of silica matrix as compared to that with larger pore size from 8.2 to 12.3 nm, although their initial HPW content are the same [148]. Eventually, the stability of HPW/*meso*-silica is also determined by the synthesis method of mesoporous silica. HPW/*meso*-silica synthesized by the one-pot synthesis method shows very stable proton conductivity under elevated temperatures and reduced RH [155]. The stability of HPW inside the mesoporous silica is most likely due to the formation of $(\equiv SiOH_2^+)(H_2PW_{12}O_{40}^-)$ species [158].

The promising application of HPW/*meso*-silica composite membranes has been demonstrated in single cell for both H_2 and alcohol fuels including methanol and ethanol [25, 143, 148, 149, 152, 155]. HPW/*meso*-silica composite membrane based fuel cells exhibit excellent cell performance under low RH, due to the prominent capability to retain water at low RH. Figure 10.17a shows that HPW/*meso*-silica based PEM with HPW loading of 80 wt% achieves power density of 171 mW cm^{-2} at 20 % RH, 80 °C, which is much higher than 33 mW cm^{-2} obtained on the Nafion membrane cell under the identical condition (Fig. 10.17b) [148]. Furthermore, only 22 % power density loss is observed for cell based on the HPW/*meso*-silica composite membrane when the RH decreases from 100 to 20 %, while it is 95 % for Nafion 115 membrane (Fig. 10.17c). Moreover, HPW/*meso*-silica composite membrane shows high stability under a constant current density of 200 mA cm^{-2} at 50 %, 80 °C for 80 h, although there is small and reversible decay in cell voltage during each circle, as shown in Fig. 10.17d.

For the application of HPW/*meso*-silica nanocomposite membrane in direct alcohol fuel cell, it shows low peak power density for direct ethanol and methanol based fuel cells at 80 °C when Pt catalyst is used in both anode and cathode. That is due to the poison effect and low electrocatalytic activity of Pt electrocatalysts for the electrochemical oxidation of ethanol and methanol at low temperatures [159]. On the contrary, when the temperature is increased to 200 °C, 25 wt% HPW/*meso*-silica based fuel cell shows peak power density of 112 mW cm^{-2} for the direct

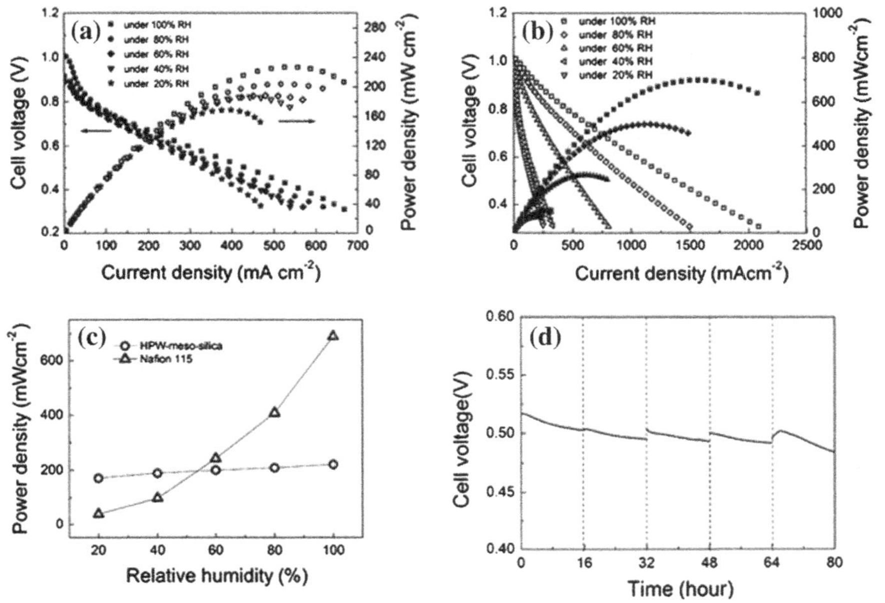

Fig. 10.17 Performance of single cells with **a** HPW/*meso*-silica (80 wt% HPW loading) composite membrane and **b** Nafion 115 membrane measured in H_2/O_2 and under 80 °C and different RH, **c** peak power density of cells with HPW/*meso*-silica composite membrane and Nafion 115 membrane against RH, **d** durability test of HPW/*meso*-silica with current load of 200 mA cm^{-2} under 50 % RH, 80 °C in H_2/O_2. Pt loading for both anode and cathode was 0.4 mg cm^{-2} (Reproduced from [148] with permission. Copyright (2011) The Royal Society of Chemistry)

ethanol fuel cell and 128.5 mW cm^{-2} for direct methanol fuel cell [155]. The outstanding cell performance of HPW/*meso*-silica based fuel cell at 200 °C indicates the elimination of the CO-poisoning effect on the Pt black catalysts at elevated high temperatures [160]. Meanwhile, according to the correlation between proton conductivity and cell performance, highest power output is expected by $Ia\bar{3}d$ HPW/*meso*-silica, as it demonstrates highest proton conductivity among the composite membranes with various silica host. And peak power density of 237 mW cm^{-2} is obtained by HPW/*meso*-silica membrane based on $Ia\bar{3}d$ mesoporous silica at 150 °C without external humidification on cathode in methanol fuel [149].

Stack cells are single cells placed in series in order to create sufficient voltage and power output. Stack cells are more critical than single cell for the practical application of PEM. However, silica based PEM is much more brittle than soft polymeric membrane such as Nafion and PBI/PA membranes [162–164]. The great challenge for HPW/*meso*-silica nanocomposite membranes based stack cells, therefore, is the fabrication of homogeneous, crack-free and mechanically strong membranes. Zeng et al. successfully fabricated relatively large HPW/*meso*-silica nanocomposite membranes with area of 38.4 cm^2 and thickness of 0.76 mm (see Fig. 10.18a) and installed the membranes into fuel cell stack (6-cells stack as example in Fig. 10.18b) [161]. HPW/*meso*-silica nanocomposite membranes were created by a modified hot-pressing process with PVDF as binder so that to increase the mechanical strength and increase the size of membrane [161]. The stack cells demonstrated the good performance in H_2 without external humidification, as shown in Fig. 10.18C. The open circuit voltage of the stack with 10 cells was 7.9 V and a maximum power output of 74.4 W was achieved at 150 °C, corresponding to a power density of 372.1 mW cm^{-2} based on the total effective cell area of 200 cm^2. The value was higher than the power output reported on alternate inorganic membrane based fuel cells such as In^{3+} doped SnP_2O_7 membrane (264 mW cm^{-2} at 250 °C) [165], NH_4PO_3 based membrane (27 mW cm^{-2} at 300 °C) [166], and CsH_2PO_4 based membrane (44 mW cm^{-2} at 220 °C) [167]. Moreover, the 10 cells stack indicated high stable performance during the 130 h on-off cycle test with almost constant power output of 35 W at cell voltage of 6.0 V, 150 °C. The high stability of stack performance demonstrated high integrity of cell components during stack operation. Overall, HPW/*meso*-silica nanocomposite membranes show the promising application of stack cell at elevated temperature and anhydrous conditions.

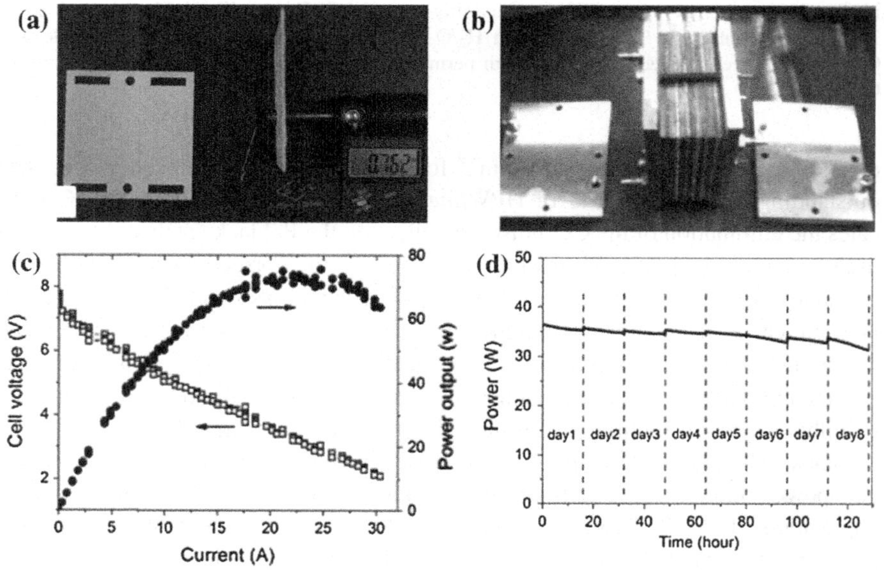

Fig. 10.18 Photographic image of HPW/*meso*-silica composite membrane (**a**) and assembled 6-cell fuel cell stack (**b**), and cell performance (**c**) and stability (**d**) of a 10-cell stack measured in H_2/O_2 (1/1 atm) without external humidification. Pt/C was used as catalyst loading of 0.4 mg cm^{-2} for both and cathode, and the effective area, and the effective area of the stack was 200 cm^2 (Reproduced from [161] with permission. Copyright (2013) Elsevier)

10.6.3 Proton Diffusion Mechanism in the HPW/Meso-Silica

Obviously, HPW loading in the mesopores of *meso*-silica material determines not only the average distances between HPWs, but also the proton conductivity and the energy barrier of proton transportation. Density functional theory (DFT) calculations using the generalized gradient approximation (CGA) reveals the relationship between the HPW loading and the proton transportation. Generally, there are two effective proton transfer pathways in the HPW functionalized mesoporous silica. One is the intramolecular proton transfer pathway, in which the proton hops within isolated HPW molecule. The other is the intermolecular proton transfer pathway, which proton hops from one HPW to the neighbouring HPW along the water-assisted hydrogen bond. The pathway with high energy barrier is the rate-determine step, and determines the overall proton conductivity of the HPW/*meso*-silica nanocomposite membrane. DFT calculation shows that the proton transfer process involves both intramolecular and intermolecular proton transfer pathways, as shown in Fig. 10.19 [168]. The energy barrier for the proton transportation in pure silica matrix is as high as 40 kJ mol^{-1}, which is consistent with the activation energy result of pure $Ia\bar{3}d$ mesoporous silica. When the adjacent HPWs are close with increase of HPW loading and less than 17 Å in distance, the calculated

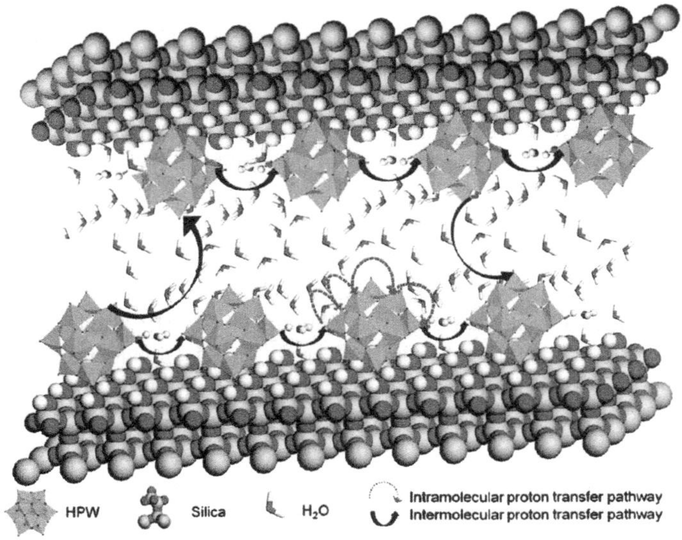

Fig. 10.19 The proton transportation was proposed to occur effectively in two ways. One is the proton-hopping on isolated phosphotungstic acid, briefly as intramolecular proton transfer pathway. The other is intermolecular proton transfer pathway, in which the proton exchange process is composed of a series of "hops" among HPW molecules and water molecules along the hydrogen bond (Reproduced from [168] with permission. Copyright (2014) American Chemistry Society)

activation energy for intramolecular proton transfer within a HPW molecule is higher (29.1–18.8 kJ mol^{-1}) than the barrier for intermolecular proton transfer along the hydrogen bond (7.8–2.5 kJ mol^{-1}). The calculation results explain why the proton conductivity keeps constant when the HPW content is higher than 67 wt%. Consequently, the overall barrier for the proton transportation in the HPW/*meso*-silica is predominated by the intramolecular proton transfer pathway.

*10.6.4 HPW/*Meso-*Silica Membrane Fabrication*

Handling and operation of mesoporous silica based PEMs require enough mechanical strength to withstand the pressure derived from the maintenance of the same RH at high temperature. However, mesoporous silica based membranes are inherently inflexible and brittle as inorganic materials are generally in the powder form. The obstacle impedes the wide application of inorganic materials in PEMFC. The HPW/*meso*-silica membrane fabrication is still one of the most critical challenges for application of the mesoporous silica based PEMs. So far, three methods have been developed to fabricate the inorganic disk by tape casting, hot press and sintering.

Tape casting combined with phase inversion is one of the simplest ways to prepare the mesoporous silica membrane [169]. In this method, mesoporous silica particles are mixed with polymer matrix such as polyethersulfone (PESf) in the organic solvent to form slurry. The slurry is tape-casted on clean substrate and then immersed into tap water to make the membrane solidified. Two layers is formed in the solidified membrane with a thin microporous layer and a thick macroporous layer because of the different precipitation rates of organic solvent occurring within the nascent membrane [170, 171]. Eventually, the solid membrane with thickness of about 0.35 mm is impregnated by HPW for proton conduction. The content of polymer matrix in the composite membrane can be reduced to as low as 25 wt% which is similar as the polymer content in the hot pressed disk, while the membrane still achieves good mechanical strength [172]. The power output of the HPW/MCM-41 membrane based single cell with 70 wt% HPW loading and 75 wt% silica loading reached up to 230 mW cm^{-2} in H_2/air at 90 °C and about 125 mW cm^{-2} in methanol/air at 150 °C without humidification for the gases. Nevertheless, the membrane is brittle under high temperature and dry conditions, although the wet membrane at low temperature shows good mechanical strength.

Hot press method increases the operation temperature of mesoporous silica based PEM up to 200 °C at low RH and even at anhydrous condition [152]. When inorganic powders are mixed with polymer binder, the membrane disk will be obtained after the hot press of composite particles at high temperature and high pressure where the polymer binder bonds the inorganic particles. For instance, phosphosilicate gel was sandwiched between two carbon papers and then membrane was obtained after hot-press at 130 °C under 11 MPa [173]. Moreover, Lu et al. used polyimide or polyethersulfone as polymer binder to glue the HPW/*meso*-silica powders into free standing PEM at 160–230 °C and 30–100 MPa [143, 147, 155]. However, the membranes fabricated by hot press technique need to be thin in the thickness range of 0.5–1.0 mm in order to have sufficient mechanical strength. However, thick membranes result in high ohmic losses in the PEMFC. Moreover, the mechanical strength of the die-pressed discs is still low and brittle in nature, making it very difficult to scale up the system to fuel cell stacks. In addition, the polyimide binder limits the thermal stability of the membrane at high temperatures.

Sintering method is effective to increase the mechanical strength of inorganic based membranes. For instance, mesoporous silica powder can form dense membranes after sintering at 650 °C with intact ordered mesoporous structure [127, 128]. The mechanical strength of sintered *meso*-silica is 51 MPa, significantly higher than ~5 MPa of hot-pressed *meso*-silica membranes [155]. The feasibility of utilization of sintered *meso*-silica based PEMs has been demonstrated by functionalization of the sintered *meso*-silica with phosphoric acid and the results indicate that the PA/sintered *meso*-silica PEM exhibits excellent conductivity under anhydrous conditions, achieving peak power density of 689 mW cm^{-2} in H_2 at 190 °C [128]. Sintering temperature is the critical parameter to make the mesoporous silica disk. The ordered structure of mesoporous silica is intact at a sintering temperature of 650 °C. However, the ordered mesoporous structure becomes disordered at 750 °C, and totally disappeared at 850 °C.

10.7 Mesoporous Materials for the Electrode Materials in PEMFC

In addition to PEMs, catalysts are also the critical component in PEMFC. Pt and Pt based alloys are employed as the catalyst in fuel cells because of its high electrocatalytic activity for hydrogen, alcohol and formic acid oxidation as well as high catalytic activity for oxygen reduction reaction (ORR). However, the scarcity and high cost of Pt impede its wide application in fuel cells. One approach is to increase the efficiency of Pt-based catalyst by employing catalyst support. Because of the interconnected pore network, high surface area, large pore volume, defined pore size distribution and tailorable surface composition, mesoporous carbon support are feasible in fuel cell applications [174]. The introduction of mesoporous structure into catalyst support not only increases the surface area of Pt-based catalyst, but also increases the dispersion and utilization of Pt NPs. The other approach is to develop non-precious metal catalyst such as metal-nitrogen-carbon materials. The non-precious metal catalyst reduces the dependence of fuel cells on precious metals and hence significantly reduces the cost of fuel cells.

10.7.1 Mesoporous Carbon Supported Catalyst

Pt/carbon black (such as Vulcan XC-72, VC) is widely used as electrocatalysts in PEMFC because of its relatively high surface area and conductivity. However, the primary micropores in carbon black are too small to be filled by the polyelectrolyte, and hence reduce the utilization of Pt NPs trapped in the micropores because of the absence of the trip-phase boundaries (TPBs) [175]. Furthermore, Pt NPs are subject to agglomerate on the surface of carbon black materials, resulting in dislodgment of Pt NPs [176]. Ordered mesoporous carbon materials (OMCs) with interconnected mesoporous structures, high surface area and narrow pore size distribution are considered as the promising catalyst support in PEMFC.

There are two strategies to synthesis mesoporous carbon materials [177, 178]. One is the hard template (nanocasting method) as developed by Ryoo's group [179], as shown in Fig. 10.20. Mesoporous materials such as mesoporous silica and aluminosilicate are employed as the hard template. Carbon precursor such as furfurylalcohol (FFA) [180], phenolic resin [175] and sucrose [181] are impregnated into the mesopores of mesoporous silica and then carbonized at inert atmosphere and high temperature. After removal of mesoporous silica template by NaOH and HF, mesoporous carbon materials are obtained. A series of ordered mesoporous carbon materials, named as CMK (carbon mesostructures from Korea), such as CMK-1 [182] and CMK-4 [183] templated from MCM-48 (*Ia3d*), CMK-2 [184] from SBA-1 (*Pm3n*), CMK-3 [185] and CMK-5 [186] from SBA-15 (*P6mm*), CMK-8 and CMK-9 [187] from KIT-6 (*Ia*$\bar{3}$*d*). Although mesoporous carbon replicates the mesoporous structure of the hard template, it is worthy to note that CMK-3 is the rod-type of *P6mm* from a complete replica of SBA-15 while CMK-5 results from incomplete filling of the SBA-15 channels with interconnected nanopipes [177].

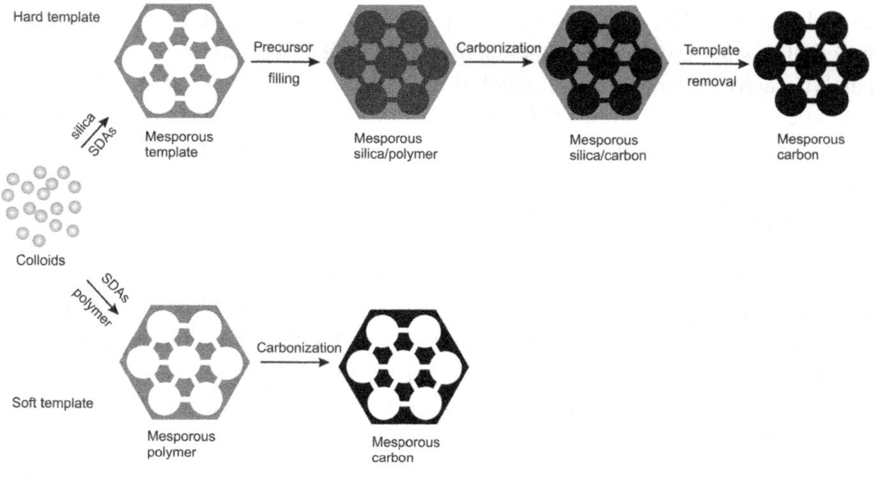

Fig. 10.20 Hard template and soft template strategies to synthesis mesoporous carbon materials, taking mesoporous carbon with *P6mm* symmetry as example

Hard template protocol shows feasible carbon sources and morphology control (rod, sphere, monolith and hollow mesoporous structure etc.). However, the pore size control in OMCs is not as easy as mesoporous silica because the pore size of OMCs is dependent on the thickness of mesoporous silica walls [188].

The other strategy to synthesize mesoporous carbon materials is the soft template method [189]. Highly ordered mesoporous structure is introduced into polymers with the assistance of SDAs, and mesoporous carbon is obtained by the carbonization of mesoporous polymer after removing the SDAs. Zhao et al. developed the system of resols/PEO-PPO-PEO, and highly ordered mesoporous polymers with different symmetries including lamellar, 2D hexagonal *P6mm*, bicontinuous cubic $Ia\bar{3}d$ and body-centred cubic $Im\bar{3}m$ were synthesized by a solvent evaporation induction self-assembly (EISA) [190, 191]. Besides PEO-PPO-PEO triblock copolymers (such as P123, F127 and F108), poly(2-vinylpyridine)-polyisopren (P2VP-PI) and poly(4-vinylpyridine)-polystyrene(P4VP-PS), and poly(ethylene oxide)-*b*-polystyrene (PEO-*b*-PS) are also used as SDAs [192]. Low molecule weight of phenolic resin colloids are self-assembled into mesoporous structure via the hydrogen-bonding interaction between block-copolymers and pyridine groups and the hydroxyl group organic precursors. Mesoporous carbon materials express the same space symmetries of mesoporous polymer, while various types of mesoporous polymers are tuned by simply varying the EO/PO rations in the template.

Pt loaded on OMCs shows significant higher performance and durability than Pt/VC for oxygen reduction reactions and methanol oxidation reaction (MOR) [176, 186]. Pt/CMK-5 showed specific activity of 433 mA mg^{-1} Pt for MOR, about twice higher than that of Pt/VC (220 mA mg^{-1} Pt) [186]. Nevertheless, the performance of OMCs is affected by the types, pore size and crystallinity of OMCs. The different pore structures of carbon support affects the surface areas of carbon supports and accordingly the electrochemical surface area (ECSA) and ORR activity of catalysts. For instance, CMK-3 and CMK-5 mesoporous carbon materials are the replica of mesoporous silica SBA-15. CMK-3 mesoporous carbon shows hexagonal structure with uniform carbon nanorods, while CMK-5 mesoporous carbon indicates hexagonal arrays of uniform carbon nanotubes. However, Pt/CMK-5 shows superior performance than Pt/CMK-3 because of higher surface area of CMK-5 than that of CMK-3 [185]. Generally, OMCs have pore size around 5 nm and it can be tuned up to 7 nm. Although the mesopores can structurally confine catalyst NPs, it shows mass transport limitations because of the blockage of mesopores by water and polyelectrolytes [181]. Large pore size mesoporous carbons developed by colloidal imprinted carbon (CIC) [193, 194] supports indicates substantial higher ORR activity than VC support [195]. Besides, Pt/CIC-15 (15 nm in pore size) demonstrates higher ORR activity than Pt/CIC-26 (26 nm in pore size), as CIC-15 has thicker walls than CIC-26, corresponding to lower electronic resistance and ORR Tafel slope [176]. For the crystallinity of carbon support, carbon nanotubes have superior corrosion resistance than amorphous carbon black because of its graphitic carbon layers. Similar as carbon black, OMCs have amorphous structure and suffer from the corrosion issues. Nevertheless, graphitic mesoporous carbons (GMPCs) show graphitic form of carbon structure, while it is much cheaper than CNTs. In comparison with Pt/VC, Pt/GMPCs with 20 wt% Pt loading shows only one third of ECSA loss during the accelerated corrosion test [196].

In comparison with OMCs spheres, hollow mesoporous carbon spheres (HMCS) shows superiority because of the unique structure of macroporous hollow core and mesoporous shell. HMCS is generally synthesized from hard template. The highly ordered mesoporous structure connects the outer surface and the inner hollow macroporous core, leading to fast mass transport for the supported catalyst via the bicontinuous network. Moreover, the hollow core can serve as reservoirs for the reactants and products, which shortens the diffusion path of mass transport. When Pt/HMCS with Pt loading of 20 wt% is employed as cathode catalyst in H_2/O_2 PEMFCs, the peak power density of the cell is 444 mW cm^{-1} at 60 °C, a 83 and 59 % higher than fuel cells with Pt/VC and commercial E-TEK Pt/VC as cathode catalyst, respectively [197]. Besides Pt NPs, Pt-based alloys such as PtPd [195] and PtRu [197] are also loaded on HMCS, showing higher performance than catalysts supported on VC. In the case of Pt/VC catalyst, the uneven dispersion of Pt on VC and the increase of the particle size of Pt with the increase of Pt loading on VC will reduce the utilization and decreases the power output of fuel cells. For instance, when Pt loading increases from 20 to 60 wt%, the particle size of Pt increases from 3.3 to 5.0 nm, resulting in 30.1 % reduction in the peak power density [198]. On the other hand, for Pt/HMCS, the particle size of Pt slightly increases from 2.5 to 3.3 nm when the Pt loading is raised from 20 to 60 wt%. Thus, the negative effect from decrease of ECSA is compensated by the positive effect from decrease of electronic resistance with reduced thickness of catalyst layer [198]. Carbon sources for HMCS also affect the properties of HMCS. HMCS derived from divinylbenzene (DVB) shows better performance than HMCS from resol [199].

Although carbon support increases the dispersion and utilization of Pt-based NPs, the scarcity of Pt also impede the application of Pt in fuel cells. Mesoporous non-noble metal catalysts have been considered as the promising catalyst for PEMFC because of its abundant resources and low cost [200, 201]. Mesoporous non-noble metal catalysts, in particular, metal-nitrogen-carbon (M-N-C) can be fabricated by heat treatment of metal and N/C precursors [202]. Transition metal salt such as iron [203] and cobalt [201] are chosen as metal sources, while polypyrrole [200], sucrose [202], furfuryl alcohol [204] and tetramethoxyphenylporphyrin (H_2TMPP) [201] are used as carbon sources. The electrocatalytic activities of M-N-C are affected by the metal sources, metal concentrations, and carbonization temperature. High concentration of metal ions in the mesoporous silica samples leads to a high catalytic activity of the resulting non-noble metal catalysts in terms of higher onset potential and current density for ORR. Although the role of metal elements in the non-noble catalyst is controversial, the absent of metal sources in the catalyst substantially decreases the catalytic activity of mesoporous M–N–C catalyst [200]. Moreover, the temperature affects the carbonization behaviour of carbon sources confined in the interconnected mesoporous network, as the geometry of mesoporous template results in different graphitization and orientation of graphene layers [205]. However, in the narrow temperature range of 800–1000 °C, the electrocatalytic activities of mesoporous M–N–C catalyst are independent on the carbonization temperature [200].

High volumetric surface area (surface area to volume ratio) is paramount important for improving the activity of non-noble metal catalysts for ORR, as high volumetric surface area corresponding to high active site density [203]. To achieve large surface area and specific porosity with tuneable pore size, unique template-assisted ultrasonic spray pyrolysis (TA-USP) technique has been developed to synthesize mesoporous M–N–C catalyst [201]. In this strategy, silica colloids with small diameter, carbon precursor and metal-N precursor such as CoTMPP and FeTMPP are homogeneously mixed into suspension, and then the suspension drops are carbonized at inert atmosphere and high temperature, as shown in Fig. 10.21a. Mesoporous M-N-C catalysts are obtained after the removal of silica template by HF or alkaline solution [202]. Large BET specific surface area of over 2000 $m^2\ g^{-1}$ is obtained for the mesoporous M–N–C catalyst synthesized by TA-USP method (Fig. 10.21b), which is much higher than that of carbon black VC (250 $m^2\ g^{-1}$) [206]. M–N–C with mesoporous structure shows improved ORR activity in comparison with conventional M–N–C catalyst derived from commercial carbon black. For instance, mesoporous Fe–N–C shows a 90 mV positive shift on peak potential than Fe-N-C (VC) for ORR, while its peak power density (64 mW cm^{-2}) is 3.4 times higher than the Fe–N–C (VC) (18 mW cm^{-2}), as shown in Fig. 10.21c [203]. Moreover, catalytic activity of mesoporous M–N–C catalyst improves with the increase of the size of silica colloid templates [206]. When the particle size of mesoporous M–N–C is the same, mesoporous M–N–C catalyst with large pore size shows a high accessible surface area of the catalyst for polyelectrolyte. For instance, when mesoporous Fe–N–C catalysts have pore size of 50, 35, and 22 nm with corresponding surface area of 2125, 1409, and 1338 $m^2\ g^{-1}$, respectively, the catalytic activity of the catalyst for ORR decreases with the decrease of pore size [206].

10.7.2 Mesoporous Metal Oxide Based Catalyst

The application of mesoporous carbon as catalyst support or non-noble catalyst reduces the Pt loading in the electrode and increases the utilization of the catalyst. However, a critical issue for carbon materials is the corrosion of carbon support at cathode during fuel cell application, as indicated by the following reaction (10.4).

$$C + 2H_2O \rightarrow CO_2 + 4H^+ + 4e^- \quad E_0 = 0.118\,\text{V versus NHE}, 25\,^{\circ}\text{C} \qquad (10.4)$$

Electrochemical corrosion of carbon materials causes the detachment of catalyst. Moreover, it also leads to the aggregation of catalyst on the surface of carbon support. Those effects result in the decrease of electrochemical surface area of the catalyst, and hence the degradation of the cell performance [174]. High electrochemical stable support instead of carbon materials, therefore, is required to be

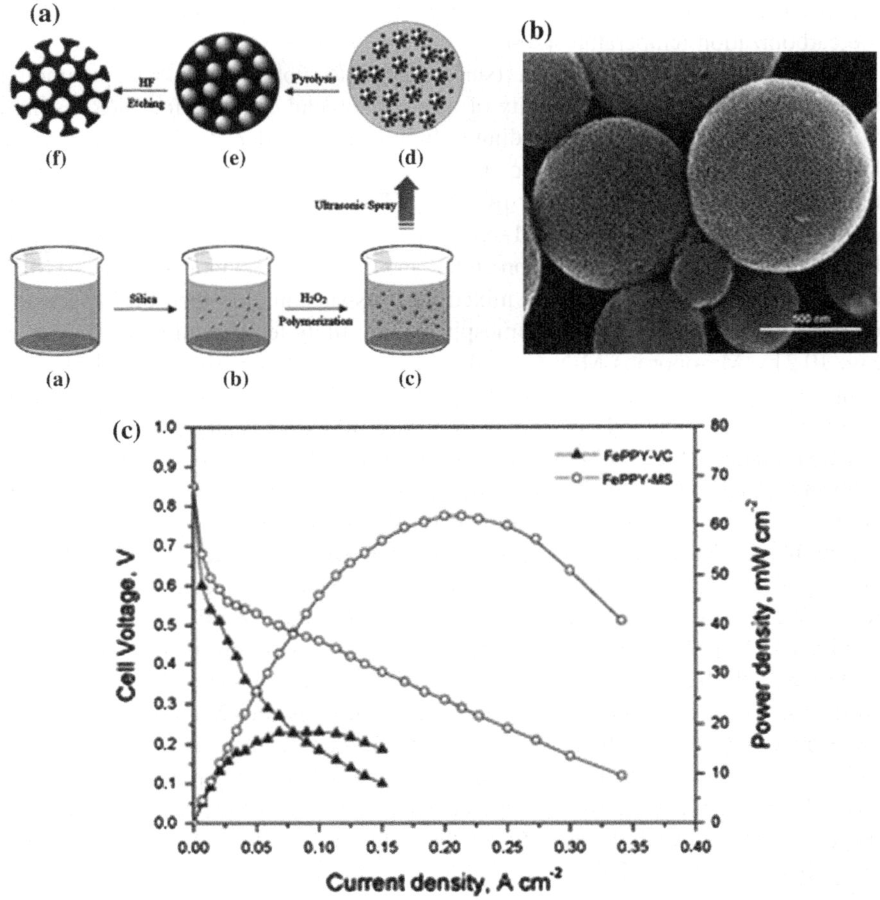

Fig. 10.21 **a** is scheme of mesoporous M-N-C catalyst via template-assisted ultrasonic spray pyrolysis technique. **b** is the accordingly SEM image of the catalyst, while **c** is the I-V plot for the fuel cell with FePPy-MS and FePPy in H_2/air fuel cell. Anode: 0.3 mg cm^{-2} PtRu (20 wt% PtRu/C, E-TEK). Cathode: 5.0 mg cm^{-2} FePPy-MS or FePPy-VC (Reproduced from [203] with permission. Copyright (2009) The Royal Society of Chemistry)

developed in fuel cell application. Mesoporous metal oxide support is one of the promising candidates. It not only inherits the merits of mesoporous carbon materials with high surface area and fast mass transportation, but also overcomes the corrosion effect of carbon materials [207].

Similar as mesoporous carbon support, mesoporous metal oxides are also synthesized by either hard template or soft template protocols. For instance, mesoporous TiO_2 [208] and NbO_2 [209] are synthesized by soft template with Pluronic P123 and poly(isoprene-*block*-ethylene oxide) (PI-*b*-PEO) as the SDAs, respectively. On the other hand, ordered mesoporous WO_{3-X} is synthesized by the hard template via calcination of HPW on the mesopores of KIT-6 mesoporous silica

[210]. Besides, NiO [211] and CeO_2 [212] based catalyst are also synthesized by hard template protocol. Mesoporous Pt with highly ordered mesoporous channels (8.5 nm in diameter and 800 $m^2\ g^{-1}$ on surface area) is also able to be fabricated by the hard template method from $Ia\,\bar{3}d$ KIT-6 mesoporous silica [213, 214].

Metal oxide support significantly increases the electrochemical stability of the support at cathode and the cell performance. For instance, the ECSA of Pt on mesoporous TiO_2 support shows decay of ~20 % after 80 h accelerated stress test (AST) at E = 1.2 V, which is much lower than the 93 % decay for Pt/carbon black [208]. The high stability of Pt/TiO_2 catalyst results from the strong metal support interaction between the Pt NPs and TiO_2 support. However, pure metal oxide materials show low electronic conductivity, leading high ohmic resistance of the electrode and then low cell performance. Doping the mesoporous metal oxide materials with carbon substantially enhance the electronic conductivity of the catalyst support. Mesoporous TiO_2 [215], Nb_2O_5 [209], and CeO_2 [212] are doped by carbon to form metal-carbon composite support, and mesoporous Pt–Pb/carbon doped Nb_2O_5(Nb_2O_5–C–Pt) shows three-times higher mass activity for formic acid oxidation than Pt–Pb/carbon black, although the reasons are not very clear [209]. Besides the increase in cell performance and durability, mesoporous oxide materials also reduce the dependence of PEMFC on RH. Pt-*meso*-CeO_2 demonstrated higher cell performance at 75 % RH, 65 °C than Pt/C catalyst due to the high water retention ability of CeO_2 [212].

10.8 Conclusions

Mesoporous materials possess periodic and interconnected channels and high surface areas. Because of the capillary condensation effect of mesoporous channels, it shows strong capability to retain water at low RH and high temperature. Moreover, water molecules confined in the mesopores induce the proton transportation by forming protonic charge carriers through hydrogen bonding. Herein, the introduction of mesoporous structure into PEMs not only facilitates proton conductivity but also alleviates the dependence of proton conductivity of PEMs on RH. Mesoporous polymer membranes, such as mesoporous sulfonated block copolymers and *meso*-Nafion membrane and mesoporous inorganic/polymer composite membranes, such as sulfonated mesoporous silica based composite membranes are directly employed as the PEMs for fuel cells. In addition, mesoporous inorganic materials such as mesoporous silica are used as the support matrix, while proton conductors such as HPW, PA, imidazole and ion liquid are impregnated into the mesopores for proton conduction. The space symmetry, pore size, porosity and ordering level of the mesoporous materials are the critical factors to obtain optimum performance as PEMs for fuel cells. Small pore size in the range of ~6 nm would be sufficient to maintain high proton conductivity and stability, while large pores will lead to a significant loss of the impregnated HPW from the HPW/*meso*-silica nanocomposites.

Functionalized mesoporous silica materials show the most promising properties as alternative PEMs for high temperature operation. The most significant aspect of the HPW/*meso*-silica nanocomposites is the formation of ($\equiv SiOH_2^+$)($H_2PW_{12}O_{40}^-$) species between the positively charged siliceous surface of mesoporous silica and negatively charged HPW in the presence of water [155]. This ensures the immobilization and high stability of impregnated or assembled HPW within a highly ordered mesoporous silica structure. The good performance and preliminary stability data of the small button cells and stacks with HPW/*meso*-silica nanocomposite membranes demonstrate that the inorganic PEMs based on highly ordered HPW/*meso*-silica nanocomposites can be applied to the fuel cell operated at temperatures up to 200 °C. Mechanical strength of the silica based membrane is also a problem and it is very difficult to fabricate MEAs with high density and adequate mechanical strength for handling and operation. Thus, it is still a great challenge to fabricate MEAs with good interface contact between the electrode and electrolyte under pressurized conditions similar to that of the polymeric material based membranes.

Mesoporous carbon support not only reduces the loading of noble metals in the electrode of PEMFC, but also facilitates the mass transportation because of its high surface area and highly ordered mesoporous channels. And the combination of metal, nitrogen and mesoporous carbon reveals the non-precious metal catalyst and diminishes the dependence of fuel cells on noble metal catalyst and reduces the cost of fuel cells. However, the corrosion issue of carbon materials at high potential limits the application of mesoporous carbon in PEMFC, while mesoporous metal oxide and carbon doped metal oxide support enhance the corrosion resistance of catalyst support in fuel cells. Mesoporous materials show limited application in high temperature SOFCs due to the high fabrication and operation temperatures associated with SOFCs, however, mesoporous materials show the promising potential in the development of PEM and electrocatalyst supports with high proton conductivity and electrocatalytic activity for low temperature PEMFCs.

Acknowledgement This work is supported by the Australian Research Council under the Discovery Project Scheme (Project Number: DP150102025).

References

1. S.P. Jiang, P.K. Shen (ed.), *Nanostructured and Advanced Materials for Fuel Cells* (CRC Press, Boca Raton, 2013)
2. J.S. Beck, J.C. Vartuli, W.J. Roth, M.E. Leonowicz, C.T. Kresge, K.D. Schmitt, C.T.W. Chu, D.H. Olson, E.W. Sheppard, A new family of mesoporous molecular sieves prepared with liquid crystal templates. J. Am. Chem. Soc. **114**, 10834–10843 (1992)
3. C.T. Kresge, M.E. Leonowicz, W.J. Roth, J.C. Vartuli, J.S. Beck, Ordered mesoporous molecular sieves synthesized by a liquid-crystal template mechanism. Nature **359**, 710–712 (1992)
4. Y. Wan, D. Zhao, On the controllable soft-templating approach to mesoporous silicates. Chem. Rev. **107**, 2821–2860 (2007)

5. Y. Ye, C. Jo, I. Jeong, J. Lee, Functional mesoporous materials for energy applications: solar cells, fuel cells, and batteries. Nanoscale **5**, 4584–4605 (2013)
6. M. Mamak, N. Coombs, G.A. Ozin, Electroactive mesoporous yttria stabilized zirconia containing platinum or nickel oxide nanoclusters: a new class of solid oxide fuel cell electrode materials. Adv. Funct. Mater. **11**, 59–63 (2001)
7. M. Mamak, N. Coombs, G. Ozin, Self-assembling solid oxide fuel cell materials: mesoporous yttria-zirconia and metal-yttria-zirconia solid solutions. J. Am. Chem. Soc. **122**, 8932–8939 (2000)
8. I.M. Hung, D.T. Hung, K.Z. Fung, M.H. Hon, Effect of calcination temperature on morphology of mesoporous YSZ. J. Eur. Ceram. Soc. **26**, 2627–2632 (2006)
9. E. Rossinyol, E. Pellicer, A. Prim, S. Estrade, J. Arbiol, F. Peiro, A. Cornet, J.R. Morante, Gadolinium doped Ceria nanocrystals synthesized from mesoporous silica. J. Nanopart. Res. **10**, 369–375 (2008)
10. G. Xiao, Z. Jiang, H. Li, C. Xia, L. Chen, Studies on composite cathode with nanostructured Ce0.9Sm0.1O1.95 for intermediate temperature solid oxide fuel cells. Fuel Cells **9**, 650–656 (2009)
11. C. Su, X.M. Xu, Y.B. Chen, Y. Liu, M.O. Tade, Z.P. Shao, A top-down strategy for the synthesis of mesoporous Ba05Sr0.5Co0.8Fe0.2O3-delta as a cathode precursor for buffer layer-free deposition on stabilized zirconia electrolyte with a superior electrochemical performance. J. Power Sour. **274**, 1024–1033 (2015)
12. I. Kivi, P. Moller, H. Kurig, S. Kallip, G. Nurk, E. Lust, Development of porous cathode powders for SOFC and influence of cathode structure on the oxygen electroreduction kinetics. Electrochem. Commun. **10**, 1455–1458 (2008)
13. J.J. Pan, H.I. Zhang, M. Pan, Self-assembly of Nafion molecules onto silica nanoparticles formed in situ through sol-gel process. J. Colloid Interface Sci. **326**, 55–60 (2008)
14. V. Ramani, H.R. Kunz, J.M. Fenton, Stabilized heteropolyacid/Nafion® composite membranes for elevated temperature/low relative humidity PEFC operation. Electrochim. Acta **50**, 1181–1187 (2005)
15. N.P. Cele, S. Sinha Ray, S.K. Pillai, M. Ndwandwe, S. Nonjola, L. Sikhwivhilu, M.K. Mathe, Carbon nanotubes based nafion composite membranes for fuel cell applications. Fuel Cells **10**, 64–71 (2010)
16. G.B. Ye, K. Li, C.A. Xiao, W. Chen, H.N. Zhang, M. Pan, Nafion (R)-titania nanocomposite proton exchange membranes. J. Appl. Polym. Sci. **120**, 1186–1192 (2011)
17. D.V. Bavykin, M. Carravetta, A.N. Kulak, F.C. Walsh, Application of magic-angle spinning NMR to examine the nature of protons in titanate nanotubes. Chem. Mater. **22**, 2458–2465 (2010)
18. E. Morgado, P.M. Jardim, B.A. Marinkovic, F.C. Rizzo, M.A.S. De Abreu, J.L. Zotin, A.S. Araujo, Multistep structural transition of hydrogen trititanate nanotubes into TiO(2)-B nanotubes: a comparison study between nanostructured and bulk materials. Nanotechnol. **18** (2007)
19. K. Li, G.B. Ye, J.J. Pan, H.N. Zhang, M. Pan, Self-assembled Nafion (R)/metal oxide nanoparticles hybrid proton exchange membranes. J. Membr. Sci. **347**, 26–31 (2010)
20. A. Vishnyakov, A.V. Neimark, Molecular simulation study of Nafion membrane solvation in water and methanol. J. Phys. Chem. B **104**, 4471–4478 (2000)
21. P. Choi, N.H. Jalani, R. Datta, thermodynamics and proton transport in Nafion: II. Proton diffusion mechanisms and conductivity. J. Electrochem. Soc. **152**, E123–E130 (2005)
22. K. Feng, L. Hou, B. Tang, P. Wu, Does thermal treatment merely make a H_2O-saturated Nafion membrane lose its absorbed water at high temperature? Phys. Chem. Chem. Phys. **17**, 9106–9115 (2015)
23. J. Lu, H. Tang, C. Xu, S.P. Jiang, Nafion membranes with ordered mesoporous structure and high water retention properties for fuel cell applications. J. Mater. Chem. **22**, 5810–5819 (2012)
24. J. Lu, S. Lu, S.P. Jiang, Highly ordered mesoporous Nafion membranes for fuel cells. Chem. Commun. **47**, 3216–3218 (2011)

25. J. Lu, H. Tang, S. Lu, H. Wu, S.P. Jiang, A novel inorganic proton exchange membrane based on self-assembled HPW-meso-silica for direct methanol fuel cells. J. Mater. Chem. **21**, 6668–6676 (2011)
26. J. Li, H. Tang, L. Chen, R. Chen, M. Pan, S.P. Jiang, Highly ordered and periodic mesoporous Nafion membranes via colloidal silica mediated self-assembly for fuel cells. Chem. Commun. **49**, 6537–6539 (2013)
27. J. Zhang, J. Li, H. Tang, M. Pan, S.P. Jiang, Comprehensive strategy to design highly ordered mesoporous Nafion membranes for fuel cells under low humidity conditions. J. Mater. Chem. A **2**, 20578–20587 (2014)
28. N.W. DeLuca, Y.A. Elabd, Polymer electrolyte membranes for the direct methanol fuel cell: a review. J. Polym. Sci., Part B: Polym. Phys. **44**, 2201–2225 (2006)
29. M.A. Hickner, H. Ghassemi, Y.S. Kim, B.R. Einsla, J.E. McGrath, Alternative polymer systems for proton exchange membranes (PEMs). Chem. Rev. **104**, 4587–4612 (2004)
30. P.V. Komarov, I.N. Veselov, P.P. Chu, P.G. Khalatur, Mesoscale simulation of polymer electrolyte membranes based on sulfonated poly(ether ether ketone) and Nafion. Soft Matter **6**, 3939–3956 (2010)
31. W.H. Choi, W.H. Jo, Preparation of new proton exchange membrane based on self-assembly of Poly(styrene-co-styrene sulfonic acid)-b-poly(methyl methacrylate)/Poly(vinylidene fluoride) blend. J. Power Sour. **188**, 127–131 (2009)
32. S.B. Darling, Directing the self-assembly of block copolymers. Prog. Polym. Sci. **32**, 1152–1204 (2007)
33. J. Won, H.H. Park, Y.J. Kim, S.W. Choi, H.Y. Ha, I.-H. Oh, H.S. Kim, Y.S. Kang, K.J. Ihn, Fixation of nanosized proton transport channels in membranes†. Macromolecules **36**, 3228–3234 (2003)
34. H.-C. Lee, H. Lim, W.-F. Su, C.-Y. Chao, Novel sulfonated block copolymer containing pendant alkylsulfonic acids: syntheses, unique morphologies, and applications in proton exchange membrane. J. Polym. Sci. Part A Polym. Chem. **49**, 2325–2338 (2011)
35. M. Ingratta, E.P. Jutemar, P. Jannasch, Synthesis, nanostructures and properties of sulfonated poly(phenylene oxide) bearing polyfluorostyrene side chains as proton conducting membranes. Macromolecules **44**, 2074–2083 (2011)
36. M.J. Park, K.H. Downing, A. Jackson, E.D. Gomez, A.M. Minor, D. Cookson, A.Z. Weber, N.P. Balsara, Increased water retention in polymer electrolyte membranes at elevated temperatures assisted by capillary condensation. Nano Lett. **7**, 3547–3552 (2007)
37. E. Helfand, Z.R. Wasserman, Block copolymer theory. 4. Narrow interphase approximation. Macromolecules **9**, 879–888 (1976)
38. M.J. Park, N.P. Balsara, Phase behavior of symmetric sulfonated block copolymers. Macromolecules **41**, 3678–3687 (2008)
39. S. Ge, X. Li, B. Yi, I.-M. Hsing, Absorption, desorption, and transport of water in polymer electrolyte membranes for fuel cells. J. Electrochem. Soc. **152**, A1149–A1157 (2005)
40. L. Rubatat, C.X. Li, H. Dietsch, A. Nykanen, J. Ruokolainen, R. Mezzenga, Structure-properties relationship in proton conductive sulfonated polystyrene-polymethyl methacrylate block copolymers (sPS-PMMA). Macromolecules **41**, 8130–8137 (2008)
41. H. Tang, Z. Wan, M. Pan, S.P. Jiang, Self-assembled Nafion-silica nanoparticles for elevated-high temperature polymer electrolyte membrane fuel cells. Electrochem. Commun. **9**, 2003–2008 (2007)
42. D. Zhao, Y. Wan, W. Zhou, *Ordered Mesoporous Materials* (Wiley-VCH Verlag GmbH & Co. KGaA, 2013), pp. 5–54
43. H. Li, M. Nogami, Pore-controlled proton conducting silica films. Adv. Mater. **14**, 912–914 (2002)
44. Y.G. Jin, S.Z. Qiao, Z.P. Xu, J.C. Diniz da Costa, G.Q. Lu, Porous silica nanospheres functionalized with phosphonic acid as intermediate-temperature proton conductors. J. Phys. Chem. C **113**, 3157–3163 (2009)

45. A. Matsuda, Y. Nono, T. Kanzaki, K. Tadanaga, M. Tatsumisago, T. Minami, Proton conductivity of acid-impregnated mesoporous silica gels prepared using surfactants as a template. Solid State Ionics **145**, 135–140 (2001)
46. Y. Jin, S. Qiao, L. Zhang, Z.P. Xu, S. Smart, J.C.D. da Costa, G.Q. Lu, Novel Nafion composite membranes with mesoporous silica nanospheres as inorganic fillers. J. Power Sour. **185**, 664–669 (2008)
47. Y. Choi, Y. Kim, H.K. Kim, J.S. Lee, Direct synthesis of sulfonated mesoporous silica as inorganic fillers of proton-conducting organic-inorganic composite membranes. J. Membr. Sci. **357**, 199–205 (2010)
48. Y. Chiba, Y. Tominaga, Poly(ethylene-co-vinyl alcohol)/Sulfonated Mesoporous Organosilicate Composites as Proton-conductive Membranes. J. Power Sour. **203**, 42–47 (2012)
49. Y. Tominaga, T. Maki, Proton-conducting composite membranes based on polybenzimidazole and sulfonated mesoporous organosilicate. Int. J. Hydrogen Energy **39**, 2724–2730 (2014)
50. K.-D. Kreuer, A. Rabenau, W. Weppner, Vehicle mechanism, a new model for the interpretation of the conductivity of fast proton conductors. Angew. Chem. Int. Ed. **21**, 208–209 (1982)
51. L. Bernard, A. Fitch, A.F. Wright, B.E.F. Fender, A.T. Howe, Proceedings of the international conference on fast ionic transport in solidsmechanisms of hydrogen diffusion and conduction in DUO2AsO4·4D2O as inferred from neutron diffraction evidence. Solid State Ionics **5**, 459–462 (1981)
52. F. Pereira, K. Vallé, P. Belleville, A. Morin, S. Lambert, C. Sanchez, Advanced mesostructured hybrid Silica –Nafion membranes for high-performance PEM fuel cell. Chem. Mater. **20**, 1710–1718 (2008)
53. H. Pu, L. Lou, Y. Guan, Z. Chang, D. Wan, Proton exchange membranes based on semi-interpenetrating polymer networks of polybenzimidazole and perfluorosulfonic acid polymer with hollow silica spheres as micro-reservoir. J. Membr. Sci. **415–416**, 496–503 (2012)
54. Y. Zhao, H. Yang, H. Wu, Z. Jiang, Enhanced proton conductivity of hybrid membranes by incorporating phosphorylated hollow mesoporous silica submicrospheres. J. Membr. Sci. **469**, 418–427 (2014)
55. X. Fang, C. Chen, Z. Liu, P. Liu, N. Zheng, A cationic surfactant assisted selective etching strategy to hollow mesoporous silica spheres. Nanoscale **3**, 1632–1639 (2011)
56. R. Deshmukh, U. Schubert, Synthesis of CuO and Cu3N nanoparticles in and on hollow silica spheres. Eur. J. Inorg. Chem. **2013**, 2498–2504 (2013)
57. G. Qi, Y. Wang, L. Estevez, A.K. Switzer, X. Duan, X. Yang, E.P. Giannelis, Facile and scalable synthesis of monodispersed spherical capsules with a mesoporous shell. Chem. Mater. **22**, 2693–2695 (2010)
58. Y. Li, J. Shi, Hollow-structured mesoporous materials: chemical synthesis, functionalization and applications. Adv. Mater. **26**, 3176–3205 (2014)
59. T. Zhang, J. Ge, Y. Hu, Q. Zhang, S. Aloni, Y. Yin, Formation of hollow silica colloids through a spontaneous dissolution-regrowth process. Angew. Chem. Int. Ed. **47**, 5806–5811 (2008)
60. Q. Zhang, T. Zhang, J. Ge, Y. Yin, Permeable silica shell through surface-protected etching. Nano Lett. **8**, 2867–2871 (2008)
61. Y. Chen, H. Chen, L. Guo, Q. He, F. Chen, J. Zhou, J. Feng, J. Shi, Hollow/rattle-type mesoporous nanostructures by a structural difference-based selective etching strategy. ACS Nano **4**, 529–539 (2010)
62. Y. Yin, W. Deng, H. Wang, A. Li, C. Wang, Z. Jiang, H. Wu, Fabrication of hybrid membranes by incorporating acid-base pair functionalized hollow mesoporous silica for enhanced proton conductivity. J. Mater. Chem. A **3**, 16079–16088 (2015)
63. K. Feng, B. Tang, P. Wu, A "H_2O donating/methanol accepting" platform for preparation of highly selective Nafion-based proton exchange membranes. J. Mater. Chem. A (2015)

64. Y.-F. Lin, C.-Y. Yen, C.-C.M. Ma, S.-H. Liao, C.-H. Lee, Y.-H. Hsiao, H.-P. Lin, High proton-conducting Nafion (R)/-SO3H functionalized mesoporous silica composite membranes. J. Power Sour. **171**, 388–395 (2007)
65. Y. Tominaga, I.-C. Hong, S. Asai, M. Sumita, Proton conduction in Nafion composite membranes filled with mesoporous silica. J. Power Sour. **171**, 530–534 (2007)
66. Z. Chen, R.S. Hsu, Nafion/acid functionalized mesoporous silica nanocomposite membrane for high temperature PEMFCs. Proton Exch. Membr. Fuel Cells **9**(25), 1151–1157 (2009)
67. J.C. McKeen, Y.S. Yan, M.E. Davis, Proton conductivity of acid-functionalized zeolite beta, MCM-41, and MCM-48: Effect of acid strength. Chem. Mater. **20**, 5122–5124 (2008)
68. S. Meenakshi, A.K. Sahu, S.D. Bhat, P. Sridhar, S. Pitchumani, A.K. Shukla, Mesostructured-aluminosilicate-Nafion hybrid membranes for direct methanol fuel cells. Electrochim. Acta **89**, 35–44 (2013)
69. L. Geng, Y. He, D. Liu, C. Lu, New organic-inorganic hybrid membranes based on sulfonated polyimide/aminopropyltriethoxysilane doping with sulfonated mesoporous silica for direct methanol fuel cells. J. Appl. Polym. Sci. **123**, 3164–3172 (2012)
70. D. Liu, L. Geng, Y. Fu, X. Dai, C. Lue, Novel nanocomposite membranes based on sulfonated mesoporous silica nanoparticles modified sulfonated polyimides for direct methanol fuel cells. J. Membr. Sci. **366**, 251–257 (2011)
71. L. Geng, Y. He, D. Liu, X. Dai, C. Lu, Facile in situ template synthesis of sulfonated polyimide/mesoporous silica hybrid proton exchange membrane for direct methanol fuel cells. Microporous Mesoporous Mater. **148**, 8–14 (2012)
72. D.X. Luu, E.-B. Cho, O.H. Han, D. Kim, SAXS and NMR analysis for the cast solvent effect on sPEEK membrane properties. J. Phys. Chem. B **113**, 10072–10076 (2009)
73. S. Zhou, D. Kim, Cross-linked aryl-sulfonated poly(arylene ether ketone) proton exchange membranes for fuel cell. Electrochim. Acta **63**, 238–244 (2012)
74. J.-R. Lee, J.-H. Won, K.-S. Yoon, Y.T. Hong, S.-Y. Lee, Multilayer-structured, SiO_2/sulfonated poly(phenylsulfone) composite membranes for proton exchange membrane fuel cells. Int. J. Hydrogen Energy **37**, 6182–6188 (2012)
75. J.-H. Won, H.-J. Lee, K.-S. Yoon, Y.T. Hong, S.-Y. Lee, Sulfonated SBA-15 mesoporous silica-incorporated sulfonated poly(phenylsulfone) composite membranes for low-humidity proton exchange membrane fuel cells: anomalous behavior of humidity-dependent proton conductivity. Int. J. Hydrogen Energy **37**, 9202–9211 (2012)
76. L. Xie, E.-B. Cho, D. Kim, Sulfonated PEEK/cubic (Im3m) mesoporous benzene-silica composite membranes operable at low humidity. Solid State Ionics **203**, 1–8 (2011)
77. V. Delhorbe, X. Thiry, C. Cailleteau, E. Mourier, M. Bathfield, L. Chikh, O. Fichet, B. Ameduri, R. Mercier, S. Vidal, L. Augier, E. Espuche, F. Gouanve, G. Gebel, A. Morin, Fluorohexane network and sulfonated PEEK based semi-IPNs for fuel cell membranes. J. Membr. Sci. **389**, 57–66 (2012)
78. D.X. Luu, D. Kim, Semi-interpenetrating polymer network electrolyte membranes composed of sulfonated poly(ether ether ketone) and organosiloxane-based hybrid network. J. Power Sour. **196**, 10584–10590 (2011)
79. S.Y. Han, J. Park, D. Kim, Proton-conducting electrolyte membranes based on organosiloxane network/sulfonated poly(ether ether ketone) interpenetrating polymer networks embedding sulfonated mesoporous benzene-silica. J. Power Sour. **243**, 850–858 (2013)
80. R. Marschall, J. Rathousky, M. Wark, Ordered functionalized silica materials with high proton conductivity. Chem. Mater. **19**, 6401–6407 (2007)
81. R. Marschall, I. Bannat, J. Caro, M. Wark, Proton conductivity of sulfonic acid functionalised mesoporous materials. Microporous Mesoporous Mater. **99**, 190–196 (2007)
82. Q. Yang, J. Liu, L. Zhang, C. Li, Functionalized periodic mesoporous organosilicas for catalysis. J. Mater. Chem. **19**, 1945–1955 (2009)
83. E.-B. Cho, D.X. Luu, D. Kim, Enhanced transport performance of sulfonated mesoporous benzene-silica incorporated poly(ether ether ketone) composite membranes for fuel cell application. J. Membr. Sci. **351**, 58–64 (2010)

84. M. Wilhelm, M. Jeske, R. Marschall, W.L. Cavalcanti, P. Tölle, C. Köhler, D. Koch, T. Frauenheim, G. Grathwohl, J. Caro, M. Wark, New proton conducting hybrid membranes for HT-PEMFC systems based on polysiloxanes and SO3H-functionalized mesoporous Si-MCM-41 particles. J. Membr. Sci. **316**, 164–175 (2008)
85. G.L. Athens, Y. Ein-Eli, B.F. Chmelka, Acid-functionalized mesostructured aluminosilica for hydrophilic proton conduction membranes. Adv. Mater. **19**, 2580–2587 (2007)
86. M. Alvaro, A. Corma, D. Das, V. Fornes, H. Garcia, Single-step preparation and catalytic activity of mesoporous MCM-41 and SBA-15 silicas functionalized with perfluoroalkylsulfonic acid groups analogous to Nafion[registered sign], Chem. Commun. 956–957 (2004)
87. F. Hoffmann, M. Cornelius, J. Morell, M. Froeba, Silica-based mesoporous organic-inorganic hybrid materials. Angew. Chem. Int. Ed. **45**, 3216–3251 (2006)
88. D. Margolese, J.A. Melero, S.C. Christiansen, B.F. Chmelka, G.D. Stucky, Direct syntheses of ordered SBA-15 mesoporous silica containing sulfonic acid groups. Chem. Mater. **12**, 2448–2459 (2000)
89. E.-B. Cho, H. Kim, D. Kim, Effect of morphology and pore size of sulfonated mesoporous benzene-silicas in the preparation of poly(vinyl alcohol)-based hybrid nanocomposite membranes for direct methanol fuel cell application. J. Phys. Chem. B **113**, 9770–9778 (2009)
90. T. Asefa, M.J. MacLachlan, N. Coombs, G.A. Ozin, Periodic mesoporous organosilicas with organic groups inside the channel walls. Nature **402**, 867–871 (1999)
91. B.J. Melde, B.T. Holland, C.F. Blanford, A. Stein, Mesoporous sieves with unified hybrid inorganic/organic frameworks. Chem. Mater. **11**, 3302–3308 (1999)
92. S. Inagaki, S. Guan, Y. Fukushima, T. Ohsuna, O. Terasaki, Novel mesoporous materials with a uniform distribution of organic groups and inorganic oxide in their frameworks. J. Am. Chem. Soc. **121**, 9611–9614 (1999)
93. C.-H. Tsai, H.-J. Lin, H.-M. Tsai, J.-T. Hwang, S.-M. Chang, Y.-W. Chen-Yang, Characterization and PEMFC application of a mesoporous sulfonated silica prepared from two precursors, tetraethoxysilane and phenyltriethoxysilane. Int. J. Hydrogen Energy **36**, 9831–9841 (2011)
94. S. Meenakshi, A. Manokaran, S.D. Bhat, A.K. Sahu, P. Sridhar, S. Pitchumani, Impact of mesoporous and microporous materials on performance of nafion and SPEEK polymer electrolytes: a comparative study in DEFCs. Fuel Cells **14**, 842–852 (2014)
95. G.L. Athens, D. Kim, J.D. Epping, S. Cadars, Y. Ein-Eli, B.F. Chmelka, Molecular optimization of multiply-functionalized mesoporous films with ion conduction properties. J. Am. Chem. Soc. **133**, 16023–16036 (2011)
96. M. Alvaro, A. Corma, D. Das, V. Fornés, H. García, "Nafion"-functionalized mesoporous MCM-41 silica shows high activity and selectivity for carboxylic acid esterification and Friedel-Crafts acylation reactions. J. Catal. **231**, 48–55 (2005)
97. S. Moghaddam, E. Pengwang, Y.-B. Jiang, A.R. Garcia, D.J. Burnett, C.J. Brinker, R.I. Masel, M.A. Shannon, An inorganic-organic proton exchange membrane for fuel cells with a controlled nanoscale pore structure. Nat. Nanotechnol. **5**, 230–236 (2010)
98. S. Fujita, A. Koiwai, M. Kawasumi, S. Inagaki, Enhancement of proton transport by high densification of sulfonic acid groups in highly ordered mesoporous silica. Chem. Mater. **25**, 1584–1591 (2013)
99. Z. Zhou, R. Liu, J. Wang, S. Li, M. Liu, J.-L. Brédas, Intra- and intermolecular proton transfer in 1H(2H)-1,2,3-triazole based systems. J. Phys. Chem. A **110**, 2322–2324 (2006)
100. S. Liu, Z. Yue, Y. Liu, Incorporation of Imidazole within the metal-organic framework UiO-67 for enhanced anhydrous proton conductivity. Dalton Trans. **44**, 12976–12980 (2015)
101. S. Scheiner, M. Yi, Proton transfer properties of imidazole. J. Phys. Chem. **100**, 9235–9241 (1996)
102. W. Munch, K.D. Kreuer, W. Silvestri, J. Maier, G. Seifert, The diffusion mechanism of an excess proton in imidazole molecule chains: first results of an ab initio molecular dynamics study. Solid State Ionics **145**, 437–443 (2001)

103. Q. Li, J.O. Jensen, R.F. Savinell, N.J. Bjerrum, High temperature proton exchange membranes based on polybenzimidazoles for fuel cells. Prog. Polym. Sci. **34**, 449–477 (2009)
104. R. He, Q. Li, J.O. Jensen, N.J. Bjerrum, Doping phosphoric acid in polybenzimidazole membranes for high temperature proton exchange membrane fuel cells. J. Polym., Sci. Part A Polym. Chem. **45**, 2989–2997 (2007)
105. R.H. He, Q.F. Li, A. Bach, J.O. Jensen, N.J. Bjerrum, Physicochemical properties of phosphoric acid doped polybenzimidazole membranes for fuel cells. J. Membr. Sci. **277**, 38–45 (2006)
106. R. Marschall, M. Sharifi, M. Wark, Proton conductivity of imidazole functionalized ordered mesoporous silica: influence of type of anchorage, chain length and humidity. Microporous Mesoporous Mater. **123**, 21–29 (2009)
107. W.L. Cavalcanti, R. Marschall, P. Tölle, C. Köhler, M. Wark, T. Frauenheim, Insight into proton conduction of immobilised imidazole systems via simulations and impedance spectroscopy. Fuel Cells **8**, 244–253 (2008)
108. K.A. Sung, K.-H. Oh, W.-K. Kim, M.-J. Choo, K.-W. Nam, J.-K. Park, Proton exchange membrane using imidazole-functionalized silica to enhance proton conductivity at lower humidity. Electrochem. Solid State Lett. **14**, B114–B116 (2011)
109. I.S. Amiinu, W. Li, G. Wang, Z. Tu, H. Tang, M. Pan, H. Zhang, Toward anhydrous proton conductivity based on imidazole functionalized mesoporous silica/nafion composite membranes. Electrochim. Acta **160**, 185–194 (2015)
110. S. Li, Z. Zhou, Y. Zhang, M. Liu, W. Li, 1H–1,2,4-triazole: an effective solvent for proton-conducting electrolytes. Chem. Mater. **17**, 5884–5886 (2005)
111. Z. Zhou, S. Li, Y. Zhang, M. Liu, W. Li, Promotion of proton conduction in polymer electrolyte membranes by 1H-1,2,3-triazole. J. Am. Chem. Soc. **127**, 10824–10825 (2005)
112. D. Basak, C. Versek, J.A. Harvey, S. Christensen, J. Hillen, S.M. Auerbach, M.T. Tuominen, D. Venkataraman, Enhanced anhydrous proton conduction in binary mixtures of 1H-imidazole-1H-1,2,3-triazole based compounds. J. Mater. Chem. **22**, 20410–20417 (2012)
113. S.J. Park, D.H. Lee, Y.S. Kang, High temperature proton exchange membranes based on triazoles attached onto SBA-15 type mesoporous silica. J. Membr. Sci. **357**, 1–5 (2010)
114. F. Yan, S. Yu, X. Zhang, L. Qiu, F. Chu, J. You, J. Lu, Enhanced proton conduction in polymer electrolyte membranes as synthesized by polymerization of protic ionic liquid-based microemulsions. Chem. Mater. **21**, 1480–1484 (2009)
115. S. Subianto, M.K. Mistry, N.R. Choudhury, N.K. Dutta, R. Knout, Composite polymer electrolyte containing ionic liquid and functionalized polyhedral oligomeric silsesquioxanes for anhydrous pem applications. ACS Appl. Mater. Interfaces **1**, 1173–1182 (2009)
116. S.S. Sekhon, J.-S. Park, E. Cho, Y.-G. Yoon, C.-S. Kim, W.-Y. Lee, Morphology studies of high temperature proton conducting membranes containing hydrophilic/hydrophobic ionic liquids. Macromolecules **42**, 2054–2062 (2009)
117. Y.-S. Ye, J. Rick, B.-J. Hwang, Ionic liquid polymer electrolytes. J. Mater. Chem. A **1**, 2719–2743 (2013)
118. I.S. Amiinu, X. Liang, Z. Tu, H. Zhang, J. Feng, Z. Wan, M. Pan, Anhydrous proton conducting materials based on sulfonated dimethylphenethylchlorosilane grafted mesoporous silica/ionic liquid composite. ACS Appl. Mater. Interfaces **5**, 11535–11543 (2013)
119. B. Lin, S. Cheng, L. Qiu, F. Yan, S. Shang, J. Lu, Protic ionic liquid-based hybrid proton-conducting membranes for anhydrous proton exchange membrane application. Chem. Mater. **22**, 1807–1813 (2010)
120. A.K. Mishra, T. Kuila, D.-Y. Kim, N.H. Kim, J.H. Lee, Protic ionic liquid-functionalized mesoporous silica-based hybrid membranes for pproton exchange membrane fuel cells. J. Mater. Chem. **22**, 24366–24372 (2012)
121. A.K. Mishra, N.H. Kim, J.H. Lee, Effects of ionic liquid-functionalized mesoporous silica on the proton conductivity of acid-doped poly(2,5-benzimidazole) composite membranes for high-temperature fuel cells. J. Membr. Sci. **449**, 136–145 (2014)

122. Y.-S. Ye, G.-W. Liang, B.-H. Chen, W.-C. Shen, C.-Y. Tseng, M.-Y. Cheng, J. Rick, Y.-J. Huang, F.-C. Chang, B.-J. Hwang, Effect of morphology of mesoporous silica on characterization of protic ionic liquid-based composite membranes. J. Power Sour. **196**, 5408–5415 (2011)
123. E. Quartarone, P. Mustarelli, Polymer fuel cells based on polybenzimidazole/H_3PO_4. Engegy Environ. Sci. **5**, 6436–6444 (2012)
124. C. Laberty-Robert, K. Valle, F. Pereira, C. Sanchez, Design and properties of functional hybrid organic-inorganic membranes for fuel cells. Chem. Soc. Rev. **40**, 961–1005 (2011)
125. S. Yu, L. Xiao, B.C. Benicewicz, Durability studies of PBI-based high temperature PEMFCs. Fuel Cells **8**, 165–174 (2008)
126. Y. Zhao, H. Yang, H. Wu, Z. Jiang, Enhanced proton conductivity of the hybrid membranes by regulating the proton conducting groups anchored on the mesoporous silica. J. Power Sour. **270**, 292–303 (2014)
127. J. Zeng, B. He, K. Lamb, R. De Marco, P.K. Shen, S.P. Jiang, Phosphoric acid functionalized pre-sintered meso-silica for high temperature proton exchange membrane fuel cells. Chem. Commun. **49**, 4655–4657 (2013)
128. J. Zeng, B. He, K. Lamb, R. De Marco, P.K. Shen, S.P. Jiang, Anhydrous phosphoric acid functionalized sintered mesoporous silica nanocomposite proton exchange membranes for fuel cells. ACS Appl. Mater. Interfaces **5**, 11240–11248 (2013)
129. J. Weber, K.-D. Kreuer, J. Maier, A. Thomas, Proton conductivity enhancement by nanostructural control of poly(benzimidazole)-phosphoric acid adducts. Adv. Mater. **20**, 2595–2598 (2008)
130. J. Lee, C.-W. Yi, K. Kim, Phosphoric acid-functionalized mesoporous silica/nafion composite membrane for high temperature PEMFCs. Bull. Korean Chem. Soc. **33**, 1397–1400 (2012)
131. S.Y. Chen, C.C. Han, C.H. Tsai, J. Huang, Y.W. Chen-Yang, Effect of morphological properties of ionic liquid-templated mesoporous anatase TiO_2 on performance of PEMFC with Nafion/TiO_2 composite membrane at elevated temperature and low relative humidity. J. Power Sour. **171**, 363–372 (2007)
132. W.H.J. Hogarth, J.C.D. da Costa, J. Drennan, G.Q. Lu, Proton conductivity of mesoporous sol-gel zirconium phosphates for fuel cell applications. J. Mater. Chem. **15**, 754–758 (2005)
133. N.K. Mal, A. Bhaumik, M. Matsukata, M. Fujiwara, Syntheses of mesoporous hybrid iron oxophenyl phosphate, iron oxophosphate, and sulfonated oxophenyl phosphate. Ind. Eng. Chem. Res. **45**, 7748–7751 (2006)
134. Z. Chai, Q. Suo, H. Wang, X. Wang, Mesoporous lanthanum phosphate nanostructures containing H_3PO_4 as superior electrolyte for PEM fuel cells. Rsc Adv. **3**, 21928–21935 (2013)
135. Z. Chai, L. Gao, C. Wang, H. Zhang, R. Zheng, P.A. Webley, H. Wang, Synthesis of mesoporous $LaPO_4$ nanostructures with controllable morphologies. New J. Chem. **33**, 1657–1662 (2009)
136. E. Rodríguez-Castellón, J. Jiménez-Jiménez, A. Jiménez-López, P. Maireles-Torres, J.R. Ramos-Barrado, D.J. Jones, J. Rozière, Proton conductivity of mesoporous MCM type of zirconium and titanium phosphates. Solid State Ionics **125**, 407–410 (1999)
137. M. Casciola, D. Bianchi, Frequency response of polycrystalline samples of α-Zr(HPO_4) $2{\cdot}H_2O$ at different relative humidities. Solid State Ionics **17**, 287–293 (1985)
138. M. Kourasi, R.G.A. Wills, A.A. Shah, F.C. Walsh, Heteropolyacids for fuel cell applications. Electrochim. Acta **127**, 454–466 (2014)
139. S. Sachdeva, J.A. Turner, J.L. Horan, A.M. Herring, in *Fuel Cells and Hydrogen Storage*, ed. by A. Bocarsly, D.M.P. Mingos (2011), pp. 115–168
140. O. Nakamura, T. Kodama, I. Ogino, Y. Miyake, High-conductivity solid proton conductors: dodecamolybdophosphoric acid and dodecatungstophosphoric acid crystals. Chem. Lett. **8**, 17–18 (1979)
141. S.P. Jiang, Functionalized mesoporous structured inorganic materials as high temperature proton exchange membranes for fuel cells. J. Mater. Chem. A **2**, 7637–7655 (2014)

142. S.P. Jiang, Functionalized mesoporous materials as new class high temperature proton exchange membranes for fuel cells. Solid State Ionics **262**, 307–312 (2014)
143. S. Lu, D. Wang, S.P. Jiang, Y. Xiang, J. Lu, J. Zeng, HPW/MCM-41 phosphotungstic acid/mesoporous silica composites as novel proton-exchange membranes for elevated-temperature fuel cells. Adv. Mater. **22**, 971+ (2010)
144. S. Choi, Y. Wang, Z. Nie, J. Liu, C.H.F. Peden, Cs-substituted tungstophosphoric acid salt supported on mesoporous silica. Catal. Today **55**, 117–124 (2000)
145. S. Lu, D. Wang, J. Lu, J. Zeng, S.P. Jiang, in *Proton Exchange Membrane Fuel Cells*, vol. 9, ed. by T. Fuller, H. Uchida, P. Strasser, P. Shirvanian, C. Lamy, C. Hartnig, H.A. Gasteiger, T. Zawodzinski, T. Jarvi, P. Bele, V. Ramani, S. Cleghorn, D. Jones, P. Zelenay (2009), pp. 1927–1933
146. S. Lu, D. Wang, Y. Xiang, S.P. Jiang, High temperature ceramic proton exchange membranes for direct methanol fuel cells. Fuel Cell Seminar **2010**(26), 269–277 (2009)
147. J. Zeng, S.P. Jiang, Characterization of high-temperature proton-exchange membranes based on phosphotungstic acid functionalized mesoporous silica nanoconnposites for fuel cells. J. Phys. Chem. C **115**, 11854–11863 (2011)
148. J. Zeng, Y. Zhou, L. Li, S.P. Jiang, Phosphotungstic acid functionalized silica nanocomposites with tunable bicontinuous mesoporous structure and superior proton conductivity and stability for fuel cells. Phys. Chem. Chem. Phys. **13**, 10249–10257 (2011)
149. J. Zeng, P.K. Shen, S. Lu, Y. Xiang, L. Li, R. De Marco, S.P. Jiang, Correlation between proton conductivity, thermal stability and structural symmetries in novel HPW-meso-silica nanocomposite membranes and their performance in direct methanol fuel cells. J. Membr. Sci. **397**, 92–101 (2012)
150. B. Grünberg, T. Emmler, E. Gedat, I. Shenderovich, G.H. Findenegg, H.-H. Limbach, G. Buntkowsky, Hydrogen bonding of water confined in mesoporous silica MCM-41 and SBA-15 studied by 1H solid-state NMR. Chem. Eur. J. **10**, 5689–5696 (2004)
151. A. Pettersson, J.B. Rosenholm, Streaming potential studies on the adsorption of amphoteric alkyldimethylamine and alkyldimethylphosphine oxides on mesoporous silica from aqueous solution. Langmuir **18**, 8447–8454 (2002)
152. H. Tang, M. Pan, S. Lu, J. Lu, S.P. Jiang, One-step synthesized HPW/meso-silica inorganic proton exchange membranes for fuel cells. Chem. Commun. **46**, 4351–4353 (2010)
153. C. Lin, T. Haolin, P. Mu, Periodic Nafion-silica-heteropolyacids electrolyte t for PEM fuel cell operated near 200 °C. Int. J. Hydrogen Energy **37**, 4694–4698 (2012)
154. C. Lin, T. Haolin, L. Junrui, P. Mu, Highly ordered Nafion-silica-HPW proton exchange membrane for elevated temperature fuel cells. Int. J. Energy Res. **37**, 879–887 (2013)
155. H. Tang, M. Pan, S.P. Jiang, Self assembled 12-tungstophosphoric acid-silica mesoporous nanocomposites as proton exchange membranes for direct alcohol fuel cells. Dalton Trans. **40**, 5220–5227 (2011)
156. R. Fan, S. Huh, R. Yan, J. Arnold, P. Yang, Gated proton transport in aligned mesoporous silica films. Nat. Mater. **7**, 303–307 (2008)
157. H. Munakata, S. Ochiai, K. Kanamura, Pore size effect on improvement of surface proton conductivity for three-dimensionally ordered macroporous silica composite membrane. J. Electrochem. Soc. **154**, B871–B875 (2007)
158. F. Lefebvre, 31P MAS NMR study of H3PW12O40 supported on silica: formation of ([triple bond, length half m-Dash]SiOH$_2$+)(H2PW12O40-). J. Chem. Soc., Chem. Commun. 756–757 (1992)
159. S.L. Gojković, T.R. Vidaković, D.R. Đurović, Kinetic study of methanol oxidation on carbon-supported PtRu electrocatalyst. Electrochim. Acta **48**, 3607–3614 (2003)
160. F. Hajbolouri, B. Andreaus, G.G. Scherer, A. Wokaun, CO tolerance of commercial Pt and PtRu gas diffusion electrodes in polymer electrolyte fuel cells. Fuel Cells **4**, 160–168 (2004)
161. J. Zeng, B. Jin, P.K. Shen, B. He, K. Lamb, R. De Marco, S.P. Jiang, Stack performance of phosphotungstic acid functionalized mesoporous silica (HPW-meso-silica) nanocomposite high temperature proton exchange membrane fuel cells. Int. J. Hydrogen Energy **38**, 12830–12837 (2013)

162. T.J. Schmidt, J. Baurmeister, Properties of high-temperature PEFC Celtec®-P 1000 MEAs in start/stop operation mode. J. Power Sour. **176**, 428–434 (2008)
163. X.-Z. Yuan, S. Zhang, H. Wang, J. Wu, J.C. Sun, R. Hiesgen, K.A. Friedrich, M. Schulze, A. Haug, Degradation of a polymer exchange membrane fuel cell stack with Nafion® membranes of different thicknesses: Part I. In situ diagnosis. J. Power Sourc. **195**, 7594–7599 (2010)
164. X.-Z. Yuan, S. Zhang, S. Ban, C. Huang, H. Wang, V. Singara, M. Fowler, M. Schulze, A. Haug, K. Andreas Friedrich, R. Hiesgen, Degradation of a PEM fuel cell stack with Nafion® membranes of different thicknesses. Part II: Ex situ diagnosis, J. Power Sources **205**, 324–334 (2012)
165. M. Nagao, A. Takeuchi, P. Heo, T. Hibino, M. Sano, A. Tomita, A proton-conducting In^{3+}-doped SnP_2O_7 electrolyte for intermediate-temperature fuel cells. Electrochem. Solid State Lett. **9**, A105–A109 (2006)
166. T. Matsui, N. Kazusa, Y. Kato, Y. Iriyama, T. Abe, K. Kikuchi, Z. Ogumi, Effect of pyrophosphates as supporting matrices on proton conductivity for NH_4PO_3 composites at intermediate temperatures. J. Power Sour. **171**, 483–488 (2007)
167. T. Matsui, T. Kukino, R. Kikuchi, K. Eguchi, Intermediate-temperature fuel cell employing CsH_2PO_4/SiP_2O_7-based composite electrolytes. J. Electrochem. Soc. **153**, A339–A342 (2006)
168. Y. Zhou, J. Yang, H. Su, J. Zeng, S.P. Jiang, W.A. Goddard, Insight into proton transfer in phosphotungstic acid functionalized mesoporous silica-based proton exchange membrane fuel cells. J. Am. Chem. Soc. **136**, 4954–4964 (2014)
169. H.J. Song, C.K. Kim, Fabrication and properties of ultrafiltration membranes composed of polysulfone and poly(1-vinylpyrrolidone) grafted silica nanoparticles. J. Membr. Sci. **444**, 318–326 (2013)
170. S.S. Madaeni, P. Moradi, Preparation and characterization of asymmetric polysulfone membrane for separation of oxygen and nitrogen gases. J. Appl. Polym. Sci. **121**, 2157–2167 (2011)
171. B. Chakrabarty, A.K. Ghoshal, A.K. Purkait, Preparation, characterization and performance studies of polysulfone membranes using PVP as an additive. J. Membr. Sci. **315**, 36–47 (2008)
172. L. Bai, L. Zhang, H.Q. He, R.K.S.A. Rasheed, C.Z. Zhang, O.L. Ding, S.H. Chan, Fabrication of phosphotungstic acid functionalized mesoporous silica composite membrane by alternative tape-casting incorporating phase inversion technique. J. Power Sour. **246**, 522–530 (2014)
173. N. Nakamoto, A. Matsuda, K. Tadanaga, T. Minami, M. Tatsumisago, Medium temperature operation of fuel cells using thermally stable proton-conducting composite sheets composed of phosphosilicate gel and polyimide. J. Power Sour. **138**, 51–55 (2004)
174. A. Garsuch, X. Michaud, C. O'Neill, P. Li, J.R. Dahn, in *Proton Exchange Membrane Fuel Cells*, vol. 8, ed. by T. Fuller, K. Shinohara, V. Ramani, P. Shirvanian, H. Uchida, S. Cleghorn, M. Inaba, S. Mitsushima, P. Strasser, H. Nakagawa, H.A. Gasteiger, T. Zawodzinski, C. Lamy (Pts 1 and 2, 2008), pp. 139–146
175. B. Fang, J.H. Kim, C. Lee, J.-S. Yu, Hollow macroporous Core/Mesoporous shell carbon with a tailored structure as a cathode electrocatalyst support for proton exchange membrane fuel cells. J. Phys. Chem. C **112**, 639–645 (2008)
176. K. Pei, D. Banham, F. Feng, T. Fuerstenhaupt, S. Ye, V. Birss, Oxygen reduction activity dependence on the mesoporous structure of imprinted carbon supports. Electrochem. Commun. **12**, 1666–1669 (2010)
177. C. Liang, Z. Li, S. Dai, Mesoporous carbon materials: synthesis and modification. Angew. Chem. Int. Ed. **47**, 3696–3717 (2008)
178. J. Liu, N.P. Wickramaratne, S.Z. Qiao, M. Jaroniec, Molecular-based design and emerging applications of nanoporous carbon spheres. Nat. Mater. **14**, 763–774 (2015)

179. S.H. Joo, S.J. Choi, I. Oh, J. Kwak, Z. Liu, O. Terasaki, R. Ryoo, Ordered nanoporous arrays of carbon supporting high dispersions of platinum nanoparticles. Nature **412**, 169–172 (2001)
180. B. Fang, J.H. Kim, M. Kim, J.-S. Yu, Ordered hierarchical nanostructured carbon as a highly efficient cathode catalyst support in proton exchange membrane fuel cell. Chem. Mater. **21**, 789–796 (2009)
181. N.P. Lebedeva, A.S. Booij, G.J.M. Janssen, in *Proton Exchange Membrane Fuel Cells*, vol. 8, ed. by T. Fuller, K. Shinohara, V. Ramani, P. Shirvanian, H. Uchida, S. Cleghorn, M. Inaba, S. Mitsushima, P. Strasser, H. Nakagawa, H.A. Gasteiger, T. Zawodzinski, C. Lamy (Pts 1 and 2, 2008), pp. 2083–2092
182. R. Ryoo, S.H. Joo, S. Jun, Synthesis of highly ordered carbon molecular sieves via template-mediated structural transformation. J. Phys. Chem. B **103**, 7743–7746 (1999)
183. M. Kaneda, T. Tsubakiyama, A. Carlsson, Y. Sakamoto, T. Ohsuna, O. Terasaki, S.H. Joo, R. Ryoo, Structural study of mesoporous MCM-48 and carbon networks synthesized in the spaces of MCM-48 by electron crystallography. J. Phys. Chem. B **106**, 1256–1266 (2002)
184. R. Ryoo, S.H. Joo, M. Kruk, M. Jaroniec, Ordered mesoporous carbons. Adv. Mater. **13**, 677–681 (2001)
185. N.-I. Kim, J. Seong, M.-J. Yang, S.-H. Cho, J.-Y. Park, H.-S. Choi, S.H. Joo, K. Kwon, in *Polymer Electrolyte Fuel Cells*, vol. 13, ed. by H.A. Gasteiger, A. Weber, K. Shinohara, H. Uchida, S. Mitsushima, T.J. Schmidt, S.R. Narayanan, V. Ramani, T. Fuller, M. Edmundson, P. Strasser, R. Mantz, J. Fenton, F.N. Buchi, D.C. Hansen, D.L. Jones, C. Coutanceau, K. SwiderLyons, K.A. Perry (2013), pp. 1277–1283
186. Z. Lei, S. Bai, Y. Xiao, L. Dang, L. An, G. Zhang, Q. Xu, CMK-5 mesoporous carbon synthesized via chemical vapor deposition of ferrocene as catalyst support for methanol oxidation. J. Phys. Chem. C **112**, 722–731 (2008)
187. T. Maiyalagan, T.O. Alaje, K. Scott, Highly stable Pt–Ru nanoparticles supported on three-dimensional cubic ordered mesoporous carbon (Pt–Ru/CMK-8) as promising electrocatalysts for methanol oxidation. J. Phys. Chem. C **116**, 2630–2638 (2012)
188. J.-S. Lee, S.H. Joo, R. Ryoo, Synthesis of mesoporous silicas of controlled pore wall thickness and their replication to ordered nanoporous carbons with various pore diameters. J. Am. Chem. Soc. **124**, 1156–1157 (2002)
189. W. Ying, Y. Haifeng, Z. Dongyuan, Biomolecular catalysis. Am. Chem. Soc. 2–48 (2008)
190. F.Q. Zhang, Y. Meng, D. Gu, Y. Yan, C.Z. Yu, B. Tu, D.Y. Zhao, A facile aqueous route to synthesize highly ordered mesoporous polymers and carbon frameworks with Ia(3)over-bard bicontinuous cubic structure. J. Am. Chem. Soc. **127**, 13508–13509 (2005)
191. Y. Meng, D. Gu, F. Zhang, Y. Shi, H. Yang, Z. Li, C. Yu, B. Tu, D. Zhao, Ordered mesoporous polymers and homologous carbon frameworks: amphiphilic surfactant templating and direct transformation. Angew. Chem. Int. Ed. **44**, 7053–7059 (2005)
192. Y. Deng, T. Yu, Y. Wan, Y. Shi, Y. Meng, D. Gu, L. Zhang, Y. Huang, C. Liu, X. Wu, D. Zhao, Ordered mesoporous silicas and carbons with large accessible pores templated from amphiphilic diblock copolymer poly(ethylene oxide)-b-polystyrene. J. Am. Chem. Soc. **129**, 1690–1697 (2007)
193. Z. Li, M. Jaroniec, Synthesis and adsorption properties of colloid-imprinted carbons with surface and volume mesoporosity. Chem. Mater. **15**, 1327–1333 (2003)
194. Z. Li, M. Jaroniec, Colloidal Imprinting: a novel approach to the synthesis of mesoporous carbons. J. Am. Chem. Soc. **123**, 9208–9209 (2001)
195. B. Ficicilar, A. Bayrakceken, I. Eroglu, Effect of Pd loading in Pd-Pt bimetallic catalysts doped into hollow core mesoporous shell carbon on performance of proton exchange membrane fuel cells. J. Power Sour. **193**, 17–23 (2009)
196. P.V. Shanahan, L. Xu, C. Liang, M. Waje, S. Dai, Y.S. Yan, Graphitic mesoporous carbon as a durable fuel cell catalyst support. J. Power Sources **185**, 423–427 (2008)
197. J.H. Kim, B. Fang, M. Kim, J.-S. Yu, Hollow spherical carbon with mesoporous shell as a superb anode catalyst support in proton exchange membrane fuel cell. Catal. Today **146**, 25–30 (2009)

198. B. Fang, J.H. Kim, M. Kim, M. Kim, J.-S. Yu, Hierarchical nanostructured hollow spherical carbon with mesoporous shell as a unique cathode catalyst support in proton exchange membrane fuel cell. Phys. Chem. Chem. Phys. **11**, 1380–1387 (2009)
199. B. Ficicilar, A. Bayrakceken, I. Eroglu, Pt incorporated hollow core mesoporous shell carbon nanocomposite catalyst for proton exchange membrane fuel cells. Int. J. Hydrogen Energy **35**, 9924–9933 (2010)
200. A. Garsuch, R. d'Eon, T. Dahn, O. Klepel, R.R. Garsuch, J.R. Dahn, Oxygen reduction behavior of highly porous non-noble metal catalysts prepared by a template-assisted synthesis route. J. Electrochem. Soc. 155, B236–B243 (2008)
201. H. Liu, C. Song, Y. Tang, J. Zhang, J. Zhang, High-surface-area CoTMPP/C synthesized by ultrasonic spray pyrolysis for PEM fuel cell electrocatalysts. Electrochim. Acta **52**, 4532–4538 (2007)
202. L. Zhang, J. Kim, E. Dy, S. Ban, K.-C. Tsay, H. Kawai, Z. Shi, J. Zhang, Synthesis of novel mesoporous carbon spheres and their supported Fe-based electrocatalysts for PEM fuel cell oxygen reduction reaction. Electrochim. Acta **108**, 480–485 (2013)
203. H. Liu, Z. Shi, J. Zhang, L. Zhang, J. Zhang, Ultrasonic spray pyrolyzed iron-polypyrrole mesoporous spheres for fuel celloxygen reduction electrocatalysts. J. Mater. Chem. **19**, 468–470 (2009)
204. M. Zhou, C. Yang, K.-Y. Chan, Structuring porous iron-nitrogen-doped carbon in a core/shell geometry for the oxygen reduction reaction. Adv. Energy Mater. **4** (2014)
205. R. Gadiou, A. Didion, S.-E. Saadallah, M. Couzi, J.-N. Rouzaud, P. Delhaes, C. Vix-Guterl, Graphitization of carbons synthesized in a confined geometry. Carbon **44**, 3348–3352 (2006)
206. L. Zhang, J. Kim, E. Dy, S. Ban, K.-C. Tsay, H. Kawai, Z. Shi, J. Zhang, Effect of template size on the synthesis of mesoporous carbon spheres and their supported Fe-based ORR electrocatalysts. Electrochim. Acta **108**, 814–819 (2013)
207. S.-Y. Huang, P. Ganesan, B.N. Popov, Electrocatalytic activity and stability of titania-supported platinum-palladium electrocatalysts for polymer electrolyte membrane fuel cell. ACS catal. **2**, 825–831 (2012)
208. S.-Y. Huang, P. Ganesan, S. Park, B.N. Popov, Development of a titanium dioxide-supported platinum catalyst with ultrahigh stability for polymer electrolyte membrane fuel cell applications. J. Am. Chem. Soc. **131**, 13898–13899 (2009)
209. M.C. Orilall, F. Matsumoto, Q. Zhou, H. Sai, H.D. Abruña, F.J. DiSalvo, U. Wiesner, One-pot synthesis of platinum-based nanoparticles incorporated into mesoporous niobium oxide–carbon composites for fuel cell electrodes. J. Am. Chem. Soc. **131**, 9389–9395 (2009)
210. E. Kang, S. An, S. Yoon, J.K. Kim, J. Lee, Ordered mesoporous WO_3-X possessing electronically conductive framework comparable to carbon framework toward long-term stable cathode supports for fuel cells. J. Mater. Chem. **20**, 7416–7421 (2010)
211. L. Samiee, F. Shoghi, A. Maghsodi, In situ functionalisation of mesoporous carbon electrodes with carbon nanotubes for proton exchange membrane fuel-cell application. Mater. Chem. Phys. **143**, 1228–1235 (2014)
212. M. Lei, T.Z. Yang, W.J. Wang, K. Huang, R. Zhang, X.L. Fu, H.J. Yang, Y.G. Wang, W.H. Tang, Self-assembled mesoporous carbon sensitized with ceria nanoparticles as durable catalyst support for PEM fuel cell. Int. J. Hydrogen Energy **38**, 205–211 (2013)
213. F. Kleitz, S. Hei Choi, R. Ryoo, Cubic Ia3d large mesoporous silica: synthesis and replication to platinum nanowires, carbon nanorods and carbon nanotubes. Chem. Commun. 2136–2137 (2003)
214. Y. Doi, A. Takai, Y. Sakamoto, O. Terasaki, Y. Yamauchi, K. Kuroda, Tailored synthesis of mesoporous platinum replicas using double gyroid mesoporous silica (KIT-6) with different pore diameters via vapor infiltration of a reducing agent. Chem. Commun. **46**, 6365–6367 (2010)
215. Y. Lei, T. Li, J. Gu, T. Yu, J. Liu, Z. Zou, Study on platinum and copper nanosheets alloys supported on mesoporous titanium dioxide doped with carbon black as electrocatalysts in pem fuel cells. Electroanalysis **24**, 699–706 (2012)

Chapter 11
Thermoelectric Nanocomposites for Thermal Energy Conversion

Yucheng Lan and Zhifeng Ren

Abstract Thermoelectric (TE) solid-state devices can convert heat into electricity. TE nanocomposites usually possess better TE performances than traditional TE bulks, such as higher figure-of-merit (*ZT*) and power output, playing a key role for industrial applications of TE devices, and thereby offering a great promise for improving energy efficiency over the traditional renewable energy systems. This chapter reviews manufacturing, TE properties and devices of TE nanocomposites. These nanocomposites are bottom-up compacted from nanoparticles that are produced by ball-milling. The produced nanocomposites possess unique microstructures to scatter phonons to decrease thermal conductivity and hold electric conductivity while maintaining or increasing the Seebeck coefficient. *ZT* of these nanocomposites are usually enhanced 10–100 %. TE devices are fabricated from these nanocomposites and their efficiencies are described. Future developments of TE nanocomposites are outlooked at the end of the chapter. The chapter will benefit material scientists, device engineers, and graduate students working in TE fields.

11.1 Introduction

Thermoelectric (TE) materials are capable of directly converting heat into electricity via the Seebeck effect [1, 2] discovered by Thomas Johann Seebeck in 1821. Likewise, a voltage gradient across a thermoelectric junction can produce a temperature gradient via the Peltier effect [3], discovered by Jean Charles Athanase Peltier in 1834. Such kinds of materials have been fabricated into various solid-state

Y. Lan (✉)
Department of Physics and Engineering Physics, Morgan State University, Baltimore, MD 21251, USA
e-mail: yucheng.lan@morgan.edu

Z. Ren (✉)
Department of Physics and TcSUH, University of Houston, Houston TX 77204, USA
e-mail: zren@uh.edu

Q. Li (ed.), *Nanomaterials for Sustainable Energy*,
NanoScience and Technology, DOI 10.1007/978-3-319-32023-6_11

thermoelectric devices. Thermoelectric devices possess unique features: no moving parts, quiet operation, low environment impact, and high reliability [4–9], attracting extensive interests for several decades. Up to now, thermoelectric devices have been widely employed in temperature measurements [5], renewable energy generators [8–10] to harvest waste heat, air-conditioners [8, 9] to lower the air temperature, and refrigerators [4, 8, 9, 11] to store foods as well as in radioisotope thermoelectric generators to power spacecraft.

One important physical parameter of TE devices is efficiency, the ability to convert energy. The efficiency of a thermoelectric device is dominated by dimensionless figure-of-merit (*ZT*) of thermoelectric materials, defined as [12]

$$ZT = (S^2\sigma/\kappa)T \tag{11.1}$$

where S, σ, k, and T are the Seebeck coefficient, electrical conductivity, thermal conductivity, and absolute temperature in Kelvin, respectively. TE generators and refrigerators are consisted of *n*-type and *p*-type TE materials. For power-generators to harvest waste heat, the energy conversion efficiency, also termed as power generation efficiency ε, can be expressed by

$$\varepsilon = \frac{T_h - T_c}{T_h}\frac{\sqrt{1+Z\bar{T}} - 1}{\sqrt{1+Z\bar{T}} + \frac{T_c}{T_h}} \tag{11.2}$$

The highest 20 % power generation efficiency is expected in PbTe nanostructures with $ZT = 2.2$ [13] to convert waste heat to electricity, lower than that of conventional electricity generation systems (around 30–40 % efficiency). For air-conditioners and refrigerators, the coefficient of performance (*COP*) can be expressed by

$$COP = \frac{T_h}{T_h - T_c}\frac{\sqrt{1+Z\bar{T}} - \frac{T_c}{T_h}}{\sqrt{1+Z\bar{T}} + 1} \tag{11.3}$$

where T_h, T_c and $\bar{T}$ are the temperature of the hot surface, the temperature of the cold surface, and the average temperature between the hot and cold surfaces of TE devices, respectively. *ZT* is sometimes defined as the modified dimensionless figure of merit of devices, which takes into consideration of the thermoelectric capacity of *n*-type and *p*-type thermoelectric materials being used in devices. Then *ZT* can be described as

$$ZT = \frac{(S_p - S_n)^2\bar{T}}{\left(\sqrt{\rho_p\kappa_p} + \sqrt{\rho_n\kappa_n}\right)^2} \tag{11.4}$$

after geometrical optimization regarding the legs sections. Here ρ is the electrical resistivity, the subscripts n and p denote the *n*-type and *p*-type thermoelectric materials, respectively. The coefficient of performance of current commercial thermoelectric refrigerators ranges from 0.3 to 0.6, much lower than that of traditional vapor-compression refrigerators.

In order to make thermoelectric devices competitive in large scale and high-power commercial applications, such as refrigeration and air-conditioning, an average *ZT* of 4 or higher in the application temperature range is required [4, 6, 7, 9, 10]. However *ZT* of all present TE materials is lower than 3.

Since the 1960s, many researches have been devoted to find new thermoelectric materials with high *ZT* [14, 15], such as Bi_2Te_3 ($ZT \sim 1$ at 400 K), PbTe ($ZT \sim 0.7$ at 700 K), $Si_{80}Ge_{20}$ ($ZT = 0.6$ at 1200 K), $CsBi_4Te_6$ ($ZT \sim 0.8$ at 225 K [16]), Tl_9BiTe_6 ($ZT \sim 1.2$ at 500 K [17]), clathrate $A^{II}_8B^{III}_{16}B^{IV}_{30}$ ($ZT = 0.7$–0.9 [18, 19], where A^{II} = Ba, Sr, B^{III} = Al, Ga, and B^{IV} = Si, Ge, Sn), lead antimony silver tellurium materials ($ZT > 1.6$ when temperature is higher than 600 K [20]), $In_4Se_{3-\delta}$ ($ZT = 1.48$ at 705 K [21]), and MgAgSb ($ZT \sim 1.4$ at 475 K [22]) and so on. More thermoelectric materials can be found in some review literatures [14, 23–25].

With the developments of nanosciences and nanotechniques, a concept of low dimensional nano-structured thermoelectric materials was proposed in the 1990s [26]. Interfaces are introduced in low-dimensional thermoelectric materials to highly scatter phonons, reducing the thermal conductivity, while not decrease the electrical conductivity significantly. It is believed that one-dimensional nano-wires, two-dimensional superlattices, and three-dimensional nanograins can increase *ZT* by decreasing thermal conductivity. Theories indicated that a large reduction in thermal conductivity is possible in TE composites of nanograins [27, 28]. Experimental data have shown that the nanostructuring approach is indeed effective in improving *ZT* [29–33]. The lattice thermal conductivity κ_L has been reduced via the increasing phonon scattering of interfaces in one-dimensional nanotubes and nano-wires [28, 34–37], two-dimensional superlattices [26, 29, 31, 32, 38–41] (such as Bi_2Te_3/Sb_2Te_3 superlattices [31], $PbSe_{0.98}Te_{0.02}/PbTe$ quantum dot superlattices [32], and SiGe/Si superlattices [40]), and three-dimensional nanocomposite bulks [22, 25, 42–54]. This chapter focuses on three-dimensional TE nanocomposite bulks.

Thermoelectric composite bulks that consist of micro- or nanograins have lower thermal conductivity. SiGe thermoelectric composites that consist of micro-grains were first fabricated in the 1960s [55]. The lattice thermal conductivity decreases with a reduction in grain size with a reduction of one third compared with single crystal materials while the Seebeck coefficient and electrical resistivity do not change with grain size [56–65]. Nanostructured thermoelectric composite materials (nanocomposites) that consist of three-dimensional nanograins were first fabricated in 2008 [42] and *ZT* was increased 40 % in Bi_2 Te_3 composites. Later *ZT* of other semiconducting thermoelectric materials have also been enhanced [25], such as *ZT* was increased 40 % from 0.93 to 1.3 in *n*-type Si-Ge alloys [45], 90 % higher to 0.95 in *p*-type SiGe alloys [44], 50 % higher to 0.8 in *p*-type half-Heuslers [48], 50 % higher to 1 in *n*-type In_4Se_{3-x} [49], and four times higher to 1.4 in MgAgSb compounds [22]. Phonons are scattered more effectively than electrons at grain boundaries in the nanocomposites, significantly reducing thermal conductivity and enhancing *ZT*s of these nanostructured materials.

This chapter introduces thermoelectric nanocomposites consisting of three-dimensional nanograins. Preparation of TE nanoparticles, including ball-mills and produced nanoparticle, will be first introduced. Then bottom-up fabrication of nanocomposites from nanoparticle will be described, followed by thermoelectric properties of TE nanocomposites and related TE devices made from nanocomposites.

11.2 High-Energy Ball-Milling and Produced Thermoelectric Nanoparticles

TE nanocomposites are usually fabricated from nanoparticles through bottom-up approaches. TE nanoparticles are first produced and then compacted into bulk nanocomposites. The thermoelectric nanoparticles can be produced by various bottom-up techniques (such as hydrothermal methods [66–72], wet chemical reaction [73, 74], and melt spinning technique [75]), or top-down approaches (such as ball-milling [42, 43, 76–79]). Among these methods, high-energy ball-milling is an effective and powerful processing method for producing large quantities of nanoparticles in a short time [80]. Here the ball-milling approach and the ball-milled TE nanoparticles are discussed.

11.2.1 Ball-Mills

The conventional ball-milling was first introduced to produce micro-sized TE particles in the 1960s [55]. The size of TE particles were gradually reduced in low energy environments until the particle size was in micrometers. The produced TE particles were compacted and *ZT*s of the composites were slightly increased. Such prepared SiGe composites have been fabricated into TE devices as radioisotope thermoelectric generators since the 1970s.

The high-energy ball-mill technique was developed in the 1970s [81] to produce superalloys with advanced properties. Homogeneous composition can be produced from ball-milled powders. Since the mid-1980s, a number of investigations have been carried out to produce a variety of materials [82], including amorphous alloys [83] and severely deformed alloys. Up to now, several kinds of high-energy ball mills have been developed to grind particles or synthesize compounds, such as shaker mills, planetary mills, and attritor mills.

Shaker mills are one kind of mill that are mostly used for laboratory investigations. Figure 11.1 shows a commercial shaker mill manufactured by SPEX Inc. Such kind of mills usually has one vial, containing particles and grinding balls, being secured in a clamp. The vial is swung energetically back and forth several thousand times per minute. The back-and-forth shaking motions impact against the particles and the ends of the vial, both milling and mixing particles. The ball

Fig. 11.1 **a** SPEX 8000 mixer/mill in the assembled condition. **b** A vial and balls

velocities are high up to 5 m/s with the amplitude of about 5 cm. Particles can be ground down to nano size and new phases can be synthesized through mechanically alloying. 10–20 g of particles can be milled at a time.

The second kind of popular high-energy mills are planetary ball mills in which a few hundred grams of particles can be milled at a time. Figure 11.2 shows a commercial planetary ball mill. Four vials are arranged on a rotating disk, rotating around their own axes like planets. The vials and the supporting disk rotate in opposite directions, causing the grinding balls to run down the inside wall of vials (friction effect) and travel freely/collide against the opposite inside wall or other balls (impact effect). The linear speed of the balls in planetary mills is higher than that in shaker mills while the frequency of impacts is much lower.

A typical ball-mill attritor consists of a rotating horizontal drum half-filled with balls to mill large quantities of powder (from 500 g to 40 kg) at a time. The balls are energized by a series of impellers to reduce particle size because of impacts and collisions. The speed of the grinding balls is about 0.5 m/s [82] when the mixture is rotating at a high speed of 250 rpm.

Table 11.1 compares these three kinds of high-energy ball-mills. Calculated velocity of balls, kinetic energy per impact, shock frequency and power for different type of mills are listed. These high-energy ball mills have high impact velocities and high impact frequencies of the grinding media, producing nanocrystalline materials [85]. The milling time depends on the energy of the mill balls and the impact frequency. It has been reported that 20 min of milling in a SPEX shaker mill is equivalent to 20 h of milling in a low-energy mill of Invicta BX 920/2 [86]. As a rule of thumb, it can be estimated that a process that takes only a few minutes in the SPEX shaker mill may take hours in an attritor and a few days in a commercial low-energy mill [82].

There are two mechanisms to process particles in high-energy ball mills. One is *mechanical milling* to grind particles. The force of an impact deforms particles and the particles are mechanically fractured because of collisions. Therefore, brittle particles are easily ground to nanoparticles. Many thermoelectric nanoparticles,

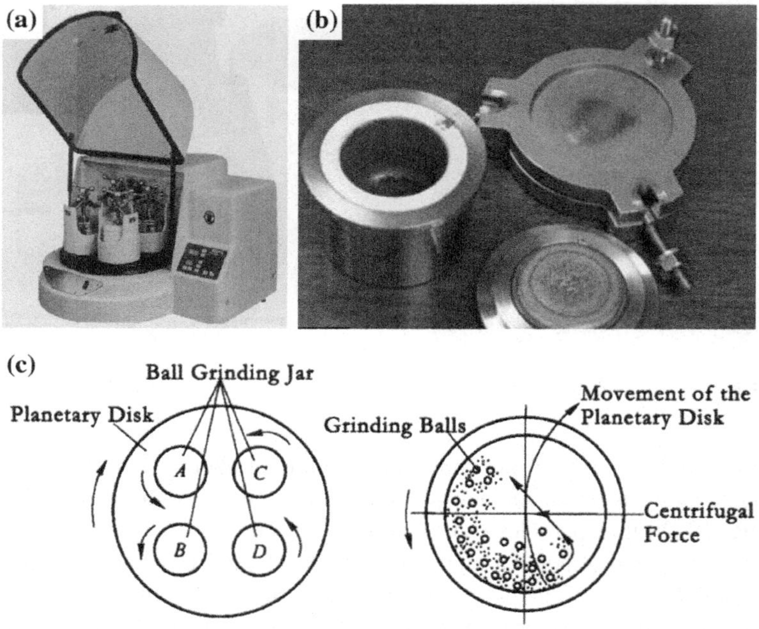

Fig. 11.2 **a** A planetary ball mill with four stations. **b** A opened ball milling jar. **c** Working principle of planetary ball mill

Table 11.1 Kinetic energy, shock frequency, and shock power for different type of high energy mills [84]

Parameter	Attritors	Shakers		Planetary ball mills		
		O[a]	SPEX mill	P5[a]	G7[a]	G5[a]
Velocity of balls (m/s)	0–0.8	0.14–0.24	<3.9	2.5–4.0	0.24–6.58	0.28–11.24
Kinetic energy (10^{-3} J/hit)	<10	3–30	<120	10–400	0.4–303.2	0.53–884
Shock frequency (Hz)	>1000	15–50	200	~100	5.0–92.4	4.5–90.7
Power (W/g/ball or rod)	<0.001	0.005–0.14	<0.24	0.01–0.8	0–0.56	0–1.604

[a]FRITSCH Pulverisette

such as *p*-type Bi_2Te_3 nanoparticles [42] and $Ga_mSb_nTe_{1.5(m+n)}$ [87] have been produced by the mechanism. However, it should be remembered that the efficiency of particle size reduction is very low, about 0.1 % in a conventional ball mill and less than 1 % in the high-energy ball milling processes of metal alloys. Destruction of long-range order or even amorphous structure of individual nanoparticles can be produced throughout the processing. The time required for the processing is short. Another mechanism is *mechanical alloying* to mill mixtures of powders together to produce new phases. The particles are repeatedly flattened, cold-welded, fractured and re-welded because of collisions. The synthesized particles are usually in microns while there are lots of defects inside the produced particles. The crystalline regions (also termed as crystalline size) of the produced particles can be in nanoscale. The mechanical alloying usually requires double the time required for mechanical milling.

11.2.2 Ball-Milled Thermoelectric Nanoparticles

High-energy ball-milling has been widely employed to produce various TE nanoparticles since 2008, such as Bi_2Te_3 nanoparticles [42–44, 88–90], SiGe alloys [45, 47], half-Heuslers [48], In_4Se_3 [49], $AgPb_{18}SbSe_{20}$ (LAST) [50], Zn_4Sb_3 [51], SnTe [52], Bi_2S_3 [53], MgAgSb [22, 91, 92], $YbAgCu_4$ [54], and skutterudites [46, 93–98].

Table 11.2 lists some typical TE nanoparticles produced by high-energy ball-milling. Ball-milling conditions and crystalline size of produced nanoparticles are also listed in the table. Some listed TE nanoparticles are directly produced through the mechanical milling processing while others are produced through the mechanical alloying.

During the mechanical milling procedure, TE ingots are first synthesized prior to ball-milling. Then, the ingots are grounded in ball-mills to produce TE nanoparticles with same chemical compositions. Figure 11.3 shows three typical kinds of ball milled nanoparticles by mechanical milling. The mechanically milled nanoparticles can be classified into two groups: single-crystalline nanoparticles and polycrystalline nanoparticles. Some TE materials are brittle, such as Bi_2Te_3 and Zn_4Sb_3. These brittle TE materials can be fragmented by ball-milling to produce nanoscaled single-crystalline particles. Figure 11.3a, b show bright-field transmission electron microscopic (BF-TEM) images of ball milled Bi_2Te_3 and Zn_4Sb_3 nanoparticles, respectively. Each individual nanoparticle is a single crystal. The crystalline size of the nanoparticles is about 20 nm or even smaller. Ductile TE materials can also be fragmented down to nanosize by high-energy milling. However the produced nanoparticles are polycrystalline. Figure 11.3c shows a typical TEM image of milled half-Heusler particles that are ductile. The size of the ball-milled nanoparticles is 50–100 nm. However selected area electron diffraction (SAED) and high-resolution transmission electron microscopic (HRTEM) images (insets in Fig. 11.3c) indicate that the produced nanoparticles are polycrystalline, plastically deformed. These nanoscaled nanoparticles constitute of crystalline subparticles. The crystalline size of ductile TE nanoparticles listed in Table 11.2 is calculated from XRD patterns or measured from HR-TEM images.

Most nanoparticles listed in Table 11.2 are produced through mechanical alloying. Mechanical alloying was first reported in 1987 [105] to synthesize nano structured materials. Highly energetic compressive impact forces repeatedly cold-weld and fracture elemental powders [106] besides blending, producing new phase nanopowders because of solid state inter-diffusion reaction [107]. Later, more

Table 11.2 Thermoelectric nanoparticles produced by ball-milling

TE materials	Ball-mill	Experimental conditions	Crystalline size[b]	References
$Si_{95}Ge_5^a$	N/A	N/A	10–30 nm	[47]
$Si_{80}Ge_{20}^a$	Pulverisette 4	90 h, β = 20: 1, argon atmosphere, 400 rpm	7 nm	[99]
$Si_{80}Ge_{20}P_2^a$	N/A	N/A	50–300 nm	[45]
$(Bi,Sb)_2\ Te_3$	N/A	Several hours	5–50 nm	[42]
$(Bi,Sb)_2\ Te_3^a$	N/A	N/A	5–10 nm	[43]
$(Bi_2\ Te3)_{0.95}$–$(Bi_2\ Se_3)_{0.05}$	Planetary	Argon atmosphere, stainless steel vial/balls, β = 15: 1, 400 rpm, 10 h	N/A	[100]
$(Bi,Sb)_2(Te,Se)_3$	Spex shaker	Ar atmosphere, steel vial/balls, β = 5:1, 1200 rpm, 6.5 h	N/A	[101]
$Co_4Sb_{12}^a$	Planetary QM-1SP2	Steel vial/balls, 15 h, 450 rpm, β = 25:1	~50 nm	[65]
Cu_2Te^a	Pulverisette P7	WC balls, 525 rpm, β = 8:1, up to 128 h	<10 nm	[76]
Cu_2Te	Pulverisette P7	WC balls, 525 rpm, β = 8:1, up to 150 h	Nanosize	[76]
$(Ni_xCo_{1-x})_4Sb_{12}^a$	N/A	ZrO_2 balls, Ar atomosphere, β = 1:1,39 h	N/A	[102]
$Fe_xCo_{1-x}Sb_{12}^a$	Attritor	100 h, Ar atomosphere	10 μm	[103]
$La_xFeCo_3Sb_{12}^a$	Planetary QM-SB	Stainless steel vial/balls, Ar atomosphere, β = 15: 1, 20 h	N/A	[104]
$Bi_{100-x}Sb_x^a$	Pulverisette 5	236 rpm, β = 10: 1, 15–22 h, Ar atomosphere	N/A	[77]
Zn_4Sb_3	SPEX	1 h	5–20 nm	[51]

β: Ball-to-powder weight ratio
[a]mechanical alloying
[b]Crystalline size is measured from TEM images or calculated from X-ray diffraction patterns

and more nanomaterials have been synthesized using the method and the synthesized nanomaterials have been summarized in literatures [108, 109]. Compared with other synthesis procedures, mechanical alloying can produce bulk quantities of nanomaterials using simple equipments and at room temperature. Similar to reported mechanical alloying [107, 110–112], *n*-type Bi_2Te_3 TE nanoparticles were mechanically alloyed [43] at room temperature from Bi and Te elemental powders in 2008. The crystalline size of produce nanoparticles is 5 nm, as shown in Fig. 11.4a. Later, $Si_{95}Ge_5$ nanopowders [45, 47] were mechanically alloyed from Si and Ge elements, MgAgSn nanoparticles [22] from Mg, Ag and Sn elemental powders, etc. Figure 11.4b, c show their TEM images respectively. Similar to mechanically milled TE nanoparticles, mechanically alloyed TE nanoparticles can be classified as single crystalline (as shown in Fig. 11.4a) and polycrystalline nanoparticles (as shown in Fig. 11.4b, c). Besides the mechanically alloyed TE nanoparticles listed in Table 11.2, more kinds of TE nanoparticles have been

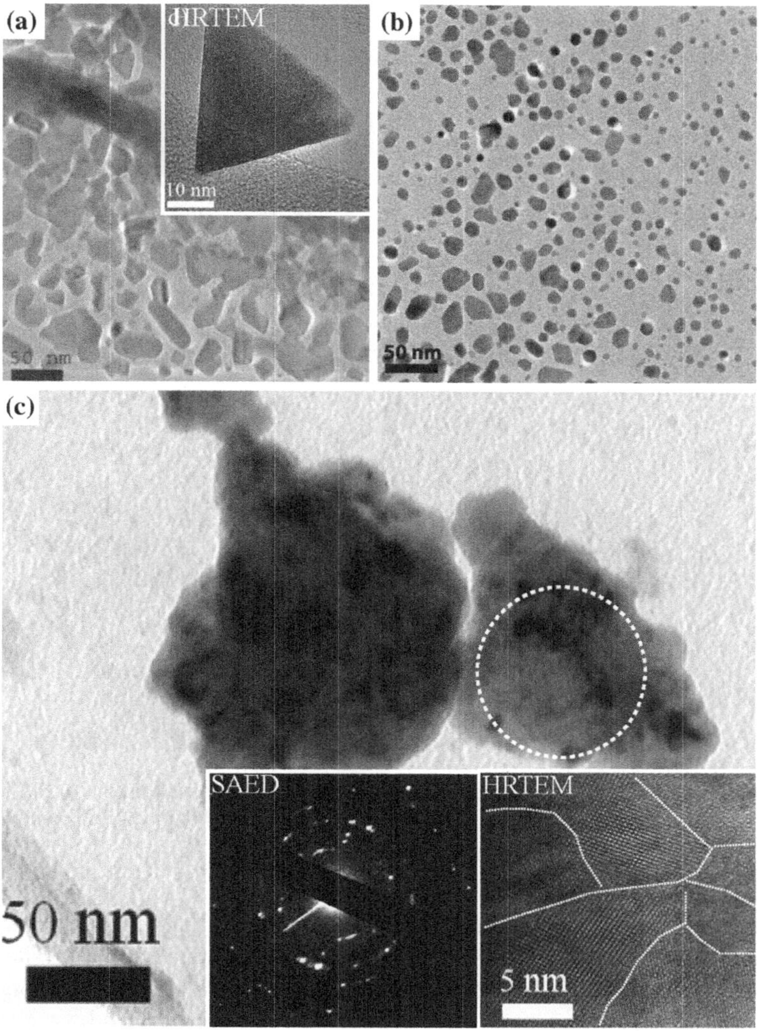

Fig. 11.3 BF-TEM images of ball-milled nanoparticles. **a** *p*-type Bi_2Te_3 [42]. **b** Zn_4Sb_3 [51]. **c** Half-Heuslers [48]. SAED and HRTEM are taken from the *circled* region

mechanically alloyed [25, 113, 114], such as $Ag_{0.8}Pb_{18+x}SbTe_{20}$ [115] and bounties [116].

Compared with mechanically milled nanoparticles, mechanically alloyed nanoparticles are usually larger. Figure 11.4d shows a scanning electron microscopic (SEM) image of mechanically alloyed Cu_2Te nanoparticles. Their geometrical size is in microns. However, XRD (not shown) indicates that the crystalline size of the micron-sized particles is below 10 nm.

Fig. 11.4 **a–c** TEM images of mechanically alloyed nanoparticles from elements. **a** $Bi_{0.4}Sb_{1.6}Te_3$ [43]. **b** $Si_{95}Ge_5$ [47]. **c** MgAlSb [22]. **d** An SEM image of Cu_2Te [76]. *Insets* SAED and HRTEM of the *circled* region

Previous research indicated that the crystalline size of ball-milled nanoparticles is significantly affected by ball-milling speed and duration. Figure 11.5 shows the crystalline size of Cu_2Te nanoparticles ball-milled for different ball-milling durations. The nanoparticles were mechanically alloyed from Cu and Te elemental powders using a planetary mill at room temperature. The crystalline size was calculated from XRD patterns. The grain size was reduced sharply with ball-milling time at first then constant with milling time.

The ball-to-powder ratio β is of minor importance [88] to affect the crystalline size of ball-milled nanoparticles. The ratio β to produce TE nanoparticles is listed in Table 11.2.

The size and shape of ball-milled nanoparticles can be directly determined by scanning electron microscopy (SEM) for relatively coarse powders or transmission electron microscopy (TEM) for fine powders. It should be mentioned that electron microscopy, especially SEM, usually shows geometrical size and not crystalline size. High-resolution TEM is a powerful tool to directly characterize crystalline size of nanoparticles.

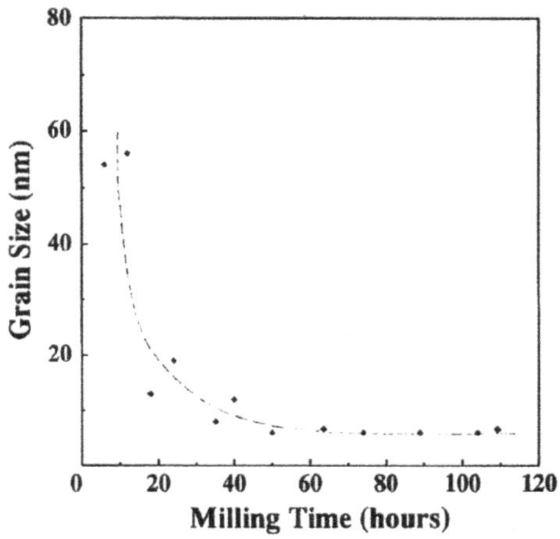

Fig. 11.5 Crystalline size of mechanical alloyed Cu_2Te nanoparticles [76]

The crystalline size of ball-milled nanoparticles can be characterized accurately from X-ray diffractions. The crystalline size and strains can be determined by measuring the Bragg peak full-width at half the maximum intensity (FWHM) and using the formula

$$B = \frac{0.9\lambda}{d\cos\theta} + \eta\tan\theta \tag{11.5}$$

where d is the crystallite size, λ is the wavelength of X-ray, η is the lattice strain, B is the peak full-width at half the maximum intensity, and θ is the Bragg angle. Thus, when $B\cos\theta$ is plotted against $\sin\theta$, a straight line is obtained with the slope as η and the intercept as 0.9 λ/d. Then, the crystalline size d and the lattice strain η can be experimentally measured from X-ray powder diffraction patterns. The X-ray peak broadening methods are most appropriate for crystalline sizes in the range of 10–100 nm.

It is important to remember that ball-milled particles are usually agglomerated and a particle may consist of several individual smaller particles. Further, an individual particle may contain a number of crystalline sub-particles defined as coherently diffracting domains. Therefore it should be careful to determine the accurate crystalline size from SEM techniques. SEM and conventional TEM usually indicate the geometric size while X-ray diffraction gives the crystalline size.

Ball-milled nanoparticles can induce high concentration of grain boundaries in TE nanocomposites. The higher concentration boundaries and defects can scatter phonons more strongly than charge carries and decrease the thermal conductivity, enhancing ZT of nanocomposite. In the next section, bottom-up fabrications of TE nanocomposites will be discussed.

11.3 Bottom-Up Techniques to Produce Thermoelectric Nanocomposites

The ball-milled thermoelectric nanoparticles can be compacted into nanocomposite bulks using various consolidation methods such as cold-pressing and sintering [62, 76, 77, 117], hot-pressing [42, 44, 73, 100, 118–121], spark plasma sintering [64, 74, 115, 122], and the extrusion method [77, 123, 124].

11.3.1 Cold Pressing

Cold pressing is the simplest bottom-up method to prepare TE nanocomposites. The nanoparticles are directly compacted at room temperature using hydraulic presses (Fig. 11.6a) with a compressive force up to 50–100 tons.

Figure 11.6b shows an SEM image of cold-pressed Cu_2Te nanocomposites. There are lots of cavities in the nanocomposites and the mass density should be much lower than that of single crystals. XRD indicated that the nanoscaled grain size (~ 10 nm) is retained without any grain growth. The nanoscale particles would scatter phonons greatly to effectively reduce thermal conductivity. According to (11.1), *ZT* of the cold-pressed nanocomposites should be increased. It is reported that the Seebeck coefficient of the cold-pressed nanocomposites depends on the cold-pressing pressure and varies with the crystalline size of the nanoparticles [76].

However, the cold-pressing method only compacts TE nanoparticles mechanically and pressed grains are only bounded together physically, resulting in poor mechanical properties and low electric conductivity. In order to enhance TE properties of cold-pressed nanocomposites, sintering usually follows cold-pressing

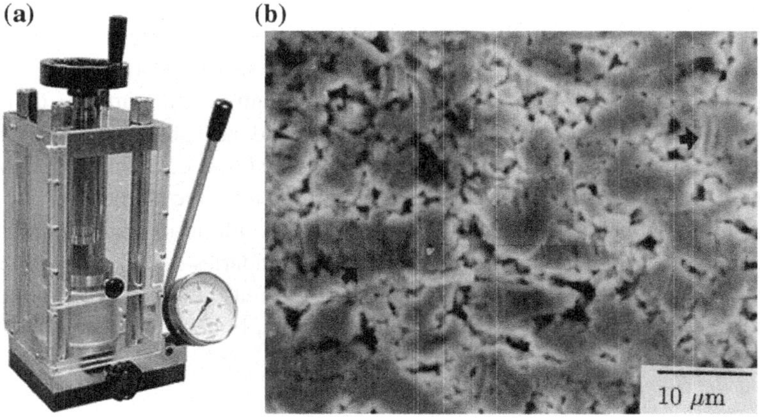

Fig. 11.6 Cold-pressing. **a** Hydraulic press. **b** An SEM image of cold-pressed Cu_2Te nanocomposites [76]. Pressure: 550 MPa. Crystalline size of nanoparticles : <10 nm (calculated from XRD)

[62, 117, 125]. Unfortunately high-temperature sintering always causes grain growth of cold pressed nanocomposites, increasing thermal conductivity and lowering *ZT*. *ZT* of cold-pressed nanocomposites is usually lower than that of ingots.

11.3.2 Hot-Pressing

Hot-pressing was employed to prepare TE composite as early as the 1960s [126]. PbSe microsized particles were hot-pressed at 750 °C and 8100 PSI for 5 min–5 h. The produced specimens possess a near-theoretical density and electrical and thermoelectric properties similar to those of the best of cast lead selenide.

Later, more kinds of thermoelectric composites were compacted by hot-pressing [55, 58, 127, 128]. Micron-sized particles were obtained from conventional ball-milling and hot-pressed into composites the 1970s. The thermal conductivity of the obtained hot-pressed composites was decreased compared with single crystals [58, 61]. The *ZT* of these composites consisting of micron-sized grains was increased by less than 20 %.

Figure 11.7 shows a typical DC hot press. TE powders are loaded into graphite dies, then compressed under high pressure. A direct current passes through the graphite dies to heat TE powders. TE powders are then compacted under high temperature (up to 1200 °C) and high pressure. The time required for this process is usually in tens of minutes because of the slow rate of heating/cooling of graphite dies. The direct current can be replaced by alternating current.

Since the invention of high-energy ball-milling, TE nanoparticles can be produced massively. The produced TE nanoparticles can be compacted into TE nanocomposites. TE grains in hot-pressed nanocomposites are in nano-size although grain growth happens during hot pressing.

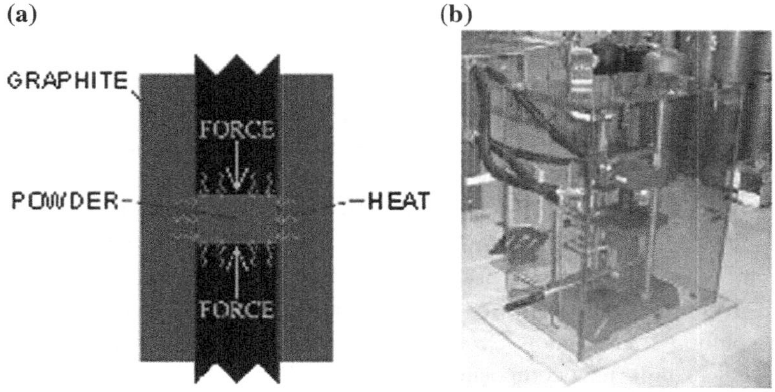

Fig. 11.7 **a** A sketch and **b** a home-made DC hot-press system

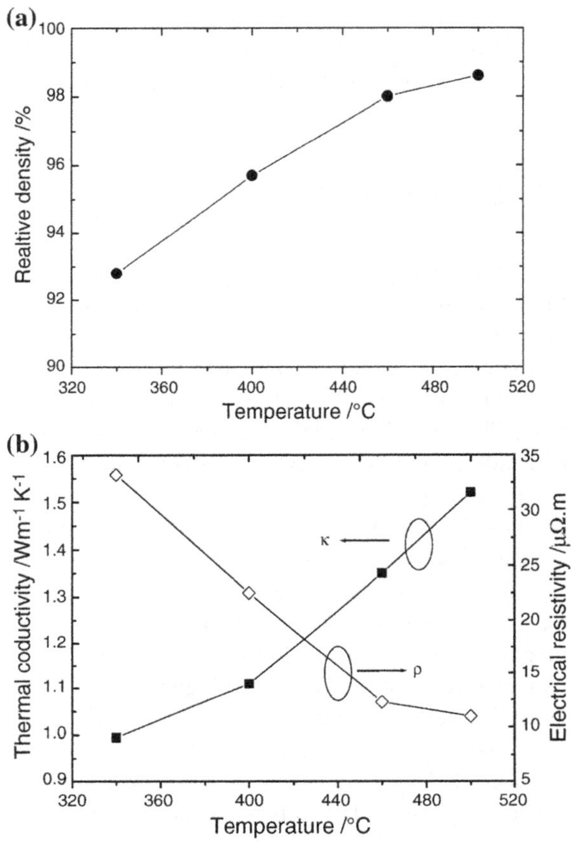

Fig. 11.8 Variation of **a** relative density and **b** carrier transport of hot-pressed *n*-type Bi_2Te_3 nanocomposites with hot-pressing temperature [100]. Hot-pressing time: 2–5 h

The physical properties of hot-pressed TE nanocomposites, such as mass density, thermal and electric conductivity, are dependent on hot-pressing procedures, especially hot-pressing temperatures. Figure 11.8a shows temperature dependence of relative density of hot-pressed Bi_2Te_3 nanocomposites. With increase of hot-pressing temperatures, relative density of TE nanocomposites increases. The density of hot-pressing nanocomposites can be 95–100 % theoretical value [25, 42]. Figure 11.8b indicates that the thermal conductivity increases while the electrical resistivity decreases with increasing hot-pressing temperature.

Grain size of hot-pressed nanocomposites is strongly affected by hot-pressing temperatures. Figure 11.9 shows grain size distribution of $Bi_{0.4}Sb_{1.6}Te_3$ nanocomposites hot-pressed at 300 °C. The grain size of the un-pressed nanograins was 8–10 nm [129], estimated from X-ray diffraction and confirmed by high-resolution transmission electron microscopy. After the nanoparticles were compacted to bulk nanocomposites by hot-pressing at 300 °C, the nanoparticles grew to several ten nanometers. Clearly, the hot pressing causes the growth of initial nanoparticles due to recrystallization.

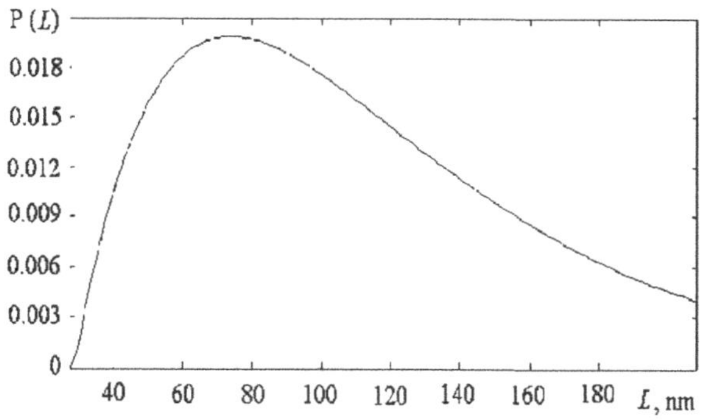

Fig. 11.9 Nanograin size distribution *P(L)* of $Bi_{0.4}Sb_{1.6}Te_3$ nanocomposites fabricated by hot-pressing at 300 °C [129]. P(L) is normalized to 1

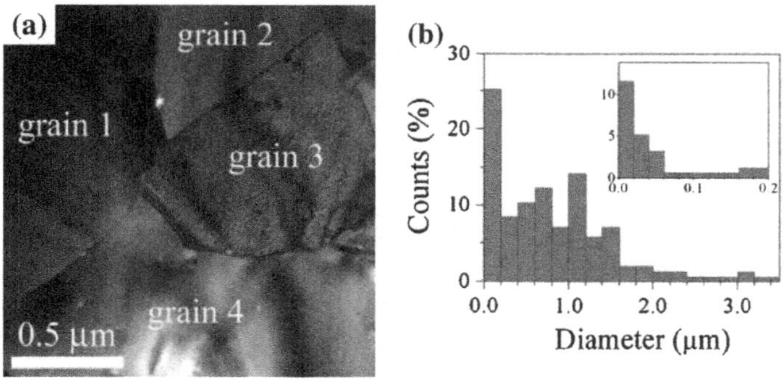

Fig. 11.10 Nanograin size distribution of $Bi_{0.4}Sb_{16}Te_3$ nanocomposites fabricated by hot-pressing at high temperature [130]. *ZT* = 1.4

Grains can grow quickly to microns once the hot-pressing temperature is high. Figure 11.10 shows the grain size of *p*-type $Bi_{0.4}Sb_{1.6}Te_3$ nanocomposites hot-pressed at high temperatures. The crystalline size can be in microns (Fig. 11.10a). More detailed analysis indicated that some nanoparticles can grow up to several microns while some nanosized nanoparticles are reserved (Fig. 11.10b).

Figure 11.11 shows a systematical study of hot-pressing temperature dependence of crystalline size of *p*-type Bi_2Te_3 nanocomposites. The average grain size increases sharply with the hot-pressing temperature once higher than 350 °C because of recrystallization.

The hot-pressing pressure also affects grain size of TE nanocomposites. Increasing pressure can slow the growth of grain size, as shown by the data in Fig. 11.11. Because of recrystallization of nanoparticles at high temperature, it is hard to produce nanocomposites with a grain size of several nanometers. It is a long way to go to control grain size of nanocomposites.

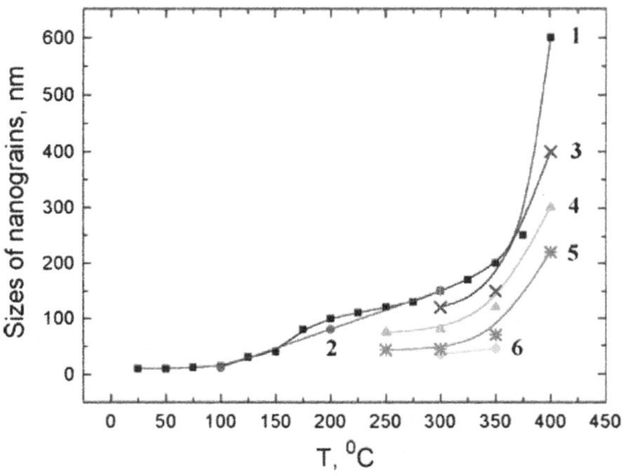

Fig. 11.11 Temperature dependence of the average nanograin size of $Bi_{0.4}Sb_{1.6}Te_3$ nanocomposites fabricated under different hot-pressing conditions [129]. ① ■: in situ heating of cold-pressed (at 25 °C and 1.5 GPa) composites. ② ●: Vacuum annealing of cold pressed (25 °C and 1.5 GPa) composites. ③ × : Hot-pressed at 35 MPa. ④ ▲: Hot-pressed at 350 MPa. Hot-pressed at 350 MPa with addition of ⑤ ※: 0.1 mass% and ⑥ ◆: 0.4 mass% MoS_2 nanoparticles

Hot-pressing time also affect grain size. Grains can grow with time. However, because the recrystallization should occur in minutes, even seconds, hot-pressing time does not affect grain size significantly [100] as the hot-pressing time is in several ten minutes even hours. Figure 11.12 shows SEM images of nanocomposites hot-pressed for different amounts of time (2–5 h). There is no any significant change in grain size. However *ZT* is affected by the heating time. A possible explanation is that the prolonging hot pressing time changes microstructures of the grains and crystalline boundaries.

Under optimal hot-pressing conditions, dense TE nanocomposites can be produced. Figure 11.13 shows an optical image of a Bi_2Te_3 nanocomposite bulk. Its mass density is close to that of crystalline ingots and thermal conductivity is

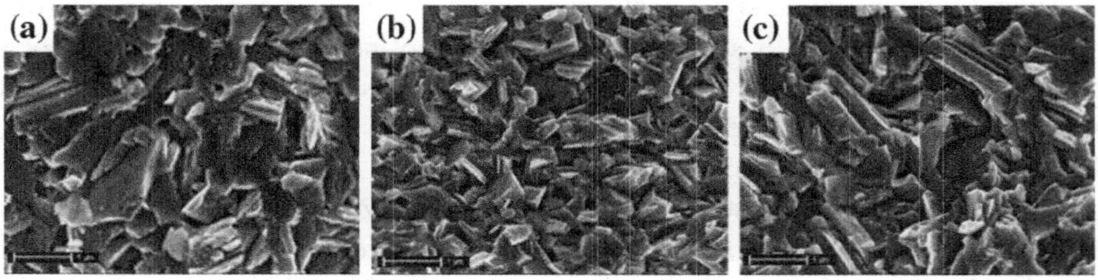

Fig. 11.12 SEM images of fractural surfaces of hot-pressed *n*-type Bi_2Te_3 nanocomposites prepared at 500 °C [100] for **a** 2 h, **b** 4 h and **c** 5 h

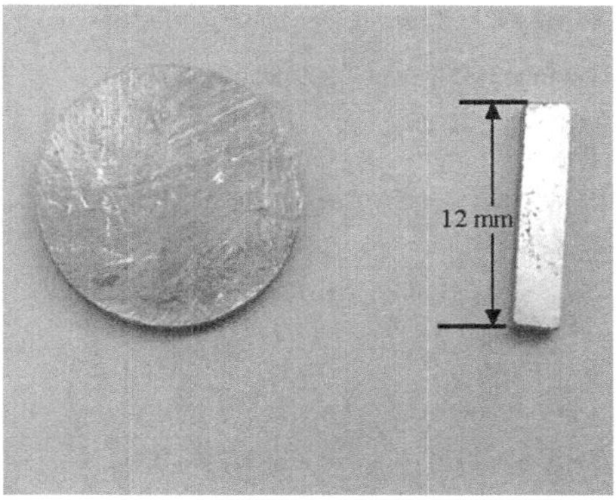

Fig. 11.13 A hot-pressed *p*-type Bi_2Te_3 thermoelectric nanocomposite (*left*) and a cut bar sample (*right*) with a peak $ZT = 1.3$ [43]

reduced while the electrical conductivity and Seebeck coefficient are almost not changed. The peak *ZT* is enhanced to 1.3 [43] while $ZT = 1.0$ for crystalline ingots.

At present the hot-pressing method is a mature technique to produce thermoelectric nanocomposites. Up to now, high-energy ball milling and DC hot pressing have been widely combined to prepare thermoelectric nanocomposites like Bi_2Te_3 [42, 43, 100, 125], SiGe alloys [44, 45, 47], skutterudite $CoSb_3$ [46, 93, 102], and TiO_x [141]. Table 11.3 lists some typical thermoelectric nanocomposites prepared by hot-pressing. Experimental conditions of hot-pressing are also listed.

11.3.3 Spark Plasma Sintering

Spark plasma sintering (SPS), an in situ sintering technique under high-pressure, is also known as the field-assisted sintering technique or pulsed electric current sintering. Figure 11.14 illustrates the working schematic. Nanoparticles are contained in graphite dies, similar to hot-pressing.

The rapid heating rates in SPS are achieved by a direct Joule heating of the graphite dies in which the TE nanoparticles are sintered.

A pulsed direct-current directly flows through nanoparticles under pressures, as well as the graphite die, internally heating nanoparticles. The pulsed direct-current with low voltages (<10 V) and high currents (typically >1000 A) can rapidly heat graphite dies and nanoparticles, facilitating a high heating or cooling rate up to 1000 °C/min [142]. Hence the sintering process generally is very fast (within a few minutes). Typically, SPS holding time of 5–10 min at sintering temperature is sufficient to densify nanoparticles during spark plasma sintering and the total processing time is less than one hour. The high heating rate of SPS ensures it to

Table 11.3 Thermoelectric properties of some typical nanocomposites produced by hot-pressing

Nanocomposites	*ZT* peak	κ at RT (W/m/K)	Experimental conditions of hot-pressing	References
PbSe	0.3 at RT	2	750 °C, 8100 psi, 5 min –5 h	[126]
$Si_{80}Ge_{20}B_x$	0.95 (0.5)	2.5 (5.0)	950–1200 °C	[44]
$Si_{80}Ge_{20}P_2$	1.3 (0.93)	2.5 (4.6)	1000–1200 °C	[45]
$Si_{63.5}Ge_{36.5}P_x$	0.128 (0.0937)	3.82 (5.07)	N/A	[58]
Si_{95} Ge_5	0.95	5(10)	N/A	[47]
SiP_x	0.7	12	>1000 °C	[131]
$Bi_{0.5}$ $Sb_{1.5}Te_3$	0.8 at RT	0.9	300–550 °C, 30 min	[125]
$(Bi,Sb)_2Te_3$	0.9 at RT	N/A	420 °C, 1.2 GPa, 1 h	[132]
$(Bi,Sb)_2(Te,Se)_3$	0.7–0.9	1.0–1.5	550 °C, 30 min	[101]
$Bi_2(Te,Se)_3$	0.5 at RT		500 °C, 4 h	[100]
$(Bi,Sb)_2Te_3$	1.4 (1.0)	1.1(1.4)	N/A	[42]
$(Bi,Sb)_2Te_3$	1.3 (1.0)	1.25 (1.4)	N/A	[43]
$CoSb_3$	<0.01 at RT	1.7	450 °C, 100 MPa	[133]
Co_4Sb_{12}	0.11	4.7	600 °C, 100 MPa, 30 min	[121]
$Yb_{0.19}$ Co_4Sb_{12}	0.26 at RT	2.6	650 °C, 26 PSI, 2 h	[134]
$(Ni_{0.09}Co_{0.91})_4Sb_{12}$	0.75 at RT	3	60–160 Mpa, 550–780 °C, 1–6 min	[93]
$(Ni_{0.08}$ $Co_{0.92})_4Sb_{12}$	0.002	5	100 MPa, 873 K	[73]
$(Ni_{0.8}$ $Co_{0.2})_4Sb_{12}$	0.1	3.0	80–160 MPa, 680–780 °C, 1–5 min	[102]
$Fe_{1.5}Co_{2.5}$ Sb_{12}	0.32 at RT	2	60 MPa, 550 °C, 2 h	[103]
$SnFe_3$ Co_5 Sb_{12}	0.15 at RT	1.5	60 MPa, 550 °C, 2 h	[135]
$La_x(Ni,Co)_3$ Sb_{12}	0.12 at RT	3.2	50 MPa, 600 °C, 2 h	[136]
$La_{1.5}$ Fe_4 Sb_{12}	0.41	2	50 MPa, 650 °C, 2 h	[104]
$La_{0.4}$ $FeCo_3$ Sb_{12}	0.04 at RT	3	70 MPa, 700 °C, 3 h	[137]
β-$FeSi_2$	0.018 at RT	4.5	80–100 MPa, 950 °C, 30 min	[138]
$Fe_{0.95}Co_{0.05}$ Si_2	0.18	4.5	50 MPa, 950 °C, 30 min	[139]
Mg_2 $Si_{0.6}$ $Ge_{0.4}$	0.21	4.3	1 GPa, 600 °C	[140]
TiO_x	<0.025 at RT	2.7–5.7	114 MPa, 1200 °C, 2 min	[141]

κ: thermal conductivity. Data in brackets is the value of single crystals
RT: room temperature

potentially densify nanoparticles without the coarsening that accompanies standard densification routes. Joule heating plays a dominant role to densify nanoparticles to achieve near theoretic density of nanocomposites [25, 42, 58, 74, 127, 143].

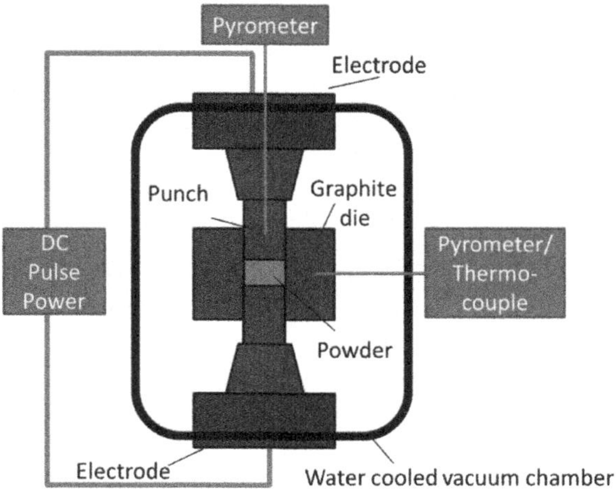

Fig. 11.14 Working schematic of a spark plasma sintering [142]

Fast consolidation involved in SPS enables the fabrication of thermoelectric nanocomposites better. Up to now, lots of thermoelectric composites have been successfully fabricated by the technique, such as $GaSb_{10}Te_{16.5}$ [87], $Ag_{0.8}Pb_{22}SbTe_{20}$ [115], Bi_2Te_3 [144–147], Co_4Sb_{12} [65, 122], La_2CuO_4 [148], PbTe [64, 149], higher manganese silicide [150], $Ca_2Co_4O_9$ [151, 152], In_2O_3 [153], and $(Zn_{1-x}Al_x)O$ [154].

Table 11.4 lists some typical thermoelectric nanocomposites produced by spark plasma sintering. SPS conditions and physical properties of produced nanocomposites are listed. Recrystallization of TE nanoparticles can usually be controlled during SPS and their thermoelectric properties are improved.

SPS has been systematically compared with hot-pressing [166]. It is widely believed that recrystallization of nanoparticles can be limited during SPS while relative density of SPS-composites is lower. *ZT* of SPS-composites is slightly higher than that of hot-pressed nanocomposites.

11.3.4 Other Bottom-Up Methods

Beside hot-pressing and spark plasma sintering, there are other kinds of bottom-up techniques to fabricate TE nanocomposites, such as hot extrusion and microwave sintering. Hot extrusion forces thermoelectric nanoparticles through a die of desired cross-sections at high temperatures [77, 124, 167, 168]. Hot extrusion usually causes preferred orientation of grains and produces heavy plastic deformations in nanocomposites, causing lots of strain. Bi_2Te_3 [169] and PbTe [170] nanocomposites have also been fabricated by microwave sintering. Here the details of these methods are not discussed.

Table 11.4 Thermoelectric nanocomposites produced by spark plasma sintering

Nano-composites	*ZT* peak	κ at RT (W/m/K)	Grain size and density	Experimental conditions of SPS	References
Cu_3SbSe_3	<0.1	0.26 (550 K)	80 nm	60 MPa, 300 °C, 5 min, HR: 300 °C/min	[155]
Cu_3SbSe_4	0.30	0.761 (550 K)	100 nm	60 MPa, 300 °C, 5 min, HR: 300 °C/min	[156]
Bi_2S_3	0.22	~0.74	200–300 nm, ρ: 96.8 %	673 K, 5 min, 40 MPa	[157]
$Bi_2Te_{2.7}Se_{0.3}$	0.55 (RT)	0.6	100 nm	40 MPa, 250 °C	[158]
$(Bi,Sb)_2Te_3$	1.15		>100 μm	60 MPa, 693 K, 5 min, HR: 50 °C/min	[144]
Sb_2Te_3	0.58	0.76	60–100 nm, ρ: 89 %	270 °C, 67 MPa, 6 min, naturally cooled	[159]
$Si_{80}Ge_{20}$	1.5	2.4	9 nm	60 MPa, 1150 °C, 3 min, HR: 400 °C/min, CR: 200 °C/min	[99]
$CrSi_2$			250 nm	1000 °C, 60 MPa	[160]
TiO_{2-x}			ρ: 99 %	1100 °C, 5 min, 34 MPa	[161]
$NaCoO_4$	0.3	~3.5	5–10 μm, ρ: 98 %	400–600 A, 50 MPa, 850 °C, 5 min, HR: 150 °C/min	[162]
$Ca_{2.7}$ $Gd_{0.25}$-Co_4O_9	0.23	2.5	2–5 μm, ρ: 98 %	400–600 A, 50 MPa, 850–1100 °C, 5 min, HR: 150 °C/min	[163]
PbTe	0.33	1.0	5–30 μm, ρ: 82–88 %	48 MPa. 5 min. 450 °C	[164]
In_4Se_3	~0.56	1.0	microns	70 MPa, 390 °C, 5 min	[165]
$GaSb_{10}Te_{16.5}$	0.6	1.2	30 nm	40 MPa	[87]

κ: thermal conductivity. Data in brackets is the value of single crystals
ρ: relative mass density, *HR:* heating rate, *CR:* cooling-rate

11.3.5 Advantages and Problems of Current Bottom-Up Techniques

There are two steps in bottom-up techniques: the production of TE nanoparticles and the fabrication of nanocomposites. TE nanoparticles can be high-energy ball-milled, chemical synthesized, or produced from other methods. Then, TE nanoparticles are

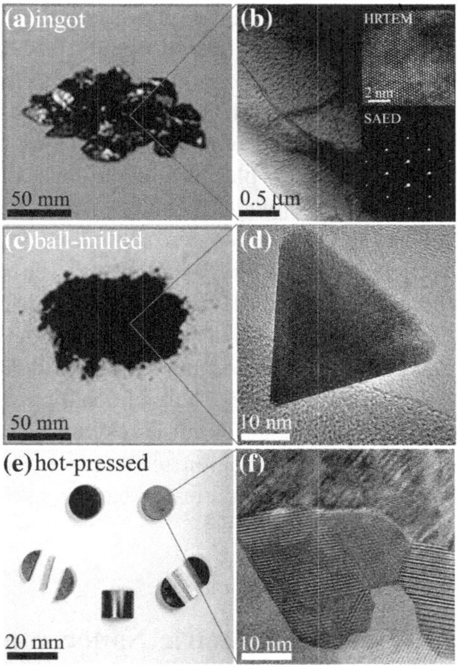

Fig. 11.15 A typical bottom-up method to produce TE nanocomposites [25]. **a** Optical image of crystalline ingots before ball-milling. **b** TEM image of a *p*-type Bi_2Te_3 ingot. *Inset* is HRTEM and SAED of the ingot. **c** Optical image of ball-milled nanoparticles. **d** HRTEM image of a typical Bi_2Te_3 nanoparticle produced by ball-milling. **e** Optical image of hot pressed bulk nanocomposites made from the ball-milled nanoparticles shown in (**c**). **f** HRTEM image of several nanograins in a *p*-type Bi_2Te_3 nanocomposite produced by the ball-milling and hot-pressing method

bottom-up-ed using cold-pressing, hot-pressing, spark-plasma-sinter, microwave sintering, or hot extrusion. Any combination of nanoparticle production and bottom-up technique is possible, providing wide feasibility to fabricate massive TE nanocomposites. Figure 11.15 shows a typical combination of ball-milling and hot-pressing. Crystalline ingots or elemental powders (Fig. 11.15a, b) are mechanically milled or mechanically alloyed into nanoparticles (Fig. 11.15c, d). The ball-milled nanoparticles are hot-pressed into nanocomposites (Fig. 11.15e). Such kind of nanocomposites are consisted of nano-sized crystalline grains (Fig. 11.15f), showing better thermoelectric properties than those of ingots. The density of the hot-pressed nanocomposites is higher up to 98 % [58] or even 100 % [42] of that of the single crystal ingots. The hot pressed nanocomposites are also thermally stable at high temperatures [42–44].

Among these bottom-up methods, cold-pressing can produce nanocomposites without recrystallization and more economically. Hot-pressing and spark-plasma-sintering are mature techniques to fabricate TE nanocomposites industrially. It usually takes several hours during hot-pressing procedures while several minutes during spark-plasma-sintering procedures to compact TE nanoparticles to TE nanocomposite bulks. Hot-pressing usually induces grain growth during long-time heating procedures while grain boundaries are crystalline, benefiting electron transportation. Spark-plasma-sintering can limit recrystallization of nanoparticles because of pulsed DC currents and short heating time. Microwave sintering and hot extrusion usually fabricate TE composites at high temperatures, inducing recrystallization of nanoparticles. Cold pressing can produce TE nanocomposites with much smaller nanograins while thermoelectric properties of cold-pressed nanocomposite are not as good as those of hot-pressed nanocomposites and spark-plasma-sintered nanocomposites.

From the commercialization point of view, the combination of high-energy ball milling and hot pressing or ball-milling and SPS probably offer the most advantages because large quantities of powders can be ball-milled and pressed in a short time. From the research point of view of thermoelectric materials, the ball-milling/hot-pressing fabrication method and ball-milling/SPS technique are superior because both methods can produce TE nanocomposites with different materials, compositions, and doping levels in a short time. The traditional procedures of melting and crystal growth take long time to prepare TE materials [80, 122, 134, 171].

A suitable fabrication technique to compact nanoparticles is still required to generate good boundaries of nanocomposites while limiting grain growth during bottom-up fabrications.

11.4 Thermoelectric Nanocomposites with Enhanced *ZT*

Experiments have shown that thermal conductivity does indeed decrease with grain size in thermoelectric nanocomposites, such as in Bi_2Te_3 composites [62, 125, 172] and PbTe composites [64]. The nanocomposites have many advantages over traditional TE crystalline materials, being widely utilized in TE devices at present. Theoretical researches also indicate that the microstructures affect the phonon transport and electron transport in TE nanocomposites, which contribute to the thermal conductivity, electrical conductivity, and Seebeck coefficient, changing *ZT*. Below some typical thermoelectric nanocomposites are discussed in details, showing how the microstructures affect thermoelectric properties.

11.4.1 Bi_2Te_3 Nanocomposites

Bi_2Te_3 is the most widely used thermoelectric material near room temperature, and has great commercial applications in refrigeration and waste heat recovery up to 200 °C. The maximum *ZT* of the commercial Bi_2Te_3 ingots remained at about 1.0 for many years. The peak *ZT* of Bi_2Te_3 nanocomposites can be enhanced to 1.3–1.4 at 100 °C [42, 43], while the nanocomposites are isotropic and thermally stable at high temperatures.

11.4.1.1 *p*-Type $Bi_xSb_{2-x}Te_3$ Nanocomposites

p-type $(Bi,Sb)_2Te_3$ nanocomposites are the first thermoelectric nanocomposites with higher *ZT* made from the ball-milling and hot-pressing method [42]. With a bottom-up approach from ball-milled nanoparticles, *ZT* increases 30–40 % compared with the starting crystalline ingots [42, 43].

The starting nanoparticles are produced by high-energy ball-milling crystalline ingots. High-energy ball milling grinds crystalline ingots into nanocrystals with size ranging from 5 to 50 nm, with an average size of about 20 nm. The produced nanoparticles are $(Bi,Sb)_2Te_3$ single phase as ingots. The milled nanoparticles are very good crystals with clean surfaces, without any nanoprecipitates or amorphous nanoregions [130].

During hot-pressing at high-temperatures, the nanoparticles would partially melt, vaporize, grow, and be compressed into solid nanocomposite bulks. The final produced nanocomposites are consisted of crystalline nanograins. Statistical analysis of thousands of nanograins shows that the majority of the nanograins are of diameters below 1.0 micron, including 12 % of nanograins with a diameter of 0–20 nm and 5 % of nanograins with a diameter of 20–40 nm. The grain boundaries scatter phonons with a mean free path comparable to or greater than the grain size, reducing the thermal conductivity. A wide grain size distribution is responsible for scattering phonon with a variety of wavelengths. There is no cavity between grains, consistent with the measured 100 % density in the nanocomposite [42]. No second phase exists between grains.

HRTEM images of grain boundaries indicate that the grain boundaries are high angle grain boundaries. Selected area electron diffraction (SAED) shows that the grains are single crystals with random orientation. The crystalline orientation randomness of the adjacent grains helps phonon scattering significantly.

In addition to the clean boundaries between grains shown in Fig. 11.16a, about 50 % of the grains in the nanocomposite are surrounded by nanometer thick interface regions (Fig. 11.16b). The interface region is slightly bismuth-rich, 1.0 ± 0.5 atomic% higher than that of the grains [130]. This bismuth-rich region builds up charges, thus increasing the hole carrier concentration in the grains. At the same time, the interface region scatters phonons. Therefore, the bismuth-rich interface region blocks phonon transport while favoring charge carrier transport.

Besides unique grain structures in the nanocomposites, each nanograin has abundant microstructural defects to scatter phonons. Usually there are many three-dimensional nanoprecipitates embedded in the nanograins shown in Fig. 11.17a. In $(Bi,Sb)_2Te_3$ nanocomposites, at least four kinds of nanoprecipitates are observed in the crystalline nanograins (Fig. 11.17a–d). A popular kind of nanoprecipitates is antimony-rich nanoprecipitates (Fig. 11.17a). There are no obvious boundaries between the nanoprecipitates and nanograin matrix. There is no clear lattice distortion between them. The lattice is continuous and the fringe space is almost same. At present, there is no any experiment to confirm if the extra Sb dopes into the lattice or not. On the HRTEM images, the nanoprecipitates are brighter than the surrounding crystalline nanograin matrix. STEM Z-contrast imaging should observe the nano precipitates more clearly. Another type of

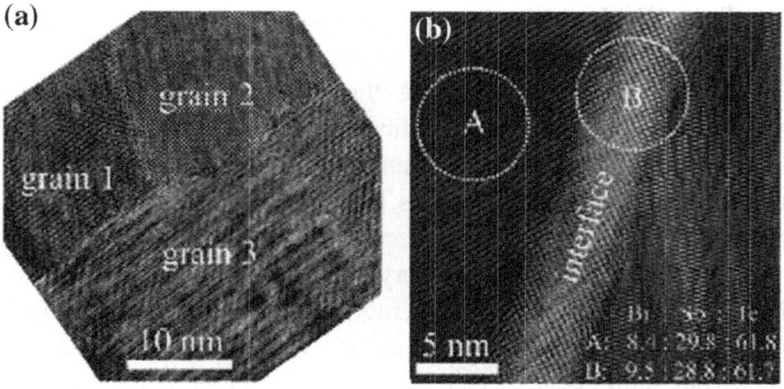

Fig. 11.16 HRTEM images of (**a**) grain boundaries and (**b**) an interface region between two grains in *p*-type Bi_2Te_3 nanocomposites. (Reprinted with permission from [130]. Copyright 2008, American Chemical Society.)

nanoprecipitates are nanodots whose orientation is twisted compared with the nanograin matrix. Their chemical composition is the same as the surrounding nanograin matrix. The third popular kind of nano precipitates (Fig. 11.17c) are also twisted with the surrounding nanograin matrix. This kind of nanoprecipitates are the combination of type one and type two nanoprecipitates. In some cases, pure tellurium nanodots are also observed (Fig. 11.17d). The polygonal tellurium precipitates are identified readily from other three kinds of irregular nanoprecipitates.

The nanoprecipitates are very similar to those observed in $Ag_{1-x}Pb_mSbTe_{m+2}$, which reduce thermal conductivity. More experiments approve that the nanometer sized precipitates embedding in crystalline hosts, like 30–40 nm Pb or Au metal in crystalline PbTe samples, can increase thermoelectric power [173] and reduce thermal conductivity [33].

The nanoprecipitates with different chemical compositions would scatter electron and phonons effectively. These precipitates would also increase the carrier concentration in the grains. Hall effect measurements at room temperature indicates that the hole carrier concentration of the nanocomposites ($n = 2.5 \times 10^{19}$ cm^{-3}) is about 39 % higher than that of the ingot ($n = 1.8 \times 10^{19}$ cm^{-3}). The increased carrier concentration should partly come from the nanoprecipitates and partly from the nano-interfaces between grains.

All of the grain boundaries, interfaces between grains, and nanoprecipitates affect the transports of electrons and phonons. More discussions of the interfacial structure effects have been reviewed recently [174].

Generally speaking, the nanoprecipitates are less than several tens of nanometers in diameter. The size distribution of these precipitates (Fig. 11.17e) measured from thousands of nanoprecipitates shows that the precipitate concentration in the nanocomposite increases rapidly with decreasing precipitate size. The volume of these nanoprecipitates accounts for about 5 % of the nanocomposites. Besides

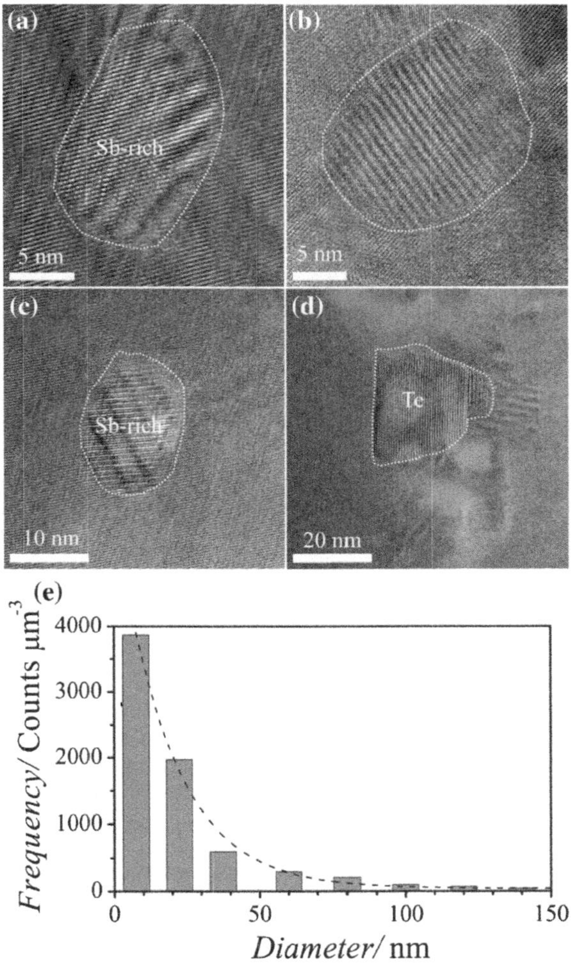

Fig. 11.17 Nanoprecipitates embedded in the grains of $(Bi, Sb)_2Te_3$ nanocomposite. **a** HRTEM image of an antimony-rich precipitate without boundary. **b** HRTEM image of a precipitate with twisted boundary while with the same chemical composition of the surrounding matrix. **c** HRTEM image of an antimony-rich precipitate with twisted boundary. **d** HRTEM image of a tellurium precipitate with high-angle boundary. **e** The precipitate size distribution. (Reprinted with permission from [130]. Copyright 2008, American Chemical Society.)

increasing the carrier concentration, the nanoprecipitates in the nanograins can scatter phonons, especially phonons with a mean free path matched with the nanoprecipitate size, thus reducing the lattice thermal conductivity.

Besides the nanoprecipitates, there are two-dimensional, one-dimensional, and zero-dimensional defects in the nanograins of the nanocomposite [130]. TEM investigations indicated that stacking faults exist in the nanograins. The threading dislocation concentration is $10^{11} cm^{-2}$ in nanograins, at least ten times higher than 5×10^9 cm^{-2} found in the crystalline ingot. The point defect concentration in nanograins is 2–3 orders of magnitude higher than that in the ingot. Structural modulations are also observed in the nanograins where the concentration is almost same as that in the ingot. All these defects would scatter phonons more effectively and further decrease the thermal conductivity of the nanocomposites.

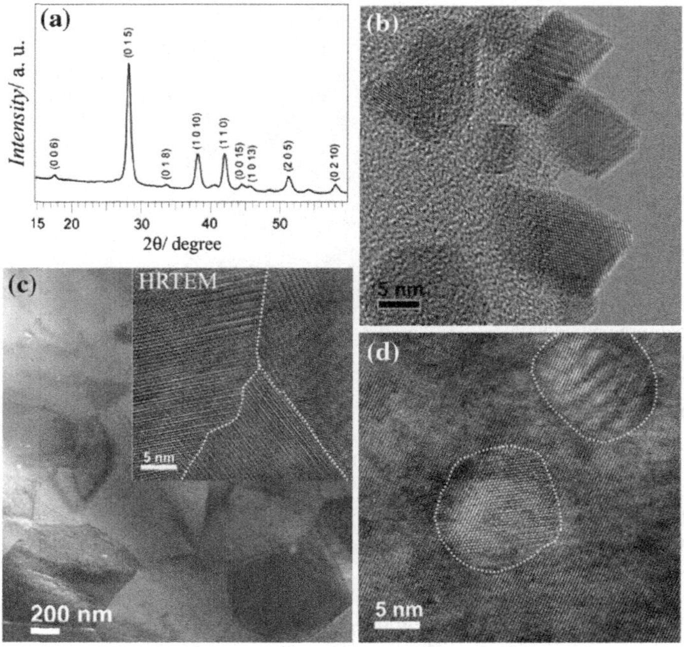

Fig. 11.18 **a** XRD pattern and **b** HRTEM image of mechanically alloyed $Bi_xSb_{1-x}Te_3$ nanopowders after ball milling from the chemical elements. **c** BF-TEM image and **d** HRTEM image of nanocomposites hot-pressed from nanoparticles shown in (**b**). The *white dots* outline the grain boundaries or nanoprecipitates embedding in the nanograins. (Reprinted with permission from [43]. Copyright 2008, American Chemical Society.)

p-type Bi_2Te_3 nano-particles can also be produced from the elements directly through mechanical alloying by the high-energy ball milling. Mechanical alloying can synthesize single-phase $(Bi,Sb)_2Te_3$ nano-particles (Fig. 11.18a) from elements Bi, Sb and Te. The size of the synthesized nano-particles is about 5–20 nm with an average size of about 10 nm. The crystalline quality of the nanoparticles mechanically alloyed from elements is almost the same as ball-milled nanoparticles from ingots (Fig. 11.18b).

It is reported that trace Sb and Te phases are observed in ball-milled Bi_2Te_3 nanopowders although $(Bi,Sb)_2Te_3$ single phase is obtained after hot-pressing [132]. The un-alloyed Sb and Te impurity should come from the insufficient milling time or insufficient transfer energy of ball-milling. If the transfer energy is not high enough, the ball-milling will only grind the elements while do not synthesize Bi_2Te_3 phase. In order to obtain single phase $(Bi,Sb)_2Te_3$ nanoparticles, the experimental conditions (like milling time, milling media, intensity of milling, ball to powder weight ratio, as well as milling temperature and milling atmosphere) should be adjusted as well.

After hot pressing, *p*-type Bi_2Te_3 nanocomposites consisting of crystalline nanograins are obtained from these mechanical alloyed nanoparticles. The microstructure of the nanocomposite is very similar to that of the nanocomposites made from ingots. The nanograin size is also in sub-micrometers and the nanocomposites are densely packed (Fig. 11.18c). The grains have excellent crystallinity and large angles between them, which explains the isotropy of the thermoelectric properties (inset in Fig. 11.18c). Compared with the starting mechanical alloyed nanopowders (Fig. 11.18b), the nanograins grow to larger sizes after hot-pressing (Fig. 11.18c). Detailed TEM studies indicate that nanoprecipitates are also formed and embedded inside the $(Bi,Sb)_2Te_3$ nanograins (Fig. 11.18d), very similar to the hot-pressed nanocomposites made from the ingots (Fig. 11.17a–d).

In both kinds of the *p*-type Bi_2Te_3 nanocomposites, these nanoprecipitates probably form during the hot-press heating and cooling processes, and are likely due to the non-uniform ball milling process during ball milling or composition fluctuation inside the grains. Nanoprecipitates have also been observed in LAST alloys and were believed to be responsible for the *ZT* enhancement in those alloys [20]. It is believed that these individual nanoprecipitates would help to enhance the scattering of the phonon with short-wavelength (high energy) and therefore contribute to the reduction of the thermal conductivity.

These unique microstructures (shown in Figs. 11.16, 11.17 and 11.18) affect the thermoelectric properties of the nanocomposites. Figure 11.19 shows the transport properties of three samples: the hot-pressed nanocomposite made from the chemical elements, the hot-pressed nanocomposite made from crystalline ingots, and the crystalline ingot itself.

Comparing with ingots, the thermal conductivities of the nanocomposites made from the elements are significantly lower than that of the ingots, as expected, and increase slower as a function of temperature (Fig. 11.19a). This indicates a smaller bipolar contribution [4] to the thermal conductivity by thermally generated electrons and holes in the nanocomposites.

Compared with the nanocomposites made from ball-milled nanoparticles from the ingots, the thermal conductivities of the nanocomposites made from the elements are systematically higher than those of the nanocomposites made from the ingots. The difference is probably due to the lack of some minor elements that were used in the ingot and some minor structural differences.

The Seebeck coefficient (Fig. 11.19b) of nanocomposites made from the elements is slightly lower than that of nanocomposites made from the ingot, both of which have a lower Seebeck coefficient than that of the crystalline ingot sample below 150 °C while higher above 150 °C. The smaller Seebeck coefficient near room temperature is due to a higher carrier concentration, while the larger Seebeck coefficient at higher temperatures is due to the suppression of the minority carrier (electron) excitation in more heavily doped samples (also due to the reduced bipolar effect). The nanocomposites made from ingots have a power factor comparable to that of the crystalline ingot below 100 °C but higher above 100 °C while the nanocomposites made from elements have a slightly lower power factor than that made from the ingots.

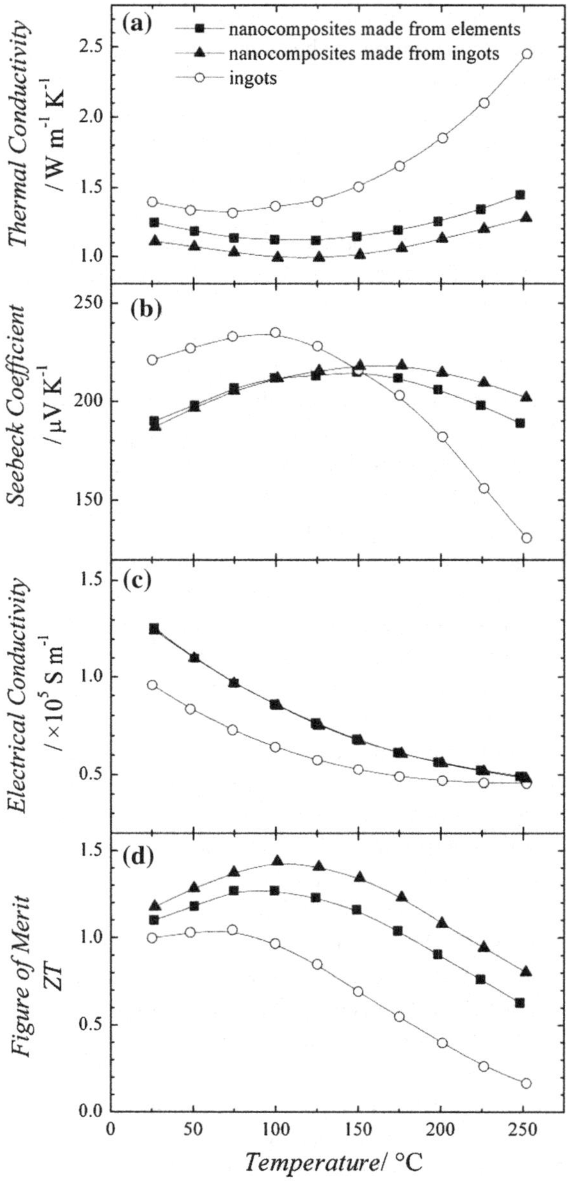

Fig. 11.19 **a** Thermal conductivity, **b** Seebeck coefficient, **c** Electrical conductivity, and **d** *ZT* dependence of temperature of hot-pressed $Bi_xSb_{1-x}Te_3$ nanocomposites made from elements (*filled squares*), $(Bi, Sb)_2Te_3$ nanocomposites made from ingots (*filled triangles*), in comparison with a commercial ingot (*open circles*). (Reprinted with permission from [43]. Copyright 2008, American Chemical Society.)

The electrical conductivity of the nanocomposites made from the elements is always higher than that of the crystalline ingots (Fig. 11.19c) because of the higher carrier concentration (2.9×10^{19} cm^{-3} for nanocomposites made from the elements and with 1.8×10^{19} cm^{-3} for the ingots), and is almost identical to the nanocomposites made from the ingots. The enhanced electrical conductivity is first observed in the thermoelectric nanocomposites. Since the 1960s, many works have been carried out to increase *ZT* through low-dimensional composites. Although the thermal conductivity of the low-dimensional composites decreases (enhancing *ZT*), the decrease of the electric conductivity weakens the *ZT* enhancement. Here the nanoprecipitates embedding in the nanograins and nano-interface between nanograins should contribute to the increase of the electric conductivity.

According to (11.1), *ZT* should be enhanced because the electrical conductivity increases while the Seebeck coefficient changes slightly when the thermal conductivity decreases.

Figure 11.19d indicates the temperature dependence of *ZT* for the hot-pressed nanocomposites made from the elements, for the hot-pressed nanocomposites made from the ingots, in comparison with that of the commercial ingots. The peak *ZT* values shift to a higher temperature and remain significantly higher than that of the ingots at all temperatures. The peak *ZT* of the hot-pressed nanocomposites made from elements is about 1.3 at 100 °C and 1.4 at 100 °C for the nanocomposites made from ingots, which are significantly higher than that of the best Bi_2Te_3-based alloy ingots ($ZT \sim 1.0$).

The *ZT* value of the ingots starts to drop above 75 °C and is below 0.25 at 250 °C whereas the nanocomposites are still higher than 0.8 at 250 °C. The superior *ZT* at high temperatures is very important for power generation applications, since there are no other materials with a similar high *ZT* in this temperature range. Such *ZT* characteristics are best for power generation applications due to a lack of available materials with high *ZT* in this temperature range.

Some research groups also prepared *p*-type $(Bi,Sb)_2Te_3$ nanocomposites [58, 125, 132, 175–179] and *p*-type $(Bi,Sb)_2(Te,Se)_3$ nanocomposites [101]. The grain size of most of these composites is in microns. The mass densities of these composites are low (80–98 % of theoretical density). *ZT* is less than 1.0. By the ball-milling and hot pressing method, *p*-type $(Bi,Sb)_2Te_3$ composites composed of 100 micron nanoparticles and 15–30 nm Bi_2Te_3 nanoparticles are also obtained ($ZT = 0.83$) [180].

The *p*-type Bi_2Te_3 nanocomposites have also been prepared by other methods, such as shear extrusion at high temperatures from ball-milled powder [181], hot pressing hydrothermally synthesized nanoparticles ($ZT = 1.47$) [72], melting spinning, and the SPS method (*ZT* increases from 0.8 to 1.35) [182].

11.4.1.2 *n*-Type $Bi_2Te_{3-Y}Se_y$ Nanocomposites

n-type Bi_2Te_3 nanocomposites are also prepared using the same method. *n*-type Bi_2Te_3 single crystalline nano-particles have been obtained by ball-milling from ingots or from elements. After hot-pressing, nanocomposites with a single phase are obtained. The microstructure of the *n*-type $Bi_2(Te,Se)_3$ nanocomposite is very similar to that of the *p*-type $(Bi,Sb)_2Te_3$ nanocomposite. The thermal conductivity is also decreased compared with that of *n*-type crystalline ingots (Fig. 11.20). *ZT* is close to 1.0 since the electric conductivity and Seebeck coefficient also decrease.

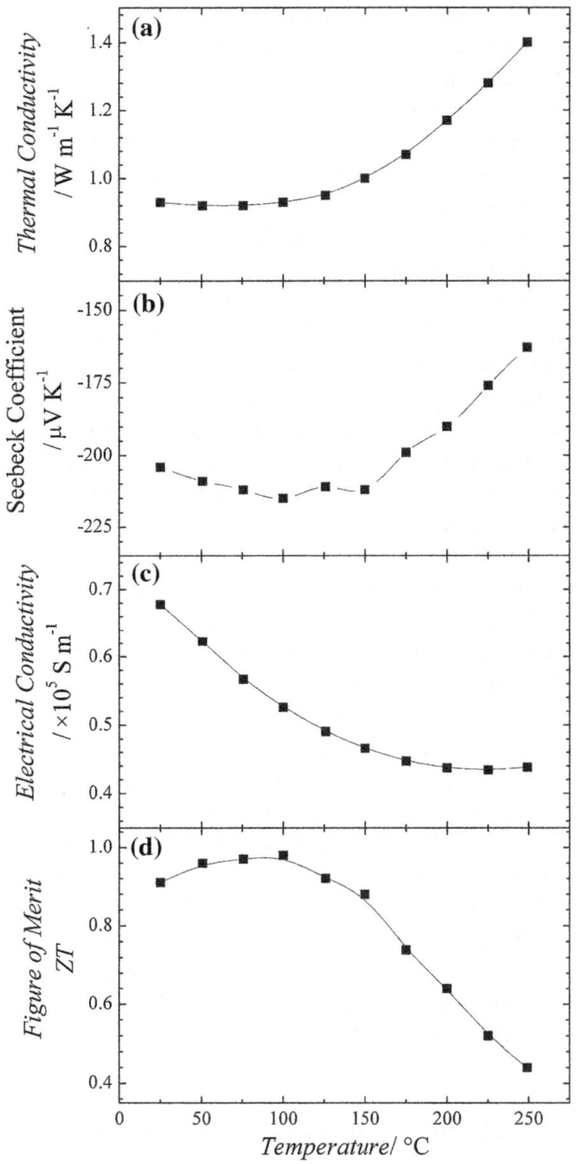

Fig. 11.20 Thermoelectric properties of hot-pressed $Bi_2Te_{3-y}Se_y$ nanocomposites [42]. **a** thermal conductivity, **b** Seebeck coefficient, **c** electrical conductivity, and **d** *ZT* dependence of temperature

More work on the *n*-type Bi_2Te_3 nanocomposites is currently under way to preserve the power factor.

n-type Bi_2Te_3 nanocomposites are also prepared by other groups using the ball milling and hot-pressing method [100, 120, 183, 184]. Grain size of these nanocomposites varies from nanometers to micron meters. *ZT* is less than 1.0.

11.4.2 SiGe Nanocomposites

Silicon-germanium alloys have been the primary thermoelectric materials in power generation devices operating in the temperature range of 600–1000 °C [6, 14], having been used in radio-isotope thermoelectric generators (RTG) for deepspace missions to convert radio-isotope heat into electricity [6, 14]. SiGe alloys also hold promise in terrestrial applications [12] such as thermoelectric generators [185] to recover waste heat.

Since the 1960s, efforts have been made to improve the *ZT* of SiGe alloys [58, 60, 185–189], with the peak *ZT* of *n*-type SiGe reaching 1 at 900–950 °C [58, 60, 185, 186, 188, 189] and *p*-type SiGe alloys at about 0.65 [60, 61, 188]. Through ball milling and hot-pressing method, a significant improvement in *ZT* with a peak value about 0.95 at 900–950 °C is achieved in *p*-type SiGe nanocomposites [44]. *ZT* is improved about 90 % over the RTG samples (*ZT* = 0.5), and 50 % over the previous highest record (*ZT* = 0.65) [60]. A peak *ZT* of about 1.3 at 900 °C was achieved [45] in *n*-type SiGe nanocomposites using the ball-milling and hot-press method.

11.4.2.1 *p*-Type $Si_{80}Ge_{20}B_x$ Nanocomposites

Using the ball-milling and hot-press method, SiGe alloy composites which consisted of micron-sized particles were prepared in the 1960s [61]. A 20 % increase in the *ZT* of *p*-type SiGe with an optimal grain-size in the 2–5 μm range was reported [61]. At present, the typical grain size is 1–10 μm in the RTG samples used for NASA space flight. In the SiGe nanocomposites, the mean free path between electrons and phonons is different: about 5 nm for electrons, and 2–300 nm for phonons in highly-doped samples at room temperature. Thus, nanocomposites can significantly reduce the phonon thermal conductivity (2.5 W/m/K of *n*-type SiGe nanocomposites while 4.6 W/m/K in RTG bulk) without significantly reducing the electrical conductivity, resulting in a higher *ZT*.

The *p*-type SiGe nanocomposites with higher *ZT* are prepared by the same technique as Bi_2Te_3 nanocomposites. *p*-type (boron doped) $Si_{80}Ge_{20}$ single phase nanoparticles can be prepared by mechanically alloying using the high-energy ball mill technique from elements [44]. The prepared SiGe nanoparticles are a single phase $Si_{80}Ge_{20}B_x$ alloy (Fig. 11.21a). The crystallinity of the mechanically alloyed SiGe nanoparticles is different from that of the ball-milled Bi_2Te_3 nanoparticles. The mechanically alloyed $Si_{80}Ge_{20}B_x$ nanoparticles are polycrystalline (Fig. 11.21c, d), consisting of several sub-nanograins (Fig. 11.21d). On the contrary, the mechanical alloyed Bi_2Te_3 nanoparticles are single crystals, as shown in Fig. 11.18b. Similar polycrystalline structure is also observed on ball-milled pure Si nanopowders. XRD and TEM indicate the crystalline region varies from 20 nm to 200 nm. The microstructure can be explained by the high-energy ball-milling theory.

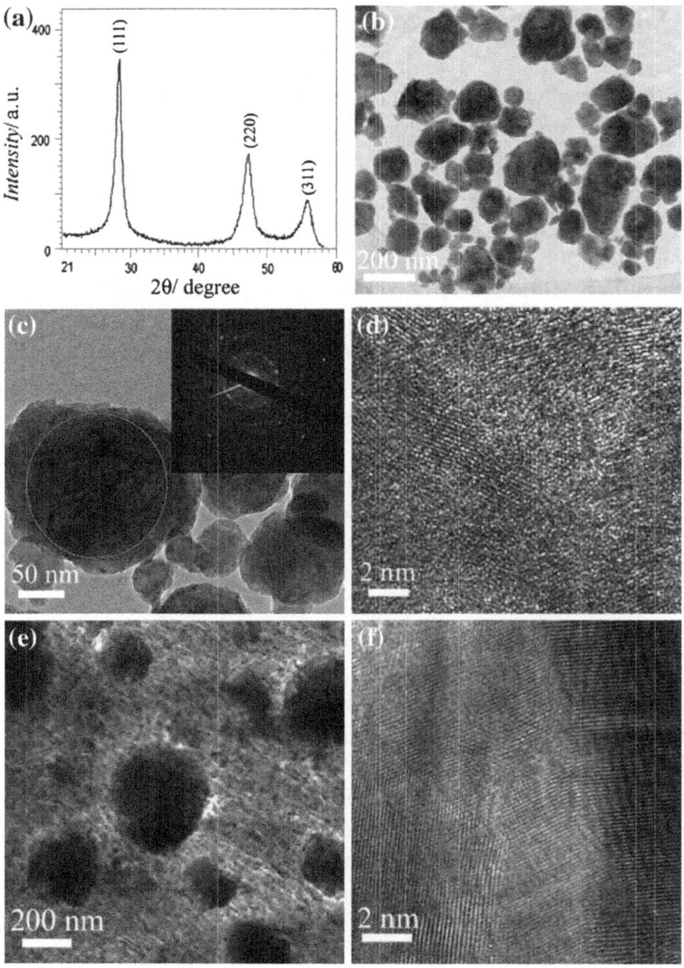

Fig. 11.21 **a** XRD pattern, **b** BF-TEM image at low-magnification, **c** TEM image at medium magnification, and **d** HRTEM image of the *p*-type $Si_{80}Ge_{20}B_x$ nanoparticles prepared by elemental chunks. **e** BF-TEM image and **f** HRTEM image of the DC hot-pressed *p*-type $Si_{80}Ge_{20}B_x$ nanocomposites. The *inset* in (**c**) is SAED of an individual grain indicating the polycrystalline nature of the grain. (Reprinted with permission from [44]. Copyright 2008, American Chemical Society.)

HRTEM image also shows that many defects exist inside an individual sub-grain (Fig. 11.21d), because the nanograins were formed by a low temperature mechanical alloying process and not by high temperature melting and solidification. There is expected to be a high strain field in the ball-milled SiGe nanoparticles. Dislocations and atomic level strains are usually produced in ball-milled nanoparticles and detailed information can be found in the articles on high-energy ball-milling [82, 111].

After hot-pressing, the grain size in the $Bi_{80}Ge_{20}B_x$ nanocomposites is up to about 20 nm, similar to the size of the initial nanopowders, indicating no significant grain growth occurring after the dc hot-press process. These nanograins are highly crystalline, completely random, closely packed (Fig. 11.21f), and have very clean boundaries, consistent with the measured high volume mass density (2.88 g/cm^3 measured by the Archimedes technique).

The Seebeck coefficient of the $Bi_{80}Ge_{20}B_x$ nanocomposites is comparable to that of RTG samples while the electrical conductivity is higher than that of RTG values in whole temperature range. More important, the thermal conductivity of the nanostructured bulk samples is much lower than that of the RTG sample over the whole temperature range up to 1000 °C, which led to a peak *ZT* of about 0.95 in the ball-milled and hot-pressed $Si_{80}Ge_{20}B_x$ nanocomposites (Fig. 11.22). Such a peak *ZT* value is about 90 % improvement over that of the *p*-type RTG SiGe alloy (*ZT* = 0.5) currently used in space missions (shown in the Fig. 11.22d as the open circles) and 50 % above that of the reported record value (*ZT* = 0.65) [60].

The significant reduction of the thermal conductivity in the nanostructured samples is mainly due to the increased phonon scattering at the numerous interfaces of the random nanostructures. The actual phonon thermal conductivity reduction is reduced at least by a factor of 2 based on the experimental data shown in Fig. 11.22a. The total thermal conductivity (k) contains contributions from both the carriers (k_e) and phonon (k_l) ($k = k_e + k_l$). The electrical conductivity of the nanocomposite is similar to that of the RTG sample.

p-type SiGe composites are also reported in other groups [55, 119, 127, 190]. The grain size of these reported composites are in microns. Although thermal conductivity is decreased, *ZT* is low because of the decrease of electrical conductivity and Seebeck coefficient.

11.4.2.2 *n*-Type $Si_{80}Ge_{20}P_2$ Nanocomposites

The microstructure of the mechanically alloyed *n*-type SiGe nanoparticles is very similar to that of the mechanically alloyed *p*-type nanoparticles. $Si_{80}Ge_{20}P_2$ single phase nanoparticles are also synthesized using the same technique (Fig. 11.23a). The particle size of the as-prepared ball-milled nanopowders is in the range of 30–200 nm (Fig. 11.23b, c). These nanoparticles are also composed of many small crystalline sub-nanograins, similar to *p*-type mechanically alloyed SiGe nanoparticles. The size of the sub-nanograins is in the range of 5–15 nm (Fig. 11.23d) with an average particle size of 12 nm.

After hot-pressing, an average grain size in the hot-pressed nanocomposites is 22 nm, indicating that the grain size is almost doubled after the hot pressing. Luckily the grain size is still very small. Compared with ball-milled nanoparticles, the strain inside the hot-pressed nanocomposites is much smaller than that of the as-prepared ball-milled nanopowders, with the strain value being ten times lower in the hot pressed samples. This smaller strain is understandable since the hot pressing

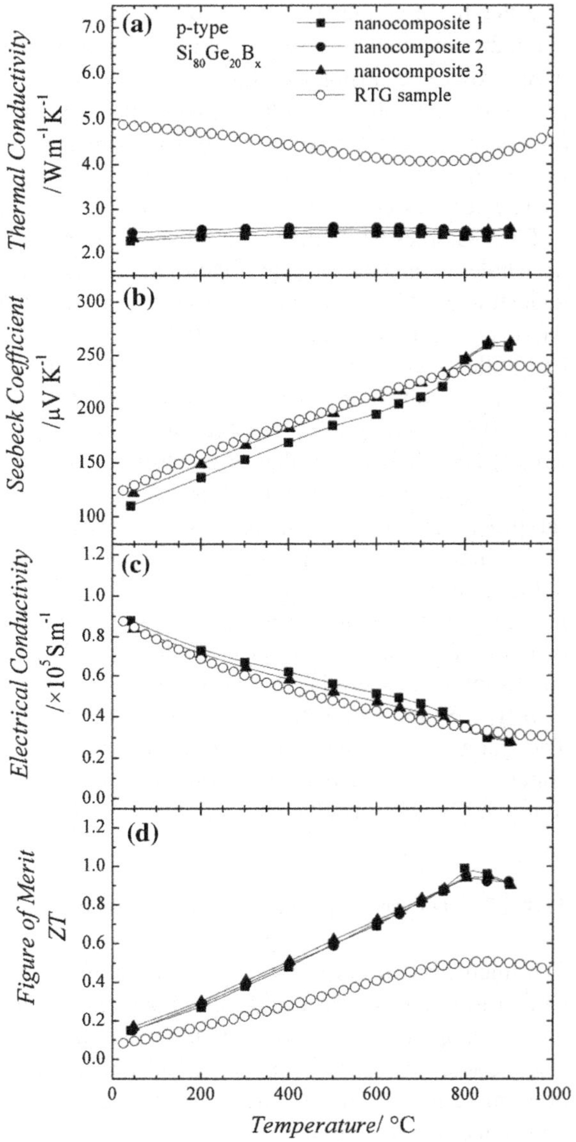

Fig. 11.22 Temperature dependence of the **a** thermal conductivity, **b** Seebeck coefficient, **c** electrical conductivity, and **d** *ZT* of three dc hot-pressed *p*-type $Si_{80}Ge_{20}B_x$ nanocomposites (*squares*, *circles*, and *triangles*) in comparison with the *p*-type SiGe bulk alloy used in RTGs for space power missions (*open circles*). Reproduced with permission from [44]. Copyright 2008, American Chemical Society

temperature is above 1000 °C, allowing the buildup stresses in the nanopowders to be relaxed during the mechanical alloy process.

These small grains with random crystalline orientations promote phonon scattering much more effectively than the micron-sized grains in bulk RTG SiGe materials.

From the comparison of the temperature dependent thermal conductivity (Fig. 11.24a), it clearly shows that the *n*-type SiGe nanocomposites have a much

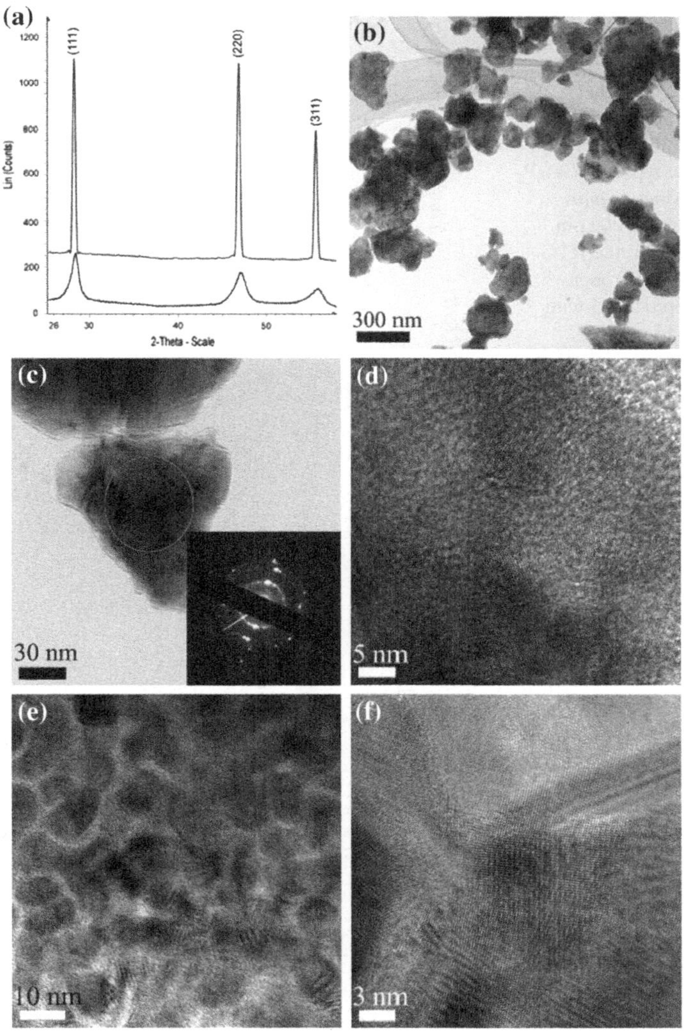

Fig. 11.23 **a** XRD pattern, **b** BF-TEM image at low-magnification, **c** BF-TEM image at medium magnification, and **d** HRTEM image of *n*-type $Bi_{80}Ge_{20}P_2$ nanoparticles. **e** BF-TEM and **f** HRTEM images of hot-pressed *n*-type $Si_{80}Ge_{20}P_2$ nanocomposites made from the nanoparticles shown in (**b**). The *inset* of (**c**) shows an SAED pattern of an individual nanoparticle. (Reprinted with permission from [45]. Copyright 2008, AIP.)

lower thermal conductivity than the RTG reference sample with a grain size of 1–10 μm. From the measured electrical conductivity of the *n*-type SiGe nanocomposites and RTG samples, $\kappa_e = 0.77$ W/m/K at room temperature with an electrical conductivity $\sigma = 1.2 \times 10^{-5}$ S/m for RTG sample, whereas for a typical nanocomposite sample $\kappa_e = 0.55$ W/m/K at room temperature for $\sigma = 0.85 \times 10^5$ S/m. By subtracting the electronic contribution κ_e from the total thermal

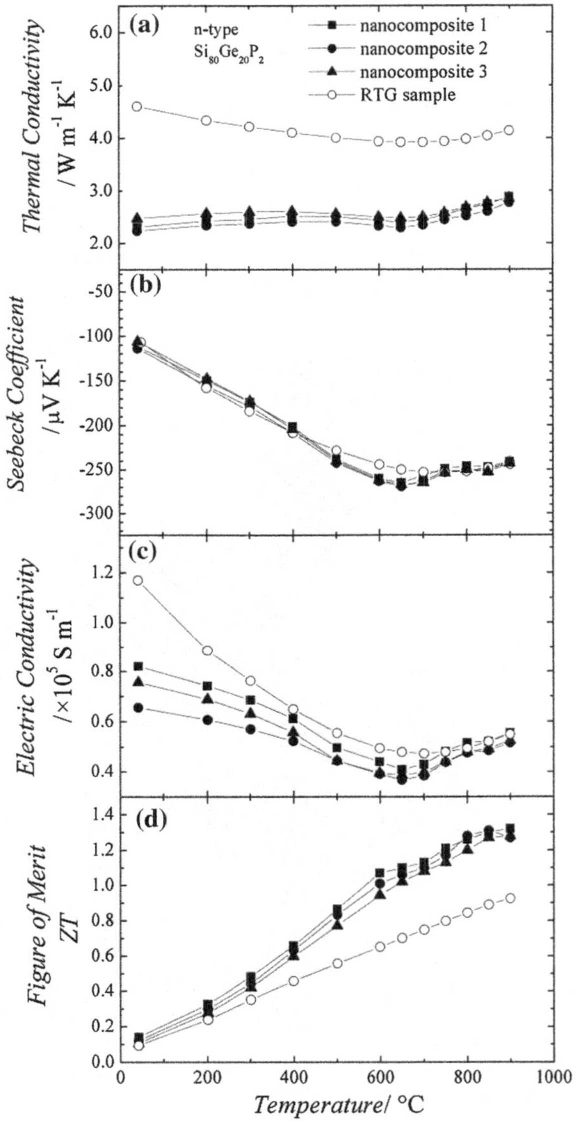

Fig. 11.24 Temperature dependence of the **a** thermal conductivity, **b** electrical conductivity, **c** Seebeck coefficient, and **d** dimensionless *ZT* of as-pressed *n*-type $Si_{80}Ge_{20}P_2$ nanocomposites (*squares*, *triangles*, and *circles*) in comparison with the RTG reference sample (*open circles*). Reproduced with permission from [45]. Copyright 2008, American Institute of Physics

conductivity κ, the lattice thermal conductivity (κ_l) of the nanocomposite samples is ~1.8 W/m/K at room temperature, which is about 47 % of the reference sample (κ_l ~ 3.8 W/m/K). The decrease of κ_l should be mainly due to a stronger boundary phonon scattering in the nanostructured samples.

The electrical conductivity of the *n*-type nanocomposites is normally lower than that of the reference in the low temperature region (Fig. 11.24b), but is similar above 750 °C, although the carrier concentrations for both types of samples are almost the same at room temperature (2.2×10^{20} cm^{-3} from the Hall effect measurement), indicating a lower electron mobility in the nanocomposite samples. The Seebeck coefficient of the nanocomposite samples is similar to that of the reference sample at below 400 °C and at above 700 °C, and is higher than that of the reference sample between 400 and 700 °C.

As a result, for the *n*-type SiGe nanocomposites, the *ZT* value shows a maximum of about 1.3 at 900 °C (Fig. 11.24d) which is about 40 % higher than that of the RTG reference (0.93). The significant enhancement of *ZT* is mainly attributed to the thermal conductivity reduction, which is strongly correlated with the nanostructure features in the *n*-type $Si_{80}Ge_{20}P_2$ nanocomposites.

11.4.2.3 Nanostructured Si and $Si_{95}Ge_5$ Nanocomposites

$Si_{95}Ge_5$ nanoparticles and pure Si nanoparticles can also be synthesized by the high-energy ball-milling from $Si_{95}Ge_5P_2$ powders and 5 atomic% GaP powders. These ball-milled nanoparticles are single phase with a size of crystalline nanoparticles in the range of 5–20 nm (obtained from the Williamson-Hall method [191]) or 20–150 nm (from TEM studies, Fig. 11.25b). Further works on SAED

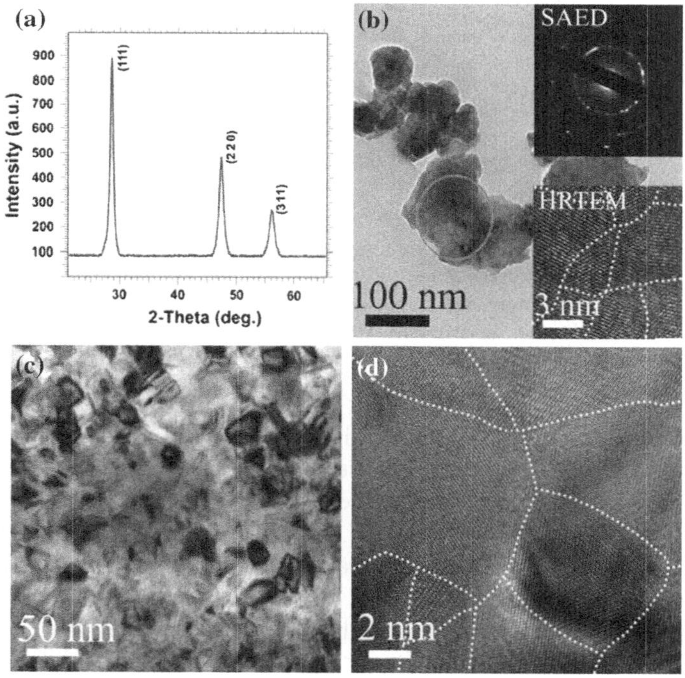

Fig. 11.25 **a** XRD pattern and **b** BF-TEM image of ball milled $Si_{95}Ge_5$ nanoparticles. **c** BF-TEM image and **d** HRTEM image of a hot-pressed $Si_{95}Ge_5$ nanocomposite. The *insets* in (**b**) are SAED pattern and HRTEM image of the *circled* region. (Reprinted with permission from [47]. Copyright 2009, AIP.)

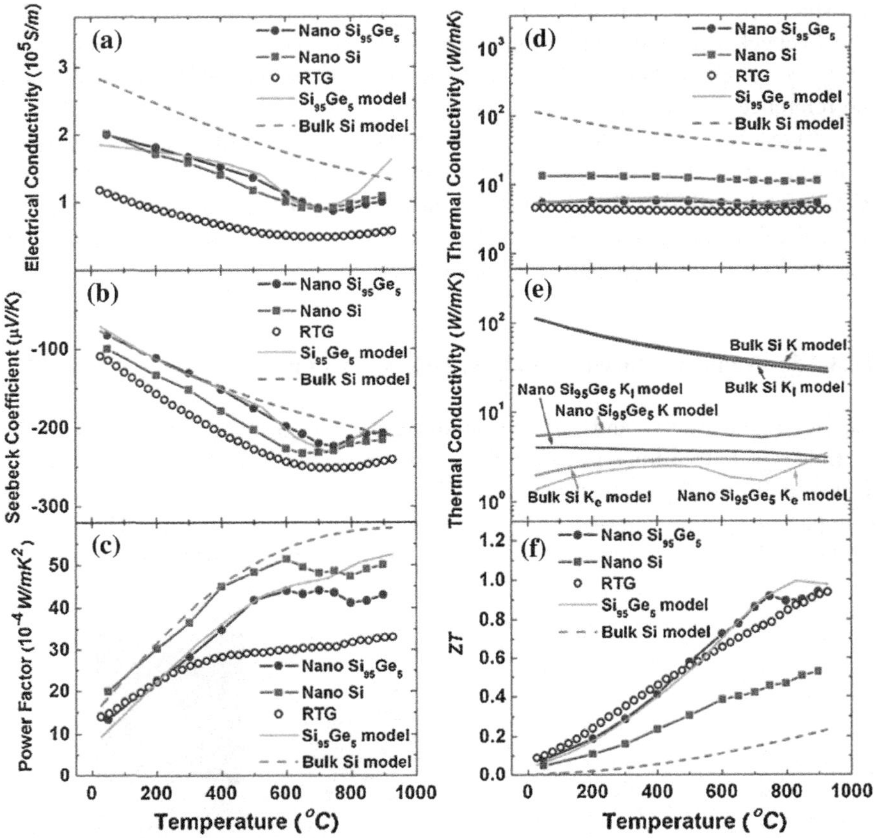

Fig. 11.26 Temperature-dependent **a** electrical conductivity, **b** Seebeck coefficient, **c** power factor, **d** thermal conductivity, **e** electron (κ_e), phonon (κ_l), and total (κ) thermal conductivity by modeling, and **f** *ZT* of nanostructured Si (*filled squares*), $Si_{95}Ge_5$ nanocomposite (*filled circles* for experiment and *solid line* for model), bulk Si model (*dashed line*), and $Si_{80}Ge_{20}$ RTG samples (*open circles*). (Reprinted with permission from [47]. Copyright 2009, AIP.)

and HRTEM (insets in Fig. 11.25b) indicate that the ball-milled nanoparticles are also consisted of sub-nanograins, as observed in the mechanically alloyed heavily doped *p*-type $Si_{80}Ge_{20}B_x$ nanoparticles (Fig. 11.21) and *n*-type $Si_{80}Ge_{20}P_2$ nanoparticles (Fig. 11.23).

After hot pressing, most of the nanograins are in the 10–30 nm range, larger than the 5–20 nm initial nanoparticle range due to the grain growth during the hot pressing process. The nanograins are highly crystalline and randomly oriented after hot pressing (Fig. 11.25d).

Figure 11.26 shows the comparative thermoelectric property measurement results for nanostructured Si, $Si_{95}Ge_5$ nanocomposite, bulk Si, and bulk $Si_{80}Ge_{20}$ alloy (RTG) that has been used by NASA for many years. Results plotted for the bulk Si are calculations using the Boltzmann transport equation under the relaxation time approximation, modified on the basis of the Vining model [60] but also considering nonparabolicity and a temperature-dependent band structure. Both the nanostructured Si and nanostructured $Si_{95}Ge_5$ samples show a higher electrical conductivity (Fig. 11.26a) but a lower absolute Seebeck coefficient (Fig. 11.26b) than that of the bulk $Si_{80}Ge_{20}$ RTG sample. This is mainly attributed to the higher solubility limit of P and the lower alloy scattering of charge carriers in nanostructured Si and nanostructured $Si_{95}Ge_5$ samples in comparison with the bulk $Si_{80}Ge_{20}$ RTG sample. The power factors for both nanostructured samples (Fig. 11.26c) are slightly lower than the values calculated for bulk materials with the same carrier concentration values as measured for the nanostructured samples. Also, due to the heavy doping in the nanostructured $Si_{95}Ge_5$ and Si samples and the activation of excess dopant (P) at high temperature during the measurement process, the nanostructured Si and nanostructured $Si_{95}Ge_5$ samples show different trends for the temperature dependent electrical conductivity and Seebeck coefficient than those of the bulk Si. In fact, the power factor of the nanostructured $Si_{95}Ge_5$ sample is much higher than that of the bulk $Si_{80}Ge_{20}$ RTG sample (Fig. 11.26c), especially at temperatures above 300 °C. Figure 11.26d shows the temperature-dependent thermal conductivity of the nanostructured Si and nanostructured $Si_{95}Ge_5$ samples in comparison with bulk Si and bulk $Si_{80}Ge_{20}$ RTG samples. The thermal conductivity of the nanostructured Si shows a significant reduction (by about a factor of 10) compared with that of the heavily doped bulk Si, which is around 100 W/m/K, at room temperature, a clear demonstration of the nanosize effect on phonon scattering. Moreover, with a 5 atomic% replacement of Si by Ge, the thermal conductivity value of the nanostructured $Si_{95}Ge_5$ is even lower, close to that of the bulk $Si_{80}Ge_{20}$ RTG sample, due to both the nanosize and point defect scattering effects in nanostructured $Si_{95}Ge_5$. Since the bulk $Si_{80}Ge_{20}$ RTG sample has 20 atomic% Ge, and the $Si_{95}Ge_5$ nanocomposite has only 5 atomic % Ge, a weaker alloy phonon scattering effect is expected in $Si_{95}Ge_5$. When the Ge concentration is increased from 5 to 20 atomic%, the thermal conductivity is decreased by another factor of 2 to about 2–3 W/m/K, but the power factor is also decreased [45] accordingly because of the reduced charge mobility due to the alloy scattering of charge carriers.

Figure 11.26e shows that the calculated thermal conductivity of $Si_{95}Ge_5$ nanocomposite matches well with the experimental results. The electron contribution to the thermal conductivity is calculated from the electrical conductivity measurement results using the Wiedemann-Franz law. The Lorenz number is calculated from the bulk model. Our modeling results show that the Lorenz number in bulk SiGe alloy varies from 1.3 to 2.2 from 25 to 1000 °C and that variation within any specific temperature is 0.2 for the range of the doping concentration in our samples. The calculated phonon thermal conductivity is below 4 W/m/K, at room temperature and reaches 3 W/m/K at 900 °C (Fig. 11.26e). The low thermal conductivity for the $Si_{95}Ge_5$ system is mainly attributed to both the enhanced boundary phonon scattering and the alloy effect. Thus, due to the significant thermal conductivity reduction without reduction of the power factor, *ZT* of the $Si_{95}Ge_5$

nanocomposites shows a maximum value of 0.95 at 900 °C, which is about the same as that of the bulk $Si_{80}Ge_{20}$ RTG sample (Fig. 11.26f).

11.4.3 PbTe Nanocomposites

Lead telluride (PbTe) is one of the best thermoelectric materials at intermediate temperatures (450–800 K). *ZT* of PbTe is 0.7 at 700 K. Recently, *ZT* was doubled to above 1.5 at 773 K in thallium doped PbTe through an enhancement of the Seebeck coefficient [192]. Based on the thallium doped PbTe, PbTe nanocomposites with high *ZT* is also achieved.

PbTe nanocomposites (95 % theoretical density) are also prepared by the spark plasma sintering method from 100–150 nm nanoparticles [74]. *ZT* of the nanocomposites is 0.1 at room temperature.

Doping affects the thermoelectric properties of PbTe nanocomposites. Experimental research indicates that the electrical conductivity at room temperature increased slightly with increasing potassium concentration, but no change was seen at high temperatures, where all of the samples showed a decrease with temperature. The thermal diffusivity was basically the same for all of the samples. The Seebeck coefficients of all of the samples changed only slightly, likely as a result of contributions from both the light and heavy holes with the high carrier concentration. As a result, the power factor increased with increasing potassium concentration and peaked at 500 K (Fig. 11.27a). The highest *ZT* value was 1.3 at 673 K for $K_{0.015}Pb_{0.985}Te$ (Fig. 11.27b), which is comparable with that of Na-doped PbTe at the same temperature [194].

Figure 11.28a shows *ZT* of Al_xPbTe nanocomposites prepared by ball-milling and hot-pressing. Aluminum doped PbTe nanocomposites have Seebeck coefficients between −100 and −200 V/K and electrical conductivities of 3.6–18 × 10^4 S/m at room temperature [195]. The maximum figure-of-merit *ZT* of 1.2 is reached

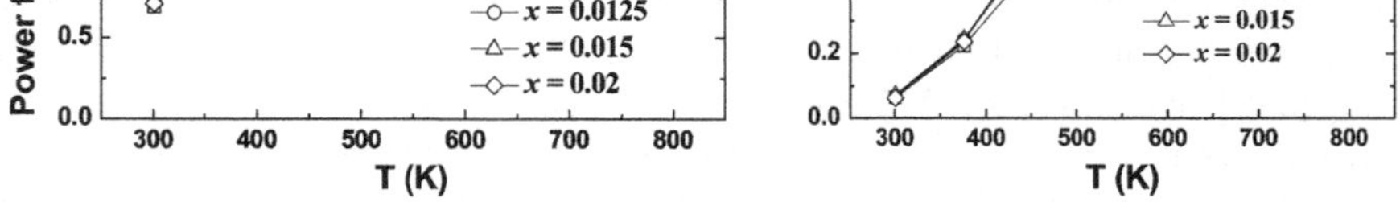

Fig. 11.27 Temperature dependence [193] of **a** power factor and **b** *ZT* for $K_xPb_{1-x}Te$ (x = 0.01, 0.0125, 0.015, 0.02)

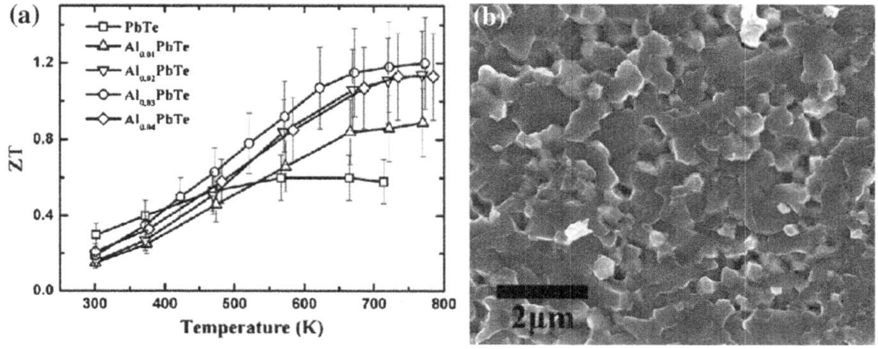

Fig. 11.28 **a** Temperature dependence of *ZT* and **b** SEM image of Al_xPbTe nanocomposites ($x = 0, 0.01, 0.02, 0.03, 0.04$) [195]

at 770 K in the $Al_{0.03}PbTe$ nanocomposites. Figure 11.28b shows an SEM image of a fractured surface Al-doped PbTe nanocomposites.

Figure 11.29 shows the temperature dependence of the thermoelectric properties of Cr-doped PbTe nanocomposites at various Cr concentrations. When Cr is doped into PbTe, the room temperature power factor increased dramatically compared with the other *n*-type PbTe alloys [197]. The thermal conductivity increased to 2.6 W/m/K at room temperature and 1.3 W/m/K at 773 K. The calculated *ZT* is 0.45 at room temperature and 0.8 at 600–773 K.

Figure 11.30 shows the temperature dependence of (a) the Seebeck coefficient, (b) the electrical conductivity, (c) the power factor, (d) the thermal diffusivity, (e) the total thermal conductivity and the lattice thermal conductivity, and (f) the *ZT* of $Cr_{0.01}Pb_{0.99}Te_{1-y}Se_y$ ($y = 0.25, 0.5, 0.75, 0.85$). The electrical conductivity of all the nanocomposites decreased with increasing temperature, consistent with the attributes of degenerate semiconductors. At fixed concentration of 1 atomic% Cr doping, the room-temperature Seebeck coefficient decreased from 211 mV/K to 157 mV/K with increasing Se concentration. The Seebeck coefficient of all the samples decreased at higher temperatures showing a bipolar transport nature and it was found that the temperature at which the bipolar effect becomes important for Te-rich samples is lower than that of Se-rich samples. The electrical conductivity increased when the Se concentration increased to 75 atomic% and then decreased above this concentration showing a higher value at an optimum Se concentration of 75 atomic %. The high electrical conductivity manifested by the Se-rich sample $Cr_{0.01}Pb_{0.99}Te_{0.25}Se_{0.75}$ resulted in a higher power factor across the entire temperature range. This high electrical conductivity is attributed to the relatively higher carrier concentration ($8.12 \times 10^{18}\ cm^{-3}$) as evidenced by the room-temperature Hall measurement. The thermal conductivity is heavily decreased compared to Cr doped PbTe nanocomposites due to phonon scattering by the point defects resulting from alloying. A minimum thermal conductivity of 0.8 W/m/K was observed at 573 K in

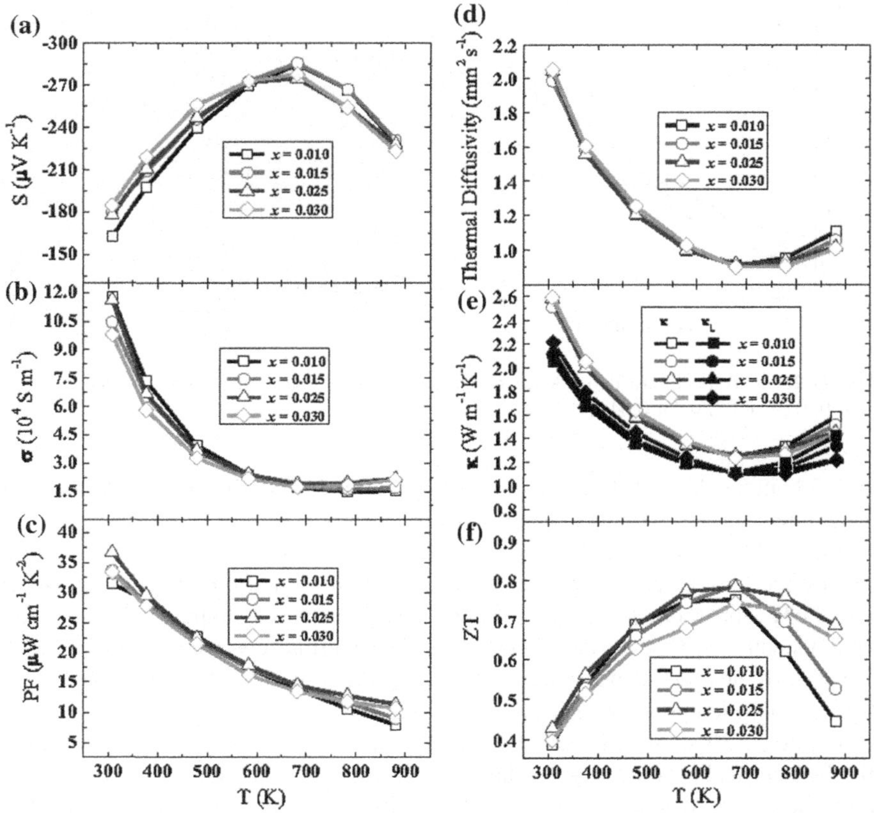

Fig. 11.29 Temperature-dependent thermoelectric properties of $Cr_xPb_{1-x}Te$ nanocomposites [196] at various dopant concentrations ($x = 0.01$, 0.015, 0.025, 0.03). **a** Seebeck coefficient, **b** electrical conductivity, **c** power factor, **d** thermal diffusivity, **e** total thermal conductivity and lattice thermal conductivity, (f) *ZT*

$Cr_{0.01}Pb_{0.99}Te_{0.5}Se_{0.5}$ with equal amount of Te and Se. This is approximately a 72 % decrease compared to PbTe for the same Cr doping level of 1 atomic%. This reduction is related to the maximum distortion of the crystal lattice due to the maximum entropic atomic mass fluctuation as confirmed by theoretical calculations and experimental studies. The increase in thermal conductivity at higher temperatures is due to the contribution of heat transport by minority carriers (holes). The highest *ZT* is 1 at approximately 573 K to 673 K in Se-rich $Cr_{0.01}Pb_{0.99}Te_{0.25}Se_{0.75}$.

The peak *ZT* temperature increased with increasing concentration of Se. The highest *ZT* of 0.6 at room temperature in Te-rich $Cr_{0.015}Pb_{0.985}Te_{0.75}Se_{0.25}$ was obtained due to a lowered thermal conductivity and enhanced power factor resulted from high Seebeck coefficient of about 220 mV/K and high Hall mobility

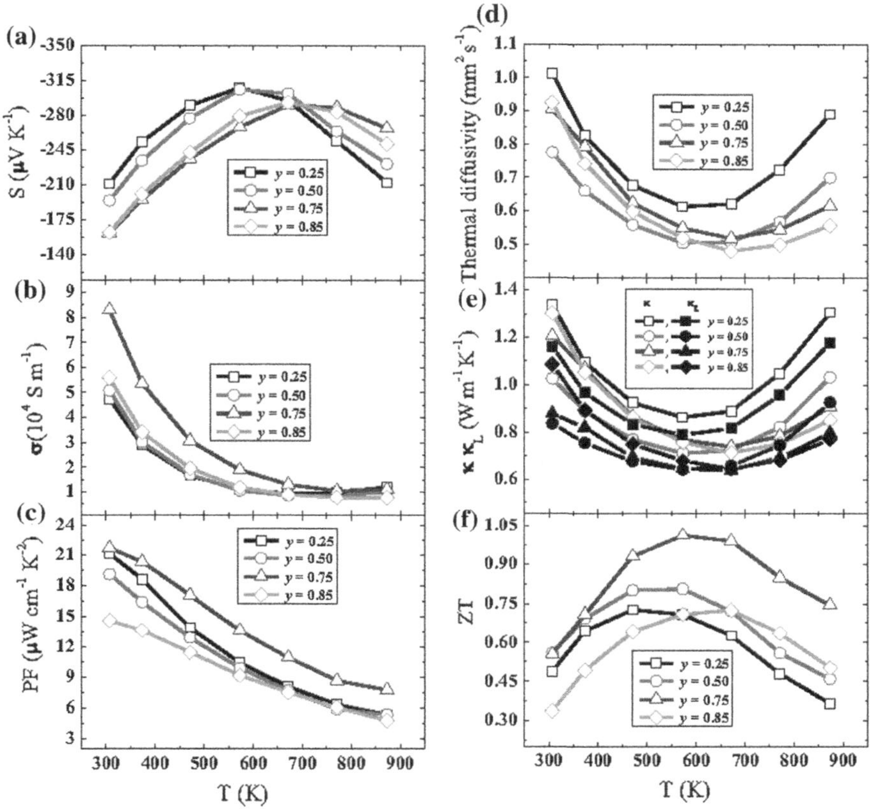

Fig. 11.30 Temperature-dependent thermoelectric properties of $Cr_{0.01}Pb_{0.99}Te_{1-y}Se_y$ (y = 0.25, 0.5, 0.75, 0.85) with fixed Cr concentration of 1 atomic% [196]. **a** Seebeck coefficient, **b** electrical conductivity, **c** power factor, **d** thermal diffusivity, **e** total thermal conductivity and lattice thermal conductivity, and **f** *ZT*

1120 $cm^2/V/s$ at room temperature. A room temperature *ZT* of 0.5 and peak *ZT* of 1 at about 573–673 K are achieved in Se-rich $Cr_{0.01}Pb_{0.99}Te_{0.25}Se_{0.75}$ nanocomposites.

11.4.4 PbSe Nanocomposites

Lead selenide is a high temperature thermoelectric material with a direct bandgap of 0.27 eV at room temperature. It's predicted that *p*-type PbSe could be competitive with its sister compound, lead telluride [198]. Several groups have reported *ZT* = 1 for doped PbSe nanocomposites [193, 196, 199–201].

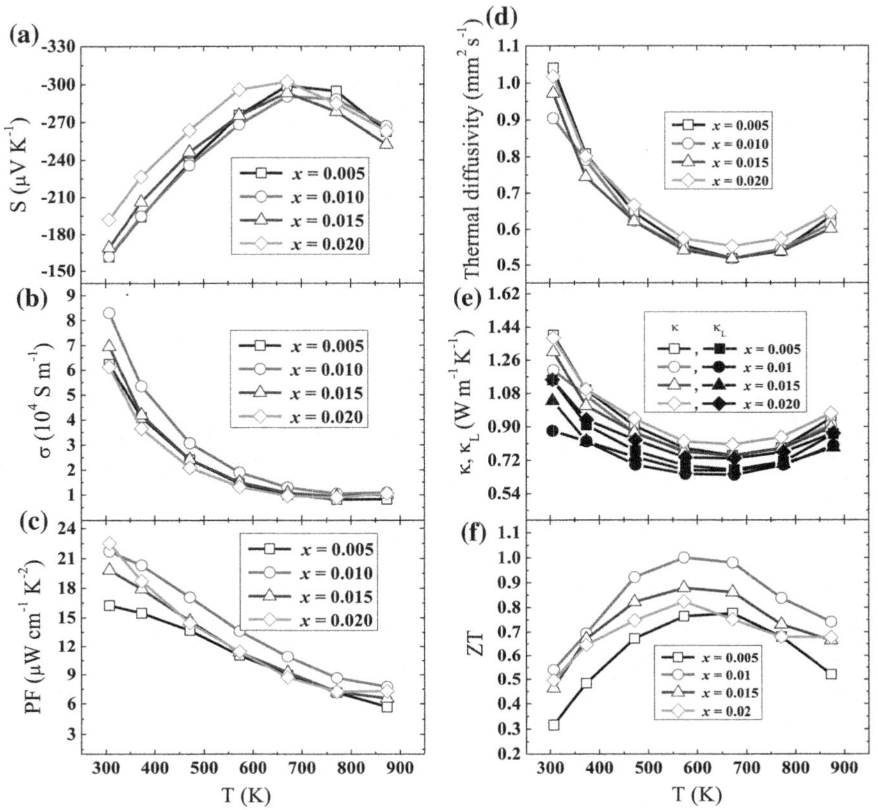

Fig. 11.31 Temperature-dependent thermoelectric properties of $Cr_xPb_{1-x}Te_{0.25}Se_{0.75}$ (x = 0.005, 0.01, 0.015, 0.02) [196]. **a** Seebeck coefficient, **b** electrical conductivity, **c** power factor, **d** thermal diffusivity, **e** total thermal conductivity and lattice thermal conductivity, and **f** *ZT*

Figure 11.31 shows the temperature-dependent thermoelectric properties of $Cr_xPb_{1-x}Te_{0.25}Se_{0.75}$ (x = 0.005; 0.01; 0.015; 0.02). The Seebeck coefficient increases slightly with increasing Cr concentration. The electrical conductivity first increases when the Cr concentration increased to a critical value of 1.5 atomic% then decreased when it exceeds this critical level. The lowest room temperature thermal conductivity is 1.25 and 0.8 W/m/K at about 673 K in $Cr_{0.01}Pb_{0.99}Te_{0.25}Se_{0.75}$, giving rise to a room temperature *ZT* of 0.5 and the highest peak *ZT* of 1 at approximately 573 K to 673 K.

Figure 11.32 shows electrical conductivity, Seebeck coefficient, power factor, thermal diffusivity, specific heat, and thermal conductivity of $K_xPb_{1-x}Se$ with different potassium concentrations (x = 0, 0.005, 0.010, 0.0125, 0.015). It is clear that for $x \geq 0.01$, the electrical conductivity increased dramatically. The electrical conductivity decreased rapidly with increasing temperature. The Seebeck coefficient goes up all the way with increasing temperatures without any sign of the

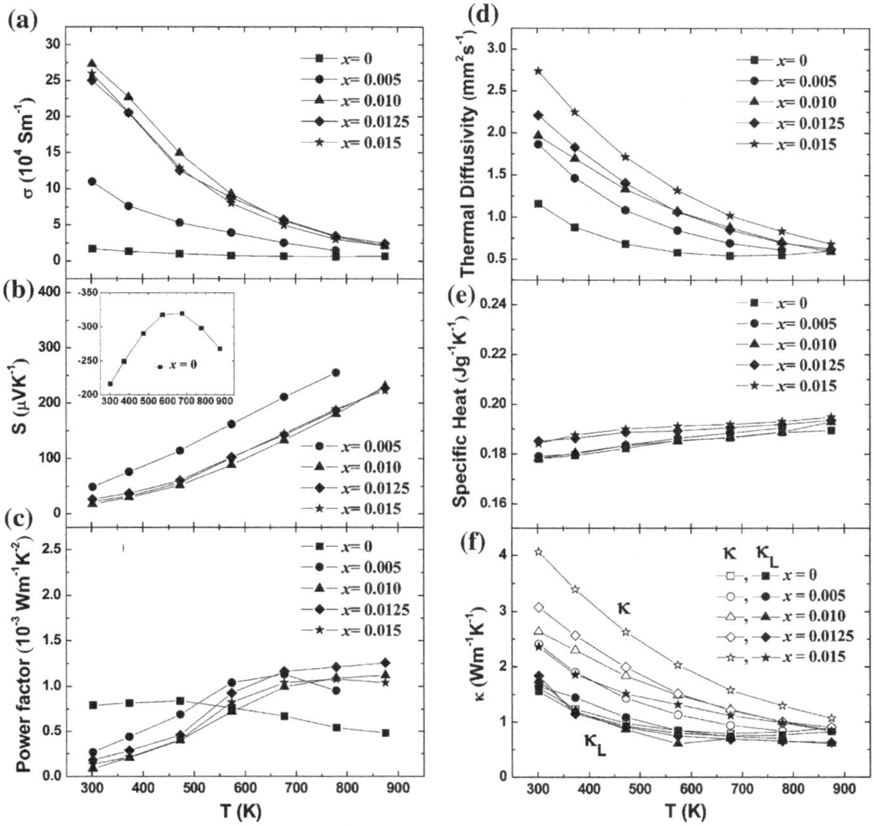

Fig. 11.32 Temperature dependences of **a** electrical conductivity, **b** Seebeck coefficient, **c** power factor, **d** thermal diffusivity, **e** specific heat, and **f** total and lattice thermal conductivities for $K_xPb_{1-x}Se$ (x = 0, 0.005, 0.010, 0.0125, 0.015) [193]

bipolar effect. Low lattice thermal conductivities of ~1.7 W/m/K at 300 K and ~0.6 W/m/K at high temperature are similar to the previously reported values [200]. The maximum *ZT* value reached was 1.2 at 873 K (Fig. 11.33).

11.4.5 Skutterudite Nanocomposites

Skutterudites have been widely studied for their promising thermoelectric properties [202, 203] and are regarded as potential candidates for next-generation thermoelectric materials for electrical power generation using either a solar source or waste heat. One of the remarkable features of this material is that the cage-like open structure can be filled with foreign atoms acting as phonon rattlers. The "rattling" of

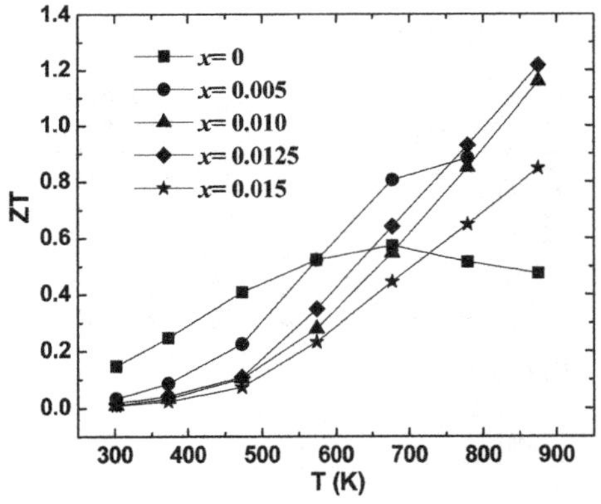

Fig. 11.33 Temperature dependence of *ZT* for $K_xPb_{1-x}Se$ (x = 0, 0.005, 0.010, 0.0125, 0.015) [193]

the filled atoms scatters phonons strongly and drastically reduces the thermal conductivity of the skutterudite compounds [202, 203].

The undoped $CoSb_3$ is *p*-type with a *ZT* of about 0.2. After doping, *ZT* increases to 0.52 at 600 K for *n*-type crystals [204]. A *ZT* of 0.7 without rattlers was achieved in $(Ni,Co)_4Sb_{12}$ nanocomposites produced by the ball-milling and hot pressing [93]. With various kinds of atoms (like Ce [205], La [206], Ca [207], Ba [208], Yb [122, 134, 171]) fill the cages, an improved *ZT* can be achieved. For example, Yb-filled *n*-type $Yb_{0.19}Co_4Sb_{12}$ with a peak *ZT* close to 1 at 373 °C [134], and $Yb_{0.15}Co_4Sb_{12}$ with *ZT* of about 0.7 at 400 °C [171] have been reported. The peak *ZT* of $Yb_xCo_4Sb_{12}$ nanocomposites made from the ball-milling and hot-pressing method was increased to 1.2 [46]. More kinds of doped Co_4Sb_{12} nanocomposites have been reported, such as $Sn_yFe_3Co_5Sb_{24}$ nanocomposites [135], rare-earth filled skutterudites Fe_4Sb_{12} nanocomposites [104], and $FeCo_3Sb_{12}$ nanocomposites [209].

11.4.5.1 *n*-Type $Yb_xCo_4Sb_{12}$ Nanocomposites

$Yb_xCo_4Sb_{12}$ nanopowders are first prepared by ball-milling from elements. Only a small portion of the ball-milled nanopowders are mechanically alloyed after high energy ball milling, regardless of the ball milling time. These same phenomena are also observed in other doped skutterudites, like $CoSb_3$ [65, 210], $(La,Ni,Co)_3Sb_{12}$ [136], $Fe_xCo_3Co_{12}$ [117, 210], $Co(Sb,Te)_3$ [211], and $FeCo_3Sb_{12}$ [209]. Similar phenomena are also observed in other systems, such as the $Ag_{0.8}Pb_{18+x}SbTe_{20}$ system [115]. This phenomenon is common in mechanical alloying. For some thermoelectric compounds like $FeSi_2$, the ball-milling process does not mechanically alloy Fe and Si elements at all [212]. However, the ball milling mixed the elements uniformly at nanometer level, favoring the chemical reaction during hot-pressing process.

After the DC hot pressing, however, the nanopowders are completely transformed into a single phase skutterudite. The average grain size is about 200–300 nm (Fig. 11.34). There are two types of nanograins in the skutterudites nancomposites:

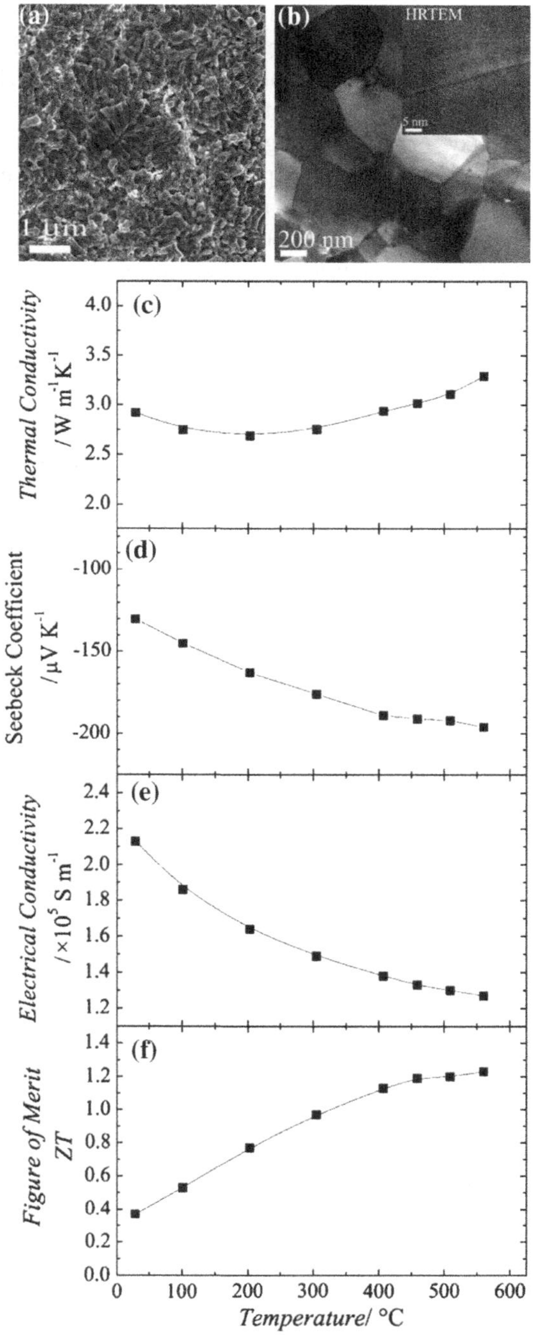

Fig. 11.34 **a** SEM image, **b** BF-TEM image, **c** thermal conductivity, **d** Seebeck coefficient, **e** electrical conductivity, **f** *ZT* of hot-pressed *n*-type skutterudite $Yb_{0.35}Co_4Sb_{12}$ nanocomposites [46]. *Inset* in (**b**) is an HRTEM image of a grain boundary

larger ones of about 1 μm in a bar shape, and smaller ones of about 0.2–0.3 μm in a spherical shape. The larger ones possibly grew from the mechanically alloyed particles while the smaller ones could form from un-alloyed powders [93]. Both types of nanograins are well crystallized with clear facets. The crystallized grains are closely packed, implying a high volume mass density. Archimedean volumetric mass density of the nanocomposites is of around 7.6 g/cm^3, which is 100 % of the theoretical density. All the nanograins in the nanocomposites are crystallized well and the large angle grain boundary is clean. The excellent crystallinity and clean grain boundaries are needed for achieving good electrical transport properties, whereas the large angle grain boundary can potentially benefit phonon scattering.

The thermoelectric properties of $Yb_{0.35}Co_4Sb_{12}$ are compared with other reported properties of nanocomposites prepared by the ball-milling and hot-pressing method. The nanocomposites produced here show a much higher electrical conductivity than those reported [122, 134], presumably due to the large electron doping effect from the high Yb concentration. The thermal conductivity of the nanocomposites (2.7–3.3 W/m/K from room temperature to 600 °C) is almost identical to these prepared by spark plasma sintering (2.4–3.9 W/m/K). The low thermal conductivity of these *n*-type $Yb_xCo_4Sb_{12}$ nanocomposites should come from the stronger phonon scattering resulting from the presence of a large concentration of rattlers and an increased concentration of grain boundaries in the nanocomposites. Therefore, the nanocomposites have a higher *ZT* value than the doped Co_4Sb_{12} crystals without rattles ($ZT = 0.52$ at 600 K [204]) reference samples as a result of the enhanced power factor and reduced thermal conductivity of the nanocomposites.

ZT increases with temperature and reaches a maximum (1.2) at around 550 °C. This is one of the highest *ZT* values reported for *n*-type filled skutterudites [93, 122, 134, 171, 213].

The thermal conductivity decreases with Yb doping content in the nanocomposites [46]. $Yb_{0.35}Co_4Sb_{12}$ has an optimized lowest thermal conductivity with a minimum of 2.7 W/m/K, which leads to the highest observed *ZT* value.

n-Type skutterudites $CoSb_3$ nanocomposites are also reported using the ball milling and hot-pressing method [103, 117, 136, 179]. $ZT = 1.1$ can be achieved in *n*-type $Ba_{0.24}Co_4Sb_{12}$ nanocomposites prepared by ball-milling, followed cold pressing and sintering [208], $ZT = 1.3$ in $Yb_xCo_4Sb_{12+y}$ nanocomposites by spark plasma sintering method [75, 122], $ZT = 1.1$ in $CoSb_{3-x}Te_x$ nanocomposites by ball-milling and spark plasma sintering method [211] while $ZT = 0.41$ in $LaxFe_4Sb_{12}$ nanocomposites [104].

11.4.5.2 *p*-Type (La,Ce,Nd)$Fe_{4-x}Co_xSb_{12}$ Nanocomposites

Rare-earth filled and Fe substituted skutterudites are *p*-type. Many *p*-type skutterudite nanocomposites can be prepared by the ball-milling and hot-pressing methods [104, 137, 202, 209, 214–216]. The ball-milled nanoparticles are mixture of Sb, skutterudites, and (Co,Fe)Sb_2 compounds. After hot-pressing, nanocomposites with a single phase are obtained. The thermal conductivity decreases significantly because of the rattling effect of rare-earth into the voids and the substitution of Fe for Co. The thermal conductivity can be reduced down to 4 W/m/K at room temperature (one third of that of $CeSb_3$ samples) [204]. The lower thermal conductivity is mainly due to the stronger phonon scattering caused by the rattling effect of rare-earth element. Lattice thermal conductivity decreases with the rare-earth filling content.

$La_x(Ni,Co)_4Sb_{12}$ nanocomposites can also be produced by ball-milling and spark plasma sintering [136]. $ZT = 0.12$ at 400 °C.

11.4.6 MgAgSb Nanocomposites

α-MgAgSb is one kind of room temperature thermoelectric material [22, 218] with a space group of I $\bar{4}$ c2 [218, 219] below 610 K. It was first reported as thermoelectric materials in 2012, with $ZT = 0.35$ at room temperature. Pure phase of α-MgAgSb nanocomposites was achieved two years later through a two-step ball milling process. MgAg phase was first mechanically alloyed from elemental powders at room temperature. Then antimony was added to the MgAg phase and ball-milled. The final powders were hot-pressed under DC currents at 575 K for 8 min to produce α-MgAgSb nanocomposites. Figure 11.35a shows an SEM image of a freshly broken surface. The grain size is 100–200 nm with fairly good uniformity and appears densely packed, consistent with its high relative density. Low magnification TEM (Fig. 11.35b), especially SAED (inset of Fig. 11.35b), shows that the grains are polycrystalline, consisting of sub-grains with crystalline size of several nanometers. HRTEM directly confirms the crystalline sub-grains, as shown in Fig. 11.35c. The very small grain size has played significantly important roles in the low thermal conductivity of the nanocomposites.

Figure 11.35d shows an HRTEM image along [0 1 $\bar{1}$] zone axis. Atom arrays of α-MgAgSb along [0 1 $\bar{1}$] orientation is shown, matching well with the HRTEM image. High-angle annular dark-field (HAADF) further indicates the local disorder of atoms in α-MgAgSb compounds, as shown in Fig. 11.35e. Figure 11.35f shows the corresponding intensity line scanning along the blue line marked in Fig. 11.35e. The intensity peaks in the linescanning should correspond to the Ag and Sb atomic columns. There is a valley in the linescanning, circled in Fig. 11.35f, corresponding to Mg atomic column. The valley is higher than the next nearest one, which indicates that other heavy atoms may move to Mg site.

Figure 11.36 shows excellent thermoelectric properties of MgAgSb nanocomposites and $MgAg_{0.97}Sb_{0.99}$ nanocomposites. Thermoelectric properties of the nanocomposites are strongly affected by Sb content in the nanocomposites. Figure 11.36a clearly shows that $MgAg_{0.97}Sb_{0.99}$ has lower thermal conductivity over the whole temperature range than MgAgSb. This can be ascribed to the strong phonon scattering due to the deficiencies at both the Ag and Sb sites and the very

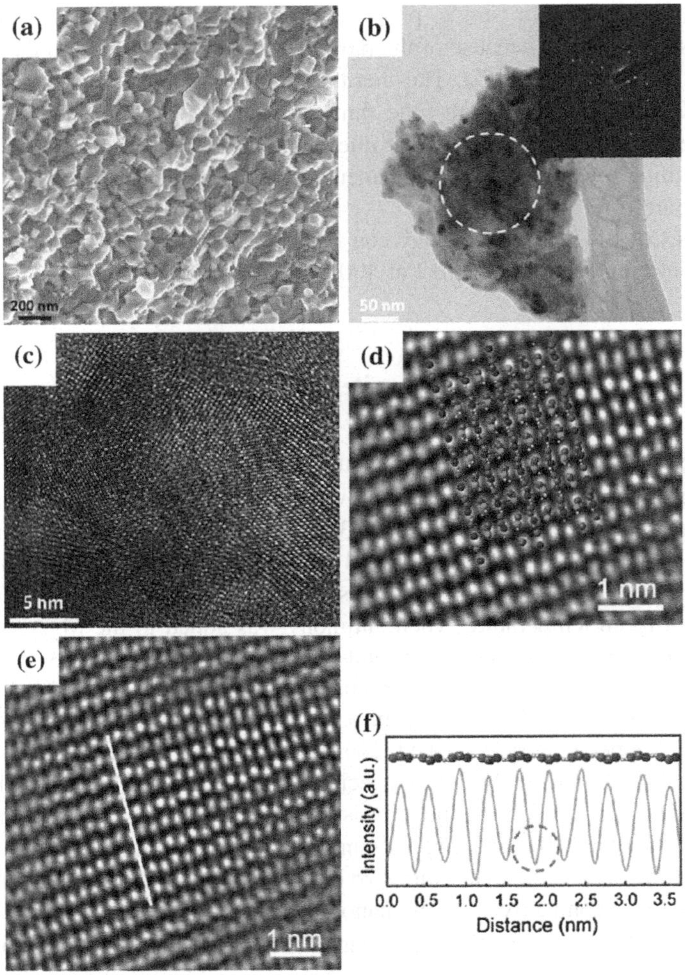

Fig. 11.35 Microstructures of MgAgSb nanocomposites. **a** SEM [22]. **b** TEM [22]. **c** HRTEM [22] and **d** structure model [217]. **e** HAADF image and **f** corresponding intensity line scanning along the *blue line* in (**e**) [217]. SAED was taken from the *circled* region

small grain size. Electric conductivity is also affected, as shown in Fig. 11.36b. The maximum Seebeck coefficient is enhanced to 2.40×10^{-4} V/K from 2.0×10^{-4} V/K (Fig. 11.36c). The power factor is enhanced around 20 % [220] over the whole temperature range when the Sb content is slightly increased from 0.99 to 1.00. Finally the deficiency of Sb enhances *ZT* peak to 1.2 from 0.8 (Fig. 11.36d).

Thermoelectric properties of MgAgSb are significantly affected by doping. Besides Sb deficiency, Ni alloying into Ag site clearly increased the electrical resistivity and the Seebeck coefficient while it decreased the thermal conductivity. A peak *ZT* of 1.4 was obtained at 450 K in $MgAg_{0.965}Ni_{0.005}Ag_{0.99}$ nanocomposites

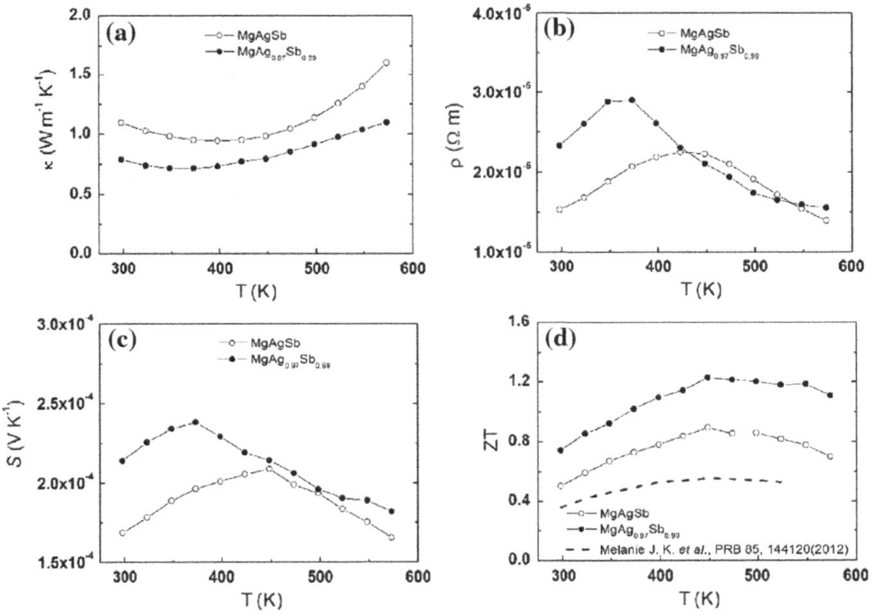

Fig. 11.36 Thermoelectric properties of MgAgSb nanocomposites [22]. **a** Thermal conductivity, **b** electrical conductivity, **c** Seebeck coefficient, and **d** *ZT*

[22]. $ZT = 1.1$ was achieved at 525 K in In-doped α-MgAgSb [221] with thermal conductivity of 1 W/m/K.

The Cu substitution of Ag in $MgAg_{0.97-x}Cu_xSb_{0.99}$ can enhance the power factor and decrease thermal conductivity [91]. The *ZT* value at room temperature is close to 1 and increases with temperature to a maximum of 1.32 at 250 °C for $MgAg_{0.963}Cu_{0.007}Sb_{0.99}$.

It is believed that α-MgAgSb is a great promising thermoelectric material for room temperature power generation because of its abundant constituent elements. It is comparable to Bi_2Te_3 based alloys, which is the best room temperature TE material.

11.4.7 $YbAgCu_4$ Nanocomposites

$YbAgCu_4$, discovered in 1987 [222], is one kind of heavy Fermion material with moderate effective mass. The compound possesses a giant Seebeck peak at 45 K which is due to Kondo scattering of conduction electrons [223], low thermal conductivity below 50 K [224], high power factor of 235 W/cm/K^2 [225], good electrical conductivity [226], and the Seebeck coefficient [223].

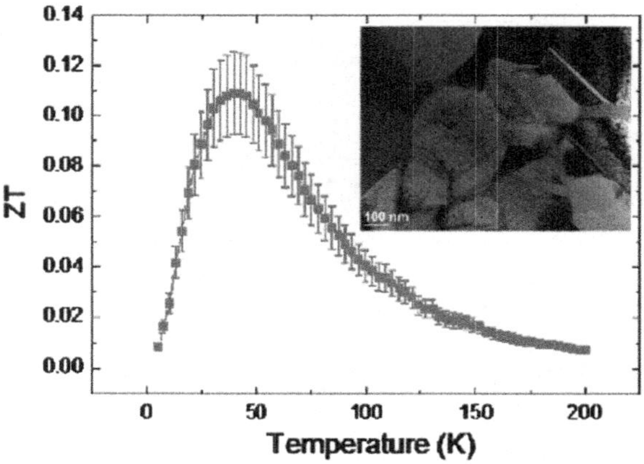

Fig. 11.37 *ZT* of YbAgCu4 nanocomposites hot pressed at 750 °C [54]. *Inset:* TEM image

$YbAgCu_4$ nanocomposites have been produced by ball milling the arc-melted ingots into nanopowders and hot pressing the nanopowders [54]. A high power factor of 131 W/cm/K^2 was obtained at 22 K. Thermal conductivity was reduced 30–50 % at 42 K through boundary scattering. Figure 11.37 shows *ZT* of $YbAgCu_4$ nanocomposites prepared by ball-milling/hot-pressing. A peak dimensionless thermoelectric figure-of-merit, *ZT*, of 0.11 was achieved at 42 K, which may find potential applications for cryogenic cooling below 77 K.

11.4.8 Other Kinds of Bottom-Up-ed Nanocomposites

Besides the previously discussed TE materials (*n*-type and *p*-type Bi_2Te_3 nanocomposites, *n*-type and *p*-type SiGe nanocomposites, *n*-type and *p*-type skutterudite Co_4Sb_{12} nanocomposites, *p*-type PbTe nanocomposites etc.), the bottom-up procedures have been widely employed to fabricate various thermoelectric nanocomposites, such as *n*-type $Bi_{85}Sb_{14}Ag$ nanocomposites [227] (power factor of 2.98×10^{-3} W/m/K^2 at 255 K, about three times of $Bi_{85}Sb_{15}$ single crystals), $Bi_{85}Sb_{13}Pr_2$ nanocomposites [228] (power factor of 3.8×10^{-3} W/m/K^2 at 235 K, about four times of $Bi_{85}Sb_{15}$ single crystals) by the ball-milling and hot-pressing methods, $Ga_mSb_nTe_{1.5+m+n}$ nanocomposites [87] (peak $ZT = 0.98$, grain size: <30 nm), $InAg_xPb_{18-x}SbTe_{20}$ nanocomposites [115] (thermal conductivity: 40 % reduced, $ZT = 1.3$ at 300 °C) by the ball-milling and spark plasma sintering methods, Zn_4Sb_3 nanocomposites [229] by cold-pressing, $Bi_{100-x}Sb_x$ nanocomposites [77] (thermoelectric conductivity is reduced by 75 %) by the ball milling and extrusion methods, Bi_2S_3 nanocomposites [53] by hot-pressing, Zn_4Sb_3 [51] and In_4Se_3 [49] nanocomposites, $AgPb_{18}SbSe_{20}$ nanocomposites [50] and SnTe nanocomposites [52] by the ball-milling and hot pressing method, etc.

In all of these thermoelectric nanocomposites, the thermal conductivity is usually decreased because of increased phonon scattering. *ZT*s are enhanced 10–20 % or even several times. More detailed information can be found in literatures [25, 113, 230].

The bottom-up procedures are also employed to produce multi-phase TE nanocomposites. It is postulated that the addition of second phase nanoparticles will work as phonon scattering centers to decrease thermal conductivity efficiently, like (1) ZnO_2 and rare-earth oxide was added into β-$FeSi_2$ nanocomposites [231] using bottom-up methods; (2) Y_2O_3 nanoparticles were dispersed in $FeSi_2$ nanocomposites by ball-milling and hot-pressing. *ZT* of *n*-type β-$FeSi_2$ nanocomposites was doubled from 0.35 to 0.63 when 2 wt% Y_2O_3 was added [232]. Thermal conductivity also decreased to 0.7 from 4.5 W/m/K; (3) TiB_2 was dispersed into β-FeSi nanocomposite using the ball-milling and hot-pressing methods [138]. Both thermal conductivity and electrical conductivity increase with increasing TiB_2 content (0–30 vol%) while power factor decreases. *ZT* decreases with TiB_2 addition; (4) TiB_2 was dispersed into B_4C nanocomposites to produce two-phase nanocomposites using the ball-milling and hot-pressing method [233, 234]. B_4C works as a thermoelectric host and TiB_2 is added to increase electrical conductivity. Both thermal conductivity and electrical conductivity decreases with TiB_2 content from 0 to 25 vol%. *ZT* decreases after addition; (5) BN and WO_3 were dispersed into *p*-type $(Bi, Sb)_2Te_3$ nanocomposites using the ball-milling and hot-press method [235]. Thermal conductivity decreases with increasing volume fraction of BN and WO_3 while the lattice thermal conductivity is constant. *ZT* increases with BN and WO_3 content at low temperature while it decreases at high temperature because of increasing electrical conductivity. In all of these nanocomposites with additions, the thermal conductivity indeed decreases because of the enhanced phonon scattering and decreases with increasing the volume fraction of a second or third phase. However, the addition of the second or third phase also affects electrical conductivity and the Seebeck coefficient. *ZT* usually decreases because of a great decreasing of electrical conductivity or the Seebeck coefficient.

11.5 Thermoelectric Devices of Nanocomposites

Compared with crystalline TE bulks, thermoelectric properties of TE nanocomposites are isotropic [25, 80] because of random orientation of nanograins. In order to confirm the isotropic property of hot-pressed TE nanocomposites, a TE nanocomposite bulk has been cut both along and perpendicular to the cold-pressing direction. Thermal conductivity, power factor and *ZT* are measured along these directions. It is found that TE properties are in agreement within 5 % [43] and the *ZTs* are the same. The isotopic properties would benefit device fabrication.

Bottom-up-ed TE nanocomposite have comparable mechanical properties with ingots. Table 11.5 lists hardness and Young's modulus of four kinds (*p*-type and *n*-type) of major thermoelectric nanocomposites. Doping can slightly affect mechanical properties. For example, hardness of *p*-type skutterudite $Ce_{0.45}Nd_{0.45}Fe_{3.5}Co_{0.5}Sb_{12}$ is 5.6

Table 11.5 Hardness and Young's modulus of TE nanocomposites [236]

Composition	Hardness (GPa)	Modulus (GPa)
p-type half-Heusler	12.8	221.0
n-type half-Heusler	9.1	186.5
p-type $Bi_{0.4}$ $Sb_{1.6}Te_3$	1.1	41.5
n-type $Bi_2Te_{2.7}Se_{0.3}$	1.2	38.8
p-type skutterudite	5.6	129.7
n-type skutterudite	5.8	136.9
n-type $Si_{0.8}Ge_{0.2}P_2$	10.8	166.3
p-type $Si_{0.8}Ge_{0.2}B_5$	10.7	155.6

p-type half-Heusler: $Hf_{0.44}Zr_{0.44}Ti_{0.12}$ $CoSb_{0.8}$ $Sn_{0.2}$
n-type half-Heusler: $Hf_{0.25}Zr_{0.75}NiSn_{0.99}Sb_{0.01}$
p-type skutterudite: $Ce_{0.45}Nd_{0.45}Fe_{3.5}Co_{0.5}Sb_{12}$
n-type skutterudite: $Yb_{0.35}Co_4Sb_{12}$

GPa while that of *n*-type skutterudite $Yb_{0.35}Co_4Sb_{12}$ is 5.8 GPa [236]. Hardness of mischmetal filled *p*-type skutterudites $(La_{0.25}Ce_{0.5}Pr_{0.05}Nd_{0.15}Fe_{0.03})_{0.9}$-$Fe_{3.1}Co_{0.9}Sb_{12}$ is 4.2 GPa [98]. Young's moduli of these three nanocomposites are 129.7 GPa, 136.9 GPa, and 116 GPa respectively.

Based on excellent physical properties of bottom-up-ed nanocomposites, various kinds of TE nanocomposites have been fabricated into TE devices.

11.5.1 Manufacturing of Thermoelectric Devices

TE nanocomposites are manufactured into devices as conventional TE materials. Bottom-up-ed nanocomposites are cut into legs (Fig. 11.38a). All legs are consisted of nanograins (Fig. 11.38b). *p*-type and *n*-type legs are arrayed thermally in parallel and electrically in series (Fig. 11.38c). The whole arrays are finally sealed as TE modules, as shown in Fig. 11.38d.

During device fabrications, *p*-type and *n*-type TE elements are soldered to electrically interconnect at high temperatures, around 300–500 °C. Interconnected materials usually diffuse into TE elements, degrading TE properties. Therefore a barrier layer, usually Ni, is deposited on TE elements before soldering. Electron microscopy indicates that interconnected materials can be blocked by Ni barrier layers. However, Ni barrier layer materials can diffuse into TE elements [237], as shown in Fig. 11.39. The rapid diffusion of barrier layers in TE elements could be an important source to degrade TE devices.

Additionally, there is a large contact resistance between traditional Ni barrier layers and TE elements (Fig. 11.40a). The high contact resistance should be related to the interface reaction of barrier materials and TE elements, wasting energy and lowering efficiency of TE devices. In order to decrease contact resistance and strengthen the bonding strength between interconnects and TE elements, a second

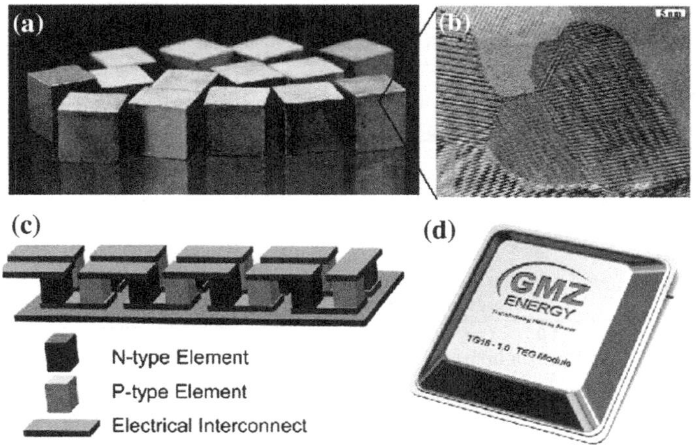

Fig. 11.38 Manufacturing of thermoelectric devices from nanocomposites. **a** Elements. **b** Microstructure of elements. **c** Structure of TE devices. **d** Module

Fig. 11.39 SEM image of a cross-sectional *n*-type $Bi_2(Te, Se)_3$ element [237]. Ni and Sn EDS line scan profiles show the element distribution across the Ni barrier layer. Inset: Ni concentration distribution in $Bi_2(Te,Se)_3$

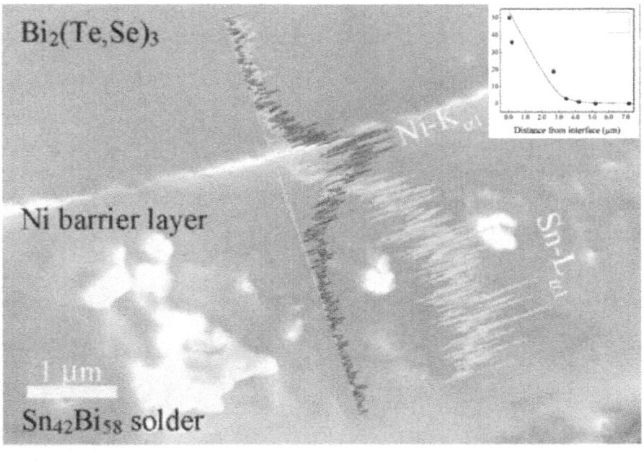

barrier layer is recently developed [238] between Ni barrier layers and TE elements. Figure 11.40b shows the contact resistances of new developed barrier layers. No voltage jump was seen in the scanning voltage probe, i.e., no significant contact resistance (<1 W cm^2). Furthermore, this new contact interface also shows enhanced bonding strength of 16 MPa as compared with that made by sputtering or electrodeposition methods (<10 MPa).

11.5.2 Thermoelectric Devices and Efficiency

Figure 11.41a shows a unicouple cooling device to test *p*-type Bi_2Te_3 nanocomposites. The unicouple used a nanostructured *p*-type Bi_2Te_3 nanocomposite as one

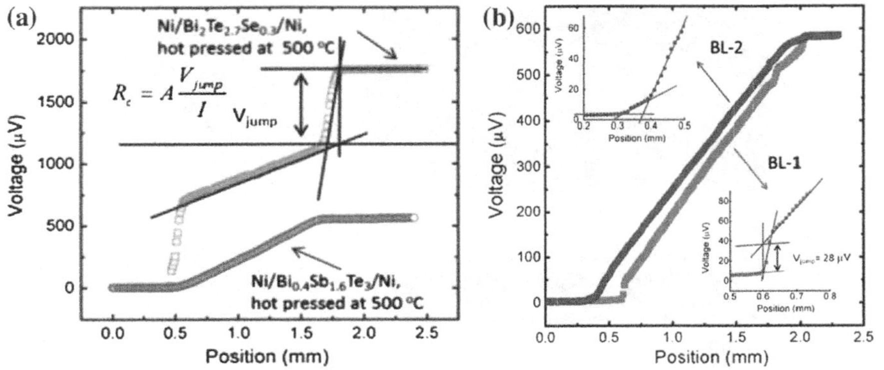

Fig. 11.40 **a** Contact resistance measurement for both n-type Ni/$Bi_2Te_{2.7}Se_{0.3}$ and p-type Ni/$Bi_{0.4}Sb_{1.6}Te_3$ [238]. The contact resistance was determined from the averaged voltage jump, where cross-section of the thermoelectric element is $A = 1.8 \times 1.8\ mm^2$ and the electrical current is $I = 0.1$ A. **b** Contact resistance measurement by the scanning probe for two n-type legs with barrier layers (BL) [238]: Ni/BL-1/$Bi_2Te_{2.7}Se_{0.3}$/BL-1/Ni and Ni/BL-2/$Bi_2Te_{2.7}Se_{0.3}$/BL-2/Ni, which is made by hot pressing at 425 °C. BL-1: 90 % $Bi_2Te_{2.7}Se_{0.3}$ + 10 % $NiSe_2$, BL-2: 1 % SbI_3 doped $Bi_2Te_{2.7}Se_{0.3}$

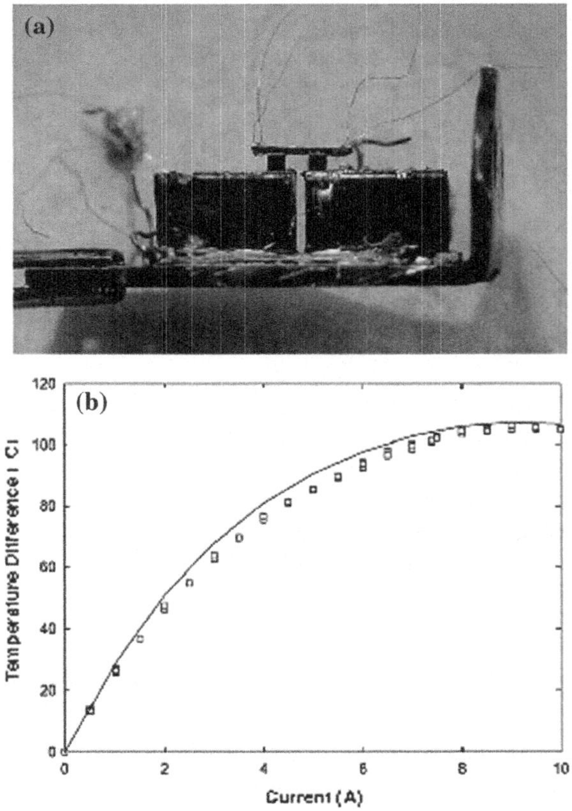

Fig. 11.41 Unicouple cooling devices [42]. **a** Experimental setup of the device cooling test. **b** Cooling test results with the hot side fixed at 100 °C

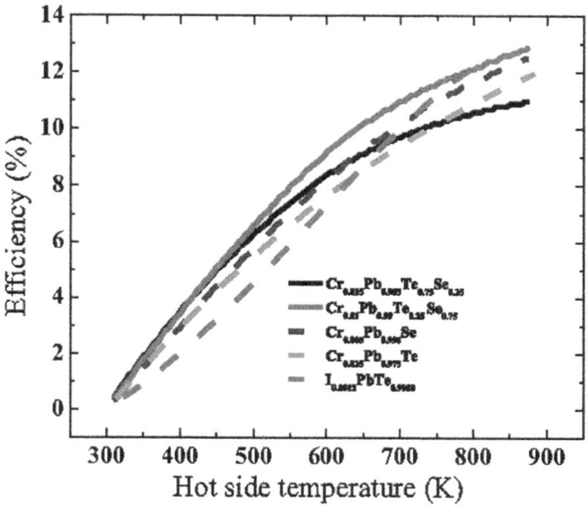

Fig. 11.42 Temperature dependence of the calculated leg efficiencies [196] of $Cr_{0.015}Pb_{0.985}Te_{0.75}Se_{0.25}$, $Cr_{0.01}Pb_{0.99}Te_{0.25}Se_{0.75}$, $Cr_{0.005}Pb_{0.995}Se$, $Cr_{0.025}Pb_{0.975}Te$, and $I_{0.0012}PbTe_{0.9988}$ with a cold side temperature at 300 K

leg and a commercially available *n*-type material as the other leg. Two thermoelectric legs were mounted onto 6.5 mm thick copper blocks and then bridged by a top copper plate. Two cold-side thermocouples were soldered into small holes drilled on both ends of the top copper plate. Hot-side thermocouples were soldered on the edge of each copper block. A temperature difference was created between the two legs when a DC current passed through the device. Figure 11.41b shows the temperature difference created by DC currents when the hot side is maintained at 100 °C.

Recently a record high thermoelectric conversion efficiency of 8.5 % has been experimentally demonstrated [239] with a single thermoelectric leg based on a recently reported *p*-type MgAgSb-based compound operating between 20 and 245 °C. The efficiency can exceed 10 % by increasing the hot side temperature to 295 °C [239].

Single leg efficiency of other TE nanocomposites has been examined over a wide range of temperatures. Figure 11.42 shows the calculated efficiency of $Cr_{0.015}Pb_{0.985}$–$Te_{0.75}Se_{0.25}$ and $Cr_{0.01}Pb_{0.99}Te_{0.25}Se_{0.75}$ nanocomposite single leg, in comparison with the efficiency of the best optimized Cr-doped PbSe ($Cr_{0.005}Pb_{0.995}Se$), Cr doped PbTe ($Cr_{0.025}Pb_{0.975}Te$), and I-doped PbTe ($I_{0.0012}PbTe_{0.9988}$). The leg efficiency is calculated accurately by discretization methods [240, 241]. An efficiency of 11 % and 13 % is obtained for the individual legs of $Cr_{0.015}Pb_{0.985}$-$Te_{0.75}Se_{0.25}$ and $Cr_{0.01}Pb_{0.99}Te_{0.25}Se_{0.75}$, respectively, when the cold side temperature is set at 300 K and hot side temperature at 873 K.

In spite of having high *ZT* exceeding 1.4 near 700 K for $I_{0.0012}PbTe_{0.9988}$, the leg efficiency is still lower than that of $Cr_{0.01}Pb_{0.99}Te_{0.25}Se_{0.75}$, especially below 700 K. This result is also comparable with the efficiency of other thermoelectric materials with a cold side temperature of 300 K and a hot side temperature of

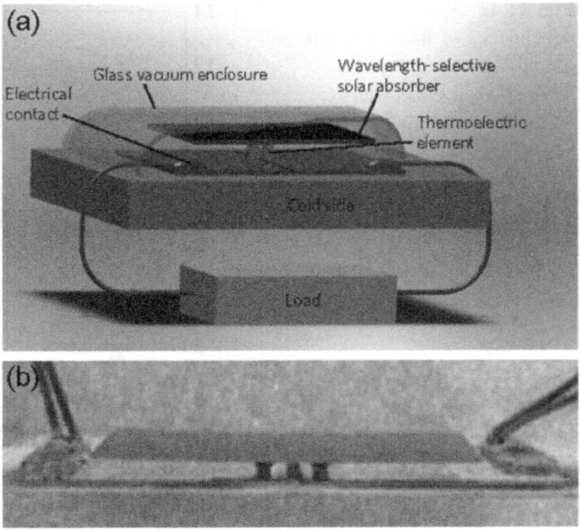

Fig. 11.43 Flat-panel solar thermoelectric generators [247]. **a** Illustration of a STEG cell. **b** Photograph of the device

773 K, such as Bi_2Te_3-Bi_2Se_3-Bi_2S_3(12.5 %) [242], PbSe: Al (9.4 %) [242, 243], half Heuslers (8.4 %) [242, 244], filled Skutterrudites (13.1 %) [243, 245], and PbTe:La (6.7 %) [242, 246].

TE nanocomposites have been fabricated into devices. Figure 11.43a illustrates a flat-panel solar thermoelectric generator (STEG) cell made of a pair of *p*- and *n*-type thermoelectric elements made of Bi_2Te_3 nanocomposites, a flat-panel selective absorber that also acts as a thermal concentrator, and two bottom electrodes that serve as heat spreaders and radiation shields. The device is surrounded by a glass enclosure maintaining an evacuated environment. The thermoelectric elements are electrically connected in series and sandwiched between the solar absorber plate and a heat sink, as shown in Fig. 11.43b. The dimensions of a typical *p*-type thermoelectric element are $1.35 \times 1.35 \times 1.65\ mm^3$ with an uncertainty of 0.01 mm, and the *n*-type elements are of similar dimensions. Each thermoelectric element is soldered onto an individual copper plate, which represents the cold side for the thermoelectric element. The copper substrate of the solar absorber serves as a heat concentrator by conducting heat laterally to the thermoelectric elements whereas the bottom copper plates serve as electrodes, heat spreaders, and radiation shields that reduce radiation losses from the rear side of the selective absorber.

The developed solar thermoelectric generators achieved a peak efficiency of 4.6 % under AM1.5G (1 kW/m^2) conditions. The efficiency is 7–8 times higher than the previously reported best value for a flat-panel STEG, and is enabled by the use of high-performance nanostructured thermoelectric materials and spectrally selective solar absorbers in an innovative design that exploits high thermal concentration in an evacuated environment.

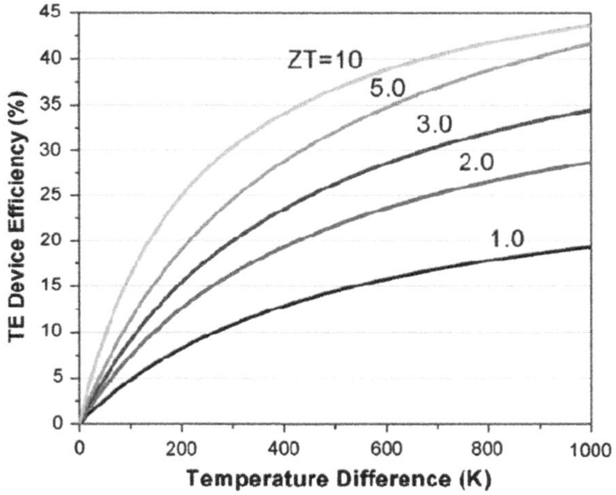

Fig. 11.44 Thermoelectric device efficiency [230]

11.5.3 Engineering Efficiency of Thermoelectric Devices

The efficiency of heat-to-electricity conversion based on the thermoelectric effect depends on the materials' *ZT*. The efficiency can usually be calculated from (11.2) and (11.3). Figure 11.44 plots device efficiency dependence at different *ZT*. Unfortunately, TE device efficiency is lower than these of solar devices, nuclear devices, coal, and even geothermal devices [248].

Recently an engineering dimensionless figure of merit $(ZT)_{eng}$ was developed [249] to accurately calculate the practical conversion efficiency of TE devices, overcoming the reporting of unrealistic efficiency using average *ZT* values. The device efficiency

$$\eta_{\max} = \eta_c \tag{11.6}$$

where η_c is Carnot efficiency. $(ZT)_{\mathrm{eng}}$ is the engineering dimensionless figure of merit defined as

$$(ZT)_{eng} = Z_{eng}\Delta T = \frac{\left(\int_{T_\in}^{T_\varepsilon} S(T)dT\right)^2}{\int_{T_\in}^{T_\varepsilon} \rho(T)dT \int_{T_\in}^{T_\varepsilon} \kappa(T)dT}\Delta T \tag{11.7}$$

Z_{eng} is the engineering figure of merit in K^{-1}, and $\hat{\alpha}$ is a dimensionless intensity factor of the Thomson effect defined as $\hat{\alpha} = S(T_h)\Delta T / \int_{T_c}^{T_h} S(T)dT$ where $S(T_h)$ is the Seebeck coefficient at the hot-side temperature T_h. The degree of Thomson effect can be identified analytically in the derived equation by $\hat{\alpha}$ enabling one to predict the relative difference of efficiency compared with numerical simulations, even though the formula does not consider the Thomson effect on heat flux evaluation. The ΔT-dependent weight factors lead to practical contributions of Joule

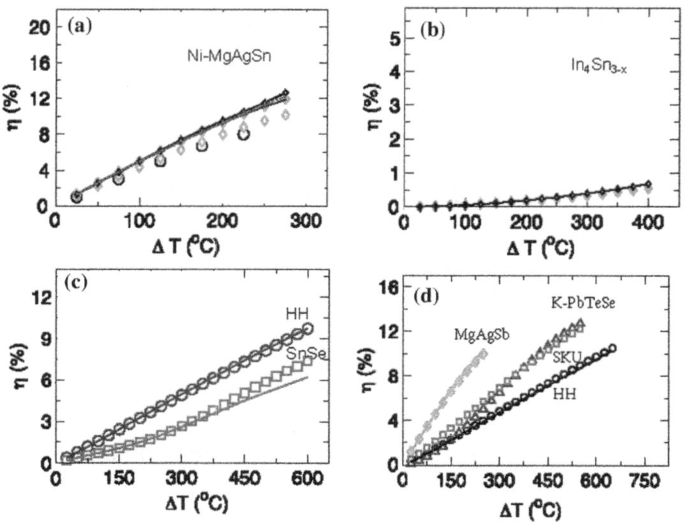

Fig. 11.45 Calculated device efficiency [249] of **a** Ni-MgAgSn at $T_c = 25$ °C, **b** In_4Se_{3-x} at $T_c = 25$ °C, **c** SnSe and HH at $T_c = 100$ °C, and **d** MgAgSb, K-PbTeSe, SKU, and HH at $T_c = 50$ °C. The cold-side temperature T_c is fixed

and Thomson heat to the heat flux at hot side at a given temperature difference rather than the concept of returning half of them each to the hot and cold sides regardless of boundary temperatures.

Figure 11.45 plots the ΔT-dependent maximum efficiency of TE devices based on (11.6), covering Ni-MgAgSn (Fig. 11.45a), In_4Se_{3-x} (Fig. 11.45b), SnSe and half-Heuslers (HH) (Fig. 11.45c), MgAgSb, K-doped PbTeSe (K-PbTeSe), skutterudites (SKU), and half-Heuslers (HH) (Fig. 11.45d). The calculated efficiency should be more reliable.

11.6 Conclusions and Outlooks

Thermoelectric nanocomposites with high *ZT* have been successfully produced by nanostructuring approaches. TE nanoparticles are first produced by ball-milling from ingots through mechanically grinding or from elements through mechanically alloying. The ball-milled nanoparticles are then bottom-up-ed by hot-pressing or SPS methods to produce nanocomposites. The produced nanocomposites consist of nanograins with structural defects. The unique nanostructures of the nanocomposites reduce the thermal conductivity by increasing phonon scattering from the high density of imperfections in the nanocomposites. Simultaneously, the high electrical conductivity and power factor of the nanocomposites are preserved while the phonon thermal conductivity is reduced significantly, resulting in a high *ZT*. In some cases, electrical conductivity and the Seebeck coefficient are increased [52] while the thermal conductivity is decreased, greatly benefiting the enhancement of *ZT*.

These nanocomposites with enhanced *ZT* have been fabricated into TE devices, having good efficiency. The nanocomposite devices will have significant commercial potential for automotive, industrial waste heat recovery, space power generation, and solar power conversion applications.

Further investigations are carried out on higher *ZT* enhancement of TE nanocomposites. A higher *ZT* of 2 can be achieved in these nanocomposites when the thermal conductivity is reduced to below 1 W/m/K and power factor is improved to a resonant level. Future thermoelectric devices fabricated from these nanocomposites will be widely used in refrigeration, air-conditioning, in large scale and high-power commercial generators, to establish a reliable, efficient, friendly, and green/clean energy supply/consumption in the global marketplace and to protect our planet's environment.

Acknowledgments The authors YL and ZFR thank the financial support by the Defense Threat Reduction Agency under Grant HDTRA122221. YL acknowledges the support from US Department of Energy under the Contract DE-NA0000720 through the program S. P. Massie Chair of Excellence and Faculty Development Funding from Morgan State University. YL is also thankful to Ms. Kara A. Green and Rola Derbeshi for their respective valuable helps on references and proofreading of the book chapter.

References

1. T.J. Seebeck, Magnetische polarisation der metalle und erze durch temperatur-differenz (magnetic polarization of metals and minerals by temperature differences), *Abhandlungen der Königlichen Akademie der Wissenschaften zu Berlin* (Treatises of the Royal Academy of Sciences in Berlin) **1825**, 265–373
2. T.J. Seebeck, Ueber die magnetische polarisation der metalle und erze durch temperatur-differenz (on the magnetic polarization of metals and minerals by temperature differences). Annalen der Physik und Chemie **6**, 133–160 (1826)
3. J.C.A. Peltier, Nouvelles expériences sur la caloricité des courants électrique (new experiments on the heat effects of electric currents). Annales de Chimie et de Physique **56**, 371–386 (1834)
4. H.J. Goldsmid, *Thermoelectric Refrigeration*. The International Cryogenics Monograph Series (Plenum Press, New York, 1964)
5. D.D. Pollock, *Thermocouples: Theory and Properties* (CRC Press, Boca Raton, 1991)
6. D.M. Rowe (ed.), *CRC Handbook of Thermoelectrics* (CRC Press, Boca Raton, 1995)
7. T.M. Tritt (ed.), *Recent Trends in Thermoelectric Materials Research III*, vol. 71 (Academic Press, San Diego, 2001)
8. F.J. DiSalvo, Thermoelectric cooling and power generation. Science **285**, 703–706 (1999)
9. L.E. Bell, Cooling, heating, generating power, and recovering waste heat with thermoelectric systems. Science **321**, 1457–1461 (2008)
10. J. Yang, T. Caillat, Thermoelectric materials for space and automotive power generation. MRS Bull. **31**, 224–229 (2006)
11. B.C. Sales, Smaller is cooler. Science **295**, 1248–1249 (2002)

12. A.F. Ioffe, *Physics of Semiconductors* (Academic Press, New York, 1960)
13. K. Biswas, J. He, I.D. Blum, C.-I. Wu, T.P. Hogan, D.N. Seidman, V.P. Dravid, M.G. Kanatzidis, High-performance bulk thermoelectrics with all-scale hierarchical architectures. Nature **489**, 414–418 (2012)
14. C. Wood, Materials for thermoelectric energy conversion. Rep. Prog. Phys. **51**, 459 (1988)
15. T.M. Tritt, M.A. Subramanian, Thermoelectric materials, phenomena, and applications: a bird's eye view. MRS Bull. **31**, 188–198 (2006)
16. D.-Y. Chung, T. Hogan, P. Brazis, M. Rocci-Lane, C. Kannewurf, M. Bastea, C. Uher, M.G. Kanatzidis, $CsBi_4Te_6$: a high-performance thermoelectric material for low-temperature applications. Science **287**, 1024–1027 (2000)
17. B. Wölfing, C. Kloc, J. Teubner, E. Bucher, High performance thermoelectric Tl_9BiTe_6 with an extremely low thermal conductivity. Phys. Rev. Lett. **86**, 4350–4353 (2001)
18. V.L. Kuznetsov, L.A. Kuznetsova, A.E. Kaliazin, D.M. Rowe, Preparation and thermoelectric properties of $A_8^{II}B_{16}^{III}B_{30}^{IV}$ clathrate compounds. J. Appl. Phys. **87**, 7871–7875 (2000)
19. J.L. Cohn, G.S. Nolas, V. Fessatidis, T.H. Metcalf, G.A. Slack, Glasslike heat conduction in high-mobility crystalline semiconductors. Phys. Rev. Lett. **82**, 779–782 (1999)
20. K.F. Hsu, S. Loo, F. Guo, W. Chen, J.S. Dyck, C. Uher, T. Hogan, E.K. Polychroniadis, M. G. Kanatzidis, Cubic $AgPb_mSbTe_{2+m}$: bulk thermoelectric materials with high figure of merit. Science **303**, 818–821 (2004)
21. J.-S. Rhyee, K.H. Lee, S.M. Lee, E. Cho, S.I. Kim, E. Lee, Y.S. Kwon, J.H. Shim, G. Kotliar, Peierls distortion as a route to high thermoelectric performance in $In_4\,Se_{3-\delta}$ crystals. Nature **459**, 965–968 (2009)
22. H. Zhao, J. Sui, Z. Tang, Y. Lan, Q. Jie, D. Kraemer, K. McEnaney, A. Guloy, G. Chen, Z. Ren, High thermoelectric performance of MgAgSb-based materials. Nano Energy **7**, 97–103 (2014)
23. G.S. Nolas, J. Poon, M. Kanatzidis, Recent developments in bulk thermoelectric materials. MRS Bull. **31**, 199–205 (2006)
24. K. Koumoto, I. Terasaki, R. Funahashi, Complex oxide materials for potential thermoelectric applications. MRS Bull. **31**, 206–210 (2006)
25. Y. Lan, A.J. Minnich, G. Chen, Z. Ren, Enhancement of thermoelectric figure-of-merit by a bulk nanostructuring approach. Adv. Funct. Mater. **20**, 357–376 (2010)
26. L.D. Hicks, M.S. Dresselhaus, Thermoelectric figure of merit of a one-dimensional conductor. Phys. Rev. B **47**, 16631–16634 (1993)
27. R. Yang, G. Chen, Thermal conductivity modeling of periodic two-dimensional nanocomposites. Phys. Rev. B **69**, 195316 (2004)
28. M.S. Dresselhaus, G. Chen, M.Y. Tang, R.G. Yang, H. Lee, D.Z. Wang, Z.F. Ren, J. P. Fleurial, P. Gogna, New directions for low-dimensional thermoelectric materials. Adv. Mater. **19**, 1043–1053 (2007)
29. G. Chen, Thermal conductivity and ballistic-phonon transport in the cross plane direction of superlattices. Phys. Rev. B **57**, 14958–14973 (1998)
30. T. Koga, S.B. Cronin, M.S. Dresselhaus, J.L. Liu, K.L. Wang, Experimental proof-of-principle investigation of enhanced $Z_{3D}T$ in (001) oriented Si/Ge superlattices. Appl. Phys. Lett. **77**, 1490–1492 (2000)
31. R. Venkatasubramanian, E. Siivola, T. Colpitts, B. O'Quinn, Thin-film thermoelectric devices with high room-temperature figures of merit. Nature **413**, 597–602 (2001)
32. T.C. Harman, P.J. Taylor, M.P. Walsh, B.E. LaForge, Quantum dot superlattice thermoelectric materials and devices. Science **297**, 2229–2232 (2002)
33. W. Kim, J. Zide, A. Gossard, D. Klenov, S. Stemmer, A. Shakouri, A. Majumdar, Thermal conductivity reduction and thermoelectric figure of merit increase by embedding nanoparticles in crystalline semiconductors. Phys. Rev. Lett. **96**, 045901 (2006)
34. M. Dresselhaus, G. Dresselhaus, X. Sun, Z. Zhang, S. Cronin, T. Koga, Low-dimensional thermoelectric materials. Phys. Solid State **41**, 679–682 (1999)

35. A.M. Rao, X. Ji, T.M. Tritt, Properties of nanostructured onedimensional and composite thermoelectric materials. MRS Bull. **31**, 218–223 (2006)
36. A.I. Hochbaum, R. Chen, R.D. Delgado, W. Liang, E.C. Garnett, M. Najarian, A. Majumdar, P. Yang, Enhanced thermoelectric performance of rough silicon nanowires. Nature **451**, 163–167 (2008)
37. A.I. Boukai, Y. Bunimovich, J. Tahir-Kheli, J.-K. Yu, W.A. Goddard III, J.R. Heath, Silicon nanowires as efficient thermoelectric materials. Nature **451**, 168–171 (2008)
38. H.Böttner, G. Chen, R. Venkatasubramanian, Aspects of thin-film superlattice thermoelectric materials, devices, and applications. MRS Bull. **31**, 211–217 (2006)
39. J.P. Heremans, C.M. Thrush, D.T. Morelli, Thermopower enhancement in lead telluride nanostructures. Phys. Rev. B **70**, 115334 (2004)
40. S.-M. Lee, D.G. Cahill, R. Venkatasubramanian, Thermal conductivity of Si-Ge superlattices. Appl. Phys. Lett. **70**, 2957–2959 (1997)
41. G.J. Snyder, E.S. Toberer, Complex thermoelectric materials. Nat. Mater. **7**, 105–114 (2008)
42. B. Poudel, Q. Hao, Y. Ma, Y. Lan, A. Minnich, B. Yu, X. Yan, D. Wang, A. Muto, D. Vashaee, X. Chen, J. Liu, M.S. Dresselhaus, G. Chen, Z. Ren, High-thermoelectric performance of nanostructured bismuth antimony telluride bulk alloys. Science **320**, 634–638 (2008)
43. Y. Ma, Q. Hao, B. Poudel, Y. Lan, B. Yu, D. Wang, G. Chen, Z. Ren, Enhanced thermoelectric figure-of-merit in *p*-type nanostructured bismuth antimony tellurium alloys made from elemental chunks. Nano Lett. **8**, 2580–2584 (2008)
44. G. Joshi, H. Lee, Y. Lan, X. Wang, G. Zhu, D. Wang, R.W. Gould, D.C. Cuff, M.Y. Tang, M.S. Dresselhaus, G. Chen, Z. Ren, Enhanced thermoelectric figure-of-merit in nanostructured *p*-type silicon germanium bulk alloys. Nano Lett. **8**, 4670–4674 (2008)
45. X.W. Wang, H. Lee, Y.C. Lan, G.H. Zhu, G. Joshi, D.Z. Wang, J. Yang, A.J. Muto, M.Y. Tang, J. Klatsky, S. Song, M.S. Dresselhaus, G. Chen, Z.F. Ren, Enhanced thermoelectric figure of merit in nanostructured *n*-type silicon germanium bulk alloy. Appl. Phys. Lett. **93**, 193121 (2008)
46. J. Yang, Q. Hao, H. Wang, Y.C. Lan, Q.Y. He, A. Minnich, D.Z. Wang, J.A. Harriman, V. M. Varki, M.S. Dresselhaus, G. Chen, Z.F. Ren, Solubility study of Yb in *n*-type skutterudites $Yb_xCo_4Sb_{12}$ and their enhanced thermoelectric properties. Phys. Rev. B **80**, 115329 (2009)
47. G.H. Zhu, H. Lee, Y.C. Lan, X.W. Wang, G. Joshi, D.Z. Wang, J. Yang, D. Vashaee, H. Guilbert, A. Pillitteri, M.S. Dresselhaus, G. Chen, Z.F. Ren, Increased phonon scattering by nanograins and point defects in nanostructured silicon with a low concentration of germanium. Phys. Rev. Lett. **102**, 196803 (2009)
48. X. Yan, G. Joshi, W. Liu, Y. Lan, H. Wang, S. Lee, J.W. Simonson, S.J. Poon, T.M. Tritt, G. Chen, Z.F. Ren, Enhanced thermoelectric figure of merit of *p*-type Half-Heuslers. Nano Lett. **11**, 556–560 (2011)
49. G.H. Zhu, Y.C. Lan, H. Wang, G. Joshi, Q. Hao, G. Chen, Z.F. Ren, Effect of selenium deficiency on the thermoelectric properties of *n*-type In_4Se_{3-x} compounds. Phys. Rev. B **83**, 115201 (2011)
50. Q. Zhang, Y. Lan, S. Yang, F. Cao, M. Yao, C. Opeil, D. Broido, G. Chen, Z. Ren, Increased thermoelectric performance by Cl doping in nanostructured $AgPb_{18}SbSe_{20-x}$ Cl_x. Nano Energy **2**, 1121–1127 (2013)
51. G. Zhu, W. Liu, Y. Lan, G. Joshi, H. Wang, G. Chen, Z. Ren, The effect of secondary phase on thermoelectric properties of Zn_4Sb_3 compound. Nano Energy **2**, 1172–1178 (2013)
52. Q. Zhang, B. Liao, Y. Lan, K. Lukas, W. Liu, K. Esfarjani, C. Opeil, D. Broido, G. Chen, Z. Ren, High thermoelectric performance by resonant dopant indium in nanostructured SnTe. Proc. Natl. Acad. Sci. USA **110**, 13261–13266 (2013)
53. W. Liu, C.F. Guo, M. Yao, Y. Lan, H. Zhang, Q. Zhang, S. Chen, C.P. Opeil, Z. Ren, Bi_2S_3 nanonetwork as precursor for improved thermoelectric performance. Nano Energy **4**, 113–122 (2014)

54. M. Koirala, H. Wang, M. Pokharel, Y. Lan, C. Guo, C. Opeil, Z. Ren, Nanostructured $YbAgCu_4$ for potentially cryogenic thermoelectric cooling. Nano Lett. **14**, 5016–5020 (2014)
55. D.M. Rowe, R.W. Bunce, The thermoelectric properties of heavily doped hot-pressed germanium-silicon alloys. J. Phys. D Appl. Phys. **2**, 1497 (1969)
56. N. Savvides, H.J. Goldsmid, Boundary scattering of phonons in fine-grained hot-pressed Ge-Si alloys. I. the dependence of lattice thermal conductivity on grain size and porosity. J. Phys. C: Solid State Phys. **13**, 4657 (1980)
57. N. Savvides, H.J. Goldsmid, Boundary scattering of phonons in finegrained hot-pressed Ge-Si alloys. II. theory. J. Phys. C: Solid State Phys. **13**, 4671 (1980)
58. D.M. Rowe, V.S. Shukla, N. Savvides, Phonon scattering at grain boundaries in heavily doped fine-grained silicon-germanium alloys. Nature **290**, 765–766 (1981)
59. C.B. Vining, A model for the high-temperature transport properties of heavily doped *n* -type silicon-germanium alloys. J. Appl. Phys. **69**, 331–341 (1991)
60. C.B. Vining, W. Laskow, J.O. Hanson, R.R. Van der Beck, P.D. Gorsuch, Thermoelectric properties of pressure-sintered $Si_{0.8}Ge_{0.2}$ thermoelectric alloys. J. Appl. Phys. **69**, 4333–4340 (1991)
61. D.M. Rowe, L.W. Fu, S.G.K. Williams, Comments on the thermoelectric properties of pressure sintered $Si_{0.8}Ge_{0.2}$ thermoelectric alloys. J. Appl. Phys. **73**, 4683–4685 (1993)
62. A.A. Joraide, Thermoelectric properties of fine-grained sintered $(Bi_2Te_3)_{25}$-$(Sb_2\ Te_3)_{75}$ *p*-type solid solution. J. Mater. Sci. **30**, 744–748 (1995)
63. C.-W. Nan, R. Birringer, Determining the Kapitza resistance and the thermal conductivity of polycrystals: a simple model. Phys. Rev. B **57**, 8264–8268 (1998)
64. K. Kishimoto, T. Koyanagi, Preparation of sintered degenerate *n*-type PbTe with a small grain size and its thermoelectric properties. J. Appl. Phys. **92**, 2544–2549 (2002)
65. W.-S. Liu, B.-P. Zhang, J.-F. Li, L.-D. Zhao, Thermoelectric property of fine-grained $CoSb_3$ skutterudite compound fabricated by mechanical alloying and spark plasma sintering. J. Phys. D Appl. Phys. **40**, 566 (2007)
66. H. Ni, T. Zhu, X. Zhao, Hydrothermally synthesized and hot-pressed $Bi_2(Te, Se)_3$ thermoelectric alloys. Physica B **364**, 50–54 (2005)
67. T. Zhu, Y. Liu, X. Zhao, Synthesis of PbTe thermoelectric materials by alkaline reducing chemical routes. Mater. Res. Bull. **43**, 2850–2854 (2008)
68. W. Wang, B. Poudel, J. Yang, D.Z. Wang, Z.F. Ren, High-yield synthesis of single-crystalline antimony telluride hexagonal nanoplates using a solvothermal approach. J. Am. Chem. Soc. **127**, 13792–13793 (2005)
69. W.Z. Wang, B. Poudel, D.Z. Wang, Z.F. Ren, Synthesis of PbTe nanoboxes using a solvothermal technique. Adv. Mater. **17**, 2110–2114 (2005)
70. B. Poudel, W.Z. Wang, D.Z. Wang, J.Y. Huang, Z.F. Ren, Shape evolution of lead telluride and selenide nanostructures under different hydrothermal synthesis conditions. J. Nanosci. Nanotechnol. **6**, 1050–1053 (2006)
71. W. Wang, X. Yan, B. Poudel, Y. Ma, Q. Hao, J. Yang, G. Chen, Z. Ren, Chemical synthesis of anisotropic nanocrystalline Sb_2Te_3 and low thermal conductivity of the compacted dense bulk. J. Nanosci. Nanotechnol. **8**, 452–456 (2008)
72. Y.Q. Cao, X.B. Zhao, T.J. Zhu, X.B. Zhang, J.P. Tu, Syntheses and thermoelectric properties of Bi_2Te_3-Sb_2Te_3 bulk nanocomposites with laminated nanostructure. Appl. Phys. Lett. **92**, 143106 (2008)
73. L. Bertini, C. Stiewe, M. Toprak, S. Williams, D. Platzek, A. Mrotzek, Y. Zhang, C. Gatti, E. Müller, M. Muhammed, M. Rowe, Nanostructured $Co_{1-x}Ni_xSb_3$ skutterudites: synthesis, thermoelectric properties, and theoretical modeling. J. Appl. Phys. **93**, 438–447 (2003)
74. J. Martin, G.S. Nolas, W. Zhang, L. Chen, PbTe nanocomposites synthesized from PbTe nanocrystals. Appl. Phys. Lett. **90**, 222112 (2007)
75. H. Li, X. Tang, X. Su, Q. Zhang, Preparation and thermoelectric properties of high-performance Sb additional $Yb_{0.2}Co_4Sb_{12+y}$ bulk materials with nanostructure. Appl. Phys. Lett. **92**, 202114 (2008)

76. K. Sridhar, K. Chattopadhyay, Synthesis by mechanical alloying and thermoelectric properties of Cu_2Te. J. Alloy. Compd. **264**, 293–298 (1998)
77. R. Martin-Lopez, A. Dauscher, H. Scherrer, J. Hejtmanek, H. Kenzari, B. Lenoir, Thermoelectric properties of mechanically alloyed Bi-Sb alloys. Appl. Phys. A **68**, 597–602 (1999)
78. J. Schilz, M. Riffel, K. Pixius, H.-J. Meyer, Synthesis of thermoelectric materials by mechanical alloying in planetary ball mills. Powder Technol. **105**, 149–154 (1999)
79. M. Zakeri, M. Allahkarami, G. Kavei, A. Khanmohammadian, M. Rahimipour, Synthesis of nanocrystalline Bi_2Te_3 via mechanical alloying. J. Mater. Process. Technol. **209**, 96–101 (2009)
80. Z.F. Ren, B. Poudel, Y. Ma, Q. Hao, Y.C. Lan, A. Minnich, A. Muto, J. Yang, B. Yu, X. Yan, D.Z. Wang, J.M. Liu, M.S. Dresselhaus, G. Chen, Enhancement of thermoelectric figure-of-merit by a nanostructure approach, in *Materials Research Society Symposium Proceedings,* ed. by J. Yang, G.S. Nolas, K. Koumoto, Y. Grin, vol. 1166, pp. 83–93 (2009)
81. J.S. Benjamin, Mechanical alloying. Sci. Am. **234**, 40–49 (1976)
82. C. Suryanarayana, Mechanical alloying and milling. Prog. Mater Sci. **46**, 1–184 (2001)
83. P. Loeff, H. Bakker, Amorphous gold-lanthanum alloy prepared by ball milling. Scr. Metall. **22**, 401–404 (1988)
84. M. Abdellaoui, E. Gaffet, The physics of mechanical alloying in a modified horizontal rod mill: mathematical treatment. Acta Mater. **44**, 725–734 (1996)
85. D. Basset, P. Matteazzi, F. Miani, Designing a high energy ball-mill for synthesis of nanophase materials in large quantities. Mater. Sci. Eng. A **168**, 149–152 (1993)
86. K. Yamada, C.C. Koch, The influence of mill energy and temperature on the structure of the TiNi intermetallic after mechanical attrition. J. Mater. Res. **8**, 1317–1326 (1993)
87. J. Cui, X. Liu, W. Yang, D. Chen, H. Fu, P. Ying, Thermoelectric properties in nanostructured homologous series alloys $Ga_mSb_nTe_{1.5(m+n)}$. J. Appl. Phys. **105**, 063703 (2009)
88. A. Kanatzia, C. Papageorgiou, C. Lioutas, T. Kyratsi, Design of ball milling experiments on Bi_2Te_3 thermoelectric material. J. Electron. Mater. **42**, 1652–1660 (2013)
89. X. Yan, B. Poudel, Y. Ma, W.S. Liu, G. Joshi, H. Wang, Y. Lan, D. Wang, G. Chen, Z.F. Ren, Experimental studies on anisotropic thermoelectric properties and structures of *n*-type $Bi_2Te_{2.7}Se_{0.3}$. Nano Lett. **10**, 3373–3378 (2010)
90. W.-S. Liu, Q. Zhang, Y. Lan, S. Chen, X. Yan, Q. Zhang, H. Wang, D. Wang, G. Chen, Z. Ren, Thermoelectric property studies on Cu-doped *n*-type $Cu_xBi_2Te_{2.7}Se_{0.3}$ nanocomposites. Adv. Energy Mater. **1**, 577–587 (2011)
91. J. Sui, J. Shuai, Y. Lan, Y. Liu, R. He, D. Wang, Q. Jie, Z. Ren, Effect of Cu concentration on thermoelectric properties of nanostructured *p*-type $MgAg_{0.97-x}Cu_x\ Sb_{0.99}$. Acta Mater. **87**, 266–272 (2015)
92. J. Shuai, H.S. Kim, Y. Lan, S. Chen, Y. Liu, H. Zhao, J. Sui, Z. Ren, Study on thermoelectric performance by Na doping in nanostructured $Mg_{1-x}Na_xAg_{0.97}Sb_{0.99}$. Nano Energy **11**, 640–646 (2015)
93. Q. He, Q. Hao, X. Wang, J. Yang, Y. Lan, X. Yan, B. Yu, Y. Ma, B. Poudel, G. Joshi, D. Wang, G. Chen, Z. Ren, Nanostructured thermoelectric skutterudite $Co_{1-x}Ni_xSb_3$ alloys. J. Nanosci. Nanotechnol. **8**, 4003–4006 (2008)
94. T. Dahal, Q. Jie, G. Joshi, S. Chen, C. Guo, Y. Lan, Z. Ren, Thermoelectric property enhancement in Yb-doped *n*-type skutterudites $Yb_xCo_4Sb_{12}$. Acta Mater. **75**, 316–321 (2014)
95. T. Dahal, Y. Lan, Q. Jie, W. Liu, K. Dahal, L. Tang, C. Guo, Z. Ren, Substitution of antimony by tin and tellurium in *n*-type skutterudites $CoSb_{2.8}Sn_xTe_{0.2-x}$. J. Miner. Met Mater. Soc. **66**, 2282–2287 (2014)
96. T. Dahal, Q. Jie, Y. Lan, C. Guo, Z. Ren, Thermoelectric performance of Ni compensated cerium and neodymium double filled *p*-type skutterudites. Phys. Chem. Chem. Phys. **16**, 18170–18175 (2014)
97. T. Dahal, Q. Jie, W. Liu, K. Dahal, C. Guo, Y. Lan, Z. Ren, Effect of triple fillers in thermoelectric performance of *p*-type skutterudites. J. Alloy. Compd. **623**, 104–108 (2015)

98. T. Dahal, S. Gahlawat, Q. Jie, K. Dahal, Y. Lan, K. White, Z. Ren, Thermoelectric and mechanical properties on misch metal filled *p*-type skutterudites $Mm_{0.9}Fe_{4-x}Co_xSb_{12}$. J. Appl. Phys. **117**, 055101 (2015)
99. S. Bathula, M. Jayasimhadri, N. Singh, A.K. Srivastava, J. Pulikkotil, A. Dhar, R.C. Budhani, Enhanced thermoelectric figure-of-merit in spark plasma sintered nanostructured *n*-type SiGe alloys. Appl. Phys. Lett. **101**, 213902 (2012)
100. J. Yang, X. Fan, R. Chen, W. Zhu, S. Bao, X. Duan, Consolidation and thermoelectric properties of *n*-type bismuth telluride based materials by mechanical alloying and hot pressing. J. Alloy. Compd. **416**, 270–273 (2006)
101. H. Kim, T. Oh, D.-B. Hyun, Thermoelectric properties of the *p*-type Bi_2Te_3-Sb_2Te_3-Sb_2Se_3 alloys fabricated by mechanical alloying and hot pressing. J. Phys. Chem. Solids **61**, 743–749 (2000)
102. Q. He, S. Hu, X. Tang, Y. Lan, J. Yang, X. Wang, Z. Ren, Q. Hao, G. Chen, The great improvement effect of pores on *ZT* in $Co_{1-x}Ni_xSb_3$ system. Appl. Phys. Lett. **93**, 042108 (2008)
103. S.-C. Ur, J.-C. Kwon, I.-H. Kim, Thermoelectric properties of Fe-doped $CoSb_3$prepared by mechanical alloying and vacuum hot pressing. J. Alloy. Compd. **442**, 358–361 (2007)
104. S. Bao, J. Yang, W. Zhu, X. Fan, X. Duan, J. Peng, Preparation and thermoelectric properties of La filled skutterudites by mechanical alloying and hot pressing. Mater. Lett. **60**, 2029–2032 (2006)
105. J.R. Thompson, C. Politis, Formation of amorphous Ti-Pd alloys by mechanical alloying methods. Europhys. Lett. **3**, 199 (1987)
106. J.S. Benjamin, T.E. Volin, The mechanism of mechanical alloying. Metallurg. Trans. **5**, 1929–1934 (1974)
107. R.B. Schwarz, C.C. Koch, Formation of amorphous alloys by the mechanical alloying of crystalline powders of pure metals and powders of intermetallics. Appl. Phys. Lett. **49**, 146–148 (1986)
108. C.C. Koch, Synthesis of nanostructured materials by mechanical milling: problems and opportunities. Nanostruct. Mater. **9**, 13–22 (1997)
109. C.C. Koch, The synthesis and structure of nanocrystalline materials produced by mechanical attrition: a review. Nanostruct. Mater. **2**, 109–129 (1993)
110. C.C. Koch, O.B. Cavin, C.G. McKamey, J.O. Scarbrough, Preparation of "amorphous" $Ni_{60}Nb_{40}$ by mechanical alloying. Appl. Phys. Lett. **43**, 1017–1019 (1983)
111. P.S. Gilman, J.S. Benjamin, Mechanical alloying. Annu. Rev. Mater. Sci. **13**, 279–300 (1983)
112. P.G. McCormick, Application of mechanical alloying to chemical refining (overview). JIM **36**, 161–169 (1995)
113. W. Liu, X. Yan, G. Chen, Z. Ren, Recent advances in thermoelectric nanocomposites. Nano Energy **1**, 42–56 (2012)
114. S. Chen, Z. Ren, Recent progress of half-Heusler for moderate temperature thermoelectric applications. Mater. Today **16**, 387–395 (2013)
115. H. Wang, J.-F. Li, C.-W. Nan, M. Zhou, W. Liu, B.-P. Zhang, T. Kita, High-performance $Ag_{0.8}Pb_{18+x}SbTe_{20}$ thermoelectric bulk materials fabricated by mechanical alloying and spark plasma sintering. Appl. Phys. Lett. **88**, 092104 (2006)
116. G. Guélou, A.V. Powell, P. Vaqueiro, Ball milling as an effective route for the preparation of doped bornite: synthesis, stability and thermoelectric properties. J. Mater. Chem. C **3**, 10624–10629 (2015)
117. J. Peng, J. Yang, T. Zhang, X. Song, Y. Chen, Preparation and characterization of Fe substituted $CoSb_3$ skutterudite by mechanical alloying and annealing. J. Alloy. Compd. **381**, 313–316 (2004)
118. C.-J. Liu, H. Yamauchi, Thermoelectric power and resistivity of $La_{1.8}Sr_{0.2}CaCu_2\ O_{6-\delta}$ and the effects of O_2 hot-isostatic-press annealing. Phys. Rev. B **51**, 11826–11829 (1995)

119. J. Harringa, B. Cook, Application of hot isostatic pressing for consolidation of n-type silicon–germanium alloys prepared by mechanical alloying. Mater. Sci. Eng., B **60**, 137–142 (1999)
120. J. Yang, R. Chen, X. Fan, S. Bao, W. Zhu, Thermoelectric properties of silver-doped *n*-type Bi_2Te_3-based material prepared by mechanical alloying and subsequent hot pressing. J. Alloy. Compd. **407**, 330–333 (2006)
121. Z. He, C. Stiewe, D. Platzek, G. Karpinski, E. Müller, S. Li, M. Toprak, M. Muhammed, Thermoelectric properties of hot-pressed skutterudite $CoSb_3$. J. Appl. Phys. **101**, 053713 (2007)
122. H. Li, X. Tang, Q. Zhang, C. Uher, Rapid preparation method of bulk nanostructured $Yb_{0.3}Co_4Sb_{12+y}$ compounds and their improved thermoelectric performance. Appl. Phys. Lett. **93**, 252109 (2008)
123. J.M. Schultz, J.P. McHugh, W.A. Tiller, Effects of heavy deformation and annealing on the electrical properties of Bi_2Te_3. J. Appl. Phys. **33**, 2443–2450 (1962)
124. M.K. Keshavarz, D. Vasilevskiy, R.A. Masut, S. Turenne, *p*-type bismuth telluride-based composite thermoelectric materials produced by mechanical alloying and hot extrusion. J. Electron. Mater. **42**, 1429–1435 (2013)
125. D.-B. Hyun, J.-S. Hwang, J.-D. Shim, T. Oh, Thermoelectric properties of $(Bi_{0.25}Sb_{0.75})_2Te_3$ alloys fabricated by hot-pressing method. J. Mater. Sci. **36**, 1285–1291 (2001)
126. J.F. Miller, R.C. Himes, Properties of PbSe prepared by powder metallurgy techniques. J. Electrochem. Soc. **107**, 915–919 (1960)
127. N. Savvides, H.J. Goldsmid, Hot-press sintering of Ge-Si alloys. J. Mater. Sci. **15**, 594–600 (1980)
128. R. Davis, C. Koch, Mechanical alloying of brittle components: silicon and germanium. Scr. Metall. **21**, 305–310 (1987)
129. L.P. Bulat, V.T. Bublik, I.A. Drabkin, V.V. Karataev, V.B. Osvenskii, Y.N. Parkhomenko, G.I. Pivovarov, D.A. Pshenai-Severin, N.Y. Tabachkova, Bulk nanostructured polycrystalline *p*-Bi-Sb-Te thermoelectrics obtained by mechanical activation method with hot pressing. J. Electron. Mater. **39**, 1650–1653 (2010)
130. Y. Lan, B. Poudel, Y. Ma, D. Wang, M.S. Dresselhaus, G. Chen, Z. Ren, Structure study of bulk nanograined thermoelectric bismuth antimony telluride. Nano Lett. **9**, 1419–1422 (2009)
131. S.K. Bux, R.G. Blair, P.K. Gogna, H. Lee, G. Chen, M.S. Dresselhaus, R.B. Kaner, J.-P. Fleurial, Nanostructured bulk silicon as an effective thermoelectric material. Adv. Funct. Mater. **19**, 2445–2452 (2009)
132. J. Yang, T. Aizawa, A. Yamamoto, T. Ohta, Thermoelectric properties of *p*-type $(Bi_2Te_3)_x(Sb_2Te_3)_{1-x}$ prepared via bulk mechanical alloying and hot pressing. J. Alloy. Compd. **309**, 225–228 (2000)
133. M.S. Toprak, C. Stiewe, D. Platzek, S. Williams, L. Bertini, E. Müller, C. Gatti, Y. Zhang, M. Rowe, M. Muhammed, The impact of nano structuring on the thermal conductivity of thermoelectric $CoSb_3$. Adv. Funct. Mater. **14**, 1189–1196 (2004)
134. G.S. Nolas, M. Kaeser, R.T. Littleton, T.M. Tritt, High figure of merit in partially filled ytterbium skutterudite materials. Appl. Phys. Lett. **77**, 1855–1857 (2000)
135. S.-C. Ur, J.-C. Kwon, I.-H. Kim, Thermoelectric properties of tin-filled skutterudites prepared by mechanical alloying process. Met. Mater. Int. **14**, 625–629 (2008)
136. J. Yang, Y. Chen, W. Zhu, S. Bao, J. Peng, X. Fan, Effect of sintering temperature on formation and thermoelectric properties of $La_{0.4}Ni_{0.4}Co_{3.6}Sb_{12}$ skutterudite by mechanical alloying and hot pressing. J. Phys. D Appl. Phys. **38**, 3966 (2005)
137. S. Bao, J. Yang, J. Peng, W. Zhu, X. Fan, X. Song, Preparation and thermoelectric properties of $La_xFeCo_3Sb_{12}$ skutterudites by mechanical alloying and hot pressing. J. Alloy. Compd. **421**, 105–108 (2006)
138. K. Cai, E. Mueller, C. Drasar, C. Stiewe, The effect of titanium diboride addition on the thermoelectric properties of β-$FeSi_2$ semiconductors. Solid State Commun. **131**, 325–329 (2004)

139. Z. He, D. Platzek, C. Stiewe, H. Chen, G. Karpinski, E. Müller, Thermoelectric properties of hot-pressed Al- and Co-doped iron disilicide materials. J. Alloy. Compd. **438**, 303–309 (2007)
140. R.B. Song, Y.Z. Liu, T. Aizawa, Thermoelectric properties of *p*-type $Mg_2Si_{0.6}Ge_{0.4}$ fabricated by bulk mechanical alloying and hot pressing. Phys. Status Solidi Rapid Res. Lett. **1**, 226–228 (2007)
141. Q. He, Q. Hao, G. Chen, B. Poudel, X. Wang, D. Wang, Z. Ren, Thermoelectric property studies on bulk TiO_x with x from 1 to 2. Appl. Phys. Lett. **91**, 052505 (2007)
142. O. Guillon, J. Gonzalez-Julian, B. Dargatz, T. Kessel, G. Schierning, J. Räthel, M. Herrmann, Field-assisted sintering technology/spark plasma sintering: mechanisms, materials, and technology developments. Adv. Eng. Mater. **16**, 830–849 (2014)
143. R. Lefever, G. McVay, R. Baughman, Preparation of hot-pressed silicon germanium ingots: Part III—vacuum hot pressing. Mater. Res. Bull. **9**, 863–872 (1974)
144. J. Jiang, L. Chen, S. Bai, Q. Yao, Thermoelectric performance of *p*-type Bi–Sb–Te materials prepared by spark plasma sintering. J. Alloys Compd. **390**, 208–211 (2005)
145. C.-H. Kuo, C.-S. Hwang, M.-S. Jeng, W.-S. Su, Y.-W. Chou, J.-R. Ku, Thermoelectric transport properties of bismuth telluride bulk materials fabricated by ball milling and spark plasma sintering. J. Alloy. Compd. **496**, 687–690 (2010)
146. S.D. Bhame, D. Pravarthana, W. Prellier, J.G. Noudem, Enhanced thermoelectric performance in spark plasma textured bulk *n*-type $BiTe_{2.7}Se_{0.3}$ and *p*-type $Bi_{0.5}Sb_{1.5}Te_3$. Appl. Phys. Lett. **102**, 211901 (2013)
147. N. Keawprak, Z. Sun, H. Hashimoto, M. Barsoum, Effect of sintering temperature on the thermoelectric properties of pulse discharge sintered $(Bi_{0.24}Sb_{0.76})_2Te_3$ alloy. J. Alloys Compd. **397**, 236–244 (2005)
148. Y. Liu, Y.-H. Lin, B.-P. Zhang, C.-W. Nan, J.-F. Li, High-temperature electrical transport behavior observed in the $La_{1.96}M_{0.04}CuO_4$ (*M*: Mg, Ca, Sr) polycrystalline ceramics. J. Am. Ceram. Soc. **91**, 2055–2058 (2008)
149. S. Yoneda, E. Ohta, H. Kaibe, I. Ohsugi, I. Shiota, I. Nishida, Grain size effect on thermoelectric properties of PbTe prepared by spark plasma sintering. Mater. Trans. **42**, 329–335 (2001)
150. T.-H. An, S.-M. Choi, W.-S. Seo, C. Park, I.-H. Kim, S.-U. Kim, Effects of spark plasma sintering temperature on thermoelectric properties of higher manganese silicide. J. Electron. Mater. **42**, 2269–2273 (2013)
151. Y.-H. Lin, J. Lan, Z. Shen, Y. Liu, C.-W. Nan, J.-F. Li, High-temperature electrical transport behaviors in textured $Ca_3Co_4O_9$-based polycrystalline ceramics. Appl. Phys. Lett. **94**, 072107 (2009)
152. T. Yin, D. Liu, Y. Ou, F. Ma, S. Xie, J.-F. Li, J. Li, Nanocrystalline thermoelectric $Ca_3Co_4O_9$ ceramics by sol-gel based electrospinning and spark plasma sintering. J. Phys. Chem. C **114**, 10061–10065 (2010)
153. Y. Liu, Y.-H. Lin, W. Xu, B. Cheng, J. Lan, D. Chen, H. Zhu, C.-W. Nan, High-temperature transport property of $In_{2-x}Ce_xO_3(0 \leq x \geq 0.10)$ fine grained ceramics. J. Am. Ceram. Soc. **95**, 2568–2572 (2012)
154. N. Ma, J.-F. Li, B. Zhang, Y. Lin, L. Ren, G. Chen, Microstructure and thermoelectric properties of $Zn_{1-x}Al_xO$ ceramics fabricated by spark plasma sintering. J. Phys. Chem. Solids **71**, 1344–1349 (2010)
155. K. Tyagi, B. Gahtori, S. Bathula, A.K. Srivastava, A.K. Shukla, S. Auluck, A. Dhar, Thermoelectric properties of Cu_3SbSe_3 with intrinsically ultralow lattice thermal conductivity. J. Mater. Chem. A **2**, 15829–15835 (2014)
156. K. Tyagi, B. Gahtori, S. Bathula, V. Toutam, S. Sharma, N.K. Singh, A. Dhar, Thermoelectric and mechanical properties of spark plasma sintered $Cu_3SbSe_3$and Cu_3SbSe_4: promising thermoelectric materials. Appl. Phys. Lett. **105**, 261902 (2014)
157. Z.-H. Ge, B.-P. Zhang, Z.-X. Yu, J.-F. Li, Effect of spark plasma sintering temperature on thermoelectric properties of Bi_2S_3 polycrystal. J. Mater. Res. **26**, 2711–2718 (2011)

158. A. Soni, Y. Shen, M. Yin, Y. Zhao, L. Yu, X. Hu, Z. Dong, K.A. Khor, M.S. Dresselhaus, Q. Xiong, Interface driven energy filtering of thermoelectric power in spark plasma sintered $Bi_2Te_{2.7}Se_{0.3}$ nanoplatelet composites. Nano Lett. **12**, 4305–4310 (2012)
159. G.-H. Dong, Y.-J. Zhu, L.-D. Chen, Microwave-assisted rapid synthesis of Sb_2Te_3 nanosheets and thermoelectric properties of bulk samples prepared by spark plasma sintering. J. Mater. Chem. **20**, 1976–1981 (2010)
160. C.-H. Lee, Fabrication and characterization of thermoelectric $CrSi_2$ compound by mechanical alloying and spark plasma sintering. J. Nanosci. Nanotechnol. **15**, 5070–5073 (2015)
161. Y. Lu, K. Sagara, L. Hao, Z. Ji, H. Yoshida, Fabrication of nonstoichiometric titanium dioxide by spark plasma sintering and its thermoelectric properties. Mater. Trans. **53**, 1208–1211 (2012)
162. I. Matsubara, Y. Zhou, T. Takeuchi, R. Funahashi, M. Shikono, N. Murayama, W. Shin, N. Izu, Thermoelectric properties of spark plasma sintered $Na_{1+x}Co_2O_4$ ceramics. J. Ceram. Soc. Jpn. **111**, 238–241 (2003)
163. I. Matsubara, R. Funahashi, T. Takeuchi, S. Sodeoka, Thermoelectric properties of spark plasma sintered $Ca_{2.75}Gd_{0.25}Co_4O_9$ ceramics. J. Appl. Phys. **90**, 462–465 (2001)
164. S.W. Finefrock, G. Zhang, J.-H. Bahk, H. Fang, H. Yang, A. Shakouri, Y. Wu, Structure and thermoelectric properties of spark plasma sintered ultrathin PbTe nanowires. Nano Lett. **14**, 3466–3473 (2014)
165. J. Cho, Y. Lim, S.-M. Choi, K. Kim, W.-S. Seo, H.-H. Park, Thermoelectric properties of spark plasma-sintered In_4Se_3-In_4Te_3. J. Electron. Mater. **40**, 1024–1028 (2011)
166. G. Delaizir, G. Bernard-Granger, J. Monnier, R. Grodzki, O. Kim-Hak, P.-D. Szkutnik, M. Soulier, S. Saunier, D. Goeuriot, O. Rouleau, J. Simon, C. Godart, C. Navone, A comparative study of spark plasma sintering (SPS), hot isostatic pressing (HIP) and microwaves sintering techniques on *p*-type Bi_2Te_3 thermoelectric properties. Mater. Res. Bull. **47**, 1954–1960 (2012)
167. J. Seo, C. Lee, K. Park, Effect of extrusion temperature and dopant on thermoelectric properties for hot-extruded *p*-type Te-doped $Bi_{0.5}Sb_{1.5}Te_3$ and *n*-type SbI_3-doped $Bi_2Te_{2.85}Se_{0.15}$. Mater. Sci. Eng. B **54**, 135–140 (1998)
168. T. Hayashi, M. Sekine, J. Suzuki, Y. Horio, H. Takizawa, Thermoelectric and mechanical properties of angular extruded $Bi_{0.4}Sb_{1.6}Te_3$ compounds. Mater. Trans. **48**, 2724–2728 (2007)
169. O. Kim-Hak, M. Soulier, P.-D. Szkutnik, S. Saunier, J. Simon, D. Goeuriot, Microwave sintering and thermoelectric properties of *p*-type $(Bi_{0.2}Sb_{0.8})_2Te_3$ powder. Powder Technol. **226**, 231–234 (2012)
170. J. Arreguin-Zavala, D. Vasilevskiy, S. Turenne, R.A. Masut, Microwave sintering of Bi_2Te_3- and PbTe-based alloys: structure and thermoelectric properties. J. Electron. Mater. **42**, 1992–1998 (2013)
171. H.Y. Geng, S. Ochi, J.Q. Guo, Solidification contraction-free synthesis for the $Yb_{0.15}Co_4Sb_{12}$ bulk material. Appl. Phys. Lett. **91**, 022106 (2007)
172. D.-H. Kim, T. Mitani, Thermoelectric properties of fine-grained Bi_2Te_3 alloys. J. Alloy. Compd. **399**, 14–19 (2005)
173. J.P. Heremans, C.M. Thrush, D.T. Morelli, Thermopower enhancement in PbTe with Pb precipitates. J. Appl. Phys. **98**, 063703 (2005)
174. D. Medlin, G. Snyder, Interfaces in bulk thermoelectric materials: a review for current opinion in colloid and interface science. Curr. Opin. Colloid Interf. Sci. **14**, 226–235 (2009)
175. J. Yang, T. Aizawa, A. Yamamoto, T. Ohta, Effect of processing parameters on thermoelectric properties of *p*-type $(Bi_2Te_3)_{0.25}(Sb_2Te_3)_{0.75}$prepared via BMA–HP method. Mater. Chem. Phys. **70**, 90–94 (2001)
176. N. Miyashita, T. Yano, R. Tsukuda, I. Yashima, Effect of excess Te on microstructure and thermoelectric properties of *p*-type $Bi_{0.40}Sb_{1.60}Te_{3.00+x}(0.00 \leq x \leq 0.40)$ materials prepared by hot pressing method. J. Ceram. Soc. Jpn. **111**, 386–390 (2003)
177. H.-J. Im, D.-H. Kim, T. Mitani, K.-C. Je, Thermal properties for hotpressed *n*-type $(Bi_{1-x}Sb_x)_2Te_3$ alloys. Jpn. J. Appl. Phys. **2004**, 43 (1094)

178. X. Fan, J. Yang, R. Chen, W. Zhu, S. Bao, Phase transformation and thermoelectric properties of *p*-type $(Bi_2Te_3)_{0.25}(Sb_2Te_3)_{0.75}$ prepared by mechanical alloying and hot pressing. Mater. Sci. Eng. A **438–440**, 190–193 (2006)
179. X.B. Zhao, S.H. Yang, Y.Q. Cao, J.L. Mi, Q. Zhang, T.J. Zhu, Synthesis of nanocomposites with improved thermoelectric properties. J. Electron. Mater. **38**, 1017–1024 (2009)
180. H. Ni, X. Zhao, T. Zhu, X. Ji, J. Tu, Synthesis and thermoelectric properties of Bi_2Te_3 based nanocomposites. J. Alloy. Compd. **397**, 317–321 (2005)
181. S. Kim, S. Yamamoto, T. Aizawa, Thermoelectric properties of anisotropy-controlled *p*-type Bi-Te-Sb system via bulk mechanical alloying and shear extrusion. J. Alloy. Compd. **375**, 107–113 (2004)
182. X. Tang, W. Xie, H. Li, W. Zhao, Q. Zhang, M. Niino, Preparation and thermoelectric transport properties of high-performance *p*-type Bi_2Te_3 with layered nanostructure. Appl. Phys. Lett. **90**, 012102 (2007)
183. I. Yashima, H. Watanave, T. Ogisu, R. Tsukuda, S. Sato, Thermoelectric properties and hall effect of $Bi_2Te_{3-x}Se_x$ polycrystalline materials prepared by a hot press method. Jpn. J. Appl. Phys. **37**, 2472 (1998)
184. J. Seo, K. Park, D. Lee, C. Lee, Thermoelectric properties of hot-pressed *n*-type $Bi_2Te_{2.85}Se_{0.15}$ compounds doped with SbI_3. Mater. Sci. Eng. B **49**, 247–250 (1997)
185. G.A. Slack, M.A. Hussain, The maximum possible conversion efficiency of silicon–germanium thermoelectric generators. J. Appl. Phys. **70**, 2694–2718 (1991)
186. B. Abeles, Lattice thermal conductivity of disordered semiconductor alloys at high temperatures. Phys. Rev. **131**, 1906–1911 (1963)
187. N.K. Abrikosov, V.S. Zemskov, E.K. Iordanishvili, A.V. Petrov, V.V. Rozhdestvenskaya, Thermoelectric properties of silicon-germanium-boron alloys. Fizikai Tekhnika Poluprovodnikov **2**, 1762–1768 (1968)
188. J.P. Dismukes, L. Ekstrom, E.F. Steigmeier, I. Kudman, D.S. Beers, Thermal and electrical properties of heavily doped Ge-Si alloys up to 1300 K. J. Appl. Phys. **35**, 2899–2907 (1964)
189. C.M. Bhandari, D.M. Rowe, Silicon–germanium alloys as high temperature thermoelectric materials. Contemp. Phys. **21**, 219–242 (1980)
190. D.M. Rowe, Theoretical optimization of the thermoelectric figure of merit of heavily doped hot-pressed germanium-silicon alloys. J. Phys. D Appl. Phys. **1974**, 7 (1843)
191. G. Williamson, W. Hall, X-ray line broadening from filed aluminium and wolfram. Acta Metall. **1**, 22–31 (1953)
192. J.P. Heremans, V. Jovovic, E.S. Toberer, A. Saramat, K. Kurosaki, A. Charoenphakdee, S. Yamanaka, G.J. Snyder, Enhancement of thermoelectric efficiency in PbTe by distortion of the electronic density of states. Science **321**, 554–557 (2008)
193. Q. Zhang, F. Cao, W. Liu, K. Lukas, B. Yu, S. Chen, C. Opeil, D. Broido, G. Chen, Z. Ren, Heavy doping and band engineering by potassium to improve the thermoelectric figure of merit in *p*-type PbTe, PbSe, and $PbTe_{1-y}Se_y$. J. Am. Chem. Soc. **134**, 10031–10038 (2012)
194. Y. Pei, A. LaLonde, S. Iwanaga, G.J. Snyder, High thermoelectric figure of merit in heavy hole dominated PbTe. Energy Environ. Sci. **4**, 2085–2089 (2011)
195. Q. Zhang, S. Yang, Q. Zhang, S. Chen, W. Liu, H. Wang, Z. Tian, D. Broido, G. Chen, Z. Ren, Effect of aluminum on the thermoelectric properties of nanostructured PbTe. Nanotechnol. **24**, 345705 (2013)
196. E.K. Chere, Q. Zhang, K. McEnaney, M. Yao, F. Cao, J. Sun, S. Chen, C. Opeil, G. Chen, Z. Ren, Enhancement of thermoelectric performance in n-type $PbTe_{1-y}Se_y$ by doping Cr and tuning Te:Se ratio. Nano Energy **13**, 355–367 (2015)
197. B. Paul, P. Banerji, The effect of chromium impurity on the thermoelectric properties of PbTe in the temperature range 100–600 K. J. Appl. Phys. **109**, 103710 (2011)
198. D. Parker, D.J. Singh, High-temperature thermoelectric performance of heavily doped PbSe. Phys. Rev. B **82**, 035204 (2010)
199. J. Androulakis, I. Todorov, J. He, D.-Y. Chung, V. Dravid, M. Kanatzidis, Thermoelectrics from abundant chemical elements: high-performance nanostructured PbSe-PbS. J. Am. Chem. Soc. **133**, 10920–10927 (2011)

200. H. Wang, Y. Pei, A.D. LaLonde, G.J. Snyder, Heavily doped *p*-type PbSe with high thermoelectric performance: an alternative for PbTe. Adv. Mater. **23**, 1366–1370 (2011)
201. H. Fan, T. Su, H. Li, B. Du, B. Liu, H. Sun, Y. Zhang, L. Li, S. Li, M. Hu, H. Ma, X. Jia, Enhanced thermoelectric performance of PbSe co-doped with Ag and Sb. J. Alloy. Compd. **639**, 106–110 (2015)
202. B.C. Sales, D. Mandrus, R.K. Williams, Filled skutterudite antimonides: a new class of thermoelectric materials. Science **272**, 1325–1328 (1996)
203. G.S. Nolas, J.L. Cohn, G.A. Slack, Effect of partial void filling on the lattice thermal conductivity of skutterudites. Phys. Rev. B **58**, 164–170 (1998)
204. T. Caillat, A. Borshchevsky, J.-P. Fleurial, Properties of single crystalline semiconducting $CoSb_3$. J. Appl. Phys. **80**, 4442–4449 (1996)
205. B. Chen, J.-H. Xu, C. Uher, D.T. Morelli, G.P. Meisner, J.-P. Fleurial, T. Caillat, A. Borshchevsky, Low-temperature transport properties of the filled skutterudites $CeFe_{4-x}Co_xSb_{12}$. Phys. Rev. B **55**, 1476–1480 (1997)
206. V. Keppens, D. Mandrus, B.C. Sales, B.C. Chakoumakos, P. Dai, R. Coldea, M.B. Maple, D. A. Gajewski, E.J. Freeman, S. Bennington, Localized vibrational modes in metallic solids. Nature **395**, 876–878 (1998)
207. M. Puyet, B. Lenoir, A. Dauscher, M. Dehmas, C. Stiewe, E. Müller, High temperature transport properties of partially filled $Ca_xCo_4Sb_{12}$ skutterudites. J. Appl. Phys. **95**, 4852–4855 (2004)
208. L.D. Chen, T. Kawahara, X.F. Tang, T. Goto, T. Hirai, J.S. Dyck, W. Chen, C. Uher, Anomalous barium filling fraction and *n*-type thermoelectric performance of Bay Co_4Sb_{12}. J. Appl. Phys. **90**, 1864–1868 (2001)
209. S. Bao, J. Yang, X. Song, J. Peng, W. Zhu, X. Fan, X. Duan, Preparation of La-filled skutterudites by mechanical alloying and hot pressing and their thermal conductivities. Mater. Sci. Eng. A **438–440**, 186–189 (2006)
210. J. Yang, Y. Chen, J. Peng, X. Song, W. Zhu, J. Su, R. Chen, Synthesis of $CoSb_3$ skutterudite by mechanical alloying. J. Alloy. Compd. **375**, 229–232 (2004)
211. W.-S. Liu, B.-P. Zhang, L.-D. Zhao, J.-F. Li, Improvement of thermoelectric performance of $CoSb_{3-x}Te_x$ skutterudite compounds by additional substitution of IVB-group elements for Sb. Chem. Mater. **20**, 7526–7531 (2008)
212. S.-C. Ur, I.-H. Kim, Phase transformation and thermoelectric properties of *n*-type $Fe_{0.98}Co_{0.02}Si_2$ processed by mechanical alloying. Mater. Lett. **57**, 543–551 (2002)
213. X. Shi, H. Kong, C.-P. Li, C. Uher, J. Yang, J.R. Salvador, H. Wang, L. Chen, W. Zhang, Low thermal conductivity and high thermoelectric figure of merit in *n*-type $Ba_xYb_yCo_4Sb_{12}$ double-filled skutterudites. Appl. Phys. Lett. **92**, 182101 (2008)
214. B.C. Sales, D. Mandrus, B.C. Chakoumakos, V. Keppens, J.R. Thompson, Filled skutterudite antimonides: electron crystals and phonon glasses. Phys. Rev. B **56**, 15081–15089 (1997)
215. X. Song, J. Yang, J. Peng, Y. Chen, W. Zhu, T. Zhang, Thermoelectric properties of La filled skutterudite prepared by mechanical alloying and hot pressing. J. Alloy. Compd. **399**, 276–279 (2005)
216. S. Bao, J. Yang, W. Zhu, X. Fan, X. Duan, Effect of processing parameters on formation and thermoelectric properties of $La_{0.4}FeCo_3Sb_{12}$ skutterudite by MA–HP method. J. Alloy. Compd. **476**, 802–806 (2009)
217. D. Li, H. Zhao, S. Li, B. Wei, J. Shuai, C. Shi, X. Xi, P. Sun, S. Meng, L. Gu, Z. Ren, X. Chen, Atomic disorders induced by silver and magnesium ion migrations favor high thermoelectric performance in *a*-MgAgSb-based materials. Adv. Funct. Mater. **25**, 6478–6488 (2015)
218. M.J. Kirkham, A.M. dos Santos, C.J. Rawn, E. Lara-Curzio, J.W. Sharp, A.J. Thompson, *Ab initio* determination of crystal structures of the thermoelectric material MgAgSb. Phys. Rev. B **85**, 144120 (2012)
219. B.R.T. Frost, G.V. Raynor, The system silver-magnesium-antimony, with reference to the theory of alloy formation. Proc. R. Soc. A **203**, 132–147 (1950)

220. Z. Liu, J. Shuai, J. Mao, Y. Wang, Z. Wang, W. Cai, J. Sui, Z. Ren, Effects of antimony content in $MgAg_{0.97}Sb_x$ on output power and energy conversion efficiency. Acta Mater. **102**, 17–23 (2016)
221. P. Ying, X. Liu, C. Fu, X. Yue, H. Xie, X. Zhao, W. Zhang, T. Zhu, High performance *a*-MgAgSb thermoelectric materials for low temperature power generation. Chem. Mater. **27**, 909–913 (2015)
222. C. Rossel, K.N. Yang, M.B. Maple, Z. Fisk, E. Zirngiebl, J.D. Thompson, Strong electronic correlations in a new class of Yb-based compounds: Yb_xCu_4 (x = Ag, Au, Pd). Phys. Rev. B **35**, 1914–1918 (1987)
223. R. Casanova, D. Jaccard, C. Marcenat, N. Hamdaoui, M. Besnus, Thermoelectric power of $YbMCu_4$ (M = Ag, Au and Pd) and $YbPd_2Si_2$. J. Magn. Magn. Mater. **90–91**, 587–588 (1990)
224. A.V. Golubkov, L.S. Parfenéva, I.A. Smirnov, H. Misiorek, J. Mucha, A. Jezowski, Heat conductivity of the heavy-fermion compound $YbAgCu_4$. Phys. Solid State **43**, 218–223 (2001)
225. G.D. Mahan, Good thermoelectrics. Solid State Phys. Academic Press **51**, 81–157 (1997)
226. T. Graf, J.M. Lawrence, M.F. Hundley, J.D. Thompson, A. Lacerda, E. Haanappel, M.S. Torikachvili, Z. Fisk, P.C. Canfield, Resistivity, magnetization, and specific heat of $YbAgCu_4$ in high magnetic fields. Phys. Rev. B **51**, 15053–15061 (1995)
227. W. Xu, L. Li, R. Huang, M. Zhou, L. Zheng, L. Gong, C. Song, Properties of Ag-doped Bi-Sb alloys as thermoelectric conversion materials for solid state refrigeration. Frontiers Energy Power Eng. China **3**, 90–93 (2009)
228. R. Huang, L. Li, W. Xu, L. Gong, Power factor enhancement in Pr substituted alloys prepared by high-pressure sintering. Solid State Commun. **149**, 1633–1636 (2009)
229. S.-C. Ur, I.-H. Kim, P. Nash, Thermoelectric properties of Zn_4Sb_3 processed by sintering of cold pressed compacts and hot pressing. J. Mater. Sci. **42**, 2143–2149 (2007)
230. Z.-G. Chen, G. Han, L. Yang, L. Cheng, J. Zou, Nanostructured thermoelectric materials: current research and future challenge. Prog. Nat. Sci.: Mater. Int. **22**, 535–549 (2012)
231. M. Ito, T. Tada, S. Katsuyama, Thermoelectric properties of $Fe_{0.98}Co_{0.02}Si_2$with ZrO_2 and rare-earth oxide dispersion by mechanical alloying. J. Alloy. Compd. **350**, 296–302 (2003)
232. M. Ito, T. Tada, S. Hara, Thermoelectric properties of hot-pressed β-$FeSi_2$ with yttria dispersion by mechanical alloying. J. Alloy. Compd. **408–412**, 363–367 (2006)
233. K.-F. Cai, C.-W. Nan, Y. Paderno, D. McLachlan, Effect of titanium carbide addition on the thermoelectric properties of B_4C ceramics. Solid State Commun. **115**, 523–526 (2000)
234. K. Cai, C. Nan, M. Schmuecker, E. Mueller, Microstructure of hot pressed B_4C-TiB_2 thermoelectric composites. J. Alloy Compd. **350**, 313–318 (2003)
235. J. Lee, T. Oh, D.-B. Hyun, Thermoelectric properties of the hot-pressed $(Bi_{0.2}Sb_{0.8})_2Te_3$ alloy with addition of BN and WO_3 powders. J. Mater. Sci. **35**, 881–887 (2000)
236. R. He, S. Gahlawat, C. Guo, S. Chen, T. Dahal, H. Zhang, W. Liu, Q. Zhang, E. Chere, K. White, Z. Ren, Studies on mechanical properties of thermoelectric materials by nanoindentation. Phys. Status Solidi A **212**, 2191–2195 (2015)
237. Y.C. Lan, D.Z. Wang, G. Chen, Z.F. Ren, Diffusion of nickel and tin in *p*-type $(Bi, Sb)_2Te_3$ and *n*-type $Bi_2(Te, Se)_3$ thermoelectric materials. Appl. Phys. Lett. **92**, 101910 (2008)
238. W. Liu, H. Wang, L. Wang, X. Wang, G. Joshi, G. Chen, Z. Ren, Understanding of the contact of nanostructured thermoelectric *n*-type $Bi_2Te_{2.7}Se_{0.3}$ legs for power generation applications. J. Mater. Chem. A **1**, 13093–13100 (2013)
239. D. Kraemer, J. Sui, K. McEnaney, H. Zhao, Q. Jie, Z.F. Ren, G. Chen, High thermoelectric conversion efficiency of MgAgSb-based material with hot-pressed contacts. Energy Environ. Sci. **8**, 1299–1308 (2015)
240. G.D. Mahan, Inhomogeneous thermoelectrics. J. Appl. Phys. **70**, 4551–4554 (1991)
241. G.J. Snyder, T.S. Ursell, Thermoelectric efficiency and compatibility. Phys. Rev. Lett. **91**, 148301 (2003)

242. W. Liu, K.C. Lukas, K. McEnaney, S. Lee, Q. Zhang, C.P. Opeil, G. Chen, Z. Ren, Studies on the Bi_2Te_3-Bi_2Se_3-Bi_2S_3 system for mid-temperature thermoelectric energy conversion. Energy Environ. Sci. **6**, 552–560 (2013)
243. Q. Zhang, H. Wang, W. Liu, H. Wang, B. Yu, Q. Zhang, Z. Tian, G. Ni, S. Lee, K. Esfarjani, G. Chen, Z. Ren, Enhancement of thermoelectric figure-of-merit by resonant states of aluminium doping in lead selenide. Energy Environ. Sci. **5**, 5246–5251 (2012)
244. G. Joshi, X. Yan, H. Wang, W. Liu, G. Chen, Z. Ren, Enhancement in thermoelectric figure-of-merit of an *n*-type half-heusler compound by the nanocomposite approach. Adv. Energy Mater. **1**, 643–647 (2011)
245. X. Shi, J. Yang, J.R. Salvador, M. Chi, J.Y. Cho, H. Wang, S. Bai, J. Yang, W. Zhang, L. Chen, Multiple-filled skutterudites: high thermoelectric figure of merit through separately optimizing electrical and thermal transports. J. Am. Chem. Soc. **133**, 7837–7846 (2011)
246. Y. Pei, J. Lensch-Falk, E.S. Toberer, D.L. Medlin, G.J. Snyder, High thermoelectric performance in PbTe due to large nanoscale Ag_2Te precipitates and La doping. Adv. Funct. Mater. **21**, 241–249 (2011)
247. D. Kraemer, B. Poudel, H.-P. Feng, J.C. Caylor, B. Yu, X. Yan, Y. Ma, X. Wang, D. Wang, A. Muto, K. McEnaney, M. Chiesa, Z. Ren, G. Chen, High-performance flat-panel solar thermoelectric generators with high thermal concentration. Nat. Mater. **10**, 532–538 (2011)
248. C.B. Vining, An inconvenient truth about thermoelectrics. Nat. Mater. **8**, 83–85 (2009)
249. H.S. Kim, W. Liu, G. Chen, C.-W. Chu, Z. Ren, Relationship between thermoelectric figure of merit and energy conversion efficiency. Proc. Natl. Acad. Sci. USA **112**, 8205–8210 (2015)

Chapter 12
Nanomaterials for Hydrogen Generation from Solar Water Splitting

Zhenhuan Zhao, Zhiming Wang and Jiming Bao

Abstract Hydrogen generation from solar water splitting has the potential to solve both environmental issues and energy crisis. This chapter begins with the discussion of the basic mechanism of photocatalytic water splitting and fundamental material requirements for particular photocatalysts. TiO_2 and alkali metal titanates are first introduced as traditional UV active photocatalysts, followed by Nb_2O_5, Ta_2O_5 and their alkali metal oxides. Metal chalcogenides and oxysulfides, nitrides and oxynitrides are then presented as lower bandgap photocatalysts than the corresponding oxides. Cu_2O and cobalt oxide nanoparticles are among very few binary oxides that can perform overall water splitting without co-catalysts under visible light. We will then show other newly developed visible light photocatalysts, namely, graphene oxide quantum dots, graphitic C_3N_4 nanostructures, and carbon-doped boron nitride nanosheets. Plasmonic nanomaterials are included as a special type of photocatalyst. The chapter concludes with discussions on current material issues, challenges and outlook for future photocatalyst development.

12.1 Background and Introduction

Our daily lives depend on a stable and affordable supply of energy. Fossil fuels have provided a major portion of the global energy, but they are not renewable, furthermore, the burning of fossil fuels has created serious environmental problems

Z. Zhao · Z. Wang · J. Bao (✉)
Institute of Fundamental and Frontier Sciences, University of Electronic Science and Technology of China, Chengdu 610054, China
e-mail: jbao@uh.edu

Z. Zhao · J. Bao
Department of Electrical and Computer Engineering, University of Houston, Houston, TX 77204, USA

Z. Wang (✉)
State Key Laboratory of Electronic Thin Films and Integrated Devices, University of Electronic Science and Technology of China, Chengdu 610054, China
e-mail: zhmwang@uestc.edu.cn

Q. Li (ed.), *Nanomaterials for Sustainable Energy*,
NanoScience and Technology, DOI 10.1007/978-3-319-32023-6_12

by producing hazardous chemicals and greenhouse gases [1, 2]. The generation of hydrogen from water using sunlight can not only provide renewable fuel to meet the world's energy demands but also mitigate climate change by reducing carbon emissions [3]. Hydrogen is a clean energy with energy density comparable to fossil fuels [4]. Energy stored in the chemical bond of hydrogen can be cleanly released either through the exothermic reaction with oxygen in a combustion engine, or through electrochemical oxidation in a fuel cell. There are three ways to generate hydrogen from water using sunlight. Electrolysis of water using the electricity from solar cells is the oldest and most mature method. Photoelectrochemical hydrogen generation is a less mature technique, it was discovered by Fujishima and Honda in 1972 where they employed TiO_2 as a photoanode and Pt as a cathode [5]. Photocatalytic water splitting utilizes particulate photocatalysts to absorb photons and decompose water directly to hydrogen and oxygen [6–9]. The last method is the least mature technique, but it has the potential to achieve a higher conversion efficiency because of the elimination of an intermediate step of conversion of solar energy to electricity. In this chapter, we will focus on the development of particulate or nanostructured photocatalysts for solar hydrogen generation.

12.2 Mechanism and Material Requirements for Solar Water Splitting

12.2.1 Mechanism of Solar Water Splitting

Figure 12.1 is a schematic illustrating the process of water splitting using a single particulate photocatalyst [10]. Upon irradiation of the photocatalyst by light with photon energy larger than its bandgap, electrons are excited to the conduction band, leaving the holes in the valence band. The photoexcited electrons and holes can migrate to the surface, initiate water oxidation to produce oxygen and water reduction to generate hydrogen. Electrons and holes can also recombine radiatively or through nonradiative channels without generating hydrogen and oxygen. To achieve a high photon to hydrogen energy conversion efficiency, electron-hole recombination has to be reduced. In addition, the sites for water reduction and oxidation should be separated to prevent the back reaction of evolved hydrogen and oxygen to water.

12.2.2 Material Requirements for Overall Solar Water Splitting

Schematic in Fig. 12.1 also illustrates the basic material requirements for overall solar water splitting. The decomposition of water into hydrogen and oxygen

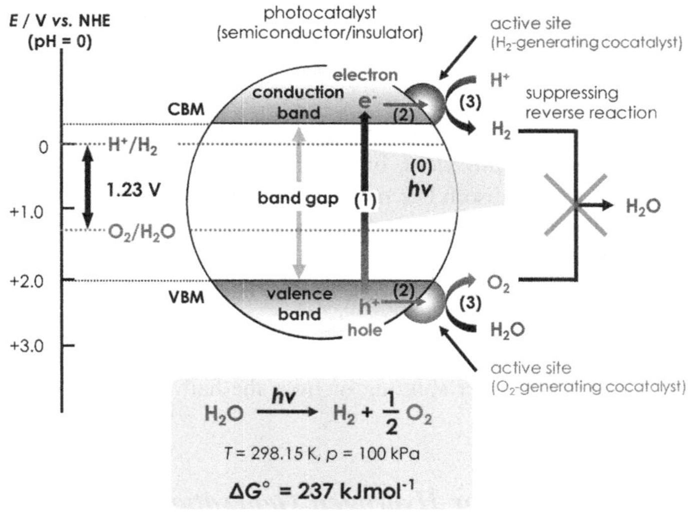

Fig. 12.1 Schematic diagram illustrating a sequence of overall water splitting by photoexcitation in a single photocatalyst: (*0*) light irradiation and absorption, (*1*) photo-excitation of electron-hole pairs, (*2*) migration of excited carriers to the surface, and (*3*) oxidation-reduction reaction on the surface. Reprint with permission from [10], Copyright 2013, Elsevier B.V

requires a minimum free energy of 237 kJ mol^{-1}, equivalent to 1.23 V between hydrogen and oxygen evolution potentials [6]. This thermodynamic limit requires a photocatalyst to have a minimum bandgap of 1.23 eV, with conduction band minimum (CBM) above the hydrogen-evolution potential and valence band maximum (VBM) below the oxygen-evolution potential. In addition to fulfilling the above thermodynamic requirement, extra energy or overpotential is needed for electrons and holes to overcome potential barriers in diffusion and to make water redox reaction proceed faster. This kinetic energy requirement raises the minimum bandgap of a photocatalyst to ~1.83 eV [11]. In order to reduce electron-hole recombination rate and enable fast diffusion of electrons and holes to the surface, a photocatalyst should also have high crystallinity and less defects [12]. It can also be seen in Fig. 12.1 that co-catalysts are typically required to provide separate sites for optimal hydrogen and oxygen evolution, and to reduce back reaction. Like many other catalysts, a good photocatalyst should be chemically stable and exhibit a strong resistance to photocorrosion, it should also be synthesized from earth-abundant elements in a green chemical way [13]. Because nanomaterials have large surface to volume ratio, and require a short distance for photoexcited electrons and holes to diffuse to surfaces, they have the potential to generate solar hydrogen at a high efficiency.

12.3 Nanomaterials for Hydrogen Generation from Solar Water Splitting

Photocatalysts with wide bandgaps are more chemically stable and have a higher photon to hydrogen quantum yield, but they are not efficient in harvesting solar energy because they only absorb UV portion of solar spectrum. It is still a challenge to identify a photocatalyst to achieve an optimal solar to hydrogen energy conversion [14]. While many semiconductors can evolve either hydrogen or oxygen from half reactions, very few photocatalysts can drive overall water splitting, especially under visible light even with the assistance of co-catalysts [15]. In this section, we mainly focus on the nanomaterials for photocatalytic hydrogen generation either from overall water splitting or from the half reaction.

12.3.1 Metal Oxides for Hydrogen Generation from Solar Water Splitting

Various semiconductor nanomaterials have been developed as photocatalysts for solar water splitting under UV light illumination [16]. UV active photocatalysts are typically wide bandgap semiconductors with d^0 and d^{10} electronic configuration [17]. For these metal oxide photocatalysts, the top of the valence band typically consists of O2*p* orbitals, located at about 3 V NHE or higher. The bottom of the conduction band is composed of *d* or *sp* orbitals of the metal cations, located between 0 and −1 V NHE [18]. Some metal oxides showing visible light activity, are also summarized in this section. Figure 12.2 summaries the band edge positions of some widely studied photocatalysts [19].

12.3.1.1 UV Active Metal Oxide Photocatalysts

TiO_2 and titanate-based nanomaterials
TiO_2 is the first material used for photoelectrochemical oxygen evolution [5]. It is an n-type semiconductor due to native oxygen vacancy. It has four crystal lattice structures: tetragonal anatase and rutile, orthorhombic brookite and monoclinic TiO_2(B) [20, 21]. Since the latter two oxides have unstable crystal lattice structures and are not easily synthesized, they are rarely used for photocatalysis. The bandgaps of anatase and rutile TiO_2 are 3.2 and 3.0 eV, respectively, with the corresponding light absorption edge of 380 and 400 nm.

The band structure of TiO_2, either rutile or anatase, is thermodynamically favorable for water splitting, but pure TiO_2 is still not able to perform overall water splitting due to the lack of catalytic activity [22]. As a result, co-catalysts such as RuO_2, Pt and Pd nanoparticles, are usually required to assist the water splitting reaction for hydrogen generation [23–25]. Sometimes, additional reagents are also

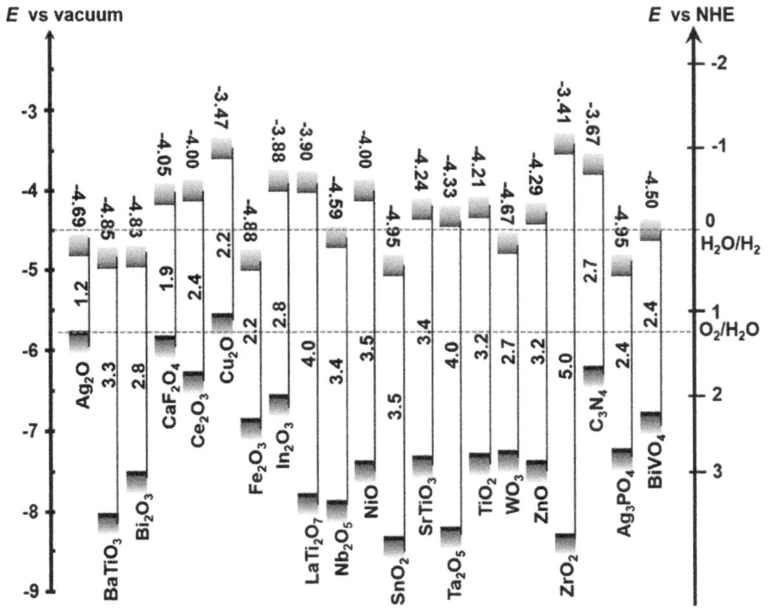

Fig. 12.2 Band edge positions with respect to vacuum level and normal hydrogen electrode (NHE) for selected semiconductors. Reprint with permission from [19], Copyright 2014, The Royal Society of Chemistry

added into water such as NaOH, Na_2CO_3 and even sacrificial reagents [26–28]. The addition of NaOH or Na_2CO_3 can facilitate the separation of photoexcited electrons and holes by forming carbonates species which cover the TiO_2 surface and consume holes to become carbonate radicals.

Recently, it is observed that anatase TiO_2 has higher electron affinity than rutile TiO_2, photogenerated electrons will diffuse from rutile to anatase as shown in Fig. 12.3. A band misalignment of about 0.4 eV significantly lowers the effective bandgap of composite materials and facilitates efficient electron-hole separation. This should lead to higher efficiency in the photocatalysis of water by rutile/anatase heterostructure [29].

Titanate-based nanomaterials, such as $SrTiO_3$, $K_2Ti_2O_5$, $K_2Ti_4O_9$, $Na_2Ti_3O_7$, $Cs_2Ti_2O_5$, $Cs_2Ti_5O_{11}$ and $Cs_2Ti_6O_{13}$ [30–36], have been investigated for photocatalytic water splitting. $SrTiO_3$, $K_2Ti_2O_5$, $K_2Ti_4O_9$, $Na_2Ti_3O_7$, $Cs_2Ti_5O_{11}$ and $Cs_2Ti_6O_{13}$ consist of six-coordination TiO_6 units while $Cs_2Ti_2O_5$ with five-coordination TiO_5 units. $Cs_2Ti_2O_5$ shows higher photocatalytic activity than $Cs_2Ti_5O_{11}$ and $Cs_2Ti_6O_{13}$ because unsaturated coordination state of the TiO_5 can act as active sites [36]. Depending on the type of intercalated cations, titanate oxides with TiO_6 unit display different photocatalytic activity [37]. $Na_2Ti_6O_{13}$, $K_2Ti_6O_{13}$ and $Rb_2Ti_6O_{13}$ show higher photocatalytic activity than $Cs_2Ti_6O_{13}$, and the activity increases in the order of Na, K, Rb and Cs [38]. More interestingly, the intercalated cations can be easily ion-exchanged with H^+, which can significantly

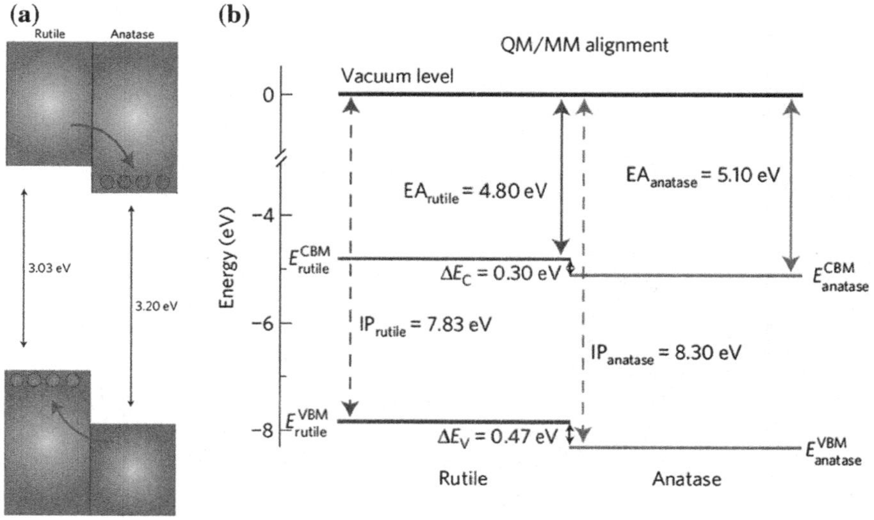

Fig. 12.3 **a** Proposed valence and conduction band alignment for the anatase/rutile interface. *Red arrows* indicate the flow of electrons (holes) in the conduction band (valence band). **b** Schematic of the quantum-mechanical/molecular-mechanical (QM/MM) alignment of rutile and anatase TiO_2. IP and EA denote ionization potential and electron affinity, respectively. E^{CBM} and E^{VBM} denote the positions of the conduction band and valence band respectively. Reprint with permission from [29], Copyright 2013, Macmillan Publishers limited

enhance the photocatalytic activity. For example, H^+ exchanged $K_2Ti_2O_5$ shows increased photocatalytic activity for H_2 generation in methanol/water solution with the quantum yield up to about 10 %. [33] Similar phenomenon is also observed in H^+ exchanged $Cs_2Ti_2O_5$ [36].

Rare earth metal based titanates such as $La_2Ti_2O_7$, $Y_2Ti_2O_7$ and $Er_2Ti_2O_7$ also show water splitting activities [39, 40]. Among them, $Y_2Ti_2O_7$ has the best photocatalytic activity for overall water splitting from pure water under UV irradiation when coupled with NiO co-catalyst. The reason is suggested to be related to the enhanced charge carrier mobility because of the corner-linked TiO_6 octahedrons. The disadvantage of rare earth metal based titanates is their high cost, making them not suitable for large-scale applications.

Nb_2O_5 and niobate-based nanomaterials

Nb_2O_5 is an n-type semiconductor with the bandgap of about 3.4 eV [41]. Bulk Nb_2O_5 itself cannot decompose water to produce hydrogen from pure water, however, mesoporous Nb_2O_5 shows excellent photocatalytic activity for hydrogen generation with the assistance of co-catalysts like CuO, NiO and Pt. Two reasons are responsible for the photocatalytic enhancement of the mesoporous Nb_2O_5: mesoporous Nb_2O_5 has higher surface area and more pore volume than bulk Nb_2O_5 for ion exchange; mesoporous Nb_2O_5 also has very thin pore walls which decrease the carrier migration distance [42]. The hydrogen evolution rate is as high as 1405, 800 and 510 μmol $h^{-1}g^{-1}$, respectively [43]. The Pt modified mesoporous Nb_2O_5

exhibits a photocatalytic activity 20 times higher than bulk Nb_2O_5 [42]. The most likely reason is suggested to be the much lower recombination rate of electrons and holes in mesoporous Nb_2O_5 than in bulk Nb_2O_5. However, these materials still cannot initiate overall water splitting.

Same with titanates, niobate based nanomaterials are also important photocatalysts for photocatalytic hydrogen generation [44]. The first example is $K_4Nb_6O_{17}$ which can promote the evolution of hydrogen. The photocatalytic performance can be further enhanced to promote both the evolution of hydrogen and oxygen by treating $K_4Nb_6O_{17}$ with H_2 at high temperature and then re-oxidization in O_2 [45]. Niobates have layered perovskite structure consisting of NbO_6 layers and metal cation interlayers. Like titanates, the cations in the interlayers can be easily ion-exchanged by other metal cations, such as Cs, Sr, Ca, Ba and Zn. For example, NiO loaded $Cs_2Nb_4O_{11}$ shows hydrogen evolution property from water under UV irradiation [46]. $Sr_2Nb_2O_7$ with (110) layered perovskite structure displays efficient photocatalytic activity with the quantum yield of about 23 %. [47] $Ca_2Nb_2O_7$ and $Ba_5Nb_4O_{15}$ gave quantum yield of about 7 and 17 % at 270 nm for water splitting when sensitized with NiO co-catalyst [47, 48], respectively.

The metal cations of the interlayer of niobates can also be ion-exchanged with H^+ ion which has been demonstrated to be beneficial for hydrogen generation. For example, the H^+ ion-exchanged $K_4Nb_6O_{17}$ shows improved hydrogen evolution performance [49]. The reason is due to the hydrated water molecules at the interlayers which increases the migration of reactants into the interlayers.

Ta_2O_5 and tantalate-based nanomaterials

The bandgap of Ta_2O_5 is as wide as 3.9 eV with the light absorption edge located at about 320 nm [50]. Bulk Ta_2O_5 can only produce trace amount of hydrogen, however, similar to Nb_2O_5, mesoporous Ta_2O_5 also displays great photocatalytic activity for overall water splitting when loaded with NiO co-catalyst under UV illumination [51]. The photocatalytic activity can be further improved by converting the amorphous pore walls into crystalline walls through high temperature calcination [52]. The rates of hydrogen and oxygen evolution reach 3.4 and 1.6 mmol h^{-1} for NiO loaded 0.3 g crystallized Ta_2O_5, respectively [53]. Therefore, nanostructure engineering is an important technique to increase the photocatalytic activity of photocatalysts.

The high photocatalytic activity of Ta_2O_5 inspired researchers to explore tantalate-based nanomaterials for photocatalytic water splitting. Alkali metal (Li, Na, K, Mg and Ba) based tantalates ($LiTaO_3$, $NaTaO_3$, $KTaO_3$, $MgTa_2O_6$ and $BaTa_2O_6$) exhibit overall water splitting without any co-catalyst [54–56]. As titanates and niobates, these tantalates typically exhibit perovskite-like structure consisting of TaO_6 octahedra with the interlayer intercalated alkali metal cations. Among them, $KTaO_3$ displays the best photocatalytic activity [54]. However, when NiO co-catalyst is loaded, $NaTaO_3$ prepared in the presence of excess alkali by solid reactions shows the best activity under UV irradiation [55]. The excess of the alkali metal cations can improve the crystallinity to suppress the defect formation and hence to increase the charge transportation and reduce electron-hole recombination. In this case, the photocatalytic activity increases in the order of $KTaO_3$, $NaTaO_3$ and $LiTaO_3$. For $BaTa_2O_6$, its activity depends on the crystalline phase with increasing order of hexagonal phase, tetragonal phase and orthorhombic phase.

Transition metal based tantalates also show activities for water splitting and some ($NiTa_2O_6$, $ZnTa_2O_6$ and $CuTa_2O_6$) display overall water splitting activity [16]. Among them, $NiTa_2O_6$ shows reasonable photocatalytic activity for overall water splitting, while $ZnTa_2O_6$ needs co-catalyst. This type of tantalates also has very wide bandgaps with the exception of $CuTa_2O_6$ whose bandgap is about 2.4 eV.

12.3.1.2 Visible Light Active Metal Oxides

Metal oxides with wide bandgaps are usually employed as UV photocatalysts. Although the quantum yield is high because of large overpotentials, their solar to hydrogen conversion efficiency is low because they cannot harvest visible light. Therefore, it is important to develop visible light active photocatalysts to maximally utilize solar energy [57, 58].

Cu_2O

Cu_2O is a p-type semiconductor with the bandgap of about 2.0−2.4 eV [59, 60], which enables Cu_2O to absorb visible light as far as 620 nm. In addition, Cu_2O is observed to split water without any co-catalyst under visible light irradiation [61]. The Cu_2O synthesized from the hydrolysis of CuCl shows quantum yield of about 0.3 % at 550 and 600 nm. Figure 12.4 shows that a stoichiometric ratio of H_2–O_2 is observed after 400 h of irradiation. Cu_2O has many advantages as a photocatalyst

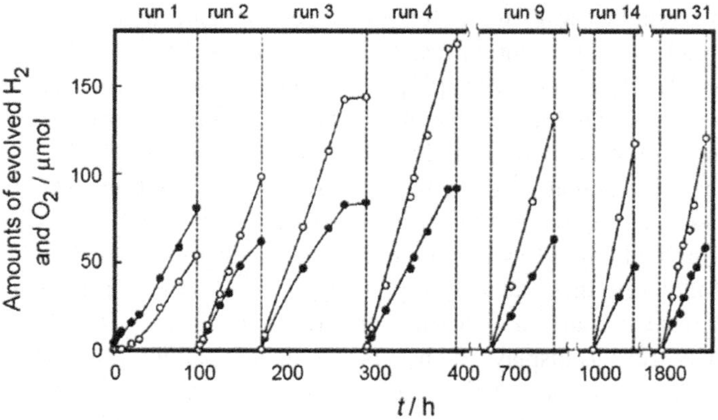

Fig. 12.4 Time courses of H_2 (*open circles*) and O_2 (*filled circles*) evolution in Cu_2O under visible light ($\lambda \geq 460$ nm) irradiation. Catalyst: 0.5 g, H_2O: 200 cm^3. The reaction system was evacuated with light irradiation after each run. Time courses in runs 5–8, 10–13 and 15–30 are omitted. Reprint with permission from [61], Copyright 1998, The Royal Society of Chemistry

for hydrogen generation such as earth abundant in nature, eco-friendly with low toxicity and available from various synthesis methods.

Cobalt oxide nanoparticles
Cobalt is also an earth-abundant transition metal. Co_3O_4 and CoO are most stable and commonly used oxides. Both oxides in bulk are not active for water splitting because their CBM is lower than hydrogen evolution potential, but nanoparticles of cobalt oxides have been recently observed to perform overall water splitting. Co_3O_4 has a bandgap of 2.1 eV. [62] Co_3O_4 is generally used as oxygen evolution photocatalyst or co-catalyst [63]. Overall water splitting oxygen by Co_3O_4 nanoparticles (quantum dots) without any co-catalyst under visible light irradiation was recently demonstrated [64]. The highly monodisperse Co_3O_4 QDs have a mean diameter of 3–4 nm, were synthesized using a reverse micelle method in a microwave-assisted solvothermal process, as shown in Fig. 12.5. The quantum confinement effect raises the CBM above the hydrogen evolution potential, making the quantum dots active for overall water splitting. The rates of hydrogen evolution

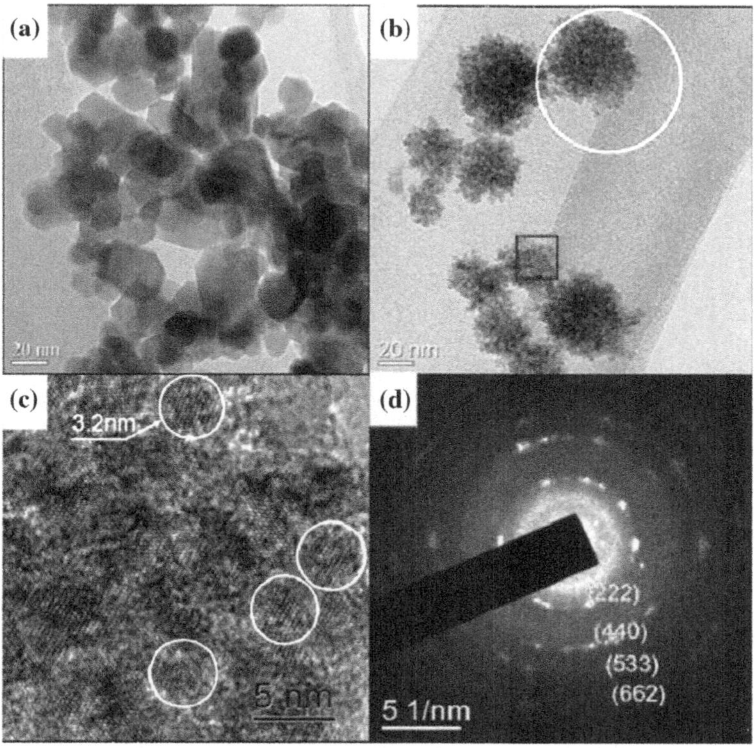

Fig. 12.5 **a**, **b** TEM images, **c** HRTEM image, and **d** selected-area electron diffraction pattern of **a** Co_3O_4-SSR (from solid-state reaction); **b**, **c** and **d** Co_3O_4-QDs (quantum dots). The squared region in (**b**) was used for HRTEM, and the circled region in (**b**) was used for selected-area electron diffraction. Reprint with permission from [64], Copyright 2002, The Royal Society of Chemistry

and oxygen evolution are reported to be 0.79 and 0.40 μmol h^{-1} with 0.01 g photocatalyst, respectively.

A higher rate overall water splitting was also observed in CoO nanoparticles (nanocrystals) [65]. CoO nanocrystals with particle size less than 10 nm (Fig. 12.6a) were fabricated by laser ablation of CoO micro-sized powders. Unlike Co_3O_4, CoO nanocrystals show no quantum confinement effect due to the relatively large size, but the band edge positions are upshifted and become aligned with water redox potentials, as shown in Fig. 12.6. CoO nanocrystals show a solar to hydrogen conversion efficiency of 5 %.

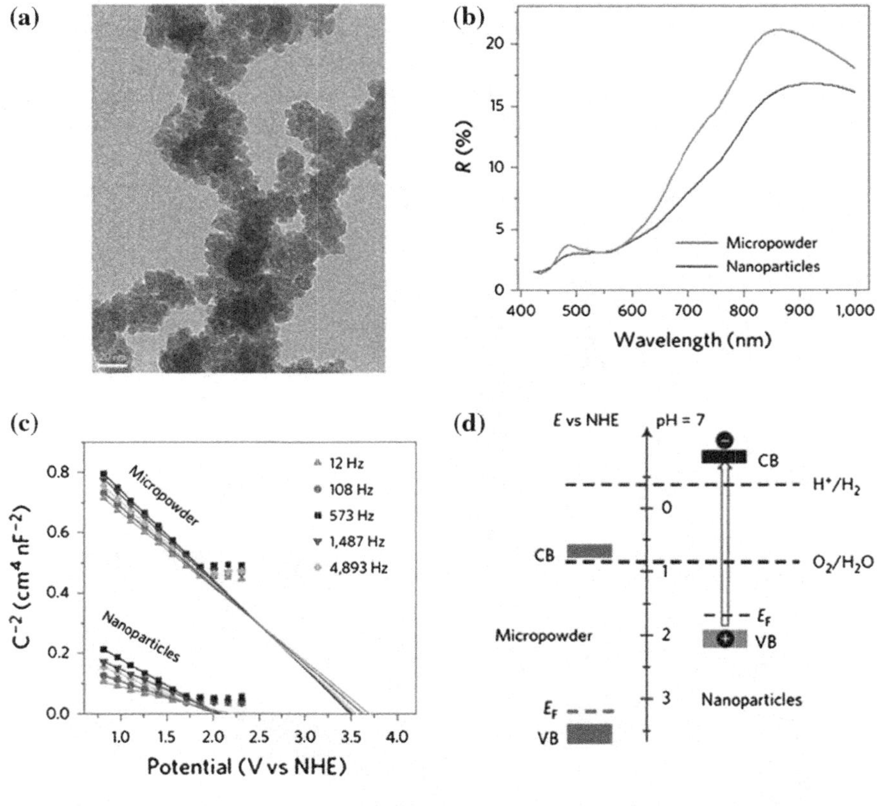

Fig. 12.6 **a** Transmission electron micrograph of CoO nanoparticles; **b** Ultraviolet-visible diffuse reflectance spectra of CoO nanocrystals and powders; **c** Mott-Schottky plots for CoO nanocrystals and micropowders according to impedance measurements. The flat-band potentials are obtained from the intercepts of the extrapolated lines. **d** Band positions of CoO nanocrystals and micropowders according to the bandgaps and flat-band potentials obtained from (**c**). *CB* conduction band; *VB* valence band. Reprint with permission from [65], Copyright 2014, Macmillan Publishers Limited

12.3.2 Metal Chalcogenide and Oxysulfide Nanomaterials

Metal chalcogenides are attractive candidates for visible light water splitting because they usually have narrower bandgaps with more negative VBM compared to that of metal oxides. Figure 12.7 summaries the band edge positions of some widely studied metal chalcogenides [19].

CdS is a well-known example of metal chalcogenide and has been extensively investigated as visible light photocatalyst for hydrogen production [66, 67]. The hydrogen evolution rate of ultrathin CdS nanosheets prepared from l-cysteine assisted exfoliation is up to 41.1 mmol h^{-1} g^{-1} under visible light irradiation [68]. However, CdS is not stable because of oxidation by photogenerated holes, and also Cd^{2+} can be dissolved in water and become a pollutant [69]. Several methods have been developed to consume photoexcited holes, including the addition of electron donors such as EDTA, sulfide and sulfite species into the reaction solution, and combination with other semiconductors [70].

MoS_2 and WS_2 are also well-studied electrocatalysts recently, but in bulk they are not suitable for photocatalytic overall water splitting because of narrow bandgaps of 1.17 and 1.35 eV, respectively, as shown in Fig. 12.7. However, monolayer MoS_2 and WS_2 prepared from chemical vapor deposition (CVD) [71, 72] or exfoliation methods [73–75] are very different from their respective bulks, they become direct bandgap semiconductors with bandgap large enough to theoretically perform overall water splitting [76, 77]. Figure 12.8 shows the CBM and VBM

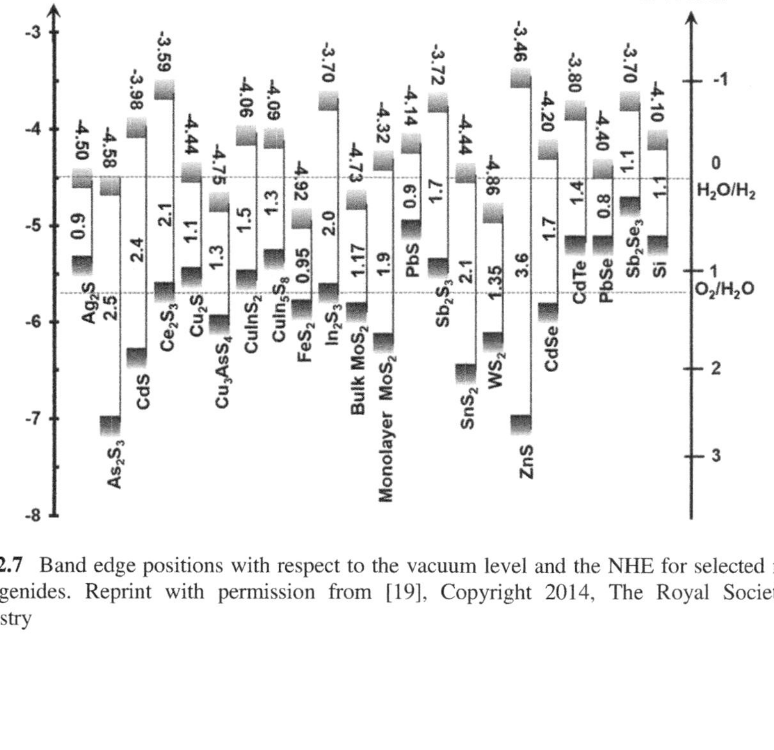

Fig. 12.7 Band edge positions with respect to the vacuum level and the NHE for selected metal chalcogenides. Reprint with permission from [19], Copyright 2014, The Royal Society of Chemistry

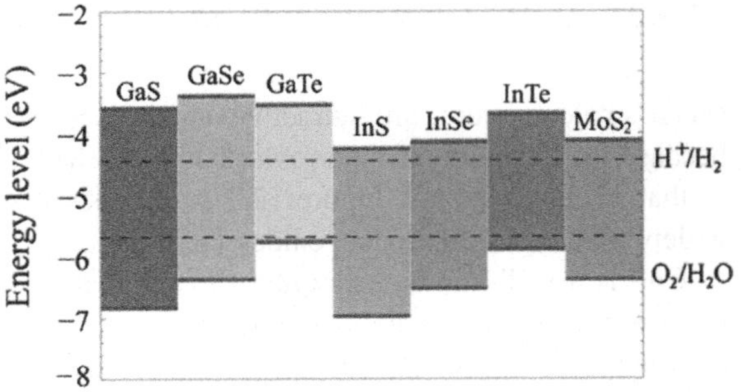

Fig. 12.8 Calculated band edge positions of monolayer metal chalcogenides relative to the vacuum level at zero strain. The band edge positions of monolayer MoS_2 and the standard redox potentials for water splitting at pH = 0 are shown for comparison [78]

energy levels with redox potentials of water splitting of monolayer group-III dichalcogenides [78]. All these monolayer chalcogenides have band edges located at energetically favorable positions for water splitting. However, there is no report on the experimental observation of hydrogen generation from solar water splitting.

Oxygen can be incorporated into the structure of metal sulfides to form oxysulfide materials. This type of materials is different from S doped metal oxides. The most typical oxysulfide is $Ln_2Ti_2S_2O_5$ (Ln: Pr, Nd, Sm, Gd, Tb, Dy, Ho, and Er) [79]. These materials can be prepared either through heat treatment of the corresponding precursors under vacuum or by sulfurization of the corresponding metal oxide powder using H_2S gas. The reason for the visible light response of these materials is because of the hybridized *O2p* and *S2p* orbitals. Similar with titanates, the above mentioned oxysulfides have layered perovskite structure with the layers composed of S–(TiO_2)–O–(TiO_2)–S double octahedra. Hydrogen can be produced if co-catalysts are loaded. For example, Pt sensitized $Sm_2Ti_2S_2O_5$ shows steady hydrogen evolution under visible light illumination [80].

12.3.3 Metal Nitride and Oxynitride Nanomaterials

Metal nitrides such as GaN, Ge_3N_4, and Ta_3N_5, and metal oxynitrides like TaON and the solid solution of $(Ga_{1-x}Zn_x)(N_{1-x}O_x)$ are also active for water splitting. [79, 81–83] For metal nitrides, the valence band consists of *N2p* orbitals and the conduction band is formed by hybridized orbitals. For metal oxynitrides, the valence band consists of hybridized *N2p* and *O2p* orbitals [84]. The conduction band is formed by hybridized orbitals like Ge *4s4p* orbitals. This electronic configuration makes metal nitrides to have wide bandgap but narrower than that of metal oxides.

Metal nitrides can promote overall water splitting. For example, Ta_3N_5 can catalyze water decomposition to hydrogen and oxygen under visible light irradiation (less than 600 nm) [85]. Oxygen has lager electronegativity than nitrogen and hence metal oxynitrides usually have a little wider bandgap than metal nitrides. The bandgap of TaON is about 2.4 eV with the light absorption edge located at 520 nm, while for Ta_3N_5 the bandgap is about 2.1 eV with the light absorption edge at about

590 nm [50]. ($Ga_{1-x}Zn_x$)($N_{1-x}O_x$) solid solution also has a narrow bandgap in the range of 2.4–2.8 eV, allowing the absorption of visible light [86]. The average apparent quantum yield of this solid solution in the range of 300–480 nm is about 0.14 % for overall water splitting [87]. Even more higher quantum yield of 2.5 % at 420–440 nm for overall water splitting can be reached by $Rh_{2-y}Cr_yO_3$ loaded ($Ga_{1-x}Zn_x$)($N_{1-x}O_x$) [88]. Stoichiometric dissociation of water into hydrogen and oxygen under visible light irradiation in the range of 400–475 nm are also realized by using InGaN nanowire arrays with the apparent quantum yield of about 12.3 % in neutral solution [89].

12.3.4 Other Newly Developed Nanomaterials

Graphite oxide and graphene oxide quantum dot

Single-phased visible light active nanomaterials are always attractive for water splitting. Recently, researchers reported graphite oxide as a new photocatalyst that can generate hydrogen under UV light and visible light irradiation [90]. Figure 12.9 shows the layered structure of the graphite oxide (GO) prepared from pristine

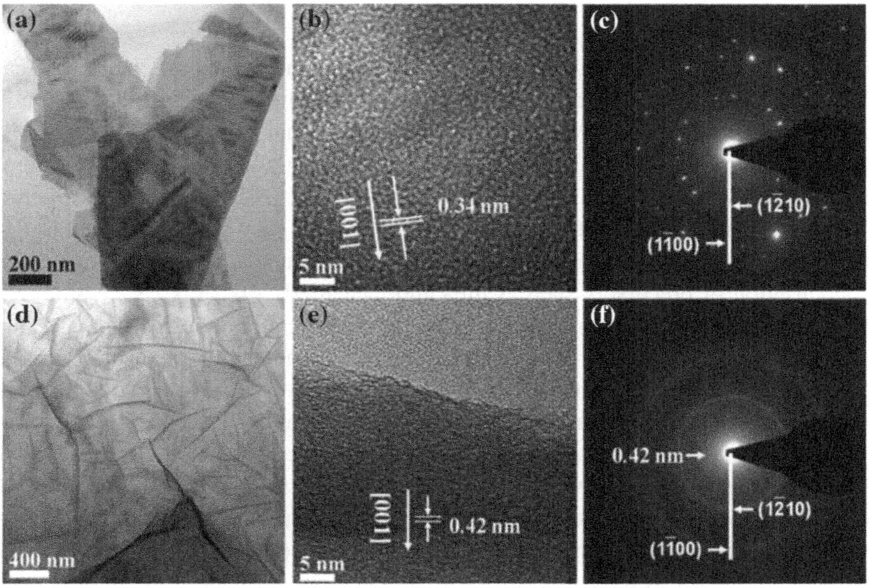

Fig. 12.9 **a** Low-magnification HRTEM image of pristine graphite (PG); **b** edge-on HRTEM image of PG, illustrating the PG interlayer spacing of 0.34 nm; **c** selected-area electron diffraction pattern of PG in panel (**b**); **d** low-magnification HRTEM image of GO (graphite oxide) winkled sheets; **e** edge-on HRTEM image of GO, illustrating the GO interlayer spacing of 0.42 nm; **f** selected-area election diffraction pattern of GO in panel (**e**). Reprint with permission from [90], Copyright 2010, Wiley-VCH

graphite (PG) by a modified Hummers method. The main feature of GO is the linked oxygen functional groups to the carbon backbone, which not only make GO highly soluble in water but also provide possibility to tune the bandgap. The synthesized GO can absorb both UV light and visible light with the absorption edge at 500 nm. The observed quantum yield of H_2 evolution was 2.7 and 0.01 % for UV light and visible light, respectively. Despite of the low quantum yield for visible light, the oxygen functional groups allow the further modification to enable GO more suitable for water splitting to generate hydrogen fuels.

Although graphite oxide has a large enough bandgap, its VBM is very close to O_2 evolution potential, so overall water splitting cannot be realized. To solve this problem, Teng and coworkers developed nitrogen-doped graphene oxide quantum dot (NGO-QDs) [91]. Graphene oxide is a monolayer graphite oxide, it itself is *p*-type due to the oxygen functional groups. Because of the different electronegativity between nitrogen and oxygen, *n*-type feature can be introduced into graphene oxide by partially replacing oxygen with nitrogen. Figure 12.10 shows TEM images of NGO-QDs with the size of about 8.1 nm prepared from the treatment of graphene oxide (GO) in NH_3 followed by oxidation. The band gap of NGO-QDs is about 2.2 eV. More importantly, the positions of band edges straddle the water redox potentials. As a result, NGO-QDs with built-in p-n structure are able to split water into hydrogen and oxygen in a stoichiometric ratio, while the nitrogen-free QDs only catalyze hydrogen evolution.

Carbon nitride

Carbon nitride is a new class of polymer consisting of carbon and nitrogen. C_3N_4 can be prepared in the form of crystalline β–C_3N_4 and graphitic carbon nitride (g–C_3N_4), as well as amorphous C_3N_4. β–C_3N_4 has a hexagonal network of tetrahedrally bonded carbon and trigonal planar nitrogen, while g–C_3N_4 is a highly ordered polymer consisting of tri-sitriazine building units. g–C_3N_4 has a layered structure with weak van der Waals force between C–N layers, very similar to graphite. g–C_3N_4 can either be exfoliated from its bulk solids or synthesized through poly-condensation of precursors such as urea and cyanamide [92]. g–C_3N_4 synthesized from the polycondensation method usually has a small amount of hydrogen because of incomplete condensation of precursors. These hydrogen atoms are usually present as amine groups on the edges, as shown in Fig. 12.11a [93]. Electron-rich nitrogen atoms and bonded hydrogen make g–C_3N_4 an excellent catalyst.

Recently, g–C_3N_4 prepared from the polycondensation method is reported to show water splitting activity [93]. The bandgap of the as-prepared g–C_3N_4 powder is about 2.7 eV, with the conduction and valence band edges higher than water redox potentials by 0.9 and 0.6 eV, and spatially separated carbon atoms and nitrogen atoms act as the active sites for reduction and oxidation respectively. It is found that in triethanolamine aqueous solution g–C_3N_4 powder shows hydrogen production under visible light illumination (>420 nm). The hydrogen evolution activity of bare g–C_3N_4 powder was not steady and varied from batch to batch.

Similar photocatalytic activity was also observed by Liu and coworkers in amorphous C_3N_4, which was prepared by heating g–C_3N_4 in an atmosphere of

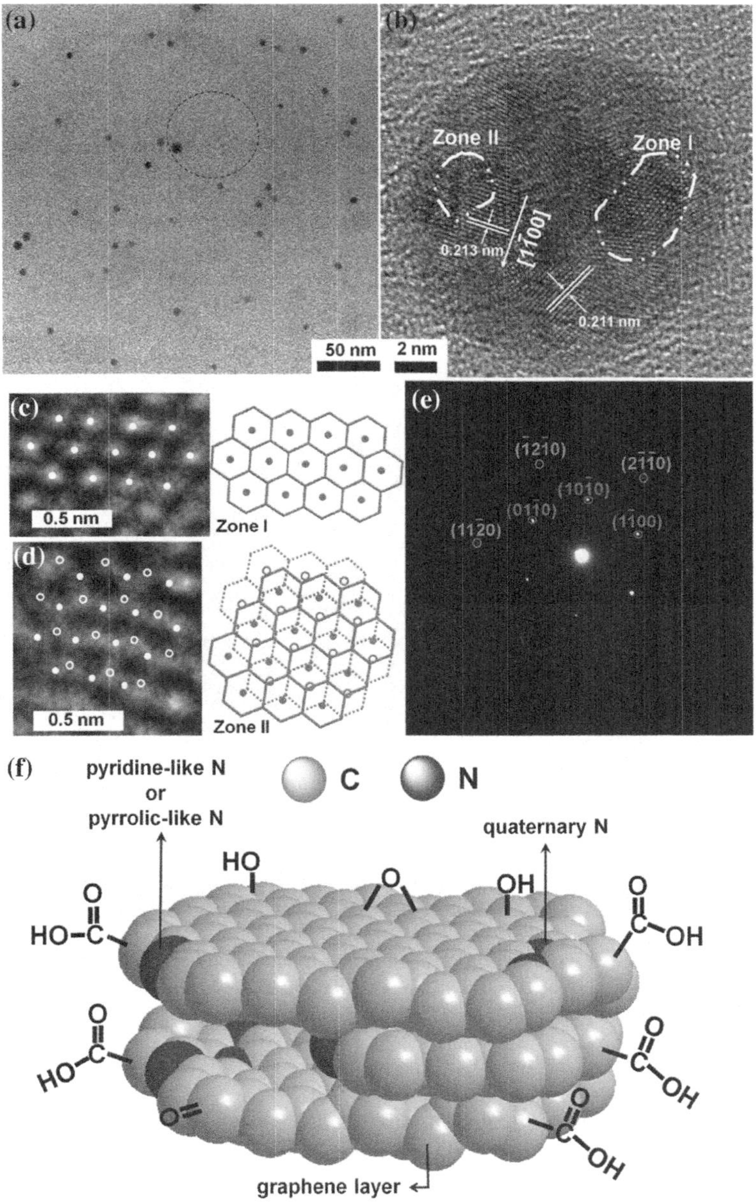
(a)
(b)
Zone II
Zone I
0.213 nm
[1$\bar{1}$00]
0.211 nm
50 nm
2 nm
(c)
0.5 nm
Zone I
(d)
0.5 nm
Zone II
(e)
($\bar{1}2\bar{1}0$)
($2\bar{1}\bar{1}0$)
($10\bar{1}0$)
($01\bar{1}0$)
($1\bar{1}00$)
($11\bar{2}0$)
(f)
pyridine-like N
or
pyrrolic-like N
C
N
quaternary N
graphene layer

◀ **Fig. 12.10** Morphology and crystal structure of NGO-QDs (nitrogen-doped graphene oxide quantum dots). **a** TEM image of NGO-QDs, also exhibiting the *circled* electron-beam spot for the high-resolution image and SAED analyses revealed in panels (**b**) and (**e**). **b** High-resolution TEM image of a NGO-QD, illustrating the graphene $\{1\bar{1}00\}$ lattice planes with a d-spacing of 0.213 nm and showing two specified zones. **c**, **d** Magnified zones I and II of the TEM image in panel **b**. The *right-hand sides* of panels **c** and **d** show the schematic graphene structure based on the *bright spots* on the TEM image. The *bright spots* correspond to the benzene-ring centers on graphene. **e** The selected-area electron diffraction pattern of the NGO-QD in panel (**b**). **f** A conceptual schematic of the NGO-QD structure. The main body of the NGO-QD consists of nitrogen-doped graphene layers stacked in the AB order, with outer surface bore with oxygen functional groups. Reprint with permission from [91], Copyright 2014, Wiley-VCH

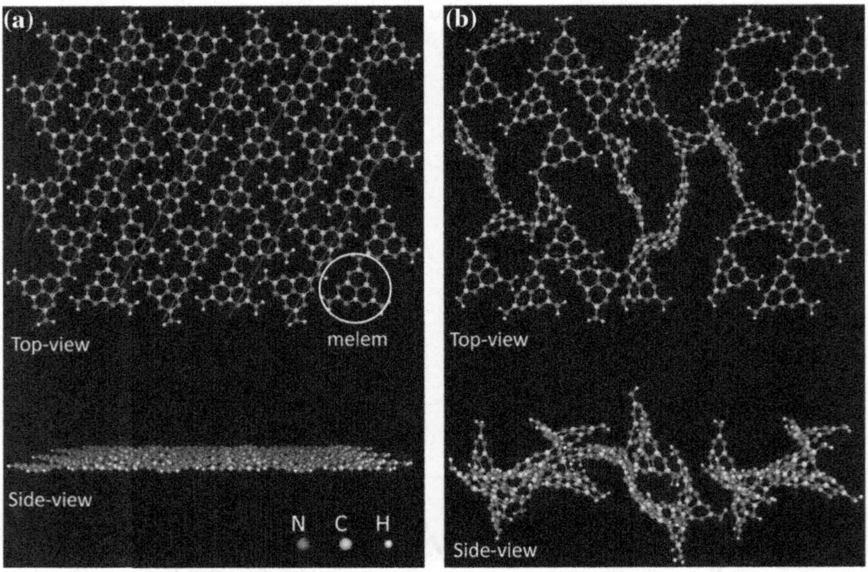

Fig. 12.11 Schematic of monolayer **a** crystalline and **b** amorphous graphitic carbon nitrides. N, C, and H atoms are represented by *light blue*, *white*, and *small yellow balls*, respectively. The melem unit is marked by the *white circle* in (**a**). Reprint with permission from [94], Copyright 2015, Wiley-VCH

argon [94]. This thermal treatment disrupted long-range atomic order between melem units, shown in Fig. 12.11b. The amorphous C_3N_4 has a smaller bandgap of about 1.90 eV, and shows an average hydrogen generation rate of 157.9 μm $h^{-1}g^{-1}$ in triethanolamine aqueous solution under visible light irradiation.

However, neither g–C_3N_4 nor amorphous C_3N_4 can realize overall water splitting. Furthermore, noble metal co-catalysts and sacrificial reagents are still needed to improve hydrogen generation rate. A recent breakthrough is the demonstration of overall water splitting by carbon dot decorated C_3N_4 composites (CDots-C_3N_4) without any co-catalysts and sacrificial reagents [95]. Figure 12.12 shows the microscopic structure of the composite. CDots are graphite particles with less than

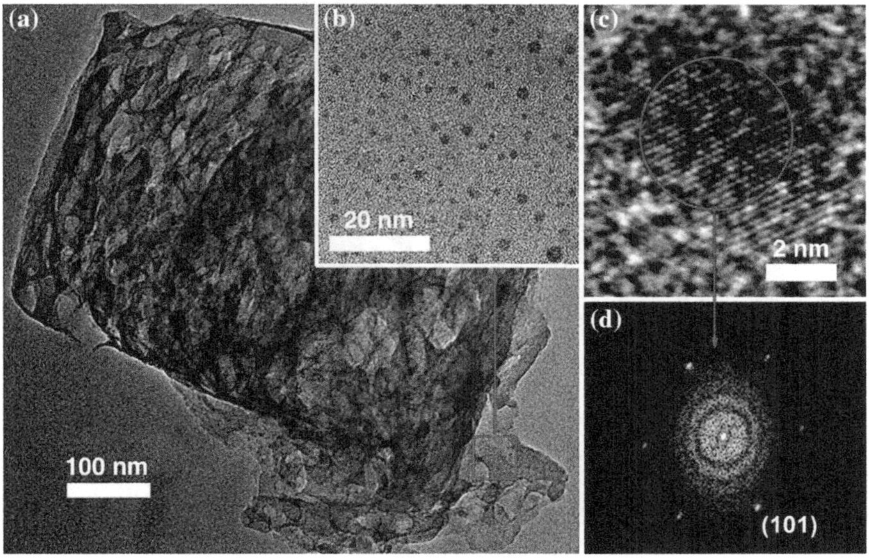

Fig. 12.12 Characterization of the physical structure of CDots-C_3N_4 composite catalyst. **a** TEM image of a grain of the composite. **b** A magnified TEM image of the region of (**a**) marked in *red*. **c** HRTEM image of a single CDot embedded in C_3N_4. **d** Corresponding FFT pattern of the lattice image in (**c**), indicating hexagonal symmetry. In all panels, the CDots concentration of the sample was 1.6×10^{-5} $g_{CDots}/g_{catalyst}$. Reprint with permission from [95], Copyright 2015, Science

10 nm in diameter, they are excellent catalyst for H_2O_2 decomposition. In addition, CDots can facilitate charge transfer and separation, even generate electrons and holes by absorbing incident photons. As a result, light absorption of the composite is significantly improved in the entire UV-vis range. Total water splitting with a high quantum yield of 16 % at 420 nm was achieved. The solar to hydrogen conversion efficiency reached as high as 1.78 % for optimized samples. It is suggested that water is first oxidized through two-electron transfer to form H_2O_2 and H_2, followed by decomposition of H_2O_2 to H_2O and O_2.

Boron nitride (BN)

Boron nitride (BN) is an important inorganic two-dimensional nanomaterial. It is widely used as transparent and flat substrate for other nanomaterials such as graphene. Hexagonal boron nitride (*h*-BN) has a large bandgap of 5.5 eV, so it is not suitable for solar energy conversion. However, by carbon doping through a pyrolysis method, the bandgap of boron nitride can be greatly reduced to 2.7 eV. The carbon-doped boron nitride (BCN) is prepared by treating the mixture of glucose, boron oxide and urea at 1250 °C in ammonia atmosphere [96]. The doping level can be controlled by tuning the weight ratio of glucose precursor to boron oxide. Figure 12.13 shows the morphology characterization of BCN-30 with glucose to boron oxide ratio of 30 %. A flake-like morphology similar to few-layer graphene can be clearly seen with a uniform distribution of B, C and N. The ternary

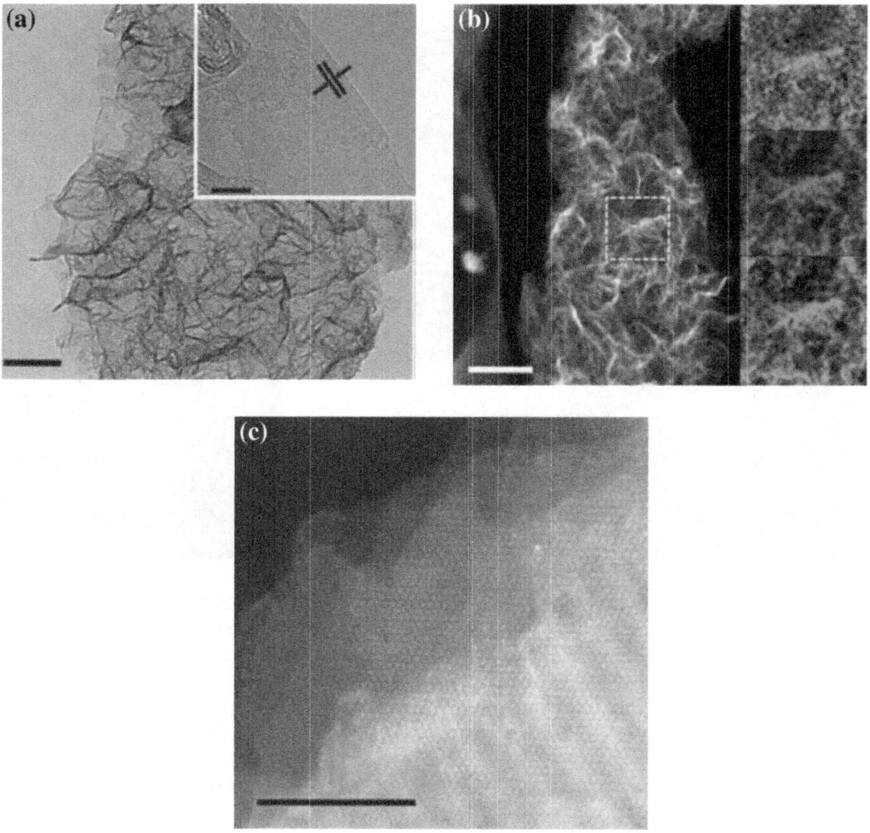

Fig. 12.13 Surface morphology of BCN-30. **a** HRTEM image of the BCN-30 sample. *Scale bar*, 100 nm, *inset scale bars*, 5 nm. **b** Typical TEM dark-field image of BCN-30 sample and the elemental mapping images of B, C and N of the enlargement of the selected-area in the picture. *Scale bar*, 300 nm. **c** High-resolution STEM of the BCN-30 sample along (002) facet. *Scale bar*, 5 nm. Reprint with permission from [96], Copyright 2015, Macmillan Publishers Limited

BCN nanosheets evolve hydrogen without co-catalysts under visible light, but suffer from deactivation under the UV light illumination in triethanolamine aqueous solution. The reason for the deactivation is suggested to the over-reduction of the –N=B-structure and the consecutive reductive disintegration of BCN.

Plasmonic nanomaterials

Plasmonic nanomaterials are often used to decorate semiconductor photocatalysts to enhance the solar energy conversion efficiency via enhanced light absorption/scattering, or resonance energy transfer [97–101]. Noble metals Au [102–104], Ag [105] and Cu [97, 106] are commonly used plasmonic nanomaterials.

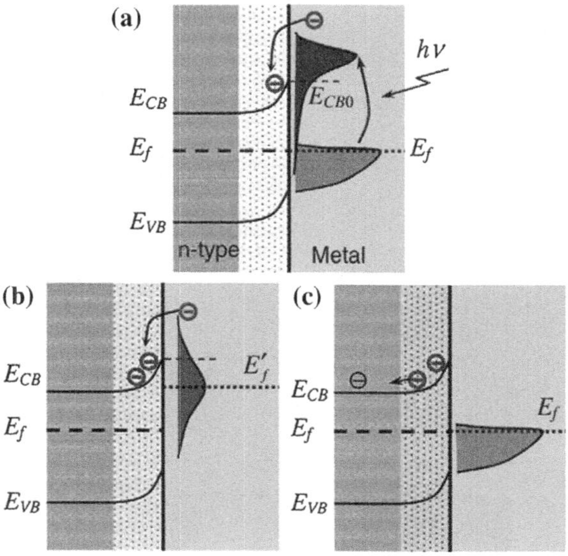

Fig. 12.14 Schematic illustration of the hot-electron injection effect: **a** excitation of the electrons from thermal equilibrium to a high-energy state upon absorbing photons and injection of the electrons to the conduction band (CB) of the semiconductor. **b** Redistribution of electron energy and formation of a Fermi-Dirac distribution at a high-temperature Fermi level after the injection of the electrons. **c** Restoration of the standard electron distribution with electrons and holes flowing to different regions in the semiconductor. Reprint with permission from [108], Copyright 2013, IOP Publishing

Recently, plasmonic nanomaterials are demonstrated to absorb light and generate electrons like semiconductor photocatalysts. Figure 12.14 shows that hot-electrons are induced by localized surface plasmon resonance (LSPR) and then spill into semiconductor co-catalysts via Schottky junction [97]. The size, morphology, and the type of plasmonic metals will all affect hot-electron generation and injection. For example, it is found that the optimal size of spherical gold nanoparticle is 10–20 nm [107].

By coupling plasmonic nanomaterials to hydrogen and oxygen evolution co-catalysts, overall water-splitting can be realized [109, 110]. Figure 12.15 shows the structure and operation of a complete water-splitting device. Au nanorods serve as light harvesting antennae to generate hot electrons, and TiO_2/Pt functions as electron transport layer and hydrogen evolution co-catalyst. A solar to hydrogen conversion efficiency of 0.1 % was achieved.

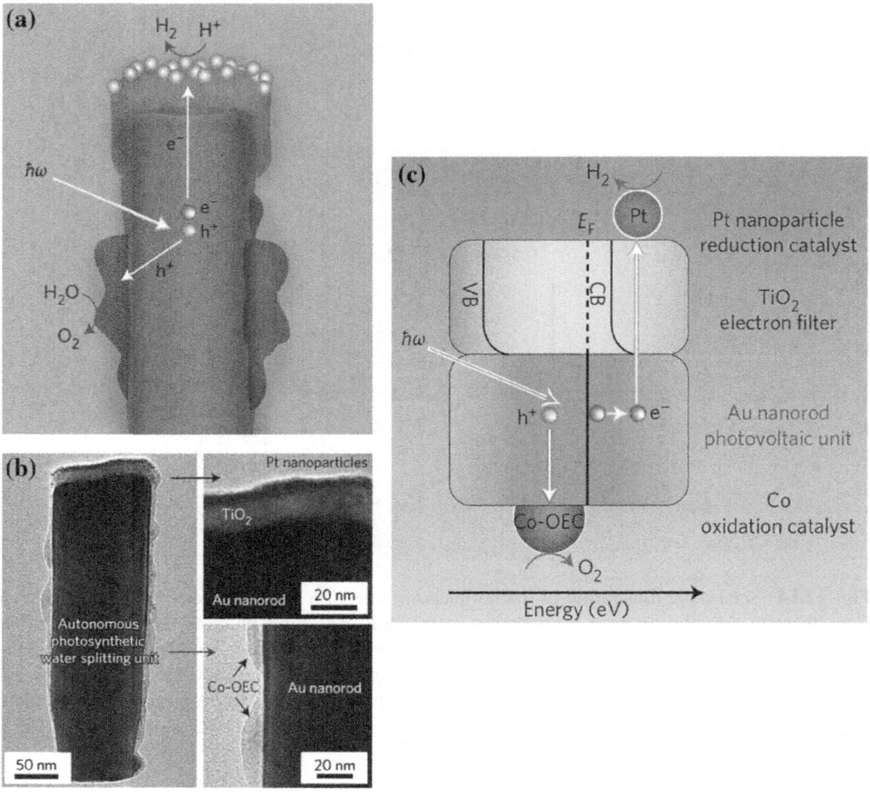

Fig. 12.15 Structure and mechanism of operation of the autonomous plasmonic solar water splitter. **a** Schematic of the cross-section of an individual photosynthetic unit showing the inner gold nanorod, the TiO_2 cap decorated with platinum nanoparticles, which functions as the hydrogen evolution catalyst, and the Co-OEC material deposited on the lower portion of the gold nanorod. **b** Corresponding transmission electron micrograph (*left*) and magnified views of the platinum/TiO_2 cap (*top right*) and the Co-OEC (*bottom right*). **c** Energy level diagram superimposed on a schematic of an individual unit of the plasmonic solar water splitter, showing the proposed processes occurring in its various parts and in energy space. *CB* conduction band; *VB* valence band; E_F Fermi energy. Reprint with permission from [110], Copyright 2013, Macmillan Publishers Limited

12.4 Conclusions and Outlook

Hundreds of photocatalysts have been identified or developed over past several decades, out of them, only very few are active and able to do water splitting under visible light, as presented in this chapter. The emergence of visible light photocatalysts is an important development toward high efficiency solar hydrogen generation, but current photocatalytic nanomaterials still suffer from many serious issues: they exhibit either low activity or short life time. For example, the efficiency of graphitic C_3N_4, Co_3O_4 quantum dots and graphene oxide quantum dots still falls below 2 %, CoO nanoparticles can only last for 1 h. A photocatalyst that can realize 10 % solar to hydrogen efficiency over long term operation is still missing. This is in a sharp contrast to photovoltaic devices such as silicon solar cells which produce solar electricity at >10 % efficiency over more than 30 years. It should be noted again that photocatalysts have to meet more stringent material requirements than those for photovoltaic cells.

Despite low performances and many challenging material problems, nanomaterials still hold the promise for low-cost clean solar fuel. This is because the efficiency of current photocatalysts is still far below the theoretical limit which is about the same as photovoltaic efficiency of the same materials. Nanomaterials such as nanoparticles have the perfect morphology for catalytic reaction owing to large surface area and high density of active sites. Nanomaterials can also allow photoexcited electrons and holes to quickly diffuse to the surface and initiate water splitting. Another advantage of nanomaterials is that they can be designed and fabricated to accommodate specific photonic and electronic properties. The efficiency of current nanomaterials is expected to improve once efficiency limiting steps are identified and overcome. New nanomaterials, such as monolayer transition metal dichalcogenides, will continue to emerge. Finally, it is expected that high efficiency photocatalysts cannot be simply implemented by a single phase material, the desired device should be a nanocomposite combining different functionalities of several nanomaterials as shown in Fig. 12.1. Certainly, they have to be synthesized by low-cost and green chemical methods.

Acknowledgments Jiming Bao acknowledges support from the Robert A Welch Foundation (E-1728). Zhiming Wang acknowledges financial support from 111 project (No. B13042). Zhenhuan Zhao acknowledges the financial support from National Science Foundation for Post-doctoral Scientists of China (2015M582538).

References

1. L. Fajersztajn, M. Veras, L.V. Barrozo, P. Saldiva, Air pollution: a potentially modifiable risk factor for lung cancer. Nat. Rev. Cancer **13**, 674–678 (2013)
2. B. Subhadra, Natural resources, renewable energy sources, GHG-emission and demographic profiles in United States: a broad analysis for developing sustainable low-carbon energy sector. Energy Sci. Technol. **5**, 36–49 (2013)
3. N. Armaroli, V. Balzani, The future of energy supply: challenges and opportunities. Angew. Chem. Int. Ed. **46**, 52–66 (2007)
4. S. Chu, A. Majumdar, Opportunities and challenges for a sustainable energy future. Nature **488**, 294–303 (2012)
5. A. Fujishima, K. Honda, TiO_2 photoelectrochemistry and photocatalysis. Nature **238**, 37–38 (1972)
6. M.G. Walter, E.L. Warren, J.R. McKone, S.W. Boettcher, Q. Mi, E.A. Santori, N.S. Lewis, Solar water splitting cells. Chem. Rev. **110**, 6446–6473 (2010)
7. Y. Tachibana, L. Vayssieres, J.R. Durrant, Artificial photosynthesis for solar water-splitting. Nat. Photon. **6**, 511–518 (2012)

8. Z.S. Li, W.J. Luo, M.L. Zhang, J.Y. Feng, Z.G. Zou, Photoelectrochemical cells for solar hydrogen production: current state of promising photoelectrodes, methods to improve their properties, and outlook. Energy Environ. Sci. **6**, 347–370 (2013)
9. T. Wang, Z. Luo, C. Li, J. Gong, Controllable fabrication of nanostructured materials for photoelectrochemical water splitting via atomic layer deposition. Chem. Soc. Rev. **43**, 7469–7484 (2014)
10. Y. Moriya, T. Takata, K. Domen, Recent progress in the development of (oxy)nitride photocatalysts for water splitting under visible-light irradiation. Coordin. Chem. Rev. **257**, 1957–1969 (2013)
11. M.J. Katz, S.C. Riha, N.C. Jeong, A.B.F. Martinson, O.K. Farha, J.T. Hupp, Toward solar fuels: water splitting with sunlight and "rust"? Coordin. Chem. Rev. **256**, 2521–2529 (2012)
12. S.C. Warren, K. Voitchovsky, H. Dotan, C.M. Leroy, M. Cornuz, F. Stellacci, C. Hebert, A. Rothschild, M. Gratzel, Identifying champion nanostructures for solar water-splitting. Nat. Mater. **12**, 842–849 (2013)
13. S. Fukuzumi, D.C. Hong, Y. Yamada, Bioinspired photocatalytic water reduction and oxidation with earth-abundant metal catalysts. J. Phys. Chem. Lett. **4**, 3458–3467 (2013)
14. A.A. Ismail, D.W. Bahnemann, Photochemical splitting of water for hydrogen production by photocatalysis: a review. Sol. Energ Mater. Sol. Cells **128**, 85–101 (2014)
15. J. Ran, J. Zhang, J. Yu, M. Jaroniec, S.Z. Qiao, Earth-abundant cocatalysts for semiconductor-based photocatalytic water splitting. Chem. Soc. Rev. **43**, 7787–7812 (2014)
16. F.E. Osterloh, Inorganic materials as catalysts for photochemical splitting of water. Chem. Mater. **20**, 35–54 (2008)
17. X. Chen, S. Shen, L. Guo, S.S. Mao, Semiconductor-based photocatalytic hydrogen generation. Chem. Rev. **110**, 6503–6570 (2010)
18. J. Zhu, D. Chakarov, M. Zäch, *Energy efficiency and renewable energy through nanotechnology* (Springer, London, 2011)
19. J.T. Li, N.Q. Wu, Semiconductor-based photocatalysts and photoelectrochemical cells for solar fuel generation: a review. Catal. Sci. Technol. **5**, 1360–1384 (2015)
20. J. Tian, Z. Zhao, A. Kumar, R.I. Boughton, H. Liu, Recent progress in design, synthesis, and applications of one-dimensional TiO_2 nanostructured surface heterostructures: a review. Chem. Soc. Rev. **43**, 6920–6937 (2014)
21. Z. Zhao, J. Tian, Y. Sang, A. Cabot, H. Liu, Structure, synthesis, and applications of TiO_2 nanobelts. Adv. Mater. **27**, 2557–2582 (2015)
22. A.L. Linsebigler, G. Lu, J.T. Yates Jr, Photocatalysis on TiO_2 surfaces: principles, mechanisms, and selected results. Chem. Rev. **95**, 735–758 (1995)
23. J. Yu, L. Qi, M. Jaroniec, Hydrogen production by photocatalytic water splitting over Pt/TiO_2 nanosheets with exposed (001) facets. J. Phys. Chem. C **114**, 13118–13125 (2010)
24. T. Kawai, T. Sakata, Photocatalytic decomposition of gaseous water over TiO_2 and TiO_2—RuO_2 surfaces. Chem. Phys. Lett. **72**, 87–89 (1980)
25. F.N. Sayed, O. Jayakumar, R. Sasikala, R. Kadam, S.R. Bharadwaj, L. Kienle, U. Schürmann, S.r. Kaps, R. Adelung, J. Mittal, Photochemical hydrogen generation using nitrogen-doped TiO_2–Pd nanoparticles: facile synthesis and effect of Ti^{3+} incorporation. J. Phys. Chem. C **116**, 12462–12467 (2012)
26. A. Galinska, J. Walendziewski, Photocatalytic water splitting over Pt-TiO_2 in the presence of sacrificial reagents. Energy Fuels **19**, 1143–1147 (2005)
27. H. Arakawa, K. Sayama, Solar hydrogen production. Significant effect of Na_2CO_3 addition on water splitting using simple oxide semiconductor photocatalysts. Catal. Surv. Jpn **4**, 75–80 (2000)
28. T. Chen, Z. Feng, G. Wu, J. Shi, G. Ma, P. Ying, C. Li, Mechanistic studies of photocatalytic reaction of methanol for hydrogen production on Pt/TiO_2 by in situ Fourier transform IR and time-resolved IR spectroscopy. J. Phys. Chem. C **111**, 8005–8014 (2007)
29. D.O. Scanlon, C.W. Dunnill, J. Buckeridge, S.A. Shevlin, A.J. Logsdail, S.M. Woodley, C.R. Catlow, M.J. Powell, R.G. Palgrave, I.P. Parkin, G.W. Watson, T.W. Keal, P. Sherwood, A. Walsh, A.A. Sokol, Band alignment of rutile and anatase TiO_2. Nat. Mater. **12**, 798–801 (2013)

30. T.K. Townsend, N.D. Browning, F.E. Osterloh, Nanoscale strontium titanate photocatalysts for overall water splitting. ACS Nano **6**, 7420–7426 (2012)
31. K. Domen, A. Kudo, T. Onishi, N. Kosugi, H. Kuroda, Photocatalytic decomposition of water into hydrogen and oxygen over nickel (II) oxide-strontium titanate ($SrTiO_3$) powder. 1. Structure of the catalysts. J. Phys. Chem. C **90**, 292–295 (1986)
32. F. Wagner, G. Somorjai, Photocatalytic and photoelectrochemical hydrogen production on strontium titanate single crystals. J. Am. Chem. Soc. **102**, 5494–5502 (1980)
33. M. Shibata, A. Kudo, A. Tanaka, K. Domen, K.-I. Maruya, T. Onishi, Photocatalytic activities of layered titanium compounds and their derivatives for H_2 evolution from aqueous methanol solution. Chem. Lett. **16**, 1017–1018 (1987)
34. M.R. Allen, A. Thibert, E.M. Sabio, N.D. Browning, D.S. Larsen, F.E. Osterloh, Evolution of physical and photocatalytic properties in the layered titanates $A_2Ti_4O_9$ (A = K, H) and in nanosheets derived by chemical exfoliation. Chem. Mater. **22**, 1220–1228 (2009)
35. J.S. Jang, S.H. Choi, D.H. Kim, J.W. Jang, K.S. Lee, J.S. Lee, Enhanced photocatalytic hydrogen production from water−methanol solution by nickel intercalated into titanate nanotube. J. Phys. Chem. C **113**, 8990–8996 (2009)
36. A. Kudo, T. Kondo, Photoluminescent and photocatalytic properties of layered caesiumtitanates, $Cs_2Ti_nO_{2n+1}$ (n = 2, 5, 6). J. Mater. Chem. **7**, 777–780 (1997)
37. J. Xing, W.Q. Fang, H.J. Zhao, H.G. Yang, Inorganic photocatalysts for overall water splitting. Chem. Asian J. **7**, 642–657 (2012)
38. S. Ogura, M. Kohno, K. Sato, Y. Inoue, Photocatalytic activity for water decomposition of RuO 2-combined $M_2Ti_6O_{13}$ (M = Na, K, Rb, Cs). Appl. Surf. Sci. **121**, 521–524 (1997)
39. H.G. Kim, D.W. Hwang, S.W. Bae, J.H. Jung, J.S. Lee, Photocatalytic water splitting over $La_2Ti_2O_7$ synthesized by the polymerizable complex method. Catal. Lett. **91**, 193–198 (2003)
40. R. Abe, M. Higashi, Z. Zou, K. Sayama, Y. Abe, Photocatalytic water splitting into H_2 and O_2 over $R_2Ti_2O_7$ (R = Y, rare earth) with pyrochlore structure. Chem. Lett. **33**, 954–955 (2004)
41. K. Sayama, H. Arakawa, K. Domen, Photocatalytic water splitting on nickel intercalated $A_4Ta_xNb_{6\text{-}x}O_{17}$ (A = K, Rb). Catal. Today **28**, 175–182 (1996)
42. X. Chen, T. Yu, X. Fan, H. Zhang, Z. Li, J. Ye, Z. Zou, Enhanced activity of mesoporous Nb_2O_5 for photocatalytic hydrogen production. Appl. Surf. Sci. **253**, 8500–8506 (2007)
43. Y.-H. Pai, S.-Y. Fang, Preparation and characterization of porous Nb_2O_5 photocatalysts with CuO, NiO and Pt cocatalyst for hydrogen production by light-induced water splitting. J. Power Sources **230**, 321–326 (2013)
44. T. Takata, A. Tanaka, M. Hara, J.N. Kondo, K. Domen, Recent progress of photocatalysts for overall water splitting. Catal. Today **44**, 17–26 (1998)
45. K. Domen, A. Kudo, A. Shinozaki, A. Tanaka, K.-i. Maruya, T. Onishi, Photodecomposition of water and hydrogen evolution from aqueous methanol solution over novel niobate photocatalysts. J. Chem. Soc. Chem. Commun. 356–357 (1986)
46. Y. Miseki, H. Kato, A. Kudo, Water splitting into H_2 and O_2 over $Cs_2Nb_4O_{11}$ photocatalyst. Chem. Lett. **34**, 54–55 (2005)
47. H. Kim, D. Hwang, Y. Kim, J. Lee, Highly donor-doped (110) layered perovskite materials as novel photocatalysts for overall water splitting. Chem. Commun. **12**, 1077–1078 (1999)
48. K. Maeda, M. Eguchi, W.J. Youngblood, T.E. Mallouk, Calcium niobate nanosheets prepared by the polymerized complex method as catalytic materials for photochemical hydrogen evolution. Chem. Mater. **21**, 3611–3617 (2009)
49. K. Domen, J. Yoshimura, T. Sekine, A. Tanaka, T. Onishi, A novel series of photocatalysts with an ion-exchangeable layered structure of niobate. Catal. Lett. **4**, 339–343 (1990)
50. W.-J. Chun, A. Ishikawa, H. Fujisawa, T. Takata, J.N. Kondo, M. Hara, M. Kawai, Y. Matsumoto, K. Domen, Conduction and valence band positions of Ta_2O_5, TaON, and Ta_3N_5 by UPS and electrochemical methods. J. Phys. Chem. B **107**, 1798–1803 (2003)

51. Y. Takahara, J.N. Kondo, T. Takata, D. Lu, K. Domen, Mesoporous tantalum oxide. 1. Characterization and photocatalytic activity for the overall water decomposition. Chem. Mater. **13**, 1194–1199 (2001)
52. T. Sreethawong, S. Ngamsinlapasathian, Y. Suzuki, S. Yoshikawa, Nanocrystalline mesoporous Ta_2O_5-based photocatalysts prepared by surfactant-assisted templating sol–gel process for photocatalytic H_2 evolution. J. Mol. Catal. A: Chem. **235**, 1–11 (2005)
53. Y. Noda, B. Lee, K. Domen, J.N. Kondo, Synthesis of crystallized mesoporous tantalum oxide and its photocatalytic activity for overall water splitting under ultraviolet light irradiation. Chem. Mater. **20**, 5361–5367 (2008)
54. H. Kato, A. Kudo, New tantalate photocatalysts for water decomposition into H_2 and O_2. Chem. Phys. Lett. **295**, 487–492 (1998)
55. H. Kato, A. Kudo, Water splitting into H_2 and O_2 on alkali tantalate photocatalysts $ATaO_3$ (A = Li, Na, and K). J. Phys. Chem. B **105**, 4285–4292 (2001)
56. J. Liu, G. Chen, Z. Li, Z. Zhang, Hydrothermal synthesis and photocatalytic properties of $ATaO_3$ and ANbO3 (A = Na and K). Int. J. Hydrogen Energy **32**, 2269–2272 (2007)
57. K. Maeda, K. Domen, Photocatalytic water splitting: recent progress and future challenges. J. Phys. Chem. Lett. **1**, 2655–2661 (2010)
58. T. Hisatomi, J. Kubota, K. Domen, Recent advances in semiconductors for photocatalytic and photoelectrochemical water splitting. Chem. Soc. Rev. **43**, 7520–7535 (2014)
59. Z.K. Zheng, B.B. Huang, Z.Y. Wang, M. Guo, X.Y. Qin, X.Y. Zhang, P. Wang, Y. Dai, Crystal faces of Cu_2O and their stabilities in photocatalytic reactions. J. Phys. Chem. C **113**, 14448–14453 (2009)
60. F.E. Osterloh, Inorganic nanostructures for photoelectrochemical and photocatalytic water splitting. Chem. Soc. Rev. **42**, 2294–2320 (2013)
61. Kondo, J., Cu_2O as a photocatalyst for overall water splitting under visible light irradiation. Chem. Commun. **3**, 357–358 (1998)
62. S. Yusuf, F. Jiao, Effect of the support on the photocatalytic water oxidation activity of cobalt oxide nanoclusters. ACS Catal. **2**, 2753–2760 (2012)
63. X.H. Deng, H. Tuysuz, Cobalt-oxide-based materials as water oxidation catalyst: recent progress and challenges. Acs Catal. **4**, 3701–3714 (2014)
64. N. Zhang, J. Shi, S.S. Mao, L. Guo, Co_3O_4 quantum dots: reverse micelle synthesis and visible-light-driven photocatalytic overall water splitting. Chem. Commun. **50**, 2002–2004 (2014)
65. L. Liao, Q. Zhang, Z. Su, Z. Zhao, Y. Wang, Y. Li, X. Lu, D. Wei, G. Feng, Q. Yu, X. Cai, J. Zhao, Z. Ren, H. Fang, F. Robles-Hernandez, S. Baldelli, J. Bao, Efficient solar water-splitting using a nanocrystalline CoO photocatalyst. Nat. Nanotechnol. **9**, 69–73 (2014)
66. B. Girginer, G. Galli, E. Chiellini, N. Bicak, Preparation of stable CdS nanoparticles in aqueous medium and their hydrogen generation efficiencies in photolysis of water. Int. J. Hydrogen Energy **34**, 1176–1184 (2009)
67. D. Jing, L. Guo, A novel method for the preparation of a highly stable and active CdS photocatalyst with a special surface nanostructure. J. Phys. Chem. B **110**, 11139–11145 (2006)
68. Y. Xu, W. Zhao, R. Xu, Y. Shi, B. Zhang, Synthesis of ultrathin CdS nanosheets as efficient visible-light-driven water splitting photocatalysts for hydrogen evolution. Chem. Commun. **49**, 9803–9805 (2013)
69. A. Kudo, Y. Miseki, Heterogeneous photocatalyst materials for water splitting. Chem. Soc. Rev. **38**, 253–278 (2009)
70. J.S. Jang, S.M. Ji, S.W. Bae, H.C. Son, J.S. Lee, Optimization of CdS/TiO_2 nano-bulk composite photocatalysts for hydrogen production from Na_2S/Na_2SO_3 aqueous electrolyte solution under visible light ($\lambda \geq 420$ nm). J. Photochem. Photobiol. A Chem. **188**, 112–119 (2007)

71. H. Schmidt, S. Wang, L. Chu, M. Toh, R. Kumar, W. Zhao, A. Castro Neto, J. Martin, S. Adam, B. Özyilmaz, Transport properties of monolayer MoS_2 grown by chemical vapor deposition. Nano Lett. **14**, 1909–1913 (2014)
72. O. Lopez-Sanchez, D. Lembke, M. Kayci, A. Radenovic, A. Kis, Ultrasensitive photodetectors based on monolayer MoS_2. Nat. Nanotechnol. **8**, 497–501 (2013)
73. M. Chhowalla, Z. Liu, H. Zhang, Two-dimensional transition metal dichalcogenide (TMD) nanosheets. Chem. Soc. Rev. **44**, 2584–2586 (2015)
74. Y. Gong, J. Lin, X. Wang, G. Shi, S. Lei, Z. Lin, X. Zou, G. Ye, R. Vajtai, B.I. Yakobson, Vertical and in-plane heterostructures from WS_2/MoS_2 monolayers. Nat. Mater. **13**, 1135–1142 (2014)
75. A. Splendiani, L. Sun, Y. Zhang, T. Li, J. Kim, C.Y. Chim, G. Galli, F. Wang, Emerging photoluminescence in monolayer MoS_2. Nano Lett. **10**, 1271–1275 (2010)
76. H.R. Gutiérrez, N. Perea-López, A.L. Elías, A. Berkdemir, B. Wang, R. Lv, F. López-Urías, V.H. Crespi, H. Terrones, M. Terrones, Extraordinary room-temperature photoluminescence in triangular WS_2 monolayers. Nano Lett. **13**, 3447–3454 (2012)
77. H. Shi, H. Pan, Y.-W. Zhang, B.I. Yakobson, Quasiparticle band structures and optical properties of strained monolayer MoS_2 and WS_2. Phys. Rev. B **87**, 155304(1–8) (2013)
78. H.L. Zhuang, R.G. Hennig, Single-layer group-III monochalcogenide photocatalysts for water splitting. Chem. Mater. **25**, 3232–3238 (2013)
79. K. Maeda, K. Domen, New non-oxide photocatalysts designed for overall water splitting under visible light. J. Phys. Chem. C **111**, 7851–7861 (2007)
80. A. Ishikawa, T. Takata, J.N. Kondo, M. Hara, H. Kobayashi, K. Domen, Oxysulfide $Sm_2Ti_2S_2O_5$ as a stable photocatalyst for water oxidation and reduction under visible light irradiation ($\lambda \leq 650$ nm). J. Am. Chem. Soc. **124**, 13547–13553 (2002)
81. T. Takata, G. Hitoki, J. Kondo, M. Hara, H. Kobayashi, K. Domen, Visible-light-driven photocatalytic behavior of tantalum-oxynitride and nitride. Res. Chem. Intermed. **33**, 13–25 (2007)
82. J. Sato, N. Saito, Y. Yamada, K. Maeda, T. Takata, J.N. Kondo, M. Hara, H. Kobayashi, K. Domen, Y. Inoue, RuO_2-loaded β-Ge_3N_4 as a non-oxide photocatalyst for overall water splitting. J. Am. Chem. Soc. **127**, 4150–4151 (2005)
83. G. Hitoki, T. Takata, J.N. Kondo, M. Hara, H. Kobayashi, K. Domen, An oxynitride, TaON, as an efficient water oxidation photocatalyst under visible light irradiation ($\lambda \leq 500$ nm). Chem. Commun. **16**, 1698–1699 (2002)
84. J. Zhu, M. Zäch, Nanostructured materials for photocatalytic hydrogen production. Curr. Opin. Colloid Interface Sci. **14**, 260–269 (2009)
85. G. Hitoki, A. Ishikawa, T. Takata, J.N. Kondo, M. Hara, K. Domen, Ta_3N_5 as a novel visible light-driven photocatalyst (LAMBDA < 600 nm). Chem. Lett. **7**, 736–737 (2002)
86. K. Maeda, K. Domen, Solid Solution of GaN and ZnO as a Stable Photocatalyst for Overall Water Splitting under Visible Light. Chem. Mater. **22**, 612–623 (2009)
87. K. Maeda, T. Takata, M. Hara, N. Saito, Y. Inoue, H. Kobayashi, K. Domen, GaN: ZnO solid solution as a photocatalyst for visible-light-driven overall water splitting. J. Am. Chem. Soc. **127**, 8286–8287 (2005)
88. K. Maeda, K. Teramura, D. Lu, T. Takata, N. Saito, Y. Inoue, K. Domen, Photocatalyst releasing hydrogen from water. Nature **440**, 295–295 (2006)
89. M.G. Kibria, F.A. Chowdhury, S. Zhao, B. AlOtaibi, M.L. Trudeau, H. Guo, Z. Mi, Visible light-driven efficient overall water splitting using p-type metal-nitride nanowire arrays. Nat. Commun. **6**, 6797 (2015)
90. T.F. Yeh, J.M. Syu, C. Cheng, T.H. Chang, H.S. Teng, Graphite oxide as a photocatalyst for hydrogen production from water. Adv. Funct. Mater. **20**, 2255–2262 (2010)
91. T.F. Yeh, C.Y. Teng, S.J. Chen, H. Teng, Nitrogen-doped graphene oxide quantum dots as photocatalysts for overall water-splitting under visible light illumination. Adv. Mater. **26**, 3297–3303 (2014)
92. J. Zhu, P. Xiao, H. Li, S.A. Carabineiro, Graphitic carbon nitride: synthesis, properties, and applications in catalysis. ACS Appl. Mater. Interfaces **6**, 16449–16465 (2014)

93. X. Wang, K. Maeda, A. Thomas, K. Takanabe, G. Xin, J.M. Carlsson, K. Domen, M. Antonietti, A metal-free polymeric photocatalyst for hydrogen production from water under visible light. Nat. Mater. **8**, 76–80 (2009)
94. Y. Kang, Y. Yang, L.C. Yin, X. Kang, G. Liu, H.M. Cheng, An amorphous carbon nitride photocatalyst with greatly extended visible-light-responsive range for photocatalytic hydrogen generation. Adv. Mater. **27**, 4572–4577 (2015)
95. J. Liu, Y. Liu, N. Liu, Y. Han, X. Zhang, H. Huang, Y. Lifshitz, S.-T. Lee, J. Zhong, Z. Kang, Metal-free efficient photocatalyst for stable visible water splitting via a two-electron pathway. Science **347**, 970–974 (2015)
96. C. Huang, C. Chen, M. Zhang, L. Lin, X. Ye, S. Lin, M. Antonietti, X. Wang, Carbon-doped BN nanosheets for metal-free photoredox catalysis. Nat. Commun. **6**, 7698 (2015).
97. P. Zhang, T. Wang, J. Gong, Mechanistic understanding of the plasmonic enhancement for solar water splitting. Adv. Mater. **27**, 5328–5342 (2015)
98. C. Clavero, Plasmon-induced hot-electron generation at nanoparticle/metal-oxide interfaces for photovoltaic and photocatalytic devices. Nat. Photon. **8**, 95–103 (2014)
99. J. Li, S.K. Cushing, F. Meng, T.R. Senty, A.D. Bristow, N. Wu, Plasmon-induced resonance energy transfer for solar energy conversion. Nat. Photon. **9**, 601–607 (2015)
100. S.K. Cushing, J. Li, J. Bright, B.T. Yost, P. Zheng, A.D. Bristow, N. Wu, Controlling plasmon-induced resonance energy transfer and hot electron injection processes in Metal@TiO_2 core–shell nanoparticles. J. Phys. Chem. C **119**, 16239–16244 (2015)
101. S.K. Cushing, N. Wu, Plasmon-enhanced solar energy harvesting. Interface **22**, 63–67 (2013)
102. F. Meng, S.K. Cushing, J. Li, S. Hao, N. Wu, Enhancement of solar hydrogen generation by synergistic interaction of $La_2Ti_2O_7$ photocatalyst with plasmonic gold nanoparticles and reduced graphene oxide nanosheets. ACS Catal. **5**, 1949–1955 (2015)
103. J. Li, S.K. Cushing, P. Zheng, T. Senty, F. Meng, A.D. Bristow, A. Manivannan, N. Wu, Solar hydrogen generation by a CdS-Au-TiO_2 sandwich nanorod array enhanced with Au nanoparticle as electron relay and plasmonic photosensitizer. J. Am. Chem. Soc. **136**, 8438–8449 (2014)
104. J. Li, S.K. Cushing, P. Zheng, F. Meng, D. Chu, N. Wu, Plasmon-induced photonic and energy-transfer enhancement of solar water splitting by a hematite nanorod array. Nat. Commun. **4**, 2651 (2013).
105. S.C. Warren, E. Thimsen, Plasmonic solar water splitting. Energy Environ. Sci. **5**, 5133–5146 (2012)
106. S. Zhang, B. Peng, S. Yang, H. Wang, H. Yu, Y. Fang, F. Peng, Non-noble metal copper nanoparticles-decorated TiO_2 nanotube arrays with plasmon-enhanced photocatalytic hydrogen evolution under visible light. Int. J. Hydrogen Energy **40**, 303–310 (2015)
107. A.O. Govorov, H. Zhang, Y.K. Gun'ko, Theory of photoinjection of hot plasmonic carriers from metal nanostructures into semiconductors and surface molecules. J. Phys. Chem. C **117**, 16616–16631 (2013)
108. X. Zhang, Y.L. Chen, R.-S. Liu, D.P. Tsai, Plasmonic photocatalysis. Rep. Prog. Phys. **76**, 046401 (2013)
109. J. Bao, Recent developments in photocatalytic solar water splitting. Mater. Today **17**, 208–209 (2014)
110. S. Mubeen, J. Lee, N. Singh, S. Kramer, G.D. Stucky, M. Moskovits, An autonomous photosynthetic device in which all charge carriers derive from surface plasmons. Nat. Nanotechnol. **8**, 247–251 (2013)

Chapter 13
Nanomaterials for Rechargeable Lithium Batteries

Sebastien Martinet

Abstract Li-Ion batteries are the technology of choice today for high energy density requirements and they are now used for most of the applications, from consumer electronics, power tools to full electric and hybrid electric vehicles. Due to their long life duration and the absence of maintenance, they are now also considered for stationary applications. In this chapter, the potentiality of nanosizing or nanostructuring the active electrode materials is discussed with examples for positive electrode compounds of low conductivity such as $LiFePO_4$ requiring carbon coating or negative electrodes such as silicon for which volume expansion during alloying with lithium is an issue. Even if going "nano" is not an answer to all the difficulties encountered in the increase of Lithium-Ion cells performances, nanotechnology may facilitate the introduction of new active materials and their commercialization.

13.1 Introduction

Li-Ion technology is based on the reversible intercalation of lithium ions into host materials at the positive and negative electrodes. Even if the first Li-Ion cells were commercialized in 1991 by Sony, the first work exhibiting the reversible intercalation of lithium into Li_xTiS_2 positive electrode was reported by Whittingham et al. for Exxon in 1976 [1]. This material was associated with lithium-aluminium alloy, but due to the cost and safety issues, this technology did not succeed to be commercialized.

Few years later, major innovations came with the discovery of $LiCoO_2$ high voltage lithiated positive electrode material by Goodenough and coworkers in 1980

S. Martinet (✉)
Laboratoire d'Innovation pour les Technologies des Energies Nouvelles et les nanomateriaux (LITEN), French Alternative Energies and Atomic Energy Commission (CEA), 17 Rue Des Martyrs BP9, 38054 Grenoble Cedex, France
e-mail: sebastien.martinet@cea.fr

Q. Li (ed.), *Nanomaterials for Sustainable Energy*,
NanoScience and Technology, DOI 10.1007/978-3-319-32023-6_13

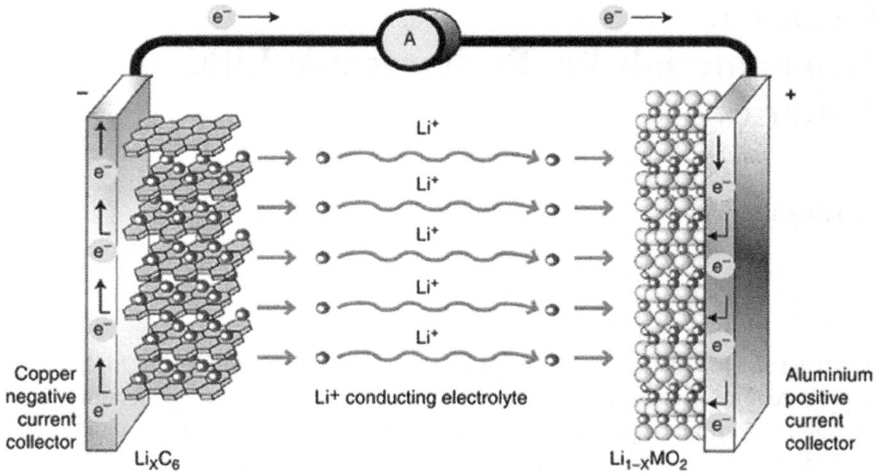

Fig. 13.1 Conventional Li-ion cell basic principle with lithium transition metal oxide at the positive electrode and graphite at the negative electrode. Reproduced with permission from [4]

[2] and the possibility to intercalate lithium into graphite at low potential by Yazami and Touzain in 1983 [3]. The Li-Ion concept was possible by the combination of these two electrodes as indicated in the scheme of Fig. 13.1 [4].

Since 1991, the lithium-ion technology developments allowed almost tripling its specific energy density from around 100 $Wh.kg^{-1}$ to close to 270 $Wh.kg^{-1}$ for the latest 18650 cells, the standardized cell for portable computers, from LG and Panasonic companies. These improvements are the result of considerable process developments that are not detailed here: (i) reduction of electrodes porosity in parallel with an increase of active material loading, (ii) reduction of thicknesses of current collectors and separator, (iii) reduction of casing weight, etc. On the active material side, new compounds were studied and introduced to overcome several drawbacks of $LiCoO_2$ (noted as LCO): (i) the specific capacity of $LiCoO_2$, is classically limited to 140 $mAh.g^{-1}$ due to the lack of stability of the lamellar structure in case of delithiation above x = 0.5 in $Li_{1-x}CoO_2$, (ii) in the charge (oxidized) state, the compound instability leads to a release of oxygen gas in case of overheating and to a potential combustion with electrolyte and (iii) the relative expansive cost of this compound linked to metallic cobalt.

Different strategies, and so developments of different technical solutions, were followed depending on the targeted application, consumer electronics or electric vehicles, for which cost, safety and energy density objectives are not always considered with the same weight. To reduce the cost, cobalt substitution by cheaper transition metals such as manganese or in a lesser extent by nickel has been and remains the subject of extensive works initiated in 1983 for spinel compounds such as $LiMn_2O_4$ [5], noted as LMO. For partial substitution, the compounds $LiNi_xMn_yCo_{(1-x-y)}O_2$ of the so-called NMC [6] family are the preferred ones with the 1/3-1/3-1/3 composition that is the most largely commercialized one.

To increase the capacity, nickel based compounds such as Li_xNiO_2 emerged in the end of the 90s [7] leading to the development and commercialization of the $LiNi_{0.8}Co_{0.15}Al_{0.05}O_2$ oxide noted as NCA in SAFT Li-Ion cells. The tendency in the 2010s is also to improve the specific capacity of NMC materials by increasing the nickel content. Additional research efforts, initiated by Amundsen and coworkers in 2002 [8] are devoted to the lithium-rich layered oxides for which specific capacities as high as 270 mAh/g are reported. Another possibility to improve energy density of Li-Ion cells came from the evidence in 1997 [9] of the electrochemical activity of the Ni^{2+}/Ni^{3+} and Ni^{3+}/Ni^{4+} couples at 4.7 V versus Li^+/Li in high voltage oxides such as the spinel oxides $LiMn_{2-x}Ni_xO_4$–LNMO.

For safety issues, LCO is preferentially substituted by LMO or NMC, but in 1997, a polyanionic compound, the olivine $LiFePO_4$ olivine—LFP, was introduced by Goodenough et al. [10], Due to the PO_4 group, oxygen is strongly linked to the structure preventing/delaying oxygen release in abuse conditions. Another strong advantage of this compound lays in the replacement of cobalt and nickel costly elements by cheap iron. Combining these 2 points makes $LiFePO_4$ of high interest for stationary applications or electric mobility.

Considering the anode negative electrode, research can be divided roughly into two different directions: (i) the increase of energy density through the replacement of graphite by high capacity compounds such as lithium-alloys (silicon, tin, and others) or conversion materials and (ii) the increase of cyclability and power ability through the use of high voltage materials less sensitive to lithium deposition reaction such as titanium based oxides. Major issues relative to the anode developments are summarized in the Fig. 13.2 [11]: Li^+ and electrons transport, structural modifications of the active material, unstability/rupture of the Solid Electrolyte Interface (SEI), reversibility of the intercalation/alloying process, long-range kinetics for Li^+ and electron all along the length of electrodes, electrode integrity and flexibility. Note that most of these issues are common to the positive electrode case.

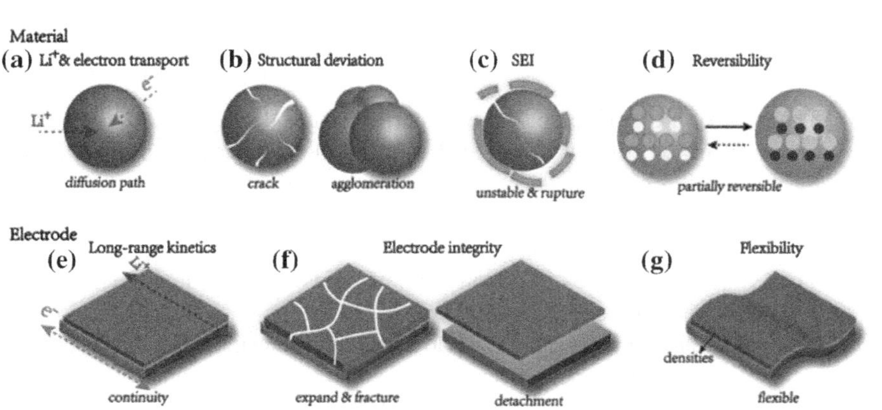

Fig. 13.2 Illustration of common issues of Li-ion batteries' anodes: **a–d** The material level, **e–g** The electrode level. Reproduced with permission from [11]

13.2 Why Nanomaterials for Batteries?

Despite the huge number of publications relative to "nano" materials for Li-Ion batteries, it is important to notice that, for several reasons that will be given below, the commercialized lithium-ion cells use typically micron-sized active materials both at the positive electrode and negative electrodes, respectively LCO, NMC, NCA, LFP or mixtures of these compounds and graphite. In the case of the positive electrode materials, especially for LFP, the micron-sized particles may be composed of primary nano-sized particles according to their synthesis process that may include steps allowing reducing particle size such as co-precipitation or mechanical milling for instance.

Going towards nanomaterials may present different benefits [12], mainly:

(i) An improved cycle life due to a better accommodation of the strain of lithium insertion/removal,
(ii) Nanomaterials usually possess a large specific surface area enabling higher charge/discharge rates through the increased electrode/electrolyte contact area
(iii) Electronic and lithium transport are improved via short path lengths, facilitating operation with low electronic conductivity or lithium diffusivity or leading to better performances at high power. In the case of lithium diffusion, one has to remind that the time constant associated to this phenomenon is proportional to the square of the diffusion length i.e. the particle size, and so switching from micro to nanosized particles significantly reduces diffusion limitations.
(iv) The opening to new reactions that are not possible with bulk materials. The case of mesoporous β-MnO_2 can be given as an example: contrary to the micronic material a reversible lithium intercalation is possible into this compound without destruction of the rutile structure [13].
(v) The thermodynamics of the reactions may be modified by the "nanosize" effect that can modify the chemical potentials for lithium ions and electrons and so result in a change of electrode potential and facilitates phase transitions [14]

Nevertheless, despite these benefits, major drawbacks have to be recalled, as they explain the difficulty to introduce the nanomaterials in the practical Li-Ion cells:

(i) The benefit of the increase of power performances through the increase of the specific surface area of the electrolyte/material interfaces is counterbalanced by undesirable electrode/electrolyte reactions leading to self-discharge, poor cycling and calendar life. For instance, manganese dissolution of positive electrode materials, electrolyte oxidation or reduction reactions.
(ii) The synthesis processes to obtain nanomaterials may in some cases be more complex and so potentially costly
(iii) Nanomaterials usually suffer from low practical densities that leads firstly to difficulties for electrode processing through the higher quantity of solvent needed to have similar viscosities of the slurries during the coating step, and secondly to reduced packing behavior of particles and so lower volumetric energy densities at electrode and cell levels.

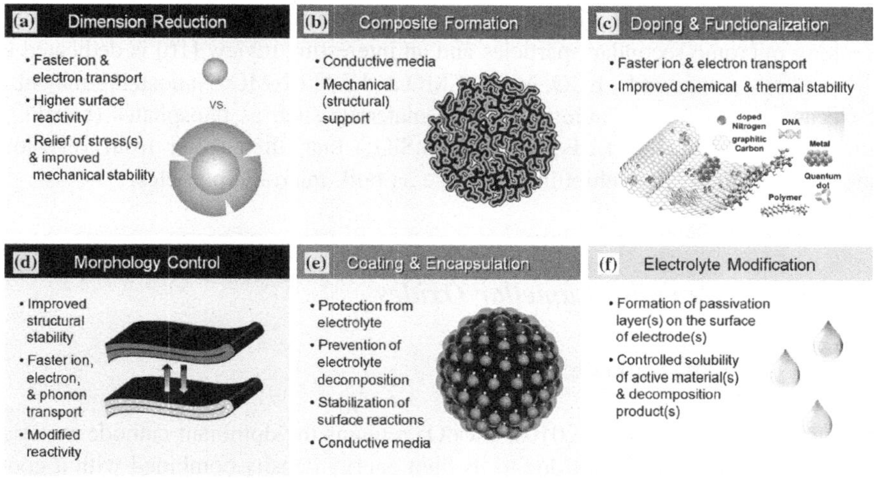

Fig. 13.3 General strategies followed at the material level to improve performances in Li-ion batteries [15]. **a** Reduction of dimensions of active materials, **b** Formation of composites, **c** Doping and functionalization, **d** Adaptation of morphology, **e** Coatings and shells around active materials and **f** Modification of electrolyte. Schematic figure in (**c**) reproduced from [120] and schematic figure of (**e**) reproduced from [107]

The design of micro/nano-heterostructures offers the possibility to overcome these issues by exploiting the nanostructure benefits in a microstructured assembly.

The different strategies that will be given in the following parts of the chapter are illustrated in Fig. 13.3 [15] such as coating, doping, size reduction and nano-structuring. The chapter is not an exhaustive review of Li-Ion battery materials developments but must be considered as a focus on the compounds for which going "nano" may have an interest.

13.3 Positive Electrode Materials

As mentioned in the introduction, despite the huge research and development relative to positive intercalation electrode material that represents the heaviest weight in the materials of the li-ion cells, the "nano" activity dedicated to size reduction is quite fewer in comparison with new negative active materials for which alloying reactions or conversion reactions may be involved, leading to huge volume variations of particles. Additionally, increasing the area of positive electrode material with electrolyte contributes to exacerbate side reactions through electrolyte oxidation and to worsen the safety behavior. In the particular case of Mn-based oxides, the rate of manganese dissolution is accelerated and negative electrode poisoning is fastened.

Nevertheless, many studies are devoted to nanostructuring the cathode materials keeping micronic secondary particles and an interesting review [16] is dedicated to 4 V cathode compounds, LCO, NMC, LMO and 5 V LNMO. Nanosizing may also have an interest for polyanionic cathode materials such as phosphates ($LiFePO_4$, $LiMnPO_4$) or silicates (Li_2FeSiO_4, Li_2MnSiO_4) that often suffer from very low intrinsic conductivities inhibiting their use as bulk micronic particles.

13.3.1 Commercial Lamellar Oxides

13.3.1.1 $LiCoO_2$ Cathode Material

Still today in the middle of 2010s, $LiCoO_2$ remains the dominant cathode material on the Li-Ion battery market due to its high energy density combined with a good cycle life and an easy synthesis.

As energy density increase is driven by the consumer electronics requirements, the tendency is to increase the end-of-charge voltage of the Li-ion cells and thus increase the specific capacity of the LCO compound with a higher delithiation ratio. One of the strategies is to use specific coatings such as oxides, fluorides, phosphates or polymers [17–21] that can increase the stability of the compound and also reduce the reactivity with electrolyte due to the high oxidative potentials up to 4.4 V versus Li^+/Li.

As an example, Atomic Layer Deposition (ALD) process can be an interesting route for the coating of the active material [21]. In this study, ALD coating of Al_2O_3 was applied to nano-sized LCO particles. As illustrated in Fig. 13.4, a significant

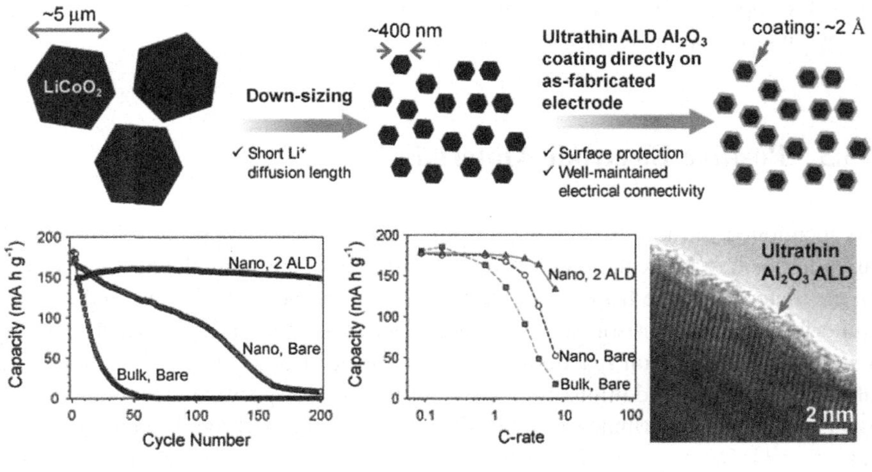

Fig. 13.4 *Top*—Preparation of ultrathin ALD Al_2O_3 coated nano-sized $LiCoO_2$. *Bottom*—Comparison of different LCO materials (bare, nano-bare and nano-ALD coated) in cycling (*left*) and at high discharge rates (*center*). *Bottom right* SEM picture of a nano-ALD coated particle. *center*. Reproduced from [21]

improvement both in cycle life and power performances has been reached for this material in comparison with classical micron-sized LCO.

13.3.1.2 NMC Cathode Materials

NMC cathode materials are of high interest for the substitution of conventional LCO. As indicated in the introduction, besides the 1/3-1/3-1/3 composition, developments are pursued to increase the nickel content or the cut-off voltage and so improve the capacity of the oxide. Nevertheless, there are several issues to overcome in terms of structural stability and safety behavior at elevated temperatures. In that scope, different approaches have been undertaken, such as protective coating [22] or spinel-layered core-shell [23] particles for instance. In another study [24], as illustrated in Fig. 13.5, a pillar layer including transition metal Ni^{3+} ions in the lithium slab is present at the periphery of the particle via a PVP functionalized precursor coating on the hydroxide precursor $Ni_{0.7}Mn_{0.15}Co_{0.15}(OH)_2$.

More recently, a new type of materials called concentration gradient cathode materials has appeared [25]. These compounds have ratio of Mn:Ni:Co that are varying through the depth of the particles. A combination of this approach with the core-shell strategy is shown in Fig. 13.6 for $Li[Ni_{0.64}Co_{0.18}Mn_{0.18}]O_2$. In this example, the nickel concentration is high in the bulk (shell) to get a high capacity. Then, a concentration gradient is present from the interface (outside layer/bulk) to the outside of the particle. Ni content gradually decreases to a minimum at the outside of the particle whereas Co and especially Mn increase to a maximum for safety and cyclability improvement. The electrochemical performances reveal an initial discharge capacity of 209 mAh.g^{-1} close to the nickel-rich Li $(Ni_{0.8}Co_{0.1}Mn_{0.1})O_2$ composition but with a much better cycling stability with a remaining capacity close to 200 mAh.g^{-1} after 50 cycles versus around 140 mAh. g^{-1} for the classical nickel-rich oxide. The stability for the concentration-gradient

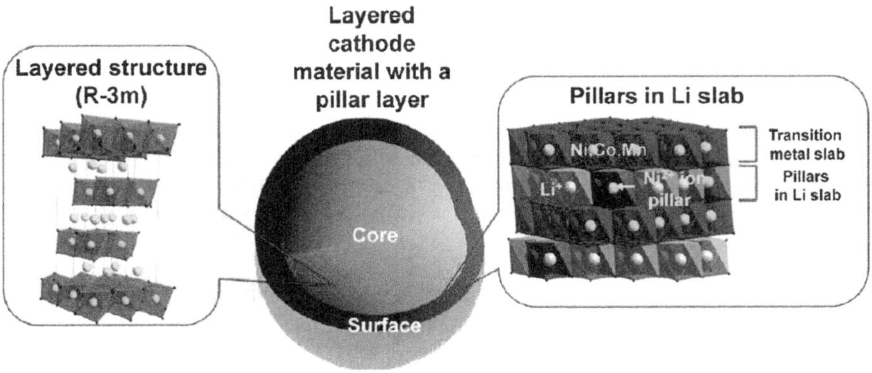

Fig. 13.5 Schematic view of the NMC layered cathode material with a pillar layer of Ni^{3+} ions in Li slab [24]

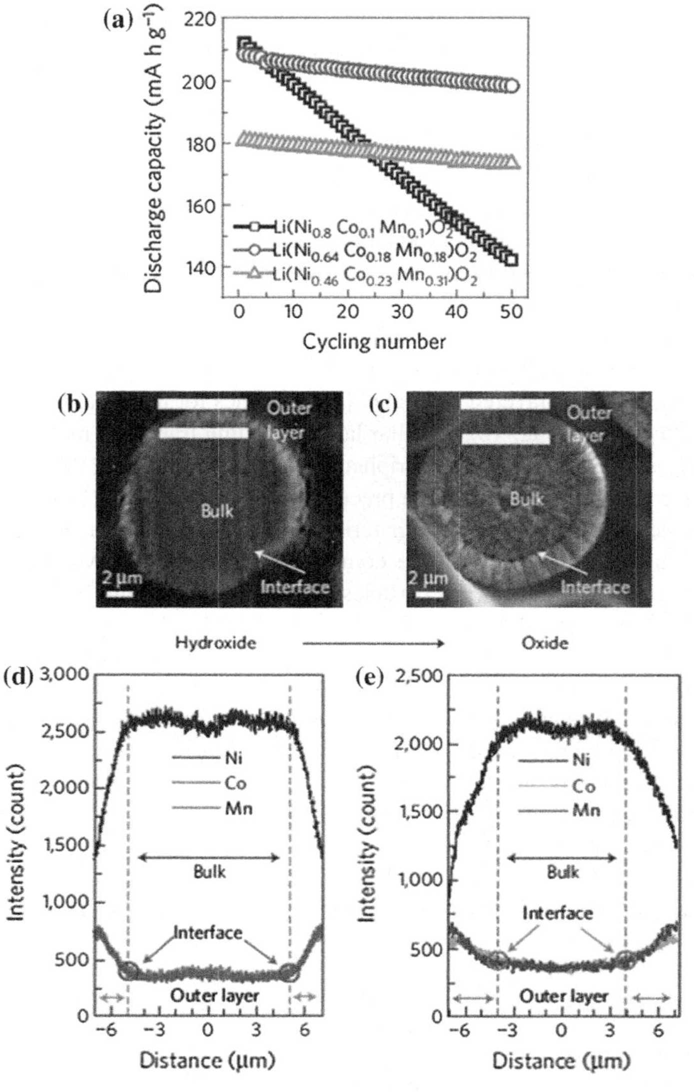

Fig. 13.6 **a** Comparison of cyclability and initial discharged capacity of NMC with different Ni:Co:Mn ratio and in blue the concentration gradient $Li[Ni_{0.64}Co_{0.18}Mn_{0.18}]O_2$. **b–e**: Scanning electron microscopy (SEM) and electron-probe X-ray micro-analysis (EPMA) results on the precursor hydroxide and final oxide $Li[Ni_{0.64}Co_{0.18}Mn_{0.18}]O_2$. Reproduced from [25]

compound is in the same order as for the $Li(Ni_{0.46}Co_{0.23}Mn_{0.31})O_2$ reference oxide of the study but this last one only provides 180 $mAh.g^{-1}$ initial discharge capacity. Additionally, differential scanning calorimetry (DSC) measurements confirmed the stability of the developed material up to 250 °C.

13.3.2 Spinel Oxides

13.3.2.1 $LiMn_2O_4$ Spinel Cathode Material

As indicated in the introduction, $LiMn_2O_4$ spinel is studied since 1983 [5] mainly for its interest in cost reduction versus conventional $LiCoO_2$ compound and for its stability in abuse conditions.

Besides these benefits, this compound faces a major issue corresponding to manganese dissolution during cycling and storage at elevated temperatures above 40 °C. This phenomenon occurs through the disproportionation of Mn^{3+} ions generating Mn^{4+} and Mn^{2+} ions, those last ones being soluble migrate to the anode side leading to manganese deposition and poisoning of negative electrode [26, 27]. This dissolution reaction is also favorited by the presence of HF acid in the electrolyte.

Many strategies have been followed to overcome this problem, such as surface coating or doping in the structure. Another approach has been followed by Kim et al. [28] who proposed to use a truncated octahedral structure for the spinel compound. As shown in Fig. 13.7, in this original structure obtained by precise control during the hydrothermal process of the hydroxide precursor (temperature, surfactant, etc.), only a small portion of the surface is orientated in the directions that support Li diffusion for power performance, and the major part of the particle

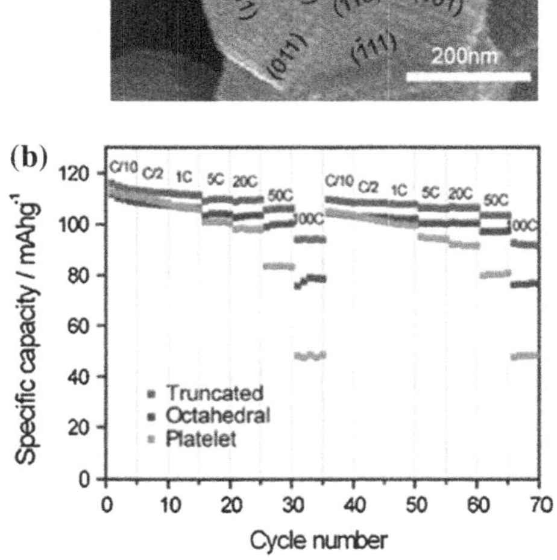

Fig. 13.7 **a** SEM image of truncated octahedral $LiMn_2O_4$ particles. **b** Comparison of performances of different types of spinel compounds at elevated discharge rates with the best performances for the octahedral material. Reproduced with permission from [28]

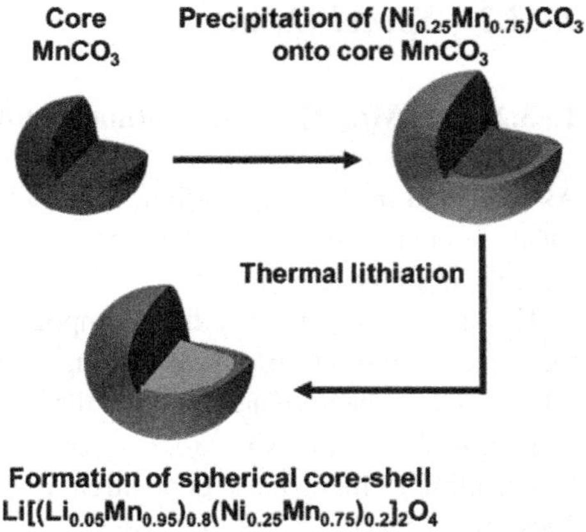

Fig. 13.8 Scheme of the formation process of the core-shell spinel structure. Reproduced from [29]

faces is aligned to the (1 1 1) direction thus reducing Mn dissolution. The rate capability up to 100C and the cycle life of this specific type of compounds exhibits a significant improvement versus other structures synthesized during the study.

In another paper from Scrosati et al. [29] the spinel compound has been nanostructured in spherical core-shell $Li[(Li_{0.05}Mn_{0.95})_{0.8}(Ni_{0.25}Mn_{0.75})_{0.2}]_2O_4$. The idea was there to protect the 4 V Li-Rich manganese spinel by a spinel $LiNi_{0.5}Mn_{1.5}O_4$ shell electrochemically inactive in the operating voltage range in which the average manganese oxidation state is Mn^{4+}. In those conditions, there is supposed to be no possibility for the disproportionation process and thus no Mn^{2+} dissolution.

The synthesis process of the core-shell compound is illustrated in Fig. 13.8, starting from a core in manganese carbonate on top of which is precipitated a mixed nickel-manganese carbonate. After thermal lithiation, the core-shell compound corresponding to a "4 V" $Li_{1.1}Mn_{1.9}O_4$ core surrounded by a "5 V" $LiNi_{0.5}Mn_{1.5}O_4$ spinel shell is obtained. The structure of the compound is confirmed by SEM of fractured particles (precursor and lithiated spinel) and XRD analysis in Fig. 13.9. The electrochemical performances of the compound are detailed in Fig. 13.10 and the clear benefit of the core-shell structure is evidenced in the graph c) where capacity retention at 60 °C after 100 cycles is only 64.7 % for the pure core compound versus 97 % for the core-shell material.

13.3.2.2 5 V Spinel $LiNi_{0.5}Mn_{1.5}O_4$ or $LiNi_{0.4}Mn_{1.6}O_4$ Cathode Material

In 1997, Amine et al. [9] and Dahn et al. [30] put in evidence the 5 V contribution in $LiMn_{2-x}M_xO_4$ spinel oxides (M = Ni, Fe with x = 0.5, 1 and M = Ni with $0<x<0.5$). In the following years, many studies were devoted to the $LiNi_{0.5}Mn_{1.5}O_4$ and variants composition [31, 32] that are of high interest for high energy density batteries based on a 20 % reduced number of "5 V" li-ion cells in series for the

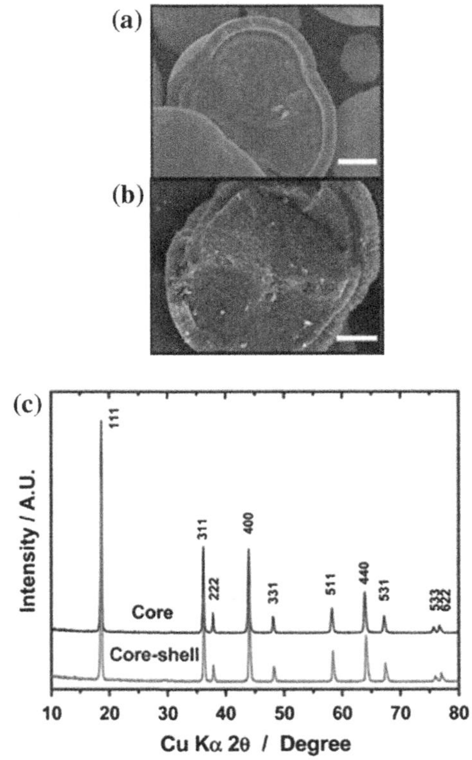

Fig. 13.9 SEM images of fractured particles (scale bar: 5 μm). **a** Core-shell precursor and **b** Lithiated spinel. **c** XRD diagrams of the core and the core-shell powders. Reproduced with permission from [29]

same total nominal voltage in comparison with conventional "4 V" cells. These 5 V compounds—with more precisely a potential at 4.7 V versus Li^+/Li and fast 3D Li^+ diffusion channels- are also perfectly suitable to be combined with the high voltage spinel $Li_4Ti_5O_{12}$ anode material, resulting in a very high power 3 V Li-Ion cell with promising characteristics for hybrid electric vehicles or other applications where fast charge or discharge rates are required. Despite the good performances obtained in this 3 V configuration [31], there is still significant improvement required concerning the interface between electrode and electrolyte due to the strong oxidative potential of the positive active material.

We do not report here all the developments (coatings, electrolyte additives, etc.) made to circumvent this issue, but we will give a specific example of structuration of the active material proposed by Lou et al. [33] who succeeded in controlling the morphology of the particles during the synthesis process via an impregnation method followed by solid-state reaction. The impregnation method allowed obtaining a homogeneous distribution of the reagents that suppressed the unwanted particle growth during the annealing step.

Different nanostructured materials were obtained through this process (Fig. 13.11), either hollow microspheres and hollow microcubes and their electrochemical performances, reported in the same figure, were promising with 97 %

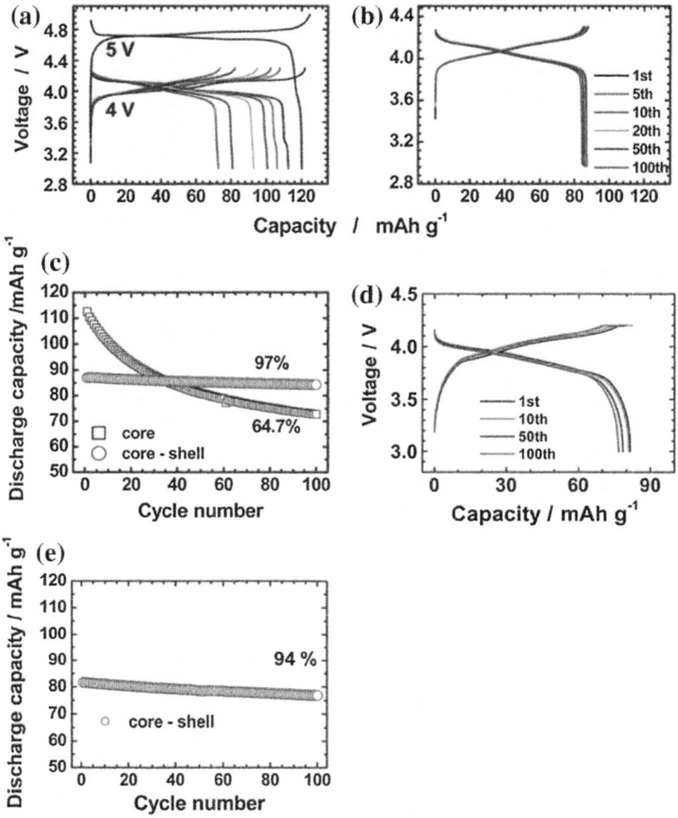

Fig. 13.10 Charge and discharge curves in coin cells at 60 °C of **a** The core and first charge and discharge curves of the shell. **b** Core-shell $Li[(Li_{0.05}Mn_{0.95})_{0.8}(Ni_{0.25}Mn_{0.75})_{0.2}]_2O_4$. **c** Cycling performances of Li/core and Li/core-shell at 60 °C. **d** Charge and discharge curves at 60 °C of C/Li $[(Li_{0.05}Mn_{0.95})_{0.8}(Ni_{0.25}Mn_{0.75})_{0.2}]_2O_4$ cells and **e** their corresponding cyclability in Al-laminate cell

capacity retained after 200 cycles at 1C, 2C or 5C rates. Nevertheless, these porous structures maybe detrimental to obtain high density electrodes and so their practical use may be limited.

13.3.3 Polyanionic Compounds

Whereas layered oxides and spinel oxides have been studied since the beginning of the 80s with a first commercialization in 1991 for $LiCoO_2$ in practical Li-Ion cells, interest in polyanionic compounds emerged later after 1997 with the first evidence of electrochemical activity of $LiFePO_4$ by Pahdi et al. [10]. These compounds are

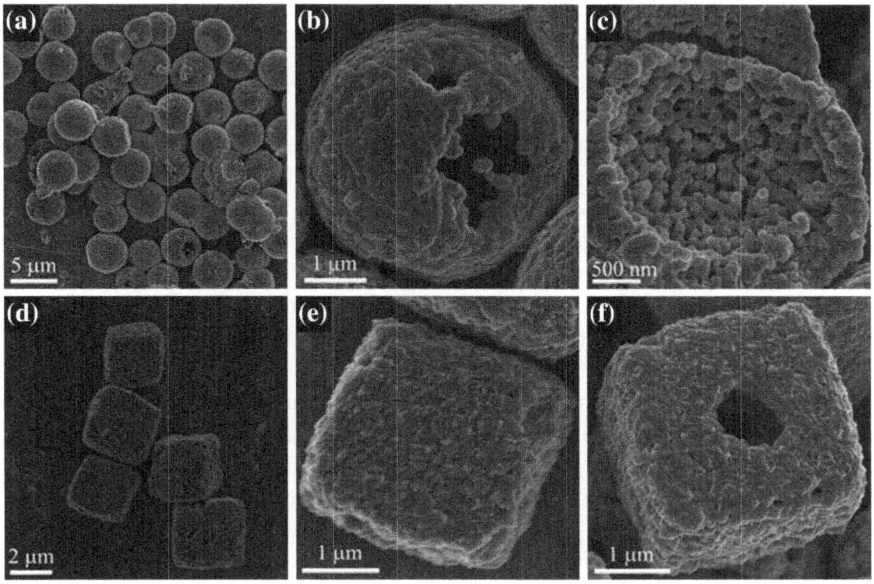

Fig. 13.11 FESEM images of $LiNi_{0.5}Mn_{1.5}O_4$ hollow microspheres (**a–c**) and microcubes (**d–f**) with nanosized subunits. Reproduced with permission from [33]

three-dimensional frameworks built on transition metals and polyanions $(XO_4)^{n-}$. C. Masquelier and L. Croguennec realized an excellent and exhaustive review of these materials [34] in which they covered phosphates, silicates, borates, sulfates, etc.

Even if polyanionic compounds suffer from a weight penalty corresponding to the replacement of O_2^{4-} by the heavy $(XO_4)^{n-}$, they present several interesting benefits:

- Their framework is very stable and provides long-term stability, resulting in improvements in cycle life but also in safety
- Thanks to the chemical nature of the polyanion and the inductive effect [35, 36], the potential of a $M^{n+}/M^{(n-1)+}$ redox couple can be monitored and increased in comparison with the corresponding "simple" oxide.
- Through the possibilities of anion and cation substitutions for a given structure, this family of materials opened a wide domain of research.

Nevertheless, most of these materials are rather poor ionic and electronic conductors and so, to circumvent this issue, many research efforts have been dedicated to their nanosizing or nanostructuration and also to conductive coatings or doping.

13.3.3.1 $LiFePO_4$

In the first work relative to the olivine $LiFePO_4$ [10], only around 0.7 Li were extracted from the compound at a very low rate of 2 $mA.g^{-1}$, i.e. roughly C/85 due to the low conductivity of the material. The olivine structure-type is built on hexagonal packing with Li and Fe atoms occupying half of the octahedral sites and P atoms occupying 1/8th of the tetrahedral sites. Li^+ and Fe^{2+} have a specific distribution within the octahedral site generating [$TeOc_2$] layers that impacts on the electronic and ionic conductivities of $LiFePO_4$. $Fe0_6$ octahedra share corners between each other and this renders electronic conductivity difficult. Ionic conductivity of Lithium is not three-dimensional and can mostly occur from $[001]_{Pnma}$ direction.

To solve these limitations and besides size reduction of active particles, carbon coating interest was shown by Armand et al. for the first time in 1999 [37], and applied to natural triphylite and synthetic $LiFePO_4$ by decomposition of a modified polycyclic aromatic exhibiting the theoretical capacity at room temperature in 2001 [38].

In Fig. 13.12, an example of a few nanometers thick carbon coating on $LiFePO_4$ that provides fast path for electron transport, is illustrated with the corresponding performances of the positive material at different C-Rates up to 40C at room temperature and 55 °C. More than 150 $mAh.g^{-1}$ are obtained at 1C rate at room temperature.

In parallel to the conductive carbon coating strategy and the size reduction of active particles via specific synthesis routes such as mechanical milling or hydrothermal synthesis [39], an alternative way was opened by a paper of Y.M. Chiang and coworkers from MIT in 2002, in which 8 orders of magnitude increase in electrical conductivity was obtained with 1 % "dopants" such as M = Zr or Nb for $Li_{1-x}M_xFePO_4$ [40] as reported in Fig. 13.13. An important academic controversy resulted from this work, but the possibility to use $LiFePO_4$ for power applications

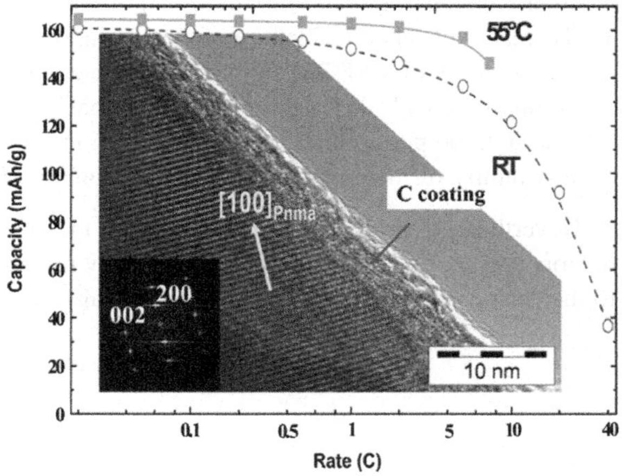

Fig. 13.12 Thin carbon coating at the surface of crystalline $LiFePO_4$ and the corresponding discharge capacities at C-rates up 40C. Reproduced with permission from [34]

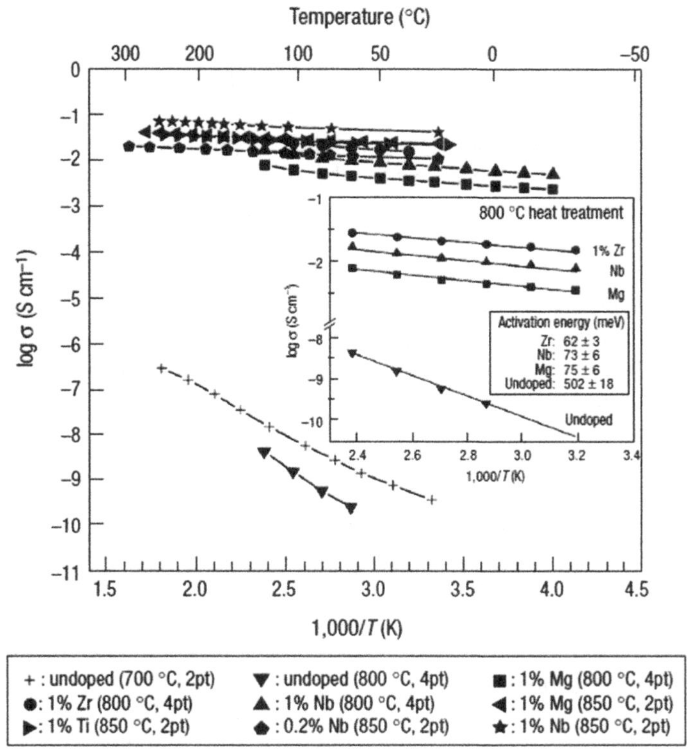

Fig. 13.13 Electrical conductivities measured between −20 °C and +150 °C for supposed "doped" $Li_{1-x}M_xFePO_4$ and undoped $LiFePO_4$. Results are for polycristals fired at 700–850 °C and measured by two points d.c. and four-point van der Pauw methods. Reproduced from [40]

was shown and A123 Start-up Company created by the MIT team confirmed the industrial interest for this material.

Nevertheless, the potential of $FePO_4/LiFePO_4$ is only 3.45 V versus Li^+/Li, resulting in a nominal 3.2 V Li-Ion cell with graphite negative electrode and so limiting the energy density of this technology. To overcome this issue total or partial substitution of iron by manganese has been and is still the purpose of many studies, as $MnPO_4/LiMnPO_4$ potential is 4.1 V versus Li^+/Li.

13.3.3.2 Other Phosphate Materials

As indicated before, the low potential of $LiFePO_4$ pushes many research towards other phosphate materials: $LiMnPO_4$ (4.1 V vs. Li^+/Li) [41], $LiCoPO_4$ (4.8 V vs. Li^+/Li) or solid-solutions such as $LiFe_{1-x}Mn_xPO_4$ [42] where potential is split into two plateau at 3.45 V for iron part and 4.1 V for manganese part in the respective proportions of the two transition metals.

The substitution of iron by manganese allows keeping the same advantage in cost reduction compared to the cobalt compound. Unfortunately, $LiMnPO_4$ has a very low intrinsic conductivity (10^{-12} S.cm^{-1}) [43], 3 orders of magnitude lower than $LiFePO_4$ (10^{-9} S.cm^{-1}), and so early work relative to this compound exhibited very poor electrochemical performances at the exception of one controversial paper [41]. A strong attention has been paid to the reduction of the synthesis temperature to control the particle morphology and limit the particle size growth. In interesting studies from Drezen and coworkers [44–46], the polyol synthesis was developed for $LiMnPO_4$. In Fig. 13.14, the effect of the particle size reduction from 270 nm to 140 nm led to an increase from 130 mAh.g^{-1} to more than 150 mAh.g^{-1} at 0.1C. Further optimization of the synthesis process allowed reaching promising practical capacities up to 113 mAh.g^{-1} at C-rate as the polyol media acts both as the solvent to dissolve the precursors and as the particle growth inhibitor. The obtained particles were platelets of only 20–30 nm thick with a reduced path length for lithium diffusion (Fig. 13.15). Despite these improvements and further works relative to doping or substitution with complex material compositions such as $LiFe_{(1-x-y-z)}Mn_xCo_yB_zPO_4$ studied by design of experiments methodology [47], up to date the only commercial Li-Ion cells using polyanionic positive active material are based on $LiFePO_4$.

13.3.3.3 Silicate Compounds

In the search for new positive-electrode materials for lithium-ion batteries, Li_2MSiO_4 silicates (M = Fe, Mn, Co) have also gained growing interest in the last few years mainly due to the theoretical possibility to exchange 2 Li and so 2 electrons per transition metal and to the low cost of materials (Fe, Mn, Si, not for Co). Their ability to reversibly exchange lithium was first evidenced by Armand et al. in 2001 for Li_2FeSiO_4 [48–50] with a practical capacity of 165 mAh.g^{-1} at 60 °C reaching the theoretical value for one electron, but the lower electronegativity of Si versus P reduces the potential of the Fe^{3+}/Fe^{2+} redox couple to around 2.8 V versus Li^+/Li which is 0.65 V below $LiFePO_4$, for similar capacity. Electrochemical performances were recorded at 60 °C due to the low intrinsic conductivity of the compound resulting from isolated FeO_4 tetrahedra and tightly bound Li^+ in tetrahedral sites.

First-principles calculations of silicate compounds [51] indicated a predicted value of 4.8 V versus Li^+/Li above the electrolyte stability limit for the second lithium exchange corresponding to Fe^{4+}/Fe^{3+} couple whereas only 4.5 V should be obtained for Mn^{4+}/Mn^{3+} opening more possibility to exchange the second electron without side reactions and reaching the total theoretical capacity (circa 330 mAh. g^{-1}) for Li_2MnSiO_4. Nevertheless, the structural instability of Li_2MnSiO_4 with a tendency to amorphize during electrochemical extraction of lithium was shown by Dominko et al. [52] both experimentally and theoretically by density functional theory (DFT) calculations. The calculations revealed that, unlike in the case of pure Li_2MnSiO_4, the solid-state solution $Li_2Mn_xFe_{1-x}SiO_4$ will not phase-separate

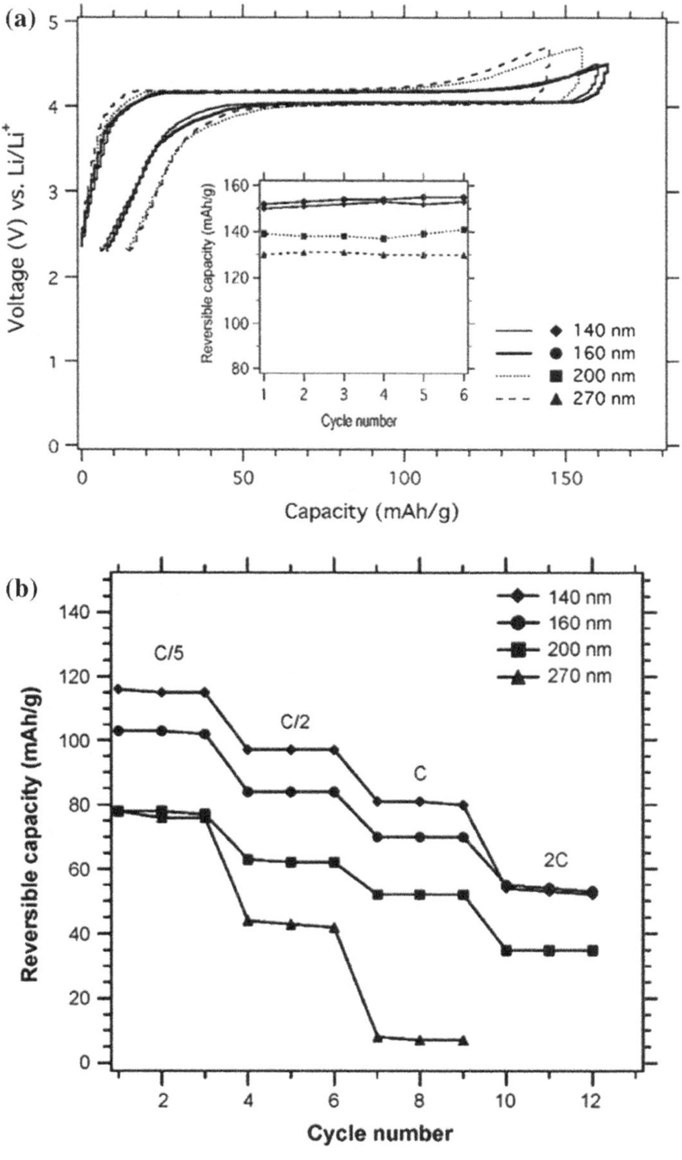

Fig. 13.14 Electrochemical performances of $LiMnPO_4$ materials with different particle sizes. **a** Galvanostatic charge-discharge at 0.1C-Rate, reproduced from [44]. **b** Rate capabilities, reproduced from [3.30]

during the delithiation and the potential required to extract more than 1 lithium per formula unit should be significantly lower than that for the Fe compound, with 0.7 V for $Li_xMn_{0.5}Fe_{0.5}SiO_4$ for instance.

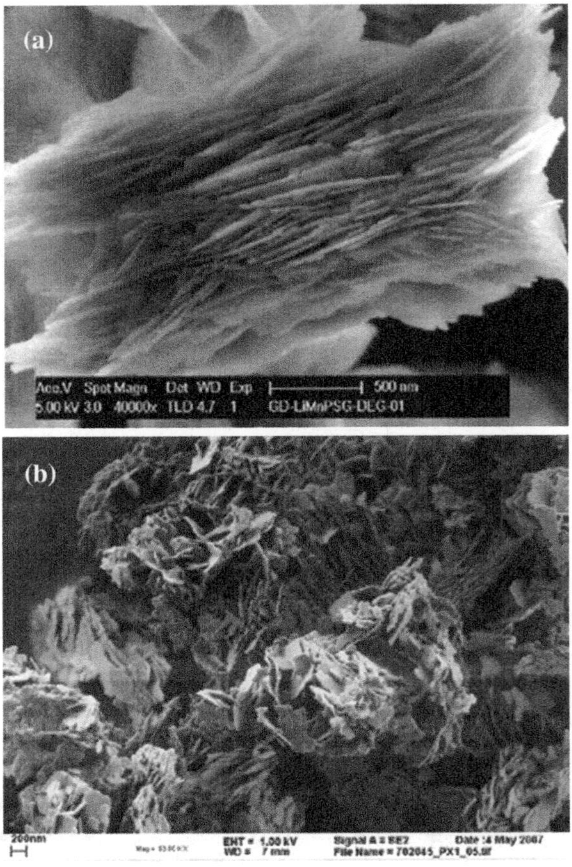

Fig. 13.15 SEM images of $LiMnPO_4$ nanoplatelets—Reproduced from [46]

Despite these findings, many works are still devoted to Li_2MnSiO_4 and especially "nano" compounds for which the nano-size effect is supposed to compensate the structural variations of the material. In a paper from Devaraju and coworkers [53], a supercritical solvothermal water/ethanol synthesis is used to obtain monodispersed nanoparticles of 15–20 nm diameter that are then coated by wet ball-milling with a conductive polymer to obtain PEDOT/Li_2MnSiO_4 nanoparticles (Fig. 13.16). The electrochemical performances are promising with high initial discharged capacities even close to the theoretical value at 40 °C with 313 mAh. g^{-1}, and moderate cyclability on 20 cycles.

In another study from Zhu et al. [54], carbon coated Li_2MnSiO_4 nanoparticles of 20–30 nm are anchored on graphene nanosheets to increase the intrinsic low power ability of the material. The benefit is significant with 210 mAh g^{-1} delivered at C-Rate (1C = 166 mA g^{-1}) and still 109 mAh g^{-1} at 10C-Rate, giving hope that there should be possibilities in the future to use the silicates in a practical Li-Ion cell.

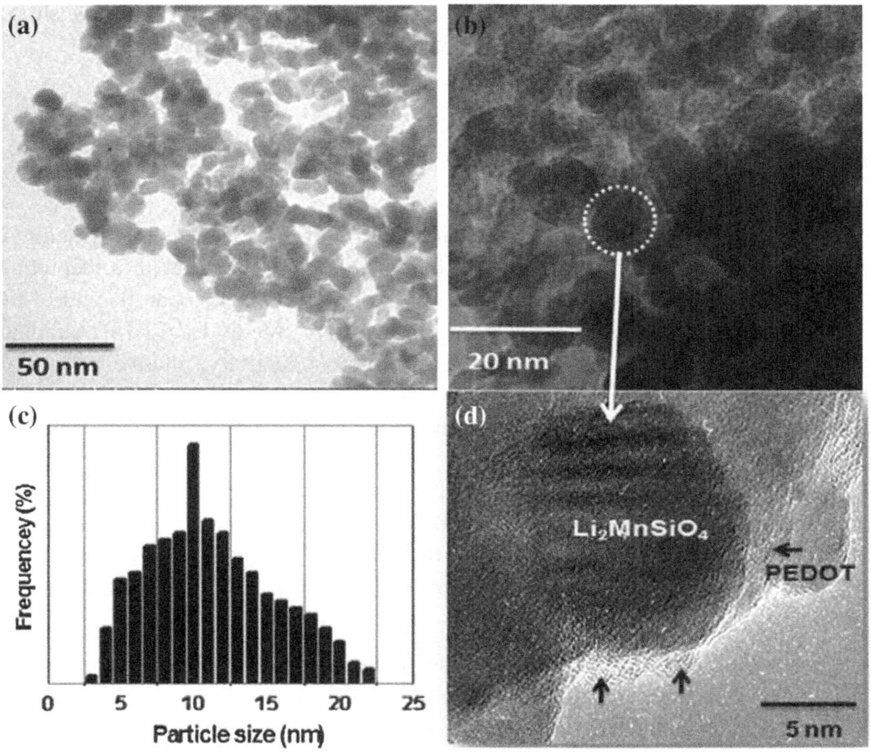

Fig. 13.16 TEM (**a**) and HRTEM (**b**) images of as-synthesized Li_2MnSiO_4 and PEDOT/Li_2MnSiO_4 nanoparticles. Size distribution (**c**) and HRTEM image of PEDOT-coated Li_2MnSiO_4 nanoparticles (**d**). Reproduced from [52]

13.4 Negative Electrode Materials

Today's Li-Ion batteries use mainly graphite or other carbon material as anode compound. To increase energy density, silicon technology is emerging by partial substitution of graphite with, for example, the commercialization of NCR18650 Li-Ion cells by Panasonic in 2015 with a nominal capacity of 3.55 Ah. Besides the energy density race, $Li_4Ti_5O_{12}$ (LTO), a high potential negative electrode, is the other commercial negative electrode alternative for extended cycle life, continuous fast charging and better safety behavior, with possible applications in SLI batteries, microHEV, HEV and even some specific EVs.

In comparison with positive electrode, many works relative to the negative electrode are concerning nanomaterials, either nano-structured or nano-sized. This point may be explained by the variety of materials available as candidates for the negative side (oxides, sulfides, nitrides, carbonaceous materials, silicon, tin and other alloying materials with lithium…) and the variety of reaction mechanisms involved (intercalation, alloying, conversion, displacement, etc.) whereas almost only intercalation materials are considered at the positive electrode.

13.4.1 Carbonaceous Materials

As indicated in the introduction, carbon in the form of three-dimensional graphite is the main negative electrode material of commercial Li-Ion cells with a theoretical capacity of 372 $mAh.g^{-1}$ corresponding to the formation of LiC_6 at the end of the Li^+ intercalation process at a potential close to 0 V versus Li^+/Li [3]. Graphite possesses advantages such as its good electronic conductivity, relatively low volume change during intercalation process (10.3 %) and good reversibility during cycling, but its volumetric capacity is rather low in reason of a theoretical density of 2.26 $g.cm^{-3}$ only.

Alternative 1D (carbon nanotubes CNT, carbon nanofibers CNF), 2D (graphene nanosheets) or 3D (mesoporous carbon, or extended graphite layer structures) carbon structures have been considered to increase the storage of lithium through different possible effects, well summarized in the review from Kaskhedikar and Maier [55].

- excess bulk storage with an increase of the graphene interlayer spacing, as evidenced in the work of Sato et al. [56] where a disordered carbon prepared from poly(p-phenylene) was able to reach the Li_2C_6 composition,
- storage in microcavities and nanopores corresponding to 3D defects
- interfacial or surface storage by adsorption of lithium of both sides of the single-layer graphene sheets in disordered carbons and filling-up of defects in carbon structures as proposed in the “house of cards” modified in “falling cards” model by Dahn and coworkers [57]

Many studies have been dedicated to the use of nanostructured carbon materials such as CNT [58–60] or CNF [61, 62] as potential anodes to replace graphite with performances varying extremely according their structure, morphology, presence of heteroatoms and surface chemistry. Nevertheless, the possible increase observed on the reversible capacity up to 600 $mAh.g^{-1}$ for example in [58] is often associated with a huge irreversible capacity coming from different phenomena such as: (i) defects (often encountered on commercial CNT [63]) or oxygen surface species presence, (ii) surface reaction with electrolyte corresponding to the formation of the Solid Electrolyte Interface (SEI) at the highly reductive potential associated with the carbon negative electrode, which is close to 100 mV versus Li^+/Li for the graphite in its fully lithiated state. This issue strongly limits their practical use in a commercial Li-Ion battery.

13.4.2 Titanium Oxides for High Voltage Anodes

The possible insertion of lithium in titanium oxides (TiO_2) was shown first by Ozhuku et al. for TiO_2 in 1979 [64], by Murphy et al. in 1983 for Li_xTiO_2 phases [65], and further studied for the spinel compounds $LiTi_2O_4$ and $Li_4Ti_5O_{12}$ by Dahn et al. in 1989 [66]. Unlike graphite, Lithium intercalation in titanium oxides such as $Li_4Ti_5O_{12}$ (LTO) or TiO_2–B occurs at an average voltage of 1.5 V versus Li^+/Li, and for that reason, these active materials were considered as cathode materials in these first studies.

In 1994, Thackeray et al. [67] proposed to take advantage of this relatively high voltage to combine some spinel oxides such as LTO with a conventional 4 V type cathode material to obtain an average 2.5 V Li-Ion cell and underlined the benefits of suppression of Li metal deposition risk at the negative electrode for safety concerns.

The theoretical capacity of these compounds is 175 mAh.g^{-1} for LTO corresponding to the reversible insertion of 3 Li in the structure ($Li_4Ti_5O_{12}$ /$Li_7Ti_5O_{12}$ couple) and 335 mAh.g^{-1} for TiO_2. The charge and discharge potential curves are indicated in Fig. 13.17 in comparison with graphite. In this example, LTO was prepared according to the patented protocol by Jouanneau et al. described in reference [68] involving alternate periods of mechanical grinding and thermal treatment under air leading to a bimodal (nm scale and μm scale) particle size distribution improving compactness of the material with a surfacic area of 10 m^2.g^{-1}. TiO_2–B (Bronze) was prepared in three steps according the protocol described by Marchand et al. [69] with a first solid-state reaction between anatase TiO_2 and KNO_3 to form $K_2Ti_4O_9$, followed by an hydrolysis with HNO_3 to exchange K^+ and a final dehydration of $H_2Ti_4O_9$,H_2o to obtain a pure microcrystalline power with surface area between 10 to 45 m^2.g^{-1}.

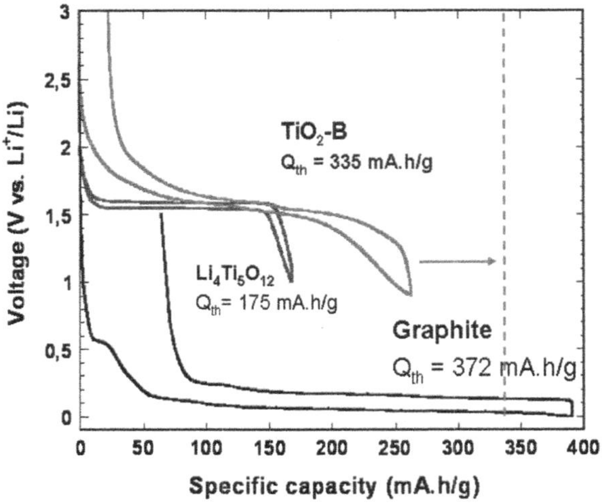

Fig. 13.17 Charge-discharge voltage of different negative electrodes versus metallic lithium. *Black* graphite, *Blue* $Li_4Ti_5O_{12}$, *Red* TiO_2–B. Reproduced with permission from [121]

13.4.2.1 $Li_4Ti_5O_{12}$ (LTO)

Despite its high potential, LTO compound has several advantages. First LTO has a spinel structure in which lithium can diffuse easily in three dimensions tunnels which is favorable for high power applications. Second, during the lithium insertion-extraction cycles only minimal expansion-contraction phenomena occurs as the oxidized and reduced phases have almost the same volume with a difference of only 0.07 % [70]. For this reason, LTO is called a zero-strain electrode material and this results in a clear advantage for extensive cycling in comparison with graphite material for which the volume expansion is around 10 %. Additionally, the high potential of LTO allows fast charging and improves safety behavior thanks to the suppression of the risk of lithium dendrites formation and also enables the use of aluminum current collector instead of copper as the formation of LiAl alloy is not possible at this working potential contrary to the graphite negative electrode.

For all these reasons, LTO is a promising anode material for Li-Ion batteries and is already integrated in commercial cells by companies like Toshiba. Nevertheless, to enable this commercialization and improve the power performances, its low intrinsic conductivity pushes many teams to work on its nanostructuration, morphology, defect chemistry, surface coating, doping and the formation of composites. An extensive review of LTO from Zhao et al. has been published in 2015 [71], covering these last points but also other issues associated with this material such as gassing problem.

In Fig. 13.18 from [71], the different types of reported LTO nano/microstructures are given: 0D nanoparticles [72], 1D nanofiber [73], nanowire [74], nanotube [75] and nanorod [76], 2D nanoplates [77]; nanosheets [78] or wave-like nanosheet [79] and 3D porous structures (detailed in Fig. 13.19), hierarchical spheres (detailed in Fig. 13.20) and arrays.

These nanostructures are obtained via synthesis routes different from the usual solid-state synthesis route. In that case, LTO is synthesized with calcination of TiO_2 and lithium precursors at a high temperature above 800 °C maintained for a long period to ensure the phase purity. This step leads to the unavoidable growth of material grains and so results in micronic particles with low specific surface area that are not well suited to power applications.

In the alternative synthesis routes to obtain nanosized particles, soft-chemical routes such as sol-gel, hydrothermal and solvothermal processes are often chosen as they allow the mixing of raw materials at the molecular level, enabling a significant decrease in the synthesis temperature and so directly reducing the grain growth. Combustion techniques like spray-pyrolysis are also a possibility to control and limit particle size. In that case, the precursors are almost instantaneously exposed at temperature that can reach several thousand degrees. For spray pyrolysis, the precursors are mixed in a solution that is sprayed through a nozzle in micrometer-sized droplets. These droplets are then directly calcined in an oven to form the active material. It is also important to notice that this technique is particularly well suited to obtain precise compositions of materials including various elements (e.g. mixture of transition metals, doping, etc.).

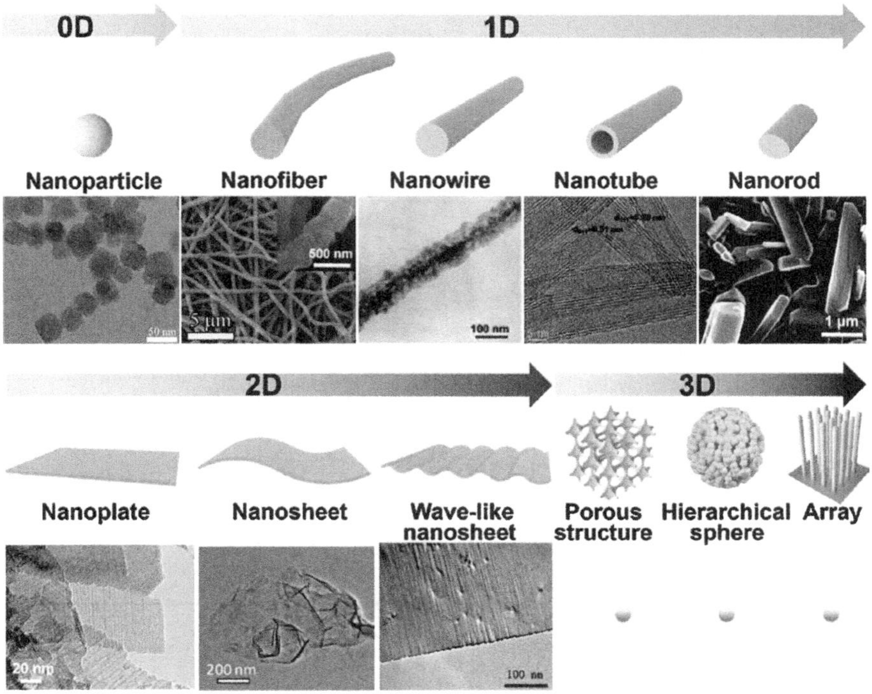

Fig. 13.18 Schematics of reported LTO nano/microstructures OD, 1D, 2D and 3D with the corresponding microscope images. Reproduced with permission from [71]

In the literature, the different routes are often combined to circumvent the drawbacks inherent to each single one (e.g. broad particle size distribution). For instance, Lu et al. [72] proposed an alternative process to prepare nearly monodisperse LTO single crystals of 40 nm by combining hydrothermal treatment of TiO_2 in LiOH to form α-Li_2TiO_3, followed by an ion-exchange with HCl and a further calcination at 400 °C. The obtained material reached 172 mAh.g^{-1}, almost the theoretical capacity, at low rate and maintained 140 mAh.g^{-1} at 5C rate.

In another example, Li et al. [4.22] synthesized LTO nanotubes through a two steps process, with first the preparation of hydrogen titanate nanotubes via a sonochemical-hydrothermal reaction of TiO_2 in concentrated NaOH solution, followed by a hydrothermal lithium ion exchange process. These nanotubes had between 100 nm to several μm length with an average 10 nm diameter.

Another possibility to increase the power performances is to develop porous 3D structures through 3 different routes illustrated in Fig. 13.19: (i) the synthesis via the use of template or pore former that generates the pores, (ii) combustion synthesis consisting mainly in generating the pores through diffusion of the gas formed by the combustion of the fuel (e.g. cellulose) with the raw materials, and (iii) the self-assembling method in which the pores result from the void space created by the special structure or the sites left by the removal of organic groups.

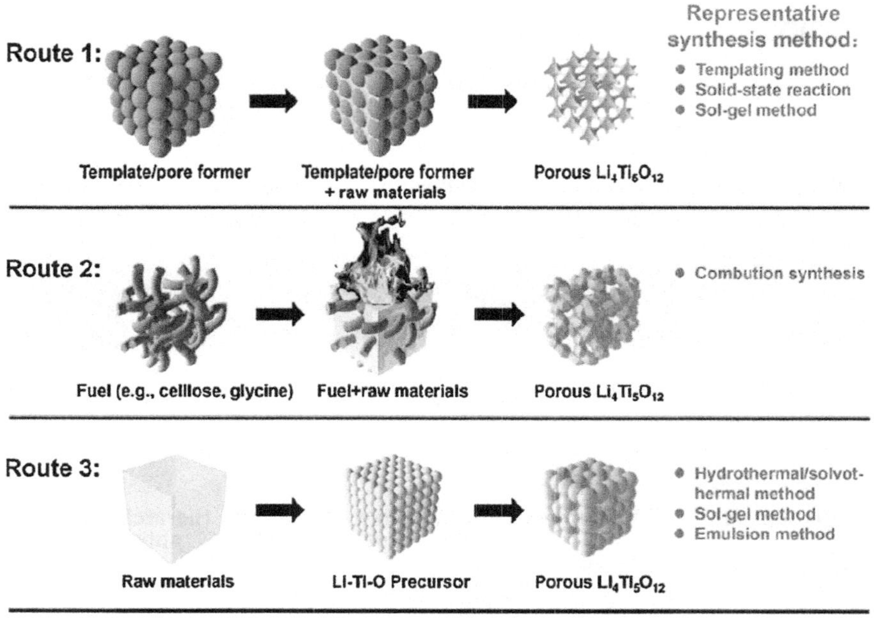

Fig. 13.19 Typical routes for the synthesis of porous LTO. Reproduced with permission from [71]

Hierarchical spheres are other 3D structures facilitating the high-rate electrochemical performances of LTO and may present better tap density than the previous porous 3D structures. In Fig. 13.20, 7 different types of those spheres are illustrated: dense and porous spheres via hydrothermal process [80], hollow spheres using nanosilica template [81], yolk-shell spheres by spray-pyrolysis [82], hollow yolk-shell spheres of carbon-Encapsulated F-Doped $Li_4Ti_5O_{12}$ via hydrothermal synthesis and subsequent solid-state lithiation at high temperature [83], flower-like spheres [84] and hollow flower-like spheres [85].

13.4.2.2 TiO_2

TiO_2 presents at least 8 polymorphs and among them 4 have been reported to intercalate electrochemically lithium, rutile, brookite and especially anatase, and TiO_2–B already mentioned in the introduction on titanium oxides. A detailed review on the nanoengineering of TiO_2 for high rate lithium storage can be found in the paper of Jiang and Zhang [86].

TiO_2–B, a metastable phase, is one of the two less compact forms with large interstitial spaces compared to the radius of lithium ions. It is composed of edge- and corner-sharing TiO_6 octahedra with perovskite-like windows between sites and this polymorph knows specific attention in the last research works on this oxide

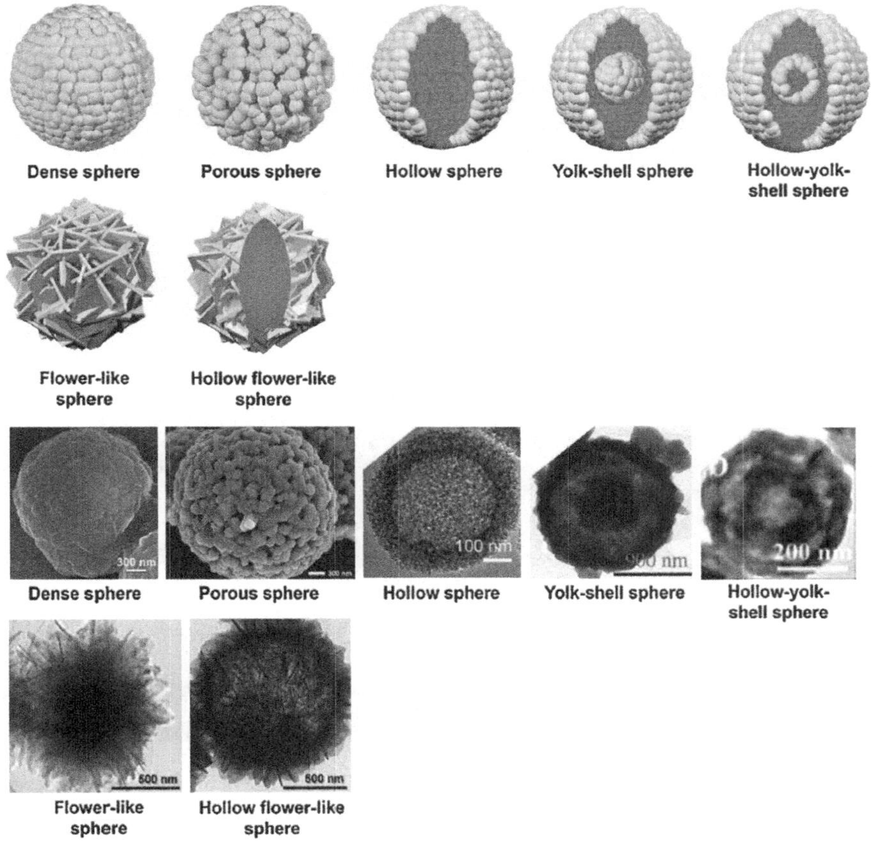

Fig. 13.20 Schematic representations of reported LTO nano/microspheres and corresponding microscope images. Reproduced with permission from [71]

family. For example, Li et al. synthesized by hydrothermal route at different temperatures from 120 to 180 °C TiO_2–B nanowires with 104–210 $m^2.g^{-1}$ BET [87] and a [88] growth direction with shorter b- and c- axis channels to fasten Li diffusion. The TEM image of the nanowires and the HRTEM image of a single nanowire are given in Fig. 13.21. The best performances were obtained for the 150 °C samples with 220 $mAh.g^{-1}$ at low rate (50 $mA.g^{-1}$) and still 140 $mAh.g^{-1}$ retained at 2 $A.g^{-1}$.

As previously shown for LTO, development of hierarchical structures is another approach in the improvement of TiO_2 performances. In a study from Wang et al. [89], TiO_2 nanocages were synthesized by facile coating of anatase TiO_2 against Cu_2O polyhedral via TiF_4 hydrolysis followed by chemical manipulation of Cu_2O via acid etching. The images of octahedral TiO_2 nanocages are presented in Fig. 13.22. The electrochemical performances of these materials were interesting with more than 110 $mAh.g^{-1}$ at C-Rate after 200 cycles.

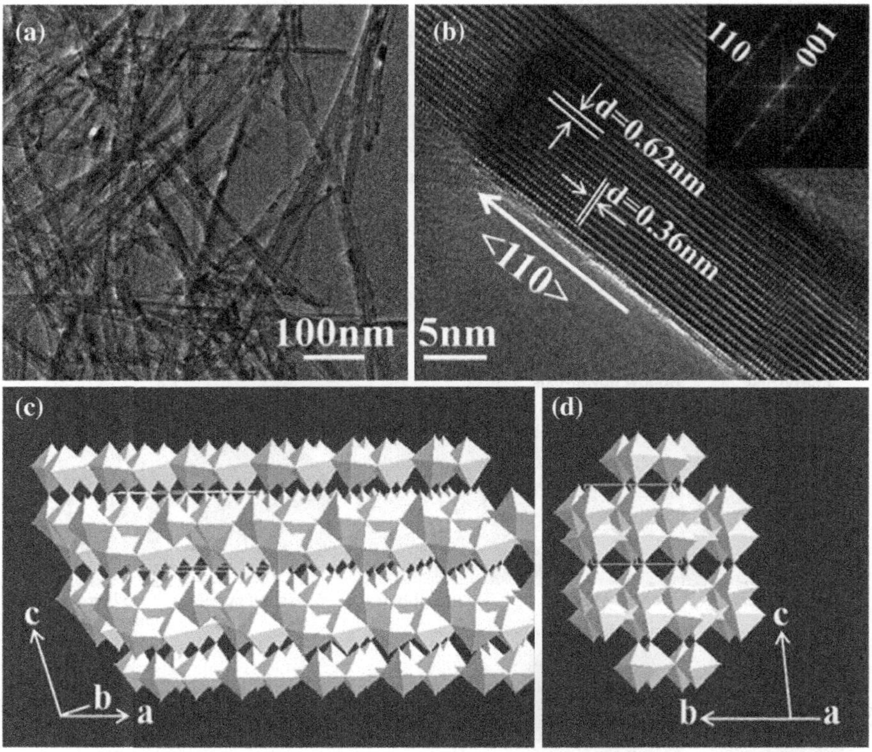

Fig. 13.21 **a** TEM image of TiO_2–B nanowires and **b** HRTEM image of a single nanowire. **c** and **d** schematic model of a single TiO_2–B nanowire composed of TiO_6 octahedra with edge-sharing along the [88] direction and corner-sharing along the (001) direction viewed from the side and cross section respectively. Reproduced with permission from [87]

13.4.3 Alloy Negative Electrodes

Contrary to the titanium oxides that are developed for high power applications, lithium alloying materials are of specific interest to significantly increase the capacity of the graphite negative electrode. In this family, we can find the elements (Al, Si, Sn, Ge, Pb, Sb, Bi) that can form electrochemically binary alloys with lithium [90–92]. After the proposal by Fujifilm in 1997 of a new TCO anode (tin composite oxide) [93], tin was the first to be integrated commercially in 2005 by Sony in Nexelion 14430 cells.

A comparison of the gravimetric and volumetric capacities of the different lithium alloying materials is given in Fig. 13.23, where one can see than silicon is the most promising compound thanks to its theoretical capacity of 3579 $mAh.g^{-1}$ corresponding to the formation of $Li_{15}Si_4$ alloy. In this figure, another important benefit from the alloys is highlighted when you compare the volumetric capacities to the commercial graphite reference electrode. Through this view point, Sn and Si

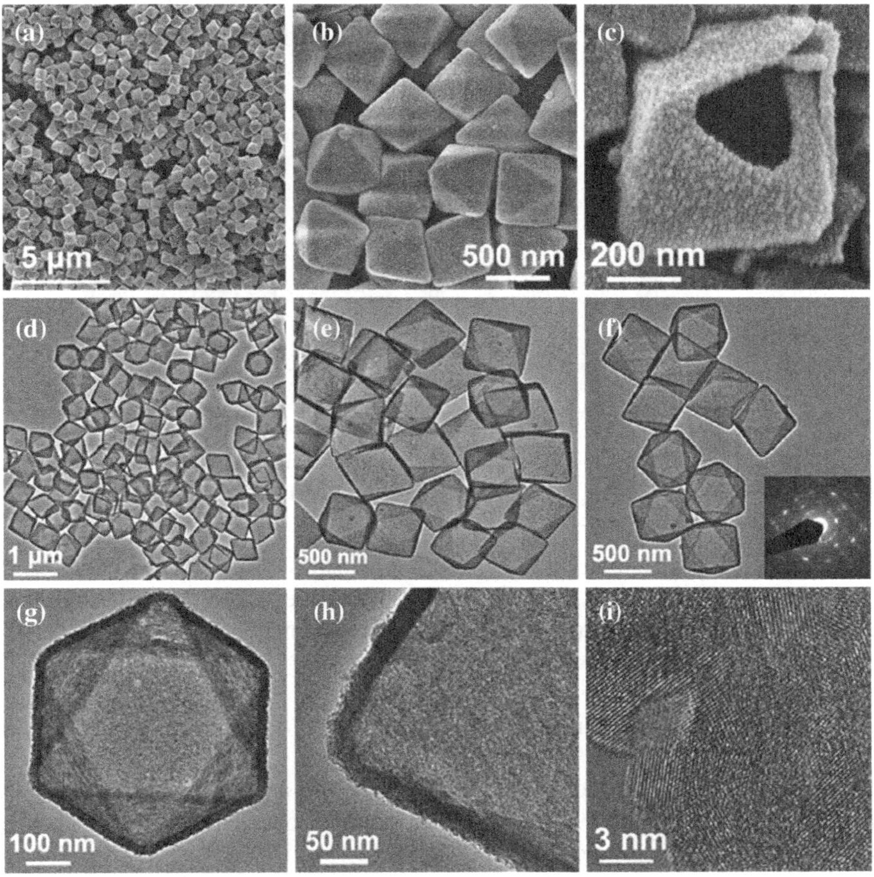

Fig. 13.22 **a**, **b** SEM images of octahedral TiO_2 nanocages with an average edge length of 500 nm; **c** A cracked nanocage; **d–f** TEM images of these nanocages; the inset in **f** Is a SAED pattern; **g** A free-standing nanocage showing each plane is an equilateral triangle; **h** Shell structure; **i** HRTEM image of a nanocage. Reproduced with permission from [89]

are quite at the same level, explaining maybe the fact that tin was studied first as volumetric energy density is a key point for portable applications. For the interested reader, M.N. Obrovac and V. I. Chevrier published a detailed review oriented toward the practical application and commercialization of alloys in LiB [94] and L. Chen and coauthors published a synthetic review on silicon/carbon composite anode in 2015 [95].

The drawback associated with the insertion of lithium during alloying reaction is the huge volume increase of the material, up to 290 % in the case of silicon at the end of the lithiation process. In the case of traditional micronic particles, the volume variations repeated during the cycling of the electrode cause a rapid degradation of the electrochemical performances through different phenomena such as decrepitation of the particles, loss of cohesion and so insulation of part of the particles,

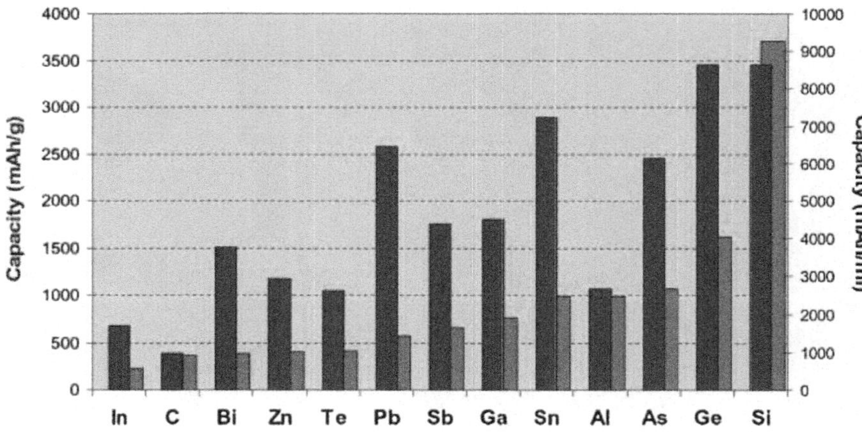

Fig. 13.23 Gravimeteric (*green*) and volumetric (*purple*) capacities for some alternative lithium alloying materials—C indicated as reference. Reproduced with permission from [96]

increase of exposed surface to the electrolyte and so irreversible lithium consumption to form SEI, etc. Much effort has been directed to the nanosizing and nanostructuration of these alloys to limit these phenomena.

13.4.3.1 Silicon Based Anodes

Lithium-silicium alloy was discovered in 1965 [97, 98] and the first use of this alloy as anode was reported in 1976 in high temperature molten-salt LiCl-KCl batteries [99, 100]. Four phases were identified, Li_2Si, $Li_{21}Si_8$, $Li_{15}Si_4$ and $Li_{22}Si_5$ at those temperatures above 400 °C with corresponding voltage plateaus on the electrochemical galvanostatic charge-discharge curve.

In 1995, Dahn et al. published one of the first papers on the utilization of silicon-carbon anode at room temperature for Li-Ion battery. They obtained by CVD nanodispersed silicon in pregraphitic carbons that exhibited reversible capacity of 600 mAh.g^{-1} with 11 % at. of silicon [101]. To better understand the research activity on silicon anode, it is necessary to have in mind the different characteristics of the different lithiated states of silicon that are summarized by Chen [95] in the Table 13.1 using data from Huggins et al. [102]. The volume expansion gradually increases with the lithium content from 60 % for LiSi phase to 290 % for $Li_{15}Si_4$ and 320 % for $Li_{22}Si_5$.

Nevertheless, the maximum lithiated phase at room temperature was demonstrated to be $Li_{15}Si_4$, by Obrovac et al. [103] who showed in 2004 that this phase crystallizes under 50 mV versus Li^+/Li in a phase unidentified in the previous studies at high temperature. This means that the real theoretical capacity at ambient temperature of silicon anode is 3590 mAh.g^{-1}, still almost 10 times higher than the value of 372 mAh.g^{-1} for graphite.

Table 13.1 Different characteristics of Si and Li–Si alloys. Reproduced with permission from [102]

Different lithiated states of silicon	Silicon atom Volume/ $Å^{-3}$	Volume/Expansion/ %	Gravimetric specific Capacity/mA•h•g^{-1}	Volumetric specific Capacity/mA•h•cm^{-3}
Si	19.6	0	0	0
LiSi	31.4	60	954	1425
$Li_{12}Si_7$	43.5	122	1635	1711
Li_2Si	51	160	1900	1741
$Li_{14}Si_6$	51.5	163	2233	2009
$Li_{13}Si_4$	67.3	247	3100	2145
$Li_{15}Si_4$	76.4	290	3590	2188
$Li_{22}Si_5$	82.4	320	4200	2357

Different strategies are followed to limit and/or control the effects of volume expansion: nanosizing of the particles, buffering with other compounds such as carbon or other matrixes such as SiO_2 for SiOx compounds, developing advanced hierarchical structures, using specific binders in the electrodes to improve the cohesion, and of course combinations of these strategies.

Silicon nanoparticles are usually prepared through Chemical Vapor Deposition (CVD) or mechanical grinding. CVD process requires pyrolysing costly and poisonous silane SiH_4 precursor limiting the possibility to use it for large scale applications. Mechanical grinding may be a valuable process to produce nano-Si in large quantities at competitive costs, with the possibility to introduce dopants or to prepare Si/C composites [104].

Nanosizing silicon does not suppress the mechanical stresses during the cycles of insertion/disinsertion of lithium, and so many researches have been dedicated to the development of Si/C composites in which carbon may both improve conductivity and act as a matrix to buffer silicon volume variations. In that case, synthesis routes such as spray-drying can be used. C. Paireau et al. obtained Si/C composites with cross-linked polyvinyl alcohol (PVA) carbon precursor [105]. According to the ratio Si/C, the cyclability can be improved and the best performances were reached for C contents between 50 to 60 % with a capacity of the composite twice as high as graphite for the 44 %Si/56 %C composite.

Recently, J. Sourice et al. [106] developed a synthesis of Silicon nanoparticles with a continuous Carbon shell based on a one-step continuous process in a novel two stages laser-pyrolysis reactor represented schematically in Fig. 13.24 in which silane is introduced at the first stage followed by ethylene in the second one. The composites exhibited a stable reversible capacity of 2400 mAh.g^{-1} at C/10 and up to 500 mAh.g^{-1} at 2C-Rate over 500 cycles. With this specific process, formation of SiO_2 at the surface of Silicon is almost suppressed, avoiding the huge irreversible capacity loss during the electrochemical lithiation of SiO_2 to form Li_2O.

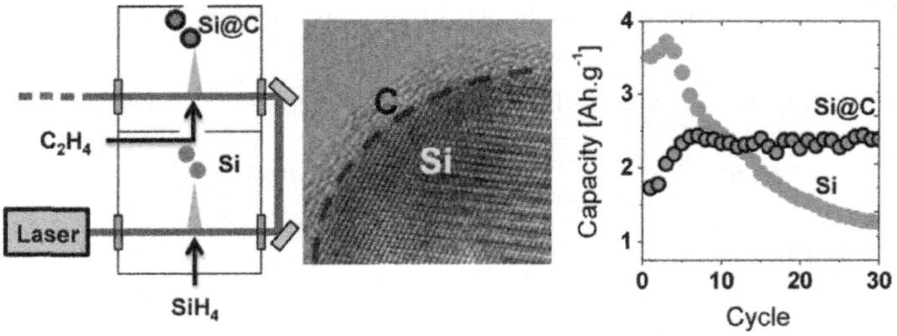

Fig. 13.24 Schematics of the double stage laser pyrolysis process (*Left*). HRTEM image of Silicon nanoparticle with carbon shell (*Center*). Comparison of the capacity of Si and Si@C nanoparticles in cycling (*Right*). Reproduced from [106]

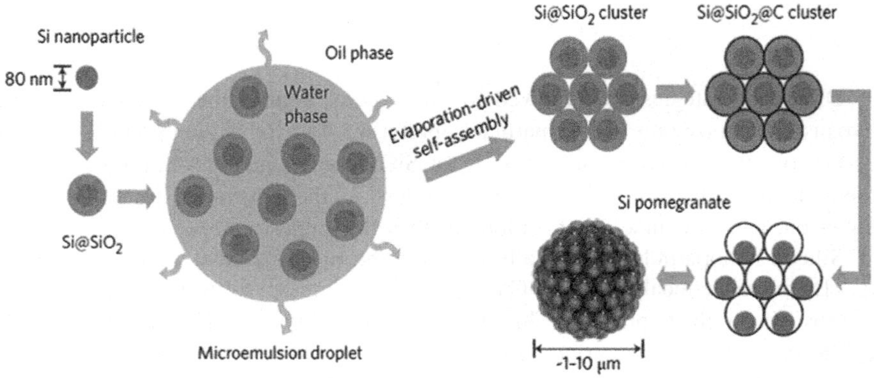

Fig. 13.25 Schematics of the process of synthesis of Pomegranate-like Si@C nanoparticles. Reproduced with permission from [107]

In 2014, Cui et al. developed a pomegranate inspired nanoscale architecture for large volume change anode materials of Lithium-Ion batteries [107] and published its application for silicon/carbon composites. The synthesis process and its different steps are illustrated in Fig. 13.25. Silicon nanoparticles are first coated with sacrificial SiO_2, then the particles are dispersed in clusters in microemulsion droplets. The self-assembly is then driven by the evaporation step and carbon coating made during pyrolysis to result in Si@SiO_2@C clusters. SiO_2 scaffold is finally removed by HF etching to obtain 1–10 μm diameter Si@C pomegranate particles. Thanks to this specific architecture, the internal void space inside the pomegranates allows a significant improvement in cycling stability over 1000 cycles as the integrity of the particles is maintained during the subsequent charge/discharge steps as shown in Fig. 13.26.

In parallel to all the work on the silicon and silicon composite materials, the role of the binder is also considered of first importance as one can easily imagine its impacts on maintaining the integrity of the composite electrode during lithiation/delithiation cycles [88, 108].

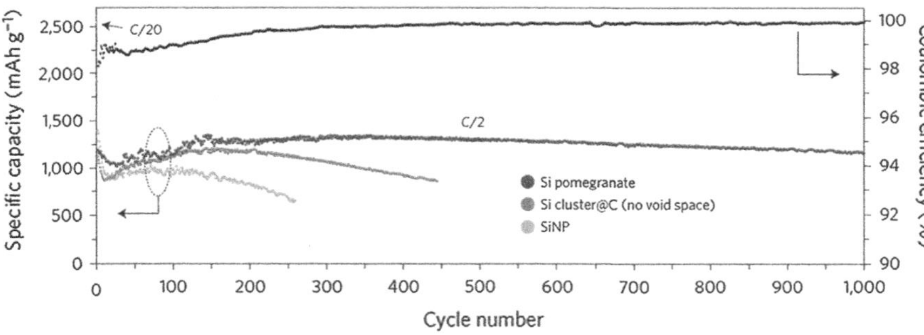

Fig. 13.26 Cycling performances of Si@C pomegranates compared with Si cluster@C without void space and Si nanoparticles. Reproduced with permission from [107]

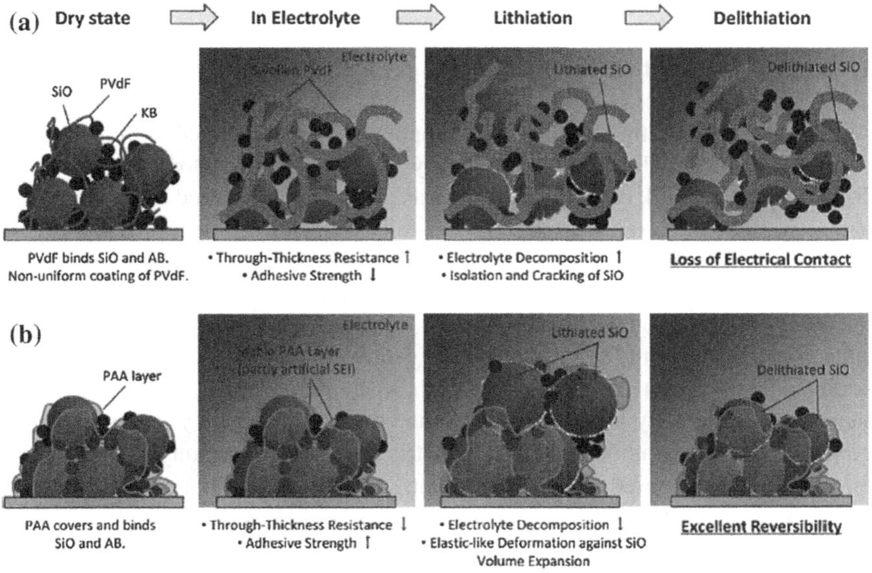

Fig. 13.27 Mechanism proposed by Komaba et al. to explain the better behavior of PAA binder compared to PVdF for SiO anode. Reproduced with permission from [88]

PVdF (Polyvinyliden fluoride) is one of the most common binder in the Li-Ion batteries, especially for positive electrodes. Nevertheless, this polymer is not well suited for silicon anodes as shown by Komaba et al. and represented in Figs. 13.27 and 13.28 where PVdF is compared with Polyacrylic acid (PAA). PVdF is a thermoplastic polymer with an average elongation that does not allow containing the volume expansion of silicon. Komaba et al. proposed using PAA as an alternative for SiO anodes and obtained 700 mAh.g^{-1} stable reversible capacity at 100 mA.g^{-1} rate. According to them, PAA makes a uniform coating on the surface

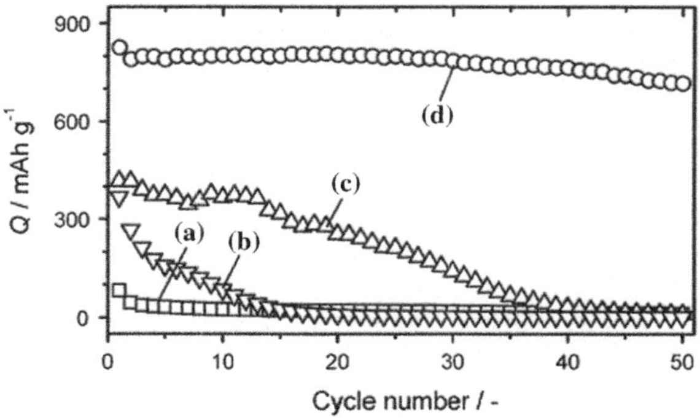

Fig. 13.28 Comparison of electrochemical performances of SiO electrodes with various binder: **a** PVdF, **b** PVA, **c** CMCNa and **d** PAA. Reproduced with permission from [88]

of SiO particles and tightly binds the particles, this contributes to the stability and also to the reduction of the irreversible capacity loss at the first cycle.

13.4.3.2 Tin Based Anodes

As indicated above, tin was the first compound to be integrated commercially in alloy anodes for Li-Ion batteries by Sony in Nexelion cells for hand-held camcorders in 2005 [109], before silicon that has only been introduced a couple of years ago (e.g. Hitachi-Maxell prismatic Li-Ion cells with SiO anode in 2011 [110]).

Many papers have been dedicated to the analysis of the performances of Nexelion 14430 cells [111–113] and the corresponding data have been compiled in the review from Obrovac and Chevrier [94] and summarized in the Fig. 13.29. The capacity retention of the cell (80 % capacity after 200–250 cycles) was moderate in comparison with conventional Li-Ion cells that usually reach 500 cycles for the same level of capacity loss, but thanks to the density of the alloy in comparison to graphite, the cells exhibited a high volumetric energy density. Sony announced that this alloy was composed of Tin and Cobalt grains dispersed in a carbon matrix. In the same review [94], the chemical and electrochemical analysis of the Nexelion

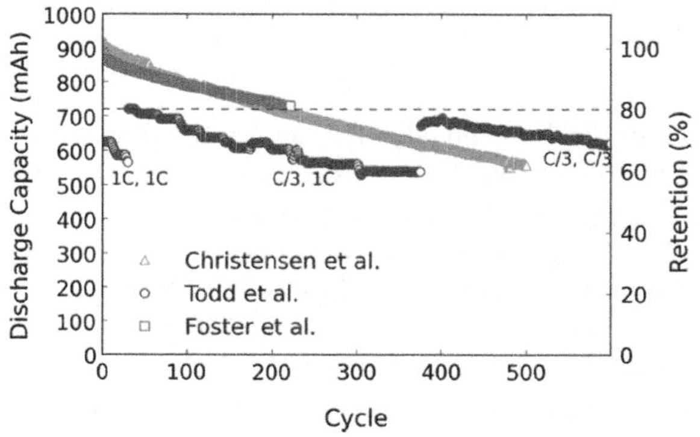

Fig. 13.29 Cycling of Nexelion 14430 cells. Reproduced with permission from [94]. Data derived from references [110–113]

cells made by various research groups are summed up indicating that the alloy has approximately 1 μm particles with a majority of Sn and Co and a small amount of Ti. The Sn-Co grains were approximately 5 nm measured by TEM analysis. Via different types of studies (sputtering, high energy ball milling and attritor milling), the composition appeared to be close to the $Sn_{30}Co_{30}C_{40}$ composition.

Similarly to silicon, binder has a strong influence on the electrochemical properties and especially on the cycling stability of tin anodes. Dahn et al. [114] published a comparison between PVdF, LiPAA and CMC binder for the alloy $Sn_{30}Co_{40}C_{40}$ prepared by mechanical attrition, with a clear evidence of the better suitability of Li-PAA, developed by 3 M company, for this electrode with a first discharged capacity close to 500 $mAh.g^{-1}$ and still almost 450 $mAh.g^{-1}$ after 100 cycles (Fig. 13.30). They suppose that LiPAA, as a polyelectrolyte, may modify the solid electrolyte interface (SEI) and so avoid continuous capacity loss during cycling. Additionally, this binder seems to enhance electronic and ionic conductivity inside the composite electrode.

In conclusion, even if the theoretical massic capacity of tin is much lower than that of silicon, due to its high density, tin remains an interesting candidate for Li-Ion cells and research activities are still on going on this compound.

13.4.4 Conversion Materials

Conversion reactions are lithiation reactions in which the active material M_zX_y (where M is a metal cation and Y an anion) is fully reduced by lithium to the metal M^0 according the (10.1):

$$M_zX_y + zLi^+ + ze^- \leftrightarrow yLi_{z/y}X + zM^0 \tag{10.1}$$

In the case of metal oxides such as the binary transition metal oxides with the rock-salt structure (CoO, CuO, NiO, FeO), there are no free voids to host lithium.

Fig. 13.30 Comparison of the cycling performance of Sn electrodes using Li-PAA, CMC and PVdF binders. Reproduced with permission from [114]

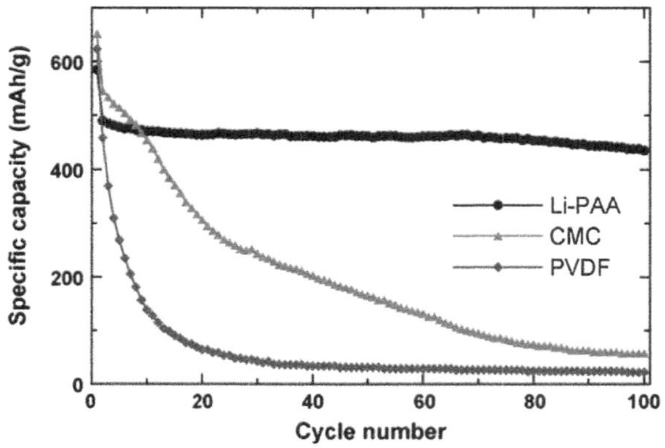

Additionally, the metallic elements Co, Cu, Ni or Fe do not form alloys with lithium but Tarascon and coworkers [115] evidenced in 2000 that they can react reversibly with lithium according to the (10.2) derived from (10.1) in the case of X = O:

$$MO + 2Li^{+} + 2e^{-} \leftrightarrow Li_2O + M^0 \qquad (10.2)$$

The full reduction of the nanometric oxide particles leads to composite materials consisting of nanometric metallic particles (1–5 nm) dispersed in an amorphous Li_2O matrix. Owing to the nanometric nature of this composite, such reactions were shown to be reversible, providing capacities up to 700 $mAh.g^{-1}$ to store lithium (twice the value of graphite materials) for 50 cycles (Fig. 13.31 where data for Co_3O_4/Li cell was added to show that the mechanism is not specific to divalent oxides). Nevertheless, one can see that this conversion process leads to a huge hysteresis in the voltage profile corresponding to the phase changes and may be associated with the energy barrier necessary to break the M–X bonds, energy barrier sensitive to the nature of the anion decreasing from around 0.9 V for oxides to 0.4 V for phosphides [116].

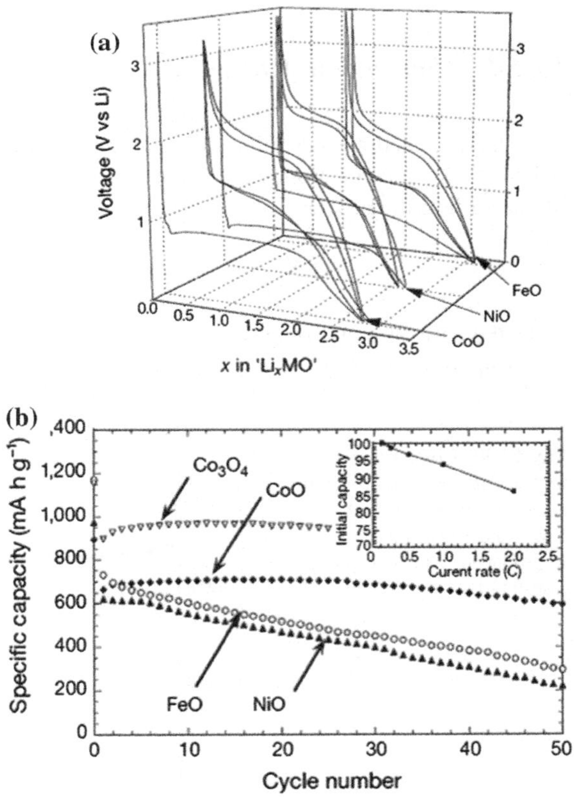

Fig. 13.31 Properties of MO/Li cells. **a** The voltage composition profile for various MO/Li cells cycled at C/5-Rate. **b** Cycling behavior for the same cells under similar conditions where data for a Co_3O_4/Li cell was added. Inset, the rate capability of a CoO electrode. Reproduced with permission from [115]

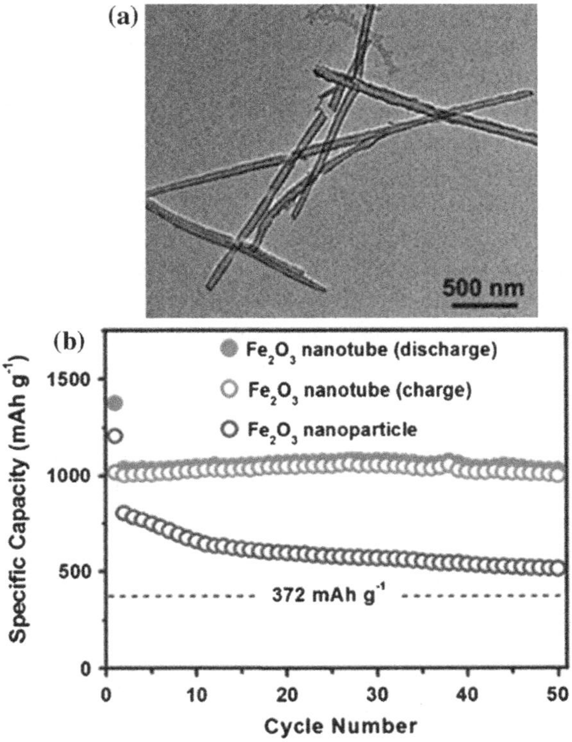

Fig. 13.32 **a** Micrographs of α-Fe_2O_3 nanotubes, and **b** electrochemical performances during cycling. *Dash line* maximum theoretical capacity of graphite indicated as a reference. Reproduced with permission from [119]

For the interested reader, J. Cabana et al. published an extended review on the electrode materials reacting through conversion reactions [117] such as transition metal oxides, phosphides, nitrides and fluorides.

Fe_2O_3 and Fe_3O_4 are potentially very attractive materials. For instance, the reduction of Fe^{3+} to Fe^0 in Fe_2O_3 corresponds to a theoretical capacity of 1007 mAh.g^{-1}. Morimoto et al. obtained α-Fe_2O_3 from α-Fe(OOH) by mechanical milling [118], reaching the theoretical capacity after the first reduction but only 800 mAh.g^{-1} for the second one. Nanostructuring is of course a possible way to improve the performances of these materials. Zhang et al. developed Fe_2O_3 nanotubes that were stable at 1000 mAh.g^{-1} for 50 cycles [119] (Fig. 13.32). Here again, the drawback of this "nano"oxides is that they suffer from a very high irreversible capacity at the first cycle, almost 400 mAh.g^{-1} in that case.

13.5 Summary and Outlook

Nanosizing and nanostructuring materials open promising routes for either the introduction of new active materials or significant improvements in materials already available. On the positive electrode side, nanocoatings allow stabilizing the material at high voltage and so increasing the energy density of the Li-Ion cells. Carbon coating was a major step in the development of low electronic conductive $LiFePO_4$ cathode material that is now available commercially for different type of applications, even those requiring high power. Regarding the negative electrode, after the first introduction of tin alloy anode, silicon is now partially substituting graphite in some standard 18650 cells. This has been possible thanks many studies and works relative to nanoalloys or nanocomposites Si-C.

Nevertheless, going "nano" presents also potential drawbacks, the major one is linked to the possible increase of the surfacic area of the compounds leading to higher electrolyte/electrode interfaces and so possible unwanted side-reactions either increasing the irreversible capacities or reducing cycle life. Nanostructuring may be a possibility to circumvent these phenomena through hierarchical structures (hollow yolk-shell spheres, pomegranates-inspired particles, etc.), but this is possible only if the synthesis process remains cost competitive. The developed nanostructures may also be of interest for other emerging battery technologies. For lithium-sulfur for instance, trapping the sulfur in hierarchical C-S composites can help in preventing polysulfides dissolution and so in improving cycling performances.

In this chapter, we focused on the positive and negative electrode materials, but other inactive compounds are and/or can be concerned by the "nano" approach: current collector nanostructuring to improve the adhesion of the active electrode layer and so increase conductivity and cohesion during cycling, ceramics coating on separator to improve thermal behavior in abuse conditions, etc.

Overall, chemistry of nanomaterials will probably enable the introduction of new battery technologies in the next years that are expected for more and more demanding applications in terms of energy and power densities.

References

1. M.S. Whittingham, Electrical energy storage and intercalation chemistry. Science **192**, 1126–1127 (1976)
2. K. Mizushima, P.C. Jones, P.J. Wiseman, J.B. Goodenough, Li_xCoO_2 (0<x<–1): a new cathode material for batteries of high energy density. Mat. Res. Bull. **15**, 783–789 (1980)
3. R. Yazami, P. Touzain, A reversible graphite-lithium negative electrode for electrochemical generators. J. Power Sources **9**, 365–371 (1983)
4. Z. Zhang, P. Ramadass, *Lithium-Ion Battery Systems And Technology, Encyclopedia of Sustainability Science and Technology* (2012, Springer, Berlin), pp 6122–6149. doi:10.1007/978-1-4419-0851-3_663, Print ISBN 978-0-387-89469-0, Online ISBN 978-1-4419-0851-3
5. M.M. Thackeray, W.I.F. David, P.G. Bruce, J.B. Goodenough, Lithium insertion into manganese spinels. Mat. Res. Bull. **18**, 461–472 (1983)
6. N. Yabuuchi, T. Ohzuku, Novel lithium insertion material of $LiCo_{1/3}Ni_{1/3}Mn_{1/3}O_2$ for advanced lithium-ion batteries. J. Power Sources **119**, 171–174 (2003)
7. M. Broussely, F. Perton, P. Biensan, J.M. Bodet, J. Labat, A. Lecerf, C. Delmas, A. Rougier, J.P. Pérès, Li_xNiO_2, a promising cathode for rechargeable lithium batteries. J. Power Sources **54**, 109–114 (1995)

8. B. Ammundsen, J. Paulsen, I. Davidson, R.S. Liu, C.H. Shen, J.M. Chen, L.Y. Jang, J.F. Leed, Local structure and first cycle redox mechanism of layered $Li_{1.2}Cr_{0.4}Mn_{0.4}O_2$ cathode material. J. Electrochem. Soc. **149**(4), A431–A436 (2002)
9. K. Amine, H. Tukamoto, H. Yasuda, Y. Fujita, Preparation and electrochemical investigation of $LiMn_{2-x}Me_xO_4$ (Me: Ni, Fe, and x = 0.5, 1) cathode materials for secondary lithium batteries. J. Power Sources **68**, 604–608 (1997)
10. A.K. Padhi, K.S. Nanjundaswamy, J.B. Goodenough, Phospho-olivines as positive-electrode materials for rechargeable lithium batteries. J. Electrochem. Soc. **144**, 1188–1194 (1997)
11. Y. Zhong, M. Yang, X. Zhou, Z. Zhou, Structural design for anodes of lithium-ion batteries: emerging horizons from materials to electrodes. Mater. Horiz. **2**, 553–566 (2015)
12. A.S. Arico, P. Bruce, B. Scrosati, J.M. Tarascon, W. van Schalkwij, Nanostructured materials for advanced energy conversion and storage devices. Nat. Mat. **4**, 366–377 (2005)
13. F. Jiao, P.G. Bruce, Mesoporous crystalline β-MnO_2__a reversible positive electrode for rechargeable lithium batteries. Adv. Mater. **19**, 657–660 (2007)
14. P. Balaya, A.J. Bhattacharyya, J. Jamnik, YuF Zhukovskii, E.A. Kotomin, J. Maier, Nano-ionics in the context of lithium batteries. J. Power Sources **159**, 171–178 (2006)
15. N. Nitta, F. Wu, J.T. Lee, G. Yushin, Li-ion battery materials: present and future. Mater. Today **18**, 252–264 (2015)
16. X. Xu, S. Lee, S. Jeong, Y. Kim, J. Cho, Recent progress on nanostructured 4 V cathode materials for Li-ion batteries for mobile electronics. Mater. Today **16**, 487–495 (2013)
17. Y. Cho, J. Eom, J. Cho, High performance LiCoO2 cathode materials at 60 °C for lithium secondary batteries prepared by the facile nanoscale dry-coating method. J. Electrochem. Soc. **157**, A617–A624 (2010)
18. Y. Oua, J. Wen, H. Xu, S. Xie, J. Li, Ultrafine $LiCoO_2$ powders derived from electrospun nanofibers for Li-ion batteries. J. Phys. Chem. Solids **74**, 322–327 (2013)
19. J. Kim, M. Noh, J. Cho, H.M. Kim, K.B. Kim, Controlled nanoparticle metal phosphates (Metal = Al, Fe, Ce, and Sr) coatings on $LiCoO_2$ cathode materials. J. Electrochem. Soc. **152**, A1142–A1148 (2005)
20. E.H. Lee, J.H. Park, J.M. Kim, S.Y. Lee, Direct surface modification of high-voltage $LiCoO_2$ cathodes by UV-cured nanothickness poly(ethylene glycol diacrylate) gel polymer electrolytes. Electrochim. Acta **104**, 249–254 (2013)
21. I.D. Scott, Y.S. Jung, A.S. Cavanagh, Y. Yan, A.C. Dillon, S.M. George, S.-H. Lee, Ultrathin coatings on Nano-$LiCoO_2$ for Li-ion vehicular applications. Nano Lett. **11**, 414–418 (2011)
22. O. Haik, N. Leifer, Z. Samuk-Fromovich, E. Zinigrad, B. Markovsky, L. Larush, Y. Goffer, G. Goobes, D. Aurbach, On the surface chemistry of $LiMO_2$ cathode materials (M = [MnNi] and [MnNiCo]): electrochemical, spectroscopic, and calorimetric studies. J. Electrochem. Soc. **157**, A1099–A1107 (2010)
23. Y. Cho, S. Lee, Y. Lee, T. Hong, J. Cho, Spinel-layered core-shell cathode materials for Li-ion batteries. Adv. Energy Mater. **1**, 821–828 (2011)
24. Y. Cho, P. Oh, J. Cho, A new type of protective surface layer for high-capacity Ni-based cathode materials: nanoscaled surface pillaring layer. Nano Lett. **13**, 1145–1152 (2013)
25. Y.K. Sun, S.T. Myung, B.C. Park, J. Prakash, I. Belharouak, K. Amine, High-energy cathode material for long-life and safe lithium batteries. Nat. Mat. **8**, 320–324 (2009)
26. J. Cho, M.M. Thackeray, Structural changes of $LiMn_2O_4$ spinel electrodes during electrochemical cycling. J. Electrochem. Soc. **146**, 3577–3581 (1999)
27. R. Benedek, M.M. Thackeray, Reaction energy for $LiMn_2O_4$ spinel dissolution in acid. Electrochem Solid State Lett. **9**, A265–A267 (2006)
28. J.S. Kim, K. Kim, W. Cho, W.H. Shin, R. Kanno, J.W. Choi, A truncated manganese spinel cathode for excellent power and lifetime in lithium-ion batteries. Nano Lett. **12**, 6358–6365 (2012)
29. S.T. Myung, K.S. Lee, D.W. Kim, B. Scrosati, Y.K. Sun, Spherical core-shell Li $[(Li_{0.05}Mn_{0.95})_{0.8}(Ni_{0.25}Mn_{0.75})_{0.2}]_2O_4$ spinels as high performance cathodes for lithium batteries. Energy Environ. Sci. **4**, 935–939 (2011)

30. Q. Zhong, A. Bonakdarpour, M. Zhang, Y. Gao, J.R. Dahn, Synthesis and electrochemistry of $LiNi_xMn_{2-x}O_4$. J. Electrochem. Soc. **144**, 205–213 (1997)
31. S. Patoux, L. Sannier, H. Lignier, Y. Reynier, C. Bourbon, F. Le Cras, S. Martinet, High voltage nickel manganese spinel oxides for Li-ion batteries. Electrochim. Acta **53**, 4137–4145 (2008)
32. S. Patoux, L. Daniel, C. Bourbon, H. Lignier, C. Pagano, F. Le Cras, S. Jouanneau, S. Martinet, High voltage spinel oxides for Li-ion batteries: from the material research to the application. J. Power Sources **189**, 344–352 (2009)
33. L. Zhou, D. Zhao, X. Lou, $LiNi_{0.5}Mn_{1.5}O_4$ hollow structures as high-performance cathodes for lithium-ion batteries. Angew. Chem. Int. Ed. **51**, 239–241 (2012)
34. C. Masquelier, L. Croguennec, Polyanionic (phosphates, silicates, sulfates) frameworks as electrode materials for rechargeable Li (or Na) batteries. Chem. Rev. **113**, 6552–6591 (2013)
35. A. Manthiram, J.B. Goodenough, Lithium insertion into $Fe_2(MO_4)_3$ frameworks: comparison of M = W with M = Mo. J. Solid State Chem. **71**, 349–360 (1987)
36. A. Manthiram, J.B. Goodenough, Lithium insertion into $Fe_2(SO_4)_3$ frameworks. J. Power Sources **26**, 403–408 (1989)
37. N. Ravet, J.B. Goodenough, S. Besner, M. Simoneau, P. Hovington, M. Armand, Meeting abstract n 127, in *196th Meeting of the Electrochemical Society*, HI, 1999
38. N. Ravet, Y. Chouinard, J.F. Magnan, S. Besner, M. Gauthier, M. Armand, Electroactivity of natural and synthetic triphylite. J. Power Sources **97**, 503–507 (2001)
39. S. Franger, F. Le Cras, C. Bourbon, H. Rouault, $LiFePO_4$ Synthesis routes for enhanced electrochemical performance. Electrochem. Solid-State Lett. **5**, A231–A233 (2002)
40. S.Y. Chung, J.T. Bloking, Y.M. Chiang, Electronically conductive phospho-olivines as lithium storage electrodes. Nat. Mater. **1**, 123 (2002)
41. G. Li, H. Azuma, M. Tohda, $LiMnPO_4$ as the cathode for lithium batteries. Electrochem. Solid-State Lett. **5**, A135–A137 (2002)
42. G. Li, H. Azuma, M. Tohda, Optimized $LiMn_yFe_{1-y}PO_4$ as the cathode for lithium batteries. J. Electrochem. Soc. **149**, A743–A747 (2002)
43. C. Delacourt, L. Laffont, R. Bouchet, C. Wurm, J.-B. Leriche, M. Morcrette, J.-M. Tarascon, C. Masquelier, Toward understanding of electrical limitations (electronic, ionic) in $LiMPO_4$ (M = Fe, Mn) electrode materials. J. Electrochem. Soc. **152**, A913–A921 (2005)
44. N.H. Kwon, T. Drezen, I. Exnar, I. Teerlinck, M. Isono, M. Graetzel, Enhanced electrochemical performance of mesoparticulate $LiMnPO_4$ for lithium ion batteries. Electrochem. Solid-State Lett. **9**, A277–A280 (2006)
45. T. Drezen, N.H. Kwon, P. Bowen, I. Teerlinck, M. Isono, I. Exnar, Effect of particle size on $LiMnPO_4$ cathodes. J. Power Sources **174**, 949–953 (2007)
46. D. Wang, H. Buqa, M. Crouzet, G. Deghenhghi, T. Drezen, I. Exnar, N.H. Kwon, J.H. Miners, L. Poletto, M. Gratzel, High-performance, nano-structured $LiMnPO_4$ synthesized via a polyol method. J. Power Sources **189**, 624–628 (2009)
47. S. Martinet, C. Bourbon, M. Amuntencei, C. Chabrol, D. Tomasi, M. Rey, S. Patoux, Design of experiment methodology to improve the energy density of lithiated metal phosphates. Solid State Ionics **268**, 247–251 (2014)
48. M. Armand, C. Michot, N. Ravet, M. Simoneau, P. Hovington, European patent EP1134 826 A1 (2001)
49. A. Abouimrane, N. Ravet, M. Armand, A. Nyten, J. O.Thomas, Abstract #350, in *IMLB 12*, Nara, Japan, 27 June–2 July 2004
50. A. Nyten, A. Abouimrane, M. Armand, T. Gustafsson, J.O. Thomas, Electrochemical performance of Li_2FeSiO_4 as a new Li-battery cathode material. Electrochem. Commun. **7**, 156–160 (2005)
51. M.E. Arroyo-de Dompablo, M. Armand, J.M. Tarascon, U. Amador, On-demand design of polyoxianionic cathode materials based on electronegativity correlations: an exploration of the Li_2MSiO_4 system (M = Fe, Mn, Co, Ni), Electrochem. Comm. **8**, 1292–1298 (2006)

52. A. Kokalj, R. Dominko, G. Mali, A. Meden, M. Gaberscek, J. Jamnik, Beyond one-electron reaction in Li cathode materials: designing $Li_2Mn_xFe_{1-x}SiO_4$. Chem. Mater. **19**, 3633–3640 (2007)
53. D.M. Kempaiah, D. Rangappa, I. Honma, Controlled synthesis of nanocrystalline Li_2MnSiO_4 particles for high capacity cathode application in lithium-ion batteries. Chem. Commun. **48**, 2698–2700 (2012)
54. H. Zhu, H. He, X. Xin, X. Ma, L. Zan, Y. Zhang, Facile synthesis of Li_2MnSiO_4/C/graphene composite with superior high-rate performances as cathode materials for Li-ion batteries. Electrochim. Acta **155**, 116–124 (2015)
55. N.A. Kaskhedikar, J. Maier, Lithium storage in carbon nanostructures. Adv. Mater. **21**, 2664–2680 (2009)
56. K. Sato, M. Noguchi, A. Demachi, N. Oki, M. Endo, A mechanism of lithium storage in disordered carbons. Science **264**, 556–558 (1994)
57. J.R. Dahn, W. Xing, Y. Gao, The "Falling cards model" for the structure of microporous carbons. Carbon **35**, 825–830 (1997)
58. F. Leroux, K. Méténier, S. Gautier, E. Frackowiak, S. Bonnamy, F. Béguin, Electrochemical insertion of lithium in catalytic multi-walled carbon nanotubes. J. Power Sources **81**, 317–322 (1999)
59. B.J. Landi, M.J. Ganter, C.D. Cress, R.A. DiLeo, R.P. Raffaelle, Carbon nanotubes for lithium ion batteries. Energy Environ. Sci. **2**, 638–654 (2009)
60. S.B. Yang, H.H. Song, X.H. Chen, A.V. Okotrub, L.G. Bulusheva, Electrochemical performance of arc-produced carbon nanotubes as anode material for lithium-ion batteries. Electrochim. Acta **52**, 5286–5293 (2007)
61. S.H. Yoon, C.W. Park, H. Yang, Y. Korai, I. Mochida, R.T.K. Baker, N.M. Rodriguez, Novel carbon nanofibers of high graphitization as anodic materials for lithium ion secondary batteries. Carbon **42**, 21–32 (2004)
62. L.W. Ji, X.W. Zhang, Fabrication of porous carbon nanofibers and their application as anode materials for rechargeable lithium-ion batteries. Nanotechnology **20**, 155705–155711 (2009)
63. J.P. Tessonnier, D. Rosenthal, T.W. Hansen, C. Hess, M.E. Schuster, R. Blume, F. Girgsdies, N. Pfänder, O. Timpe, D.S. Su, R. Schlögl, Analysis of the structure and chemical properties of some commercial carbon nanostructures. Carbon **47**, 1779–1798 (2009)
64. T. Ohzuku, Z. Takehara, S. Yoshizawa, Nonaqueous lithium/titanium dioxide cell. Electrochim. Acta **24**, 219–222 (1979)
65. D.W. Murphy, R.J. Cav, S.M. Zahurak, A. Santoro, Ternary Li_xTiO_2 phases from insertion reactions. Solid State Ionics **9**, 413–418 (1983)
66. K.M. Colbow, J.R. Dahn, R.R. Haering, Structure and electrochemistry of the spinel oxides $LiTi_2O_4$ and $Li_{4/3}Ti_{5/3}O_4$. J. Power Sources **26**, 397–402 (1989)
67. E. Ferg, R.J. Gummow, A. de Kock, M.M. Thackeray, Spinel anodes for lithium-ion batteries. J. Electrochem. Soc. **141**, L147–L150 (1994)
68. S. Jouanneau, F. Le Cras, C. Bourbon, H. Lignier, Commissariat à l'Energie Atomique, FR 2 874 603; WO 2006/027449
69. R. Marchand, L. Brohan, M. Tournoux, TiO_2 (B), a new form of titanium dioxide and the potassium octatitanate $K_2Ti_8O_{17}$. Mat. Res. Bull **15**, 1129–1133 (1980)
70. T. Ohzuku, A. Ueda, N. Yamamoto, Zero-strain insertion material of $Li[Li_{1/3}Ti_{5/3}]O_4$ for rechargeable lithium cells. J. Electrochem. Soc. **142**, 1431–1435 (1995)
71. B. Zhao, R. Ran, M. Liu, Z. Shao, A comprehensive review of $Li_4Ti_5O_{12}$-based electrodes for lithium-ion batteries: the latest advancements and future perspectives. Mater. Sci. Eng. R **98**, 1–71 (2015)
72. J. Lu, C. Nan, Q. Peng, Y. Li, Single crystalline lithium titanate nanostructure with enhanced rate performance for lithium ion battery. J. Power Sources **202**, 246–252 (2012)
73. H. Xu, X. Hu, W. Luo, Y. Sun, Z. Yang, C. Hu, Y. Huang, Electrospun conformal $Li_4Ti_5O_{12}$/C fibers for high-rate lithium-ion batteries. ChemElectroChem **1**, 611–616 (2014)

74. J.R. Li, Z.L. Tang, Z.T. Zhang, Controllable formation and electrochemical properties of one-dimensional nanostructured spinel $Li_4Ti_5O_{12}$. Electrochem. Commun. **7**, 894–899 (2005)
75. S.C. Lee, S.M. Lee, J.W. Lee, J.B. Lee, S.M. Lee, S.S. Han, H.C. Lee, H.J. Kim, Spinel $Li_4Ti_5O_{12}$ nanotubes for energy storage materials. J. Phys. Chem. C **113**, 18420–18423 (2009)
76. W. Wang, Y. Guo, L. Liu, S. Wang, X. Yang, H. Guo, Gold coating for a high performance $Li_4Ti_5O_{12}$ nanorod aggregates anode in lithium-ion batteries. J. Power Sources **245**, 624–629 (2014)
77. Y. Sha, B. Zhao, R. Ran, R. Cai, Z. Shao, Synthesis of well-crystallized $Li_4Ti_5O_{12}$ nanoplates for lithium-ion batteries with outstanding rate capability and cycling stability. J. Mater. Chem. A **1**, 13233–13243 (2013)
78. N. Li, G. Zhou, F. Li, L. Wen, H.-M. Cheng, A self-standing and flexible electrode of $Li_4Ti_5O_{12}$ nanosheets with a N-doped carbon coating for high rate lithium ion batteries. Adv. Funct. Mater. **23**, 5429–5435 (2013)
79. J. Liu, X. Wei, X.-W. Liu, Two-dimensional wavelike spinel lithium titanate for fast lithium storage. Sci. Rep. **5**, 9782 (2015)
80. Y.S. Lin, J.G. Duh, Facile synthesis of mesoporous lithium titanate spheres for high rate lithium-ion batteries. J. Power Sources **196**, 10698–10703 (2011)
81. L. Yu, H.B. Wu, X.W. Lou, Mesoporous $Li_4Ti_5O_{12}$ hollow spheres with enhanced lithium storage capability. Adv. Mater. **25**, 2296–2300 (2013)
82. K.M. Yang, Y.N. Ko, J.-Y. Yun, Y.C. Kang, Preparation of $Li_4Ti_5O_{12}$ yolk-shell powders by spray pyrolysis and their electrochemical properties. Chem. Asian J. **9**, 443–446 (2014)
83. Y. Ma, B. Ding, G. Ji, J.Y. Lee, Carbon-encapsulated F-doped $Li_4Ti_5O_{12}$ as a high rate anode material for Li+ batteries. ACS Nano **7**, 10870–10878 (2013)
84. J. Chen, L. Yang, S. Fang, Y. Tang, Synthesis of sawtooth-like $Li_4Ti_5O_{12}$ nanosheets as anode materials for Li-ion batteries. Electrochim. Acta **55**, 6596–6600 (2010)
85. Z. Zhang, G. Li, H. Peng, K. Chen, Hierarchical hollow microspheres assembled from N-doped carbon coated $Li_4Ti_5O_{12}$ nanosheets with enhanced lithium storage properties. J. Mater. Chem. A **1**, 15429–15434 (2013)
86. C. Jiang, J. Zhang, Nanoengineering titania for high rate lithium storage: a review. J. Mater. Sci. Technol. **29**, 97–122 (2013)
87. J. Li, W. Wan, H. Zhou, J. Li, D. Xu, Hydrothermal synthesis of TiO_2(B) nanowires with ultrahigh surface area and their fast charging and discharging properties in Li-ion batteries. Chem. Commun. **47**, 3439–3441 (2011)
88. S. Komaba, K. Shimomura, N. Yabuuchi, T. Ozeki, H. Yui, K.J. Konno, Study on polymer binders for high-capacity SiO negative electrode of Li-Ion batteries. J. Phys. Chem. C **115**, 13487–13495 (2011)
89. Z. Wang, X.W.D. Lou, TiO_2 nanocages: fast synthesis, interior functionalization and improved lithium storage properties. Adv. Mater. **24**, 4124–4129 (2012)
90. A.N. Dey, Electrochemical alloying of lithium in organic electrolytes. J. Electrochem. Soc. **118**, 1547–1549 (1971)
91. D. Fauteux, R. Koksbang, Rechargeable lithium battery anodes: alternatives to metallic lithium. J. Applied Electrochem. **23**, 1–10 (1993)
92. R.A. Huggins, Lithium alloy negative electrodes. J. Power Sources **81**, 13–19 (1999)
93. Y. Idota, T. Kubota, A. Matsufuji, Y. Maekawa, T. Miyasaka, Tin-based amorphous oxide: a high-capacity lithium-ion-storage material. Science **276**, 1395–1397 (1997)
94. M.N. Obrovac, V.L. Chevrier, Alloy negative electrodes for Li-ion batteries. Chem. Rev. **114**, 11444–11502 (2014)
95. F. Luo, B. Liu, J. Zheng, G. Chu, K. Zhong, H. Li, X. Huang, L. Chen, Review—nano silicon/carbon composite anode materials towards practical application for next generation Li-ion batteries. J. Electrochem. Soc. **162**, 2509–2528 (2015)

96. D. Larcher, S. Beattie, M. Morcrette, K. Edström, J.C. Jumas, J.M. Tarascon, Recent findings and prospects in the field of pure metals as negative electrodes for Li-ion batteries. J. Mat. Chem. **17**, 3759–3772 (2007)
97. H. Axel, H. Schäfer, A. Weiss, Angew. Chem. Int. Ed. **4**, 358 (1965)
98. H. Axel, H. Schäfer, A. Weiss, Zur Kenntnis der Phase $Li_{22}Si_5$. Zeitschrift für Naturforschung—Section B J. Chem. Sci. **21**, 115–117 (1966)
99. S.-C. Lai, Solid lithium-silicon electrode. J. Electrochem. Soc. **123**, 1196–1197 (1976)
100. R.A. Sharma, R.N. Seefurth, Thermodynamic properties of the lithium-silicon system, J. Electrochem. Soc 123, 1763 (1976)
101. A.M. Wilson, B.M. Way, J.R. Dahn, T. van Buuren, Nanodispersed silicon in pregraphitic carbons. J. Appl. Phys. **77**, 2363–2369 (1995)
102. B.A. Boukamp, G.C. Lesh, R.A. Huggins, All-solid lithium electrodes with mixed conductor matrix. J. Electrochem. Soc. **128**, 725–729 (1981)
103. M.N. Obrovac, L. Christensen, Structural changes in silicon anodes during lithium insertion/extraction. Electrochem. Solid-State Let. **7**, A93–A96 (2004)
104. W.R. Liu, Z.Z. Guo, W.S. Young, D.T. Shieh, H.C. Wu, M.H. Yang, N.L. Wu, Effect of electrode structure on performance of Si anode in Li-ion batteries: Si particle size and conductive additive. J. Power Sources **140**, 139–144 (2005)
105. C. Paireau, S. Jouanneau, M.-R. Ammar, P. Simon, F. Béguin, E. Raymundo-Piñero, Si/C composites prepared by spray drying from cross-linked polyvinyl alcohol as Li-ion batteries anodes. Electrochim. Acta **174**, 361–368 (2015)
106. J. Sourice, A. Quinsac, Y. Leconte, O. Sublemontier, W. Porcher, C. Haon, A. Bordes, E. De Vito, A. Boulineau, S. Jouanneau, Si Larbi, N. Herlin-Boime, C. Reynaud, One-step synthesis of Si@C nanoparticles by laser pyrolysis: high-capacity anode material for lithium-ion batteries. ACS Appl. Mater. Interfaces **7**, 6637–6644 (2015)
107. N. Liu, Z. Lu, J. Zhao, M.T. McDowell, H.-W. Lee, W. Zhao, Y. Cui, A pomegranate-inspired nanoscale design for large-volume-change lithium battery anodes. Nat. Nanotechnol. **9**, 187–192 (2014)
108. B.P.N. Nguyen, S. Chazelle, M. Cerbelaud, W. Porcher, B. Lestriez, Manufacturing of industry-relevant silicon negative composite electrodes for lithium ion-cells. J. Power Sources **262**, 112–122 (2014)
109. Sony's New Nexelion Hybrid, Lithium ion batteries (2005), http://www.sony.net/SonyInfo/News/Press/200502/05-006E/. Accessed 9 Jan 2016
110. R. Nagai, F. Kita, M. Yamada, H. Katayama, Development of highly reliable high-capacity batteries for mobile devices and small- to medium-sized batteries for industrial applications. H. Hitachi Rev. **60**, 28–32 (2011)
111. A.D.W. Todd, P.P. Ferguson, M.D. Fleischauer, J.R. Dahn, Tin-based materials as negative electrodes for Li-ion batteries: combinatorial approaches and mechanical methods. Int. J. Energy Res. **34**, 535–555 (2010)
112. D. Foster, J. Wolfenstine, J. Read, J.L. Allen, Performance of sony's alloy based li-ion battery (ARL-TN-0319), 1–14 (2008)
113. L. Christensen, D.B. Le, J. Singh, M.N. Obrovac, in *216th Meeting of the Electrochemical Society* (Vienna, 2009)
114. J. Li, D.-B. Le, P.P. Ferguson, J.R. Dahn, Lithium polyacrylate as a binder for tin–cobalt–carbon negative electrodes in lithium-ion batteries. Electrochim. Acta **55**, 2991–2995 (2010)
115. P. Poizot, S. Laruelle, S. Grugeon, L. Dupont, J.-M. Tarascon, Nano-sized transition-metaloxides as negative-electrode materials for lithium-ion batteries. Nature **407**, 496–499 (2000)
116. P.G. Bruce, B. Scrosati, J.-M. Tarascon, Nanomaterials for rechargeable lithium batteries. Angew. Chem. Int. Ed. **47**, 2930–2946 (2008)
117. J. Cabana, L. Monconduit, D. Larcher, M. Rosa Palacin, Beyond intercalation-based li-ion batteries: the state of the art and challenges of electrode materials reacting through conversion reactions. Adv. Mater. **22**, E170–E192 (2010)

118. H. Morimoto, S.I. Tobishima, Y. Iizuka, Lithium intercalation into α-Fe_2O_3 obtained by mechanical milling of α-FeOOH. J. Power Sources **146**, 315–318 (2005)
119. Z.Y. Wang, D.Y. Luan, S. Mahdavi, C.M. Li, X.W. Lou, α-Fe_2O_3 nanotubes with superior lithium storage capability. Chem. Commun. **47**, 8061–8063 (2011)
120. W.J. Lee, U.N. Maiti, J.M. Lee, J. Lim, T.H. Han, S.O. Kim, Nitrogen-doped carbon nanotubes and graphene composite structures for energy and catalytic applications. Chem. Commun. **50**, 6818–6830 (2014)
121. S. Jouanneau, S. Patoux, Y. Reynier, S. Martinet, High energy and high power Li-ion cells-practical interest/limitation of nanomaterials and nanostructuration, in *Nanomaterials for Lithium-Ion Batteries: Fundamentals and Applications*, ed. R. Yazami (Pan Stanford Ed., 2013)
122. M. Winter, J.O. Besenhard, in *Lithium Ion Battery: Fundamentals and Performance*, eds. by M. Wahihara, O. Yamamoto (Wiley-VCH, Weinheim, 1998)

Chapter 14
Self-organized Chiral Liquid Crystalline Nanostructures for Energy-Saving Devices

Zhigang Zheng and Quan Li

Abstract Energy saving, as important as energy harvesting and storage in sustainable energy endeavor, would greatly decrease the energy production and will help in alleviating the increasing pressure on nonrenewable energy sources, which will not only lengthen their exhaustion but also would provide sufficient time to explore alternative potential sources of sustainable energy. Hitherto, extensive efforts have been devoted to developing energy saving technology to meet environmental and economic restrictions. In this context, liquid crystal (LC) based energy-saving devices employing chiral LC phases as the active materials have been demonstrated. As unique LCs, chiral LC phases possess many excellent characteristics on molecular arrangement and optical properties due to their periodic modulated structure, which enables their promising applications in energy-saving devices. Here we discuss different approaches adopted to develop LC devices with energy saving characteristics. The devices are based on the chiral LCs, i.e., chiral nematic LCs and blue phase LCs, however the use of chiral nematic LCs is highlighted in this chapter. The bistable states of chiral nematic LCs present different optical properties in absence of external field for potential applications in low energy consumption displays, image storage, and light shutters. Furthermore, their selective reflection exhibits the high efficiency on solar energy tuning as a smart window or adaptor of photovoltaic devices. By doping photosensitive chiral switches or motors into the LC, light tunability on reflection band of chiral LCs has been achieved. In this sense, recent developments of delicate light driven energy-saving photonic devices will be also discussed.

Z. Zheng · Q. Li (✉)
Liquid Crystal Institute and Chemical Physics Interdisciplinary Program, Kent State University, Kent, OH 44242, USA
e-mail: qli1@kent.edu

Q. Li (ed.), *Nanomaterials for Sustainable Energy*,
NanoScience and Technology, DOI 10.1007/978-3-319-32023-6_14

14.1 Introduction

Accelerated developments of industry, accompanied by a large consumption of traditional energy resources, result in the appearance of energy problem, which is gradually emerging as a serious global issue. These issues have alerted people to realize the efficient use of energy, reduce its consumption as well as gradually get rid of or significantly reduce the dependence on traditional energy sources, and find new sustainable energy sources to secure an energy sufficient future. Thus, energy-saving devices are highly desirable and have been attracting special attention similar to energy harvesting and storage. More energy-saving materials and devices are in continuous development, and liquid crystal (LC) display (LCD) can be regarded as prototype representatives in this challenging undertaking.

LC is a kind of functional material with the fluidity of ordinary liquids and anisotropy of crystalline solids. It perfectly reflects the blend of characteristics of fluids and crystals, exhibits the properties of soft matter, and thus, is sometimes called as soft crystal. Upon applying a weak external field, the response of LC leads to a rearrangement of molecules, thereby producing the distinct variations on optical, electric and mechanical properties associated to LC. The discovery of LC phase can be traced back to the end of 19th century, whereas its real practical applications started from 1971 when the first twisted nematic (TN) display mode was developed by Schadt, Helfrich and Fergason [1, 2]. Currently, LC has been widespread utilized in diverse electric devices, especially the thin-film-transistor LCD (TFT-LCD). The emergence of low energy consumption TFT-LCD rapidly replaces the conventional cathode ray tube (CRT) display, and thus brings a revolution on energy. New LC display modes and LC applications beyond display were proposed and investigated, promoting the advancements of LC devices.

The majority of LC devices use nematic LCs, i.e., LC phases with only orientational ordering of the molecules, as the active materials. The nematic phase is not chiral. However, when the nematic phase is transformed into chiral LCs (CLCs) such as chiral nematic phase and blue phases, the resulting CLCs exhibit fascinating properties compared to the parent nematic phases. The chiral nematic LC phase is a self-organized helical superstructure whereas the blue phase is the cubic structure made of double twisted cylinders. The self-organized helical structure of chiral nematic phase enables it to exhibit different optical textures caused by the coupling between elastic free energy and the surface anchoring [3], and thereby leads to different transmission properties with a light scattering for a focal conic texture (helical axis aligns randomly) while a transparent for a planar oily streaks texture (helical axis aligns vertical to the substrates). If the helical pitch length of chiral nematic LC is comparable to light wavelength, the selective Bragg reflection occurs in the case that oily streaks texture is presented, i.e., the circular polarized light (CPL) with the same handedness as the helix can be reflected. This quality qualifies this CLC as a self-organized one dimensional (1D) photonic crystal. The reflection wavelength (λ) is determined by the pitch length (P), having a relationship as $\lambda = \langle n\rangle P$, in which, $\langle n\rangle$ is the average refractive index of LC. Generally, chiral nematic LCs can be prepared conveniently by doping a certain amount of chiral dopant into the conventional achiral nematic LCs, and the concentration of chiral dopant determines the helical pitch length, thus influences their stable states, consequently presenting different optical properties. It means that chiral nematic LC possesses several stable arrangement states corresponding to different optical properties, while the stable states are influenced by the content of dopant. Most importantly, these CLCs present bistability, meaning that more than one of the aforementioned optical states is stable on the absence of applied external field. Consequently, once an optical state is set upon the material (such as a reflective state), no change occurs, which presents huge impacts on energy consumption of chiral nematic LCs based devices. Blue phases occur in chiral nematic LCs with high chirality. Since the cubic lattice dimensions of blue phases are of the order of several hundred nanometers, these dielectric mediums are able to reflect light in all three directions. Accordingly, these CLCs have been regarded as self-organized three dimensional (3D) photonic crystals.

Inspired by the excellent characteristics of CLC, many applications focusing on energy-saving were proposed and the relevant devices were designed and developed in the past decades. This chapter gives an overview of several kinds of typical energy-saving devices based on self-organized CLC nanostructure from the materials, mechanisms, devices designing and fabrications to their performances. We start our presentations from bistable CLC devices, including the generation of bistability and materials; followed by some specific application of CLC for solar energy manipulation in virtue of selective reflection, and subsequently, a new generation of LCDs driven by photovoltaic devices incorporated into the display architecture are introduced; finally, novel light-driven CLC photonic devices capable of exploiting solar energy for operation are demonstrated, focusing on the wide tunable Bragg reflector and the judiciously designed and fabricated micro-size integrated devices through micro-fluidic technique.

14.2 Bistability in Self-organized Chiral Liquid Crystals

Twisted nematic (TN) mode has been widely adopted in high definition displays at present, especially the well-known TN-90 mode, i.e. LC molecules rotate 90° across the cell thickness on going from one substrate to the other [1, 2]. Device structure of TN-90 is shown in Fig. 14.1. Alignment directions on the two substrates are perpendicular with each other, which forces the direction of LC director near one substrate to be perpendicular to that close to the other, therefore forming the 90° twisted arrangement of LC. It should be noted that to lift the double degeneracy of 90° twisted arrangement, the nematic LC host is doped with a small amount (generally < 1.0 wt%) of chiral agent with weak twisting power. Such TN cell is inserted between a pair of crossed polarizers where the polarization direction is either parallel (i.e., E mode) or perpendicular (i.e., O mode) to alignment direction of one of the substrates. In absence of external driving field (OFF-state), polarization direction of the impinging light rotates 90° and transmits through the

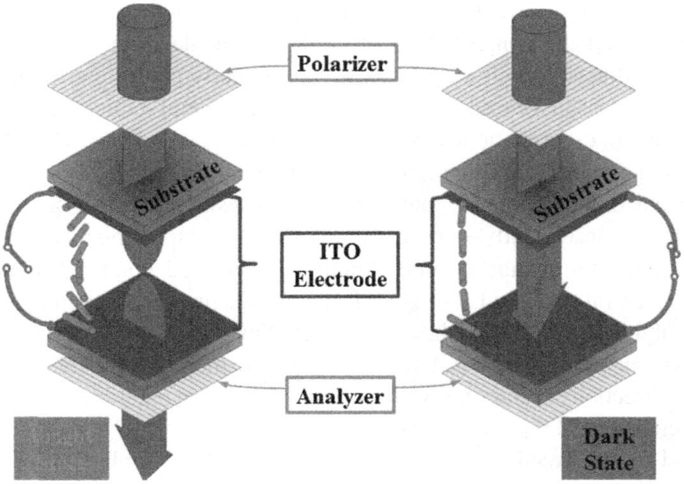

Fig. 14.1 Illustration of the mechanism of TN-90 LCD. Transmission axes of polarizer and analyzer are orthogonal to each other. Bright state is obtained in absence of applied voltage, i.e., normal white mode; however, if the transmission axes of polarizer and analyzer are parallel to each other, dark state is obtained in absence of applied voltage, which is called normal dark mode

device, generating the bright state due to the waveguide effect of the LC layer. When a driving field is applied (ON-state), the molecules uniformlyalign along the direction of the field because of Fréedericksz transition, thereby giving rise to the disappearance of twisted arrangement and waveguide effect, consequently blocking the transmission light, and thus producing the dark state. Although TN mode facilitates the fabrication processing of LCD and possesses good optical performances, nonetheless the main drawback is that every grey level should be maintained by a corresponding applied field, i.e., the field is held at ON-state, consequently resulting in high power consumption of the devices. This is a significant undesirable aspect of otherwise efficient display devices.

Bistability, which was first observed by Bigelow et al. [4] 40 years ago, is a common phenomenon existing in CLC system owing to its special helical topological structure. Just as its name implies, there exist two possible molecular arrangements at OFF-state, normally presenting the distinct optical property, such as the transmission intensity, phase retardation or the color. For the reason that the free energy of the two arrangements is almost equal to each other, which ensures the stability at these two states without any external supplementary energy, bistable LC devices are recognized as the representative energy-saving devices and show promise in next generation LCD and beyond. Currently, such bistability can be induced by many approaches, such as the coupling effects of external electric-field and judiciously considered surface anchoring conditions, light irradiation of photosensitive LC system, and appropriate optimization of the dielectric property of materials.

In this section, bistability of the self-organized CLC devices is introduced from material composition, mechanism, structural design and processing points of view.

Since bistable CLC has been widely adopted in displays and photonics, and a review on this aspect is rare and not systematic, herein, we focus our discussion on bistability of different CLC systems, while neglect bistable state in nematic and smectic LCs which fall outside the scope of this chapter. Three aspects will be presented as follows: (i) bistability generated by the coupling effect of applied electric-field and surface anchoring, (ii) bistable device based on some of specific LCs, and (iii) light-driven bistable LC devices.

14.2.1 Bistability Enabled by Device Optimizations

Surface anchoring determines the arrangement of LC molecules in a LC cell. If such anchoring conditions, i.e., alignment direction, anchoring strength, surface geometry, and symmetry of anchoring, are changed or broken by applied electric-field, bistable state can be induced in some systems.

A prior work showed that bistability was achieved through the asymmetrically surface anchoring treatment [5]. SiO was evaporated onto the surface of two substrates with the incidence of 75° and was grazed gently to form a uniform orientation of LC molecules. The thickness of SiO layers on top and bottom substrates were controlled in 25 and 30 Å respectively to produce a slight difference of surface anchoring strength. If the surface critical electric-fields corresponding the two surfaces were E_{c1} and E_{c2}, and $E_{c1} < E_{c2}$, the bistability was observed when a short square pulse, E_0, was applied on the cell. As $E_{c1} < E_0 < E_{c2}$ was satisfied, a quasi-uniform alignment of LC molecules was appeared; while the other stable 180°-twisted arrangement generated in the case of $E_0 > E_{c2}$. Such bistable arrangement was caused by the hydrodynamic breaking of anchoring under the coupled effect of applied electric-field and asymmetric surface anchoring. The 180°-twisted state was formed as two anchoring were broken (i.e., LC molecules near two substrates are driven by the external field), however the quasi-uniform state was obtained as only one anchoring was broken (i.e., LC molecules near one of the substrates are driven by the field). A similar work reported the bistability obtained using the short pulse direct-current (DC) voltage to drive LC layer confined in a hybrid cell (i.e., the top substrate was treated with vertical alignment by coating ortho-decyl-dimethyl-[3-(trimethoxysilyl)-propyl] ammonium chloride thin film, and the other was evaporated with SiO layer to achieve the planar alignment) [6]. The stable twisted and homeotropic alignments were realized, and can be switched fast between the two states by high electric-field within 100 μs.

An expected bistability appeared as well by applying the non-uniform distributed electric-field on the cell. On the basis of this, a pioneer work on bistable TN (BTN) LCD was carried out by Berreman and Heffner [7]. They used the specially shaped short electric pulse generated by judiciously designed electrodes to drive LC layer, and consequently, two stable states with 0 and 2π twists were observed. Numerical solutions based on Oseen-Frank equations and Leslie-Ericksen equations confirmed the existence of bistability. However, as a display device, the contrast ratio (CR) of BTN is lower, therefore reflective BTN was proposed as an efficient mode to improve the CR as well as diminish the energy consumption, for the reason that the rear polarizer in transmission BTN does not required, and subsequently, a bistability between −36° and 324° twist states was reported [8]. However, the tested CR reached only 6:1. Although such ratio showed to be slightly higher than that in transmission mode, it was much lower than the expectation. So that the parameter optimizations, focusing on twisting angle, phase retardation (i.e., cell thickness and birefringence of LC material) and cell thickness-to-pitch ratio (*d*/*P*), were carried out [9]. In addition, to make a better matching between polarization direction of reflected light and alignment direction of LC close to the rear substrate, a quarter waveplate was inserted between the reflector and LC layer and the matching parameter was optimized by means of a parameter space approach. Finally, the CR was improved more than 16 times [10].

With particular design of driving manner, the other type of bistability between splay and twist states can be achieved facilely. The splay arrangement was generally formed in planar cell for a CLC system with weak twisted energy. Splay arrangement transited to the dynamic stable bend state as a certain voltage, higher than the critical value, was applied across the cell. The bend arrangement would quickly transit to twist in the case of removing the voltage due to backflow of LC molecules. The stability of twist arrangement was improved as *d*/*P* value increased, however the driving voltage went up concurrently, therefore optimization of *d*/*P* was necessary [11].

One obvious shortage was that the twist cannot be driven back to the splay except for the thermodynamic recovery over a much longer time. Taking into account of this problem, a three-terminal electrode structure, containing one common electrode on one substrate and two electrodes on the other substrate, was proposed, thus vertical and horizontal electric-field can be applied through different pair of electrodes. As shown in Fig. 14.2, the original splay state can be driven by the vertical electric-field to a high bend state, passing through a low bend state; the high bend state transformed to twist state after removing the vertical voltage as aforementioned; as the horizontal electric-field was applied, the twist was quickly driven to the original splay state [12]. Based on such three-terminal electrode driving manner, a viewing-angle controlled LCD and reflective dual-mode low-energy-consumption LCD were developed. The switching between twist (bright state) and bend (dark state) showed the viewing-angle less than 45°, but a dramatic extension to 175° was obtained by in-plane switching of splay state (Fig. 14.3). Therefore, different viewing-angle can be selected by manipulating the device between two stable states, splay and twist [13]. Similarly, by adjusting the angle between polarization and alignment directions to 108°, and maintaining *d*/*P* ratio at 0.25, the initial dark π-twist state can be driven to a dynamic stable bright bend state with vertical electric-field or to a memory stable bright splay state with horizontal electric-field (Fig. 14.4). Since that one polarizer was removed in the reflective LCD and two stable no-energy-consumption states corresponding to dark and bright states respectively, such reflective LCD reached the requirements of green and low-energy-consumption devices [14]. Apart from the bistability between

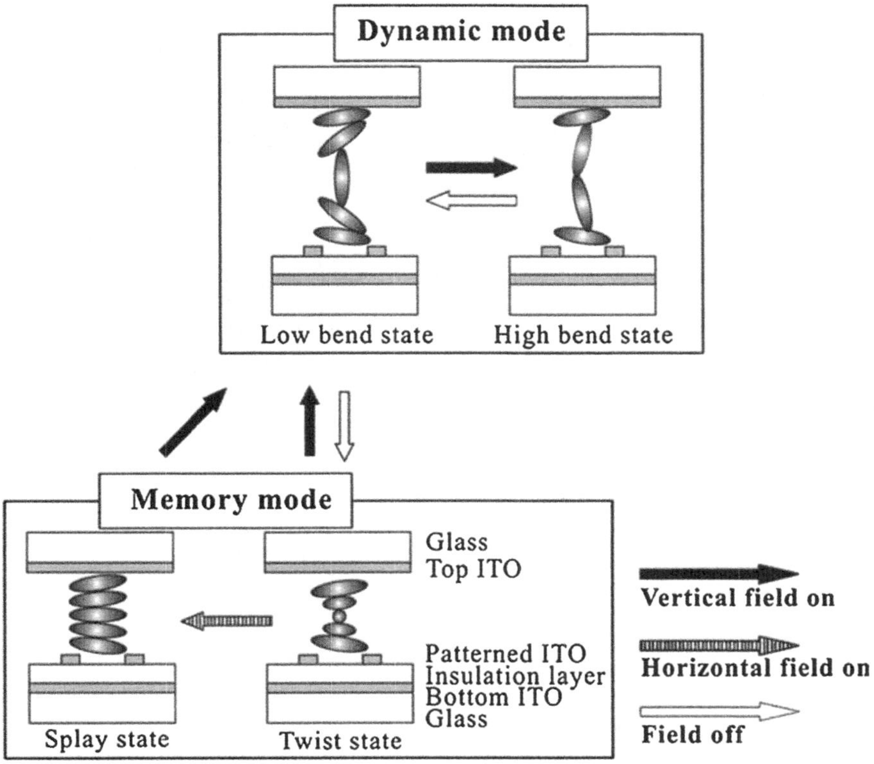

Fig. 14.2 Mechanism of three-terminal driving. The splay state can be driven by vertical electric-field to the high bend state, passing through the low bend state. As such vertical field is removed, the high bend state gradually recovered to low band state, finally to the stable twist state. Such twist state can be driven back to original splay state through the horizontal field [12]. Copyright 2006 American Institute of Physics

splay and twist, the bistable $+\pi/2$ and $-\pi/2$ twist states was achieved as well through the three-terminal electrode driving recently [15].

Recently, a very interesting transmissive and reflective mode switching LCD was achieved by stacking a homeotropically aligned LC layer and a bistable long-pitch CLC layer [16]. The homeotropic LC layer, formed by negative nematic LC, generated the half-wave phase retardation as it was driven by a vertical saturation electric-field to form a homogeneous planar arrangement. Such layer was inserted between two crossed polarizers. The long-pitch CLC layer presented the optical transparent planar state at the initiate, while such state transformed to a scattering focal conic state by applying a certain short pulse voltage (8–11 V in the reported experiment). The recovery from focal conic to planar state can be obtained by appropriately raising the voltage of short pulse (14 V in the reported experiment). To realize mode switching LCD, two configurations were designed: (i) the homeotropic LC layer was placed upon the bistable CLC (Fig. 14.5a). In reflective mode (Fig. 14.5b), the bistable layer was triggered to scattering focal conic state;

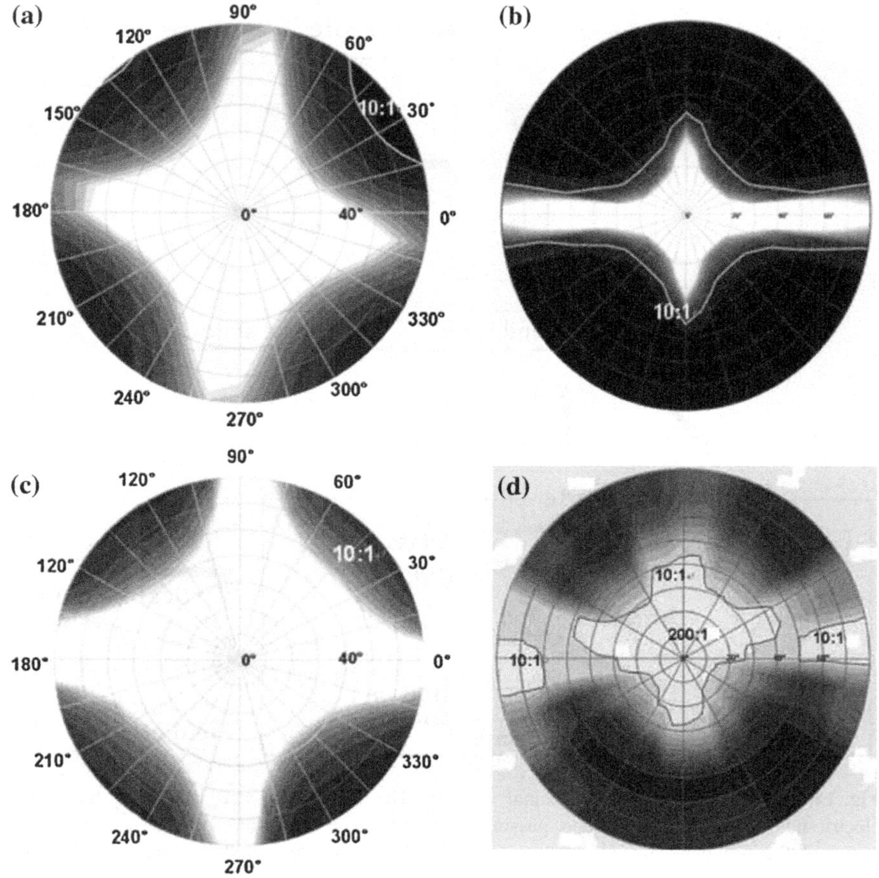

Fig. 14.3 Numerically-calculated (**a**, **b**) and measured (**c**, **d**) viewing angle characteristics. Viewing angle characteristics of splay state with horizontal switching (**a**, **c**), which shows wide viewing angle characteristics; viewing angle characteristics of 180°-twisted state with vertical switching (**b**, **d**), which shows narrow viewing angle characteristics. Reproduced with permission from [13]. Copyright 2008 Optical Society of America

however, in the transmissive mode, planar state was required for the bistable CLC layer. The dark and bright states of transmissive and reflective mode were controlled by manipulating the homeotropic LC layer with vertical electric-field; (ii) the bistable CLC layer was placed upon the homeotropic LC layer. In reflective mode, the vertical alignment of LC in homeotropic LC layer was maintained. The switch between dark and bright state was obtained by driving the bistable CLC to planar (no reflection in visible band) and focal conic states. In transmissive mode, the bistable CLC at upper part was driven to planar state, the lower homeotropic LC layer was switched between homeotropic and homogeneous alignment for achieving the dark and bright state. The CRs at reflective and transmissive modes were 20:1 and 700:1 respectively.

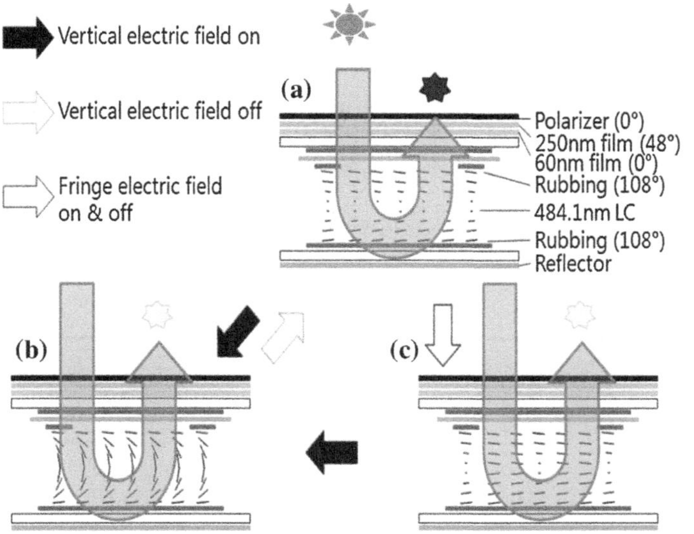

Fig. 14.4 Dual-mode operation of the reflective LCD. **a** Stable dark at π-twist state; **b** bright at bend state driven by vertical switching, **c** stable bright at planar state driven by horizontal switching. Reproduced with permission from [14]. Copyright 2012 Optical Society of America

Some kinds of specific alignment materials, with tunable properties of their anchoring strength and direction by light, temperature and electric-field, were used to fabricate bistable LC devices. The typical one is photosensitive polymer alignment film, known as photoalignment film, which possesses an anchoring strength dependence on light irradiation time. Therefore the asymmetric anchoring of LC cell can be readily achieved by exposing two substrates with different time, and thus the BTN device with two stable twist states of −22.5° and 157.5° were developed [17]. Anchoring strength of such photoalignment material (Fig. 14.6a) was increased with the prolonging of exposure time. Not only the anchoring strength, but the alignment direction can be controlled by linearly polarized light for some other photoalignment materials. A photorewritable alignment material (Fig. 14.6b), mixed by three kinds of azobenzene based photoisomerizable polymers, was adopted to form the patterned alignment [18]. By switching with two pairs of electrodes, generating the orthogonal horizontal electric-field, the devices can be transformed between the stable homogeneous and twist arrangements (Fig. 14.6c) [19]. Recently, a very interesting alignment film, with switching between vertical and parallel anchoring by whatever temperature or electric-field, was studied and recognized as a novel promising alignment material for bistable devices. Such material, named as CYTOP (perfluoropolymer poly [perfluoro(4-vinyloxy-1-butene)]), showed vertical alignment property originally. As the temperature rose to 49.5 °C, alignment property transited from vertical to parallel, however the transition from parallel to vertical during cooling occurred at 44 °C. That means a transition hysteresis within the temperature range from 44 to 49.5 °C where the bistability generated [20]. When a

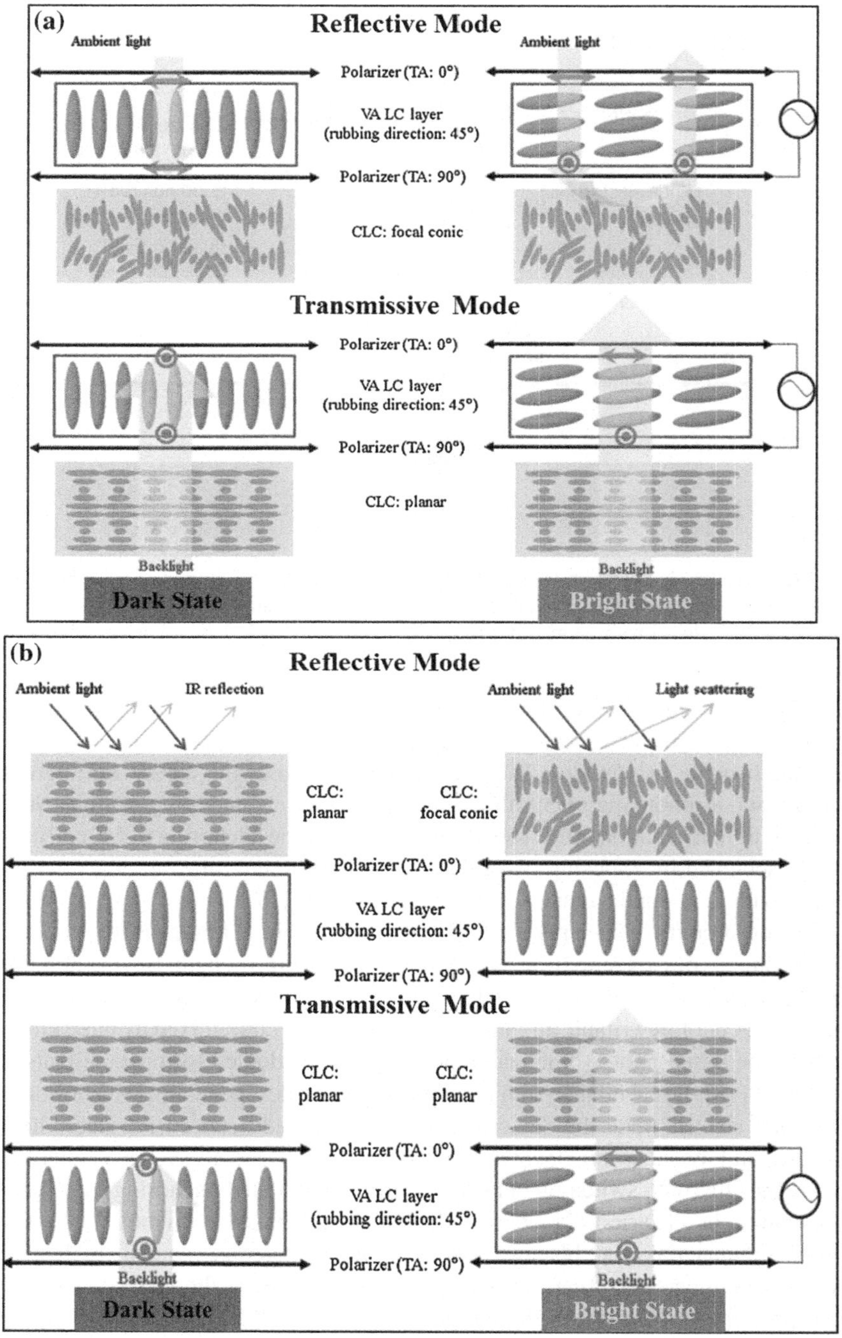

Fig. 14.5 Display mode switching between reflection and transmission. Two configurations are shown: **a** VA LC layer upon CLC layer; **b** CLC layer upon VA LC layer. Reproduced with permission from [16]. Copyright 2010 Optical Society of America

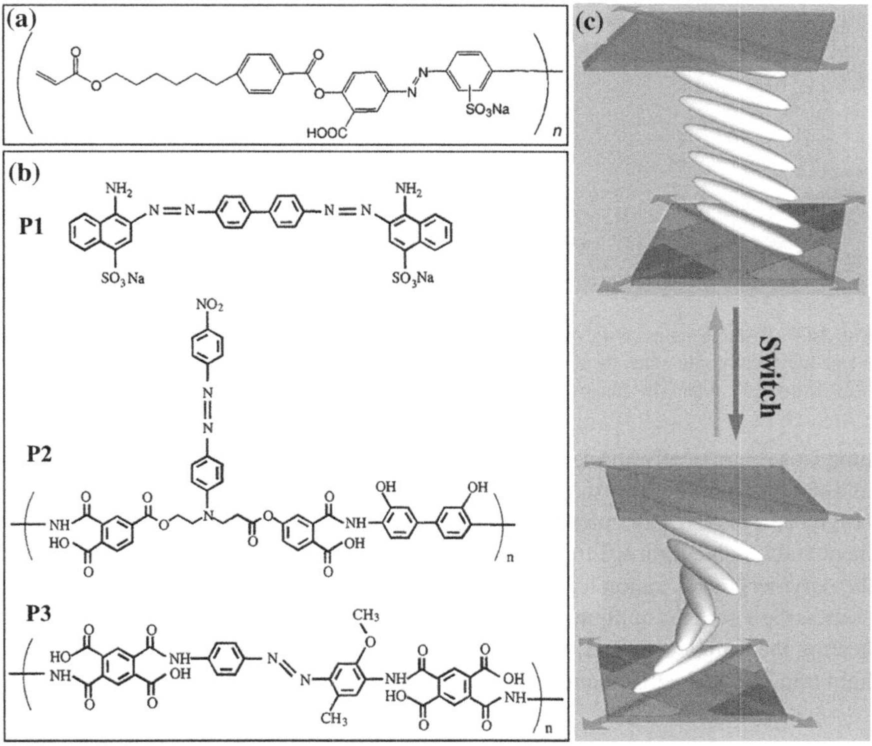

Fig. 14.6 Molecular structure of photoalignment materials and illustration of two horizontal electric-field driving. **a** Photoalignment film used in [17]; **b** rewritable photoalignment film, blended by three kinds of azobenzene based polymer, P1, P2 and P3 [18]; **c** bistability induced by a pair of in-plane electrodes [19]. Reproduced with permission from [19]. Copyright 2003 American Institute of Physics; Copyright 1998 Publication Board, Japanese Society of Applied Physics; Copyright 2008 American Institute of Physics

vertical electric-field within a certain range was applied across the LC cell, the original homeotropic alignment transformed to a stable homogeneous alignment, and the recovery can be achieved by cooling the sample [21]. Therefore, bistable device was controlled by temperature solely or combining the applied electric-field and temperature.

Bistability also commonly appears in polymer stabilized LC systems due to the significant hysteresis resulted from strong anchoring of polymer network, which hinders the rotation of LC molecules under the effect of external field. The bistable property is generally depended on the concentration of polymer, the higher polymer content leads to the larger hysteresis. In addition, the existence of polymer plays a positive role in avoiding the degradation and improving the switching fatigue. A bistability between the reflective colorful planar CLC and transparent fingerprint CLC was observed in a polymer stabilized fingerprint CLC system [22]. The CR of such device was 2.5 times higher than the conventional CLC device and the response

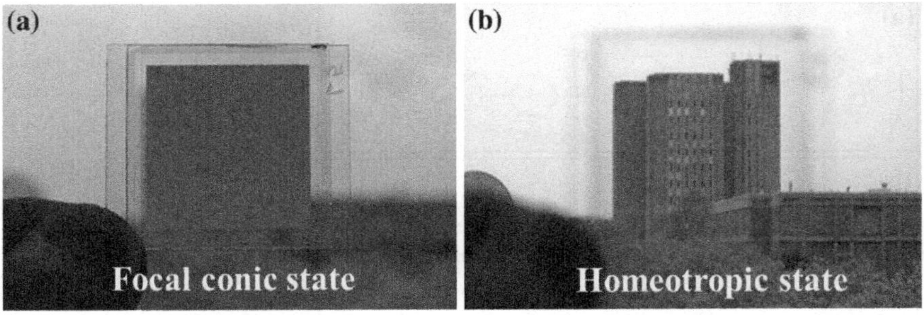

Fig. 14.7 Bistable states of polymer stabilized chiral liquid crystal in absence of applied voltage. **a** opaque focal conic state; **b** transparent homeotropic state. Reproduced with permission from [23]. Copyright 2009 The Japan Society of Applied Physics

time was dramatically shortened to 10 ms, which was significantly shorter than that of conventional device, usually about 230 ms. Similarly, the other kind of bistable polymer stabilized CLC presented a stable transparent states of homeotropic alignment and a stable light scattered focal conic state by doping proper polymer [23]. As the polymer concentration was 5.1 wt%, the focal conic state can be triggered to stable homeotropic state by applying a short-pulse voltage and the recovery was obtained by heating the sample to a high temperature and followed by a successive cooling. The light transmission performance of such sample was confirmed as shown in Fig. 14.7.

14.2.2 Bistability Enabled by Novel Chiral Liquid Crystals

Some kinds of CLC possess either the special molecular polarizability or the arrangement, enabling bistability with the stimulation of external field. Among these CLC materials, dual-frequency CLC (DFCLC) is a typical one. Molecules in such CLC exhibits positive dielectric property when driven by low-frequency (normally <5 kHz) electric-field, i.e., molecules align parallel to the direction of electric-field; in contrast, if a high-frequency (several tens or hundreds of kilohertz) electric-field is applied, CLC shows negative dielectric constant and the molecules align perpendicular to electric-field. Such frequency-dependent dielectric property enables a much faster switching between two arrangements of LC. By injecting a low twisted DFCLC into a planar cell with high pretilt angle (about 70°–80°) on the surface of substrates and driving with a low frequency electric-field, the DFCLC transformed from the initial tilted twisted (tT) arrangement to the bias homeotropic (bH) arrangement, and followed by the transition from bH to tilt homeotropic (tH) arrangement after removing the electric-field. The high surface pretilt angle was achieved by mixing the homogeneous and homeotropic alignment materials with a certain weight ratio, and spin coated on the substrates followed by a high temperature baking and a rubbing. The transformation from tH to tT arrangement can be reached by applying a low frequency electric-field to drive the molecule arrangement from tH to bH firstly, and subsequently applying a high frequency electric-field, thereby forcing the bH to transit to bias twisted (bT) arrangement because of backflow effect. As the bias voltage was removed, bT arrangement recovered to the original tT. Consequently, the bistable transition between two stable states, tT and tH, without the need of external bias, was achieved by DFCLC [24]. Such bistability is dependent on *d*/*P* ratio, i.e., tH cannot be maintained in the case of larger *d*/*P*, while tT is absent as this ratio is too small. The crucial *d*/*P* influence on the stability of DFCLC was studied from Erickson-Leslie-Parodi theory, and the optimized range of *d*/*P* ratio for exhibiting a better bistability indicated that the ratio in the range 0.59–1.16 when the pretilt angle was 62°; the range narrowed to 0.95–1.28 when the pretilt angle increased to 80° [25].

When a similar DFCLC was confined in a cell with parallel alignment, the bistability between splay and 180°-twist arrangement was realized (Fig. 14.8). The initial splay arrangement transited to bend state when the material was driven by low frequency electric-field. After removing such electric-field, the arrangement changed to 180°-twist due to the backflow effect. As 180°-twist arrangement was driven successively by a low frequency electric-field and high-frequency electric-field, it transformed to bend state at first, and followed by the transition from bend to the initial splay [26].

Two stable states, focal conic and planar, are observed commonly in DFCLC (Fig. 14.9a, b). The planar state with reflection center at near infrared (NIR) was obtained when the material was confined in a cell treated with homogeneous alignment. The transition from transparent planar state to light scattered focal conic

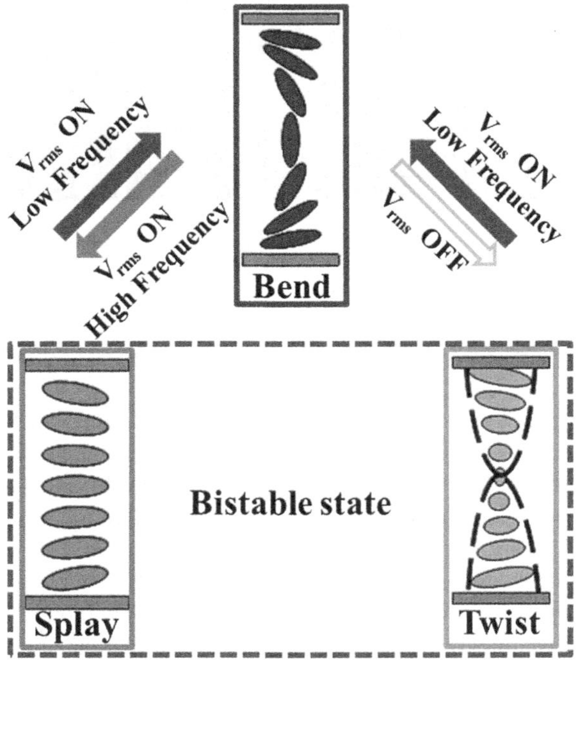

Fig. 14.8 Bistability of splay and π-twist configuration based on dual-frequency CLC. Adapted with permission from [26]. Copyright 2009 American Institute of Physics

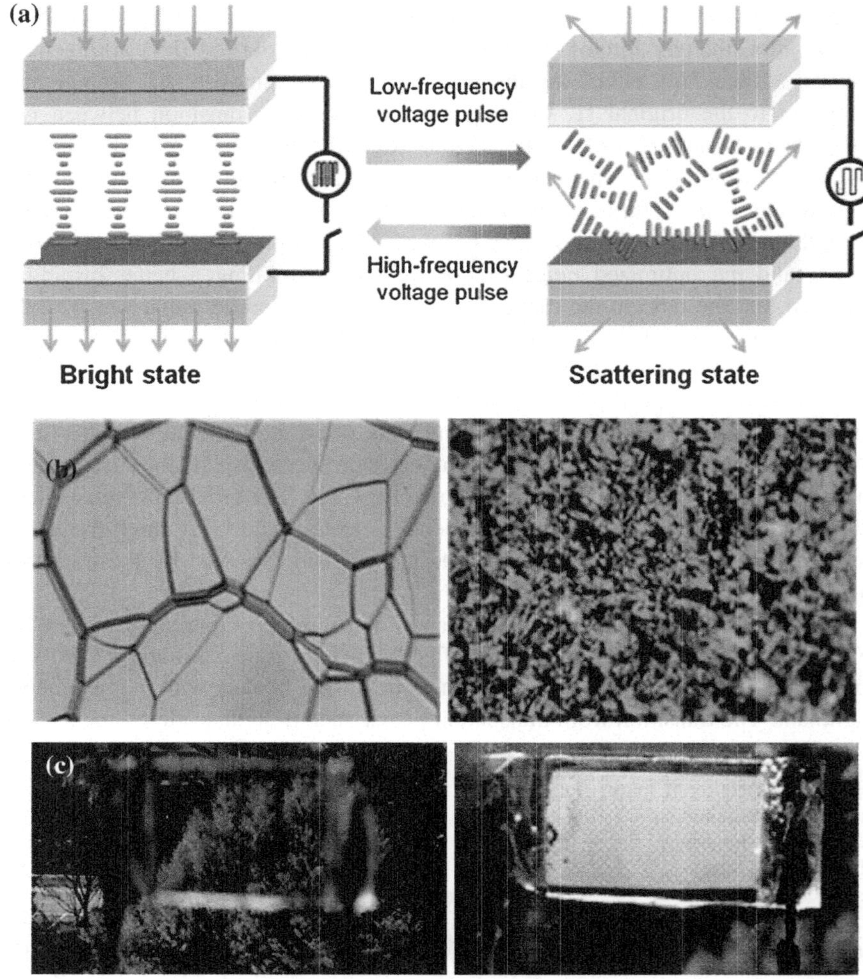

Fig. 14.9 Switching between transparent planar state and opaque focal conic state. **a** Schematically shown the driving manner between two states [27]; **b** the corresponding POM micrographs of planar state and focal conic state [27]; **c** transmission performance of the planar state and focal conic state [28]. Reproduced with permission from [27, 28]. Copyright 2011 Optical Society of America; Copyright 2012 Optical Society of America

state was achieved by applying low frequency (1 kHz) voltage pulse. When the sample was driven by a high frequency (100 kHz) voltage pulse, the optical transparent planar state reappeared. The driving voltage was determined by the pitch length of DFCLC, i.e., the concentration of chiral dopant contained in the system. Owing to the dual-frequency property, the switching between planar and focal conic states was as fast as 10 ms [27]. Stable planar and focal conic states can be formed in DFCLC confined in a cell without any alignment. The optical and dynamic response properties were similar as those of DFCLC injected into a homogeneous cell as shown in Fig. 14.9c. Such bistability is related to the balance between the elastic free energy and the anchoring [28].

Often, polymer stabilization is adopted as a facile way to improve the stability of arrangement in DFCLC. As an enough strong low frequency electric-field was applied to form an electric-field-induced homeotropic alignment of LC molecules, the sample was irradiated by a UV source to stabilize such state. Subsequently, a high frequency voltage pulse drove the homeotropic state to a focal conic state; while a low frequency voltage pulse drove the sample back to the initial homeotropic alignment. Both of the focal conic and homeotropic states were stable in absence of external field because of the memory effect caused by polymer [29]. Polymer always leads to other new effects apart from the arrangement stabilizations. When a planar state was stabilized by polymer in a homogeneous cell, the extension of reflection band was observed after applying a low frequency voltage pulse instead of the light scattering commonly observed in DFCLC without polymer stabilization. The reflection band width was narrowed after the sample was driven by a high frequency voltage pulse. In addition, light transmission of the samples was related to polymer concentration. High transmission usually obtained in the case of low polymer concentration, and vice versa. Such bistability between two planar states with different width of reflection band can be used as a transflective light shutter [30].

Based on bistability between tH and tT arrangements [31, 32], splay and twist arrangements [33] or focal conic and planar states of DFCLC [34], tunable one dimensional photonic crystals (1D PC) were fabricated by sandwiching such material between two substrates periodically coated with two kinds of inorganic films (Ta_2O_5 and SiO_2), forming a refractive index modulation along a single direction (Fig. 14.10a). By successively applying the low frequency and high frequency voltage pulses, the distinct arrangements of bistable DFCLC led to the tunability for mode field of such sandwiched 1D PC, such as the wavelength, reflectance and number of modes, as shown in Fig. 14.10b. Similarly, a guest-host (GH) mode DFCLC with stable tH and tT arrangements was developed by doping a very small amount of dichromic black dyes into DFCLC [35]. The light transmission in tT state was higher than that in tH state for the reason that the absorbance of dye in the former was larger than that in the latter. These two stable states can be switched with the same manner as aforementioned, the CR reached 6.5 which was higher than that of conventional GH mode LCD.

In addition to DFCLC, blue phase LC (BPLC) exhibits obvious bistability. BPLC is a specific kind of LC phase existing between isotropic and chiral nematic phases in high chirality system. LC molecules arrange as the double twisted cylinders (DTCs) in such phase. BPLC has three sub-phases. During cooling process from isotropic state, the DTCs first arranged randomly, presenting a fluid fog-like optical texture (BPIII); and followed successively by an orderliness simple cubic (BPII) and body center cubic (BPI) arrangements, presenting the colorful platelets texture. BPLC possesses a significant temperature hysteresis, i.e., the temperature range of BP in heating process is different with that in cooling process, thus there will be a certain temperature range, where it exhibits distinct sub-phases in heating and cooling. Therefore, bistability between two sub-phases of BP can be controlled by temperature. Besides temperature,

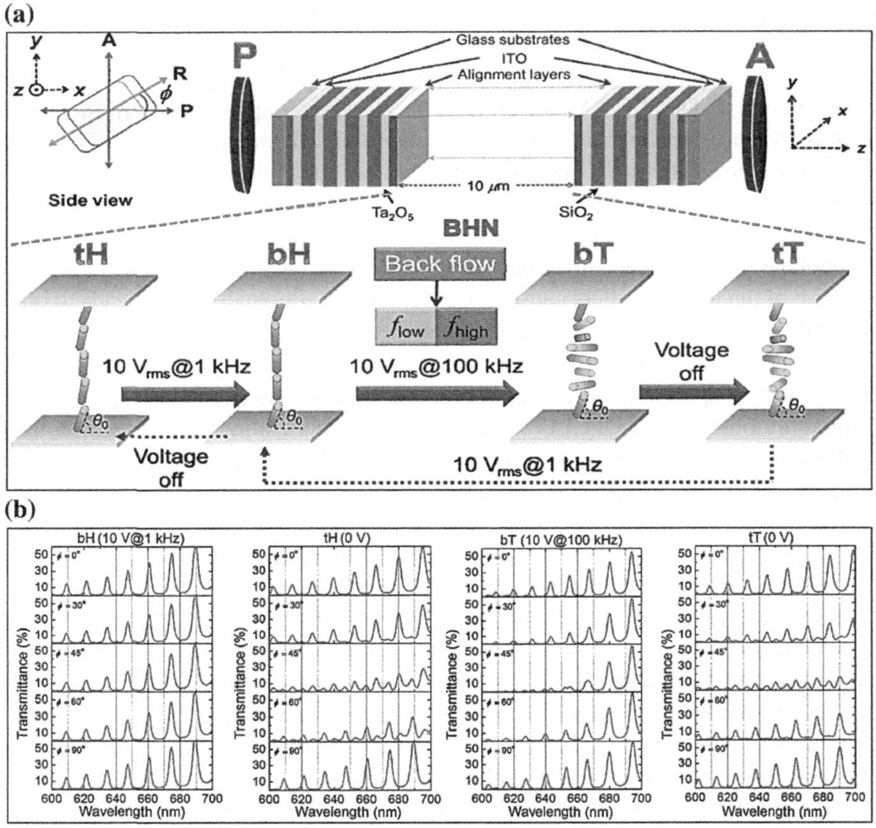

Fig. 14.10 Driving mechanism of 1D PC and the corresponding transmission spectra. **a** the device structure of 1D PC and the arrangement change between two stable states, tH and tT; **b** transmission spectra at different arrangement states. Reproduced with permission from [32]. Copyright 2013 Optical Society of America

electric-field controlled bistable BPLC were found recently [36]. Bistability of BPLC, controlled with electric-field, in a very narrow temperature range (47–50 °C), appeared in a BPLC with BP range about 16 °C (50.1–34.5 °C). At the temperature of 48 °C, the arrangement of BPI with orientation denoted by Miller indices (h, k, l), BPI (2,0,0), was observed at the initiate. This state passed through a planar state and transformed to a flow state due to the electrohydrodynamic effect as the voltage was increased to 80 V. When the voltage slowly decreased from 80 V, the flow state successively transformed to planar state and recovered to the initial BPI (2,0,0); however, if this voltage was suddenly removed, the flow state quickly transited to another BPI with a different orientation, BPI (1,1,0), which can be further transited to BPI (2,0,0) by applying a voltage of 25–35 V, or to planar state directly by applying a larger voltage of 35–80 V. If a voltage pulse of 35 V was applied, BPI (2,0,0) was maintained after this pulse; nevertheless, BPI (1,1,0) formed as the voltage increased

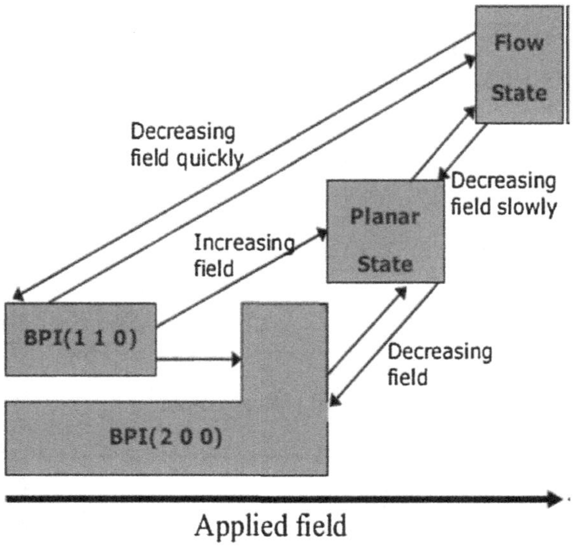

Fig. 14.11 Phase transition of BP with different driving manners. Reproduced with permission from [37]. Copyright 2010 American Institute of Physics

to 80 V after the pulse. Figure 14.11 depicts phase transition as applying a voltage [37]. Recently, the theoretical works indicated that the bistable defect structure of BP can be obtained by applying different voltages on the sample at the fixed boundary conditions [38].

It is noteworthy that the bistability of BPLC only can exist in a narrow temperature range as mentioned above, thus polymer stabilization was adopted to improve such drawback. The experimental results indicated that BP transited to chiral nematic phase (N*) firstly, and then to an unwound homeotropic alignment when the electric-field slowly increased to 5.3 V/μm. The difference was that the homeotropic state went back to BP if the electric-field was removed suddenly, however if the electric-field was decreased slowly, N* was obtained consequently (Fig. 14.12). Such bistability resulted from the coupling effect among electric-field energy, elastic free energy and the polymer anchoring energy. For generating the bistability, an appropriate polymer concentration should be considered. One attentive aspect was the temperature range that existed bistability by polymer stabilization was significantly widened to about 55 °C (i.e., −18.4–37.2 °C) [39].

The bistable in smectic A (SmA) LC was found by doping a small amount of ionic liquid into the system. Under the effect of low frequency (<1 kHz) voltage pulse, the transparent SmA transformed to light scattered focal conic phase because of the ion-moving induced electrohydrodynamic effect. The recovery from focal conic state to homeotropic arrangement can be achieved by applying a high frequency (>1 kHz) voltage pulse [40]. In addition, considering the flexibility of the bistable device, polymer walls were fabricated at some certain regions by UV irradiation under the controlling of electric-field based on Kelvin polarization force. However, the CR of the devices was only 3–6 [41]. A feasible way to increase CR is mixing a small amount chiral agent with SmA LC host, thereby enhancing the light scattering of the sample at focal conic state.

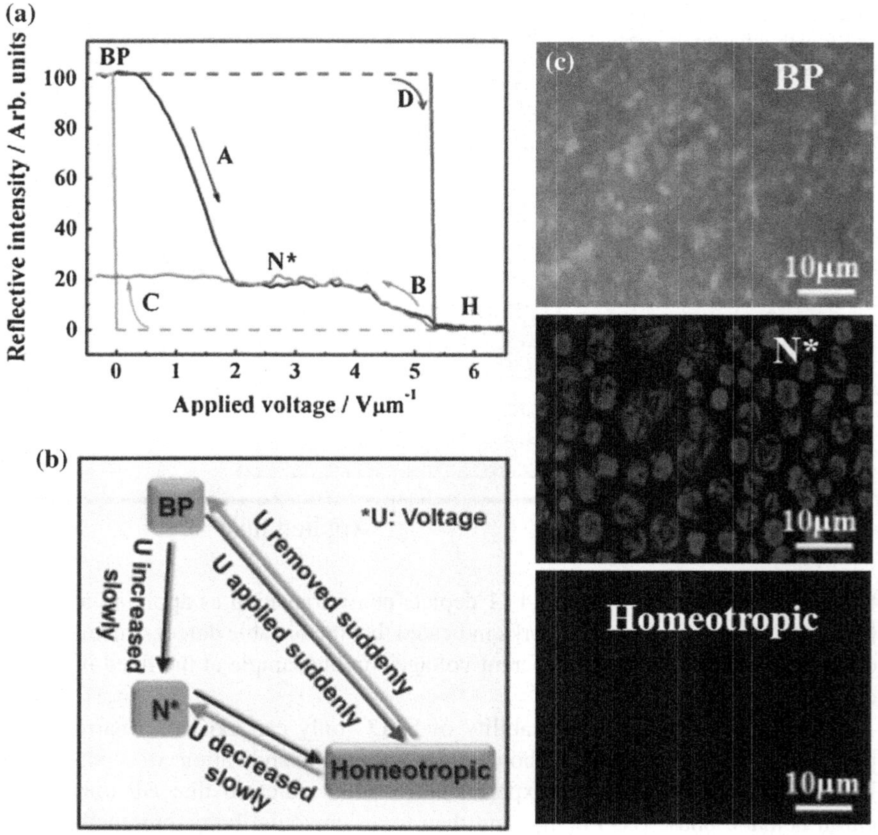

Fig. 14.12 Bistability of polymer stabilized BPLC. **a** voltage dependent reflection intensity during driving; **b** phase transitions with different driving manner; **c** corresponding optical texture of BP, chiral nematic phase, N*, and homeotropic alignment. Reproduced with permission from [39]. Copyright 2012 Optical Society of America

14.2.3 Light Driven Bistable Chiral Liquid Crystal Devices

The bistabilities mentioned above are all driven by electric-field. However, light is another efficient external stimulus adopted as the driving source. A judiciously structure design for a photo addressable bistable reflective LCD was carried out based on the photosensitive Methyl Red (MR) doped system, and its mechanism was illustrated in Fig. 14.13a. Such reflective LCD contained only one polarizer, a dye-doped homogeneous aligned LC layer, a quarter waveplate (QWP) and a transflective mirror (TFM), which were set successively from the front to the backside. The transmission axis of polarizer was parallel to the orientation of LC molecules, while it had an angle of 45° with respect to the fast axis of QWP. One of the substrates was treated with homogeneous alignment while the other was coated

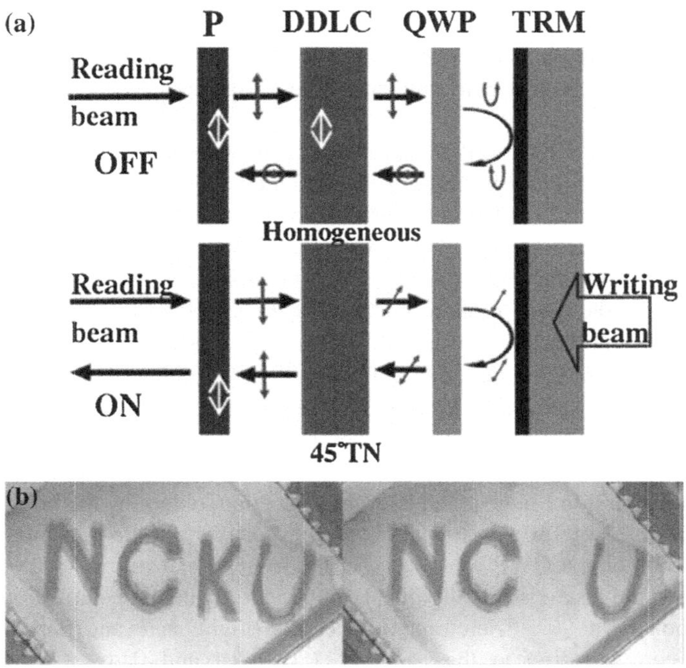

Fig. 14.13 Photoaddressablebistable reflective LCD. **a** Mechanisms of the reflective display. *P* Polarizer; *DDLC* MR dye-doped liquid crystal; *QWP* Quarter waveplate; *TRM* Transflective mirror. **b** Display performance of the reflective LCD before and after the letter 'K' was erased. Reproduced with permission from [42]. Copyright 2006 American Institute of Physics

by a non-rubbed optically transparent polymer film to generate a weak anchoring. At the OFF state, one reading beam impinged from the front polarizer side, passed through the LC layer and transformed to a CPL; after reflecting by the TFM, the rotation of CPL changed to the opposite direction and transformed to a linear polarized light with the polarization perpendicular to the orientation of LC molecules after re-passing through the QWP. Such reflection light was stopped by the polarizer. As a writing polarized beam (polarization direction was parallel to fast axis of QWP) impinged from the backside, MR molecules underwent photoisomerization and adsorbed on the surface of substrate covered with polymer, leading to a homogeneous orientation of LC molecules along the direction perpendicular to the polarization direction of writing beam, while with an angle of 45° to the alignment on the other substrate, thus forming a 45°-TN layer. Therefore, the reading beam can pass through the polarizer, thereby achieving the ON state. Simultaneously, the writing beam was stopped by polarizer because its polarization direction was perpendicular to the transmission axis after passing through the 45°-TN layer, so the reading beam will not be washed out by the writing beam. CR of such reflective LCD was estimated to exceed 10 [42]. The performance is shown in Fig. 14.13b. A similar structure layout of light driven reflective LCD was developed as well by doping another kind of azo dye, SD-2, into LC material. The ON state was obtained by the original 45°-TN dye-doped LC layer. Phase transition from LC to isotropic state upon photo irradiation yields the OFF state of the device [43].

Another kind of light driven bistable LC devices was realized by doping a photosensitive binaphthyl-based azobenzene chiral molecular switch into the conventional nematic LC, obtaining the photoresponsive CLC. Helical pitch length can be elongated and compressed by ultra-violet (UV) and visible light irradiation respectively. The bistable transmissive and reflective LCD were fabricated respectively on the basis of pitch modulation of such chiral switch. At the initiate, the 180°-twist arrangement was formed in a homogeneous (0°-twist) cell by adjusting the d/P ratio in the range of 0.25–0.75. UV irradiation made the d/P ratio decreased to less than 0.25, leading to a homogeneous arrangement. When the photoresponsive LC layer was sandwiched between the crossed polarizers with the transmission axis of one of the polarizers parallel to the alignment direction of the cell, the transmission can be tuned by light irradiation, obtaining the bright and dark state in the cases of 180°-twist and 0°-twist respectively in the transmissive mode. In reflective mode, a QWP was placed at the front of rear mirror with an angle of 45° between the fast axis and alignment direction to rotate the polarization direction of reflection light 90°. Therefore, a reflective bright state and dark state were obtained corresponding to the cases of 180°-twist and 0°-twist arrangements, respectively [44]. By injecting similar photoresponsive CLC into the cell without any surface treatment, a focal conic state was observed at the initiate and was triggered to planar state with the UV light irradiation due to the elongation of helical pitch, thereby exhibiting the bistability between focal conic and planar states. Figure 14.14 displays the letters recorded by light irradiation [45]. In addition, a photosensitive chiral molecular switch was mixed with the conventional chiral dopant with a certain weight ratio and doped into the nematic LC to form a

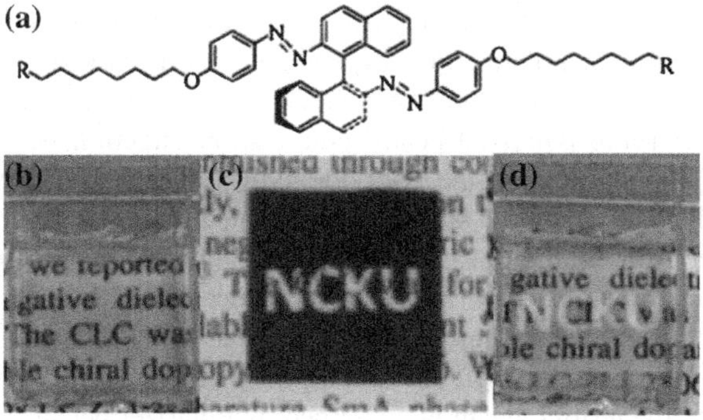

Fig. 14.14 Bistability in photoresponsive liquid crystal. **a** Molecular structure of binaphthyl photosensitive chiral switch; **b** transparent state as helical pitch length is long after irradiated by UV light; **c** photomask with 'NCKU' transmissive letters; **d** displayed 'NCKU' when the sample was irradiated through photomask. Reproduced with permission from [45]. Copyright 2013 Optical Society of America

light driven handedness inversion CLC. Such material was confined in a cell with orthogonal alignments on the two substrates, forming the 90°-twist arrangement. Two stable states of 90°-twist (right handed) and 90°-twist (left handed) were controlled by light irradiation with different wavelength, and a π-phase shifter was achieved based on this performance [46].

14.3 Solar Energy Related Applications of Chiral Liquid Crystal Bistability

One unique characteristic of CLC is selective reflection determined by the helical pitch. Such reflection performances, including band-shift (red-shift or blue-shift), band widening or narrowing and the reflectance, can be tailored by judiciously designing the arrangement of LC molecules and by driving with external field, thereby enabling a facile tunability for the impinged sunlight from the reflection wavelength range to the transmission intensity, i.e., the manipulation of solar energy. For the purpose of effective utilization of solar energy, some kinds of CLC devices have been developed and studied in the past decade.

Among these devices, smart window is one of the most attractive aspects, which can be adopted to manipulate the sunlight transmitting into the room according to ambient condition. The majority of smart windows developed and utilized at present are based on electrochromic metal oxides, polymer electrochromics and all-solid-state electrochromics [47]. Such kind of smart window has many advantages, such as good electro-optical performance, easy processing and low cost; however the instability of the performance caused by the non-uniform diffusion of metal oxides in the glass and the degradation of polymer is a bottleneck problem. Besides, the intelligence of such kind of smart window seems not sufficient, i.e., the device only responds to the electric-field [47]. LC shows diverse responsiveness to not only the electric-field, but also the light and temperature, rendering a novel kind of promising material which is suitable to be applied in smart windows. The most common LC smart windows are based on two kinds of polymer/LC composites, one is polymer dispersed LC (PDLC) and the other is polymer stabilized cholesteric texture (PSCT, i.e., polymer stabilized CLC). The mechanism of these windows is switching between the light scattering and optical transparent states under an applied electric-field. The voltage-off state seems opaque due to the light scattering by randomly distributed LC droplets, while the voltage-on is transparent for the reason of refractive index matching between the uniform aligned LC and polymer. This means that an optical transparency should be maintained by a sustaining applied voltage, and consequently the power consumption is very high. In addition, such smart window can only be used for stopping or passing through the sunlight however without any more intelligent tunable performances, which causes a less efficient solar energy tunability. Apart from the smart window, some conventional CLC devices driven with the help of sustainable solar energy by integrating the photovoltaic elements or materials have attracted more attention recently. The combination of LC devices with the common solar energy devices, enabling adaptiveness of solar energy devices, indicates the significance and importance of such devices for energy efficient future.

This section will be presented as the following parts: (i) adaptive infrared reflective smart window; (ii) photovoltaic driven LCD and other LC modulated solar energy devices, focusing the developments and study progresses on these two aspects at present. We will not discuss the conventional PDLC and PSCT smart windows herein, instead of those, adaptive infrared reflective CLC smart window which shows more intelligent and better performance in the context of energy saving is focused.

14.3.1 Adaptive Infrared Reflective Smart Window

Sunlight has a wide band spectral distribution and the majority energy ($\sim$60 %) is concentrated in the range of near infrared (NIR) band (from 700 to 1100 nm) [47]. However, for single CLC layer, the reflection band width is only 50–100 nm, due to the limited birefringence of LC material, which weaken the capability of CLC for reflecting the solar energy. Therefore CLC smart window with a wide reflection band is desired [48].

A simple and efficient way was proposed in the early days, which involves stacking of two CLC layers one on the top of another. Two different kinds of CLC materials were injected into the homogenous cell to form the planar arrangements. The reflection band of each CLC layer was modulated to NIR band, and simultaneously, the central reflection position of one of the CLC layers should be overlapped with the edge of reflection band of the other CLC layer. With such stacked configuration, a reflection band of about 460 nm has been achieved [49]. Such stacked structure is very simple theoretically, however, it has some technical difficulties. On one hand, the two CLC layer should be sufficiently parallel to each other to maintain the light impinges on each layer with the same incident angle; on the other hand, to eliminate the Fresnel reflection on the interface of two CLC layer, a refractive index matching layer is necessary to be inserted between them. In addition, from the economical point of view, the cost of processing for such stacked structure is higher. Recently, a much more convenient and economical processing of CLC film was proposed, which adopted several kinds of CLC polymers with different helical pitches and homogeneously mixed them together to generate the pitch gradient in a single film. To enhance the mechanical properties and cohesiveness as well as improve the flexibility of such film, the ethylene-vinyl acetate (EVA) copolymer which is widely utilized in building glasses, vehicle windshields, etc., was added with a certain weight ratio. The optimized content of EVA was investigated, indicating a satisfied reflection performance as EVA was 50 wt% (optical texture is shown in Fig. 14.15a). It was difficult to fabricate the film in the case of low EVA content, while phase separation imposed a problem if the EVA

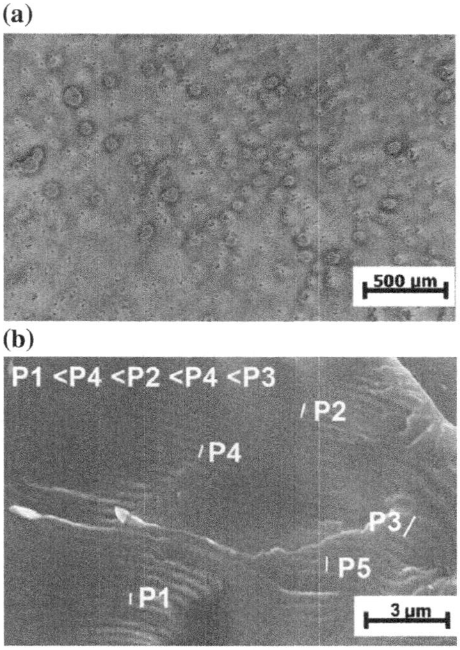

Fig. 14.15 Optical performance and helical structure of the reflective film. **a** Optical texture taken by POM; **b** TEM photograph of the film. Reproduced with permission from [50]. Copyright 2015 The Royal Society of Chemistry

content is too high. The morphological, optical and energy-saving performances as a smart window film were studied. An evident pitch gradient structure was observed under the transmission electron microscope (TEM) (Fig. 14.15b), and the energy conservation efficiency reached 40.4 % [50].

Another efficient way, commonly utilized to widen the reflection band, is forming the pitch gradient across the CLC layer. Such structure can be obtained through many different methods. A small amount of light absorber was doped into the mixture of CLC and photopolymerizable acrylate-based monomers, thereby leading to an attenuation of the intensity of exposure light as the light penetrate into the mixture during polymer stabilization, thus polymer density on the side closing to the irradiation source was stronger than that on the side away from the source, and consequently a pitch gradient formed due to the non-uniform anchoring strength between polymer network and LC molecules. The helical pitch is small as the anchoring strength was stronger and vice versa. In the experiment, the exposure source was UV light with the wavelength of 365 nm, and the light absorber, which was appropriate to UV, was selected as Irgacure 907, and in the meanwhile, another cell filled into pure LC was used as the UV intensity attenuator due to the UV absorption of the pure LC. Reflection band width before UV irradiation was about 80 nm, however after the irradiation, it widened to 220 nm caused by the formation of pitch gradient (Fig. 14.16a) which was increased almost two times. If the sample was simultaneously irradiated from both sides of CLC layer (i.e., symmetrical irradiation), the tendency of pitch gradient was weakened (Fig. 14.16b), therefore

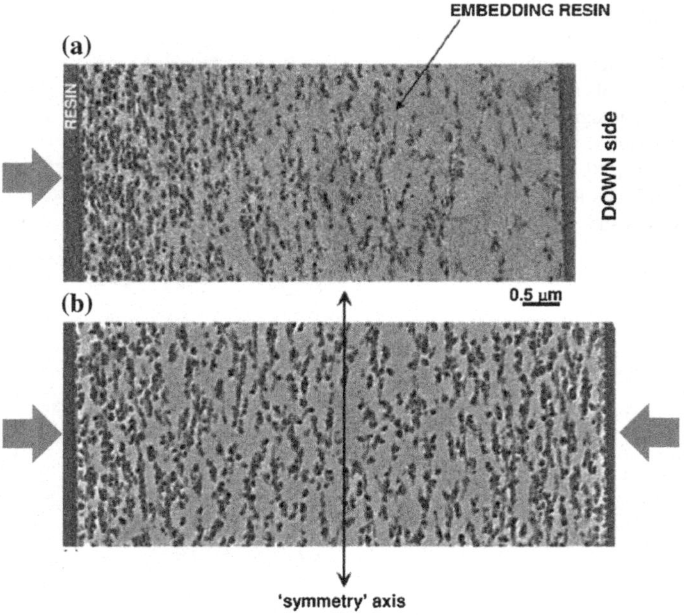

Fig. 14.16 TEM photographs for the broad reflective liquid crystal gel. **a** UV irradiation from one side to the other as shown by *violet arrow*; **b** UV irradiation from both sides of the film. Reproduced with permission from [51]. Copyright 2006 American Institute of Physics (Color figure online)

resulting in the decrease of reflection band width to 160 nm which was 37.5 % narrowed compared with the former, 220 nm, irradiated from one side to the other (i.e., asymmetrical irradiation) [51]. In order to obtain a wider band width, some of physical parameters, such as thickness of the layer, content of light absorber, irradiation intensity and exposure time, need to be optimized. Similar work has been carried out with the same light absorber as well and obtained the reflection band width of 240 nm. The transmission tunability was tested by driving with an applied external voltage, the wide reflection band disappeared and was driven to a low transmission scattering state as the voltage was 25 V, while as the voltage was increased to 218 V, an optical transparent state was induced due to the electric-field-induced uniform alignment of LC molecules along the direction of electric-field [52]. Very recently, an electrically switchable smart window was achieved by using aforementioned methods and its energy-saving performance was estimated according to the medium size office building model through Trnsys simulation software. Three different climates were chosen to comprehensive evaluate the application potentiality of such smart window in the distinct environmental conditions. Consequently, it was predicted that more than 12 % of energy can be saved after adopting such smart windows [53].

In addition, some specific chiral dopants were adopted in the preparation of wide reflection band CLC smart window. A temperature-sensitive chiral dopant, **CM**

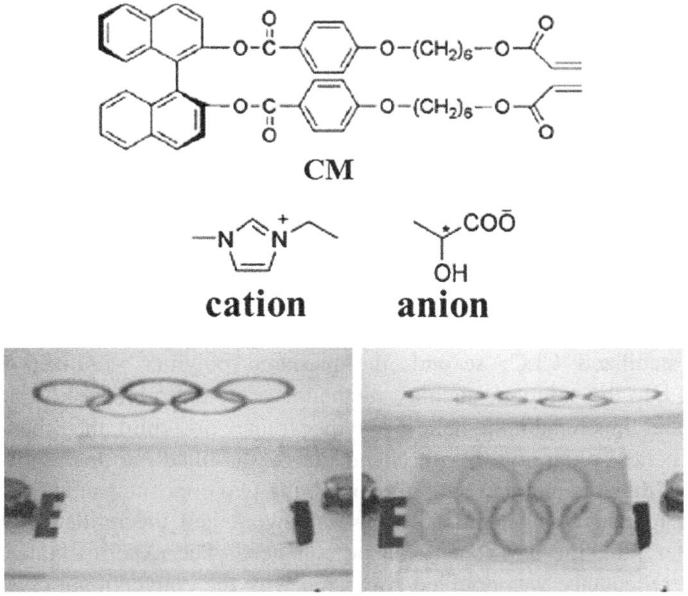

Fig. 14.17 Molecular structures of temperature-sensitive chiral dopant CM [54] and ionic dopants, and optical performances of CLC doped with such ionic chiral dopant. Reproduced with permission from [55]. Copyright 2009 Elsevier Ltd.; Copyright 2010 The Royal Society of Chemistry

(Fig. 14.17), which contains two photopolymerizable end groups and possesses a significant change of helical twisting power (HTP) with temperature, was homogeneously doped into the nematic LC to form the temperature sensitive CLC. Before polymerization, reflection band shifted about 300 nm as the temperature rose from 15 to 60 °C. Therefore, pitch gradient can be easily produced by polymerizing the material and simultaneously raising the temperature gradually. Finally, a much wider reflection band with the width over 500 nm was achieved [54]. Another kind of ionic liquid chiral dopant (Fig. 14.17) was used to develop the pitch gradient as well. As a high frequency (100 kHz) AC voltage was applied on the sample, the anions and cations contained in such ionic liquid chiral dopant moved toward the anode and cathode, respectively, thus forming concentration gradient of chiral dopant across the cell which led to the pitch gradient. Polymer stabilization was carried out as the AC voltage of 40 V was applied to obtain a stable pitch gradient structure, and the reflection band from 400 to 850 nm, covering the whole visible spectrum, was obtained. The transmission performance can be switched by driving with a DC voltage between the opaque light scattering state and the light reflective state, as shown in Fig. 14.17 [55]. It was reasonable that the reflection band can be modulated to NIR band just by appropriately doping the ionic liquid chiral dopant, and thus the wide NIR reflection CLC can be achieved as well. It is interesting that the broad reflection can be generated in polymer stabilized CLCs with ions trapped into the polymer network. Upon driving with DC voltage, the movement of ions leading to the motion of polymer network which stretches the helical pitch of CLC on one side while compress the pitch on the other side, therefore forming a gradient of helical pitch, thus enabling the broad reflection band [56].

The reflectance of CLC is the other aspect which should be further improved. Generally, CLC only reflects the CPL with the same handedness sense, which means only less than 50 % of reflectance for the non-polarized sunlight can be reflected. In order to break through the barrier of natural property of conventional CLC material, a film containing both of the right-handed and left-handed helical CLC was achieved. Such film was prepared through the following steps: first, the mixture of conventional CLC and monomers was photopolymerized to form the polymer stabilized CLC; second, the prepared polymer stabilized CLC was immersed into the solutions such as, methanol, acetone, to remove LC in polymer network, and thus remain the helical polymer framework; third, the other CLC with the same helical pitch but opposite handedness was filled into the empty polymer framework, thus the unpolarized light containing two opposite handed CPLs can be efficiently reflected by helical polymer framework and the refilled CLC respectively. Therefore, the reflectance can be enhanced. The experimental result indicated that reflectance was doubled compared with the conventional CLC [57].

By combining all of aforementioned techniques together, a wide reflection band with high reflectance was achieved in a CLC film, and the energy consumption of such film applied as a smart window was calculated through Trnsys simulation software. The wide reflection band was obtained by doping about 1 wt% of UV-absorber, Tinuvin 328, and the reflectance was enhanced by stacking two layers of CLCs with opposite handedness. Consequently, the reflection band was widened from 100 to 277 nm, while the reflectance was enhanced from less than 50 % to more than 95 %. As a potential application of smart window, such CLC film can produce a net energy-saving about 15 % on average [58].

In addition, by doping a small amount of black dyes into CLC can modulate the transmission intensity as well as the reflection band. At the planar state (voltage-off), the reflection band was located in NIR, thus the majority energy of sunlight was reflected; simultaneously, the absorption of black dyes in that case was high which modulated the transmission intensity into the room. As a driving voltage was applied, the planar arrangement transformed to homeotropic alignment, leading to the disappearance of NIR reflection band and the increasing of transmission intensity [59].

An adaptive smart window, which can modulate its transmission in accordance with the environmental condition, was developed very recently. Such adaptiveness was achieved by doping a small amount of photosensitive binaphthyl based azobenzene chiral molecular switch into the nematic LC. In the daylight, the UV contained in sunlight triggered the photoisomerization of chiral switch, leading to the decrease of HTP and lengthening the helical pitch, thus reflecting the NIR component of sunlight while transmitting visible spectral band; the helical pitch recovered back to the original state at the night, so the NIR transmitted into the room and maintain the temperature of the room. The visible light was absorbed by the dye sensitized solar cell (DSSC) embedded at the front or back of the photoresponsive CLC layer and transformed to the electric energy as the driving source for the CLC layer [60]. In addition, to decrease the energy consumption of driving, the photoconductive zinc phthalocyanine particles were dispersed in ionic liquid and mixed within the LC layer to increase the conductivity of such layer and thus decreased the strength of the applied driving field, so that the net energy consumption can be cut [61]. Similarly, the transparent silver nanowire was proposed to replace the conventional transparent conductive electrode, indium tin oxide (ITO). Experimental results indicated that a significant decrease of sheet resistance from 100 to 20 Ω/□ as the density of nanowire increased from 10 to 40 mg/m^2, however the spectrum transmission can be maintained above 90 %. The driving voltage was reduced about 15 V, while the transmission at ON state was enhanced more than 10 % after using the silver nanowire as the transparent electrode [62].

14.3.2 Photovoltaic Driven LCD and Other LC Modulated Solar Energy Devices

Currently, majority of LCD is driven by an embedded voltage-generator, thus the energy consumption of the conventional LCD is high, although such consumption has been improved compared to the early CRT display. In addition, as the size of LCD increases gradually, its energy consumption is raised dramatically. The average power consumption of a 50-in. LCD is more than 100 W at present, which is clearly considerable high at the current energy shortage situation. In the past decades, studies on photovoltaic materials and devices have made remarkable achievements. If such devices can be integrated in LCDs and the LCDs can be driven with photovoltage, the power consumption can be decreased significantly, which is always desired and preferred. A polarizing organic photovoltaics (ZOPV) was proposed and adopted in the conventional TN LCD with a judicious design on the structure of the device. Light with the polarization direction perpendicular to the alignment of molecular chain transmits through ZOPV, while the light energy is absorbed if its polarization direction is parallel to molecular alignment. Generally, ZOPV can be realized by uniaxial orientation of the conjugated polymer molecules. In the reported work, the common p-type organic photovoltaic polymer, poly (3-hexylthiophene) (P3HT), was coated on the substrate and unidirectional rubbed at 150 °C. Such ZOPV not only has an obvious light extinction effect (Fig. 14.18a, b), but also can be utilized as a photovoltage generator, and therefore a TN LCD sandwiched in between two ZOPV devices was developed. A high ratio of short-circuit current density, 3.54, was achieved consequently [63].

By unidirectional rubbing the P3HT layer, a polarizing solar cell (PSC) integrated reflective TN LCD was developed. An ITO glass, used as the anode of PSC, was successively coated with a hole-collecting layer, poly (3,4-ethylenedioxythiophene): poly(styrene sulfonate) (PEDOT: PSS), a PV layer of regioregular P3HT and [6,6]-phenyl C_{61}-butyric acid methyl ester (PCBM) which was unidirectionally rubbed subsequently, and an aluminum thin layer as the reflective as well as the cathode layer. A TN cell was sandwiched between a polarizer and the PSC as illustrated in Fig. 14.18c. The transmission axis of polarizer was perpendicular to the alignment direction of PV layer. At OFF state, the polarization direction of impinged light was rotated 90° which is parallel to the alignment direction of PV layer, and thus the light was absorbed and transformed into electric energy. At the ON state, the polarization direction was unchanged and perpendicular to the alignment direction. The absorption of PV layer in this case reached the minimum, and the light transmitted through the layer was reflected back by aluminum reflector, thereby obtaining the bright state. CR of such reflective TN LCD was tested as 1.7, indicating that the device was suitable to be used in character display. Figure 14.18d presents its performance [64]. A normal white (i.e., bright at OFF state) mode of reflective TN LCD was achieved with the similar device structure but adjusted the alignment direction of PV layer paralleling to the transmission axis of polarizer. CR was increased more than an order of magnitude due to the better dark state in normal white mode [65].

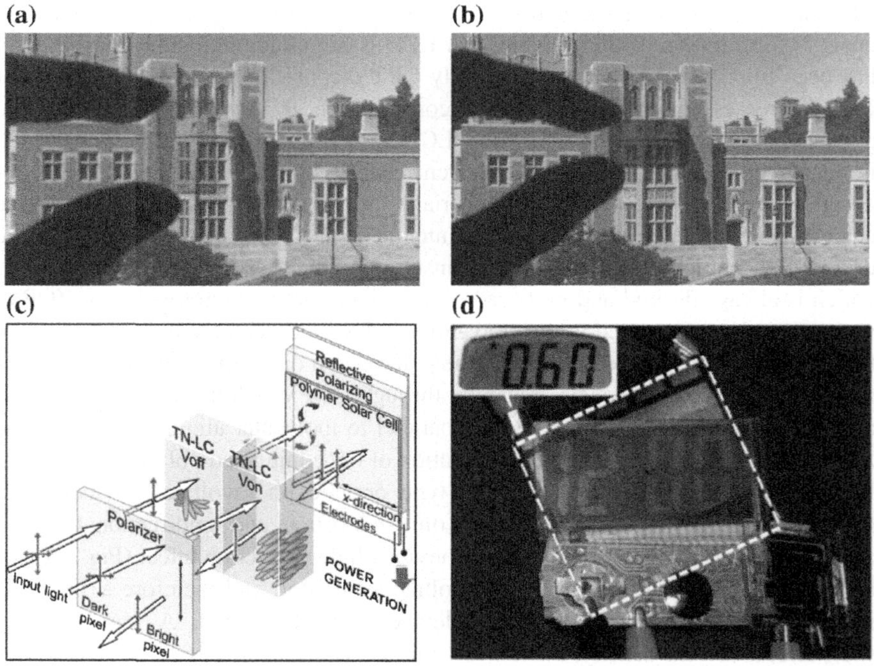

Fig. 14.18 Photovoltaic driven reflective LCD. **a** Transmission of ZOPV; **b** transmission of ZOPV after rotating the sample for 90°; **c** device structure and mechanism of photovoltaic driven reflective LCD; **d** display performance of the reflective LCD; the inset showed the voltage generated by polarizing PSC. Reproduced with permission from [63, 64]. Copyright 2011 WILEY-VCH; Copyright 2012 Optical Society of America

Apart from the aforementioned novel photovoltaic driven LCD, some other solar energy devices incorporated with LC device as an energy modulator have been proposed recently. For instance, the TN LC device used commonly in display was adopted as a light modulator to adjust the light intensity irradiated on the surface of an a-Si:H solar cell. Considering the durability of the device, two polymer-based polarizers used in conventional TN LCD were replaced by the sub-wavelength metal grating polarization beam splitter with a stable optical performance. Tested results indicated that the power conversion efficiency of solar cell can be tuned from 0.209 to 0.417 % [66]. Similarly, the fluorescent dye-doped LC was used as the solar energy modulator as well. The doped dye can absorb the impinged light, and simultaneously, the dye emitted fluorescence after absorbing the energy of incident light. The fluorescence was guided by a slab waveguide and irradiated to a photovoltaic device placed aside, thereby generating the electric energy which can be used to drive the LC molecules and modulate the transmission intensity again [67].

14.4 Light Driven Chiral Liquid Crystal Photonic Devices with Wide Tuning Range

In the development of LC devices, electric-field is always utilized widely as a convenient driving force, but it also results in high power consumption of LC devices. However, light can be employed as a convenient and new driving force. Compared with the electric-field, light simplifies the processing of device fabrications and realizes the remote control of the devices. To achieve light-driven capability, photosensitive LCs are necessary either by covalently embedding a photosensitive group into the molecular structures or doping some amount of photosensitive dye, such as azobenzene based materials, into the LC host. Generally, the latter is preferred due to the complicated molecular design, synthesis and purification of photosensitive LC over the former approach. The irradiation of light with certain wavelength (determined by the absorption band of photosensitive materials) triggers the reversible photoisomerization, resulting in the arrangement changes of the LC. Sunlight possesses a wide spectrum range covered from UV, visible to infrared band (0.15–4 μm) and provides sufficient energy as a light stimuli, therefore enabling the driving of LC with such a kind of sustainable energy, which is the most prominent advantage of light driven LC devices [68].

Sackmann carried out the pioneer research on light driven LC material in 1971. A light tunable reflection color in non-halide based azobenzene doped cholesteric mixture from 560 to 610 nm was demonstrated [69]. Although the tuning range was not wide, this work opened a new pathway of photo responsive LC. Subsequently, further elaborate studies focused on molecular design of the photosensitive chiral dopant with good solubility in LC, high initial HTP, large variation of HTP range and good fatigue resistance during light stimulation. A series of azobenzenophane based planar chiral dopants were synthesized [70]. The free rotation of naphthalene moiety contained in molecules is restricted, thereby giving rise to planar charality.

In addition, some systematically studies indicated that the initial HTP of photosensitive chiral dopant is usually related to the number of chiral group contained in the molecules [71, 72]. More chiral groups cause higher initial HTP, however the variation of HTP range is decreased in the meanwhile. In the recent years, some kinds of photosensitive chiral dopants with both high initial HTP and large variation of HTP range were synthesized, such as the isosorbide or isomannide-based [73] and a series of binaphthyl-based compounds [74].

Inspired by the developments on photosensitive chiral dopant, and combining with the specific optical properties result from the refractive periodic structure of CLC, several kinds of novel light driven photonic devices with wide tunable range were produced successively, mainly concentrated on tunable Bragg reflector, micro-patterned optical devices, and micro-fluidic active and passive optical elements. In this section, the presentations focus on the recent developments of the three aforementioned light driven photonic devices, from their mechanisms, and structures to optical performances.

14.4.1 Light Tunable CLC Bragg Reflector

Selective Bragg reflection is one of the most important characteristics of CLC which can be directly utilized in optical filters, reflective colorful displays and temperature sensors. The central wavelengths of reflection band are determined by pitch length of the CLCs. Band-shift of the conventional CLC during heating is a common phenomenon caused by the HTP variation of chiral dopant dissolved in LC. However, such temperature induced shifting is not significant, generally less than tens of nanometers, limited by the small changes on the solubility of chiral dopant in LC. As aforementioned, by doping the photosensitive chiral dopant, possessing wide HTP changes, thus the wide band-shift can be achieved by light irradiation. White et al. reported a dramatic wide band-shift of light tunable CLC reflector by doping an appropriate binaphthyl-based azobenzene axially chiral molecular switch (**QL1**) with a large initial HTP (53 μm^{-1}) in the nematic LC host and good solubility with LC (Fig. 14.19a). After irradiating the CLC containing 10 wt% of such chiral dopant for several hundreds of seconds, a red-shift of reflection band over 2000 nm, covering the visible red-color to NIR and even to the short-wave IR, was achieved. For accelerating the restoration process, polymer stabilization was adopted which significantly shortened the recovery time from several days to a few minutes [75].

Li's group carried out a series of molecular designs of the chiral molecular switch to improve the solubility of such materials with LC host, increase initial HTP values and widen variation of HTPs. It can be anticipated that good solubility and high initial HTP will reduce the amount of chiral dopant mixed into the LC, while the large variation of HTP will enable the wide tuning range of reflection band. Accordingly, another kind of binaphthyl-based chiral molecular switch was synthesized by replacing the flexible long alkyl groups of **QL1** to two rigid mesogenic groups (**QL2**, Fig. 14.19a). HTP of such chiral switch at initial state was 90 μm^{-1} in E7, and the

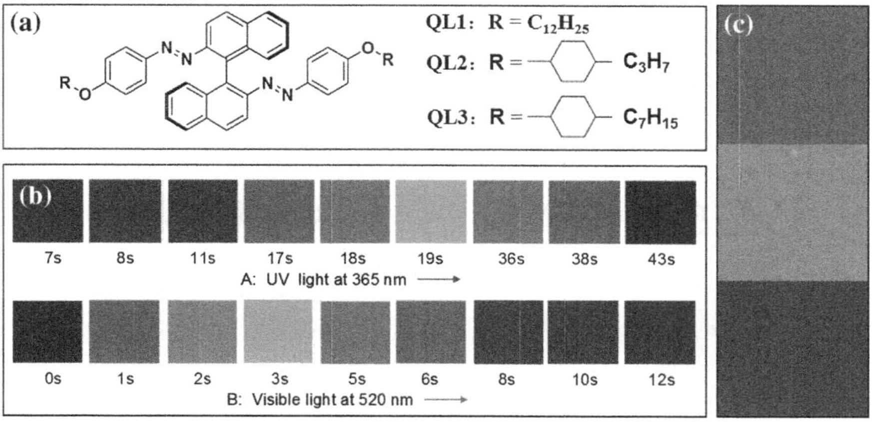

Fig. 14.19 Light driven Bragg reflector based on binaphthyl photosensitive chiral molecular switches. **a** Molecular structure of chiral switches; **b** RGB full color range tuned by UV and visible lights; **c** photostationary *red*, *green* and *blue* colors by light irradiation with 440, 450 and 550 nm wavelengths, respectively [75–77]. Reproduced with permission from [76, 77]. Copyright 2009 WILEY-VCH; Copyright 2010 The Royal Society of Chemistry; Copyright 2011 WILEY-VCH (Color figure online)

further experimental results indicated that a wide reversible band-shift, covering the full-color range, was realized by stimulating the photoresponsive CLC containing only 6.5 wt% of chiral switch with UV (365 nm) and visible (520 nm) light alternatively (Fig. 14.19b). The whole band-shift was completed within 1 min. Moreover, there was no detrimental effect of chiral switch on the stability and electric-tuning performance. A reflective image recorded by photomask exposure can be erased by applying an appropriate voltage pulse and reappeared by enhancing the voltage. The tuning among red, green and blue (RGB) reflections enables the color LCD driven by light [76]. However such RGB reflections were dynamic states, i.e., such states can't be maintained for sufficient long time, which is obviously one of the major problems for display. In view of this, further improvement and synthesis of chiral switch were conducted by lengthening the flexible alkyl group (**QL3**, Fig. 14.19a). After doping 6.0 wt% of **QL3** into nematic LC, E7, an exciting phenomenon was observed that RGB static reflections can be achieved by stimulating the mixture with different wavelengths. The red, green and blue reflections can be obtained by the irradiation of 440, 450 and 550 nm light, respectively (Fig. 14.19c) [77].

One of the most crucial problems of azobenzene based chiral switch is the spontaneous thermal relaxation, which causes the instability of reflection for the CLC reflector. Such problem can be resolved by using a kind of dithienylcyclopentene based axially chiral switch ((***S***,***S***)**-1**, Fig. 14.20a), which undergoes the photoisomerization between the ring-open and ring-closed isomers by UV (310 nm) or visible (550 nm) light irradiation. Both of the two photoisomers exhibit excellent fatigue resistance without thermal relaxation, enabling the stability of LC arrangement at any state in the process of light stimulation. In addition, the

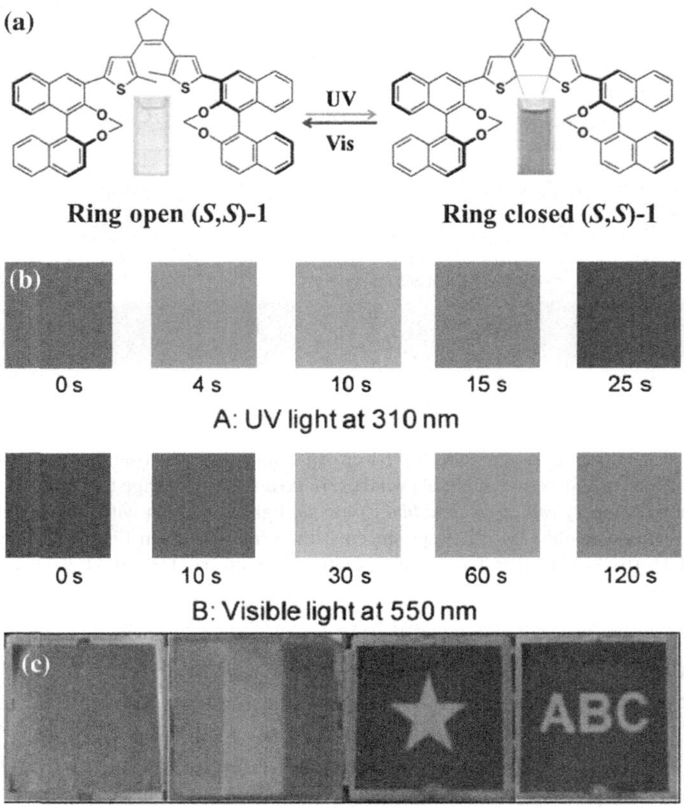

Fig. 14.20 Light tunable full color range based on thermo-stable chiral molecular switch, (*S*,*S*)-**1**. **a** Molecular structure of (*S*,*S*)-**1**; **b** full color reflection tuning by UV and visible light; **c** light pattern storage. Reproduced with permission from [78]. Copyright 2012 American Chemical Society

high HTP values at initial state and two photostationary states, PSS_{UV} and PSS_{vis}, assure the band-shift, covering full-color range (Fig. 14.20b), just by doping a small amount of chiral switch. The image storage was achieved as well owing to the lack of thermal relaxation of the material (Fig. 14.20c) [78].

Except for the light stimulation by UV and visible lights, LC also can be driven by NIR source after specific treatment of the materials. NIR light possesses a good penetrability but low irradiation energy, however utilizing NIR to trigger the photoisomerization of chiral switch is a challenging work [79–81]. A breakthrough was achieved very recently by loading a very small amount of upconversion nanoparticles (UCNPs) with core-shell nanostructure (Fig. 14.21a, b) in the photoresponsive CLC mixture composed of nematic LC E7 and chiral switch **QL3**. Upon the irradiation of NIR light, UCNPs absorbed NIR photons and transformed the upconverted excitation energy to chiral switch through the Förster energy transfer. The wavelength of upconverted emission was determined by the

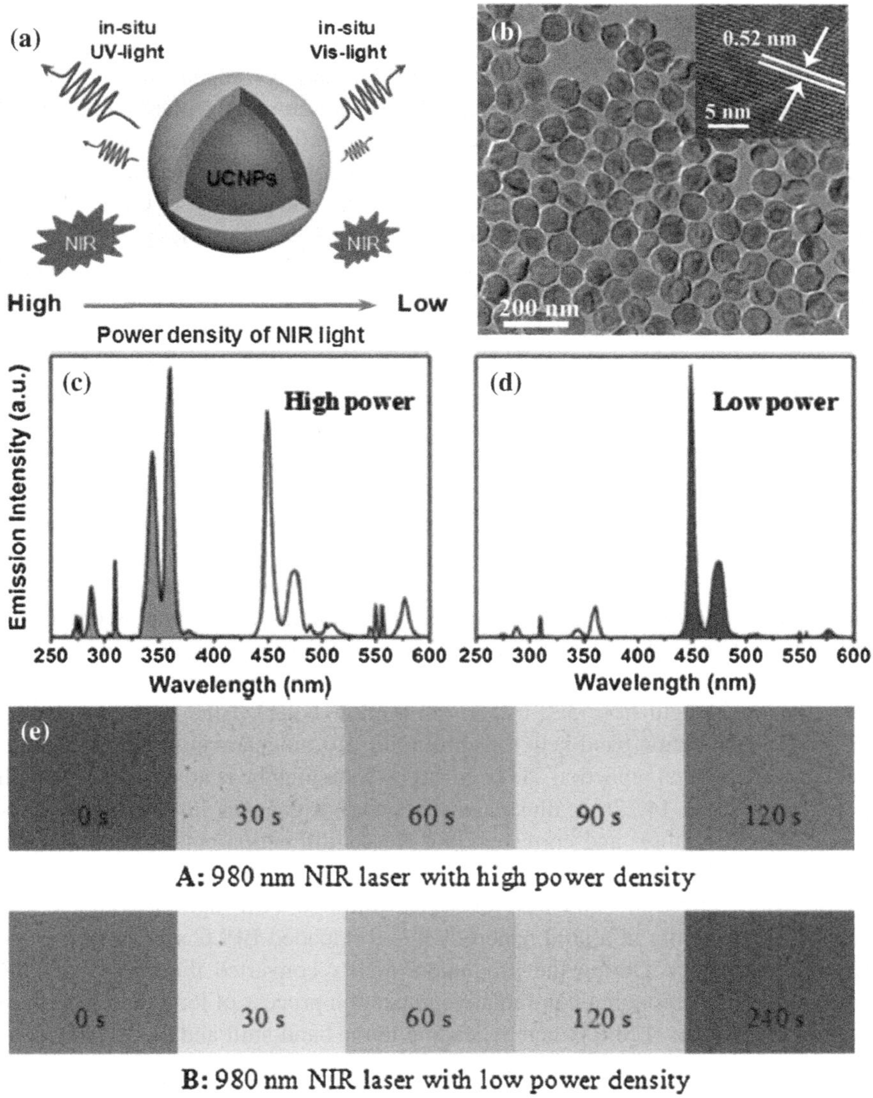

Fig. 14.21 NIR light tunable CLC reflector. **a** Illustration of the structure of UCNP; **b** TEM and high resolution TEM (the *inset*) images of the UCNPs; **c**, **d** emission spectra of UCNPs upon irradiation of 980 nm laser at high output power (2.0 W/mm^2, **c**) and low output power (0.15 W/mm^2, **d**); **e** full color reflection tunability by 980 nm laser with high power and low power. Reproduced with permission from [79]. Copyright 2014 American Society of Chemistry

irradiation energy of NIR (Fig. 14.21c, d). The experimental results showed that the reflection band blue-shifted from 435 nm (blue color) to 625 nm (red color) under the irradiation of 980-nm laser with high power density, however it was interesting that the blue-shift of reflection band occurred only by attenuating the power density

(Fig. 14.21e) [79]. That means the full-color reversible tuning can be achieved by only one light source with tunable power intensity.

In a high chiral system, LC molecules arrange in a three dimensional (3D) cubic nanostructure by self-organization, i.e., BP. BP presents the typical narrow band Bragg reflection as the photonic crystal due to its regular 3D structure with a period about several hundreds of nanometers. Such reflection band can be shifted into the visible band by modulating the concentration of chiral dopant. Generally, the reflection color purity of BP is higher than that of CLC, thus the BPLC reflector attracts more attentions. However, the tunability of the reflection of BP encountered great difficult because of the thermo-instability of such phase. The cubic structure is vulnerable to the disturbance from external field. Although a great many of efforts have been made to obtain the reflection color tuning of BPLC, the tuning range was still less than 100 nm. Li's group achieved a breakthrough on RGB full color tunable reflection by light stimulation. The chiral switch denoted as **QL4** (3.5 wt%) (Fig. 14.22a) was homogeneously mixed with conventional chiral dopants, S811 (29.5 wt%) and R811 (13.5 wt%), and nematic LC, E48 (53.5 wt%) and filled into a 7.5 μm-thick planar cell. As shown in Fig. 14.22b, the sample was maintained at 40 °C, exhibiting BPII with the reflection band at UV. After irradiating with 408 nm light for 2 s, the reflection band red-shifted to blue color, however the BPII phase was invariable confirmed through Kössel diffraction. Continuous irradiation led to the red-shift of reflection band to 520 nm, presenting green color, and followed by the phase transition from BPII to BPI accompanied by a jump of reflection wavelength from 520 to 560 nm (yellow color). After such wavelength jumping, the reflection band kept red-shifting to 710 nm, showing bright red color. The recovery of such reflection can be tuned by changing the irradiation wavelength to 532 nm. Figure 14.22b-A illustrated the structure changes in light tuning process, while the textures and corresponding Kössel diffraction patterns were shown respectively in Fig. 14.22b-B, b-C [82].

The wide reflection band-shift of BPLC through NIR light irradiation was realized very recently in a gold nanorods (GNRs) loaded BPLC system due to the photothermal effect. During the irradiation, GNRs converted the energy of NIR photons into heat through a non-radiative relaxation process of longitudinal surface plasmon resonances (LSPR), thereby leading to the band-shift and phase transitions from low temperature BPI to high temperature BPII, consequently achieving the wide reflection tuning by NIR source [83].

14.4.2 Light Manipulated Micro-patterned and Micro-fluidic Photonic Devices

Taking advantage of the light controllable reflection, various multi-color micro-structures can be conveniently recorded on the sample by photomask exposure. Kurihara et al. synthesized a kind of isomannide-based chiral compound

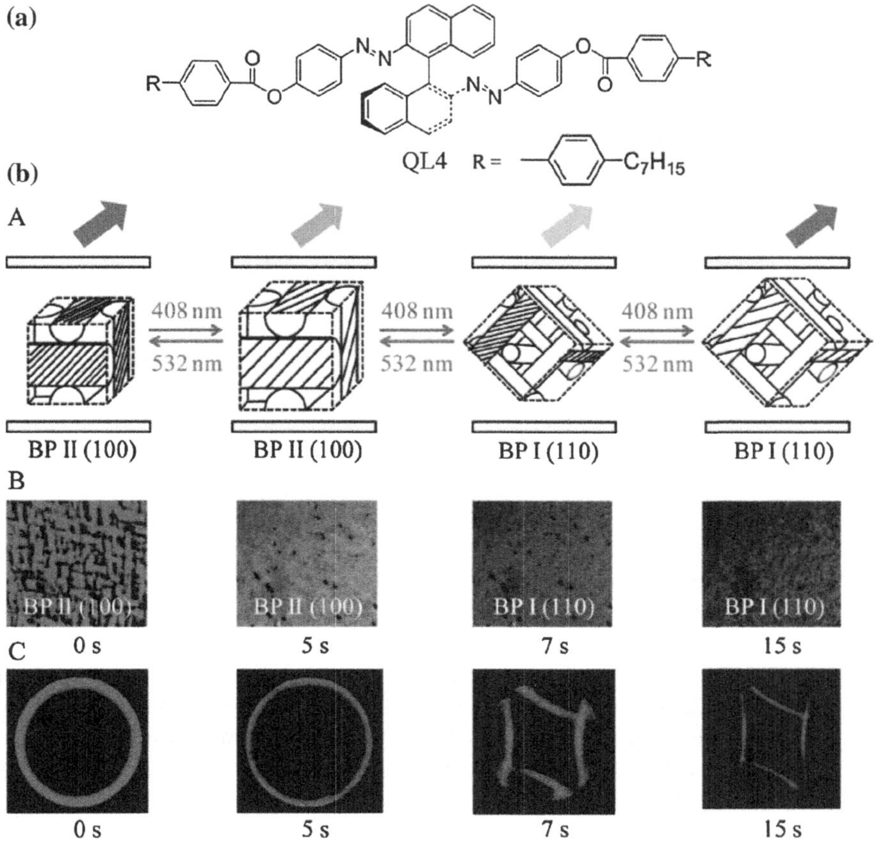

Fig. 14.22 Light tunable 3D cubic LC nanostructure. **a** Molecular structure of **QL4**; **b** illustration of structure changes (*A*) and corresponding optical texture and Kössel diffraction patterns (*B* and *C*) during light stimulation. Reproduced with permission from [82]. Copyright 2013 WILEY-VCH

with plural azobenzene groups (Fig. 14.23a) and mixed it with the other photo-insensitive chiral compound as the chiral agent doped into the conventional nematic LC, E44. The sample was confined in a planar cell, maintained at 25 °C and exposed with the UV lamp for different time, red, green, blue reflection colors can be obtained (Fig. 14.23b). Irradiating the sample through a patterned grey mask (inset of Fig. 14.23c) for 10 s, a RGB rectangle pattern (Fig. 14.23c) was recorded. The red, green and blue reflection colors were resulted from the different irradiation intensity after UV light passing through the grey pattern. Higher intensity caused longer reflection wavelength. The resolution of the color patterns, determined by the properties of materials, was estimated to be 70–100 μm, which was very close to the size of sub-pixel of TFT-LCD, therefore such RGB micro-pattern reflector was promising to be applied as a reflective color filter in LCD [73].

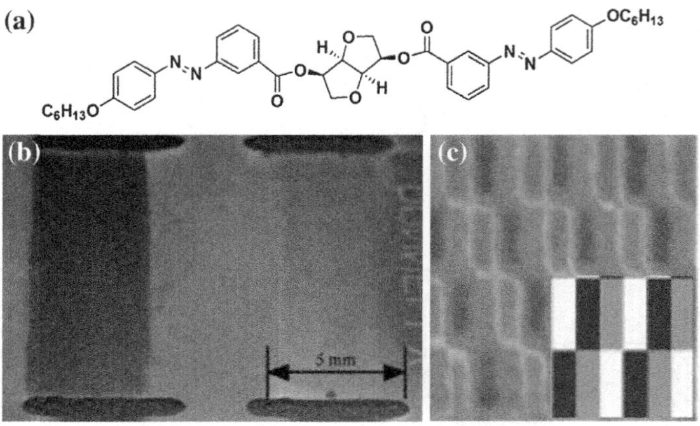

Fig. 14.23 Micro-pattern recording by light. **a** Molecular structure of chiral dopant; **b** R, G, B reflections at different areas; and **c** corresponding RGB micro-patterns exposed by *grey* photomask (*inset*). Reproduced with permission from [73]. Copyright 2005 WILEY-VCH

Similarly, as the size of pattern was decreased to several microns, some obvious interesting optical phenomenon can be observed. An elegant periodic amplitude-modulated stripe-pattern with the period of 10 μm was achieved by convenient holographic recording. As a probe beam impinged along the normal direction of the sample, the diffraction patterns can be received at both sides of the sample due to the formation of transmission and reflection gratings simultaneously. In the experiment, the photoresponsive CLC composed by 61.3 wt% nematic LC, E7, 38.1 wt% right-handed chiral dopant, CB15, and 0.6 wt% photosensitive dye, DR1, was confined in a 10 μm-thick planar cell. Herein, the content of chiral dopant should be carefully adjusted to make the transmission intensity of a circular polarized probe beam to be sensitive to the intensity of holographic beams. A 532 nm laser, used as the holographic exposure source, was divided into two beams with the equal intensity of 0.56 W/cm^2, impinged on the sample with the incident angle of 1.5° and −1.5°, respectively. In the meanwhile, a 632 nm He–Ne laser was used as the probe beam, incident along the normal direction of the sample as shown in Fig. 14.24a. An interference intensity modulation with the period of 10 μm was generated during the holographic exposure, leading to a higher transmittance at interference bright area, while a lower transmittance at interference dark area, and therefore forming an intensity periodic modulation of the probe beam at both of the transmission and reflection sides. Consequently, the diffraction was observed in both sides of the sample, i.e., forming the transmission and reflection grating simultaneously as illustrated in Fig. 14.24b [84].

Very recently, a charming selective diffraction grating based on nematic LC side-chain polymer doped with photosensitive chiral agent (Fig. 14.25a) was developed by irradiating the polymer film through periodic photomask. Helical pitch of the area undergone light exposure was elongated while at the area without exposure, the pitch length was maintained, and thus forming a periodic pitch modulation.

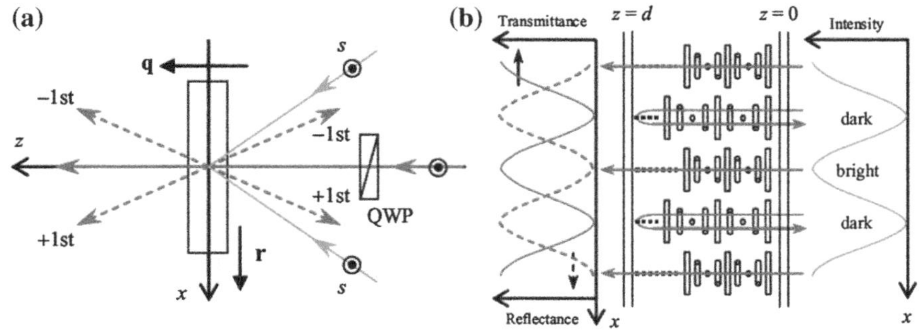

Fig. 14.24 Light induced transmission and reflection grating. **a** Optical setup of grating recording and detecting. '**q**' and '**r**' denote the direction of wavevector and rubbing direction, respectively; the polarization direction of recording and detecting beam are vertical to paper; 'QWP' denotes quarter waveplate; **b** mechanism for the formation of grating. Reproduced with permission from [84]. Copyright 2009 American Institute of Physics

The reflection band at the masked region was maintained at the original position, while reflection band at exposed region red-shifted to the other position without overlap with the original one. As a light with the wavelength located in either of the reflection band, a light amplitude modulation formed and thus the diffraction appeared (Fig. 14.25b). Band-shift at exposed region can be controlled by adjusting the exposure time, and therefore a light tunable wavelength-selective diffraction grating was achieved. Figure 14.25c shows the dual-reflection color stripe patterns and the corresponding selected diffraction. In addition, the structure of such light tunable grating was very stable due to the adoption of LC polymer [85].

Recently, functionalized LC microdroplets or microshells confined in a polymer spherical shell and uniformly dispersed in a solution seems promising for a novel micro-scale photonic application [86, 87]. Monodispersed photoresponsive CLC microdroplets were fabricated by dispersing CLC into a 10 wt% poly(vinyl alcohol) (PVA) aqueous solution and appropriately controlling the flow rates of PVA solution and CLC. The CLC droplets with the diameter about 50 μm were collected in a 125 μm-thick transparent glass cell. Chiral switch used herein was **(*S*,*S*)-a** (Fig. 14.20a) as aforementioned. As these microdroplets were far apart from each other, some reflection spots caused by central reflection of each droplet were observed (Fig. 14.26a). The color of these spots can be reversibly tuned with the irradiation of UV and visible light, and blue (Fig. 14.26a), green (Fig. 14.26b), red (Fig. 14.26c) reflection colors were obtained due to the variation of HTP during the irradiation. When a single layer of microdroplets were packed densely to form a hexagonal pattern, a peculiar circular arranged 'lit fire-cracker' was exhibited on account of the photonic cross-communication between two adjacent microdroplets (Fig. 14.26d). If these microdroplets were densely packed to two layers as depicted (Fig. 14.26e), different reflection pattern was observed under microscope, moreover, such pattern was related to the incident direction of the light, i.e., from top to bottom or opposite (top–bottom, Fig. 14.26f; bottom–top, Fig. 14.26g) [87].

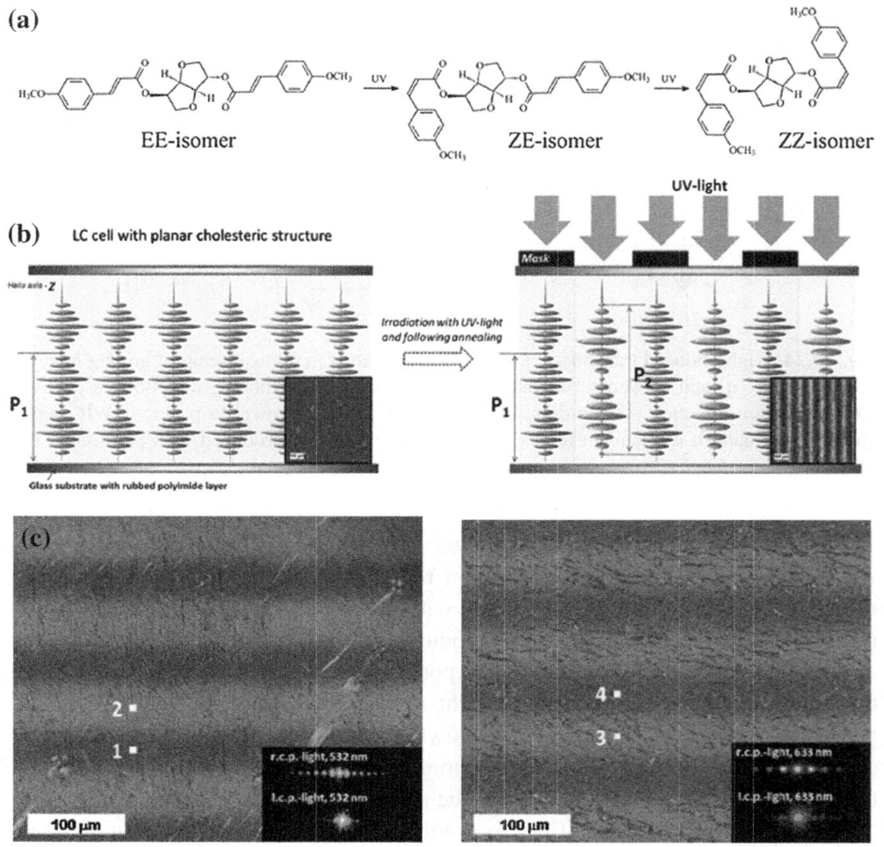

Fig. 14.25 Light tunable color selective grating. **a** Molecular structure of chiral dopant; **b** illustration for the formation of grating; picture of *insets* are optical textures at uniform planar state and the planar state with periodic changes of pitch length; **c** optical texture of gratings which is applicable to *green light* and *red light* respectively; their corresponding diffraction patterns are shown in the insets. Reproduced with permission from [85]. Copyright 2015 American Chemistry Society (Color figure online)

CLC also can be adopted to generate the mirror-less laser after mixing with a small amount of fluorescent dye due to its periodic photonic structure. If the reflection band of CLC matches with the emission spectrum of the dye, laser emission at the reflection band edge is observed owing to the maximum density of state of photons at this position. The reflection band can be tuned by external field, such as electric-field, temperature and the light irradiation, which enables the tunability of laser wavelength, forming the tunable laser. Generally, three manners are always adopted to realize the wavelength tuning: (i) electric-field manipulation; (ii) temperature manipulation; and (iii) mechanical manipulation. Actually, the following two consume electric energy as well due to the temperature controller and

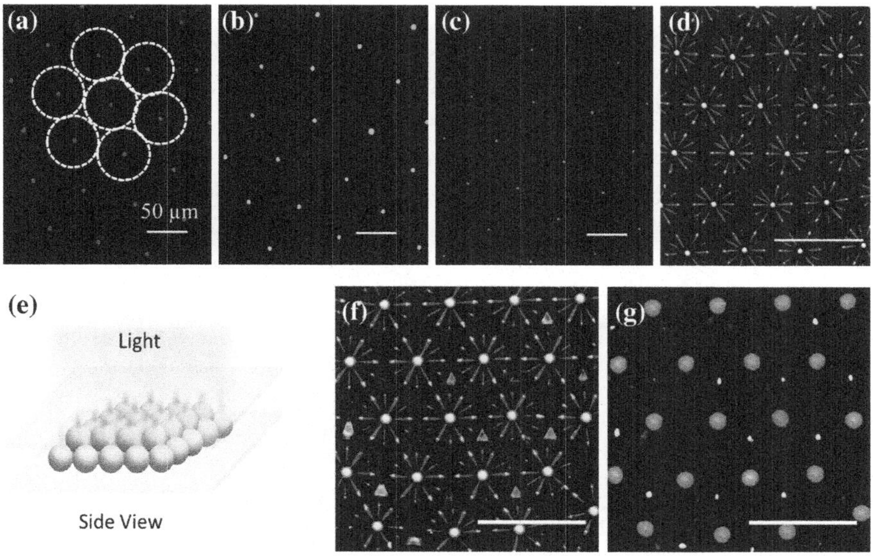

Fig. 14.26 Light controllable LC microdroplets and their photonic cross-communication. **a–c** Hexagonal reflection *dots arrays* showing RBG full color controlling; **d** photonic cross-communication between droplets when their distance is closer; **e** double-layer hexagonal packing of the microdroplets; **f**, **g** reflection patterns as the light irradiates from up-layer to the bottom (**f**) and from the bottom-layer to the up (**g**). Adapted with permission from [87]. Copyright 2015 WILEY-VCH

the mechanical devices which work by applying electric-field. Therefore, the electric-energy saving light tunable laser is always desirable.

By virtue of microfuidic technology, the CLC with a small amount of laser dye, 4-dicyanomethylene-2-methyl-6-(*p*-(dimethylamino)styryl)-4H-pyran (DCM), was used to fabricate microshells in a PVA solution. The direction of helical axis was along the radius of the shell owing to the planar anchoring effect of PVA (Fig. 14.27b). As mentioned above, the reflection band-gap can be shifted under the light stimulation since the existence of photosensitive chiral switch, **QL5** (Fig. 14.27a), in the laser dye-doped CLC. When the reflection band-edge overlapped with the fluorescence emission range of laser dye, under the pumping of nanosecond-pulsed laser (λ = 532 nm), a band-edge laser was emitted along the helical axes, thus forming the omnidirectional lasing as the red outer rings around the microshell shown in the micrograph taken through the microscope (Fig. 14.27c). Furthermore, the wavelength of emitted laser can be reversibly tuned in the range from ~590 to ~640 nm [88].

In addition, a convenient and economic fabrication of CLC band-gap laser was demonstrated very recently. A homogeneous mixture contained with photorespon-sive CLC, laser dye (DCM) and 15 wt% PVA aqueous solution was painted on a smooth glass with the doctor blade to form the wet-film with CLC droplets randomly dispersed in PVA matrix. The wet-film was maintained in room temperature to evaporate the water, accompanied by the shrinkage of the film thickness which

Fig. 14.27 Light tunable laser emission in CLC microshells. **a** Molecular structure of chiral switch **QL-5**; **b** structure illustration of photoresponsive CLC microshell as a water/oil/water double emulsion; **c** CLC microshells observed using cross polarized confocal microscopy. Reproduced with permission from [88]. Copyright 2014 WILEY-VCH

enforced the random dispersed CLC uniformly arranged with their helical axes vertical to the substrate, thus forming the dry-film with the reflection color shown in the droplets (Fig. 14.28a). Such reflection color can be tuned reversibly with light (Fig. 14.28b, c). Similar as the aforementioned, when the reflection band-gap

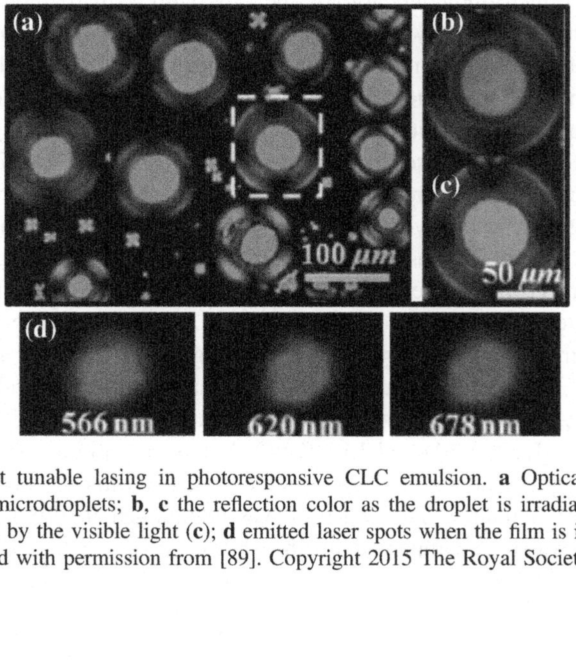

Fig. 14.28 Light tunable lasing in photoresponsive CLC emulsion. **a** Optical texture of the dispersed CLC microdroplets; **b**, **c** the reflection color as the droplet is irradiated by UV light (**b**) and followed by the visible light (**c**); **d** emitted laser spots when the film is irradiated by UV light. Reproduced with permission from [89]. Copyright 2015 The Royal Society of Chemistry

overlapped with the emission band of laser dye, lasing at band-edge occurred by pumping with a second-harmonic Q-switched Nd:YAG pulsed source (output wavelength: 532 nm). Contemporarily, a wide wavelength tunability of the emitted laser from 566 to 678 nm was achieved by the irradiation of UV light. Figure 14.28d showed the emitted laser spots with the wavelength of 566, 620 and 678 nm. Apart from the glass substrate, such material can be painted on Poly(methyl methacrylate) (PMMA), forming a flexible tunable laser device [89].

14.5 Conclusions and Outlook

This chapter provides a glimpse of devices for energy-saving applications involving self-organized chiral liquid crystals. Around this topic, an overview of several typical energy-saving chiral liquid crystal devices are given, more specifically focusing on (i) bistable chiral liquid crystal devices; (ii) chiral liquid crystal devices for solar energy manipulation and photovoltaic driven chiral liquid crystal devices; and (iii) wide tunable novel chiral liquid crystal photonic devices, and demonstrated from the materials, mechanisms, optical performances to their potential applications. Liquid crystal is a smart functional material and exhibits a strong compatibility with other electronic or photonic elements. Due to their extremely high sensitivity in response to weak external forces, the liquid crystal devices have always been regarded as typical energy-saving devices and attract more attentions. In addition, chiral liquid crystal possesses a strong self-organization, which causes the fabrication of devices easier, more economic and convenient. The above advantages enable the higher efficient, better performing, more intelligent and lower cost energy-saving chiral liquid crystal devices.

However, some prominent drawbacks of the present chiral liquid crystal devices are impeding their further widespread application. The ambient temperature, UV-absorption of molecules and the flow disturbance caused by fluidity lead to obvious degradation of the devices. Through the developments in the recent decade, many efforts have been contributed to improve these drawbacks and make sure better performance of devices, and undeniably, these works are really effective, but some problems still persist. Looking forward, the role of chiral liquid crystal devices on energy-saving endeavor is self-evident, and therefore their diverse applications in the future can be expected. The advancements of chiral liquid crystal devices toward energy-saving are significant and exciting for the application on next generation displays and beyond. The demonstrations of energy-saving liquid crystal devices show that there is ample scope for further developments. The gradual applications of such kind of smart, low energy consuming devices will continually promote the future relevant investigations go deeper and may create numerous possibilities in an energy efficient world.

Acknowledgements The preparation of this chapter benefited from the support to Quan Li by US Air Force Office of Scientific Research (AFOSR), US Department of Energy (DOE), US Army

Research Office (ARO), US Department of Defense Multidisciplinary University Research Initiative (DoD MURI), US National Aeronautics and Space Administration (NASA), US National Science Foundation (NSF), Ohio Third Frontier, and CSC Scholarship to Zhigang Zheng.

References

1. M. Schadt, W. Helfrich, Voltage-dependent optical activity of a twisted nematic liquid crystal. Appl. Phys. Lett. **18**, 127–128 (1971)
2. J.L. Fergason, T.R. Taylor, T.B. Harsch, Liquid crystals and their applications. Electron. Tech. **1**, 41–50 (1970)
3. D. Demus, J.W. Goodby, G.W. Gray, H.W. Spiess, V. Vill (eds.), *Handbook of liquid crystals: low molecular weight liquid crystals I* (Wiley-VCH, Weinheim, 1998)
4. J.E. Bigelow, R.A. Kashnow, Observations of a bistable twisted nematic liquid-crystal effect. IEEE Trans. Electron Devices **22**, 730–733 (1975)
5. I. Dozov, M. Nobili, G. Durand, Fast bistablenematic display using monostable surface switching. Appl. Phys. Lett. **70**, 1179–1181 (1997)
6. R. Barberi, M. Giocondo, J. Li, R. Bartolino, I. Dozov, G. Durand, Fast bistablenematic display with grey scale. Appl. Phys. Lett. **71**, 3495–3497 (1997)
7. D.W. Berreman, W.R. Heffner, New bistable cholesteric liquid-crystal display. Appl. Phys. Lett. **37**, 109–111 (1980)
8. Z.L. Xie, H.S. Kwok, Reflective bistable twisted nematic liquid crystal display. Jpn. J. Appl. Phys. **37**, 2572–2575 (1998)
9. Z.L. Xie, H.J. Gao, S.Y. Xu, H.S. Kwok, Optimization of reflective bistable twisted nematic liquid crystal displays. J. Appl. Phys. **86**, 2373–2376 (1999)
10. Z.L. Xie, H.S. Kwok, Optimization of reflective bistable twisted nematicdisplaays with retardation compensation. J. Appl. Phys. **88**, 1718–1721 (2000)
11. S.H. Lee, K.H. Park, T.H. Yoon, J.C. Kim, Bistable chiral-splay nematic liquid crystal device using horizontal switching. Appl. Phys. Lett. **82**, 4215–4217 (2003)
12. C.G. Jhun, C.P. Chen, U.J. Lee, S.R. Lee, T.H. Yoon, J.C. Kim, Tristable liquid crystal display with memory and dynamic operating modes. Appl. Phys. Lett. **89**, 123507 (2006)
13. J.S. Gwag, Y.J. Lee, M.E. Kim, J.H. Kim, J.C. Kim, T.H. Yoon, Viewing angle control mode using nematic bistability. Opt. Express **16**, 2663–2669 (2008)
14. T. Kim, J.H. Lee, T.H. Yoon, S.W. Choi, Reflective dual-mode liquid crystal display possessing low power consumption and high contrast ratio under ambient light. Opt. Express **20**, 15522–15529 (2012)
15. D.H. Song, J.H. Lee, S.J. Lee, S. Kim, S. Lim, S.T. Shin, J.C. Kim, T.H. Yoon, Bistable switching of twist direction in a twisted-nematic liquid crystal cell. Appl. Phys. Lett. **97**, 063501 (2010)
16. K.H. Kim, H.J. Kim, D.H. Song, B.H. Cheong, H.Y. Choi, S.T. Shin, J.C. Kim, T.H. Yoon, Switching of liquid-crystal devices between reflective and transmissive modes using long-pitch cholesteric liquid crystals. Opt. Lett. **35**, 3504–3506 (2010)
17. F.S.Y. Yeung, H.S. Kwok, Truly bistable twisted nematic liquid crystal display using photoalignment technology. Appl. Phys. Lett. **83**, 4291–4293 (2003)
18. B. Park, Y. Jung, H.H. Choi, H.K. Hwang, Y. Kim, S. Lee, S.H. Jang, M. Kakimoto, H. Takezoe, Thermal and optical stabilities of photoisomerizable polyimide layers for nematic liquid crystal alignments. Jpn. J. Appl. Phys. **37**, 5663–5668 (1998)
19. J.I. Niitsuma, M. Yoneya, H. Yokoyama, Contact photolithographic micropatterning for bistable nematic liquid crystal displays. Appl. Phys. Lett. **92**, 241120 (2008)
20. J.K. Kim, F. Araoka, S.M. Jeong, S. Dhara, K. Ishikawa, H. Takezoe, Bistable device using anchoring transition of nematic liquid crystals. Appl. Phys. Lett. **95**, 063505 (2009)

21. J.K. Kim, K.V. Le, S. Dhara, F. Araoka, K. Ishikawa, H. Takezoe, Heat-driven and electric-field-driven bistable devices using dye-doped nematic liquid crystals. J. Appl. Phys. **107**, 123108 (2010)
22. M.H. Lu, Bistable reflective cholesteric liquid crystal display. J. Appl. Phys. **81**, 1063–1066 (1997)
23. R. Bao, C.M. Liu, D.K. Yang, Smart bistable polymer stabilized cholesteric texture light shutter. Appl. Phys. Express **2**, 112401 (2009)
24. J.S. Hsu, B.J. Liang, S.H. Chen, Bistable chiral tilted-homeotropicnematic liquid crystal cells. Appl. Phys. Lett. **85**, 5511–5513 (2004)
25. B.J. Liang, C.L. Lin, Crucial influence on d/p range in bistable chiral tilted-homeotropicnematic liquid crystal cells. J. Appl. Phys. **102**, 124504 (2007)
26. I.A. Yao, C.L. Yang, C.J. Chen, J.P. Pang, S.F. Liao, J.H. Li, J.J. Wu, Bistability of splay and π twist states in a chiral-doped dual frequency liquid crystal cell. Appl. Phys. Lett. **94**, 071104 (2009)
27. Y.C. Hsiao, C.Y. Tang, W. Lee, Fast-switching bistable cholesteric intensity modulator. Opt. Express **19**, 9744–9749 (2011)
28. P. Kumar, S.W. Kang, S.H. Lee, Advanced bistable cholesteric light shutter with dual frequency nematic liquid crystal. Opt. Mater. Express **2**, 1121–1134 (2012)
29. J. Ma, L. Shi, D.K. Yang, Bistable polymer stabilized cholesteric texture light shutter. Appl. Phys. Express **3**, 021702 (2009)
30. C.Y. Huang, K.Y. Fu, K.Y. Lo, M.S. Tsai, Bistable transflective cholesteric light shutters. Opt. Express **11**, 560–565 (2003)
31. C.Y. Wu, Y.H. Zou, I. Timofeev, Y.T. Lin, V.Y. Zyryanov, J.S. Hsu, W. Lee, Tunable bi-functional photonic device based on one-dimensional photonic crystal infiltrated with a bistable liquid-crystal layer. Opt. Express **19**, 7349–7355 (2011)
32. Y. C. Hsiao, Y. H. Zou, I. V. Timofeev, V. Ya. Zyrayanov, W. Lee, Spectral modulation of a bistable liquid-crystal photonic structure by the polarization effect. Opt. Mater. Express **3**, 821–828 (2013)
33. H.T. Wang, I.V. Timofeev, K. Chang, V. Ya, Zyryanov, and W. Lee, Tunable narrow-bandpass filter based on an asymmetric photonic bandgap structure with a dual-mode liquid crystal. Opt. Express **22**, 15097–15103 (2014)
34. Y.C. Hsiao, C.Y. Wu, C.H. Chen, V.Y. Zyryanov, W. Lee, Electro-optical device based on photonic structure with a dual-frequency cholesteric liquid crystal. Opt. Lett. **36**, 2632–2634 (2011)
35. Y.H. Lee, K.C. Huang, W. Lee, C.Y. Chen, Low-power displays with dye-doped bistable chiral-tilted homeotropic nematic liquid crystals. J. Disp. Tech. **10**, 1106–1109 (2014)
36. C.T. Wang, H.C. Jau, T.H. Lin, Bistable cholesteric blue phase liquid crystal using thermal hysteresis. Opt. Mater. **34**, 248–250 (2011)
37. C.T. Wang, H.Y. Liu, H.H. Cheng, T.H. Lin, Bistable effect in the liquid crystal blue phase. Appl. Phys. Lett. **96**, 041106 (2010)
38. A. Tiribocchi, G. Gonnella, D. Marenduzzo, E. Orlandini, F. Salvadore, Bistable defect structures in blue phase devices. Phys. Rev. Lett. **107**, 237803 (2011)
39. Z.G. Zheng, D. Zhang, X.W. Lin, G. Zhu, W. Hu, D. Shen, Y.Q. Lu, Bistable state in polymer stabilized blue phase liquid crystal. Opt. Mater. Express **2**, 1353–1358 (2012)
40. N. Gheorghiu, J.L. West, A.V. Glushchenko, M. Mitrokhin, Patterned field induced polymer walls for smectic a bistable flexible displays. Appl. Phys. Lett. **88**, 263511 (2006)
41. E.A. Buyuktanir, N. Gheorghiu, J.L. West, M. Mitrokhin, B. Holter, A. Glushchenko, Field-induced polymer wall formation in a bistable smectic-A liquid crystal display. Appl. Phys. Lett. **89**, 031101 (2006)
42. T.H. Lin, H.C. Jau, S.Y. Hung, H.R. Fuh, A.Y.G. Fuh, Photoaddressable bistable reflective liquid crystal display. Appl. Phys. Lett. **89**, 021116 (2006)
43. C.T. Wang, H.C. Jau, T.H. Lin, Optically controllable bistable reflective liquid crystal display. Opt. Lett. **37**, 2370–2372 (2012)

44. C.T. Wang, Y.C. Wu, T.H. Lin, Photo-switchable bistable twisted nematic liquid crystal optical switch. Opt. Express **21**, 4361–4366 (2013)
45. A.Y.G. Fuh, Z.H. Wu, K.T. Cheng, C.K. Liu, Y.D. Chen, Direct optical switching of bistable cholesteric textures in chiral azobenzene-doped liquid crystals. Opt. Express **21**, 21840–21846 (2014)
46. C.W. Chen, C.C. Li, H.C. Jau, C.H. Lee, C.T. Wang, T.H. Lin, Bistable light-driven π phase switching using a twisted nematic liquid crystal film. Opt. Express **22**, 12133–12138 (2014)
47. R. Baetens, B.P. Jelle, A. Gustavsen, Properties, requirements and possibilities of smart windows for dynamic daylight and solar energy control in buildings: a state-of -the-art review. Sol. Energy Mater. Sol. Cells **94**, 87–105 (2010)
48. M. Mitov, Cholesteric liquid crystals with a broad ligght reflection band. Adv. Mater. **24**, 6260–6276 (2012)
49. C. Binet, M. Mitov, M. Mauzac, Switchable broadband light reflection in polymer-stabilized cholesteric liquid crystals. J. Appl. Phys. **90**, 1730–1734 (2001)
50. Y. Gao, W. Yao, J. Sun, H. Zhang, Z. Wang, L. Wang, D.K. Yang, L.Y. Zhang, H. Yang, A novel soft matter composite material for energy-saving smart windows: from preparation to device application. J. Mater. Chem. A **3**, 10738–10746 (2015)
51. S. Relaix, C. Bourgerette, M. Mitov, Broadband reflective liquid crystalline gels due to the ultraviolet light screening made by the liquid crystal. Appl. Phys. Lett. **89**, 251907 (2006)
52. M. Mitov, E. Nouvet, N. Dessaud, Polymer-stabilized cholesteric liquid crystals as switchable photonic broad bandgaps. Eur. Phys. J. E **15**, 413–419 (2004)
53. H. Khandelwal, R.C.G.M. Loonen, J.L.M. Hensen, M.G. Debije, A.P.H.J. Schenning, Electrically switchable polymer stabilised broadband infrared reflectors and their potential as smart windows for energy saving in buildings. Sci. Rep. **5**, 11773 (2015)
54. R. Guo, K. Li, H. Cao, X. Wu, G. Wang, Z. Cheng, F. Wang, H. Zhang, H. Yang, Chiral polymer networks with a broad reflection band achieved with varying temperature. Polymer **51**, 5990–5996 (2009)
55. W. Hu, L. Zhang, H. Cao, L. Song, H. Zhao, Z. Yang, Z. Cheng, H. Yang, L. Guo, Electro-optical study of chiral nematic liquid crystal/chiral ionic liquid composites with electrically controllable selective reflection characteristics. Phys. Chem. Chem. Phys. **12**, 2632–2638 (2010)
56. H. Nemati, S. Liu, R.S. Zola, V.P. Tondiglia, K.M. Lee, T.J. White, T.J. Bunning, D.K. Yang, Mechanism of electrically induced photonic band gap broadening in polymer stabilized cholesteric liquid crystals with negative dielectric anisotropies. Soft Matter **11**, 1208–1213 (2015)
57. S. Relaix, M. Mitov, The effect of geometric and electric constraints on the performance of polymer-stabilized cholesteric liquid crystals with a double-handed circularly polarized light reflection band. J. Appl. Phys. **104**, 033539 (2008)
58. H. Khandelwal, R.C.G.M. Loonen, J.L.M. Hensen, A.P. H.J. Schenning, M.G. Debije, Application of broadband infrared reflector based on cholesteric liquid crystal polymer bilayer film to windows and its impact on reducing the energy consumption in buildings. J. Mater. Chem. A **2**, 14622–14627 (2014)
59. B.H. Yu, J.W. Huh, K.H. Kim, T.H. Yoon, Light shutter using dichroic-dye-doped long-pitch cholesteric liquid crystals. Opt. Express **21**, 29332–29337 (2015)
60. H.K. Kwon, K.T. Lee, K. Hur, S.H. Moon, M.M. Quasim, T.D. Wilkinson, J.Y. Han, H. Ko, I. K. Han, B. Park, B.K. Min, B.K. Ju, S.M. Morris, R.H. Friend, D.H. Ko, Optically switchable smart windows with integrated photovoltaic devices. Adv. Energy Mater. **5**, 1401347 (2015)
61. D. Cupelli, F.P. Nicoletta, S. Manfredi, M. Vivacqua, P. Formoso, G.D. Filpo, G. Chidichimo, Self-adjusting smart windows based on polymer-dispersed liquid crystals. Sol. Energy Mater. Sol. Cells **93**, 2008–2012 (2009)
62. H.H. Khaligh, K. Liew, Y. Han, N.M. Abukhdeir, I.A. Goldthorpe, Silver nanowire transparent electrodes for liquid crystal-based smart windows. Sol. Energy Mater. Sol. Cells **132**, 337–341 (2015)

63. R. Zhu, A. Kumar, Y. Yang, Polarizing organic photovoltaics. Adv. Mater. **23**, 4193–4198 (2011)
64. Y.H. Huh, J.C. Shin, Y.C. Kim, B. Park, Reflective type solar-LCDs by using polarizing polymer solar cells. Opt. Express **20**, A278–A286 (2012)
65. Y.H. Huh, B. Park, Power generating reflective-type liquid crystal displays using a reflective polariser and a polymer solar cell. Sci. Rep. **5**, 11558 (2015)
66. C.Y. Chen, Y.L. Lo, Integration of a-Si: H solar cell with novel twist nematic liquid crystal cell for adjustable brightness and enhanced power characteristics. Sol. Energy Mater. Sol. Cells **93**, 1268–1275 (2009)
67. M.G. Debije, Solar energy collectors with tunable transmission. Adv. Funct. Mater. **20**, 1498–1502 (2010)
68. H.K. Bisoyi, Q. Li, Light-directing chiral liquid crystal nanostructures: From 1D to 3D. Acc. Chem. Res. **47**, 3184–3195 (2014)
69. E. Sackmann, Photochemically induced reversible color changes in cholesteric liquid crystals. J. Am. Chem. Soc. **93**, 7088–7090 (1971)
70. M. Mathews, N. Tamaoki, Planar chiral azobenzenophanes as chiroptic switches for photon mode reversible reflection color control in induced chiral nematic liquid crystals. J. Am. Chem. Soc. **130**, 11409–11416 (2008)
71. N. Tamaoki, Cholesteric liquid crystals for color information technology. Adv. Mater. **13**, 1135–1147 (2001)
72. M. Brehmer, J. Lub, P. van de Witte, Light-induced color change of cholesteric copolymers. Adv. Mater. **10**, 1438–1441 (1998)
73. T. Yoshioka, T. Ogata, T. Nonaka, M. Moritsugu, S.N. Kim, S. Kurihara, Reversible-photon-mode full-color display by means of photochemical modulation of a helically cholesteric structure. Adv. Mater. **17**, 1226–1229 (2005)
74. Y. Wang, Q. Li, Light-driven chiral molecular switches or motors in liquid crystals. Adv. Mater. **24**, 1926–1945 (2012)
75. T.J. White, R.L. Bricker, L.V. Natarajan, N.V. Tabiryan, L. Green, Q. Li, T.J. Bunning, Phototunableazobenzene cholesteric liquid crystals with 2000 nm range. Adv. Funct. Mater. **19**, 3484–3488 (2009)
76. J. Ma, Y. Li, T. White, A. Urbas, Q. Li, Light-driven nanoscale chiral molecular switch: reversible dynamic full range color phototuning. Chem. Comm. **46**, 3463–3465 (2010)
77. Q. Li, Y. Li, J. Ma, D.K. Yang, T.J. White, T.J. Bunning, Directing dynamic control of red, green, and blue reflection enabled by a light-driven self-organized helical superstructure. Adv. Mater. **23**, 5069–5073 (2011)
78. Y. Li, A. Urbas, Q. Li, Reversible light-directed red, green, and blue reflection with thermal stability enabled by a self-organized helical superstructure. J. Am. Chem. Soc. **134**, 9573–9576 (2012)
79. L. Wang, H. Dong, Y. Li, C. Xue, L.D. Sun, C.H. Yan, Q. Li, Reversible near-infrared light directed reflection in a self-organized helical superstructure loaded with upconversion nanoparticles. J. Am. Chem. Soc. **136**, 4480–4483 (2014)
80. K. Gutierrez-Cuevas, L. Wang, C. Xue, G. Singh, S. Kumar, A. Urbas, Q. Li, Near infrared light-driven liquid crystal phase transition enabled by hydrophobic mesogen grafted plasmonic gold nanorods. Chem. Comm. **51**, 9845–9848 (2015)
81. L. Wang, H. Dong, Y. Li, R. Liu, Y.F. Wang, H.K. Bisoyi, L.D. Sun, C.H. Yan, Q. Li, Luminescence-driven reversible handedness inversion of self-organized helical superstructures enabled by a novel near-infrared light nano transducer. Adv. Mater. **27**, 2065–2069 (2015)
82. T.H. Lin, Y. Li, C.T. Wang, H.C. Jau, C.W. Chen, C.C. Li, H.K. Bisoyi, T.J. Bunning, Q. Li, Red, green and blue reflections enabled in an optically tunable self-organized 3D cubic nanostructured thin film. Adv. Mater. **25**, 5050–5054 (2013)
83. L. Wang, K.G. Gutierrez-Cuevas, H.K. Bisoyi, J. Xiang, G. Singh, R.S. Zola, S. Kumar, O.D. Lavrentovich, A. Urbas, Q. Li, NIR light-directing self-organized 3D photonic suuperstructures loaded with anisotropic plasmonic hybrid nanorods. Chem. Comm. **51**, 15039–15042 (2015)

84. T. Sasaki, A. Emoto, T. Shioda, H. Ono, Transmission and reflection phase gratings formed in azo-due-doped chiral nematic liquid crystals. Appl. Phys. Lett. **94**, 023303 (2009)
85. A. Ryabchun, A. Bobrovsky, Y. Gritsai, O. Sakhno, V. Shibaev, J. Stumpe, Stable selective gratings in LC polymer by photoinduced helix pitch modulation. ACS Appl. Mater. Interfaces **7**, 2554–2560 (2015)
86. J.H. Noh, H.L. Liang, I. Drevensek-Olenik, P.F. Jan, Lagerwall, Tunable multicoloured patterns from photonnic cross-communication between cholesteric liquid crystal droplets. J. Mater. Chem. C **2**, 806–810 (2014)
87. J. Fan, Y. Li, H.K. Bisoyi, R.S. Zola, D.K. Yang, T.J. Bunning, D.A. Weitz, Q. Li, Light-directing omnidirectional circularly polarized reflection from liquid-crystal droplets. Angew. Chem. Int. Ed. **54**, 2160–2164 (2015)
88. L. Chen, Y. Li, J. Fan, H.K. Bisoyi, D.A. Weitz, Q. Li, Photoresponsivemonodisperse cholesteric liquid crystalline microshells for tunable omnidirectional lasing enabled by a visible light-driven chiral molecular switch. Adv. Opt. Mater. **2**, 845–848 (2014)
89. Z.G. Zheng, B.W. Liu, L. Zhou, W. Wang, W. Hu, D. Shen, Wide tunable lasing in photoresponsive chiral liquid crystal emulsion. J. Mater. Chem. C **3**, 2462–2470 (2015)

Chapter 15
Nanomaterials for the Production of Biofuels

Sudipta De and Rafael Luque

Abstract Decreasing fossil fuel resources has forced the mankind to explore alternative sources to make up the gap between energy demand and its production. Lignocellulosic biomass is currently considered as one of the best solutions for this consequence, since it is highly abundant and has no competition with food crops. This chapter discloses some efficient catalytic approaches for the conversion of lignocellulose feedstocks into liquid hydrocarbon fuels via deoxygenation processes. Different catalytic factors of nanomaterials including composition, functionalities, and effect of catalyst support in the deoxygenation process are critically discussed.

15.1 Introduction

The world demand for energy is expected to double from 2000 to 2050, due to a $\sim$50 % increase in world population and a comparable increase in energy consumption per capita [1, 2]. The growing production of natural gas, coal and unconventional oil (e.g., tar sands) will compensate for the declining supply of oil but is not expected to satisfy the growth in energy demand throughout the rest of the century. Therefore, renewable energies will be required to cover about half of demand at around 2050 and about two-thirds of it by 2100. Biomass is a renewable resource at a human timescale, unlike oil, coal and gas which are formed over millions of years. The implementation of biomass-derived liquid fuels, i.e., "biofuels", is much easier. For instance, biofuels are presently being introduced in the European and US market by blending biofuel components into the fossil fuel pool. In developed countries there is a growing trend toward employing modern technologies and efficient bioenergy conversion using a range of biofuels, which are becoming cost competitive with fossil fuels (Puhan et al. 2005). Beyond the transportation sector, biomass is also a promising feedstock for the chemical industry which accounts for 5–10 % of today's

S. De (✉) · R. Luque
Departamento de Quimica Organica, Universidad de Cordoba,
Campus de Rabanales, Edificio Marie Curie, E-14014 Cordoba, Spain
e-mail: q62alsor@uco.es

Q. Li (ed.), *Nanomaterials for Sustainable Energy*,
NanoScience and Technology, DOI 10.1007/978-3-319-32023-6_15

oil and gas consumption. It is expected that the biofuel economy will grow rapidly during the 21st century and with this, biofuel economy and its associated biorefineries will be shaped accordingly.

15.1.1 Types of Biofuels

We can currently distinguish between "first-generation" biofuels derived from plant storage organs, which are already present on the market, and "second-generation" biofuels which are produced from the entire plant including the lignocellulosic part. Bioethanol as biogasoline and fatty acid alkyl esters as biodiesel are now commercialized on a large scale. However, these first-generation biofuels have problems such as low energy density and low stability. Besides, the first generation biofuels are produced from sugars and vegetable oils, and thus competing with the food chain for their feedstock. Therefore, on exceeding demand the available volumes first-generation biofuels may no longer be a sustainable option. Hence the mankind has to turn towards second-generation biofuels that exploit cheaper and more abundant lignocellulosic feedstock. In this chapter, we will discuss the critical role of heterogeneous nanomaterials used in the catalytic production of second generation biofuels from biomass feedstocks.

15.1.2 Chemistries in Biofuel Production

Selective cleavage of C–O bonds is a key transformation for the production of renewable fuels from biomass-derived platform compounds. Other processes associated to the utilization of biomass feedstocks are hydrolysis, dehydration, decarbonylation, decarboxylation etc. Since this chapter discusses the production of fuel components, we will focus our discussion mainly on the deoxygenation processes that effectively reduce oxygen content in the platform compounds. Unfortunately, C–O bond cleavage is not an easy reaction to achieve. The average energy of C–O bonds is $\sim$358 kJ/mol, which is well above that of most other C–X single bonds (except C–H and C–F bonds). This makes the C–O bond tolerant (and hence inert) to many reagents and catalysts, and due this reason compounds like diethyl ether, tetrahydrofuran and dioxane used as solvents in many reactions. In classical organic chemistry, the C–O bond in ethers may be cleaved in the presence of concentrated HBr and BBr_3, but is stable in catalytic hydrogenolysis reactions, unless activated (e.g., as in allylic and benzylic ethers) or strained (e.g., as in ethylene oxide) [3].

In the energy sector, current research is focused on activating/functionalizing the various hydrocarbons available, and their downstream transformations since fossil fuels are composed of largely hydrocarbons. In contrast, the chemistry of biomass, and specifically that of catalysts to transform biomass, was left behind due to many reasons. Compared to fossil fuel components which are mostly unfunctionalized, biomass feedstocks are over-functionalized since they contain a large amount of oxygen. Different methods including dehydration, hydrogenolysis, hydrogenation, decarbonylation, decarboxylation have been reported to remove oxygen functionalities (Scheme 15.1).

Hydrodeoxygenation (HDO) is currently considered as the most effective method for platform molecules upgrading which improves the effective H/C ratio, eventually leading to hydrocarbons. The key challenge of HDO processes is to achieve a high degree of oxygen removal with minimum hydrogen consumption, for which catalysts need appropriate and careful designed. To date, several catalytic materials have been reported for HDO, with various advantages and disadvantages [4]. The active phase of those materials is indeed a metal center that acts as a hydrogen carrier. In this consequence, both precious metals (e.g., Pd, Pt, Re, Rh, and Ru) and non-precious metals (e.g., Fe, Ni and Cu) have exhibited good activities in the HDO process. However, in terms of process cost, transition metals are more advantageous than noble metals and sometimes exhibit comparable HDO efficiency. The industrial catalysts based on Co–Mo–Ni formulations provide comparatively superior HDO performance but they undergo rapid deactivation due to coke formation and water poisoning [5].

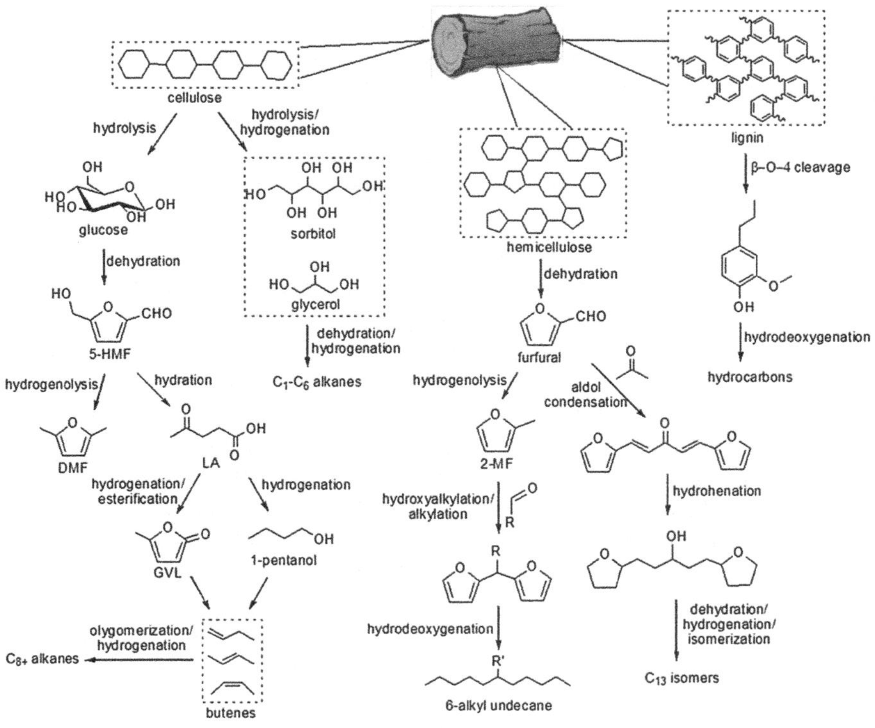

Scheme 15.1 Different types of fuels from biomass-derived platform chemicals

One of the major problems in upgrading biomass-derived platform chemicals is that, these chemicals generally contain various different functionalities that have different probabilities of chemical transformations. Selectivity is the most important issue here, since HDO reactions generally require high pressures of H_2 and higher H_2 pressure can reduce both C=O and C=C bonds simultaneously. Taking into account a highly useful platform molecule, 5-hydroxymethylfurfural (HMF) which contains four active functional groups, often undergoes unwanted transformations in HDO process, leading to the formation of several by-products with low DMF yield. Therefore, the central challenge in converting biomass into hydrocarbon fuels relates to an efficient cleavage of ubiquitous ether and alcoholic C–O linkages within the feedstock molecules to reduce both the oxygen content and degree of polymerization. And for this, a selective HDO catalyst is highly desirable in order to prevent complete hydrogenation of unsaturated compounds as well as to prevent over-utilization of expensive hydrogen.

15.2 Fuels Derived from Furfural and 5-Hydroxymethylfurfural

Among the different biomass-derived products, 2, 5-dimethylfuran (DMF) is one of the most attractive furan derivatives that can be considered as an attractive gasoline blend. DMF has a boiling point of 367 K and an energy content of 31.5 MJ L^{-1} similar to that of gasoline (35 MJ L^{-1}) and 40 % higher than ethanol (23 MJ L^{-1}) [6]. DMF is not soluble in water and can be used as a blender in transportation fuels. DMF can be obtained by reduction of both the formyl and hydroxyl groups of the HMF using supported metal catalysts in organic solvents such as butanol. The process consists of the production of HMF starting from fructose (in a biphasic reactor with and acid catalyst) followed by hydrogenation over a carbon supported copper/ruthenium (Cu–Ru/C) catalyst. The initial hydrogenation reactions were carried out in the presence of copper chromite ($CuCrO_4$) catalysts. This catalyst was however shown to be easily deactivated by chloride ions even at ppm level. The low melting point and high surface mobility of Cu (I) chloride species were observed to accelerate Cu catalysts sintering. To overcome this problem, a chloride-resistant carbon-supported bimetallic copper–ruthenium (Cu–Ru/C) catalyst was developed. Copper has a comparably lower surface energy to that of ruthenium and their combination generates a two-phase system in which the copper phase coats the ruthenium surface as confirmed by electron spectroscopy. The designed Carbon-supported Ru catalyst is resistant to deactivation by chloride ions, while Cu shows the predominant role in hydrogenolysis over Ru. Cu–Ru/C catalyst consequently exhibits copper-like hydrogenolysis behavior combined with ruthenium-like chlorine resistance. The liquidphase hydrogenation of HMF using Cu–Ru/C catalyst provided quantitative conversion of HMF with a maximum 71 % DMF selectivity under 6.8 bar H_2 in 1-butanol.

Owing to the different functionalities in HMF, several by-products are typically formed in HMF conversion processes which lead to a low DMF yield and increase the costs of product purification. One of the proposed approaches to improve DMF yield relates to the exploration of bimetallic catalysts that can selectively convert HMF to DMF. In bimetallic systems, the incorporation of a second metal creates a number of possibilities to modify the surface structure and composition of metal catalysts. The properties of bimetallic catalysts significantly differ from their monometallic analogs due to geometric and electronic effects between the two metals [7]. The best advantage of these bimetallic systems is that the catalytic activity can be easily tuned by changing their size and composition. Tsang et al. reported a novel approach of making PtCo bimetallic nanoparticles that were effective catalysts for selective hydrogenation of C=O bond in the presence of C=C bonds [8]. Incorporation of Co effectively minimizes the unselective low coordination sites and optimizes the electronic environment of Pt nanoparticles. Pt (111) surfaces preferentially adsorb α,β-unsaturated aldehyde in a terminal di-σ_{CO} mode which leads to a preferential terminal aldehyde reduction to unsaturated alcohol. In contrast, low coordination sites favor π interactions with C=C bonds, which account for unselective products. In a very recent work, Wang et al. devised a bimetallic system consisting of PtCo nanoparticles encapsulated in hollow carbon spheres via soft-templating method [9]. Pt nanoparticles were incorporated inside the spheres by in situ impregnation during the carbonization process, and subsequently the Co nanoparticles were introduced using as-synthesized Pt@HPS. The final PtCo@HCS materials were obtained upon Pt@HPS-Co^{+2} pyrolysis under H_2/Ar atmosphere. The catalytic activity of the catalyst was tested in the HDO of HMF. At 120 °C, 2, 5-di(hydroxymethyl)furan was obtained as main product (70 %), which indicates the effectiveness of the catalyst for the selective hydrogenation of formyl groups in HMF. Upon raising the temperature to 160 °C, almost quantitative DMF yield (96 %) was obtained, confirming that catalyst can also catalyze the hydrogenolysis of the hydroxyl group at high temperature. A rapid increase of DMF yield (from 44 to 95 % after 6 min) at this temperature indicates that the hydrogenolysis is the rate-determining step for the formation of DMF. The role of the second metal (here Co) was clearly understood when activated carbon-supported platinum (Pt/AC) and graphitized carbon-supported platinum (Pt/GC) catalysts were used under the same reaction conditions, only 9 and 56 % DMF yields were respectively obtained. While the modification of these catalysts with Co dramatically increased the HMF conversion and DMF yield to 100 and 98 % respectively.

Later on, Ebitani et al. reported a carbon supported PdAu bimetallic catalyst (Pd_xAu_y/C) for the selective hydrogenation of HMF to DMF [10]. Bimetallic Pd_xAu_y/C catalyst exhibited significantly higher activity as compared to monometallic Pd/C and Au/C catalysts due to the existence of a charge transfer phenomenon from Pd to Au atoms as proved by XPS and X-ray absorption near-edge structure (XANES) analyses. In another recent report by Nakagawa et al., a remarkable synergy between Pd and Ir has been observed in bimetallic Pd–Ir alloy particles supported on SiO_2 for the hydrogenation of furfural and HMF in water [11]. Incorporation of Ir showed a remarkably higher TOF than the monometallic Pd catalysts with similar particle size, particularly for C=O hydrogenation. Ir atoms on the surface were found to promote the adsorption at C=O site, whereas Pd surface strongly interacts with the furan ring.

Non-precious Ni and Ni–Fe bimetallic catalysts supported on SiO_2 were investigated for the conversion of furfural [12]. Furfuryl alcohol and furan were the primary products over monometallic Ni/SiO_2, resulting from hydrogenation and decarbonylation of furfural. In contrast, 2-methylfuran (2-MF) yield was greatly increased with reduced yield of furan and C_4 products using Fe–Ni bimetallic catalysts. Results proved that the addition of Fe suppressed the decarbonylation activity of Ni while promoting C=O hydrogenation (at low temperatures) and C–O hydrogenolysis (at high temperatures). DFT analysis was conducted to understand possible surface species on mono- and bimetallic surfaces, which proved that the selectivity differences displayed by these two catalysts were dependent on the stability of η^2-(C, O) surface species (Scheme 15.2). These η^2-(C, O) species were found to be comparatively more stable on Ni–Fe to those on pure Ni. Furfural could then be readily hydrogenated to furfuryl alcohol and subsequently hydrogenolyzed to 2-MF. The strong interaction between O (from the carbonyl group) and the oxyphilic Fe atoms supports a preferential hydrogenolysis reaction on the bimetallic alloy. On the other hand, the Ni surface initiates the decomposition of η^2-(C, O) species to produce furan and CO.

So far we have discussed the hydrogenation reactions which have used gaseous hydrogen as reducing agent. Therefore, in most cases the hydrogenation needs to be performed at high pressure. Hydrogen donor solvents (e.g., formic acid, alcohols, etc.) have been comparably utilized in replacement of molecular H_2 as hydrogenating agent in HMF conversion to DMF. A successful one-pot conversion of sugar into DMF was developed using formic acid (FA) as hydrogen carrier and Pd/C catalyst [13]. FA was used as dehydrating agent for the conversion of fructose to HMF and also played key roles in subsequent steps of hydrogenation and hydrogenolysis in presence of Pd/C. This method produced 51 % DMF yield from fructose and 95 % DMF yield from HMF. Ru/C used in the same process produced 30 and 37 % DMF from fructose and HMF respectively, which proves that Pd/C is a more effective hydrogenation catalyst [14]. This is to note here that, using a neutral support (like carbon) DMF is produced as main product. When the support

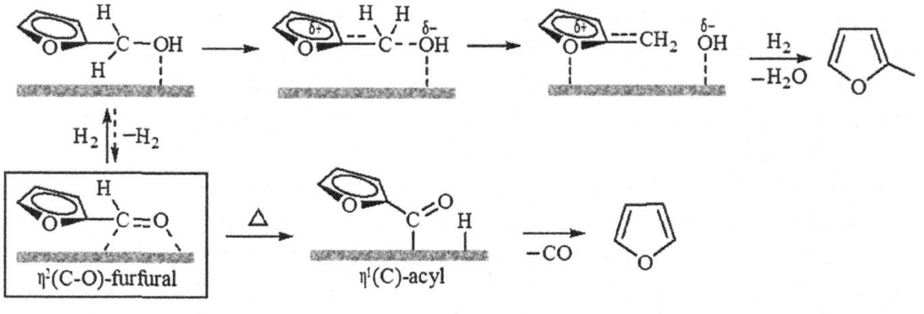

Scheme 15.2 Possible species on the catalyst surface during the conversion of furfural [12]

is changed from neutral to acidic, ring opening also takes place in addition to hydrogenation. The hydrogenolysis of HMF in formic acid over a Pd/ZrP catalyst produced 1, 6-hexanediol as major product with very slight amount of DMF [15]. It is observed that the deoxygenation of the furan ring (C–O bond cleavage) is favored by Brønsted-acid catalytic sites, and hydrogenation of C=O and C=C bonds is favored by the metal catalytic sites. High Brønsted/Lewis acid ratio of ZrP effectively cleaves the C–O bonds in furan and Pd dissociates formic acid to form in situ hydrogen for hydrogenation (Table 15.1).

An alternative approach for the selective hydrogenation of HMF is catalytic transfer hydrogenation (CTH), in which alcohols are used as both hydrogen donor and reaction medium. Catalytic systems used in this process must be effective in generating hydrogen from alcohols. Hansen et al. reported a Cu-doped porous metal oxide (Cu-PMO) catalyst for HMF hydrogenation in supercritical methanol that produced a mixture of products including dimethylfuran (DMF), dimethyltetrahydrofuran (DMTHF) and 2-hexanol in good yields [22]. However, the critical temperature of methanol is very high in this process (300 °C) and the selectivity of DMF is very low (*ca.* 34 %) at this temperature. This can be overcome by replacing methanol by isopropyl alcohol and a high DMF yield (81 %) can be achieved at 190 °C using a Ru/C catalyst [23]. Isopropyl alcohol can also be employed as hydrogen donor in the sequential transfer hydrogenation/hydrogenolysis of furfural and HMF over in situ reduced, Fe_2O_3-supported Cu, Ni, and Pd catalysts [24]. Pd/Fe_2O_3 exhibits an extraordinary activity in comparison to Cu and Ni catalysts but ring-hydrogenation and decarbonylation compounds are observed as reaction side products. Pd nanoparticles strongly coordinate with the π-system of the furfural ring which causes ring hydrogenation. However, the formation of ring-hydrogenated

Table 15.1 Selective conversion of HMF and furfural over different HDO catalysts

HDO catalyst	H_2 source	Solvent. T (°C)	Substrate	Product	Yield (%)	Reference
Cu–Ru/C	H_2 (6.8 bar)	1-BuOH, 220	HMF	DMF	71	[6]
Pd/C	H_2 (62 bar)	[EMIM]C1, 120	HMF	DMF	15	[16]
Pd/C	H_2 (10 bar)	sc CO_2–H_2O, 80	HMF	DMF	100	[17]
PtCo@HCS	H_2 (10 bar)	1-BuOH, 180	HMF	DMF	98	[9]
Ni-W_2C/AC	H_2 (40 bar)	THF, 180	HMF	DMF	96	[18]
Pd–Au/C	H_2 (1 bar)	THF, 60	HMF	DMF	100	[10]
Pd/C	Formic acid	THF, 150/70	HMF	DMF	95	[13]
Ru/C	Formic acid	THF, 150/75	HMF	DMF	37	[14]
Cu-PMO	MeOH	MeOH, 260	HMF	DMF	48	[22]
Ru/C	*i*-PrOH	*i*-PrOH, 190	HMF	DMF	81	[23]
Pd/Fe_2O_3	*i*-PrOH	*i*-PrOH, 180	HMF	DMF	72	[24]
Cu chromite/AC	H_2	Vapor phase, 230	Furfural	2-MF	95	[19]
Cu/Zn/Al/Ca/Na (59:33:6:1:1)	H_2	Vapor phase, 250	Furfural	2-MF	87	[20]
Ni–Fe/SiO_2	H_2	Vapor phase, 250	Furfural	2-MF	39	[12]
Mo_2C	H_2	Vapor phase, 150	Furfural	2-MF	4.5	[21]

products can be reduced by decreasing Pd loading (specific activity decreases due to an increased metal dispersion). The observed higher hydrogenolysis activity of Pd/Fe_2O_3 catalysts could be correlated to the morphology and size of Pd particles and their strong interaction with the hematite (Fe_2O_3) support. The oxophilic nature of Fe along with the intimate contact between Pd and Fe species promotes the activation of O–H bonds which readily undergoes hydrogenolysis in the presence of active hydrogen generated over Pd surfaces.

15.3 Long Chain Hydrocarbons via C–C Coupling

Furfural and HMF can be converted relatively easily to larger molecules by employing the reactive aldehyde group or furan ring. C–C coupling reactions are considered as excellent strategies to upgrade HMF and furfural into liquid alkane fuels using external carbonyl-containing molecules followed by HDO processes. Carbon number can be increased by aldol condensation of furan aldehydes (HMF or furfural) with external carbonyl containing molecules having α-hydrogen (e.g., ketones) in the presence of a base or an acid catalyst (Scheme 15.3). Further hydrogenation and dehydrodeoxygenation of aldol products can produce high-quality diesel range hydrocarbons [25, 26]. Base catalyzed condensation of HMF with acetone produces C_9 and C_{15} unsaturated intermediates depending on utilized HMF/acetone molar ratio. Aldol compounds are subsequently subjected to hydrogenation/dehydration/ring opening processes in the presence of bifunctional catalysts such as Pd/Al_2O_3 and $Pt/NbOPO_4$, producing a mixture of linear C_9 and C_{15} alkanes in high yields. Acidic supports are favored in producing high yield of hydrocarbons by enhancing the rate of dehydration and ring opening steps. The use of $TaOPO_4$ instead of $NbOPO_4$ increased the C–O dissociation activity slightly. However, C–C dissociation reactions also became more prevalent. Solid acid catalysts composed of silicon and aluminum oxides such as SiO_2–Al_2O_3 and H-zeolites have also been used as co-catalysts for total hydrodeoxygenation, especially in early studies. The first study of the total hydrodeoxygenation of furan-based compounds in 2005 used Pt/SiO_2–Al_2O_3 catalysts [25], although the same group later reported that $Pt/NbOPO_4$ is a superior catalyst, as discussed above. The total hydrodeoxygenation by Pt/SiO_2–Al_2O_3 usually requires a pre-reduction step of furanic substrates with a simple hydrogenation catalyst such as Pd/Al_2O_3 at a low temperature [27, 28]. Excellent yields of *n*-tridecane (72.6 %) and *n*-dodecane (15.6 %) can be obtained from the hydrogenated 2:1 aldol adduct [see (1), Scheme 15.3] of furfural and acetone.

Zhang et al. investigated the total hydrodeoxygenation of furanic compounds with various zeolite-supported noble-metal catalysts in a fixed-bed reactor [29]. The total hydrodeoxygenation of the 1:1 aldol adduct of furfural and 2-pentanone [see (2), Scheme 15.3, R = H, R′ = C_3H_7] was performed over various supported Pd catalysts (Fig. 15.1). Much higher yields of C_9 and C_{10} alkanes were obtained over

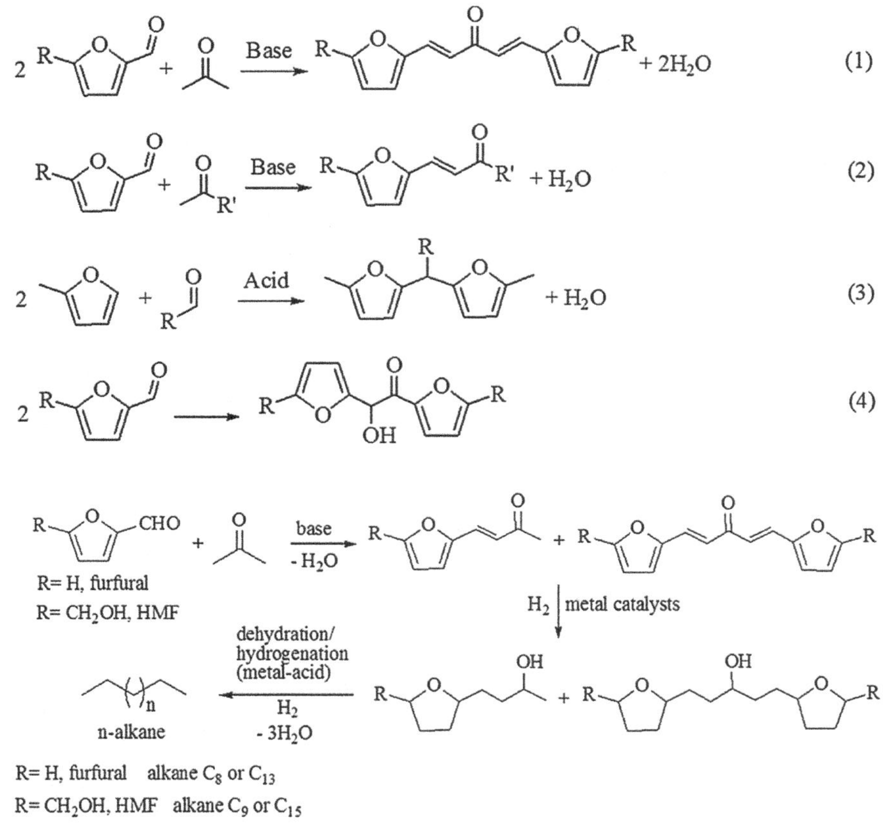

Scheme 15.3 Coupling of furfural or HMF with acetone and subsequent dehydroxygenation to alkanes

Pd/H-ZSM-5 (90 %) than Pd/SiO_2–Al_2O_3. The higher activity of the zeolite-supported catalyst can be explained by the higher acid strength and larger amount of Brønsted acid sites indicated by NH_3 temperature-programmed desorption (TPD) and FTIR spectroscopy of adsorbed pyridine. The same group also reported a Ru/H-ZSM-5 catalyst for the total hydrodeoxygenation of 1-(2-furanyl)-5-methyl-1-hexen-3-one [1:1 aldol adduct of furfural and methyl isobutyl ketone; [see (2), Scheme 15.3], R = H, R′ = *i*-Bu] [30]. Intense C–C dissociation reactions occurred and a wide range of alkane products were obtained at 573 K. The C_8–C_{11} range products (68.4 % yield) were mainly composed of isoundecane and isodecane.

Hydroxyalkylation-alkylation (HAA) combined with HDO is a comparatively promising route for the synthesis of renewable high-quality diesel or jet fuel (Scheme 15.4). Taking advantage of this combined process, 2-MF (Sylvan) can be used in the Sylvan diesel process where it serves as starting material [31–33]. HAA process combines two Sylvan molecules with an aldehyde or a ketone to yield oxygenated intermediate molecule [see (3), Scheme 15.3] which undergoes HDO

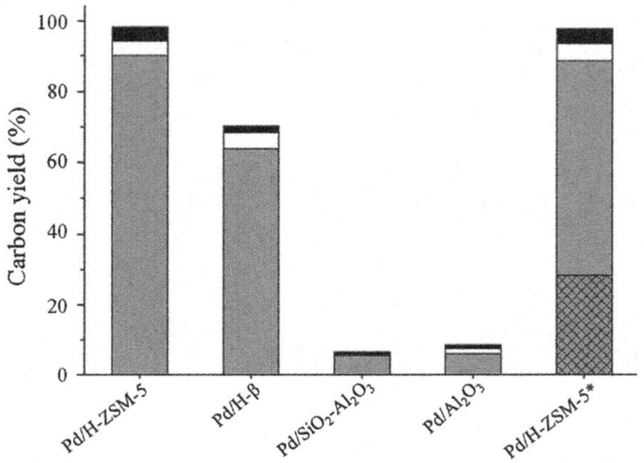

Fig. 15.1 Hydrodeoxygenation of the 1:1 aldol adduct of furfural and 2-pentanone [see (2), Scheme 15.3, R = H, R′ = *i*-Pr) over various Pd catalysts. Black, white, gray, and hatched bars represent C_1–C_4, C_5–C_8, C_9 and C_{10}, and C_{11} and C_{12} alkanes, respectively. The asterisk indicates the use of the aldol adduct of furfural and 2:1 mixture of 2-pentanone and 2-heptanone as the substrate. Conditions: catalyst (Pd 5 wt%, 1.8 g), substrate (0.04 $mLmin^{-1}$), H_2 (6 MPa, 120 $mLmin^{-1}$), 533 K [29]

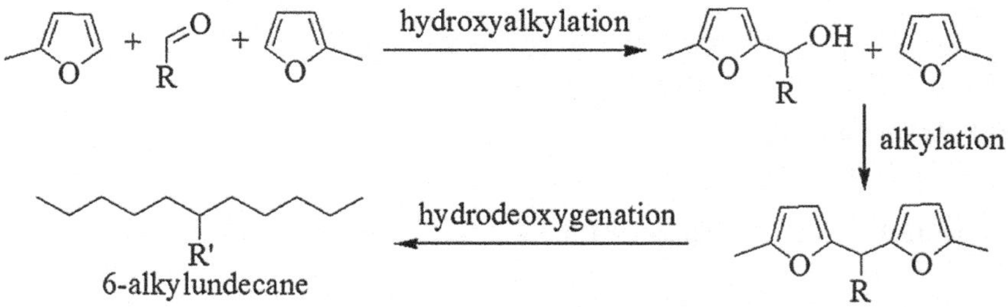

aldehyde (RCHO)	R	R' (after hydrodeoxygenation)
butanal	$-CH_2CH_2CH_3$	$-CH_2CH_2CH_3$
5-hydroxymethylfurfural	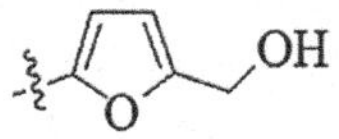	$-CH_2CH_2CH_2CH_2CH_3$
5-methylfurfural	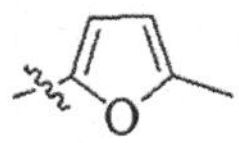	$-CH_2CH_2CH_2CH_2CH_3$

Scheme 15.4 Different steps in Sylvan process for the production of 6-alkylundecane [31]

reaction at moderate temperatures and high H_2 pressures to yield diesel fuel. Different types of resins (such as, Nafion, Amberlyst etc.) are used to couple 2-MF and furfural, and the HDO steps are performed using metal catalysts such as Pd/C, Pt/C and Ni–W_xC/C. Catalysts containing acidic support always produce higher

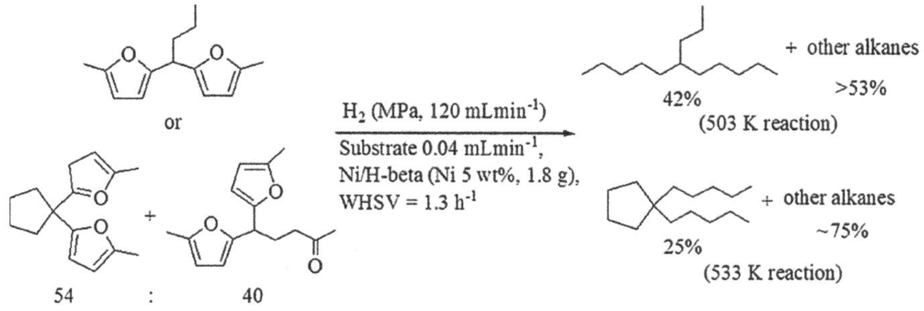

Scheme 15.5 Conversion of furanic derivatives into branched alkanes [35]

carbon yield. For example, a 94 % carbon yield of diesel and 75 % carbon yield of C_{15} hydrocarbons (with 6-butylundecane as major component) can be achieved using a 4 % Pt/ZrP catalyst. However, sometimes it is observed that the strongly acidic supports could initiate the dissociation of C–C bond to produce lower alkanes. In contrast, weakly acidic supports such as C, TiO_2, or Al_2O_3 can suppress C–C dissociation, although a higher reaction temperature (~623 K) is required than in cases that use strong acids (typically ≤573 K).

The extended applications of the Sylvan process have been reported in recent studies using different catalytic systems. The total hydrodeoxygenation of alkylated furans using a Ni/H-beta catalyst was performed which produced high alkane yields from 1,1-bis(5-methyl-2-furanyl)butane, an alkylation product of 2-methylfuran with butanal [see (3), Scheme 15.3, R = *n*-Pr] [34]. Among various M/H-beta catalysts (M = Pt, Pd, Ru and Ni), Ni/H-beta was selected because of the lower cost of Ni than noble metals. The Ni/H-beta catalyst was tested on another substrate which is the alkylation product of 2-methylfuran with cyclopentanone [35]. From both substrates, the corresponding alkane was obtained in ~40 % yield without isomerization or C–C dissociation (Scheme 15.5).

15.4 Levulinic Acid-Based Fuels

Similar to HMF and furfural, levulinic acid (LA) derived from lignocellulosic biomass has the potential to be utilized as a platform intermediate molecule in the production of renewable liquid fuels for the transportation sector. Recently, levulinic acid has been identified by Bozell and Petersen as one of the top 10 platform molecules for the production in biorefineries of value-added chemicals and liquid transportation fuels [36]. In this section, we will discuss the role of levulinic acid as a functional intermediate for the production of liquid hydrocarbon fuels through its hydrogenation and C–C coupling.

15.4.1 Hydrogenation of Levulinic Acid

Levulinic acid can be transformed into its derivatives such as ethyl levulinate (EL), γ-valerolactone (GVL), and methyltetrahydrofuran (MTHF) which have shown fuel applications [37]. GVL itself can be used as a fuel additive to current fuels derived from petroleum due to combustion energy similar to ethanol (35 MJ L^{-1}) [38]. GVL was also identified as a potential intermediate for the production of fuels and chemicals based on renewable feedstocks. GVL is generally produced from levulinic acid via two main routes: (i) hydrogenation of levulinic acid to γ-hydroxyvaleric acid followed by an intramolecular esterification through cyclization to produce GVL and (ii) acid catalyzed dehydration of levulinic acid to angelica-lactone followed by hydrogenation. In the 1950s, Quaker Oats reported a continuous process for the vapor-phase commercial-scale production of GVL via LA hydrogenation [39]. Quantitative yields to GVL could be achieved using a mixture of metal oxide catalysts (CuO and Cr_2O_3) at 200 °C. Later on, hydrogenation of LA has been typically performed in the presence of H_2 using various metal catalysts such as Ru, Pd, Pt, Ni, Rh, Ir, Au on different supports. Among the precious metal catalysts, Ru based catalysts have shown high performance to reduce levulinic acid or its esters to GVL [40]. XPS studies revealed that a higher extent of Ru^0 species in case of carbon supported Ru could account for its higher hydrogenation activity as compared to Ru on other supports.

An alternative route to produce GVL is the catalytic transfer hydrogenation (CTH) of levulinic acid through the Meerwein–Ponndorf–Verley (MPV) reaction using secondary alcohols as hydrogen donors and solvents. This process does not require expensive noble metal catalysts. Different heterogeneous metal oxides including ZrO_2, MgO/Al_2O_3, MgO/ZrO_2, $CeZrO_x$ and γ-Al_2O_3 were tested for the hydrogenation of levulinic acid and its esters, among which ZrO_2 was most active (92 % GVL yield) using 2-butanol at 150 °C [41]. GVL can be produced directly from furfural through sequential CTH and hydrolysis reactions catalyzed by zeolites with Brønsted and Lewis acid sites [42]. In the first step, furfural is converted into furfuryl alcohol and butyl furfuryl ether via CTH promoted by a Lewis acid catalyst. Furfuryl alcohol and butyl furfuryl ether are subsequently converted into LA and butyl levulinate through hydrolytic ring opening reactions using a Brønsted acid, which finally undergo a second CTH step to produce 4-hydroxypentanoates followed by lactonization to GVL.

15.4.2 Upgrading of Levulinic Acid into Hydrocarbon Fuels

Levulinic acid can also be transformed into hydrocarbon fuels by different catalytic routes. Unlike furfural and HMF, LA is not coupled with any external molecules directly. The multistep process involves hydrogenation of LA into GVL which serves as the starting material for the production of hydrocarbons through C–C coupling reactions. The Dumesic group extensively worked on the conversion of GVL to kerosene- and diesel-range hydrocarbons (Serrano-Ruiz and Dumesic 2011) [43]. A series of catalytic approaches were developed to convert aqueous solutions of LA into different types of liquid hydrocarbon transportation fuels. The catalytic pathways involve oxygen removal via dehydration/hydrogenation and decarboxylation reactions combined with C–C coupling processes through ketonization, isomerization, and oligomerization that are required to increase the molecular weight as well as to adjust the structure of the final hydrocarbon product. Aqueous levulinic acid is firstly hydrogenated to water-soluble GVL over non-acidic catalysts (e.g., Ru/C) at low temperatures. Water soluble GVL was subsequently upgraded to liquid hydrocarbon fuels following two main pathways: C_9 route and C_4 route (Scheme 15.6).

In the C_9 route, GVL is converted to 5-nonanone via pentanoic acid over a water-tolerant multifunctional Pd/Nb_2O_5. Subsequently, 5-nonanone is transformed into its corresponding alcohol that is further converted to C_9 alkanes through hydrogenation/dehydration cycles using the same bifunctional Pt/Nb_2O_5 catalyst. Comparatively, GVL is first decarboxylated in the C_4 route using a silica/alumina catalyst at elevated pressure to give butene followed by oligomerization over acidic catalysts (e.g., H-ZSM5, Amberlyst 70), resulting in different C_{12} alkanes.

A stepwise pathway to produce branched C_7–C_{10} gasoline-like hydrocarbons in high yields has also been recently reported by Mascal et al. [44]. The three-step process proceeds through the formation of an angelica lactone dimer which serves as a novel feedstock for hydrodeoxygenation. LA is converted using a solid acid

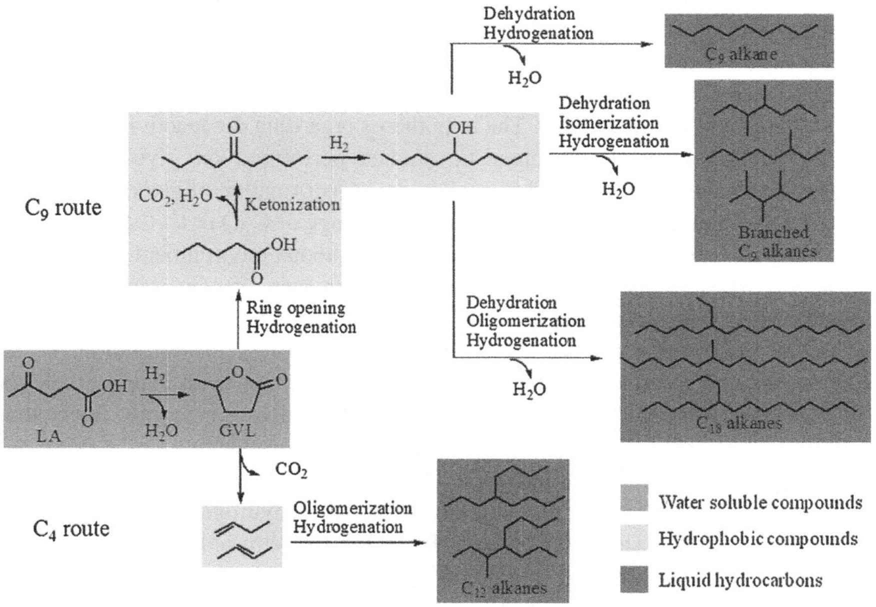

Scheme 15.6 Catalytic routes for the conversion levulinic acid into liquid hydrocarbon fuels via GVL intermediate [43]

catalyst (e.g., montmorillonite clay, K10) into angelica lactone, which dimerises in the presence of catalytic amounts of K_2CO_3. This dimer product is eventually hydrodeoxygenated to gasoline range hydrocarbons using a combination of oxophilic metal and noble metal catalysts under mild conditions. Different catalysts were screened in HDO reactions of angelica lactone dimers. Ir–ReO_x/SiO_2 catalyst exhibited the highest activity, with quantitative conversion producing 88 % total hydrocarbon yield. Pt–ReO_x/C catalysts were also effective in providing analogous hydrocarbon yields but their C_{10} hydrocarbon selectivity was comparatively inferior to that of Ir–ReO_x/SiO_2.

15.5 Fuels from Sugar Alcohols

Sugar alcohols are straight-chain molecules with the general formula CH_2OH–$(CHOH)_n$–CH_2OH in which all of the carbon atoms have one OH functional group. These alcohols can be obtained easily by the hydrogenation of sugars such as glucose to sorbitol (C_6), and xylose to xylitol (C_5) [45]. The C_3 sugar alcohol, glycerol, is manufactured on a large scale as a byproduct of soap or biodiesel fuel from triglycerides [46]. Therefore, sugar alcohols with a carbon number less than 6 are very important platform molecules, but due to the low carbon numbers, the total hydrodeoxygenation systems of sugar alcohols should suppress C–C dissociation reactions as much as possible. In this consequence, sorbitol can be widely used as the substrate because of its availability, although sorbitol has a more complex structure than other biomass-derived sugar alcohols.

The transformation of sorbitol into hydrocarbons that can be used in the gasoline group is currently considered to be a promising technology for the production of second-generation biofuels [47]. The hypotheses regarding the reactivity of sorbitol are most often based on mechanistic studies performed on glycerol. Several mechanisms have been identified to explain the transformation of glycerol into diols and are presented in reactions that lead to the cleavage of C–O or C–C bonds. These transformations are generally carried out in an aqueous environment, at low temperatures (120–150 °C), most often in a closed batch-type reactor (or in flow reactor sometimes) with hydrogen supply. The main reason of using water as solvent is to lower the viscosity of the sugar alcohols and to suppress the formation of carbonaceous solid during hydrodeoxygenation. The reaction temperature employed for the total hydrodeoxygenation of sorbitol is generally significantly lower than that for furanic substrates probably because of the higher reactivity of sorbitol and its susceptibility to C–C dissociation. This reaction pathway involves four key classes of reactions: (1) C–C bond cleavage through dehydrogenation, followed by decarbonylation and dehydrogenation, followed by retro-aldol condensation at metal sites; (2) C–O cleavage through dehydration at acid sites; (3) hydrogenation at metal sites; and (4) isomerization at metal–acid sites. These basic mechanisms can also be extended to larger polyols (4–6 carbons).

Supported metal catalysts are generally used when the target product is hydrogen; while bifunctional catalysts composed of a metal phase and an acid phase are more often chosen to produce hydrocarbons. Catalyst support has an influence on the catalytic activity, as shown for example by a comparative study of the ethylene glycol conversion on platinum catalysts on different supports in aqueous phase [48]. In particular, acidic sites favor dehydration and basic sites favor retro-aldol reaction. Platinum supported catalysts can be classified according to the ethylene glycol conversion in decreasing order as follows: activated carbon ≈ TiO_2 > Al_2O_3 > SiO_2–Al_2O_3 ≈ ZrO_2 > platinum black > CeO_2 > SiO_2 > ZnO. The most selective for alkanes are, by decreasing selectivity: activated carbon > SiO_2–Al_2O_3 > TiO_2 > ZrO_2. Different zeolites supports such as H-ZSM-5, H-mordenite and HY have been found to be active for the aqueous phase transformation of sorbitol [49, 50] or glycerol [51] into hydrocarbons. However in many cases, these catalysts are shown to be unstable under hydrothermal conditions (liquid water between 150 and 250 °C) [52]. Niobium oxide (Nb_2O_5) and its derivatives (phosphated derivatives in particular) are highly effective for the conversion of sorbitol in aqueous phase process. When compared the activity of Pt/silica-alumina and Pt on niobium-based supports (niobic acid, niobium phosphate, phosphated niobic acid) it is observed that better alkanes yield could be achieved with niobium-based supports [53]. The presence of stronger acid groups in niobium-based catalysts promotes the dehydration reaction due to the strong interaction between acid sites and alcohol groups of polyols. Although, better yield of alkanes is observed using acidic support, the selectivity towards straight chain alkanes is lower. Experiment shows that a Ni/H-ZSM-5 catalyst produces *iso*-rich C_5 and C_6 alkanes from sorbitol [54]. The addition of MCM-41 to the Ni/H-ZSM-5 catalyst increases the selectivity to hexanes substantially [55]. FTIR spectroscopy of adsorbed pyridine shows that the addition of MCM-41 decreases the number of strong Brønsted acid sites of H-ZSM-5 by extracting the Al atoms from the framework of H-ZSM-5.

Huber et al. studied the aqueous-phase hydrodeoxygenation of sorbitol over two fundamentally different types of bifunctional catalysts: Pt/Zr-P and Pt–ReO_x/C [56]. Products obtained were classified into three major categories: (1) light gases (CO_2 and C_1–C_4 alkanes), (2) gasoline-range products (mainly C_5–C_6 alkanes), and (3) aqueous-phase products (isosorbide, mannitol, sorbitan, hexanetriol, C_2–C_6 diols, glycerol, and methanol). The differences in product selectivity between the Pt/Zr-P and Pt–ReO_x/C can be related to the different acid–metal interaction of these catalysts. The acid sites on the Pt–ReO_x catalyst were able to isomerize sorbitol into mannitol, whereas no mannitol was observed with the Pt/Zr-P catalyst. This suggests that the isomerization occurs over bifunctional metal–acid sites that are mixed at the atomic level. The Pt–ReO_x/C catalyst had a higher selectivity toward secondary alcohols than the Pt/Zr-P catalyst, which produced all primary alcohols. The Pt/Zr-P had a higher C_6 product selectivity and lower CO_2 selectivity than Pt–ReO_x/C, suggesting that Pt/Zr-P has a higher relative rate of C–O versus C–C bond cleavage than Pt–ReO_x/C (Table 15.2).

Table 15.2 Total hydrodeoxygenation of sugar alcohols using different HDO catalysts

Catalyst (loading amount [wt%])	Substrate	Conv. (%)	Main products (C-yield [%])	Reference
Pt(4)/SiO_2–Al_2O_3	Sorbitol	92	Total alkanes (77), hexanes (42, >95 % *n*), pentanes (19)	[57]
Pd(3)/SiO_2–Al_2O_3	Sorbitol	>97	Total alkanes (66), hexanes (37), pentanes(18)	[57]
Pt(4)/niobic acid	Sorbitol	–	Total alkanes (49), oxygenates (18)	[53]
Pt(4)/$NbOPO_4$	Sorbitol	–	Total alkanes (61), oxygenates (7)	[53]
Pt(4)/zirconium phosphate	Sorbitol	74	Total alkanes (59), hexanes (37), pentanes (8), CO_2 (8)	[58]
Pt(5)-Re(4.9)/C	Sorbitol	100	Total alkanes (44), hexanes (3.4), pentanes (5), CO_2 (40)	[56]
Pt(0.86)-Re(1.2)/TiO_2	Sorbitol	99.8	Total alkanes (82), hexanes (5.4), pentanes (4.9), CO_2 (13)	[59]
Ni(10)/H-ZSM-5	Sorbitol	–	Total alkanes (47.6), *i*-hexane (28.3), *n*-hexane (10.9), *i*- pentane (4.2), *n*-pentane (4.2)	[54]
Ni(2)/ H-ZSM-5 + MCM-41	Sorbitol	67	Total alkanes (72), hexanes (66)	[55]
Pt(2.2)/ZrO_2 + TiO_2-WO_x.	Sorbitol	100	Total alkanes (19), *n*-hexane (7), other *n*-alkanes (12), CO_2 (10)	[60]
Ir(4)-Re(3.9)O_x/ SiO_2 + H-ZSM-5	Sorbitol	>99.9	Total alkanes (98.9), *n*-hexane (95.3)	[61]
Ir(4)-Re(3.9)O_x/SiO_2	Sorbitol	>99.9	Total alkanes (52.7), *n*-hexane (50.5), 2-hexanol (7.7), 3- hexanol (35.6)	[61]
Rh(4)-Re(3.6)O_x/ SiO_2 + H-ZSM-5	Sorbitol	>99.9	Total alkanes (88.1), *n*-hexane (66 0), ≤ C_5 alkanes (21.6)	[61]
Ru(5)/C + H-ZSM-5	Sorbitol	99.6	Total alkanes (51.2), *n*-hexane (13 7), ≤ C_5 alkanes (35.1)	[61]
Skeletal Ni + H-ZSM-5	Sorbitol	85.1	Total alkanes (7.6), *n*-hexane (0.3), ≤ C_5 alkanes (7.3)	[61]
Pt(4)/zirconium phosphate	Xylitol	76	Total alkanes (67), pentane (48), CO_2 (6)	[58]
Ir(4)-Re(3.9)O_x/ SiO_2 + H-ZSM-5	Xylitol	>99.9	Total alkanes (98.5), *n*-pentane (95.9)	[61]
Ni(2)/H-ZSM-5	Xylitol	94	Total alkanes (91; *iso*/*n* = 0.32), pentanes (90)	[62]
Ni(2)/MCM-22	Xylitol	97	Total alkanes (97; *iso*/*n* = 0.15), pentanes (91)	[62]
Ir(4)-Re(3.9)O_x/ SiO_2 + H-ZSM-5	Erythritol	>99.9	Total alkanes (96.0), *n*-butane (94.8)	[61]
Ir(4)-Re(3.9)O_x/ SiO_2 + H-ZSM-5	Glycerol	>99.9	Total alkanes (98.8), propane (98.0)	[61]
Pt(5)/H-ZSM-5	Glycerol	–	Total alkanes (51), propane (28), C_4–C_7 alkanes (1.1), ethane (19), methane (2.6), CO_2 (29)	[51]

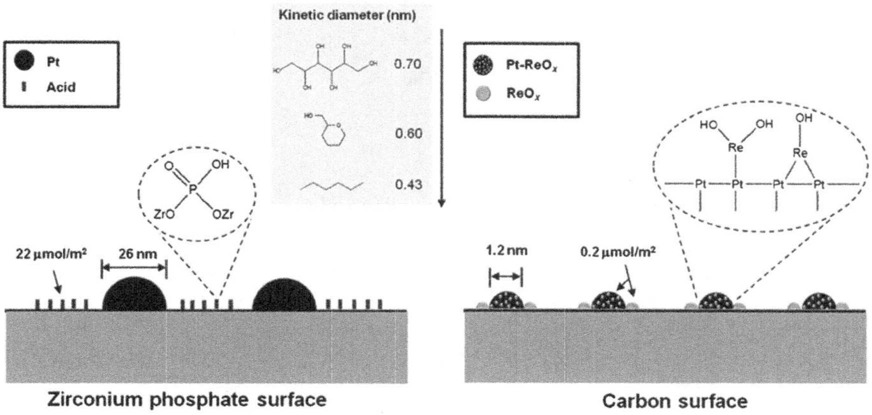

Fig. 15.2 Probable structures of Pt/Zr-P and Pt–ReO$_x$/C under reaction conditions [56]

The acid site relative to the metal site is important in determining the catalyst performance. Figure 15.2 shows the plausible structure of Pt/Zr-P and Pt–ReO$_x$/C under reaction conditions. The acid sites on the Pt–ReO$_x$ are in intimate contact with the metal Pt sites, whereas those on the Pt/Zr-P catalysts are atomically separated from the metal Pt sites. Pt particles on the Pt–ReO$_x$ catalyst are covered with or contain ReO$_x$ species that can interact directly on Pt metal particles. In contrast there is no proof of Pt and Zr-P forming a single phase which suggests that the Zr-P acid sites and the metal sites on the Pt/Zr-P catalyst are in two different locations. The amount of acid sites based on the surface area is 110 times higher for Pt/Zr-P than for Pt–ReO$_x$/C, however this is of no use due to the higher (21.7 times) Pt particle size in Pt/Zr-P than in Pt–ReO$_x$/C. Sorbitol has a kinetic diameter below 0.7 nm very similar to the size of Pt–ReO$_x$ particles. The oxophilic Re species may be able to strongly adsorb the hydroxyl groups of reactants near Pt metal sites. The adjacent acid sites on the Pt–ReO$_x$ catalyst can be affected by the metal with high coverage under reaction conditions, resulting in enhanced activity for acid-catalyzed reactions such as isomerization and hydrogenolysis.

Two reaction routes are proposed for the total hydrodeoxygenation of sorbitol (Fig. 15.3). One is the route via sorbitan and isosorbide, which are dehydration products of sorbitol, as intermediates. This route is proposed for the systems that use strong acids such as SiO_2–Al_2O_3. The other route is the step-by-step removal of all OH groups. This route is likely with metal-catalyzed C–O dissociation catalysts such as Ir-ReO$_x$/SiO_2.

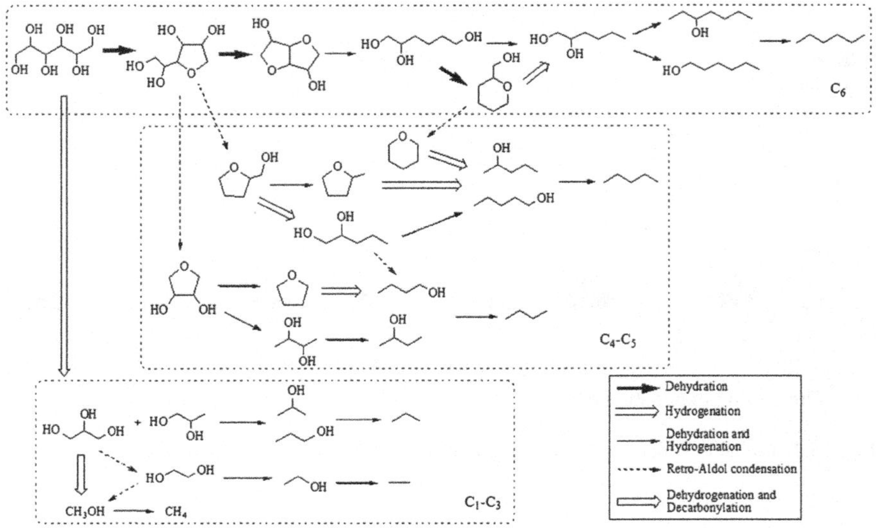

Fig. 15.3 Transformation of sorbitol into fuel units over Pt/SiO_2–Al_2O_3 catalyst [56]

15.6 Lignin-Based Fuels

Lignin is a natural aromatic polymer and a main constituent of lignocellulosics biomass; however, it has received much less biorefining efforts than plant polysaccharides due to its complexity, biological recalcitrance and relative thermal stability. To improve carbon utilization and economic competitiveness of biomass refineries, biomass-derived lignin can be partially utilized for the production of fuels and chemicals. Lignin is a rich source for phenolic bio-oils and has a high energy density than cellulose and hemicelluloses [63]. But due to the large oxygen content, phenolic bio-oils, consisting of phenolic molecules such as phenol, guaiacol, syringol, and their derivatives, requires significant deoxygenation to convert into conventional transport alkane fuels. Therefore, phenolic compounds are generally regarded as important model compounds for bio-oil; whereas, hydrodeoxygenation is used for bio-oil upgrading.

Research groups of Gates and Resasco have extensively studied HDO chemistries to upgrade different model compounds from lignin-derived bio-oils including anisole, guaiacol, vanillin, eugenol, phenol and cresol [64–66]. Their findings indicate that noble metals (e.g., Pt, Pd, Ru etc.) in combination with an acidic support (such as Al_2O_3, SiO_2, zeolites) can offer most effective catalytic systems for selective HDO processes. For example, Pt/γ-Al_2O_3 as catalyst was used for the conversion of different lignin model compounds as well as lignin-derived bio-oils. The proposed bifunctional system served two different roles in the reaction; the metallic function offered enhanced HDO kinetics, while the acidic support played a key role in the transalkylation reaction for the effective cleavage of ether linkages from the lignin structure. Another bifunctional catalyst consisting of Ru supported on zeolite H-ZSM-5 exhibited an excellent HDO activity towards the conversion of lignin-derived phenolic monomers and dimers to cycloalkanes in aqueous solution at 150 °C [67]. Initially, a series of noble metals supported on H-ZSM-5 (Si/Al = 38) were tested in the aqueous-phase hydrodeoxygenation of phenol at 150 °C. Ru was shown to be most active and selective for the production of cyclohexane as compared to Pd and Pt. The protocol discloses the removal of oxygen functionalities through C–O bond cleavage in phenolics, followed by an integrated metal- and acid-catalyzed hydrogenation and dehydration. The separate role of Brønsted acid sites from the zeolite (promotes dehydration reactions) and Ru (catalyzes hydrogenation processes) make this system ideal for alkanes formation from lignin-derived phenolics. In addition to metallic sites, the Si/Al ratio had a crucial role in determining the acid strength as well as the catalyst hydrophobicity. Although phenol conversions did not depend on Si/Al ratios and topology of the zeolite, the selectivity to cyclohexane remarkably increased with decreased Si/Al ratios in H-ZSM-5. Experiments revealed that Ru/H-ZSM-5 with the lowest Si/Al ratio in H-ZSM-5 (Si/Al = 25) was most selective to cycloalkanes production. These findings indicate that the presence of a larger concentration of acid sites in the zeolite favored cyclohexanol dehydration during HDO, which leads to a higher selectivity to hydrocarbons, in good agreement with recent studies showing that the integration of acid functionality with noble metal catalysts can provide useful bifunctional catalytic systems to achieve fast oxygen removal [68].

As per requirement of the acidic properties of a catalyst support acidic zeolites of the type H-Beta and H-ZSM-5 are very effective for designing bifunctional catalysts to convert monomeric lignin compounds [69, 70]. A bifunctional Ni/H-ZSM-5 catalyst (Si/Al = 45 and Ni = 20 wt%) exhibits high activity and selectivity for the hydrodeoxygenation of various C–O and C=O bonds in furans, alcohols, ketones, and phenols. The same catalyst is also able to convert a series of alkyl-, ketone-, or hydroxysubstituted phenols and guaiacols, alkyl-substituted syringol to produce cycloalkanes (73–92 %) as major products along with some aromatics (5.0–15 %) and methanol (0–17 %).

Understanding the interaction between bio-oils (or raw lignin) with the catalyst surface as well as the design of optimum catalytic surfaces is essential in order to achieve high conversion of lignin-derived bio-oils to fuels via HDO. The alcoholic fractions of lignin bio-oils are water soluble while alkylated phenolic compounds lead to water/oil emulsions. An easily recoverable catalytic system that simultaneously stabilizes emulsions is highly advantageous for HDO technologies in a biphasic reaction set-up. Hybrid catalytic system consisting of deposited Pd nanoparticles on a carbon nanotube–inorganic oxide (SiO_2) hybrid can stabilize water–oil emulsions and catalyze reactions at the liquid/liquid interface [66]. The hybrid solid nanoparticles were reported to be capable of catalyzing reactions in both aqueous and organic phases. Pd deposited on the hydrophilic interface catalyzes aqueous reactions, whereas its deposition on its hydrophobic counterpart favors reactions in the organic solvent. Different bimetallic systems including noble

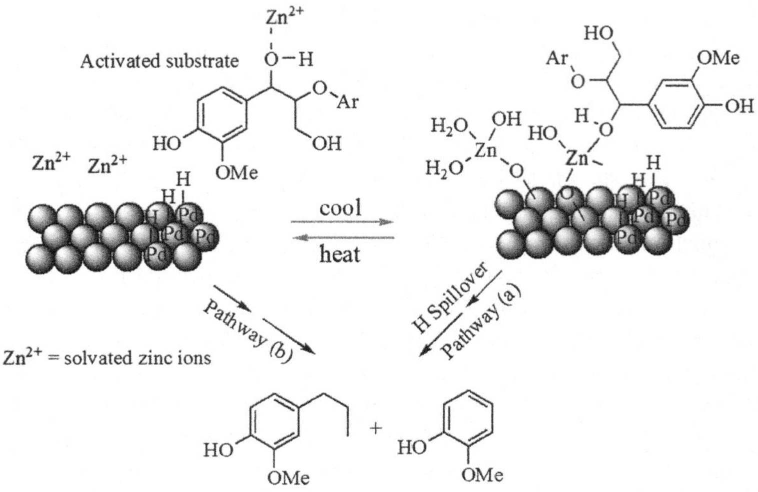

Fig. 15.4 Proposed pathways for HDO of β–O–4 lignin substrate over Zn/Pd/C catalyst. Pathway **a** involves substrate binding to Zn^{2+} sites and hydrogen spillover from Pd sites. Pathway **b** represents desorption of Zn ions into solution, activation of substrate via binding, and it is the Zn bound substrate that reacts with Pd sites on the catalyst's surface [71]

metals combined with a transition metal (e.g., Fe, Ni, Cu, Zn or Sn) have also been identified as highly selective for oxygen removal even under mild HDO conditions.

Bimetallic Zn/Pd/C system is another good example of synergistic effect applied in the selective HDO of lignin compounds [71]. The β–O–4 linkages found in dimeric lignin model complexes and synthetic lignin polymers were successfully cleaved by this system, producing HDO products in high yield and maintaining aromatic functional groups. Controlled experiments suggested that the single use of Pd/C or Zn^{2+} was unable to promote HDO, indicating a synergistic effect between Pd/C and Zn^{2+} as represented in a mechanistic approach (Fig. 15.4). X-ray absorption spectroscopy (EXAFS) confirmed that there was no alloy formation occurred between Pd and Zn in the system.

15.7 Conclusions and Outlook

The proposed contribution has been aimed to provide an overview of the possibilities to design a range of processes and technologies including nanomaterials for the production of biofuels from biomass-derived platform chemicals (i.e., furanics, levulinic acid, etc.) and lignocellulosic fractions including lignin and sugars. The preparation of a range of heterogeneous catalysts for such processes remains a significant challenge that needs to be addressed in the future, in terms of process/catalyst compatibility (e.g., stability, activity, selectivity) as well as to be able to process complex biomass feedstocks. In this regard, metal-based nanomaterials are of special interest due to their easily controllable physical properties like size, shape and composition. Biomass conversion requiring efficient catalytic activity and selectivity thus fits well with these advantages of nano-scale metal catalysts. Currently, the conversion of lignincellulose, lipids, and chitin has been effectively catalyzed by nano metal catalysts. Despite this well-recognized demand of nano-catalysis, knowledge of the prevailing mechanisms at such small scales is necessary. A detailed study of the mechanistic aspect of these catalytic processes to develop a scientific basis is needed which will help to tune and tailor the new catalyst system. We believe the optimization and design of such nanomaterials can be of utmost importance for a more sustainable society and we hope this contribution can further stimulate research efforts in the area leading to future developments in biofuels production from biomass using heterogeneous nanocatalysts.

References

1. World Business Council for Sustainable Development: *Facts and Trends to 2050—Energy and Climate Change* (2004), http://www.wbcsd.org
2. US Department of Energy, Energy Information Administration: *International Energy Outlook* (2006), http://www.eia.doe.gov
3. S. Nishimura (ed.), *Handbook of Heterogeneous Catalytic Hydrogenation for Organic Synthesis* (Wiley, 2001)
4. Z. He, X. Wang, Hydrodeoxygenation of model compounds and catalytic systems for pyrolysis bio-oils upgrading. Catal. Sustainable Energy Prod. **1**, 28–52 (2012)
5. M.V. Bykova, D.Y. Ermakov, V.V. Kaichev, O.A. Bulavchenko, A.A. Saraev, M.Y. Lebedev, V.A. Yakovlev, Ni-based sol–gel catalysts as promising systems for crude bio-oil upgrading: guaiacol hydrodeoxygenation study. Appl. Catal. B: Environ. **113–114**, 296–307 (2012)
6. Y. Roman-Leshkov, C.J. Barrett, Z.Y. Liu, J.A. Dumesic, Production of dimethylfuran for liquid fuels from biomass-derived carbohydrates. Nature **447**, 982–985 (2007)
7. F. Tao, S.R. Zhang, L. Nguyen, X.Q. Zhang, Action of bimetallic nanocatalysts under reaction conditions and during catalysis: evolution of chemistry from high vacuum conditions to reaction conditions. Chem. Soc. Rev. **41**, 7980–7993 (2012)
8. S.C. Tsang, N. Cailuo, W. Oduro, A.T.S. Kong, L. Clifton, K.M.K. Yu, B. Thiebaut, J. Cookson, P. Bishop, Engineering preformed cobalt-doped platinum nanocatalysts for ultraselective hydrogenation. ACS Nano **2**, 2547–2553 (2008)
9. G.H. Wang, J. Hilgert, F.H. Richter, F. Wang, H.J. Bongard, B. Spliethoff, C. Weidenthaler, F. Schuth, Platinum–cobalt bimetallic nanoparticles in hollow carbon nanospheres for hydrogenolysis of 5-hydroxymethylfurfural. Nat. Mater. **13**, 293–300 (2014)
10. S. Nishimura, N. Ikeda, K. Ebitani, Selective hydrogenation of biomassderived 5-hydroxymethylfurfural (HMF) to 2,5-dimethylfuran (DMF) under atmospheric hydrogen pressure over carbon supported PdAu bimetallic catalyst. Catal. Today **232**, 89–98 (2014)
11. Y. Nakagawa, K. Takada, M. Tamura, K. Tomishige, Total hydrogenation of furfural and 5-hydroxymethylfurfural over supported Pd–Ir Alloy catalyst. ACS Catal. **4**, 2718–2726 (2014)
12. S. Sitthisa, W. An, D.E. Resasco, Selective conversion of furfural to methylfuran over silica-supported Ni–Fe bimetallic catalysts. J. Catal. **284**, 90–101 (2011)
13. T. Thananatthanachon, T.B. Rauchfuss, Efficient production of the liquid fuel 2,5-dimethylfuran from fructose using formic acid as a reagent. Angew. Chem. Int. Ed. **49**, 6616–6618 (2010)

14. S. De, S. Dutta, B. Saha, One-pot conversions of lignocellulosic and algal biomass into liquid fuels. ChemSusChem. **5**, 1826–1833 (2012)
15. J. Tuteja, H. Choudhary, S. Nishimura, K. Ebitani, Direct Synthesis of 1, 6-Hexanediol from HMF over a Heterogeneous Pd/ZrP catalyst using formic acid as hydrogen source. ChemSusChem **7**, 96–100 (2014)
16. M. Chidambaram, A.T. Bell, A two-step approach for the catalytic conversion of glucose to 2, 5-dimethylfuran in ionic liquids. Green Chem. **12**, 1253–1262 (2010)
17. M. Chatterjee, T. Ishizaka, H. Kawanami, Hydrogenation of 5-hydroxymethylfurfural in supercritical carbon dioxide/water: a tunable approach to dimethylfuran selectivity. Green Chem. **16**, 1543–1551 (2014)
18. Y.-B. Huang, M.-Y. Chen, L. Yan, Q.-X. Guo, Y. Fu, Nickel–tungsten carbide catalysts for the production of 2,5-dimethylfuran from biomass-derived molecules. ChemSusChem **7**, 1068–1072 (2014)
19. L.W. Burnett, I.B. Johns, R.F. Holdren, R.M. Hixon, Production of 2-methylfuran by vapor-phase hydrogenation of furfural. Ind. Eng. Chem. **40**, 502–505 (1948)
20. H.-Y. Zheng, Y.-L. Zhu, B.-T. Teng, Z.-Q. Bai, C.-H. Zhang, H.-W. Xiang, Y.-W. Li, Towards understanding the reaction pathway in vapour phase hydrogenation of furfural to 2-methylfuran. J. Mol. Catal. A: Chem. **246**, 18–23 (2006)
21. W.-S. Lee, Z. Wang, W. Zheng, D.G. Vlachos, A. Bhan, Vapor phase hydrodeoxygenation of furfural to 2-methylfuran on molybdenum carbide catalysts. Catal. Sci. Technol. **4**, 2340–2352 (2014)
22. T.S. Hansen, K. Barta, P.T. Anastas, P.C. Ford, A. Riisager, One-pot reduction of 5-hydroxymethylfurfural *via* hydrogen transfer from supercritical methanol. Green Chem. **14**, 2457–2461 (2012)
23. J. Jae, W. Zheng, R.F. Lobo, D.G. Vlachos, Production of dimethylfuran from hydroxymethylfurfural through catalytic transfer hydrogenation with ruthenium supported on carbon. ChemSusChem **6**, 1158–1162 (2013)
24. D. Scholz, C. Aellig, I. Hermans, Catalytic transfer hydrogenation/hydrogenolysis for reductive upgrading of furfural and 5-(hydroxymethyl)furfural. ChemSusChem **7**, 268–275 (2014)
25. G.W. Huber, J.N. Chheda, C.J. Barrett, J.A. Dumestic, Production of liquid alkanes by aqueous-phase processing of biomass-derived carbohydrates. Science **308**, 1446–1450 (2005)
26. R.M. West, Z.Y. Liu, M. Peter, J.A. Dumesic, Liquid alkanes with targeted molecular weights from biomass-derived carbohydrates. ChemSusChem **1**, 417–724 (2008)
27. R. Xing, A.V. Subrahmanyam, H. Olcay, W. Qi, G.P. van Walsum, H. Pendse, G.W. Huber, Production of jet and diesel fuel range alkanes from waste hemicellulose-derived aqueous solutions. Green Chem. **12**, 1933–1946 (2010)
28. H. Olcay, A.V. Subrahmanyam, R. Xing, J. Lajoie, J.A. Dumesic, G.W. Huber, Production of renewable petroleum refinery diesel and jet fuel feedstocks from hemicellulose sugar streams. Energy Environ. Sci. **6**, 205–216 (2013)
29. J. Yang, N. Li, S. Li, W. Wang, L. Li, A. Wang, X. Wang, Y. Cong, T. Zhang, Synthesis of diesel and jet fuel range alkanes with furfural and ketones from lignocellulose under solvent free conditions. Green Chem. **16**, 4879–4884 (2014)
30. B. Pholjaroen, N. Li, J. Yang, G. Li, W. Wang, A. Wang, Y. Cong, X. Wang, T. Zhang, Production of renewable jet fuel range branched alkanes with xylose and methyl isobutyl ketone. Ind. Eng. Chem. Res. **53**, 13618–13625 (2014)
31. A. Corma, O. de la Torre, M. Renz, N. Villandier, Production of high-quality diesel from biomass waste products. Angew. Chem. Int. Ed. **50**, 2375–2378 (2011)
32. A. Corma, O. de la Torre, M. Renz, Production of high quality diesel from cellulose and hemicellulose by the Sylvan process: catalysts and process variables. Energy Environ. Sci. **5**, 6328–6344 (2012)
33. G. Li, N. Li, Z. Wang, C. Li, A. Wang, X. Wang, Y. Cong, T. Zhang, Synthesis of high-quality diesel with furfural and 2-methylfuran from hemicelluloses. ChemSusChem **5**, 1958–1966 (2012)

34. G. Li, N. Li, J. Yang, L. Li, A. Wang, X. Wang, Y. Cong, T. Zhang, Synthesis of renewable diesel range alkanes by hydrodeoxygenation of furans over Ni/Hβ under mild conditions. Green Chem. **16**, 594–599 (2014)
35. G. Li, N. Li, X. Wang, X. Sheng, S. Li, A. Wang, Y. Cong, X. Wang, T. Zhang, Synthesis of Diesel or Jet Fuel Range Cycloalkanes with 2-Methylfuran and Cyclopentanone from Lignocellulose. Energy Fuels **28**, 5112–5118 (2014)
36. J.J. Bozell, G.R. Petersen, Technology development for the production of biobased products from biorefinery carbohydrates—the US Department of Energy's "Top 10" revisited. Green Chem. **12**, 539–554 (2010)
37. F.M.A. Geilen, B. Engendahl, A. Harwardt, W. Marquardt, J. Klankermayer, W. Leitner, Selective and flexible transformation of biomass-derived platform chemicals by a multifunctional catalytic system. Angew. Chem. Int. Ed. **49**, 5510–5514 (2010)
38. I.T. Horvath, H. Mehdi, V. Fabos, L. Boda, L.T. Mika, γ-Valerolactone–a sustainable liquid for energy and carbon-based chemicals. Green Chem. **10**, 238–242 (2008)
39. A.P. Dunlop, J.W. Madden, Process of preparing gamma-valerolactone. US Patent 2,786,852 (1957)
40. A.M. Hengne, N.S. Biradar, C.V. Rode, Surface species of supported ruthenium catalysts in selective hydrogenation of levulinic esters for biorefinery application. Catal. Lett. **142**, 779–787 (2012)
41. M. Chia, J.A. Dumesic, Liquid-phase catalytic transfer hydrogenation and cyclization of levulinic acid and its esters to γ-valerolactone over metal oxide catalysts. Chem. Commun. **47**, 12233–12235 (2011)
42. L. Bui, H. Luo, W.R. Gunther, Y. Roman-Leshkov, Domino reaction catalyzed by zeolites with brønsted and lewis acid sites for the production of cvalerolactone from furfural. Angew. Chem. Int. Ed. **52**, 8022–8025 (2013)
43. J.C. Serrano-Ruiz, J.A. Dumesic, Catalytic routes for the conversion of biomass into liquid hydrocarbon transportation fuels. Energy Environ. Sci. **4**, 83–99 (2011)
44. M. Mascal, S. Dutta, I. Gandarias, Hydrodeoxygenation of the angelica lactone dimer, a cellulose-based feedstock: simple, high-yield synthesis of branched C_7–C_{10} gasoline-like hydrocarbons. Angew. Chem. Int. Ed. **53**, 1854–1857 (2014)
45. A. Corma, S. Iborra, A. Velty, Chemical routes for the transformation of biomass into Chemicals. Chem. Rev. **107**, 2411–2502 (2007)
46. C.-H. Zhou, J.N. Beltramini, Y.-X. Fan, G.Q. Lu, Chemoselective catalytic conversion of glycerol as a biorenewable source to valuable commodity chemicals. Chem. Soc. Rev. **37**, 527–549 (2008)
47. G.W. Huber, S. Iborra, A. Corma, Synthesis of transportation fuels from biomass: chemistry, catalysts, and engineering. Chem. Rev. **106**, 4044–4098 (2009)
48. J.W. Shabaker, G.W. Huber, R.R. Davda, R.D. Cortright, J.A. Dumesic, Aqueous-phase reforming of ethylene glycol over supported platinum catalysts. Catal. Lett. **88**, 1–8 (2003)
49. Q. Zhang, K. Qiu, B. Li, T. Jiang, X. Zhang, L. Ma, T. Wang, Isoparaffin production by aqueous phase processing of sorbitol over the Ni/HZSM-5 catalysts: effect of the calcination temperature of the catalyst. Fuel **90**, 3468–3472 (2011)
50. M. Banu, S. Sivasanker, T.M. Sankaranarayanan, P. Venuvanalingam, Hydrogenolysis of sorbitol over Ni and Pt loaded on NaY. Catal. Commun. **12**, 673–677 (2011)
51. K. Murata, I. Takahara, M. Inaba, Propane formation by aqueous-phase reforming of glycerol over Pt/H-ZSM5 catalysts. Reaction Kinetics Catal. Lett. **93**, 59–66 (2008)
52. R.M. Ravenelle, F. Schuber, A. D'Amico, N. Danilina, J.A. van Bokhoven, J.A. Lercher, C. W. Jones, C. Sievers, Stability of zeolites in hot liquid water. J. Phys. Chem. C **114**, 19582–19595 (2010)
53. R.M. West, M.H. Tucker, D.J. Braden, J.A. Dumesic, Production of alkanes from biomass derived carbohydrates on bi-functional catalysts employing niobium-based supports. Catal. Commun. **10**, 1743–1746 (2009)
54. Q. Zhang, T. Wang, B. Li, T. Jiang, L. Ma, X. Zhang, Q. Liu, Aqueous phase reforming of sorbitol to bio-gasoline over Ni/HZSM-5 catalysts. Appl. Energy **97**, 509–513 (2012)

55. Q. Zhang, T. Jiang, B. Li, T. Wang, X. Zhang, Q. Zhang, L. Ma, Highly selective sorbitol hydrogenolysis to liquid alkanes over Ni/HZSM-5 catalysts modified with pure silica MCM-41. ChemCatChem **4**, 1084–1087 (2012)
56. Y.T. Kim, J.A. Dumesic, G.W. Huber, Aqueous-phase hydrodeoxygenation of sorbitol: a comparative study of Pt/Zr phosphate and Pt–ReO_x/C. J. Catal. **304**, 72–85 (2013)
57. G.W. Huber, R.D. Cortright, J.A. Dumesic, Renewable alkanes by aqueous-phase reforming of biomass-derived oxygenates. Angew. Chem. Int. Ed. **43**, 1549–1551 (2004)
58. N. Li, G.A. Tompsett, G.W. Huber, Renewable high-octane gasoline by aqueous-phase hydrodeoxygenation of C_5 and C_6 carbohydrates over Pt/zirconium phosphate catalysts. ChemSusChem **3**, 1154–1157 (2010)
59. J. Duan, Y.T. Kim, H. Lou, G.W. Huber, Hydrothermally stable regenerable catalytic supports for aqueous-phase conversion of biomass. Catal. Today **234**, 66–74 (2014)
60. L. Vilcocq, A. Cabiac, C. Especel, S. Lacombe, D. Duprez, New insights into the mechanism of sorbitol transformation over an original bifunctional catalytic system. J. Catal. **320**, 16–25 (2014)
61. K. Chen, M. Tamura, Z. Yuan, Y. Nakagawa, K. Tomishige, One-pot conversion of sugar and sugar polyols to n-alkanes without C–C dissociation over the Ir-ReOx/SiO2 catalyst combined with H-ZSM-5. ChemSusChem **6**, 613–621 (2013)
62. T. Jiang, Q. Zhang, T. Wang, Q. Zhang, L. Ma, High yield of pentane production by aqueous-phase reforming of xylitol over Ni/HZSM-5 and Ni/MCM-22 catalysts. Energy Convers. Manage. **59**, 58–65 (2012)
63. J. Zakzeski, P.C. Bruijnincx, A.A.L. Jongerius, B.M. Weckhuysen, The catalytic valorization of lignin for the production of renewable chemicals. Chem. Rev. **110**, 3552–3599 (2010)
64. R.C. Runnebaum, T. Nimmanwudipong, D.E. Block, B.C. Gates, Catalytic conversion of compounds representative of lignin-derived bio-oils: a reaction network for guaiacol, anisole, 4-methylanisole, and cyclohexanone conversion catalysed by Pt/γ-Al_2O_3. Catal. Sci. Technol. **2**, 113–118 (2012)
65. M. Saidi, F. Samimi, D. Karimipourfard, T. Nimmanwudipong, B.C. Gates, M.R. Rahimpour, Upgrading of lignin-derived bio-oils by catalytic hydrodeoxygenation. Energy Environ. Sci. **7**, 103–129 (2014)
66. S. Crossley, J. Faria, M. Shen, D.E. Resasco, Solid nanoparticles that catalyze biofuel upgrade reactions at the water/oil interface. Science **327**, 68–72 (2010)
67. W. Zang, J. Chen, R. Liu, S. Wang, L. Chen, K. Li, Hydrodeoxygenation of lignin-derived phenolic monomers and dimers to alkane fuels over bifunctional zeolite-supported metal catalysts. ACS Sustainable Chem. Eng. **2**, 683–691 (2014)
68. C. Zhao, J. He, A.A. Lemonidou, X. Li, J.A. Lercher, Aqueous-phase hydrodeoxygenation of bio-derived phenols to cycloalkanes. J. Catal. **280**, 8–16 (2011)
69. C. Zhao, J.A. Lercher, Upgrading pyrolysis oil over Ni/HZSM-5 by cascade reactions. Angew. Chem. Int. Ed. **51**, 5935–5940 (2012)
70. C. Zhao, J.A. Lercher, Selective hydrodeoxygenation of lignin-derived phenolic monomers and dimers to cycloalkanes on Pd/C and HZSM-5 catalysts. ChemCatChem **4**, 64–68 (2012)
71. T.H. Parsell, B.C. Owen, I. Klein, T.M. Jarrell, C.L. Marcuum, L.J. Hupert, L.M. Amundson, H.I. Kenttamaa, F. Ribeiro, J.T. Miller, M.M. Abu-Omar, Cleavage and hydrodeoxygenation (HDO) of C–O bonds relevant to lignin conversion using Pd/Zn synergistic catalysis. Chem. Sci. **4**, 806–813 (2013)

Index

Q. Li (ed.), *Nanomaterials for Sustainable Energy*,
NanoScience and Technology, DOI 10.1007/978-3-319-32023-6

The manufacturer's authorised representative in the EU is Springer Nature Customer Service Centre GmbH, Europaplatz 3, 69115 Heidelberg, Germany. If you have any concerns regarding our products, please contact ProductSafety@springernature.com

Printed and bound by CPI Group (UK) Ltd, Croydon, CR0 4YY
30/09/2026
02225583-0020